Changing Global Perspectives on Horseshoe Crab Biology, Conservation and Management

Ruth H. Carmichael • Mark L. Botton
Paul K.S. Shin • Siu Gin Cheung
Editors

Changing Global Perspectives on Horseshoe Crab Biology, Conservation and Management

Springer

Editors
Ruth H. Carmichael
Dauphin Island Sea Lab
University of South Alabama
Dauphin Island, AL, USA

Mark L. Botton
Department of Natural Sciences
Fordham University
New York, NY, USA

Paul K.S. Shin
Department of Biology and Chemistry
City University of Hong Kong
Kowloon, Hong Kong

Siu Gin Cheung
Department of Biology and Chemistry
City University of Hong Kong
Kowloon, Hong Kong

ISBN 978-3-319-19541-4 ISBN 978-3-319-19542-1 (eBook)
DOI 10.1007/978-3-319-19542-1

Library of Congress Control Number: 2015954955

Springer Cham Heidelberg New York Dordrecht London

Printed on acid-free paper

Springer International Publishing AG Switzerland is part of Springer Science+Business Media (www.springer.com)

We dedicate this book to the memory of departed colleagues:

Since the first International Symposium on the Science and Conservation of Horseshoe Crabs and the resulting publication of Biology and Conservation of Horseshoe Crabs *in 2009 (Springer), the field of horseshoe crab research and conservation lost three renowned pioneers, including Dr. Robert Barlow, Dr. Anne Rudloe, and Dr. Koichi Sekiguchi. We thought it only appropriate to give tribute to these great mentors who shaped the field we know today. Additional information, including selected key publications and personal reflections on each colleague, are provided by Dr. Carl N. Shuster Jr.*

Dr. Robert B. Barlow (1939–2009)
Pioneer investigator into the neural basis of horseshoe crab behavior and vision

Credit: Author's collection (Courtesy G. Kreamer, 2005)

Bob Barlow was an internationally distinguished vision scientist who applied research on the compound eye of Limulus polyphemus *to advance knowledge of human visual processes. Dr. Barlow had a B.S. in Physics from Bowdoin College (1961) and received a Ph.D. at Rockefeller University (1967) under Nobel Laureate Dr. H. K. Hartline. For three decades he was a professor at Syracuse University and summer researcher at the Marine Biological Laboratory in Woods Hole, MA. In the late 1990s he became director of the Center for Vision Research and a professor of ophthalmology, biochemistry, and molecular biology at SUNY Upstate Medical University. Dr. Barlow was an engaging educator who combined stories from his personal research experiences with his passion and appreciation for horseshoe crabs to teach at all levels, from the university to teacher workshops and public venues. Dr. Barlow's dedication to the cause of horseshoe crab conservation on Cape Cod, sense of humor, and openness were integral to his character and, like his scientific advancements, remain a part of his legacy.*

Dr. Anne E. Rudloe (1947–2012)
A pioneer in horseshoe crab ecology and an early advocate for conservation

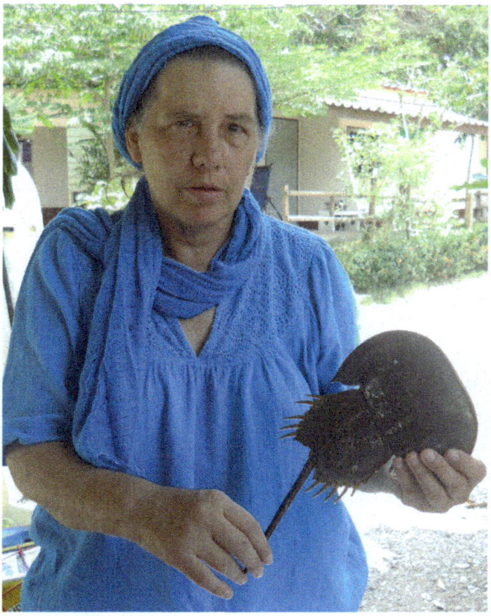

Credit: Courtesy Jack Rudloe

Anne Rudloe was a noted pioneering marine biologist, educational activist, and author. She was the managing director (retired) of the Gulf Specimen Marine Lab and Aquarium in Panacea, FL. She had a B.S. from Mary Washington College (1972) and received her M.S. (1972) and Ph.D. (1978) from Florida State University under Prof. W. F. Herrnkind.

She was highly respected for her research on horseshoe crabs, sea turtles, electric rays, and mysid shrimp, as well as for her tireless efforts to educate and advocate for stewardship and conservation of Florida wetlands. Dr. Rudloe also was dedicated to the art of Zen meditation. Anne's unique and compelling combination of science and spirituality helped make her an extraordinary researcher, educator, and conservationist. In addition to her many publications, Dr. Rudloe shared her insights in an essay blog, Of Birds and the Rainbow.

Dr. Koichi Sekiguchi (1920–2012)
Biogeographer of the Indo-Pacific species and editor/coauthor of the first comprehensive book on the Biology of Horseshoe Crabs

Credit: *Author's collection*

Dr. Sekiguchi led the first extensive distribution survey of the Indo-Pacific species of horseshoe crabs in 1975. His monumental book, The Biology of Horseshoe Crabs, *was the first comprehensive book on the subject and remains a basic resource for scholars and researchers worldwide. He received his Ph.D. in 1957 from Tokyo Kyoiku University under Dr. Hidemiti Oka (1902–1982). Dr. Sekiguchi assumed the study begun by Dr. Oka on the embryology of the Japanese horseshoe crab and wrote more than 50 publications, many with students. In 1978 he founded and was president of the Japanese Society of Preservation of Horseshoe Crabs until 2003 (and later became the first honorary president). Dr. Sekiguchi was made an honorary fellow of the Arthropodan Embryological Society of Japan and the Japanese Society of Systematic Ecology. In retirement he was tireless in organizing local horseshoe crab preservation groups and lecturing to various audiences. He is remembered as one of the true giants in the field of horseshoe crab biology and ecology and remains a source of inspiration to those who followed him.*

Citations: Passaglia CL (2010) In memoriam Robert B. Barlow Jr. J Exp Mar Biol Ecol 213:1397; The Wakulla News (2 May 2012, http://www.thewakullanews.com/content/anne-rudloe-obituary); Makioka T (2012) Obituary. Proc Arthrop Embryol Soc Jpn 47:1–4

Memories and Reflections on Drs. Barlow, Rudloe, and Sekiguchi

By Dr. Carl N. Shuster Jr.

During the past 20 years or so—owing to increasing interest in and awareness of horseshoe crabs and the various roles they play in ecosystems and the human health field—the scope of research focus on these animals, and resulting pulse of publications relating to them, has grown much more steady and substantial than at any previous period of time. It would probably not be too great an approximation to estimate that this recent great research surge has resulted in more scientists studying horseshoe crabs than during any other comparable period in history and that horseshoe crabs are recognized as a major species in some ecosystems.

*Skipping much history, here the focus is on remembering Drs. Robert Barlow, Anne Rudloe, and Koichi Sekiguchi for their important leadership roles in horseshoe crab research. Their contributions began in the 1970s, including participation in meetings. At the conference at Woods Hole in 1979, on the biomedical application of horseshoe crabs (see Cohen et al. 1979), the opening session featuring ten papers on the biology and ecology of horseshoe crabs included those by Rudloe and Sekiguchi. The following year, they were participants in the "*Limulus *Expedition" lecture series at the Duke Marine Laboratory (Bonaventura et al. 1982).*

In time, Delaware Bay attracted all three experts. The first to arrive were Anne Rudloe and her husband Jack who were then in the process of preparing their outstanding article for National Geographic (1981). I imagine them characteristically barefooted among the spawners stopping occasionally to observe the behavior of the crabs and comparing it to their crabs in Florida. Dr. Sekiguchi, as did the Rudloes, rendezvoused at my home in Virginia before we drove to Delaware Bay. Sekiguchi-san visited on two separate occasions: the first in

1989 with former students. In a later year he brought a scientific crew to film the spawning of Limulus; *they highlighted the deposition of the eggs by the use of an optical fiber.*

Dr. Anne Rudloe speaks to Dr. Sekiguchi and others at the Gulf Specimen Aquarium circa late 1980s (*top panel*) and Dr. Rudloe and a group of young learners at the Gulf Specimen Marine Lab (*bottom panel*) (*Courtesy Jack Rudloe*)

In 1994, Dr. Mark Botton (Fordham University) arranged a research trip to Japan, coordinated by Drs. Sekiguchi and Hiroaki

Sekiguchi (alert and amused by Shuster's dozing) on the Delaware Bay ferry between Cape May, New Jersey, and Cape Henlopen, Delaware (*top*). Dr. Sekiguchi and colleagues (his former students) on Highs Beach, Delaware Bay, New Jersey, in 1989 (*bottom*: Hiroaki Sugita, Carl N. Shuster Jr., Koichi Sekiguchi, Tomio Itow, and Fumio Shishikura). *Author's collection*

Sugita. There, among field trips, meetings with fisheries and education specialists and horseshoe crab conservationists and research on amplexus of the Japanese species, Tachypleus tridentatus, *resulted in a coauthored paper (Botton et al. 1996). By the 1970s all four living species had been observed in their natural habitat as well as studied at Shimoda Marine Research Center, University of Tsukuba. Live specimens of all four were also available for study in 1994.*

Dr. Barlow's natural athletic ability contributed to how he approached research. Known as a strong, long-distance swimmer, Bob made use of this attribute and SCUBA gear to survey the extent and nature of the domain of the crabs on Cape Cod where he conducted field studies on their vision. It was also an opportunity for Dr. Barlow to visualize what Limulus *had to contend with in the water environment and to account for those effects (Widener and Barlow 1999).*

*When it became obvious that the time was ripe for a scientific book on the American horseshoe crab, Dr. Barlow hosted a planning session at Woods Hole, followed by a visit to a company that produced LAL (*Limulus *amebocyte lysate) on Cape Cod.*

Meeting at Associates of Cape Cod (ACC), MA, to discuss writing a book on American horseshoe crabs (*The American Horseshoe Crab,* Harvard Press) (*seated, left to right*: Bob Barlow, Mark Botton, Steve Chamberlain, Jane Brockmann; (*standing*) Michael Dawson (ACC), Maureen Powers, Stanley Watson (ACC; this was his office), Bob Loveland, Carl N. Shuster Jr. *Author's collection*

Although Dr. Barlow focused most of his research at Woods Hole on Cape Cod, he visited Delaware Bay several times during the decades of his research, the most recent being 2002 to share his knowledge at a Green Eggs & Sand teacher workshop. He participated in many of these and other educational workshops throughout his career.

Dr. Bob Barlow, Carl N. Shuster. Jr., and Brian Harrington at the Green Eggs & Sand teacher workshop at Wellfleet, Massachusetts, in 2005. *Author's collection* (Courtesy of Michael Oates)

In these and many other ways, Drs. Barlow, Rudloe, and Sekiguchi are viewed as pioneers in horseshoe crab research, most notably to field biologists for their observational and ecological approaches to their studies of the animal. All three were also known and admired for their depth of understanding of the behavior of Limulus, *their teaching abilities, and their dedication to the conservation of the species. In this brief salute we respectfully share our memories of these three colleagues who were at the forefront of the current "surge" of scientific and conservation-based studies on horseshoe crabs. Gratefully we acknowledge their research accomplishments and pay tribute to the level of grace, passion, and dedication they brought to doing it.*

References *(Also see the selected key publications listed below)*

Bonaventura J, Bonaventura C, Tesh S (eds) (1982) Physiology and biology of horseshoe crabs: studies on normal and environmentally stressed animals. Alan R. Liss, Inc., New York

Botton ML, Shuster CN Jr, Sugita H et al (1996) Amplexus and mating behavior in the Japanese horseshoe crab, Tachypleus tridentatus. *Zool Sci 13:151–159*

Cohen E (ed) (1979) Biomedical applications of the horseshoe crab (Limulidae). Alan R. Liss, Inc., New York

Shuster CN Jr, Barlow RB, Brockmann HJ (eds) (2003) The American horseshoe crab. Harvard University Press, Cambridge, MA

Selected Key Publications *(Author's annotations)*

Dr. Robert Barlow

***Barlow RB**, Kaplan E (1971)* Limulus *lateral eye: properties of receptor units in the unexcised eye. Science 174:1027–1029 (Barlow's first major discovery as a principle investigator; the physiological properties of the eye change dramatically when retina is excised)*

***Barlow RB**, Bolanowski SJ, Brachman ML (1977) Efferent optic nerve mediate circadian rhythms in the* Limulus *eye. Science 197:86–89.* ***Barlow RB**, Chamberlain SC, Levinson JZ (1980)* Limulus *brain modulates the structure and function of the lateral eye. Science 210:1037–1039 (Second major discovery;*

in vitro, the structure and function of the Limulus *lateral eye shows diel modulation by efferent optic nerve signals from a circadian clock)*

Barlow RB*, Ireland LC, Kass L (1982) Vision as a role in* Limulus *mating behavior. Nature 296:65–66 (Third major discovery;* Limulus *uses its lateral eye to find mates)*

Powers MK, ***Barlow RB****, Kass L (1991) Visual performance of horseshoe crabs day and night. Vis Neurosci 7:179–189 (Male crabs see crablike objects of widely different contrast during day and night despite a relatively simple visual system)*

Widener JW, ***Barlow RB*** *(1999) Decline of a horseshoe crab population on Cape Cod. Biol Bull 197:300–301*

Barlow RB *(2009) Vision in horseshoe crabs. In: Tanacredi JT, Botton ML, Smith DR (eds) Biology and conservation of horseshoe crabs. Springer Science+Business Media LLC, New York, pp 223–235*

Dr. Anne Rudloe

Rudloe A*, Herrnkind WF (1976) Orientation of* Limulus polyphemus *in the vicinity of breeding beaches. Mar Behav Physiol 4:75–89 (Adults orient to wave surge and other stimuli while moving to and from spawning sites)*

Rudloe A *(1979) Locomotor and light responses of larvae of the horseshoe crab,* Limulus polyphemus *(L.). Biol Bull 157:494–505 (Reported lunar, tidal, and circadian responses)*

Rudloe A *(1980) The breeding behavior and pattern of movement of horseshoe crabs,* Limulus polyphemus, *in the vicinity of breeding beaches in Apalachee Bay, Florida. Estuaries 3:177–183*

Rudloe A *(1981) Aspects of the biology of juvenile horseshoe crabs,* Limulus polyphemus. *Bull Mar Sci 32:125–133*

Rudloe A*, Rudloe J (1981) The changeless horseshoe crab. Natl Geogr 159:562–572 (Well-illustrated article featuring a two-page view that testifies to the unusually great abundance of spawners at Delaware Bay circa 1980)*

Rudloe A *(1982) Man's influence as an environmental threat to* Limulus. *In: Bonaventura J (ed) Physiology and biology of horseshoe crabs: studies on normal and environmentally stressed animals. Alan R. Liss, New York, pp 297–300 (Worldwide exploitation of horseshoe crabs exemplifies a call to revere and protect Mother Earth)*

Rudloe A *(1983) The effect of heavy bleeding on mortality of the horseshoe crab,* Limulus polyphemus, *in the natural environment. J Invert Pathol 42:167–176 (A proactive conservation project to test bleeding effects; considered not a significant impact on the crabs)*

Dr. Koichi Sekiguchi

Sekiguchi K *(1973) A normal plate of the development of the Japanese horseshoe crab,* Tachypleus tridentatus. *Sci Rep Tokyo Kyoiku Daigaku Sec B 15:152–162*

Sekiguchi K, *Nakamura K (1979) Ecology of the extant horseshoe crabs. In: Cohen E (ed) Biomedical applications of the horseshoe crabs (Limulidae). Alan R. Liss, New York, pp 39–45*

Sekiguchi K, *Sugita H (1980) Systematics and hybridization in the four living species of horseshoe crabs. Evolution 34:712–718*

Sekiguchi K, *Yamamichi Y, Costlow JD (1982) Horseshoe crab development studies I. Normal embryonic development of* Limulus polyphemus *compared with* Tachypleus tridentatus. *In: Bonaventura J, Bonaventura C, Tesh S (eds) Physiology and biology of horseshoe crabs: studies on normal and environmentally stressed animals. Alan R. Liss, New York, pp 53–73*

Sekiguchi K, *Seshimo H, Sugita H (1988) Post-embryonic development of the horseshoe crab. Biol Bull 174:337–345*

Sekiguchi K *(1988) Biology of horseshoe crabs. Science House Co., Ltd., Tokyo (Essentially a review of the research accomplished under Dr. Sekiguchi's direction)*

Sekiguchi K, *Shuster CN Jr (2009) Limits on the global distribution of horseshoe crabs (Limulacea): lessons learned from two lifetimes of observations: Asia and America. In: Tanacredi JT, Botton ML, Smith DR (eds) Biology and conservation of horseshoe crabs. Springer Science+Media, LLC, New York, pp 5–24 (Large-scale, global biotic, and environmental parameters help to explain the distribution or lack thereof of the Limulacea)*

Foreword

As a keen young naturalist growing up in the UK, I had a childhood with no direct experience of horseshoe crabs. To me, horseshoe crabs were exotic creatures from faraway seas, featured in the many natural history books that I owned.

It was not until I lived in the USA from 2001 to 2006 that I became personally acquainted with these remarkable species. In May 2002 I first visited the mass spawning congregation of horseshoe crabs in Delaware Bay, with many thousands of animals laying millions of eggs. The timing of this spawning event to coincide with the northward migration of a huge number of wading birds provides them with a much-needed, abundant food source as they travel to their breeding grounds; and it provides us with one of the world's great wildlife spectacles as horseshoe crabs and birds crowd together in a restricted space. It also demonstrates the broader ecological importance of these species.

Horseshoe crabs hold a further fascination beyond their extraordinary breeding behavior. They look like nothing else on earth! And they are a truly ancient group, dating back 500 Ma and scarcely changed over the last 150 Ma. Few species can compete for the title of "ultimate living fossil" as convincingly as the horseshoe crabs. They are also very unusual physiologically (e.g., their blood has some remarkable properties), and this has made them very important in biomedical research.

In recent years there has been a welcome growth in interest in the conservation and management of horseshoe crabs. The *International Workshop on the Science and Conservation of Asian Horseshoe Crabs*, held in Hong Kong in June 2011, was a seminal development that led directly to the establishment of the Horseshoe Crab Specialist Group within the IUCN Species Survival Commission (SSC) later that year. The SSC Horseshoe Crab Specialist Group has rapidly become very active in raising the profile of the conservation of these species worldwide, and it has been very exciting for me to see the increased attention being given to animals that, until recently, were being ignored by the international conservation community. The successful 2011 meeting was followed by *Comparative Approaches to Horseshoe Crab Ecology and Conservation in North America and Southeast Asia*, a Special Session at the November 2013 meeting of the Coastal and Estuarine Research Federation in San Diego, California.

Four species of horseshoe crab are extant today, three in the seas of Asia (from India to Japan) and one in eastern North America (from Maine to the Yucatan). Despite the resilience of these species over millions of years, they all now face a wide range of anthropogenic threats with which they never had to contend through their evolutionary history. These new challenges include the impacts of commercial fisheries and the loss of spawning habitats. Horseshoe crab eggs (and legs) are considered a delicacy in parts of Asia, and the species are also captured to be used as bait in certain types of fishery. In the past there was a massive harvest of animals in the Delaware Bay (up to 1.2 million crabs per year between the 1870s and the 1920s). The animals were processed into fertilizer or used to feed livestock. This fishery proved to be unsustainable and fortunately has now ended. But fisheries remain in Asia, where loss of spawning habitat due to land reclamation is also a growing problem, especially with the burgeoning economies in this region. The SSC Horseshoe Crab Specialist Group is currently in the process of reassessing the IUCN Red List status of all four species.

At the end of the day, conservation is about leadership. Horseshoe crab conservation has accelerated over the last few years because of the dynamism and vision of Mark Botton of Fordham University, New York, and Paul Shin at the City University of Hong Kong. Mark and Paul are the cochairs of the SSC Horseshoe Crab Specialist Group, and I am extremely grateful to them for all that they have done over the last 3 years. I also congratulate them and the chapter authors for helping to produce *Changing Global Perspectives on Horseshoe Crab Biology, Conservation and Management*. This outstanding book gives chapter and verse on the conservation of these species, and this is extremely timely given the growing threats to these remarkable animals. I believe and hope that it leads to greatly increased attention to horseshoe crab conservation worldwide.

Simon N. Stuart
IUCN Species Survival Commission

Preface

Since the first International Symposium on Science and Conservation of Horseshoe Crabs in New York, USA, in 2007, we have driven and witnessed a revolution in horseshoe crab study, including substantially increased international participation in and awareness of research, conservation, and management of horseshoe crabs. We overcame cultural and language barriers to enhance communication and enable collaborative interactions and emerging opportunities for data and concept sharing. The expanding commercial uses for horseshoe crabs, coupled with the continuing degradation of their essential spawning and nursery grounds, make these efforts essential for the vitality of natural populations in North America and Asia, where the four living species of horseshoe crabs reside.

The second international meeting, held in Hong Kong in 2011, reaffirmed the value of international collaborations through scientific studies, public education, and outreach. Further attention was drawn to research and conservation needs by participants who attended a Special Session, Comparative Approaches to Horseshoe Crab Ecology and Conservation, at the biennial meeting of the Coastal and Estuarine Research Federation in San Diego, CA, USA, in 2013. The enthusiastic participation of scientists from around the world inspired the idea for this new book to report on the significant progress of scientific studies on horseshoe crabs and new issues related to their conservation.

The content of this book, therefore, is a blend of primary research on topics from *Evolution, Adaptation, and Genetics* (Part 1, Chaps. 1, 2, 3, 4, 5, and 6) to *Ecology and Population Dynamics* (Part 2, Chaps. 7, 8, 9, 10, 11, 12, 13, and 14) and *Biology and Physiology* (Part 3, Chaps. 15, 16, 17, 18, 19, and 20), balanced by contributions on the assessment of conservation threats, status, and needs in *Conservation and Management* (Part 4, Chaps. 21, 22, 23, 24, 25, 26, 27, 28, 29, 30, 31, 32, 33, and 34). In many cases, findings documented in this book were part of presentations first given at the 2011 and 2013 meetings and were expanded and refined from those initial reports. In other cases entirely new works are presented. Supplemental material is available online to support some contributions.

Part 4 also includes a summary of motivations for establishing an International Union for Conservation of Nature (IUCN) Species Survival Commission Horseshoe Crab Specialist Group (another major success from the 2011 meeting). The need for working through the IUCN for conservation of horseshoe crabs and the current status of these efforts are described in detail at the beginning of Part 4 (Chap. 21). The book closes with a visually stunning example of global grassroots horseshoe crab outreach education with the contribution, *Young Voices: Through the Arts, Future Environmental Stewards Have a Global Voice* (Part 4, Chap. 34). We also included a special dedication to our departed colleagues Drs. Robert Barlow, Anne Rudloe, and Koichi Sekiguchi, including reflections by another pioneer in horseshoe crab research, Dr. Carl N. Shuster Jr.

Overall, the book contains the work of 84 contributors, 44 % from the USA (11 Atlantic and Gulf of Mexico coast states from Maine to Alabama) and 56 % from other countries including Australia, Canada, China, India, Japan, Malaysia, New Zealand, Poland, and Singapore. This book is part of an ongoing global effort to increase data and concept sharing to support basic research and advance conservation for these remarkable animals. We look forward to the Third International Workshop on Science and Conservation of Horseshoe Crabs at Sasebo City, Japan, in 2015, where we will extend the international horseshoe crab network to include participants from Indonesia, the Philippines, Thailand, and Vietnam and continue discussions on research and the future work of the IUCN Horseshoe Crab Specialist Group.

Dauphin Island, AL, USA — Ruth H. Carmichael
New York, NY, USA — Mark L. Botton
Kowloon, Hong Kong SAR — Paul K.S. Shin
Kowloon, Hong Kong SAR — Siu Gin Cheung

Acknowledgements

On the road to compiling any document, there are many contributors beyond the authors and editors who make the work possible. The editors are grateful to everyone who provided advice, input, and support for producing this book.

Dr. Carl N. Shuster Jr. was an enormous source of inspiration to each of the editors and to many of the contributors to this book. We owe him a huge debt of gratitude. In addition to contributing directly to chapters within the book, Dr. Shuster provided background information, images, and reflections from his personal experiences to prepare the dedication for the book. We recognize and thank the funders, conveners, and attendees of the second International Workshop on Science and Conservation of Asian Horseshoe Crabs (Hong Kong 2011) and the Special Session at the Coastal and Estuarine Research Federation (CERF) biennial meeting (USA 2013). Funders included the City University of Hong Kong (both meetings), Ocean Park Conservation Foundation and the Agriculture, Fisheries and Conservation Department of the Hong Kong SAR Government (Hong Kong meeting), and Fordham University (CERF meeting). We also enthusiastically thank Ms. Elizabeth Hieb at the Dauphin Island Sea Lab who acted as copyeditor for each contribution and assisted with organizing and compiling the many components of the book. Her organization, attention to detail, and patient diligence were invaluable to timely completion of the work. Dr. Jack Rudloe generously contributed images of his wife, Anne, for the dedication. We must also thank Alexandrine Cheronet, senior publishing editor, for her outstanding support of our book concept and facilitating completion of the final product with Springer.

Ruth H. Carmichael thanks Dr. I Valiela and the Friends of Pleasant Bay who first introduced her to this unique animal and concerns that prompted the demand for science to support management initiatives on Cape Cod, MA. C Shuster, R Barlow, M Botton, J Brockmann, J Levin, G Kreamer, G Gauvry, D Smith, T Novitsky, J Harrington, and N Wainwright helped to jump-start her applied horseshoe crab research program, and D Rutecki, E Gaines, students, and collaborators at the University of Maine, Machias, and Dauphin Island Sea Lab have kept the work going. Development funding from the Dauphin Island Sea Lab allowed participation in international meetings and writing associated with producing this book. Mark L. Botton would also like to thank his many students and collaborators, including B Loveland, D Smith, J Tanacredi, H Sugita, T Itow, the late K Sekiguchi, M Hamilton, J Mattei, M Beekey, C Colón, T Jacobsen, and the late J Ropes. Fordham University provided travel support to attend the 2011 Hong Kong workshop and awarded a Faculty Fellowship during fall of 2013 to organize the horseshoe crab session at the San Diego CERF meeting and establish the foundation for this book. Paul K.S. Shin appreciates the dedicated work on horseshoe crab research from many students under co-supervision with S.G. Cheung and is grateful to the continuous support from Suzanne Gendron and her team from the Ocean Park Conservation Foundation. SGC would like to acknowledge his Ph.D. supervisor, Professor Brian Morton, who introduced him to this animal and its amazing biology. His pioneering work on this animal in Hong Kong sets the stage for a more integrated approach to its conservation.

No scientific publication would be possible without the dedicated efforts of the many peer reviewers who donated their time to ensure the high quality of contributions, thank you:

Dr. Doris W.T. Au (City University of Hong Kong, Hong Kong)
Mr. Allen Aven (Dauphin Island Sea Lab, University of South Alabama, Alabama, USA)
Dr. Mark Beekey (Sacred Heart University, Connecticut, USA)
Dr. Punyasloke Bhadury (Indian Institute of Science Education and Research, Kolkata, India)
Mr. Gregory Breese (Delaware Bay Estuary Project, Delaware, USA)
Dr. H. Jane Brockmann (University of Florida, Florida, USA)
Dr. Lesley Cartwright-Taylor (Nature Society, Singapore)
Dr. Benny K.K. Chan (Academia Sinica, Taiwan)
Prof. Laurie H.M. Chan (University of Ottawa, Canada)
Dr. Jill M.Y. Chiu (Hong Kong Baptist University, Hong Kong)
Prof. Loke Ming Chou (National University of Singapore, Singapore)
Prof. Ka Hou Chu (The Chinese University of Hong Kong, Hong Kong)
Dr. Alok Das (Siksha O Anusandhan University, Bhubaneswar, India)
Prof J.L. Ding (National University of Singapore, Singapore)
Dr. Peter Funch (Aarhus University, Denmark)
Dr. Kevin Gao (Ocean University of China, China)
Mr. Glenn Gauvry (ERDG Inc., Delaware, USA)
Dr. Sara P. Grady (Massachusetts Bays National Estuary Program, Massachusetts, USA)
Prof. Eric M. Hallerman (Virginia Polytechnic Institute and State University, Virginia, USA)
Dr. Yumiko Iwasaki (Fordham University, New York, USA)
Dr. Nancy L. Jackson (New Jersey Institute of Technology, New Jersey, USA)
Dr. Mary-Jane James-Pirri (University of Rhode Island, Rhode Island, USA)
Dr. Jessica Jones (US Food and Drug Administration, Alabama, USA)

Mr. Gary Kreamer (Delaware Division of Fish and Wildlife, Delaware, USA)
Dr. Michael Lau (World Wide Fund For Nature Hong Kong, Hong Kong)
Mr. Kevin Laurie (Hong Kong Coast Watch, Hong Kong)
Prof. Joe S.Y. Lee (Griffith University, Australia)
Dr. Jack Levin (University of California School of Medicine, California, USA)
Ms. Alison Leschen (Waquoit Bay National Estuarine Research Reserve, Massachusetts, USA)
Dr. Jennifer H. Mattei (Sacred Heart University, Connecticut, USA)
Dr. Michael J. Millard (US Fish and Wildlife Service, USA)
Dr. Jayant K. Mishra (Pondicherry University, India)
Dr. Faridah BT Mohamad (Universiti Malaysia Terengganu, Malaysia)
Dr. Shin Nishida (University of Miyazaki, Japan)
Dr. Thomas J. Novitsky (Horseshoe Crab Conservation Association, Massachusetts, USA)
Dr. James Pierce (LaSalle University, Pennsylvania, USA)
Dr. David Smith (US Geological Survey, Leetown Science Center, West Virginia, USA)
Dr. Stephen A. Smith (VA/MD Regional College of Veterinary Medicine, Virginia, USA)
Dr. John T. Tanacredi (Molloy College, New York, USA)
Dr. Ling Ming Tsang (National Taiwan Ocean University, Taiwan)
Dr. Winsor Watson III (University of New Hampshire, New Hampshire, USA)
Dr. David W.H. Wong (State University of New York at Oneonta, USA)

Contents

Part II Ecology and Population Dynamics

Part III Biology and Physiology

Contributors

Y.M. Abdel-Hadi Department of Aquaculture, Faculty of Agriculture, Universiti Putra Malaysia, Serdang, Selangor, Malaysia

Nik Hafiz Abdullah School of Fundamental Sciences, Horseshoe Crab Research Group (HCRG), Universiti Malaysia Terengganu, Kuala, Terengganu, Malaysia

Amirrudin Ahmad School of Marine and Environmental Sciences, Horseshoe Crab Research Group (HCRG), Universiti Malaysia Terengganu, Kuala, Terengganu, Malaysia

Amirul Asyraf Alia'm Institute of Marine Biotechnology, Horseshoe Crab Research Group (HCRG), Universiti Malaysia Terengganu, Kuala, Terengganu, Malaysia

Aaron W. Aunins Cherokee Nation Technology Solutions, Leetown Science Center, Kearneysville, WV, USA

Allen M. Aven University Programs, Dauphin Island Sea Lab, Dauphin Island, AL, USA

Department of Marine Sciences, University of South Alabama, Mobile, AL, USA

Jacy D. Baggett Murphy High School, Mobile, AL, USA

Muhammad Farhan Saiful Bahri School of Marine and Environmental Sciences, Horseshoe Crab Research Group (HCRG), Universiti Malaysia Terengganu, Kuala, Terengganu, Malaysia

B. Bal Bioengineering Laboratory, Centre of Biotechnology, Siksha O Anusandhan University, Bhubaneswar, Odisha, India

Mark A. Beekey Department of Biology, Sacred Heart University, Fairfield, CT, USA

Satyaranjan Behera Odisha Biodiversity Board, Regional Plant Resource Centre Campus, Bhubaneswar, Odisha, India

Punyasloke Bhadury Integrative Taxonomy and Microbial Ecology Research Group, Department of Biological Sciences, Indian Institute of Science Education and Research Kolkata, Nadia, West Bengal, India

Tiffany Black Department of Marine Fisheries Biology, Florida Fish and Wildlife Conservation Commission, Senator George Kirkpatrick Marine Laboratory, Cedar Key, FL, USA

Błażej Błażejowski Polish Academy of Sciences, Institute of Paleobiology, Warsaw, Poland

Mark L. Botton Department of Natural Sciences, Fordham University, New York, NY, USA

Angela J. Brady Marine Sciences, University of Maine-Machias, Machias, ME, USA

H. Jane Brockmann Department of Biology, University of Florida, Gainesville, FL, USA

Bonnie L. Brown Department of Biology, Virginia Commonwealth University, Richmond, VA, USA

Ruth H. Carmichael Dauphin Island Sea Lab, University of South Alabama, Dauphin Island, AL, USA

Department of Marine Sciences, University of South Alabama, Mobile, AL, USA

Lesley Cartwright-Taylor Nature Society (Singapore), The Sunflower, Singapore

Christopher C. Chabot Department of Biological Sciences, Plymouth State University, Plymouth, NH, USA

Chang-Po Chen Biodiversity Research Center, Academia Sinica, Taipei, Taiwan, Republic of China

Helen Cheng Department of Biological Sciences, University of New Hampshire, Durham, NH, USA

Siu Gin Cheung Department of Biology and Chemistry, City University of Hong Kong, Kowloon, Hong Kong

State Key Laboratory in Marine Pollution, City University of Hong Kong, Kowloon, Hong Kong

B.C. Choudhury Wildlife Trust of India, Noida, Uttar Pradesh, India

A. Christianus Institute of Bioscience, Universiti Putra Malaysia, Serdang, Selangor, Malaysia

Department of Aquaculture, Faculty of Agriculture, Universiti Putra Malaysia, Serdang, Selangor, Malaysia

Christina P. Colón Department of Biological Sciences, Kingsborough Community College, Brooklyn, NY, USA

A.P. Das Bioengineering Laboratory, Centre of Biotechnology, Siksha O Anusandhan University, Bhubaneswar, Odisha, India

Jeak Ling Ding Department of Biological Sciences, Faculty of Science, National University of Singapore, Singapore, Singapore

Michael S. Eackles U.S. Geological Survey, Leetown Science Center-Aquatic Ecology Branch, Kearneysville, WV, USA

H.T. Eng Department of Aquaculture, Faculty of Agriculture, Universiti Putra Malaysia, Serdang, Selangor, Malaysia

Maurice G. Estes Jr. Atmospheric Science, University of Alabama in Huntsville, Huntsville, AL, USA

Biological and Environmental Sciences, Alabama A&M University, Normal, AL, USA

M.I.M. Faizul Institute of Bioscience, Universiti Putra Malaysia, Serdang, Selangor, Malaysia

Glenn Gauvry Ecological Research & Development Group, Inc., Little Creek, DE, USA

Izzatul Huda Abdul Ghaffar School of Fundamental Sciences, Horseshoe Crab Research Group (HCRG), Universiti Malaysia Terengganu, Kuala, Terengganu, Malaysia

Eric Hallerman Department of Fish and Wildlife Conservation, Virginia Polytechnic Institute and State University, Blacksburg, VA, USA

Elizabeth E. Hieb University Programs, Dauphin Island Sea Lab, Dauphin Island, AL, USA

Bow Ho Department of Microbiology, Yong Loo Lin School of Medicine, National University of Singapore, Singapore, Singapore

Hwey-Lian Hsieh Biodiversity Research Center, Academia Sinica, Taipei, Taiwan, Republic of China

Menghong Hu Department of Biology and Chemistry, City University of Hong Kong, Hong Kong, Hong Kong

College of Fisheries and Life Science, Shanghai Ocean University, Shanghai, China

Noraznawati Ismail Institute of Marine Biotechnology, Horseshoe Crab Research Group (HCRG), Universiti Malaysia Terengganu, Kuala, Terengganu, Malaysia

Chikako Iwaoka Kujukushima Aquarium, Sasebo, Nagasaki, Japan

Sheri L. Johnson Department of Zoology, University of Otago, Dunedin, New Zealand

David R. Jones The New Jersey Aquaculture Innovation Center at Rutgers University, Cape May, NJ, USA

Mohd Mustakim Mohd Kasturi School of Fundamental Sciences, Horseshoe Crab Research Group (HCRG), Universiti Malaysia Terengganu, Kuala, Terengganu, Malaysia

Oh Sheng Khai Institute of Marine Biotechnology, Horseshoe Crab Research Group (HCRG), Universiti Malaysia Terengganu, Kuala, Terengganu, Malaysia

Tim L. King U.S. Geological Survey, Biological Resources Division, Leetown Science Center, Aquatic Ecology Branch, Kearneysville, WV, USA

Hiroko Koike The Kyushu University Museum, Fukuoka, Fukuoka, Japan

Gary Kreamer Delaware Division of Fish and Wildlife, Smyrna, DE, USA

Sharon W. Kreamer The Tatnall School, Wilmington, DE, USA

Kenji Kuroyanagi Minamichita Beach Land Aquarium, Chita-Gun, Aichi, Japan

Billy K.Y. Kwan Department of Biology and Chemistry, City University of Hong Kong, Hong Kong, Hong Kong

Brenda J. Landau Pendoley Environmental Pty Ltd, Booragoon, WA, Australia

Peng Li Centre for Biomedical and Life Sciences, Singapore Polytechnic, Singapore, Singapore

Robert E. Loveland Department of Ecology, Evolution & Natural Resources, Rutgers University, New Brunswick, NJ, USA

Peter D.M. Macdonald Mathematics and Statistics, McMaster University, Hamilton, ON, USA

P.S. Mahapatra Bioengineering Laboratory, Centre of Biotechnology, Siksha O Anusandhan University, Bhubaneswar, Odisha, India

Azwarfarid Manca School of Marine and Environmental Sciences, Horseshoe Crab Research Group (HCRG), Universiti Malaysia Terengganu, Kuala, Terengganu, Malaysia

Jennifer H. Mattei Department of Biology, Sacred Heart University, Fairfield, CT, USA

John McFadyen Biology, Dauphin Island Sea Lab, Dauphin Island, AL, USA

Conor P. McGowan U.S. Geological Survey, Alabama Cooperative Fish and Wildlife Research Unit, Auburn University, Auburn, AL, USA

Michael J. Millard U.S. Fish and Wildlife Service, Northeast Fishery Center, Lamar, PA, USA

Anita Mishra Department of Ocean Studies and Marine Biology, Pondicherry University (Brookshabad Campus), Port Blair, Andaman & Nicobar Islands, India

Jayant K. Mishra Department of Ocean Studies and Marine Biology, Pondicherry University (Brookshabad Campus), Port Blair, Andaman & Nicobar Islands, India

Mohamad Faridah School of Marine and Environmental Sciences, Horseshoe Crab Research Group (HCRG), Universiti Malaysia Terengganu, Kuala, Terengganu, Malaysia

Shin Nishida Biology, Science Education, Faculty of Education and Culture, University of Miyazaki, Miyazaki, Miyazaki, Japan

Chihiro Nishimura Department of Regional Policy, Faculty of Economics, University of Nagasaki, Sasebo, Nagasaki, Japan

Thomas J. Novitsky Horseshoe Crab Conservation Association, Orleans, MA, USA

Sixto Portilla Center for Environmental Research and Coastal Oceans Monitoring (CERCOM), Molloy College, Rockville Center, NY, USA

Muhammad Zul Fayyadh Azizo Rahman School of Marine and Environmental Sciences, Horseshoe Crab Research Group (HCRG), Universiti Malaysia Terengganu, Kuala, Terengganu, Malaysia

Daniel Sasson Department of Biology, University of Florida, Gainesville, FL, USA

Paul K.S. Shin Department of Biology and Chemistry, City University of Hong Kong, Kowloon, Hong Kong

State Key Laboratory in Marine Pollution, City University of Hong Kong, Kowloon, Hong Kong

Carl N. Shuster Jr. Biological Oceanographer (retired), Arlington, VA, USA

K. Sivakumar Department of Endangered Species Management, Wildlife Institute of India, Dehradun, Uttarakhand, India

David R. Smith U.S. Geological Survey, Leetown Science Center, Leetown, WV, USA

Matthew D. Smith Department of Biology, University of Florida, Gainesville, FL, USA

Muhd Fawwaz Afham Mohd Sofa School of Marine and Environmental Sciences, Horseshoe Crab Research Group (HCRG), Universiti Malaysia Terengganu, Kuala, Terengganu, Malaysia

John A. Sweka U.S. Fish and Wildlife Service, Northeast Fishery Center, Lamar, PA, USA

John T. Tanacredi Center for Environmental Research and Coastal Oceans Monitoring (CERCOM), Molloy College, Rockville Center, NY, USA

Basudev Tripathy Malacology Division, Prani Vigyan Bhawan, M-Block, Zoological Survey of India, Kolkata, West Bengal, India

Youji Wang Department of Biology and Chemistry, City University of Hong Kong, Hong Kong, Hong Kong

College of Fisheries and Life Science, Shanghai Ocean University, Shanghai, China

Winsor H. Watson III Department of Biological Sciences, University of New Hampshire, Durham, NH, USA

Yasmin Department of Ocean Studies and Marine Biology, Pondicherry University (Brookshabad Campus), Port Blair, Andaman & Nicobar Islands, India

Chester B. Zarnoch Baruch College, City University of New York, New York, NY, USA

Part I
Evolution, Adaptation, and Genetics

Chapter 1
The Oldest Species of the Genus *Limulus* from the Late Jurassic of Poland

Błażej Błażejowski

Abstract This paper describes the discovery of new, exceptionally well preserved Late Jurassic (Late Tithonian) horseshoe crabs belonging to the genus *Limulus*, which are considered to be the oldest known representatives of the genus. The three-dimensionally preserved fossil horseshoe crabs, belonging to two species: *Limulus darwini* and *Crenatolimulus* sp. nov., recently were discovered in the Owadów-Brzezinki Quarry in Central Poland. The state of preservation of the fossil horseshoe crabs is described and the evolutionary lineage of both extinct and extant limulids is discussed. The palaeoenvironment inhabited by the new species and the trophic relationships of living horseshoe crabs are investigated in an attempt to determine the potential relationships between Mesozoic and recent xiphosurans.

Keywords Limulus darwini • Jurassic • Mesozoic • Fossil • Stabilomorphism • Owadow-Brzezinki Quarry • Poland

1.1 Introduction

The recently discovered paleontological site of the Sławno limestones at the Owadów-Brzezinki Quarry in Central Poland (Fig. 1.1) has yielded exceptionally well-preserved fossils of Late Jurassic horseshoe crabs. The discovery of new, more or less three-dimensionally preserved Late Jurassic Xiphosurida arthropods adds significantly to our understanding of a group that has a stratigraphic range throughout almost the entire Phanerozoic period.

The uniqueness of these new finds lies in their very close stratigraphical and geological relationship to one of the most famous *Fossil-Lagerstätte* localities in the world – Solnhofen-Eichstätt and Nusplingen, in southern Germany, where the best-known genus of the Jurassic horseshoe crab, *Mesolimulus,* is found. Horseshoe

B. Błażejowski (✉)
Polish Academy of Sciences, Institute of Paleobiology,
Twarda 51/55, 00-818 Warsaw, Poland
e-mail: bblazej@twarda.pan.pl

R.H. Carmichael et al. (eds.), *Changing Global Perspectives on Horseshoe Crab Biology, Conservation and Management*, DOI 10.1007/978-3-319-19542-1_1

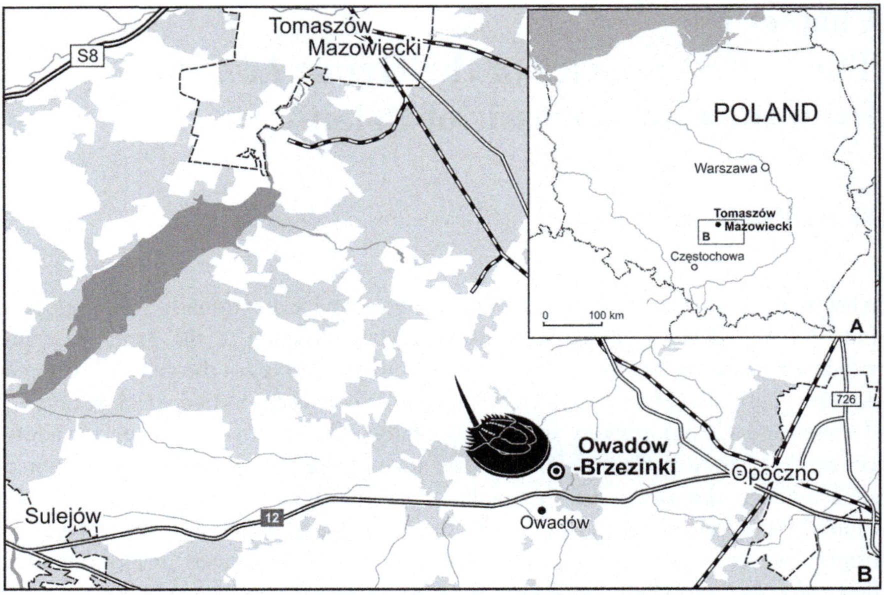

Fig. 1.1 Road map with the location of Owadów-Brzezinki Quarry and its proximity to Tomaszów Mazowiecki in Central Poland

crabs lived during the Late Jurassic at Nusplingen (Late Kimmeridgian, Beckeri Zone), Solnhofen-Eichstätt (Early Tithonian, Hybonotum Zone) and in Owadów-Brzezinki (Late Tithonian = Middle Volgian, Zarajskensis Subzone of the Scythicus Zone), under very similar environmental conditions. As such, it presents an unrivalled opportunity for studying their evolution and rate of speciation.

1.2 Geological and Paleontological Settings

The exposed carbonate sequence of the Sławno limestones belongs to the Kcynia Formation (Kutek 1994) and is divided into four successive units. In general, first three units (I, II and III) appear to represent a continuous transition from an offshore to nearshore, perhaps lagoonal settings, whereas unit IV shows evidence of a return to more open marine conditions. The Kcynia Formation is underlain by yellowish marls and marly clays of the Pałuki Formation.

The lowermost part of the unit III (Fig. 1.2) is very fossiliferous. Numerous specimens of horseshoe crabs were found in association with an enormously rich assemblage of the soft-shelled bivalves *Corbulomima* sp., remains of various fish and marine reptiles, rare ammonites, decapod crustaceans (Feldmann et al. 2015), land insects (dragonflies, beetles, grasshoppers) and pterosaurs (Kin and Błażejowski 2012; Kin et al. 2012, 2013). There are many similarities between the Owadów-

Fig. 1.2 Panoramic view of the highest level of exploitation in Owadów-Brzezinki Quarry, Poland (i.e. unit III and most fossiliferous '*Corbulomima* horizon' occurring in the middle of the quarry wall)

Brzezinki and Solnhofen-Nusplingen faunas, which allow for comparative paleontological studies at a previously unattainable level of taxonomic resolution. The Owadów-Brzezinki Quarry may be regarded as a new 'taphonomic window' into the living world of the latest Jurassic, being the first near-contemporary paleontological 'supplement' to a previously known *Fossil-Lagerstätten*.

1.3 Materials and Methods

The studied fossils of horseshoe crabs were collected by the author during field work in the Owadów-Brzezinki Quarry (lat. 51°37.4238′, lon. 20°13.6343′) in 2012–2014. The collected materials are housed at the Institute of Paleobiology, Polish Academy of Science in Warsaw (ZPAL X.1/O-B).

1.4 The Finds of Late Jurassic *Limulus* Fossils

After 6 years of field work, limestones of this unit III have yielded more than 150 well-preserved specimens of horseshoe crabs, including 11 almost completely articulated exoskeletons (Figs. 1.3 and 1.4). Tables of biometric measurements of selected specimens *Limulus darwini* are given in Kin and Błażejowski (2014). The preservation of ventral side in some specimens (Fig. 1.5) is one of the most promising aspects of the limuline material from Owadów-Brzezinki. This material is currently the subject of detailed study and will undoubtedly provide new information on the evolution of the horseshoe crabs.

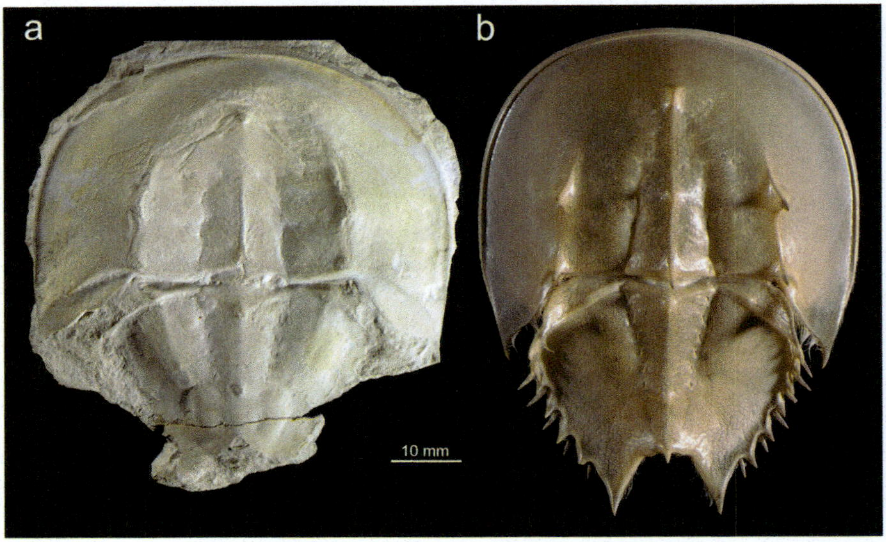

Fig. 1.3 (**a**) Late Jurassic horseshoe crab species *Limulus darwini*, Owadów-Brzezinki, Poland (holotype ZPAL X.1/O-B/XA 1, Kin and Błażejowski 2014), (**b**) juvenile specimen of the extant Atlantic horseshoe crab species *Limulus polyphemus*, USA (ZPAL X.1/JB-NY)

The ventral presentation of the fossil limuline provides an opportunity to study the appendages and mouthparts, and comparison of these extant forms will probably allow us to determine the evolutionary pathways of these features in far more detail than previously.

The vast majority of these finds are incomplete; they are preserved as isolated fragments of the prosoma and opisthosoma, and represent all four stages of disarticulation (Fig. 1.6) as observed by Babcock et al. (2000). Specimens of *L. darwini* (Kin and Błażejowski 2014) constitute the first Late Jurassic representatives of the order Xiphosurida to be recorded from Poland. Previously, the only known limulid was a single specimen of lower Pliensbachian *Limulitella* cf. *liasokeuperinus* (Braun 1860) from a brackish-marine stratum of the Gielniów Formation at Skarżysko-Kamienna, south-central Poland (Karaszewski 1975). This particular specimen was lost during World War II. Because extinct horseshoe crabs are extremely rare in the fossil record, it is virtually impossible to establish an exact phylogenetic relationship between extinct and extant families and genera.

After a detailed analysis of three-dimensionally preserved Late Jurassic limulids from Owadów-Brzezinki, Kin and Błażejowski (2014) showed that there are no significant morphological differences between specimens from Owadów-Brzezinki and extant juvenile individuals of the genus *Limulus* (Müller 1785). Clearly, modern horseshoe crabs of that genus are more closely related to Late Jurassic forms than previously thought (Riek and Gill 1971; Sekiguchi and Shuster 2009) (Fig. 1.3). Indeed, the morphological features of the opisthosoma of Late Jurassic *Limulus* from Poland are very similar to the holotype of *Limulus coffini* (Reeside and Harris

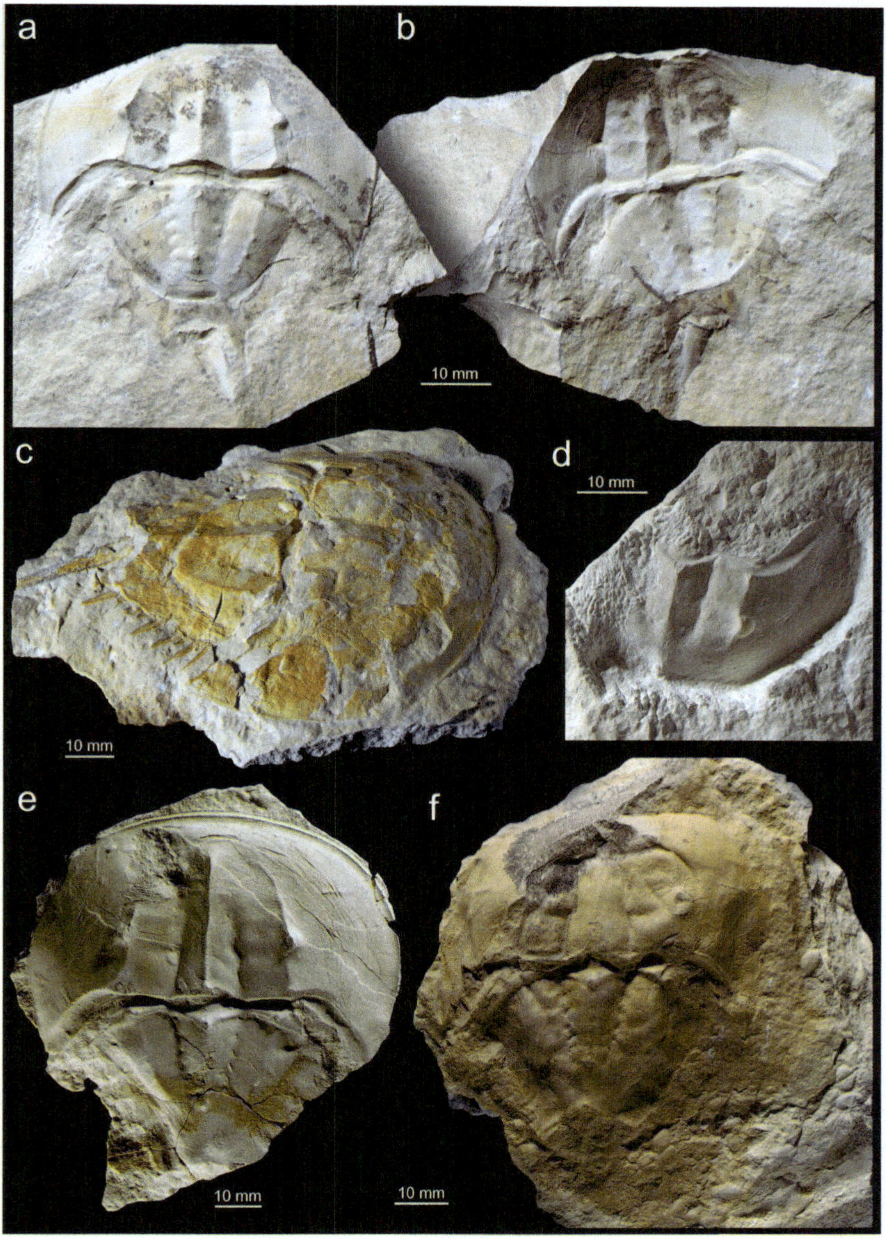

Fig. 1.4 Three-dimensionally preserved representatives of Late Jurassic horseshoe crabs from Owadów-Brzezinki, Poland. *Crenatolimulus* sp. nov. (**a**) Negative (external mold) and (**b**) positive (rock slab with imprint) (ZPAL X.1/O-B/14.1), (**c**) *Crenatolimulus* sp. nov. (ZPAL X.1/O-B/XA 13.B), (**d**) *Limulus darwini* prosoma with exceptionally well-preserved left compound eye (ZPAL X.1/O-B/14.2), (**e**) nearly complete exuvia of *Crenatolimulus* sp. nov. (ZPAL X.1/O-B/XAC 5n) (**f**) *Crenatolimulus* sp. nov. (ZPAL X.1/O-B/XAC 2), no telson preserved, exhibits depressed and beaded occipital band

Fig. 1.5 The ventral side of the late Jurassic horseshoe crab form Owadów-Brzezinki, Poland (ZPAL X.1/O-B/XA 14.4)

1952) from the Upper Cretaceous of Colorado. This clearly indicates a close relationship, while *L. darwini* is almost certainly an early link both with *L. coffini* and the extant *Limulus polyphemus* (Linnaeus 1758). Thus, the genus *Limulus* existed at least 148 million years ago and has survived to the present day in an almost unchanged form (Kin and Błażejowski 2014).

1.5 Comparison of Late Jurassic *Limulus* with Extinct and Extant Limulines

Much of the general paleontological background to the horseshoe crabs may be found in several earlier publications, especially in Anderson and Selden (1997), Anderson and Shuster (2003), and Rudkin and Young (2009), and numerous references cited therein.

The fossil record of the basic xiphosurid horseshoe crab body plan certainly extends back as far as the Ordovician, which indicates an origin which lies outside of the paraphyletic ‘synziphosurine’ (Rudkin and Young 2009). Thus Xiphosurans have existed for some 480 Mya (Lamsdell 2013), with the earliest unequivocal representatives found in the Upper Ordovician of Manitoba, Canada (Rudkin et al. 2008), followed by further Xiphosurida finds from the Lower Ordovician of Morocco (Van Roy et al. 2010). Fossil species from the earliest Palaeozoic

Fig. 1.6 Late Jurassic horseshoe crabs from Owadów-Brzezinki, Poland; different stages of disarticulation (ZPAL X.1/O-B/XAC 8.3)

(Cambro-Silurian) have a segmented opisthosoma characteristic of the order Synziphosurida, whereas, for the Xiphosurida, the midpart of the body was essentially a solid piece of fused segments.

Of all known extinct Mesozoic and Cenozoic horseshoe crabs (Størmer 1952; Kin and Błażejowski 2014) *Limulus darwini* seems to be almost identical with the previously mentioned Late Jurassic limuline from Germany (Hauschke and Wilde 1991) and unusually well preserved early Cretaceous limulines from the Glen Rose Formation, north-central Texas (Feldmann et al. 2011).

Certainly the new species of Late Jurassic horseshoe crabs presented here shows some features in common with representatives of the genus *Mesolimulus*, in particular with *M. walchi* (Desmarest 1822) from the late Kimmeridgian pelitic limestones of Brunn (Röper et al. 1996) and Nusplingen (Schweigert and Dietl 2002; Briggs et al. 2005), and the early Tithonian lithographic limestones of Solnhofen (Münster 1840; Frickhinger 1994). A narrower furrow along the inner margin of the genal angles on the prosoma and the presence of short articular processes with an absence of distinct dorsal furrows on the opisthosoma are the main features that distinguish *Limulus darwini* from *M. walchi* (Kin and Błażejowski 2014). However, the material from Germany gives no clues as to whether these differences are merely a function of the state of preservation of the significantly less compressed *L. darwini* specimens. In respect to some similarities to the extant genus *Tachypleus*, the Jurassic

Mesolimulus walchi is referenced by many researchers as *Mesotachypleus* (Anderson and Shuster 2003).

The most important morphological features that distinguish the Lower Cretaceous (Albian) *Crenatolimulus paluxyensis* (Feldman et al. 2011) from *Limulus darwini* (Kin and Błażejowski 2014) and all other post-Palaeozoic horseshoe crabs are the beaded rims of the posterior region on the prosoma and flanks with two prominent ridges defining a tripartite surface on the opisthosoma. The other exoskeleton features are generally similar to *L. darwini*, with the particular exception of the proportionally much longer genal spines on the prosoma and the clearly pronounced axial ridge on the opisthosoma (Kin and Błażejowski 2014). Significantly, as well as *L. darwini,* other specimens belonging to the genus *Crenatolimulus* sp. nov. (Figs. 1.4 and 1.6) have recently been found. These are the first records of the genus in Europe. They obviously represent a species new to science and are the subject of ongoing studies.

With respect to the abdominal morphology, it is important to compare *L. darwini* with another extinct horseshoe crab *Limulus coffini*, which has a three-dimensionally preserved complete opisthosoma. Both show similar morphological characteristics. Among the most important features are six fixed spines terminated posteriorly by two large marginal spines occurring on both sides of the relatively deep and wide posterior margin, six distinct apodomes on either side of the axis and large muscle scars separating anterior and posterior opisthosomal parts. The morphological features that distinguish *L. darwini* from *L. coffini* are the lack of a pronounced axial ridge, the occurrence of three (not two) tubercles along the axial ridge and slightly larger articular processes (Kin and Błażejowski 2014).

The separation of the larger anterior part and abdominal posterior part of the opisthosoma in *L. darwini* is similar in comparably sized individuals, i.e. juveniles of *L. polyphemus* (compare Babcock et al. 2000). Other similar morphological features include the median cardiac node and the shape of the occipital bands on the prosoma and the size and shape of the articular processes on the opisthosoma (Kin and Błażejowski 2014).

Unfortunately, there are no specimens comparable in size with mature individuals of *L. polyphemus* in the recent collection from Owadów-Brzezinki, and the absence of fully-grown exoskeletons known from the late Jurassic remains unexplained.

According to Barthel (1974), all specimens of *M. walchi* were presumably exoskeletons (molts) of only young animals. The finds of giant limulid trackways (i.e. *Kouphichnium lithographicum,* Oppel 1862) in the Upper Jurassic of Germany (Kolb 1963; Frickhinger 1994; Schweigert 1998) and France (Gaillard 2010), which are attributed to much larger individuals, seem to support an assumption that only juveniles can be preserved in the fossil state. This puzzling feature could possibly be explained by the fact that subadult horseshoe crabs molted less often than juveniles, as is the case in modern limulids (Barthel 1974; Carmichael et al. 2003; Chiu and Morton 2004). Shuster and Sekiguchi (2003) suggest that *Limulus polyphemus* takes 9–11 years and 19 or fewer molts to reach maturity. The shallow water in which the Owadów-Brzezinki sediments accumulated may thus be interpreted a "limulid nursery," where young individuals spend their first years rapidly increasing its size, prior to venturing out into deeper waters (Carmichael et al. 2003; Chiu and Morton 2004).

The poor fossil record is, of course, the main obstacle for clear understanding of how exactly the living horseshoe crabs derived from Mesozoic lineages (Fisher 1982, 1984). Molecular studies on extant population (Avise et al. 1994; Xia 2000) have provided some clues to their genealogical relationships, but many unsolved questions still concern their patchy distribution.

1.6 Summary and Conclusions

Because Xiphosurida arthropods are extremely rare in the fossil record, the recent finds have an exceptional importance. It is, indeed, particularly fortunate that the horseshoe crabs are found in such an excellent state of preservation. Furthermore, *L. darwini* from Owadów-Brzezinki is regarded as the oldest known representative of the genus (Kin and Błażejowski 2014). Possibly, these new discoveries have only a limited bearing on genealogical relationships within the group, but they certainly throw a great deal of new light on several aspects of evolution and adaptation processes. Evidently, this extraordinary group of organisms has survived more than one major extinction event and global biotic crises (Raup and Sepkoski 1982) in the history of our planet, including P/Tr (Permian–Triassic) when trilobites became extinct, and K/T (Cretaceous–Tertiary) which 65 million years ago turned dinosaurs and ammonites into museum exhibits. All horseshoe crabs found in Owadów-Brzezinki were located in strata with abundant soft-shelled bivalves, within so-called '*Corbulomima* horizon', what may be related to their potential feeding habits. The diet of modern Atlantic *Limulus polyphemus* is highly diverse and consists of variety of small marine fauna including bivalves (Smith 1953; Botton 1984). According to Botton and Ropes (1989), bivalves of the genus *Corbula* form part of the diet of *L. polyphemus*, and the prolific *Corbulomima* sp. at Owadów-Brzezinki acted as possible food attractor for *L. darwini.* Thus the depositional area of '*Corbulomima* horizon' could have been both "nursery" and feeding zone for *L. darwini*.

Kin and Błażejowski (2014) suggested that the relative evolutionary conservatism noticeable in *Limulus* (and, presumably, all known representatives of Xiphosurida), may be related to a unique formula of adaptative strategy. In order to systematize this phenomenon, Kin and Błażejowski (2014) introduced the term 'stabilomorphism', which is *a relative morphological stability of organisms in time and spatial distribution, the taxonomic status of which does not exceed genus level.* The definition refers exclusively to genera that have survived at least one of the great mass extinctions. By this definition, the resulting morphological structure of the presumed stabilomorph must have been virtually unchanged for more than 65 million years (e.g. *Limulus*). A high effectiveness of adaptation significantly reduces the need for differentiated phenotypic variants in response to environmental changes and provides for long-term evolutionary success (Kin and Błażejowski 2014).

Acknowledgements I would like to sincerely thank Peter Walsh (University of Silesia) for many useful suggestions and correction of English. I appreciate the help provided in the field of Robert Siuda (Owadów-Brzezinki Quarry, Nordkalk Company). And last but not least, I wish to acknowledge Kevin Laurie (independent horseshoe crab researcher, Hong Kong), Mark Botton (Department of Natural Sciences, Fordham College at Lincoln Center, New York) and one anonymous reviewer for his critical review and very helpful comments that improved the manuscript. This work was supported by the Polish National Science Centre (grant no.2012/07/B/ST10/04175).

References

Anderson LI, Selden PA (1997) Opisthosomal fusion and phylogeny of Palaeozoic Xiphosura. Lethaia 30:19–31

Anderson LI, Shuster CN Jr (2003) Throughout geologic time: where have they lived? In: Shuster CN, Barlow RB, Brockman HJ (eds) The American horseshoe crab. Harvard University Press, Cambridge, MA, pp 189–224

Avise JC, Nelson WS, Sugita H (1994) A speciational history of "living fossils": molecular evolutionary patterns in horseshoe crabs. Evolution 48:1986–2001

Babcock LE, Merriam DF, West RR (2000) *Paleolimulus*, an early limuline (Xiphosurida), from Pennsylvanian—Permian Lagerstatten of Kansas and taphonomic comparison with modern *Limulus*. Lethaia 33:129–141

Barthel KW (1974) *Limulus*: a living fossil. Horseshoe crabs and interpretation of an Upper Jurassic environment (Solnhofen). Nature 61(10):428–433

Botton ML (1984) Importance of predation by horseshoe crabs, *Limulus polyphemus*, to an intertidal sand flat community. J Mar Res 42:139–161

Botton ML, Ropes JW (1989) Feeding ecology of horseshoe crabs on the continental shelf, New Jersey to North Carolina. Bull Mar Sci 45:637–647

Braun CF (1860) Die Thiere in den Pflanzenschiefern der Gegend von Bayreuth. 1771, Zool. Fundamenta

Briggs DEG, Moore RA, Shultz JW et al (2005) Mineralization of soft–part anatomy and invading microbes in the horseshoe crab *Mesolimulus* from the Upper Jurassic Lagerstatte of Nusplingen, Germany. Proc R Soc B Biol Sci 272:627–632

Carmichael RH, Rutecki D, Valiela I (2003) Abundance and population structure of the Atlantic horseshoe crab *Limulus polyphemus* in Pleasant Bay, Cape Cod. Mar Ecol Prog Ser 246:225–239

Chiu HMC, Morton B (2004) The behaviour of juvenile horseshoe crabs, *Tachypleus tridentatus* (Xiphosura), on a nursery beach at Shui Hau Wan, Hong Kong. Hydrobiologia 523(1–3):29–35

Desmarest AG (1822) Les Crustacés prominent dits. In: Brongniart A, Desmarest AG (eds) Historie naturelle des crustacés, sous les rapports zoologiques. Chez F.G. Levrault Libraire, Strasbourg, pp 67–142

Feldmann RM, Schweitzer CE, Dattilo B et al (2011) Remarkable preservation of a new genus and species of limuline horseshoe crab from the Cretaceous of Texas, USA. Palaeontology 54(6):1337–1346

Feldmann RM, Schweitzer CE, Błażejowski B (2015) A new species of lobster (Glypheoidea: Mecochiridae) from the Late Jurassic (late Tithonian) Lagerstätte from central Poland. N Jahrb Geol Paläont Abh 275:107–114

Fisher DC (1982) Phylogenetic and macroevolutionary patterns within the Xiphosurida. Proc N Am Paleontol Convent 1:175–180

Fisher D (1984) The Xiphosurida: archetypes of bradytely? In: Eldredge N, Stanley SM (eds) Living fossils. Springer, New York, pp 196–213

Frickhinger KA (1994) The fossils of Solnhofen. Korb Goldschneck-Verlag, Ostfildern, pp 1–336
Gaillard C (2010) A giant limulid trackway (*Kouphichnium lithographicum*) from the lithographic limestones of Cerin (Late Kimmeridgian, France): ethological and environmental implications. Swiss J Geosci 103:1–16
Hauschke N, Wilde V (1991) Zur Verbreitung und Okologiemesozoischer Limuliden. N Jahrb Geol Paläon Abh 183:391–411
Karaszewski W (1975) Obecność rzadkiej skamieniałości–limulusa w liasie świętokrzyskim. Prz Geol 23(10):500–501. [in Polish]
Kin A, Błażejowski B (2012) Polskie Solnhofen. Prz Geol 60(7):375–379. [in Polish]
Kin A, Błażejowski B (2014) The horseshoe crab of the genus *Limulus*: living fossil or stabilomorph? PLoS One 9(10):e108036. doi:10.1371/journal.pone.0108036
Kin A, Błażejowski B, Binkowski M (2012) 'Polish Solnhofen': a long-awaited alternative? Geol Today 28(3):91–94
Kin A, Gruszczyński M, Martill D et al (2013) The paleoenvironment and taphonomy of a Late Jurassic (Late Tithonian) Lagerstädtte from central Poland. Lethaia 46:71–81
Kolb A (1963) Riesige *Limulus*-Fährte aus den lithographischen Schiefern bei Pfalzpaint. Geol Blätter Nordost-Bayern 13:73–78
Kutek J (1994) The *Scythicus* zone (Middle Volgian) in Poland: its ammonites and biostratigraphic subdivisions. Acta Geol Pol 44:1–33
Lamsdell JC (2013) Revised systematics of Palaeozoic 'horseshoe crabs' and the myth of monophyletic Xiphosura. Zool J Linn Soc 167:1–27
Linnaeus C (1758) Systema naturae per regna tria naturae, secundum classes, ordines, genera, species, cum characteribus, differentiis, synonymis, locis, vol 1, Regnum animale. Editio decima, reformata. Laurentii Salvii, Stockholm
Müller OF (1785) Entomostraca seu Insecta Testacea. Lipsiae et Hafniae, pp 1–135
Münster G (1840) Über die fossilen Arten *Limulus* in den lithographischen Schiefern von Bayern. In: Münster G (ed) Beiträge zur Petrefacten-Kunde 3:26–27
Oppel A (1862) Ueber Fährten lithographischen Schiefer (*Ichnites lithographicus*). Paläeontologische Mitteilungen aus dem Museum des Koeniglich Bayerischen Staates 1, pp 121–125
Raup DM, Sepkoski JJ Jr (1982) Mass extinctions in the marine fossil record. Science 215:1501–1503
Reeside JB Jr, Harris DV (1952) A Cretaceous horseshoe crab from Colorado. J Wash Acad Sci 42(6):174–178
Riek EF, Gill ED (1971) A new xiphosuran genus from lower cretaceous freshwater sediments at Koonwarra, Victoria, Australia. Paleontology 14(2):206–210
Röper M, Rothgaenger M, Rothgaenger K (1996) Die Plattenkalke von Brunn (Landkreis Regensburg). Eichendorf Verlag, Eichendorf, pp 1–102
Rudkin DM, Young GA (2009) Horseshoe crabs – an ancient ancestry revealed. In: Tanacredi JT, Botton M, Smith DR (eds) Biology and conservation of horseshoe crabs. Springer, New York, pp 25–44
Rudkin DM, Young GA, Nowlan GS (2008) The oldest horseshoe crab: a new xiphosurid from late Ordovician konservat-lagerstätten deposits, Manitoba, Canada. Palaeontology 51:1–9
Schweigert G (1998) Die Spurenfauna des Nusplinger Plattenkalks (Oberjura, Schwäbische Alb). Stuttg Beitr Nat B 262:1–47
Schweigert G, Dietl G (2002) Miscellanea aus dem Nusplinger Plattenkalk (Ober-Kimmeridgian, Schwäbische Alb), 4. Limuliden Jber Mitt Oberrhein Geol Ver NF 84:323–331
Sekiguchi K, Shuster CN Jr (2009) Limits on the global distribution of horseshoe crab (Limulacea): lessons learned from two lifetimes of observations: Asia and America. In: Tanacredi JT, Botton ML, Smith DR (eds) Biology and conservation of horseshoe crabs. Springer, New York, pp 5–24

Shuster CN Jr, Sekiguchi K (2003) Growing up takes about ten years and eighteen stages. In: Shuster CN Jr, Barlow RB, Brockmann HJ (eds) The American horseshoe crab. Harvard University Press, Cambridge, MA, pp 103–133

Smith OR (1953) Notes on the ability of the horseshoe crab *Limulus polyphemus* to locate soft-shell clams *Mya arenaria*. Ecology 34:636–637

Størmer L (1952) Phylogeny and taxonomy of fossil horseshoe crabs. J Paleontol 26:630–640

Van Roy P, Orr PJ, Botting JP et al (2010) Ordovician faunas of Burgess shale type. Nature 465:215–218

Xia X (2000) Phylogenetic relationship among horseshoe crab species: effect of substitution models on phylogenetic analyses. Syst Biol 49:87–100

Chapter 2
The Delaware Bay Area, U.S.A.: A Unique Habitat of the American Horseshoe Crab, *Limulus polyphemus*

Carl N. Shuster Jr.

Abstract Continuing the theme developed by Shuster and Botton (Estuaries 8:363–372, 1985) on the dynamics of the Delaware Bay population of *Limulus polyphemus* and by Sekiguchi and Shuster (Limits on the global distribution of horseshoe crabs (Limulacea): lessons learned from two lifetimes of observations: Asia and America. In: Tanacredi JT, Botton ML, Smith DR (eds) Biology and conservation of horseshoe crabs. Springer, New York, pp 5–24, 2009) on global limits on the distribution of horseshoe crabs and by Shuster and Sekiguchi (Basic habitat requirements of the extant species of horseshoe crabs (Limulidae). In: Tanacredi JT, Botton ML, Smith DR (eds) Biology and conservation of horseshoe crabs. Springer, New York, pp 115–129, 2009) on local habitat requirements, this paper considers why the Delaware Bay is the epicenter of *Limulus polyphemus* abundance. The emphasis is on how its geologic and geographic characteristics are advantageous to *Limulus*. Although the species ranges along the east coast of North America from Yucatan, Mexico to Maine, USA (21°N to 44.5°N), only at Delaware Bay (38°40′ N to 39°20′ N) has it produced a population of millions of adult horseshoe crabs, unmatched anywhere. Why? It appears most likely that it was the interaction between the ecological generalist, *Limulus polyphemus*, and the exceptionally favorable environment that developed in the Delaware Bay area after the past great ice age. However, this productive relationship is short-lived. Chronologically the bay will be shorter-live with a probable existence of some 85,000 years, while *Limulus polyphemus* may have already existed some 135 million years.

Keywords *Limulus polyphemus* • Delaware Bay • Horseshoe crab epicenter • Multiple spawning beaches

---

C.N. Shuster Jr. (✉)
Biological Oceanographer (retired), 3733 North 24th Street, Arlington, VA 22207-5011, USA
e-mail: carlshuster@verizon.net

R.H. Carmichael et al. (eds.), *Changing Global Perspectives on Horseshoe Crab Biology, Conservation and Management*, DOI 10.1007/978-3-319-19542-1_2

2.1 Uniqueness of Habitats and Horseshoe Crab Populations

If there ever was another place where a species of horseshoe crabs existed so prominently, it must have been comparable to Delaware Bay and the adjoining continental shelf. Although there is a wide range in the uniqueness of its habitats and of its adaptability, the American species *Limulus polyphemus*, of all the extant and probably many of the extinct species, has developed the largest numbers in any population at Delaware Bay. Why? This paper offers one interpretation.

Shuster (1955, 1979) interpreted differences in temperature tolerance at two widely separated populations (Florida and Woods Hole; Mayer 1914, later confirmed by Fraenkel 1960), and morphometric data collected on 16 populations of *Limulus polyphemus*, from Florida to Maine during surveys in 1953 and 1954, as evidence of physiological races. Since then several studies, e.g., by Riska (1981) and by King et al. (2005), reached similar conclusions regarding differences between populations. Therefore, a reasonable conclusion is that all populations of horseshoe crabs are unique, as seen in variations in their habitat and/or in their anatomy, genetics, and behavior. However, regardless of the four extant species considered, some horseshoe crab populations are more unique than others; among these are:

***Tachypleus tridentatus* in the Seto Inland Sea, Japan** Once numerous (Sekiguchi 1988; Shuster and Sekiguchi 2009), this species may have approached the great numbers of *Limulus* seen in Delaware Bay. However, early in the twentieth century the reclamation of land by dikes and polders eliminated most spawning areas in the Seto Sea.

Southern-most Populations, Yucatan, Mexico Several populations of *Limulus* inhabit lagoons around the Yucatan Peninsula, from Campeche Sound on the west around to Bahia de la Ascension on the east coast (Zaldivar-Rae et al. 2009). The geology of the Gulf of Mexico must have been responsible for the ultimate isolation of these populations from all others for a long time – evidenced by tissue samples from northeastern Yucatan crabs, which are highly genetically differentiated from *Limulus* populations in the United States, suggesting a taxonomic revision of the genus may be due (King et al. 2005).

Indian River Lagoon (near Cape Canaveral), Florida, U.S.A. This, a shallow water, hyper-saline habitat lacks significant tides. As a result the reproductive ecology and early life history of *Limulus* differ markedly from populations inhabiting tidal areas with the usual estuarine salinities (Ehlinger and Tankersly 2009).

A Long-Spined Variant, Georgia, U.S.A. The presence of a prominent extra spine on the opercular ridges during the mid-juvenile stages of *Limulus* in the coastal waterway behind the barrier islands of Georgia, first reported by Say (1818), was rediscovered in 2005 at the aquarium maintained at the Oceanographic Institute, Skidaway, Georgia. All of the dorsal spines are greatly enlarged during development of the juveniles as well as smaller spines arising from the nodules on the opercular ridge (Fig. 2.1). This characteristic is being studied (H.J. Brockman,

Fig. 2.1 An exuviae of a juvenile *Limulus polyphemus australis* (Say 1818), prosomal width 9 cm (Specimen courtesy of Dodie Sanders, University of Georgia Aquarium; photograph by Shuster; the coin is 23 mm in diameter)

personal communication). Actually, of all the dorsal spines, only the spine midway along each opercular ridge is the one described by Say (1818).

Northernmost Populations, Maine, U.S.A. In virtually every measurement of quantity, these populations are among the smallest. Within Taunton Bay, lesser bays, Egypt Bay and Hog Bay, with home ranges of 64.1 ha and 61.4 ha, respectively, support discreet populations where the adult mean prosomal width is 15.4 cm in the males and 18.0 cm in the females (Moore and Perrin 2007). The numbers of *Limulus* spawning at Hog Bay, from 2001 to 2005, averaged 1,082 per year, with a range of 741–1,525 individuals. Lack of any populations to the north of Taunton Bay may be due to the cold, more-northern climate that precluded the establishment of any reproducing *Limulus* population (Moore and Perrin 2007).

Restricted distributions also exist at Great Bay, New Hampshire (Schaller et al. 2010) and Pleasant Bay, Massachusetts (James-Pirri 2010). These populations apparently rarely utilize resources on the continental shelf, acting as if they were land-locked, except for such influences as those of salinity and the tidal amplitudes. It also appears that it is the stress of the seasonal length of cold weather that keeps population numbers low. Otherwise their local movements are similar to those of other horseshoe crabs except on a much smaller scale. Within their bays they exhibit the annual migration from deeper waters to the shallow waters and spawning beaches, and then back to deeper waters during over-wintering. They differ from nearby populations, at Plum Island Sound, Massachusetts (Baptist et al. 1957) and at Cape Cod Bay (Shuster 1950). The habitats at these locations are more open to deeper waters, and the crabs moved freely from them over relatively long areas. Yet, the dimensions of the annual migrations of the lesser populations and the closeness and sufficient abundance of all the usual resources demonstrate that reproducing populations of *Limulus* can exist within relatively small reaches of habitats essential in the life cycle of the species. However, the northern-most crabs are probably living at the edge of extinction. It is the combination of the several stresses that puts these populations at a greater local risk (Botton et al. 2010). Otherwise, even a spawning site that is only a small patch of sand might be the start of a population (see Colorplate 14 in Shuster et al. 2003).

2.2 A Changing Coastline

All estuaries are ephemeral geological events, rare in the history of the earth. (Amos 1979)

The coastline in the vicinity of the present Delaware Bay has been ever-changing and so has the ecology of the area. For example, before the last ice age, ocean waters covered most of the area of present-day Delmarva Peninsula as evidenced by oyster beds that were buried there some 35,000–50,000 years ago by sediments eroded from the uplands that were transported seaward (Fig. 2.2). This is only one of many fossil-bearing geologic formations in the mid-Atlantic coastal areas, e.g., at Calvert Cliffs on Chesapeake Bay or that revealed by the excavation of the Delaware Chesapeake Canal (Shuster 1960a).

Fast-forward to some 20,000 years ago when the earth's climate began an unusually long interglacial interval that is now more than half over (Pielou 1991). In 13,000–11,000 years ago the mass of ice that covered the northeastern portion of North America as far south as Long Island and the northern part of

Fig. 2.2 Excavation of a reservoir and drainage ditches near Laurel, Delaware, revealed that ancient oyster shells were the most abundant fossils in a layer of blue clay that varied from 1 to 2 ft at a depth of 8 ft below the surface and just above the reservoir water level. At *left*, Henry Hutchinson lowers a black bucket to Shuster to collect shell samples; on *right*, a portion of the oyster-laden blue clan layer (Shuster 1960a, photographs by Dr. Frank Daiber)

New Jersey began to melt and recede northward. At that time the edge of the continent and the Atlantic Ocean met at the edge of the present continental shelf some 175 km due east of Delaware. Sea level rose rapidly during the initial melting of the ice shield and the ocean moved closer to Delaware. By 7,000–5,000 years ago a long and narrow embayment, the proto-Delaware Bay, was approaching a recognizable shape (Custer 1984). John Kraft and colleagues at the University of Delaware have studied the history of development of the bay including coastline partitioning and physical and chemical attributes that further defined the estuarine nature of the bay (e.g., Chrzastowski 1986; Knebel et al. 1988; Fletcher et al. 1992). If other estuaries were developing at about the same rate along the Atlantic coast, as was the Georgia coast some 5,000 years ago (Henry 2009), then at least most of the coastal area to the south was also nearing environmental conditions in which *Limulus* could exist. Then, sea level rise escalated 2,000 years ago and has continued (Fletcher et al. 1993; John and Pizzuto 1995).

2.2.1 Dimensions and Related Features of Delaware Bay

Certain features of the physical configuration of the bay (Shuster 1959) are important in the ecology of *Limulus* (Table 2.1, see also Fig. 2.3). The bay occupies a relatively large area and has a quantity of habitats favorable to horseshoe crabs, including: (a) extensive tidal marshes that contribute to the base of the food web, (b) many lengthy sandy beaches that support large spawning populations and the incubation of the eggs, (c) shallow-water areas that provide ample nursery areas for the early life stages, (d) prey species are abundant, and (e) the NNW-SSE axis of the channel and the southerly flow of the Delaware River were probably other formative factors. When the interaction of the river and the rising ocean formed the bay, the result was a southern entrance, protected by widely separated capes that provide a less turbulent access to and egress from the bay depending upon the force from westerly or easterly winds.

In essence, tidal streams, from some eight drainage basins along the Delaware Bay shore (Tiner et al. 2011: New Jersey has a similar geography) break up the shoreline into long sandy stretches. This might divide and double the beaches between the major streams of each basin in Delaware into sub-beaches if different chemical and physical changes occurred at each end (Fig. 2.4). This creates an interesting pattern with the north and south extremes of each beach possibly more influenced by the adjacent stream. If there are effluents in such a stream or the streams and the along-shore currents create a topographical difference in the sub-beaches would these sub-beaches be equally noticed by *Limulus* as a local homing signal? Regardless, these streams might isolate a discrete spawning population at each sub-beach. This and the tendency of the crabs to spawn at one beach, that between two tributaries, has been the basis for beach-by-beach studies to obtain: (a) the total spawning effort of the bay (Finn et al. 1991), (b) its

Table 2.1 Dimensions of Delaware Bay

[a]**Estuarine wetlands**
Delaware = 308 km^2
[b]New Jersey = 210 km^2
[c]**Dimensions of the Bay**
Surface area: 1,862 km^2
Max. effective length = 75 km
Max. effective width = 25 km
Max. depth = 46 m (mean depth 10 m)
About 83 % of the bay area is less than 9 m in depth
[c]**Shoreline**
Beach shoreline: DE = 68 km, NJ = 62 km
Mainly sandy: DE = 89 km, NJ = 100 km
[d]**Suitability of shoreline for horseshoe crabs**
(Total shoreline: DE = 91 km, NJ = 148 km)
Optimal habitat: DE = 31 km, NJ = 26 km
Suitable habitat: DE = 11 km, NJ = 5 km
Less suitable: DE = 29 km, NJ = 49 km
Unsuitable: DE = 20 km, 67 km

From [a]Tiner et al. (2011)
[b]Chan and Schulte (2008) – Manomet Center for Conservation Sciences. See also Fig. 2.3
[c]Shuster (1959)
[d]Lathrop et al. (2006)

spawning index (Smith et al. 2002a, b), (c) egg densities (Pooler et al. 2003), or (d) the number of the nests of eggs in the beaches (Weber and Carter 2009) of at least several beaches to establish either a trend or an overview of the activity.

An atlas (NOAA 1980) depicts, on the small scale of about 37.5 mm = 100 statute miles, 13 maps on the physical environments of the Atlantic coast from Florida to Maine showing, on separate maps, that the bay area has: (a) an entrance about 80 miles from the edge of the continental shelf, (b) sedimentary rocks of the Quaternary, (c) mudflats and marshes, (d) a southerly surface drift of ocean currents in January and north and south during July, (e) sea surface temperature in February at 5.0 °C and 22.5 °C in July, (f) prevailing winds in January to the SE and those in July to the NE, and (g) the probability that hurricane force winds of >73 mph will occur 0–4 % a year. The shores of Delaware Bay are lined with non-forested wetlands.

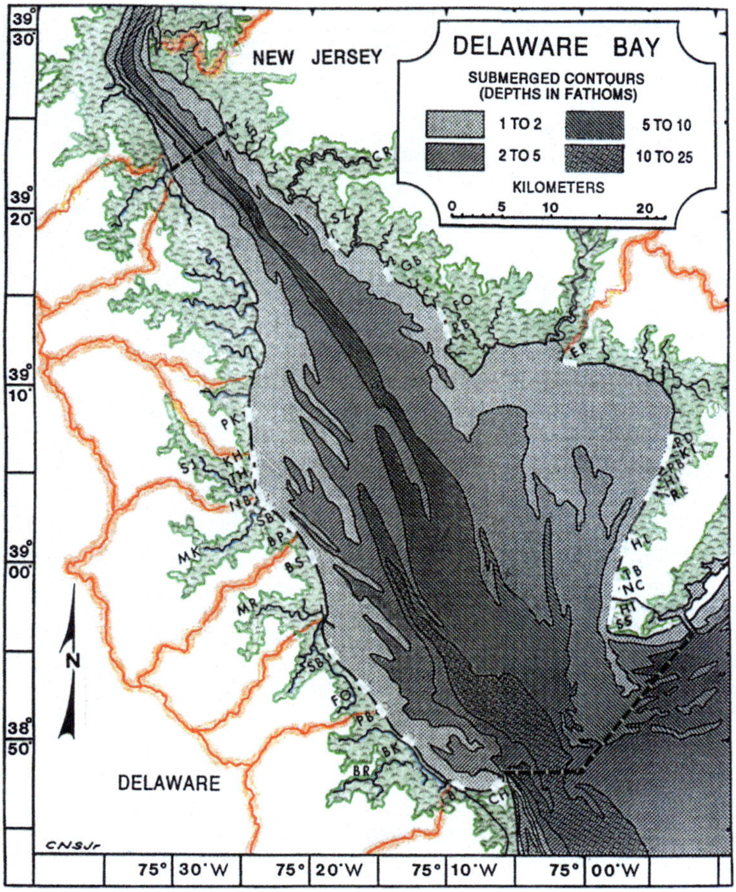

Fig. 2.3 Areas along Delaware Bay contributing to the shoreline habitats where *Limulus* is generally found: Submerged contours of the bay (After Shuster 1959); outlines of the watersheds (*inland boundary in orange*), marshes (*inland boundary in green*), and tidal streams (*in undulating black lines*) have been added. The data summarized in Table 2.1 are for the bay area between the heavy *dashed lines*. **Delaware** (DNREC 2001): Rivers – Little (LC), St. Jones (SJ), Murderkill (MK), Mispillion (MR), and Broadkill (MR). **New Jersey**: the state has five watershed regions in its Watershed Management Program: the Cohansey (CR) and Maurice (MA) are within the Lower Delaware Region and Cape May County is within the Atlantic Coastal Region (Shinn 2000). The following beaches (those indicated by the *white rectangles* on the shoreline are potential annual sampling sites for the U.S. Geological Index of Spawning Activity (ISA) – Smith et al. 2002a, b). **Delaware**: Pickering (PK), Kitts Hummock (KH), Ted Harvey (TH), No. Bowers (BN), So. Bowers (BS), Bennetts Pier (PB), Big Stone (BG), Slaughter (SB), Fowlers (FO), Primehook (PB) Broadkill (BK), Lewes (LB), and Cape Henlopen (CH). **New Jersey**: Sea Breeze (SZ), Gandys (GB), Fortesque (FO), Raybins (RB), East Point (EP), Reeds (RD), Kimbles (KI), Pierces Point (PB), Highs B (HI), South Cape Shore (RL = Rutgers Lab), Norburys Landing (NL), Town Bank (TB), North Cape May (NC), Higbee (HI), and Sunset (SS)

Fig. 2.4 A diagrammatic representation of a portion of the western shore of Delaware Bay (*wave symbol*) showing four drainage basins. Each basin, in sequence from the bay inland consists of sandy beaches (the *white stripes* along the bay), tidal streams (*black*), tidal marshes (shown by *clumps of marsh grass*), the four drainage basins (one *white* and three differently *shaded*), and beyond the upland (*tree symbols*). This depiction is not to scale, north is to the *right*

2.2.2 Significance of the Orientation of Delaware Bay

Due to the southerly axis of Delaware Bay and its shores, prevailing storms usually affect one shore more than the other shore that was more shielded from the winds and higher waves. Whether *Limulus* moves to the protected shore under such circumstances, Cook (1857) noted that when the crabs were scarce along the Cape May shores they were correspondingly abundant on Delaware shores. Further, it appears that other features may facilitate the coming and going of the "goes with the flow *Limulus*": (a) the flooding tides on the New Jersey side arrive sooner and stronger than on the Delaware side (NOAA 1987), (b) the wide entrance to the bay, and (c) the along-shore currents. Crabs tagged to the north enter the bay on the New Jersey side usually arrive earlier on Cape May shores, than those tagged south of Delaware Bay that tend to enter the bay on the Delaware side (Swan 2005). The benthic topography of the bay, with several channels from the shores heading toward the main channel, also may serve in orienting the movements of the crabs. Horseshoe crabs tend to move in the direction of benthic currents (observed by author in the clear waters of Florida and Massachusetts during distribution surveys during the summers of 1949–1950 and 1943–1944). This directional movement was first confirmed by Rudloe and Herrnkind (1976) who released float-tagged adults near breeding beaches. The tagged crabs moved most closely in the direction of the wave surge; even in the absence of a surge they usually moved in relatively straight paths. Similar results were obtained during tests in a large wave-controlled tank (Rudloe and Herrnkind 1980). Moving with the flow was also demonstrated in test runs using large juvenile crabs in a flume (Luckenbach and Shuster 1997) and in observing adults using optics on a benthic sled (Michael Oates, personal communication).

2.2.3 *Considering Climate/Weather*

Apparently Xiphosura have mostly inhabited warm-water habitats, tropical and temperate, based on two ancient populations (Barthel et al. 1990; Selden and Nudds 2004) and the present geographic ranges of the extant species (Sekiguchi and Shuster 2009). Two species of Xiphosura existed in the fauna of Mazon Creek, Illinois (ca. 323 million years ago (mya); Shabica and Hay 1997) and one, *Mesolimulus walchi* (see Fig. 2.8) at Solnhofen, Bavaria, Germany (ca. 150 mya; Barthel et al. 1990).

The three Indo-Pacific species straddle the equator but are limited in their northward range (Yamasaki et al. 1988; Sekiguchi and Shuster 2009). During the last ice age it seems likely that *Limulus* probably existed only in the Gulf of Mexico area and the eastern coast from Florida to Georgia, in that they served as warm water refuges (deduced from the study by Saunders et al. 1986).

At Delaware Bay, data on the hydroclimate has been accumulating ever since 1927 when the Rutgers oyster research laboratory was established on Cape May, New Jersey. Five years of that data from oyster growing area at Maurice River Cove, a midway area of the bay on the New Jersey side, were selected to construct hydroclimagraphs (Shuster 1960b). These also serve to illustrate a short-term trend in variations in the hydroclimate of horseshoe crabs at mid-bay (Fig. 2.5).

2.2.4 *Increasing Beach Erosion*

Over the years, after the author's first trip to Delaware Bay in 1949 to study *Limulus*, there has been a noticeable sea level rise and significant erosion of the shoreline with several beaches having been swept clean of the sandy deposits by high storm tides, as during hurricane Sandy in October 2010. Whatever the causative factor, erosion of a beach exposed previously covered tidal marshes and usually moved the sand landward covering/eliminating more tidal marshland, e.g., at a severely eroded stretch of beach (Fig. 2.6).

Notable changes, especially during heavy storms, are usually associated with certain locations. Other types of change also occur. There has been an increasing

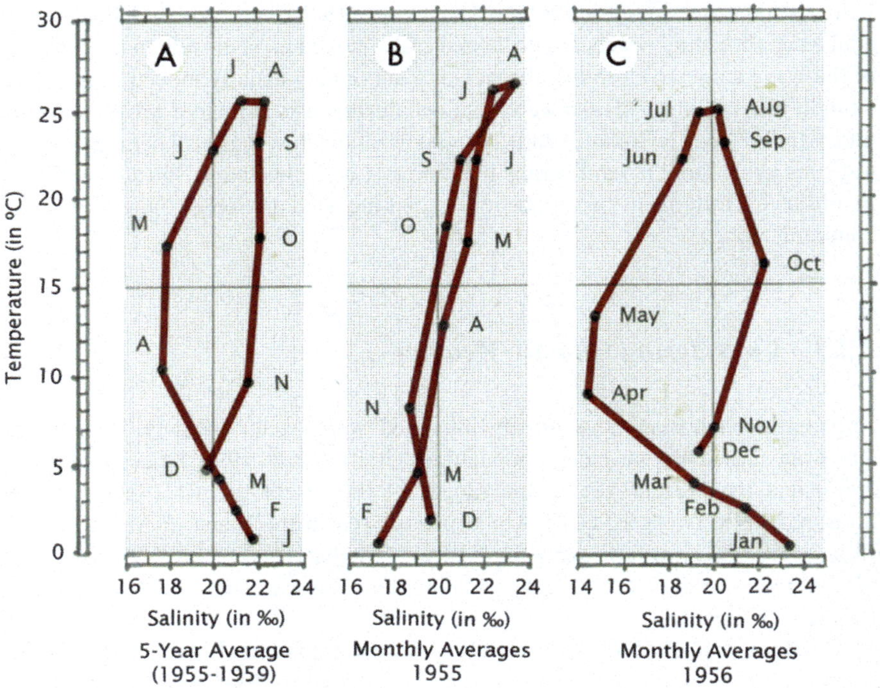

Fig. 2.5 Although water temperatures and salinities are plotted together, they are independent variables, with changes in one not affecting the other. Nevertheless, during the spring and fall seasons, the salinity does seem to vary according to temperature. This variation is due to the difference in evaporation and precipitation as well as to the relative amounts of freshwater and seawater in the bay and to their temperatures (Shuster 1960b). Extremes in salinity, narrow in 1955 (**b**) and wide in 1956 (**c**), are "hidden" in the 5-year average of 1955–1959 (**a**)

accumulation of soupy clay-silt in a relatively quiet water area at least since 1961 (personal observation) off the northern portion of Slaughter Beach (also known as Cedar Beach) where a back-eddy has been created by a mile-long jetty at Mispilion River Inlet. Excessive die-offs of *Limulus* frequently occur in this area. One instance was after a mass tagging effort (Swan 2005); another was a summer die-off in June 1979 (cause unknown, Shuster and Botton 1985). In 2003 (personal observation) finely ground detritus mixed with the silty-clay created an adverse habitat during the spawning season for the crabs (Fig. 2.7).

2.2.5 *Defining Environmental Spawning Parameters*

After World War II more attention was devoted to two major coastal resources of Delaware Bay, at a time when impacts upon beaches and spawning populations of horseshoe crabs were becoming more obvious. Action revved up in 1992 when the spawning population decreased to a level probably at least 40 % less than in the

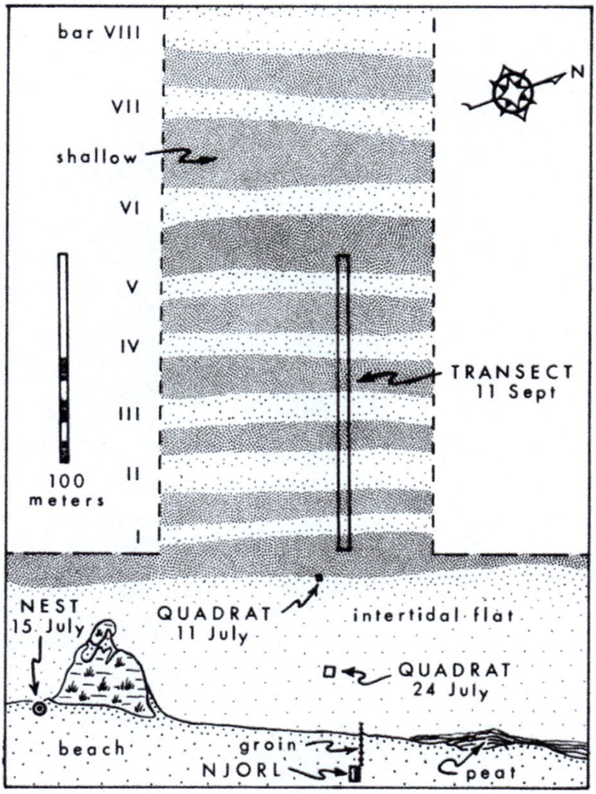

Fig. 2.6 Exposure of intertidal flats at Cape May, New Jersey on 11 September 1951 in the vicinity of the Rutgers oyster research lab (NJORL), showing several collection sites of early life stages of *Limulus* (nest, quadrats, and transect). The marshy nub extending from the beach developed on a discarded pile of oyster shells (Shuster 1979). In 2014 the entire beach area delineated at the *bottom* of the picture is now flooded at high tide

previous year. Decreases in the annual stop-over of migratory shorebirds that fed on *Limulus* eggs soon followed. The effect was most intense in the migratory Red Knot, and this sparked a marked increase in the research and budget considerations at Delaware Bay. One consideration led to the mapping and classification of the suitability of beaches totaling 239 km in length as *Limulus* spawning habitats (Lathrop et al. 2006). Of 31 km out of 91 km (Delaware) and 26 km of 148 km (New Jersey), 24 % received the highest rating ("optimal") (see Table 2.1 and Fig. 2.3). Only an additional 7 % of the total shoreline was "suitable," 33 % was "less suitable," 31 % was "avoided" by the crabs and 5 % was physically "disturbed" habitat (see Table 2.1). Even so, the available spawning habitat (a combined 64 km of optimal, suitable, and less suitable) was either the most suitable in any single embayment, worldwide, or horseshoe crabs at Delaware Bay were the most efficient spawners.

Fig. 2.7 (**a**) Horseshoe crabs have difficulty in surviving and in spawning when mud and finely-ground detritus cover a beach, as in 1999 at the northern portion of Slaughter Beach (sometimes referred to as Cedar Beach), Delaware (Photographs by Shuster). This portion of the beach was covered with a mat of fine organic detritus that contained fine silt (almost a clay – a whitish coating on the crabs). Underneath the mat of detritus the beach was "muddy." The faint shoreline in the upper photograph is the mile-long Mispillion River that juts out into Delaware Bay creating a back-eddy to the south where fine sediments are collected, (**b**) In this close up of a portion of the beach (Fig. 2.7a), the crabs are emerging from the covering of detritus; most of the crabs are coated with dried clay

Table 2.2 Environmental variables measured at the mean high water mark

Environmental variable	Range of measurements	Peak suitability
Depth of sand over peat	8–16 in.	16 in.
Sediment moisture	1–18 %	3–4 %
Beach slope	0–14 %	7 %
Grain size	0.0–2.5 mm	0.6–0.8 mm

Based on Brady and Schrading (1997)

Attention was also directed toward beach restoration. In the late 1980s, Karen Day, US Fish & Wildlife Service, and Shuster developed some preliminary data on beach characteristics that were later researched in detail by Brady and Schrading (1997) as part of a larger study on the effect of beach replenishment upon the suitability/quality of *Limulus* spawning beaches (Table 2.2). Besides the ranges of these physical parameters, the bay also has the full range of salinity, from 32 to 8 ppt, within which *Limulus* can survive (the area within the dashed lines shown on Fig. 2.3).

In Delaware the Department of Natural Resources and Environmental Control program of beach restoration and protection of saltwater tidal marshes, sediment replenishment on the beaches (after extensive storm/erosion damage to beaches), did not appear to affect the spawning indices and egg counts of horseshoe crabs (Weber and Carter 2009; Kalasz and Weber 2010; Michels et al. 2010).

Recent studies on the effect of beach dynamics on horseshoe crabs spawning and on distribution of their eggs have expanded an important dimension to research on the environment of *Limulus*. Jackson et al. (2005) explored the relative impacts of horseshoe crab nest-building versus wave action in redistributing sediments and eggs, with the activities of the crabs generally the more important. On wave and swash transport and redistribution of *Limulus* eggs, Nordstrom et al. (2006) reported that, because vegetative beach wrack traps the eggs locally, more eggs remain on the surface during flooding tides and when wave energy is low than during ebb tides when more eggs are available in the swash zone.

2.2.6 *Concerning Beach "Fidelity"*

Beach fidelity–that horseshoe crabs always spawn at a certain beach (their natal beach) – is most fully expressed by the behavior of the females during the 5–7 days of high tides during a new or full moon. Most female horseshoe crabs rarely completely spawn on one high tide; most often returning on a few more succeeding highs. Acoustic-tag data (Brousseau et al. 2004; Smith et al. 2010) indicate females only move a short distance from the beach and return on a subsequent high tide, usually to a somewhat different spot on the same beach, usually in a up-bay direction (allied to along-shore currents?). Could this short-term/short-distance be interpreted, alternatively to beach fidelity, as conservation of energy?

After completing spawning all of her mature eggs (up to around 100,000 at Delaware Bay), how long does a female remain in the vicinity? Does she actually mature and spawn another full clutch of eggs the same season? If so, remaining in the vicinity would be logical. However, she could leave the bay during the spawning season (presumably after spawning) as has been suggested via benthic sled/video evidence on a large female (Michael Oates, personal observation during an exploratory cruise at the entrance to Delaware Bay on the *F/V Maggie S. Myers*, Captain Frank "Thumper" Eicherly IV).

But then, how to explain erratic spawning behavior associated with specific sites, as when *Limulus* tries to spawn in a pile of oyster shells? Did they just tire out when they happened to reach a beach covered with shells or were the shells covering a former spawning site? Or why do they sometimes try to spawn on a beach composed of peat?

Further, tagging data (Swan 2005) demonstrate that adult *Limulus* wander all around the in-shore continental shelf in the vicinity of Delaware Bay (mostly between Atlantic City, New Jersey, and Toms Cove, Virginia). Also, crabs tagged outside of Delaware Bay are recovered within the bay. How the crabs are guided back to "their native beach" may be as much dependent upon where they enter the bay, as on the first flooding currents that enter the bay on the Cape May side (NOAA 1987) and then follow certain benthic currents within the bay.

Overall, the postulation of rigorous spawning "fidelity" to a beach seems to be an overstatement and contradictory to the concept that horseshoe crabs are ecological generalists. Also, it is inconceivable that *Limulus* followed the exactly same-beach conditions northward in establishing populations from Florida to Maine after the last great ice age.

2.2.7 Wave-Protected Beaches and Shallow Waters Essential

Of the environmental components vital to the existence of horseshoe crabs, the weakest links are the intertidal areas suitable for spawning (including incubation of the eggs) and the associated adjacent shallow-water areas suitable as nurseries of the early juvenile stages. However, although observations confirm that sandy beaches have been those largely populated during spawning season, *Limulus* has spawned in shell heaps along the shore, among gravel and cobble stones, in silty-sand, in muddy detritus, and often are found on marsh banks (Shuster, personal observations).

2.3 *Limulus polyphemus*

> Apparently, horseshoe crabs long ago came up with a body plan that works well, and have simply stuck tenaciously to it. Yet, internally, their molecular genetic clocks have kept on ticking. (Avise 2002)

In Sect. 2.2 the focus was on the habitat with references to *Limulus*, here the emphasis is on the adaptations of the species to and relationships with environmental conditions in the Delaware Bay area.

2.3.1 A Brief Geologic History of Horseshoe Crabs

Horseshoe crabs are ecological/environmental generalists (Eldredge 1991; Loveland et al. 1997) that move freely between oceanic and estuarine environments. This may have been during most of their geologic history. Apparently, millions of years ago, as evidence at least at two coastal environments (Selden and Nudds 2004) – at Mazon Creek (Shabica and Hay 1997) and Solnhofen (Hecht et al. 1985; Barthel et al. 1990) – were inhabited by species of Xiphosura.

Considering the geologic age of *Limulus* it may have been subjected to adverse conditions during at least two major extinction events that occurred between the Cretaceous and Tertiary Periods some 65 mya and the Pleistocene and the Recent Epochs (ca. 0.01 mya) (Hallam and Wignall 1997). *Limulus* certainly is a candidate as a survivor.

Mesolimulus walchi was fossilized about 150 mya in the Solnhofen formation, Bavaria, Germany, during the Jurassic Period. If not the ancestral species, it so closely resembles the anatomy of the extant species that when alive could be mistaken for a contemporary species (Fig. 2.8).

The most recent fossilized connection with the American species is *Limulus coffini* – found in a Cretaceous formation in Colorado, USA, dated at about 80 million years ago (Reeside and Harris 1952). It is represented by a nearly complete dorsal mold of a horseshoe crab opisthosoma, almost identical with that of *Limulus polyphemus*.

Just when *Limulus* occupied the Delaware Bay area has not been ascertained, but this bay probably has not been "a *Limulus* habitat" much more than a few thousand years. During the ocean level rise, *Limulus* must have moved from Florida waters northward, incrementally (as suggested by the study by Saunders et al. 1986), when each embayment became environmentally hospitable. It seems reasonable to also assume that each successive population produced a population large enough to spin off sufficient "wandering" members to form the next more northward population.

2.3.2 Delaware Bay as a Habitat and Limulus as a Successful Colonizer

Delaware Bay, after Chesapeake Bay, is the second largest on the east coast of the U.S. Both bays are geologically young, having developed since the end of the last ice age. Both are inhabited by *Limulus* but the later bay, with more extensive sandy

Fig. 2.8 Internal dorsal mold of *Mesolimulus walchi* (Specimen VIII(10a) at the Senckenberg Museum, Frankfurt-am-Main, Germany: courtesy of Dr. Heinz Malz (photograph by Shuster))

shorelines, is more favorable for horseshoe crabs. Essentially, Delaware Bay was shaped when sea level began rising some 13,000 years ago, when the last great glaciers began to melt and ocean waters began to push into sediments that had been transported seaward from upland erosion. Certain geologic results – size of the bay, its general north-to-south orientation, and the partitioning of the bay shoreline by tidal streams into many beaches – have been of major importance in the spawning and the distribution of the horseshoe crabs. And *Limulus* was ready, as an ecological generalist produced by a lengthy evolutionary lineage, to take advantage of the resultant favorable environment.

2.3.3 Life Stage as a Factor in Distribution

At Delaware Bay there is a clear pattern of distribution with the life cycle stages/age of the crabs tracked from the beaches to throughout the bay with many crabs on the continental shelf (Botton et al. 2010). Earlier, in related studies, Botton and Loveland (2003) and Botton et al. (2003a) had also recently added an important part of that information – that plankton tows revealed the newly hatched larvae were most numerous within the immediate shoreline area and the next stage, the first-tailed stage, settled down, also near the shore. As soon as the digestive tract of the hatchling became functional, the next instar (the first tailed stage) of the life stages initiated, albeit slowly, the trek toward deeper water, instar by instar. I do not recall the source of information – that chelicerates, at least spiders, tend to be cannibalistic and disperse once the hatchlings have a functional digestive tract. Is this a partial answer to the dispersion of *Limulus*?

Early stages of *Limulus*, usually those in their second and third years, have been relatively common on the intertidal bars and in the sloughs at low tide. The smaller, first-year instars are not always prominent, possibly due to storm dispersal before they become evident. But, when prevalent as at low tide on 11 September 1951, a total of slightly over 2,000 instars, stages IV–VI, were collected from exposed intertidal bars, 1 through 5; instar 6 is the stage usually reached during the year it hatched (Shuster 1979).

The collection efficiency of a suction-dredge (greater) was compared with a trawl (less) in a study of the distribution and abundance of small (mostly first-year instars) juveniles at seven sites (four at Delaware, three at New Jersey) in the upper portion of Delaware Bay (Burton et al. 2009). Sampling was along three parallel transects from 1 to 4 m from shore to about 925 m. In July, juveniles had prosomal widths predominately at 7, 10, and 15 mm; by September/October they were at 13.5, 17, and 24 mm. Densities of the juveniles were greatest at the near-shore transect in July, but by September/October most had gone past the offshore transect into deeper waters.

Beginning in 1992 and continuing through 2008, annual surveys by trawls equipped with appropriate mesh size, depending on the age class to be examined, have compiled data on the distribution of the larger juveniles and adults throughout Delaware waters (Jordan Zimmerman, Delaware Division of Fish & Wildlife, personal communication). Clearly, the early instars increasingly make their way into the deeper waters of the bay and, later, are distributed throughout the area with the mid-sized to almost adult-sized instars common on the continental shelf. This was observed personally off Chincoteague, Virginia in the 1990s on Captain Eustler's trawler when Steve Doctor, a Maryland fisheries biologist, was conducting a survey of horseshoe crabs and on an exploratory mission with Jim Berkson when a large number of juveniles were trawled. Adults notably congregate during feeding and spawning; otherwise they are widely and unevenly distributed, at least on the continental shelf off Delaware Bay (Botton and Haskin 1984; Botton and Ropes 1987, 1989; Captain Jeff Eutsler, Ocean City, Maryland, personal observation).

2.3.4 *Management of the* Limulus *Resource*

In 2000 the Horseshoe Crab Management Committee of the Atlantic State Marine Fisheries Commission (ASMFC) voted to investigate the protection of horseshoe crabs on the continental shelf. This was due to the abundance of three age classes of *Limulus*: (a) large adult-sized juvenile females that would molt in the fall and join the spawning migration the next year, (b) those that were first-year spawners, and (c) those older but still active females that would spawn for several more years. The result was the establishment of a large refuge off the mouth of Delaware Bay, 3 miles off the shores of New Jersey, Delaware, and Maryland (NOAA 2001).

2.3.5 Diet as a Factor in Distribution

The diet of *Limulus* within the bay and on the continental shelf was an early, important contribution of Mark Botton in his numerous studies on the American horseshoe crab (e.g., Botton 1982, 1984a, b; Botton and Haskin 1984; Botton and Ropes 1987, 1989; Botton et al. 2003b). The ecological importance of the bay/continental shelf to *Limulus* is evident from the numbers of adult horseshoe crabs on the continental shelf off the entrance to Delaware Bay (Hata and Berkson 2003), where surf clams and whelks are important commercial fisheries, and where tagged *Limulus* move in and out of the bay (Swan 2005). The shelf not only provides an ample source of food but also an avenue for distribution and a wintering-over area. For example, major concentrations of hard clams, *Mercenaria mercenaria*, occur in Delaware Bay while the ocean quahog, *Arctica islandica*, and the surf clam, *Spisula solidissima*, are abundant on the continental shelf in the mid-Atlantic region (NOAA 1980).

2.3.6 Abundance and Distribution

Most recently the enormity of the population of horseshoe crabs within Delaware Bay was calculated from a massive, bay-wide tagging/recapture study in 2003 when over 17,500 adult horseshoe crabs were tagged within Delaware Bay (Smith et al. 2006). A total of 7,221 crabs were obtained for tagging by trawl during pre-spawning (26 March to 8 May) and 10,322 crabs from 28 to 30 May prior to peak spawning. Recaptured tags were obtained on 29 May, 31 May, and 2 June 2003, during the annual spawning survey. Resultant estimates were about 20 million within the bay (90 % confidence interval: 13–28 million) of which 6.25 million (90 % confidence interval: 4.0–8.8 million) were females. Earlier, lesser numbers but still impressive numbers of horseshoe crabs were collected by trawl within a 2,912 km^2 study area on the continental shelf off Delaware Bay (Hata and Berkson 2003). In their survey the mean abundance for all daytime collections was 6.81 million horseshoe crabs with a confidence interval of 2.29–11.33 million; night-time sampling was greater – 11.4 million with a confidence interval of 5.95–16.85 million.

A preliminary survey of the spawning population of Delaware Bay (Shuster and Botton 1985) was conducted by the senior author in 1977, 1979, and 1980. Realistically this one-man survey was weak in the amount of data collected, so submission was withheld until more was available. By the time of submission of the manuscript to Estuaries in 1984 a different but substantial kind of information was available. Botton had completed most of his studies on the diet of *Limulus* in the Delaware Bay area. This enabled the authors to expand the paper to consider population dynamics. The added feature that *Limulus* fed on the surf clam (*Spisula solidissima*) on the shelf strongly suggested that diet was probably the answer. The source of the crabs also seemed obvious – that the shelf crabs were part of the Delaware Bay spawning population, Thus, the authors designated that portion of the population within the bay as one cohort, that on the shelf as another cohort. This was reinforced by a 15-year tagging program (Swan 2005) that showed adults moved in and out of the bay. All of the above illustrated an enormous population of *Limulus* and a great amount (and range) of movement. Today this situation is considered to be a fact – that the bay is an enormous incubator producing more horseshoe crabs than can be fed during their life cycle. Migration is the relief value – large numbers of the juveniles move out of the bay onto the shelf where they mature.

Thus, one interpretation of the numbers of the large juvenile and adult horseshoe crabs on the continental shelf is that Delaware Bay is inadequate to support such an enormous population of crabs. Mapping of the distribution of fisheries species such as clams (NOAA 1980) and their harvests strongly indicate that large sources of species eaten by large juveniles and adult horseshoe crabs (Botton and Haskin 1984; Botton and Ropes 1989) are abundant on the shelf. Another reason for the shelf distribution could be that the crabs appear to spread out away from each other (Captain Jeff Eustler, personal observation) or the belief that the crabs do not congregate except when feeding or breeding or at least stop in certain areas to "rest". Observations also suggest that distribution in the bay and shelf populations may be age-related. For example, since old males tend to be more numerous during the early periods of spawning at Fortesque Beach, is this because they have remained in the upper bay during the cold-water months (Michael Oates, personal communication)? Also, it would be reasonable to expect that more of the younger, more active adults would be on the shelf due to the initial dispersal of the juveniles and to the return of the young adults in search of prey.

Crabs tagged during spawning usually have short-term (i.e., within days) and local movements (Swan 2005). Over longer time periods (months and years) after spawning they show increasing distances and changes in locations (Swan 2005).

2.4 Delaware Bay: A Multiplicity of Optimal Conditions

The larger-sized crabs found from Georgia to Cape Cod compared with elsewhere within their geographical range suggest that the mid-Atlantic geography and environmental conditions favored *Limulus*. At Delaware Bay a number of drainage basins and their tidal streams separate the shoreline into numerous, lengthy sandy beaches, each capable of supporting a large spawning aggregation of *Limulus* (see Figs. 2.3 and 2.4). These beaches extend along a favorable salinity gradient from

32 ppt at the entrance to the bay to 8–10 ppt up-bay, usually the lowest tolerance level of *Limulus*. The distance from the capes to the last major spawning area upbay is about 50 km.

2.5 Conclusions

It was and is the juxtaposition of the ecological generalist and consummate survivor behavior of *Limulus* with the geologic evolution and characteristics of the bay, and the geographic location of Delaware Bay within the mid-range of the species that have contributed to the great success of *Limulus* at Delaware Bay. It is this juxtaposition that has resulted in an ecological situation that out-produces other populations. How? In essence, each beach supports a seasonally discrete spawning population that contributes to the total production of the bay – with a lesser yield or failure of some beaches and greater success of others that would compensate for the lesser results. This conclusion is supported by the extensive research that has been conducted over the past 30 years at Delaware Bay.

While many changes in the shoreline have probably occurred during the few thousand years since *Limulus* arrived at Delaware Bay, including the times after the arrival of the early Dutch and Swedish colonists, the species still thrives at this bay. But, in the long run, the ocean will have inundated Delmarva Peninsula, covering both Chesapeake and Delaware Bays within the next 75,000 years (Fig. 2.9). If changes in the shoreline proceed at a pace that include development of shallow-water coastal embayments with sandy beaches within the migration ranges of some populations, *Limulus* will probably persist. But sea-level rise is not always constant as transgressive facies changes in peat, mud, and sand and muddy sand in tidal wetlands in Delaware suggest a rapid rise 2,000 years ago (Fletcher et al. 1993; John and Pizzuto 1995). The contrast between the persistence of a landform (from about 7,000 years before present to 75,000 years in the future) and the some 135 million years that the species has existed, is more than suggestive that *Limulus polyphemus* is indeed a survivor.

Acknowledgements If no other detailed natural history/ecological studies had been conducted on *Limulus polyphemus* at Delaware Bay, the contributions by Drs. Mark L. Botton (Fordham University), Robert E. Loveland (Rutgers University), and their colleagues would be notable as being among the first and most significant. The influx of federal scientists during the past score of years, led by Drs. David R. Smith (US Geological Survey) and Michael J. Millard (US Fish and Wildlife Service), that supported the Atlantic States Marine Fisheries Commission Horseshoe Crab Management Program, also added to an understanding of the activities of *Limulus* at Delaware Bay. Thanks are also due to the hundreds of volunteers who have participated in spawning surveys, in tagging programs, and in development of individual and community-based horseshoe crab sanctuaries and other conservations efforts (in the "Just Flip 'Me" and community-based horseshoe crab sanctuaries organized by Glenn Gauvry, the Ecological Research and Development Group Inc.), as well as to watermen who shared their opinions on *Limulus*.

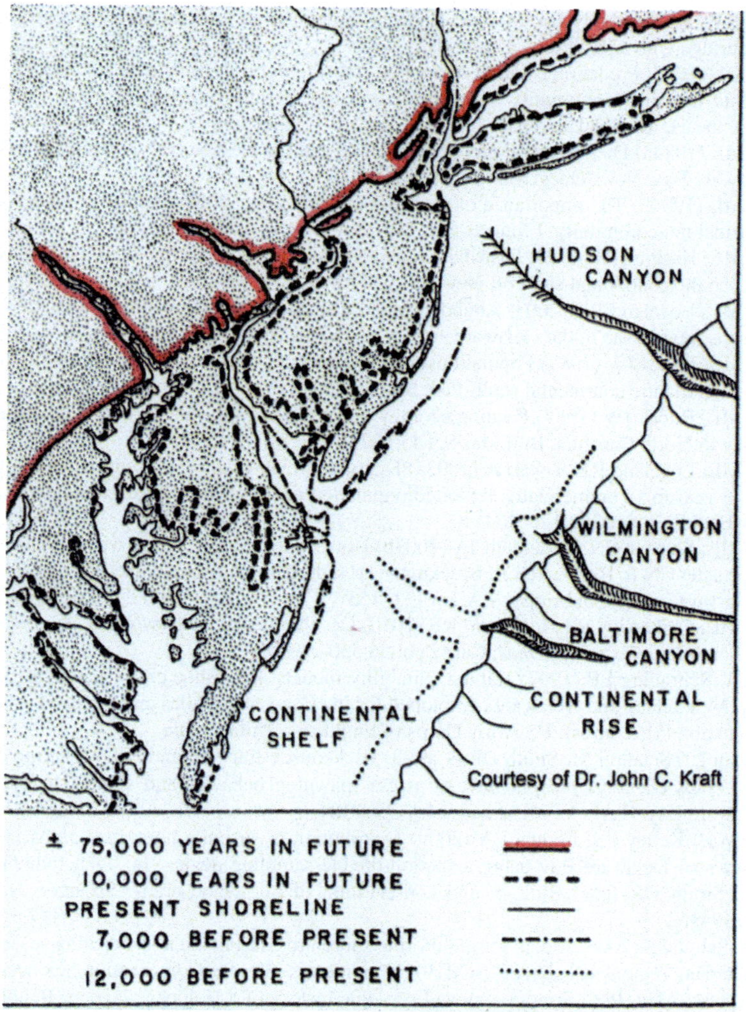

Fig. 2.9 Long before the next major ice age, seawater will bathe much of the Atlantic coast (Courtesy of Dr. John C. Kraft). Whether the retreating shoreline will also have habitat suitable for *Limulus polyphemus* is unknown but since this species can spawn in only a patch of sand (Shuster et al. 2003: COLORPLATE 14) and has already existed some 140 million years (Shishikura et al. 1982) it probably will survive whatever occurs within the next 75,000 years

References

Amos WH (1979) The infinite river: a biologist's vision of the world of water. Random House, New York

Avise JC (2002) Genetics in the wild. Smithsonian Institution Press, Washington, DC

Baptist JP, Smith OR, Ropes JW (1957) Migrations of the horseshoe crab, *Limulus polyphemus*, in Plum Island Sound, Massachusetts. U.S. Fish & Wildlife Service, Special Scientific Report Fisheries No. 220

Barthel KW, Swinburne NHM, Morris SC (1990) Solnhofen: a study in Mesozoic palaeontology. Cambridge University Press, Cambridge

Botton ML (1982) Predation by adult horseshoe crabs, *Limulus polyphemus* (L.), and its effect on benthic intertidal community structure of breeding beaches in Delaware Bay, New Jersey. Ph.D. thesis, Rutgers University

Botton ML (1984a) Diet and food preferences of the adult horseshoe crab *Limulus polyphemus* in Delaware Bay, New Jersey. Mar Biol 81:199–207

Botton ML (1984b) The importance of predation by horseshoe crabs, *Limulus polyphemus*, to an intertidal flat community. J Mar Res 42:139–161

Botton ML, Haskin HH (1984) Distribution and feeding of the horseshoe crab, *Limulus polyphemus*, on the continental shelf off New Jersey. Fish Bull 82:383–389

Botton ML, Loveland RE (2003) Abundance and dispersal potential of horseshoe crab (*Limulus polyphemus*) larvae in the Delaware estuary. Estuaries 26:1472–1479

Botton ML, Ropes JW (1987) Populations of horseshoe crabs, *Limulus polyphemus*, on the northwestern Atlantic continental shelf. Fish Bull 85:805–812

Botton ML, Ropes JW (1989) Feeding ecology of horseshoe crabs on the continental shelf, New Jersey to North Carolina. Bull Mar Sci 45:637–647

Botton ML, Loveland RE, Tiwari A (2003a) Distribution, abundance, and survivorship of young-of-the-year in a commercially exploited population of horseshoe crabs, *Limulus polyphemus*. Mar Ecol Prog Ser 265:175–184

Botton ML, Shuster CN Jr, Keinath JA (2003b) Horseshoe crabs in a food web who eats whom? In: Shuster CN Jr, Barlow RB Jr, Brockmann HJ (eds) The American horseshoe crab. University of Harvard Press, Cambridge, MA, pp 133–153

Botton ML, Tankersley RA, Loveland RE (2010) Developmental ecology of the American horseshoe crab *Limulus polyphemus*. Curr Zool 56:550–562

Brady JT, Schrading EP (1997) Habitat suitability models: horseshoe crab (spawning) Delaware Bay, New Jersey and Delaware (developed for the Cape May, Villas and Reeds Beach Habitat Evaluation Procedures). US Army Corps of Engineers, Philadelphia

Brousseau LJ, Sclafani M, Smith DR et al (2004) Acoustic-tracking and radio-tracking of horseshoe crabs (*Limulus polyphemus*) to assess spawning behavior and subtidal habitat use in Delaware Bay. N Am J Fish Manag 24:1376–1384

Burton WH, Kelley FS, Franks EA (2009) Distribution of juvenile horseshoe crabs in subtidal habitats of Delaware Bay using a suction-dredge sampling device. In: Tanacredi JT, Botton ML, Smith DR (eds) Biology and conservation of horseshoe crabs. Springer, New York, pp 149–162

Chan S, Schulte S (2008) A plan for monitoring shorebirds during the non-breeding season in bird monitoring region New Jersey–BCR 30. Manoment Center for Conservation Sciences

Chrzastowski MJ (1986) Stratigraphy and geologic history of a Holocene lagoon: Rehoboth Bay and Indian River Bay, Delaware. Ph.D. thesis, University of Delaware

Cook GH (1857) Geology of Cape May, NJ, p 105 (see Fowler HW (1908) The king crab fisheries in Delaware Bay. In: 1907 Annual Report New Jersey State Museum, Trenton, NJ, pp 111–119 + plates 59–65

Custer JF (1984) Delaware prehistoric archaeology, an ecological approach. University of Delaware Press, Newark

DNREC (2001) Inland Bays/Atlantic Ocean Basin assessment report. Delaware Department of Natural Resources and Environmental Control, Dover. 40-01/01/01/02

Ehlinger GS, Tankersley RA (2009) Ecology of horseshoe crabs in microtidal lagoons. In: Tanacredi JT, Botton ML, Smith DR (eds) Biology and conservation of horseshoe crabs. Springer, New York, pp 149–162

Eldredge N (1991) Fossils: the evolution and extinction of species. Harry N. Abrams, Inc., New York

Finn JJ, Shuster CN Jr, Swan BL (1991) *Limulus* spawning activity on Delaware Bay shore. Finn Tech Industries, Cape May Court House

Fletcher CH III, Knebel HJ, Kraft JC (1992) Holocene depocenter migration and sediment accumulation in Delaware Bay: submerging marginal marine sedimentary basin. Mar Geol 103:165–183

Fletcher CH, Pizzuto JE, John S et al (1993) Sea-level rise acceleration and the drowning of the Delaware Bay coast at 1.8 ka. Geology 21:121–124

Fraenkel G (1960) Lethal high temperatures for three marine invertebrates: *Limulus polyphemus*, *Littorina littorea* and *Pagurus longicarpus*). Oikos 11:171–182

Hallam A, Wignall PB (1997) Mass extinctions and their aftermath. Oxford University Press, Oxford

Hata D, Berkson J (2003) Abundance of horseshoe crabs (*Limulus polyphemus*) in the Delaware Bay area. Fish Bull 101:933–938

Hecht MK, Ostrom JH, Viohl G, Wellnhofer P (eds) (1985) The beginnings of birds. Freunde des Jura-Museums, Eichstätt

Henry VJ (2009) Geology of the Georgia coast. The New Georgia encyclopedia. University of Georgia Press, Athens

Jackson NL, Nordstrom KF, Smith DR (2005) Influence of waves and horseshoe crab spawning on beach morphology and sediment grain-size characteristics on a sandy estuarine beach. Sedimentology 52:1097–1108

James-Pirri MJ (2010) Seasonal movement of the American horseshoe crab *Limulus polyphemus* in a semi-enclosed bay on Cape Cod, Massachusetts (USA) as determined by acoustic telemetry. Curr Zool 55:575–586

John SJ, Pizzuto JE (1995) Accelerated sea level rise 2,000 years BP in the Delaware Bay: stratigraphic evidence from the Leipsic River valley, Delaware, USA. J Coastal Res 11(3):573–582

Kalasz KS, Weber RG (2010) 2005–2010 Delaware horseshoe crab (*Limulus polyphemus*) egg survey project. Report to the Atlantic States Marine Fisheries Commission's Horseshoe Crab Technical Committee, Washington, DC

King TL, Eackles MS, Spidle AP et al (2005) Regional differentiation and sex-based dispersal among populations of the horseshoe crab *Limulus polyphemus*. Trans Am Fish Soc 134:441–465

Knebel HJ, Fletcher CH, Kraft JC (1988) Late Wisconsin Holocene paleogeography of Delaware Bay: a large coastal plain estuary. Mar Geol 83:115–133

Lathrop R, Allen M, Love A (2006) Mapping and assessing critical horseshoe crab habitats in Delaware Bay. Rutgers University Center for Remote Sensing and Spatial Analysis. http://deathstar.rutgers.edu/projects/delbay/

Loveland RE, Botton ML, Shuster CN Jr (1997) Life history of the American horseshoe crab (*Limulus polyphemus*) in Delaware Bay and its importance as a commercial resource. In: Farrell J, Martin C (eds) Proceedings of the horseshoe crab forum status of the resource. University Delaware Sea Grant Program, Lewes

Luckenback M, Shuster CN Jr (1997) Preliminary test runs on the behavior of juvenile *Limulus* in a flume (unpublished notes and photographs)

Mayer AG (1914) The effects of temperature upon tropical marine animals. Papers Tortugas Laboratory, Carnegie Institute Washington, Publ 183(6):1–24

Michels S, Smith D, Bennet S (2010) Horseshoe crab spawning activity in Delaware Bay 1999–2009. Report to the Atlantic States Marine Fisheries Commission's Horseshoe Crab Technical Committee, Washington, DC

Moore S, Perrin S (2007) Seasonal movement and resource-use patterns of resident horseshoe crab (*Limulus polyphemus*) populations in a Maine, USA estuary. Estuar Coast 30:1016–1026

NOAA (1980) Eastern United States coastal and ocean zones data atlas. National Oceanic and Atmospheric Administration, Washington, DC

NOAA (1987) Delaware river and bay: tidal circulation and water level forecast atlas. U.S. Department of Commerce, Washington, DC

NOAA (2001) National Marine Fisheries Service, 50 CFR Part 697, Atlantic Coastal Fisheries Cooperative Management Act Provisions; Horseshoe Crab Fishery; Closed Area. Fed Regist 66(23):8906–8911

Nordstrom KF, Jackson NL, Smith DR (2006) Transport of crab eggs by waves and swash on an estuarine beach: implications for foraging shorebirds. Estuar Coast Shelf Sci 70:438–448

Pielou EC (1991) After the ice age: the return of life to glaciated North America. University Chicago Press, Chicago

Pooler PS, Smith DR, Loveland RE et al (2003) Assessment of sampling methods to estimate horseshoe crab (*Limulus polyphemus*) egg density in Delaware Bay. Fish Bull 101:698–703

Reeside JB Jr, Harris DV (1952) A cretaceous horseshoe crab from Colorado. J Wash Acad Nat Sci 42:174–187

Riska B (1981) Morphological variation in the horseshoe crab *Limulus polyphemus*. Evolutionary 35:647–658

Rudloe A, Herrnkind W (1976) Orientation of *Limulus polyphemus* in the vicinity of breeding beaches. Mar Behav Physiol 4:75–89

Rudloe A, Herrnkind W (1980) Orientation by horseshoe crabs, *Limulus polyphemus*, in a wave tank. Mar Behav Physiol 7:199–211

Saunders NC, Kessler LG, Avise JC (1986) Genetic variation and geographic differentiation in mitochondrial DNA of the horseshoe crab, *Limulus polyphemus*. Genetics 112:613–627

Say T (1818) An account of the Crustacea of the United States. J Acad Nat Sci Phila Part II 1(5):423–458

Schaller SW, Chabot CC, Watson WH (2010) Seasonal movements of American horseshoe crabs *Limulus polyphemus* in the Great Bay estuary, New Hampshire (USA). Curr Zool 56:587–598

Sekiguchi K (1988) Horseshoe crabs of the Japanese coast. In: Sekiguchi K (ed) Biology of horseshoe crabs. Science House, Tokyo, pp 39–45

Sekiguchi K, Shuster CN Jr (2009) Limits on the global distribution of horseshoe crabs (Limulacea): lessons learned from two lifetimes of observations: Asia and America. In: Tanacredi JT, Botton ML, Smith DR (eds) Biology and conservation of horseshoe crabs. Springer, New York, pp 5–24

Selden P, Nudds J (2004) Evolution of fossil ecosystems. Manson Publishing Ltd., London

Shabica CW, Hay AA (eds) (1997) Richardson's guide to the fossil fauna of Mazon Creek. Northeastern Illinois University, Chicago

Shinn RC (2000) Watershed management and estuary programs: perfect together. In: Kosko K (ed) The Jersey shore line, New Jersey Marine Science Consortium & Sea Grant Special Edition 19(03):5–6

Shishikura F, Nakamura S, Takahashi K et al (1982) Horseshoe crab phylogeny based on amino acid sequences of the fibrino peptide-like C. J Exp Zool 223:89–91

Shuster CN Jr (1950) Observations on the natural history of the American horseshoe crab, *Limulus polyphemus*. In; 3rd report on investigations of methods of improving the shellfish resources of Massachusetts. Woods Hole Oceanographic Institute, Woods Hole, pp 18–23

Shuster CN Jr (1955) On morphometric and serological relationships within the Limulidae, with particular reference to *Limulus polyphemus*. Ph.D. Biology dissertation, New York University

Shuster CN Jr (1959) Biological evaluation of the Delaware River Estuary. In: Smith JG, Haber RA, Kaplosky AJ, Simpson CO (coordinating committee) 1959 State of Delaware Intrastate Water Resources Survey, Dover, pp 21–173

Shuster CN Jr (1960a) Oysters in Delaware waters. Estuar Bull Univ Del 5(3):1–15

Shuster CN Jr (1960b) Hydroclimate on the bar grounds. Estuar Bull Univ Del 5(1):7–11

Shuster CN Jr (1979) Distribution of the American horseshoe "crab", *Limulus polyphemus (L.)*. In: Cohen E (ed) Biomedical applications of the horseshoe crab (Limulidae). Alan R. Liss, Inc., New York, pp 3–26

Shuster CN Jr, Botton ML (1985) A contribution to the population biology of horseshoe crabs, *Limulus polyphemus*, in Delaware Bay. Estuaries 8:363–372

Shuster CN Jr, Sekiguchi K (2009) Basic habitat requirements of the extant species of horseshoe crabs (Limulidae). In: Tanacredi JT, Botton ML, Smith DR (eds) Biology and conservation of horseshoe crabs. Springer, New York, pp 115–129

Shuster CN Jr, Barlow RB, Brockmann HJ (eds) (2003) The American horseshoe crab. Harvard University Press, Cambridge, MA

Smith DR, Pooler PS, Swan BL et al (2002a) Spatial and temporal distribution of the horseshoe crab (*Limulus polyphemus*) spawning in Delaware Bay: implications for monitoring. Estuaries 25(1):115–125

Smith DR, Pooler PS, Loveland RE et al (2002b) Horseshoe crab (*Limulus polyphemus*) reproductive activity on Delaware Bay beaches: implications for monitoring. J Coast Res 18(4):730–750

Smith DR, Millard MJ, Eyler S (2006) Abundance of adult horseshoe crabs (*Limulus polyphemus*) in Delaware Bay estimated from a bay-wide mark-recapture study. Fish Bull 104:456–464

Smith DR, Brousseau LJ, Mandt MT et al (2010) Age and sex specific timing, frequency, and special distribution of horseshoe crabs spawning in Delaware Bay: insights from a large-scale radio telemetry array. Curr Zool 56(5):563–574

Swan BL (2005) Migrations of adult horseshoe crabs, *Limulus polyphemus*, in the Middle Atlantic Bight: a 17-year tagging program. Estuaries 28:28–40

Tiner RW, Biddle AD, Jacobs AD et al (2011) Delaware wetlands: status and changes from 1992 to 2007. Cooperative National Wetlands Inventory Publication, US Fish & Wildlife Service, Northeast Region, Hadley, MA and Delaware Department of Natural Resources & Environmental Control, Dover

Weber RG, Carter DB (2009) Distribution and development of *Limulus* egg clusters on intertidal beaches in Delaware Bay. In: Tanacredi JT, Botton ML, Smith DR (eds) Biology and conservation of horseshoe crabs. Springer, New York, pp 249–266

Yamasaki T, Makioka T, Saito J (1988) External morphology. In: Sekiguchi K (ed) The biology of horseshoe crabs. Science House, Tokyo, pp 69–132

Zaldivar-Rae J, Sapién-Silva RE, Rosales-Raya M et al (2009) American horseshoe crabs, *Limulus polyphemus*, in Mexico: open possibilities. In: Tanacredi JT, Botton ML, Smith DR (eds) Biology and conservation of horseshoe crabs. Springer, New York, pp 97–113

Chapter 3
Sea Level Rise in Delaware Bay, U.S.A.: Adaptations of Spawning Horseshoe Crabs (*Limulus polyphemus*) to the Glacial Past, and the Rapidly Changing Shoreline of the Bay

Robert E. Loveland and Mark L. Botton

Abstract Horseshoe crabs have proven adept at locating suitable areas of sandy beach spawning habitat throughout their long geological history. Paleogeographic studies have shown that the most recent period of sea level rise (SLR) has been occurring in Delaware Bay for at least 6,000 years. Comparison of aerial photographs from the 1930s with contemporary satellite imagery clearly indicates a landward movement of the shoreline along the New Jersey coastline of Delaware Bay. Habitat for horseshoe crab spawning has been adversely impacted over this period of time by the loss or degradation of spawning beaches, which to some extent has been offset by the deposition of this sand in "marginal habitats" such as tidal creeks and sandy deltas. The well-documented natural landward movement of a beach-marsh system in a time of SLR has been compromised in some locations by the hardening of the coastline through construction of bulkheads, groins and jetties. This directly reduces the productivity of these beaches for horseshoe crabs, and, consequently, their use by shorebirds. The response to SLR and storms in the recent past has emphasized the protection of coastal property; however, there has been some effort to restore beach ecosystems through nourishment. Given that SLR is an ongoing process, beach nourishment projects to protect a developed shoreline will require a long-term commitment at considerable cost. From the perspective of horseshoe crab conservation and habitat preservation, we suggest that consideration be given to the strategy of property buy-outs and abandonment, thus enabling a more natural beach response to SLR.

Keywords Delaware Bay • Limulus • Sea level rise • Beach erosion • Paleogeology • Geomorphology • Spawning habitat

---

R.E. Loveland (✉)
Department of Ecology, Evolution & Natural Resources, Rutgers University, New Brunswick, NJ 08903, USA
e-mail: robert.loveland@gmail.com

M.L. Botton
Department of Natural Sciences, Fordham University, 113 West 60th Street, New York, NY 10023, USA

R.H. Carmichael et al. (eds.), *Changing Global Perspectives on Horseshoe Crab Biology, Conservation and Management*, DOI 10.1007/978-3-319-19542-1_3

3.1 Introduction

An elevated chunk of salt marsh cordgrass (*Spartina alterniflora*) once formed a circular mini-marsh of approximately one acre (ca. 4,000 m^2), just off the beach of the Rutgers University Cape Shore Laboratory on the Delaware Bay in Cape May County. Beginning in the early 1960s, the marsh became noticeably smaller over the years. Eventually one summer, it was gone – not a trace of the marsh could be found. By the beginning of the 1980s, it was indeed rare to find any remnants of a salt marsh sitting on the flats of Delaware Bay along the coast of the Cape May peninsula in New Jersey. After the demise of the mini-marsh, a gently sloping beach became important spawning habitat for the American horseshoe crab *Limulus polyphemus*. Although both of the authors studied the population biology of horseshoe crabs along the East coast of North America since the 1970s, and conducted research at the very site where the mini-marsh once stood, it never dawned on us until recently that we had witnessed an event which has become commonplace along the coast of Delaware Bay. At about that same time, we started to observe dying stands of forests near the banks of the small creeks along Delsea Drive (New Jersey Route 47), the main road connecting small towns that dot the New Jersey shore of Delaware Bay. Not only are salt marshes being fractured and "pushed" further upland, but salt water is now intruding farther into the creeks, leading to the demise of the trees. The very beach itself is being eroded, overwashed and moved inland; areas of peat – remnants of earlier salt marsh vegetation – are being exposed for the first time in thousands of years. Although not discussed much in the early 1980s, we now realize that a global phenomenon known as Sea Level Rise (SLR) continues to impact our study sites. It has become common knowledge that SLR can be observed across the world's oceans, posing a potential threat to coastal communities everywhere.

When we began our collaborative research on *Limulus polyphemus* in the mid-1980s, horseshoe crabs were extremely abundant along the New Jersey coast of Delaware Bay. However, it was not always that way. The Delaware Bay population of horseshoe crabs was decimated for fertilizer and animal feed in the late 1880s (Shuster 2003; Kreamer and Michels 2009); populations declined until the beginning of the Great Depression. A period of harvesting of such magnitude, just to sustain the fertilizer industry, nearly extirpated the local horseshoe crabs of lower Delaware Bay. That near-extinction event created a bottleneck that lasted for over four decades, from the early 1880s to the middle of the 1930s (Shuster and Botton 1985). With the development of chemical fertilizers, harvesting of horseshoe crabs ceased, and the population began a slow comeback. However, the crabs were no longer "useful"; in fact, during the summer spawning season many local residents along the beachfront considered the crabs to be a nuisance.

In the backwaters of Delaware Bay, around 1990 the local fishermen discovered that horseshoe crabs were an excellent bait for catching eels. Although most Americans do not eat eels, they are in great demand in Asia. The commercial value of horseshoe crabs soared as the market value for eels and conchs increased (Kreamer and Michels 2009). The race was on to harvest as many horseshoe crabs as possible for the bait industry. Unregulated harvesting has contributed to further reduction in population size, and although the fishery is now regulated by the Atlantic States Marine Fisheries Commission (ASMFC 1998), the bait fishery continues to be a contentious factor in protecting *Limulus polyphemus*.

How human activity affects the population level and distribution of horseshoe crabs is important because *Limulus polyphemus* plays a central role in the dynamics of shorebird migration in Delaware Bay (Mizrahi and Peters 2009), and because blood of the horseshoe crab is used to produce LAL (*Limulus* amoebocyte lysate), necessary for the US Food and Drug Administration (FDA) approved deterministic test for pyrogenic infections and contamination (Levin et al. 2003). Current decisions which relate to beach maintenance and erosion directly influence the quality of the beach as spawning habitat for *Limulus polyphemus*. How horseshoe crabs respond to the rapidly changing environment of Delaware Bay is the subject of this chapter. Knowledge of the geologic history and a contemporary view of SLR are necessary to understand the challenge for survival in a somewhat chaotic and entropic environment. Fortunately, *Limulus polyphemus* has confronted such elements in the past, and has done reasonably well, given that we can now trace this species back to the upper-Jurassic (148 million years ago).

3.2 A Brief Geological History of South Jersey and Delaware Bay

In 1925, it was Alfred Wegener who first described and documented the phenomenon of "continental drift" (Wegener 1929). A belt of Cretaceous fossils stretching from the Highlands of Sandy Hook to Camden in New Jersey has a remarkable similarity to Cretaceous fossils found north of London, England. There was no doubt in Wegener's mind that the only explanation for the distribution of such Cretaceous fossils is that New Jersey must have been adjacent to what is now Western Europe, since the European Plate was once in contact with the North American Plate. An explanation of *how* the continents drifted apart emerged only after geophysicists discovered alternating bands of extruded magma along the Mid-Atlantic Ridge. Although Wegener, at the beginning of the twentieth Century, was not able to explain the mechanism of Continental Drift, he explained an incredible amount of the findings of the other Alfred, namely Alfred Wallace, and incidentally Charles Darwin (Winchester 2003). The scientific community in the early 1960s *finally* accepted his evidence for the concept of continental drift. We now understand that it is Plate Tectonics which is the driving force, causing continents to move as they float on a semi-liquid mantle beneath individual plates.

Now, why is the above brief synopsis very important in a contemporary discussion of horseshoe crabs? *Limulus polyphemus* is abundant and the only species of horseshoe crab in the Western Atlantic, presently distributed along the East Coast of North America from Maine to the Yucatan peninsula. In Asia, illustrations of horseshoe crabs can be found dated as early as 800 AD (Sekiguchi 1988). In fact, a species of "kingcrab" was first mentioned by Rudyard Kipling (1902) in his "Just So" stories; since Kipling had lived in India, it is likely that he was referring to *Tachypleus gigas*. Although there are few early references to the American horseshoe crab, there is some indication that Native Americans and early European settlers may have used horseshoe crabs for food and a variety of other purposes (reviewed by Kreamer and Michels 2009). The sharp, spiny telsons may have been used as spear tips, while dried horseshoe crabs were used to fertilize corn and other farm crops. The famous painting (ca. 1590) by John White, entitled "The Method of Fishing of the Inhabitants of Virginia" (reproduced as Fig. 3.2 in Kreamer and Michels 2009) clearly depicts Native Americans in the process of harvesting horseshoe crabs with long spears from a dugout canoe.

We are very aware of the modern American horseshoe crab, even though not much documentation exists about *Limulus polyphemus* prior to its widespread use as fertilizer in the nineteenth century (Shuster 2003). Thus, we must look further back in time.

3.2.1 Importance of the Fossil Record

Recent fossil evidence suggests that horseshoe crabs have always lived in shallow coastal areas (reviewed by Rudkin and Young 2009), and may even have spawned intertidally in a manner similar to living horseshoe crabs (Diedrich 2011). Curiously, fossils of *Limulus polyphemus* have never been found in North America, largely because the type of lithographic rock where horseshoe crab fossils occur is not found along the Northeast coast of the US. So, how long ago did *Limulus polyphemus* arrive on the shores of North America? *Mesolimulus*, found in Triassic limestone deposits in Germany, used to be considered the closest relative of the American horseshoe crab. That species was of rather small size relative to *Limulus polyphemus*; adult American horseshoe crabs are large, especially the female (Delaware Bay females: 25 cm; males: 20 cm). Even today there are remnants of a population in which both males and females are rather small – namely the horseshoe crabs that are resident in embayments at the northern extent of the range, such as Great Bay, New Hampshire. The fact that most horseshoe crab fossils are small simply suggests that it was the young and smaller individuals that fossilized, not the adults. While most discussion of the fossil record of horseshoe crabs has been largely speculative relative to *Limulus polyphemus* (Shuster and Anderson 2003), a most amazing discovery was recently reported by the young paleontologist from Poland, Blazej Blazejowski. While digging in a lithographic Owadow-Brzezinki Quarry for ancient clams and microfossils, Blazej and his colleagues came across a small collection of extremely rare horseshoe crab fossils, so well preserved that he was able to reconstruct three-dimensional models of the specimens (Kin et al. 2013; see also Chap. 1 by Blazejowski in this book). The fossils were small and strongly resembled *Limulus polyphemus*, so Blazej decided to compare his upper-Jurassic fossils (148 million years old) with extant juvenile specimens of New Jersey horseshoe crabs. Analysis of many morphological features of the fossils matched the extant exuviae (the molted carapace) of *Limulus polyphemus* (Fig. 3.1).

Fig. 3.1 Comparison of cast of juvenile *Limulus polyphemus* with Upper-Jurassic fossil of *Limulus darwini* (Courtesy of B. Blazejowski)

The external appearance of the Polish fossils was similar enough to suggest that they were ancestral to *Limulus polyphemus*. A few minor differences in spination led Blazej to assign the species *Limulus darwini* to the newly discovered fossil. Blazejowski and Kin have proposed a new concept which explains the incredible similarity of *Limulus polyphemus* to a creature that moved among the calcareous silty inlets of the Jurassic Sea. That is, the American horseshoe crab is a stabilomorph, not a "living fossil" as it is so commonly referred to in the literature (Kin and Blazejowski 2014). Błażejowski (2015) would attribute the constancy in external appearance of *Limulus* for at least the last 148 million years as an indicator of extreme stabilomorphy. As also argued by Shuster and Anderson (2003), the morphology of *Limulus polyphemus* needs not respond to an entropic world because it has proven itself to be completely adapted to whatever environment it exploits. So, even though the absolute position of sandy shorelines changes over time as a function of sea level, horseshoe crabs have proven to be quite capable of tracking this habitat, behavior that may strongly relate to the morphology of *Limulus*.

3.2.2 Limulus polyphemus *Took a Ride*

How long ago did *Limulus polyphemus* arrive on the shores of North America? Thanks to Alfred Wegener, we now know that there was a time when the European Plate and the North American Plate were contiguous (e.g. Fig. 3.13 in Garrison 1993). About 200 million years ago, roughly at the beginning of the Triassic, the tectonic plates began to separate due to the advent of sea floor spreading along the mid-Atlantic rift. Most of the European Plate drifted to the east, and what was to become the North American Plate drifted west – but the cleavage was not exact. It is known that due to a bit of randomness in Continental Drift, chunks of the European Plate drifted to the coast of North America. Several examples come to mind: it has been shown that parts of two very famous islands are of European origin in the nature of their country rock, namely Mt. Desert Island in Maine, and Staten Island in New York (McPhee 1982). Wegener had already pointed out that certain rocks of Scotland and Ireland were very similar to those that can be found in New Brunswick and Newfoundland, Canada. Well, if entire blocks of country rock were transported to what is now Eastern North America, couldn't an early version of *Limulus polyphemus* also be moved from the waters of Europe, albeit slowly?

3.2.3 Why Did **Limulus polyphemus** *"Prefer" North America to Europe?*

If we examine the current distribution of *Limulus polyphemus*, we find that it is most abundant in the mid-Atlantic states. Although there are populations of horseshoe crabs along the coast of Long Island, and even along the coast of Cape Cod, those were regions that were heavily impacted by recurrent glacial periods. North of Long Island, New York, there are few estuaries that become wide, shallow bays near the ocean, which also are characterized by extensive sandy beaches. But south of Long Island, there are many estuaries that terminate in shallow, sandy bays with well-developed beaches such as Raritan Bay, Delaware Bay, and Chesapeake Bay. Why did horseshoe crabs not remain in European waters? Simply put, there are few if any estuaries in Europe that have the requisite topography necessary for horseshoe crab spawning habitat. Clearly, such habitat must have existed in the Jurassic, but it may have disappeared once continental drift created a completely different world by the time of the Cretaceous. The truth is, we simply do not know why *Limulus polyphemus* moved to the mid-Atlantic states of North America. But we do know that *Limulus* went extinct in Europe.

As the Atlantic Ocean continued to expand, continental drift largely settled down to a snail's pace (it still exists, moving New Jersey to the west by a few centimeters per year). New Jersey was roughly where it is now, but the ocean was not. During the Cretaceous, which ended about 60 million years ago, much of the eastern part of North America was an inland sea. Somewhere in the middle of the Miocene, the ocean was retreating from the land; however, most of southern New Jersey was still under water. As illustrated in the very fine work of David P. Harper (2013, p. 154), the Atlantic coastline was roughly along the current boundary of the Cretaceous deposits. We assume that *Limulus polyphemus* was around then, and most likely was adapted to the retreating coastline. After all, horseshoe crabs belong to an extremely old group of Arthropods, dating back to mid-Ordovician, about 450 million years ago. About two million years ago, during the beginning of the Pleistocene

Ice Ages, continental glaciers came down from the north and covered much of what is now Canada and the northern United States. With a periodicity of about 80,000 years (US Army Corps of Engineers 2004), the glaciers cycled between their southernmost transgression, and their retreat back toward the North Pole. Of the many transgressions, only three glaciers moved as far south as New Jersey. Their influence on coastal geomorphology and ocean level was profound. Each time the glaciers advanced to their terminal moraine, the level of the ocean dropped to an unfathomable depth relative to the level seen along the coast today. In fact, land was 'created' during the advance of a glacier and the shoreline was many kilometers east of the present coastline of South Jersey. We suspect that *Limulus polyphemus* simply adjusted to the dropping level of the ocean, as did the Lenape natives who were around at least during the last glacial transgression and the subsequent regression.

3.3 Sandy Beach Habitat in Delaware Bay: Reconstructing the Past and Predicting the Future

As shown in a seminal paper by Knebel et al. (1988), Delaware Bay has undergone a remarkable geological transformation within the past 18,000 years. At that time, with sea level much lower than at present, the long stretches of sandy beach that now comprise the main horseshoe crab spawning areas along the lower bay were completely absent (Fig. 3.2a). Sea level began rising at the end of the Pleistocene (Fig. 3.2b) and small areas of beach began to develop near the present bay mouth, though most of the shoreline still consisted of tidal wetlands. The main channel widened, and the extent of sandy beach continued to increase as sea level rose (Fig. 3.2c, d), presumably giving horseshoe crabs access to more suitable spawning locations within the estuary. Were there horseshoe crabs in Delaware Bay 18,000 years ago, before there were extensive sandy beaches? Unfortunately, direct evidence is unobtainable as this predates historical records, and no archaeological evidence has yet been uncovered. We would suggest, however, that the numerous small tidal creek areas (Fig. 3.2a) could well have had sand bar systems, and these sites could well have been refugia for spawning horseshoe crabs at times when expansive beaches were scarce.

Contemporary rates of SLR have been the subject of much discussion. As noted earlier, we first became aware of the local effects of SLR through some of our own casual observations of salt marsh loss, beach erosion and peat exposure, and salt water intrusion into uplands. There is now voluminous evidence linking SLR to global climate change. Since 1900, the rate of global SLR is some 1.7 mm/year, with the most rapid rate of increase (3.2 mm/year) occurring from 1993 to 2010 (IPCC 2013). Sea level is rising because warming temperatures contribute to the melting of glaciers and the thermal expansion of sea water. It is not only the increase in ocean volume that is significant, because land subsidence in southern New Jersey exacerbates the increase in relative sea level. NOAA records for the past 100 years

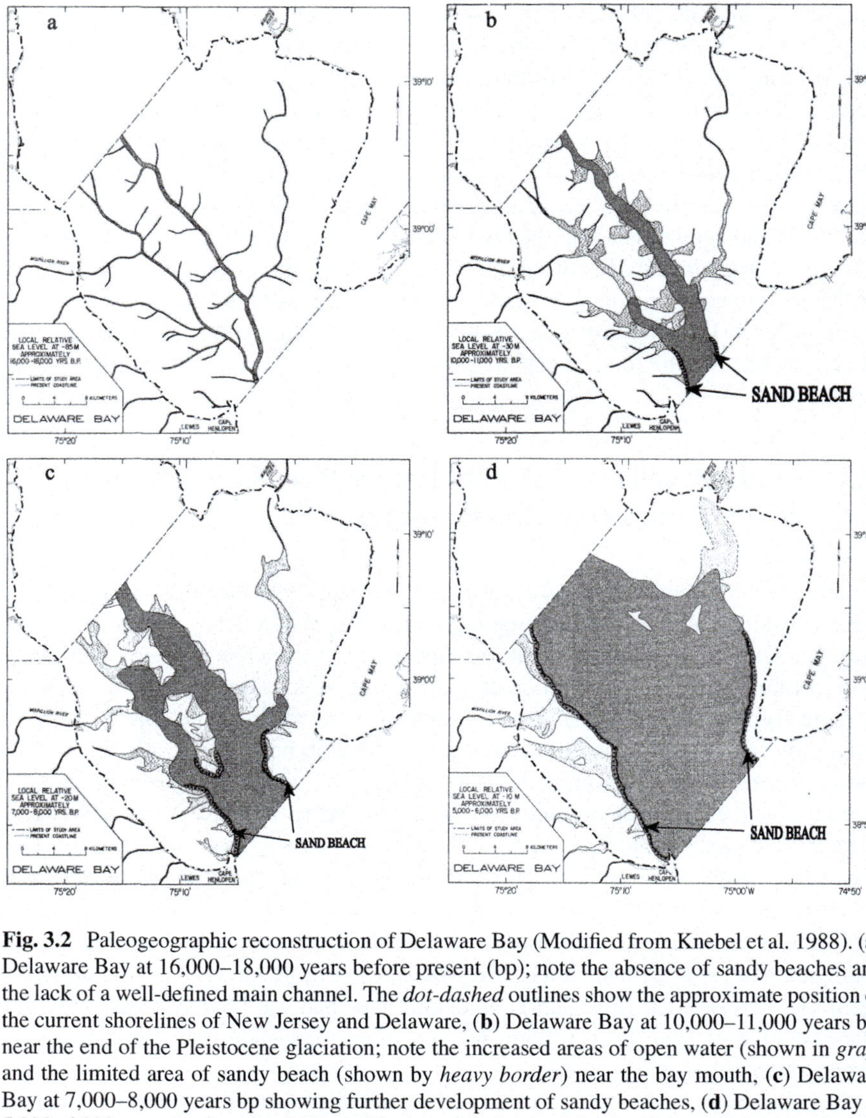

Fig. 3.2 Paleogeographic reconstruction of Delaware Bay (Modified from Knebel et al. 1988). (**a**) Delaware Bay at 16,000–18,000 years before present (bp); note the absence of sandy beaches and the lack of a well-defined main channel. The *dot-dashed* outlines show the approximate position of the current shorelines of New Jersey and Delaware, (**b**) Delaware Bay at 10,000–11,000 years bp, near the end of the Pleistocene glaciation; note the increased areas of open water (shown in *gray*) and the limited area of sandy beach (shown by *heavy border*) near the bay mouth, (**c**) Delaware Bay at 7,000–8,000 years bp showing further development of sandy beaches, (**d**) Delaware Bay at 5,000–6,000 years bp showing further widening of the bay and the development of sandy beaches along the New Jersey and Delaware shores

show that sea level in the region has risen at an average rate of 3.99 mm/year (Fig. 3.3), much higher than the global average (IPCC 2013).

Direct evidence of the effects of SLR on the Delaware Bay can be seen by comparing aerial photographs taken in 1933 and 1990 (Fig. 3.4). We show three different locations for purposes of illustration. At East Point, New Jersey, the sandy beach was completely lost in less than 60 years (Fig. 3.4a1, a2), and the shoreline was hardened to protect homes (Fig. 3.4a3). The only suitable spawning habitat in the

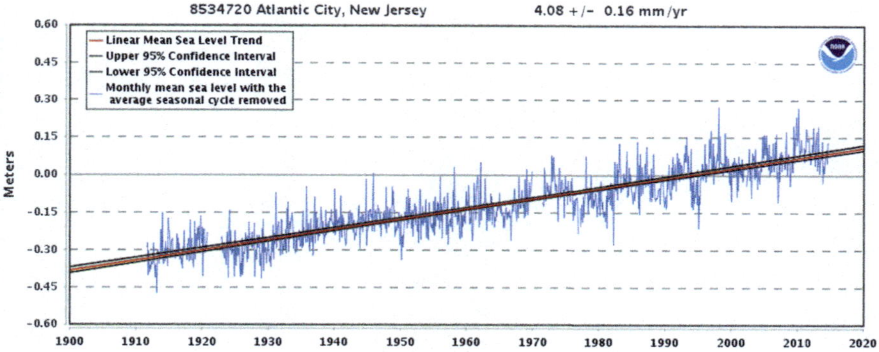

Fig. 3.3 Changes in sea level at station Atlantic City, New Jersey during the past century as recorded by National Oceanic and Atmospheric Administration (NOAA) (http://tidesandcurrents.noaa.gov/sltrends/sltrends_station.shtml?stnid=8534720). The mean sea level trend is 3.99 mm/year with a 95 % confidence interval of ±0.18 mm/year

immediate vicinity consists of a few pocket beaches, which may be found between areas of salt marsh and/or peat (Fig. 3.4a4). At Kimbles Beach, New Jersey, part of the Cape May National Wildlife Refuge, there are no houses on the shoreline so the beach has been "allowed" to migrate landward in a more natural way (Fig. 3.4b1, b2). Areas of overwashed sand are now common along the open bay shore, while a system of sand bars and spits (Fig. 3.4b3) has developed into a locally important horseshoe crab spawning and shorebird feeding location (Botton et al. 1994). Lastly, Reeds Beach demonstrates the combined effects of SLR and jetty construction on the movement of sand (Fig. 3.4c1, c2). The jetty traps sand moving northward with the longshore drift; consequently sand is accreting near the jetty (Fig. 3.4c3) while the southern portion of town is severely eroded (Fig. 3.4c4). It is important to emphasize that these are long-term changes in SLR that can be documented since (at least) the early 1900s. It is simply disingenuous to attribute the erosion of beaches and loss of property along the Delaware Bay shore of New Jersey to rare and catastrophic events such as Hurricane Sandy. This distinction must be kept in mind when we consider how best to respond to the situation.

3.3.1 Sea Level Rise and Horseshoe Crab Habitat

There is abundant evidence that the level of the ocean along New Jersey stopped dropping for the "last time" around 10,000 years ago. As the ocean retreated, landward sediments washed down and over the salt marshes, onto the mudflats, and eventually covered both the marshes and beaches. So what happened to the sandy beaches, which are necessary as spawning habitat for horseshoe crabs? The answer comes from the fact that quartzite (the major component of beach sand) has always been around, even in some of the oldest deposits ever found. Sand is a product of the

Fig. 3.4 Comparison of aerial photographs of Delaware Bay in 1933 and 1990, with ground truth photographs of selected regions. The *black line* superimposed on the coastline is a fixed contour giving perspective to the extent of beach erosion along the bay shore. (**a**) East Point in 1993 (**a1**) and 1990 (**a2**); (**a3**) shows the armored portion of the town of East Point, (**a4**) is a typical "pocket beach" to the east of town, (**b**) Kimbles Beach in 1933 (**b1**) and 1990 (**b2**); note the extensive

slow destruction of mountains and the erosion which follows. Sand is readily moved by water currents, and is deposited downstream of where it originated. Given enough time, sand can cover vast areas both on the land and the sea. Thus, sand was never a limiting factor in the movement of the coastline of New Jersey, particularly in the region of the coastal plain. Some beaches were no doubt covered by advancing terrestrial sediments. Much of the sand seems to have moved seaward as the level of the water dropped. But what happened when the ocean began to rise, as it did when the Wisconsin Glacier (the last to reach New Jersey) started its journey back to the Arctic? Simply put, the East Coast began to experience a period of Sea Level Rise (SLR), where the edge of the ocean advanced landward at an inexorable pace about 10,000 years ago. Glaciers, of course, begin their retrenchment when the average temperature of the earth's surface increased. The many cycles of advance and retreat of glaciers are highly correlated with changing temperature. With the loss of glacial mass, the amount of water in the ocean increased. With the rising temperature, the volume of the ocean increased, and consequently the land adjacent to the ocean was inundated.

But there are other factors which change the rate of relative SLR, not the least of which is subsidence, or sinking of the land. South Jersey is an example of a sinking coastline. One of the reasons relates to the work of Alfred Wegener done so long ago, but given little regard until recently. He proposed the concept of isostatic adjustment. This phenomenon is easy to understand from the point of view of compression. As an example, consider the rising coastline of Alaska, which requires that piers and bulkheads be moved seaward periodically. The land is actually rebounding from a time when the sheer mass of the glaciers, which covered that land, literally pushed the land downward. When the Wisconsin Glacier retreated, the land referred to as New England arose. What is now understood is that isostasy requires that for every mass of land that sinks, some other land must rise! Thus, land that was compressed under the glacier had to sink below its previous level, and the land adjacent (south) of the glacier actually elevated in order to balance the distribution of continental landmass. As glaciers retreat, the once elevated land to the south of the glacier sinks. There are many reasons why land sinks, but it is a fact that the shores of Delaware Bay are sinking.

The rate of SLR is not constant. The rate is influenced by both geophysical and climatological changes (both up and down). It turns out that recent evidence of samples taken from the exposed marshes along the shore of Delaware Bay indicates that SLR took a statistical turn upward around 6,000 years ago. The people of the Lenape tribe must have recognized this change, since they were forced to move inland, again, by the rising waters. In fact, if one digs deeply in the marshes or beaches of today, one might encounter artifacts of that ancient civilization.

Fig. 3.4 (continued) setback of the beach and the areas of sand overwash in 1990; (**b3**) demonstrates the intensive spawning by horseshoe crabs on sandy tidal creek delta areas along Delaware Bay, (**c**) Reeds Beach in 1993 (**c1**) and 1990 (**c2**). The long jetty seen at the top of (**c2**) was constructed to maintain the navigational channel at Bidwell's Ditch; it has resulted in sand accretion at the northern section of town near the jetty (**c3**) and the loss of sand from the southern section (**c4**)

Eventually, those deposits and artifacts will once again see the light of day. We are not sure why the rate of SLR changed 6,000 years ago, but based on the fossil foraminifera preserved in cores, it appears that temperature began to increase slightly faster than previously. After all, the glaciers were in retreat and the influence of the tropical waters of the Gulf Stream may have begun to affect the coast, much as the Gulf Stream off the coast of New Jersey still does today. We know enough about the coastal geology of southern New Jersey, including land subsidence as well as contemporary rates of SLR (Fig. 3.3) that it is possible to model the advance of the coast with rising sea level (Cooper et al. 2008).

What did the coast look like as the ocean moved onto the shelf and into river valleys? Beside the fact that the coast was further out on the continental shelf, its appearance was not much different than today. There were mudflats and marshes, and yes, there were sandy beaches. Ancient beaches are not too difficult to validate in the geological deposits of the past. As an expert in fossil conodonts, the geopaleontologist Anita Harris was able to identify ancient sand beaches in the sediments of the Delaware Water Gap (McPhee 1982). Some of the eroded ancient sands of the Gap may have moved far down the Delaware River becoming the contemporary beaches of Cape May, also giving rise to the famous Cape May "diamonds" so abundant on Sunset Beach. How did SLR influence the spawning behavior of horseshoe crabs? Probably not much. As SLR advances, so do the marshes and the beaches. Adjacent upland forests are lost as sea water invades the land, causing major changes in the types of vegetation along the coast. In light of the apparent 'destruction' of habitat, one sees a balance in the 'creation' of new habitat. As the authors have publically suggested, new habitats resulting from SLR may appear initially as being "marginal." With time, these new habitats become the status quo of the coast. Accordingly, we hypothesize that shifting habitat for horseshoe crab spawning has always been a hallmark of the advancing coastline. Under most conditions, *Limulus polyphemus* is able to find the new habitat and exploit it for the purpose of reproduction.

What do we mean by "marginal habitat"? This term has both temporal and spatial implications. As an ecosystem, coastal marshes and beaches are notoriously dynamic – they can be created, or destroyed, by overnight storms. At some locations such as East Point, New Jersey, the sandy beach was completely lost in less than 60 years due to the combined effects of erosion, sea level rise and hurricanes such as Gloria in 1985. Large portions of a beach may transmute into a mudflat if the sand is removed during an onshore storm. Nearshore sandbars may appear suddenly if an erosion event deposits sand parallel to the beach. Sandy beaches may develop at the mouth of tidal creeks, replacing peaty banks. These are examples of rapidly changing habitats. On a slower scale, SLR may contribute to the transgression of a sandy beach, exposing ancient peat bogs and former salt marsh. Wind may literally blow sand from a beach onto the adjacent, higher elevated salt marsh, thus creating an overwash beach; such an event may form the basis of a future, more stable sandy beach which constantly moves to higher, inland ground. Our research indicates that horseshoe crabs seem to exploit new habitats, while at the same time abandoning former high-quality, sandy beaches that were deeper but are now eroding at a fast pace. As we found in the late 1980s, horseshoe crabs are repelled by toxic materials emanating from exposed ancient peaty bogs on eroding beaches, particularly hydrogen sulfide which is a byproduct of anaerobic decomposition of marsh material (Botton et al. 1988; Draxler 1993). On the other hand, horseshoe crabs are strongly attracted to newly formed offshore sandbars. While *Limulus polyphemus* successfully spawns on beaches where the sand is deep (greater than 25 cm) and somewhat intertidal, this species seems prone to spawn in great numbers on sandbars and spits that are largely submerged by the tide most of the day. Periodically, when the bar is exposed at low tide, one can observe many spawning "nests" where the horseshoe crabs laid their eggs. One also can observe great flocks of shorebirds feeding on the green eggs of horseshoe crabs; these are washed out of the nest by wave action and are an important part of the diet of many species of migratory shorebirds (Botton et al. 1994).

On a larger, spatial scale, habitat formation is relatively infra-dispersed in Delaware Bay. This presents a problem for spawning horseshoe crabs – they must locate small patches of sand which might provide a place to reproduce. *Limulus polyphemus* has an uncanny sense of finding a nesting habitat. We have observed spawning crabs at the base of upland forests, in tidal sloughs, on the surface of salt marshes, on newly formed overwash beaches and on small "pocket beaches". The crabs make nests along tidal creeks, under houses propped up by pilings, and among rocky fill areas with pockets of sand even within highly disturbed estuaries (Botton et al. 2006). Over the course of time, sand accumulates along the front of salt marshes and constitutes an emerging spawning beach. However, *Limulus polyphemus* appears to abandon highly erosional beaches, particularly where human activity has armored the beach with debris, bulkheads, or houses (Botton et al. 1988; Jackson and Nordstrom 2009). There is a simple explanation for this, which is spatially related. We previously alluded to the fact that SLR is responsible for the movement of the entire coastal system landward. This includes the beaches of Delaware Bay, as well as the barrier island system on the Atlantic side of the coastal plain of New Jersey. However, humans cannot seem to tolerate the advancing beaches. This advancement is slow and progressive, but it is a geophysical reality. The placement of barriers may temporarily stop the ocean from progressing landward, but the cost of developing a fixed coastline is extraordinary in time, space and resources. Are horseshoe crabs affected? Yes, but only to the extent that a barrier to the movement of a beach, or the loss of sand from a developed beach, will reduce the net habitat available to the crabs for the purpose of spawning. Irrespective of the existence of horseshoe crabs, socioeconomic reasons often trump environmental issues. During the past four decades that we have been doing research on horseshoe crabs in Delaware Bay, we have witnessed the demise of the small towns of Moores Beach, Thompsons Beach, and Seabreeze, which were formally viable communities of summer residents, local fishermen, vacationers, boaters, and beach walkers. SLR ultimately 'won out' because the roads (often through elevated berms on the surface of a salt marsh) that led to the towns became impassable, particularly at high tide and storms. The towns became isolated from their necessary infrastructural needs. Eventually, local and state resources helped the residents to abandon the town, and the buildings were removed. Beach rubble, used to stabilize the encroaching bay water, was removed; in time, the beaches were restored by natural forces of tides, currents, and storms. Now these once rough-hewn beaches have reverted to spawning habitat for horseshoe crabs. Of particular interest is Moores Beach where through the efforts of many conservation groups the beach has been completely restored, and appears to attract both horseshoe crabs and shorebirds (discussed in Sect. 3.4.4).

As of this writing, there are towns along Delaware Bay that are experiencing severe impacts from both SLR and coastal storms such as Hurricanes Irene in 2011 and Sandy in 2012. In the following section, we describe some of the changes that have taken place in four of our study sites along the Delaware Bay and summarize our observations on habitat quality for horseshoe crabs in these locations.

3.4 Case Studies of SLR Along the New Jersey Coast of Delaware Bay

3.4.1 Norburys Landing

Norburys Landing has been one of our southern-most sampling stations since 1985. The landing is a minor promontory into the bay at the end of Millmann Road (County Road 642). The beach to the north is sand-starved and has been eroding for a long time. In the mid-1980s, we used to observe enormous numbers of horseshoe crabs spawning, and numerous migratory shorebirds were to be seen, particularly at high tide when horseshoe crabs were spawning. We sampled the beach at Norburys Landing from the late 1980s until 2001. Our data for egg abundance both in the surface sediments and at a depth of 20–25 cm indicate a severe drop in egg density over time. In other words, *Limulus polyphemus* was no longer using this beach for spawning, largely due to the lack of deep sand and the increase of muddy sediments along the present narrow beach. Total live horseshoe crab eggs dropped from 553,998 eggs/m^2 in June of 1990 to 104,712 eggs/m^2 in June of 2001. Norburys Landing went from a relatively robust site for spawning and egg development, to a place that horseshoe crabs no longer visited – all a result of beach erosion. No attempt has been made to restore this beach.

3.4.2 Kimbles Beach

Kimbles Beach is several miles up-bay from Norburys Landing; it includes an open bay beach and three tidal creek systems (Fig. 3.4b). When we first sampled Kimbles, the open beach was wide, gently sloped and of excellent quality sand for the development of horseshoe crab eggs. Prior to 2001, the density of horseshoe crab eggs was always quite high; it was not unusual to find densities of 10^6 eggs/m^2 (see

Table 3.1 Mean densities of horseshoe crab eggs/m^2 at three habitats in the vicinity of Kimbles Beach, New Jersey

Date	Open beach	Tidal creek	Sand bar
15 May 1997	488,689	149,821	948,079
29 May 1997	879,832	1,284,051	873,348
12 June 2001	0	172,653	80,327

Eggs were sampled using a series of replicate 5 cm cores to a depth of 20 cm (See Smith et al. 2002 for detailed description of sampling procedure)

Table 3.1 for 1997). Consequently, there were large numbers of migratory shorebirds feeding on the eggs at the water's edge. Ruddy Turnstones (*Arenaria interpres*) would dig up the nests of horseshoe crabs in search of eggs. Great Black Backed Gulls (*Larus fuscus*) could be seen feeding on overturned or stranded mated pairs of *Limulus polyphemus*. By the end of the 1990s, we noticed that the beach was not only losing sand due to erosion, but the beach was moving inland. Emerging marshes contained the roots and rhizomes of *Phragmites* and *Spartina* that had been buried for nearly 5,000 years. By 2001, we found many mated pairs of horseshoe crabs spawning, not on the beach, but on a sand bar that had formed off the mouth of the creek. We also found that mated pairs were entering the tidal creek from the bay, making their way to the newly formed sandy plots along the edge of the tidal creek. Samples taken from the open beach, the sandbar, and the tidal creek indicated that the crabs showed preference for the newly formed "marginal" habitats (see Table 3.1 for 2001).

We conclude, therefore, that the loss of pristine beach habitat over a short period may result in the development of a very rich habitat elsewhere. From the perspective of a mated pair of horseshoe crabs, not all is lost if a particular beach is no longer suitable for spawning due to beach erosion.

3.4.3 *Reeds Beach*

Reeds Beach, about a half mile north of Kimbles Beach, is probably the best known beach in the world for observing spawning horseshoe crabs and large flocks of migratory shorebirds, which feed on the abundant eggs of the crabs. The beach at Reeds is relatively narrow to non-existent in the south, but very expansive to the north (Fig. 3.4c). Why? Because there is a navigable creek at the north end of town, called Bidwells Ditch. To prevent the silting in of the creek, a large jetty was constructed that parallels the creek well out into the bay. The jetty has also acted as a block to the natural movement of sand from the south. Thus, at Reeds the southern beaches are eroding landward and the northern beaches are expanding bayward. Recently, fewer horseshoe crabs use Reeds Beach for spawning. The beach profile at the north end of Reeds has remained constant for many years; similarly, the composition of the sand is the same now as it was during the height of migratory

Table 3.2 Mean and maximum densities of horseshoe crab eggs/m^2 at the open beach at Reeds Beach, New Jersey

Date	Mean	Maximum
8 May 1986	516,956	1,197,720
21 May 1986	1,282,517	2,563,862
5 Jun 1986	1,913,395	3,979,886
14 Jun 1986	2,630,165	10,709,594
9 May 2001	325,716	708,373
23 May 2001	584,781	1,045,542
6 Jun 2001	278,489	720,128
20 Jun 2001	58,497	169,977

shorebirds and mating horseshoe crabs. It is improbable that beach quality has resulted in reduced spawning. More likely, there are simply fewer horseshoe crabs around due to a severe reduction in population density since the early 1990s. One of our earliest sampling forays onto the beach at the jetty yielded incredibly high counts of eggs in the sand (see Table 3.2 for 1986). By 2001, we found that egg density has dropped by at least one order of magnitude (see Table 3.2 for 2001). Observations of Reeds beach near the jetty in 2014 suggest that there are simply not enough crabs spawning there to sustain the level of eggs to attract shorebirds.

3.4.4 Moores Beach

Moores Beach, located about 1.5 miles (2.4 km) from the town on Route 47 in Cumberland County, was once a thriving community. During the mid-1980s, over 25 cottages (including double wide trailers) were resident on the beach. Over time, the occupants began to protect the cottages from the ravages of the encroaching bay. Our observations and sampling of the beach began before 1990; it was apparent even then that erosion was systematically destroying the town, first by damaging the most bayward cottages, then destroying the paved road, then eroding a temporary dirt road, then finally washing the most landward cottages onto the marsh (Fig. 3.5a–c). Sometime in the late 1990s to early 2000s, Moores Beach was abandoned and inaccessible. The loss of Moores Beach parallels the abandonment of other beachfront towns along the New Jersey shoreline of Delaware Bay, including Thompsons Beach and Sea Breeze, since the late 1980s.

Due to the effort of a number of agencies, and inspired by the leadership of Dr. Larry Niles, Moores Beach is in the process of being restored, with the use of funds that were recently appropriated for Superstorm Sandy recovery. The results have been striking – a person standing on today's beach (Fig. 3.5d) would see no evidence that the beach was once home to numerous summertime dwellers. Much of the added sand appears to have moved north toward the entrance of the tidal creek at the north end of Moores Beach, thereby forming sandbars and a larger beach along the creek. In the summer of 2014, we observed a fair number of horseshoe

Fig. 3.5 Composite images of Moores Beach, New Jersey showing changes between 1987 and 2014. (**a**) October 1987, showing the extensive amount of rubble fill used to protect dwellings. The trailer with the red roof seen here is the same as the one in the backgrounds of Fig. 3.5b, c, (**b**) May 1992, note the pilings in the bay where houses once stood; also note in the foreground a remnant of the older blacktopped road which was washed away, and the reddish colored earthen road to the left which became inundated and impassable, (**c**) June 1995, the town was essentially abandoned; note the extensive peat formations, a further indicator of beach erosion, (**d**) July 2014, after the removal of all debris followed by beach nourishment

crabs on the upper beach along the tidal creek. Small flocks of returning shorebirds were seen in the shallows adjacent to the sandbars. It appears that the effort to restore Moores Beach to a *natural* beach has merit. Even without beach nourishment, it is likely that some aspect of the beach will prevail even as the level of the bay rises inexorably.

3.5 Responding to Future SLR

As we have shown in the preceding sections, horseshoe crabs have successfully tracked the available shallow water marine habitats throughout geological time. Sea level rise is nothing new, and in our view, SLR itself is not a threat to horseshoe crabs. They have adapted and found suitable spawning habitats wherever they are, regardless of the position of the shoreline. Problems for horseshoe crabs are caused by man-made alterations of the coast that are engineered to "stabilize" beaches and protect coastal properties (Botton et al. 1988; Botton 2001; Berkson et al. 2009; Hsieh and Chen 2009).

By combining sea level rise scenarios with a digital elevation model, Cooper et al. (2008) showed that virtually all of the Delaware Bay coastline of New Jersey would be susceptible to inundation with as little as 0.61 m of sea level rise. Moreover, there are only very slight differences in elevation between beaches, marshes, and uplands. The amount of landward shoreline retreat is much greater than the vertical rise in sea level (Bruun 1962), and shorelines along the east coast of the US have moved inland some 23.8 m for each 0.3 m of sea level rise over the last century (Zhang et al. 2004). These studies suggest that the narrow sandy beaches on Delaware Bay, particularly those which are inhabited, are at great risk from sea level rise, even without factoring in the possible impacts of catastrophic storm events.

Moving forward, what can be done to best conserve beaches that are of importance to horseshoe crabs? There are three basic strategies that can be considered:

1. Armoring – harden the shoreline through the use of structures such as bulkheads, revetments, or groins;
2. Beach nourishment – maintain the present position of coastal dwellings through the periodic addition of massive quantities of sand;
3. Shoreline retreat – either abandon coastal property and infrastructure, or move it landward in anticipation of future SLR.

Shoreline armoring is probably most viable when the coastal zone is very highly developed and economically valuable, and where retreat is not practical, e.g. New York City. Seawalls and other such structures tend to give property owners a false sense of security, encouraging further development and artificially lowering the perceived risks of living in the coastal zone (Kousky 2014). Once in place, armored shorelines generally have less ecological functionality than the sandy beaches that they replace (e.g. Dugan et al. 2011), and in the specific case of horseshoe crabs, bulkheads or revetments may limit access to portions of the intertidal spawning habitat (Jackson and Nordstrom 2009; Jackson et al. 2010) and reduce the amount of suitable habitat (Botton et al. 1988; Lathrop et al. 2006). Given the relatively modest economic value of Delaware Bay beach front communities, when balanced against their ecological value for horseshoe crabs and shorebirds, we would not recommend this method of shore protection.

Beach nourishment is used throughout the coastal US for shore protection. It is expensive and must be done regularly in order to maintain its effectiveness. For example, the beach at Cape May, New Jersey was nourished 12 times between 1962 and 1999 at a total cost of over US $119 million; in the most recent project, over US $102 million was spent to nourish 3.06 km of beach (http://beachnourishment.wcu.edu/). The region from Cape May Inlet to Lower Township (along the Cape May Peninsula) has been nourished 14 times from 1991 to 2012 at a total cost of over US $74 million. The most recent project added sand to 4.1 km of beach at a cost of US $9.1 million, or about US $2.2 million per km (http://beachnourishment.wcu.edu/). As with shoreline armoring, beach nourishment encourages development in vulnerable coastal areas and requires enormous public subsidies when the systems fail; the cost is absorbed through Federal Emergency Management Agency (FEMA) grants and other government programs (Maly and Ishikawa 2013). And while beach nourishment may provide protection under current conditions, it is prudent to factor in SLR in any future cost/benefit scenarios (McGuire and Lynch 2013). When considering the potential benefits of beach nourishment for horseshoe crabs, it is important to choose source sediments that have the proper texture and organic carbon content (Avissar 2006) because it may prove difficult to modify the sediments once they are in place (Jackson et al. 2007). An ongoing effort to nourish several Delaware Bay beaches in the aftermath of Hurricane Sandy (Bauers 2014) employs sand trucked in from nearby sand quarries rather than beach sand; the long term suitability of these sediments for horseshoe crab spawning and egg development is not yet known. These nourished beaches may contribute sand to the mouths of nearby tidal creeks and create suitable habitat for spawning.

As sea level rises, it may become increasingly prudent to consider the third option, retreat from the coastline. Certainly, this poses numerous challenges. Many owners of coastal property will be reluctant to leave; they have come to expect that publically funded beach nourishment and/or armoring projects will continue to maintain their property values, and that flood insurance policies will be made available to them at rates that do not truly reflect the risks of living in the coastal zone (Kousky 2014). When their property is damaged, the homeowner's response is often to rebuild-in-place rather than to consider moving away from the coast. The developed Delaware Bay shoreline of New Jersey could serve as a model for considering the coastal retreat option because it has very high ecological value and is rather sparsely settled with modest property values when compared to houses along the Atlantic coast barrier beaches.

As an illustration of the comparative economics of beach nourishment and buyouts, we examined the southern section of Reeds Beach, New Jersey (Table 3.3). In order to approximate the one-time costs of a buyout, we used the most recent tax assessment data from the township to estimate of the value of the land and property, and we also factored in the demolition costs. In estimating the costs of beach nourishment, we assumed that costs would be comparable to a recent project in Lower Township-Cape May, US $2.2 million per km (http://beachnourishment.wcu.edu/) and that nourishment of the 907 m of beach would need to take place once per decade. Maintaining the existing housing would also require continued services, including roads, electricity and telephone services, but we did not include these in our analysis. Based on our estimates, the one-time costs of a buyout would be

Table 3.3 Comparison of the costs (in US $) of beach nourishment and buyout scenarios for South Reeds Beach

Category	Total land value[a]	Total dwelling value[a]	Total value[a]
1. Buyout scenario			
All lots (n=30)	$2,862,100	$1,309,600	$4,171,700
Occupied lots (n=22)	$2,848,700	$1,309,600	$4,158,300
Vacant lots (n=8)	$13,400	$0	$13,400
Demolition and removal of debris[b]			$250,000
Total costs			$4,421,700
2. Nourishment scenario			
Length of beach = 907 km			
Cost per km = $2,200,000[c]			
Cost = $2,000,000 per event, repeated at 5–10 year intervals for an indefinite period of time			
Additional costs: maintain roads, electricity, telephone lines, etc.			

[a]Data on property values is taken from Middle Township, New Jersey for tax year 2013 (http://www.middletownship.com/Assessor/yearly%20assessment%20listing%202013.pdf, accessed 6 July 2014)

[b]Estimated at $10,000 for demolition and removal of debris from existing homes, plus $30,000 for removal of pilings and bulkheads from vacant lots

[c]Estimated at $2,200,000 per km for the most recent nourishment project at Cape May, New Jersey (http://beachnourishment.wcu.edu/, accessed 6 July 2014)

approximately US $4.4 million, versus approximately US $2.0 million for the first nourishment project, and approximately US $2.4 million for the second nourishment project 10 years later (assuming an inflation rate of 2 % per annum). In other words, beyond 10 years, the economics shift in favor of a buyout. Interestingly, there is renewed interests on the part of residents of Delaware Bay shore communities to consider buyouts due to the damage caused by Hurricane Sandy in 2012 (Nutt 2013).

3.6 Summary and Conclusions

The geological record informs us that there have been many fluctuations of sea level during the 445 million years that horseshoe crabs have existed. We suggest that these animals have always been able to track the available sandy beaches which they then used for spawning. Today, concern for the state of coastal beaches in Delaware Bay is a contentious issue because humans, horseshoe crabs and migratory shorebirds co-occur. Many beachfront communities that are now experiencing SLR (whether they believe it or not!) are in trouble due to the inexorable landward movement of the intertidal zone. Recent storms, such as Hurricane Sandy, have certainly accelerated the problem, but the erosion of Delaware Beaches has been occurring for a much longer period of time (Fig. 3.4). Several options, among many, are most common to temporarily stabilize the land where dwellings occur. The most popular, and expensive, method is to stay put and restore the dwelling by rebuilding and/or elevating the building to conform to federal standards. Another method of protection is to build higher berms or vegetated dunes, often in association with beach nourishment. A third solution, but not final, is to remove the dwellings from the beach and then restore habitat through a beach nourishment program. It is a costly procedure to "buy out" an entire community, or even a portion thereof. After the community is abandoned, infra-structural elements must be removed and sand must be brought to the beach by truck or dredge. None of these approaches provides a lasting buffer against SLR. However, we also are aware that horseshoe crabs can benefit by the third solution above. The case of Moores Beach is a good example of returning a beach area to its original state, before it was occupied by settlers. The reason that we are cautious regarding any of the above actions is that *the beach will still move inland as SLR proceeds, and sand will have to be periodically renewed on the restored area.* Over extensive coastal regions of Delaware Bay, new habitat in the form of overwash beaches, sandbars, and sand beaches along tidal creeks is emerging in areas that are not at all influenced by human activity. This is the condition that has challenged spawning horseshoe crabs for thousands of years, and they have readily adapted to the changing environment spurred on by SLR.

Acknowledgements Principal support for our work in Delaware Bay has been provided by funds from New Jersey Sea Grant, Public Service Enterprise Group, US Army Corps of Engineers Philadelphia District, New Jersey Audubon Society, New Jersey Department of Environmental Protection, Fordham University Research Council, and New Jersey Chapter of the Nature Conservancy. Our research used the facilities of the Haskin Shellfish Laboratory (Rutgers University) Cape Shore Laboratory, and we are grateful to the directors and staff for their cooperation. Lastly, we thank our many student assistants for their crucial help in the field and laboratory.

References

Army Corps of Engineers (2004) Coastal geology. University Press of the Pacific, Honolulu

Atlantic States Marine Fisheries Commission (ASMFC) (1998) Interstate fishery management plan for horseshoe crab. Atlantic States Marine Fisheries Commission, Fishery Management Report No. 32. Washington, DC

Avissar NG (2006) Modeling potential impacts of beach replenishment on horseshoe crab nesting habitat suitability. Coast Manag 34:427–441

Bauers S (2014) Projects to restore habitat post-Sandy begin. http://articles.philly.com/2014-04-04/news/48838956_1_red-knot-moores-beach-nongame-species-program. Accessed 19 Aug 2014

Berkson J, Chen CP, Mishra J et al (2009) A discussion of horseshoe crab management in five countries: Taiwan, India, China, United States, and Mexico. In: Tanacredi JT, Botton ML, Smith DR (eds) Biology and conservation of horseshoe crabs. Springer, New York, pp 465–474

Błażejowski B (2015) The oldest species of the genus *Limulus* from the Late Jurassic of Poland. In: Carmichael RH, Botton M, Shin PKS, Cheung SG (eds) Changing global perspectives on horseshoe crab biology, conservation and management. Springer, Cham pp 3–14

Botton ML (2001) The conservation of horseshoe crabs: what can we learn from the Japanese experience? In: Tanacredi JT (ed) *Limulus* in the limelight. Kluwer Academic/Plenum Publishing, New York, pp 41–51

Botton ML, Loveland RE, Jacobsen TR (1988) Beach erosion and geochemical factors: influence on spawning success of horseshoe crabs (*Limulus polyphemus*) in Delaware Bay. Mar Biol 99:325–332

Botton ML, Loveland RE, Jacobsen TR (1994) Site selection by migratory shorebirds in Delaware Bay, and its relationship to beach characteristics and the abundance of horseshoe crab (*Limulus polyphemus*) eggs. Auk 111:605–611

Botton ML, Loveland RE, Tanacredi JT et al (2006) Horseshoe crabs (*Limulus polyphemus*) in an urban estuary (Jamaica Bay, New York), and the potential for ecological restoration. Estuar Coasts 29:820–830

Bruun P (1962) Sea level rise as a cause of shore erosion. J Waterw Harb Div 88:117–130

Cooper MJP, Beevers MD, Oppenheimer M (2008) The potential impacts of sea level rise on the coastal region of New Jersey, USA. Clim Chang 90:475–492

Diedrich CG (2011) Middle Triassic horseshoe crab reproduction areas on intertidal flats of Europe with evidence of predation by archosaurs. Biol J Linn Soc 103:76–105

Draxler AFJ (1993) The influence of sediment biogeochemistry on soft-bottom marine benthos. Ph.D. thesis, Rutgers University

Dugan JE, Airoldi L, Chapman MG et al (2011) Estuarine and coastal structures: environmental effects, a focus on shore and nearshore structures. Treat Estuar Coast Sci 8:17–41

Garrison T (1993) Oceanography. Wadsworth Publishing Co., Belmont

Harper D (2013) Roadside geology of New Jersey. Mountain Press Publishing Co., Missoula

Hsieh H-L, Chen C-P (2009) Conservation program for the Asian horseshoe crab *Tachypleus tridentatus* in Taiwan: characterizing the microhabitat of nursery grounds and restoring spawning grounds. In: Tanacredi JT, Botton ML, Smith DR (eds) Biology and conservation of horseshoe crabs. Springer, New York, pp 417–438

IPCC (2013) Climate change 2013. The physical science basis. Working Group I Contribution to the Fifth Assessment Report of the Intergovernmental Panel on Climate Change Summary for policymakers. http://www.climatechange2013.org/images/uploads/WGI_AR5_SPM_brochure.pdf. Accessed 6 July 2014

Jackson NL, Nordstrom KF (2009) Strategies to conserve and enhance sandy barrier habitat for horseshoe crabs (*Limulus polyphemus*) on developed shorelines in Delaware Bay, USA. In: Tanacredi JT, Botton ML, Smith DR (eds) Biology and conservation of horseshoe crabs. Springer, New York, pp 399–416

Jackson NL, Smith DR, Tiyarattanachai R et al (2007) Evaluation of a small beach nourishment project to enhance habitat suitability for horseshoe crabs. Geomorphology 89:172–185

Jackson NL, Nordstrom KF, Smith DR (2010) Armoring of estuarine shorelines and implications for horseshoe crabs on developed shorelines in Delaware Bay. In: Shipman H, Dethier MN, Gelfenbaum G, Fresh KL, Dinicola RS (eds) Puget sound shorelines and the impacts of armoring–proceedings of a state of the science workshop, May 2009. US Geological Survey Scientific Investigations Report 2010–5254, pp 195–202

Kin A, Blazejowski B (2014) The horseshoe crab of the genus *Limulus*: living fossil or stabilomorph? PLoS ONE 9(10):e108036. doi:10.1371/journal.pone.0108036

Kin A, Gruszczynski M, Martill M et al (2013) Palaeoenvironment and taphonomy of a Late Jurassic (Late Tithonian) Lagerstätte from central Poland. Lethaia 46:71–81

Kipling R (1902) Chapter 10: the crab that played with the sea. In: Just so stories for little children. Macmillan & Co., New York

Knebel HJ, Fletcher CH III, Kraft JC (1988) Late Wisconsinian-Holocene paleogeography of Delaware Bay: a large coastal plain estuary. Mar Geol 83:115–133

Kousky C (2014) Managing shoreline retreat: a US perspective. Clim Chang 124:9–20

Kreamer G, Michels S (2009) History of horseshoe crab harvest on Delaware Bay. In: Tanacredi JT, Botton ML, Smith DR (eds) Biology and conservation of horseshoe crabs. Springer, New York, pp 299–313

Lathrop RG Jr, Allen M, Love A (2006) Mapping and assessing the critical horseshoe crab spawning habitats of Delaware Bay. Rutgers University, Center for Remote Sensing and Spatial Analysis. http://crssa.rutgers.edu/projects/coastal/hcrab/report/ALS_DelBay_hcrab_report_20060718.pdf

Levin J, Hochstein HD, Novitsky TJ (2003) Clotting cells and *Limulus* amoebocytes lysate: an amazing analytical tool. In: Shuster CN Jr, Barlow RB, Brockmann HJ (eds) The American horseshoe crab. Harvard University Press, Cambridge, MA, pp 310–340

Maly E, Ishikawa E (2013) Land acquisition and buyouts as disaster mitigation after Hurricane Sandy in the United States. In: Proceedings of international symposium on City Planning 2013, City Planning Institute of Japan, Sendai. http://www.cpij.or.jp/com/iac/sympo/13/ISCP2013-8.pdf. Accessed 6 July 2014

McGuire CJ, Lynch D (2013) Thinking ahead: the impacts of sea level rise on coastal landscape protections. Nat Resour Environ 27(4):1–5

McPhee J (1982) In suspect terrain. Farrar, Straus, and Giroux, New York

Mizrahi DS, Peters KA (2009) Relationships between sandpipers and horseshoe crabs in Delaware Bay: a synthesis. In: Tanacredi JT, Botton ML, Smith DR (eds) Biology and conservation of horseshoe crabs. Springer, New York, pp 65–87

Nutt AE (2013) State to buy back Hurricane Sandy-damaged homes in Cumberland County. http://www.nj.com/news/index.ssf/2013/12/christie_to_buy_homes_in_cumberland_county_damaged_by_sandy.html. Accessed 6 July 2014

Rudkin DM, Young GA (2009) Horseshoe crabs–an ancient ancestry revealed. In: Tanacredi JT, Botton ML, Smith DR (eds) Biology and conservation of horseshoe crabs. Springer, New York, pp 25–44

Sekiguchi K (1988) Biology of horseshoe crabs. Science House, Tokyo

Shuster CN Jr (2003) King crab fertilizer: a once-thriving Delaware Bay industry. In: Shuster CN Jr, Barlow RB, Brockmann HJ (eds) The American horseshoe crab. Harvard University Press, Cambridge, MA, pp 341–357

Shuster CN Jr, Anderson LI (2003) A history of skeletal structures: clues to relationships among species. In: Shuster CN Jr, Barlow RB, Brockmann HJ (eds) The American horseshoe crab. Harvard University Press, Cambridge, MA, pp 154–183

Shuster CN Jr, Botton ML (1985) A contribution to the population biology of horseshoe crabs, *Limulus polyphemus* (L.), in Delaware Bay. Estuaries 8:363–372

Smith DR, Pooler PS, Loveland RE et al (2002) Horseshoe crab (*Limulus polyphemus*) reproductive activity on Delaware Bay beaches: interaction with beach characteristics. J Coast Res 18:730–740

Wegener A (1929) The origin of continents and oceans. Translated by John Biram, 1962. Dover Publications, Inc., Mineola

Winchester S (2003) Krakatoa. HarperCollins Perennial Publications, New York

Zhang K, Douglas BC, Leatherman SP (2004) Global warming and coastal erosion. Clim Chang 64:41–58

Chapter 4
Conservation Genetics of the American Horseshoe Crab (*Limulus polyphemus*): Allelic Diversity, Zones of Genetic Discontinuity, and Regional Differentiation

Tim L. King, Michael S. Eackles, Aaron W. Aunins, H. Jane Brockmann, Eric Hallerman, and Bonnie L. Brown

Abstract Extensive studies of genetic variation of Atlantic horseshoe crab *Limulus polyphemus* populations have revealed the presence of considerable allelic diversity and population structuring that appear to reflect the actions of various evolutionary processes. We have expanded on our previous efforts to gain a more refined understanding of *L. polyphemus* population structure by surveying 792 additional animals distributed among 12 additional spawning aggregations. Here we report on variation at 13 microsatellite DNA markers for 1,684 horseshoe crabs sampled from 33 spawning assemblages from northern Maine to the Yucatan Peninsula, Mexico. Average unbiased heterozygosity (uH_E) was high (0.74±0.01), the number of private alleles was low (0.06±0.04), effective population size (N_e) ranged from 22 to 187, inbreeding (F) ranged from −0.07 to 0.07, and tests for genic differentiation among populations indicated shallow but statistically significant differentiation within regions and highly significant differences among regions ($P<0.005$). Current findings are consistent with previous research by this group in suggesting a series of genetic discontinuities across the species' range that could indicate regional adap-

T.L. King (✉) • M.S. Eackles
U.S. Geological Survey, Biological Resources Division, Leetown Science Center, Aquatic Ecology Branch, 11649 Leetown Road, Kearneysville, WV 25430, USA
e-mail: tlking@usgs.gov

A.W. Aunins
Cherokee Nation Technology Solutions, Leetown Science Center, 11649 Leetown Road, Kearneysville, WV 25430, USA

H.J. Brockmann
Department of Biology, University of Florida, Gainesville, FL 32611-8525, USA

E. Hallerman
Department of Fish and Wildlife Conservation, Virginia Polytechnic Institute and State University, Blacksburg, VA 24061, USA

B.L. Brown
Department of Biology, Virginia Commonwealth University, 1000 W. Cary Street, Richmond, VA 23284, USA

R.H. Carmichael et al. (eds.), *Changing Global Perspectives on Horseshoe Crab Biology, Conservation and Management*, DOI 10.1007/978-3-319-19542-1_4

tive significance or reflect vicariant geographic events. Additional collections allowed improved delineation of structuring (as reflected by two new zones of genetic discontinuity) along the southeast Atlantic coast as well as identification of previously undetected shallow but significant structuring along the Florida Gulf coast. Regional groupings may warrant management unit recognition based on the patterns observed among multiple genetic metrics. The integration of this information with previously identified genetic variation and ecological data is essential to developing an ecologically and evolutionarily sound conservation management strategy.

Keywords *Limulus polyphemus* • American horseshoe crab • Microsatellite DNA • Genetic diversity • Differentiation • Genetic discontinuity • Management units

4.1 Introduction

The American horseshoe crab, *Limulus polyphemus*, is a resource of economic and biological significance along North America's Atlantic coast, particularly in the mid-Atlantic region of the United States (Shuster et al. 2003). *Limulus polyphemus* eggs are a primary food source for migratory shorebirds such as the western Atlantic red knot, *Calidris canutus rufa* (Baker et al. 2004), and other life stages serve as a food source for many species of finfish and the protected Atlantic loggerhead sea turtle (*Caretta caretta*). Horseshoe crabs are commercially harvested in large numbers as bait for American eel (*Anguilla rostrata*) and whelk (*Busycon* spp.) fisheries (Walls et al. 2002). In addition, *L. polyphemus* are harvested for non-destructive use in the production of *Limulus* amoebocyte lysate (LAL, Levin and Bang 1968), the standard reagent used to detect bacterial endotoxins in patients, injectable drugs, intravenous devices, and other medical applications.

As a direct result of perceived declines in the abundance of *L. polyphemus*, the Atlantic States Marine Fisheries Commission (ASMFC 1998) adopted a fishery management plan (FMP) for the species. The FMP (as amended) makes stock identification a principal research need in regional management of the species. Heritable genetic information offers an objective means of defining management units and provides an evolutionary framework from which to develop and evaluate conservation priorities (Ryder 1986; Moritz 1994). Molecular genetics occupies an important place in contemporary conservation biology as a robust tool for identifying fine-scale population structure, determining the degree of reproductive isolation among populations, and identifying the presence and extent of metapopulation structure (Hallerman 2003).

A range of molecular genetic techniques has been utilized in attempts to assess population structure in *L. polyphemus*. A survey of allozyme variation among four collections suggested that Atlantic Ocean and Gulf of Mexico populations of *L. polyphemus* were genetically differentiated (Selander et al. 1970). A subsequent study of mitochondrial DNA (mtDNA) variation identified a major genetic discontinuity distinguishing northern from southern populations, with the phylogeographic break occurring around Cape Canaveral, Florida (Saunders et al. 1986). Pierce et al. (2000) reported little evidence of gene flow between Delaware and Chesapeake Bay *L. polyphemus* populations as reflected by sequence variation in the mtDNA COI region, although variation at randomly amplified polymorphic DNA (RAPD) markers was uniform, implying that gene flow may be sex-biased. Simple-sequence repeat loci, often termed microsatellites, have become the standard markers for identification of population structure due to their hypervariability and codominant expression (Ellegren 2004). These markers have provided previously unrealized utility at the individual level, including genetic tagging, assignment (or allocation) to population or collection of origin (Cornuet et al. 1999), assessment of relatedness (Brockmann et al. 1994, 2000), and demonstrations of sex-biased dispersal (Goudet et al. 2002). Most recently, King et al. (2005) surveyed selectively neutral genetic variation at 14 microsatellite DNA markers of 892 *L. polyphemus* sampled at 21 locations from northern Maine to the Yucatan Peninsula, Mexico. This extensive intraspecific examination of the nuclear genome revealed the presence of considerable allelic diversity and population structuring that appeared to reflect various evolutionary processes. Their findings suggested the presence of similar levels of genetic diversity and variation among the collections, punctuated by a series of genetic discontinuities of varying "depth" across the species' range that could indicate demographic independence, regional adaptation, and/or reflect vicariant geographic events. Gender-specific estimates of population differentiation (F'_{ST}) and assignment to collection of origin suggested the presence of male-biased dispersal throughout the study area by virtue of observing higher differentiation among females across populations than for males. Moreover, patterns of population relatedness were consistent with the observations that populations at both ends of the species' range are more differentiated from proximal populations than those in the middle, the zone of greatest abundance. Faurby et al. (2010) applied Bayesian coalescent-based methods to these microsatellite data to infer the historic demography of *L. polyphemus* populations. Their results showed strong declines in population sizes throughout the species' distribution except in the geographically isolated southernmost Mexico population, where a strong increase in population size was observed. Analyses suggested that demographic changes in the core of the distribution occurred within the last 150 years and thus were likely caused by anthropogenic effects including past overharvest of the species for fertilizer, and current use of the animals as bait and for biomedical testing (i.e., for production of *Limulus* amebocyte lysate or LAL). Declines of the peripheral northern and southern populations that occurred during the "Little Ice Age" are suggested to most likely have been climatically driven (Faurby et al. 2010).

A key short-term goal for horseshoe crab management is to sustain populations demographically, whereas the long-term goal is to conserve adaptively important genetic variation to maintain the species' evolutionary potential. The management process must begin with a definition of the biologically appropriate units of conservation. Against this background, we here expand on the previous surveys of microsatellite DNA variation among spawning aggregations (predictable gatherings of adults with the specific purpose of reproducing) of *L. polyphemus* (King et al. 2005; Faurby et al. 2010) to identify previously undetected population, phylogeographic, and evolutionary relationships. We report on the screening of 13 polymorphic markers from 1,684 animals collected throughout the species' range from 33 spawning aggregations to estimate allelic diversity, characterize allelic patterns within and among collections, identify phylogeographic structure, and assess the demographic status of each collection. This research, which includes collections from 12 previously untested populations distributed from Connecticut to Florida's Gulf Coast (Table 4.1; Fig. 4.1), provides a refined view of the nature and extent of neutral (assumed) genetic variation in the southeastern portion of the *L. polyphemus* range. The integration of this information with previously identified allozyme (Selander et al. 1970), mitochondrial DNA (Saunders et al. 1986; Pierce et al. 2000), microsatellite DNA (King et al. 2005; Faurby et al. 2010), and ecological data is essential to developing an ecologically and evolutionarily sound conservation management strategy.

4.2 Methods

4.2.1 Sample Collections, DNA Extraction, and Microsatellite Genotyping

Limulus polyphemus were sampled in 32 spawning aggregations along the United States Atlantic coast from Franklin, Maine to St. Joseph Bay, Florida (Table 4.1; Fig. 4.1). A collection also was obtained from the north coast of the Yucatan Peninsula in the Ria Lagartos estuary, Mexico. Crabs were hand collected from shallow waters. Non-lethal somatic tissue samples were obtained by either clipping a 5 mm section from the terminus of a locomotive appendage or collecting a 100 μL sample of hemolymph via syringe. Solid samples were preserved in 95 % ethanol and stored at 4 °C; hemolymph was spotted onto two quadrants of an FTA card (Whatman, GE Health Care Life Sciences) and allowed to air-dry. After documenting sex and measuring prosomal width, each specimen was returned live to the location from which it was collected. Genomic DNA from somatic tissue was extracted with one of two extraction methods: the Puregene DNA extraction kit (Gentra Systems, Inc., Minneapolis, MN, used according to manufacturer's guidelines), or using glass milk according to the method outlined by Höss and Pääbo (1993). DNA isolated with the Puregene kit was resuspended in 100 μl of 10 mM Tris-HCl, pH 8.0, 1 mM EDTA before use in PCR. The supernatant resulting from the glass milk extraction was not diluted and used directly in PCR. For hemolymph samples preserved on FTA cards, DNA was extracted from the card following the manufacturer's instructions.

Thirteen polymorphic microsatellite loci (King et al. 2005) were used for the present analysis. The PCR reaction mixtures consisted of 100–200 ng of genomic

Table 4.1 Abbreviation, general location, and sample size for 33 spawning collections of Atlantic horseshoe crabs *Limulus polyphemus* genotyped at 13 microsatellite DNA loci

Abbreviation	Spawning collection site	Sample size
MEH	Hog Bay, Franklin, Maine	47
MET	Thomas Point Beach, Maine	45
MEM	Middle Bay, Brunswick, Maine	48
NHS	Chadman's Landing, Squamscott River, New Hampshire	48
MAP	Pleasant Bay, Massachusetts	48
RIN	Green Island, Narragansett Bay, Rhode Island	48
CTH	***Housatonic River, Milford Point, Connecticut***	***48***
NYP	Great Peconic Bay, Long Island, New York	48
NJF	Fortescue Beach, New Jersey	48
NJR	Reeds Beach, New Jersey	48
NJH	Highs Beach, New Jersey	49
DKH	Kitts Hummock Beach, Delaware	36
DBS	Big Stone Beach, Delaware	31
DFB	Fowler Beach, Delaware	47
MDT	Turkey Point, Chesapeake Bay, Maryland	30
MDF	Flag Pond State Park, Chesapeake Bay, Maryland	29
MD5	***Ocean City, Maryland – 2005***	***48***
MD6	***Ocean City, Maryland – 2006***	***48***
VAC	***Chincoteague, Virginia***	***48***
NCS	Shackleford Banks, North Carolina	55
SBB	Bulls Bay, South Carolina	53
SBE	***Beaufort, South Carolina***	***48***
GSA	***Savannah, Georgia***	***48***
GSI	***Sapelo Island, Georgia***	***32***
FIR	Indian River, Florida (Atlantic coast)	46
FBB	***Biscayne Bay***	***20***
FMI	***Tiger Tail Beach, Marco Island, Florida (Gulf coast)***	***81***
FCH	***Charlotte Harbor, Florida***	***51***
FTB	***Tampa Bay, Florida***	***141***
FCK	Seahorse Key, Cedar Keys NWR, Florida	132
FAP	***Alligator Point, Apalachicola Bay, Florida***	***92***
FSJ	St. Joseph Bay, Florida	23
MXY	Ria Lagartos and San Felipe, Yucatan, Republic of Mexico	20
	Total	1,684

This presentation expands on the previous surveys of microsatellite DNA variation among populations of *L. polyphemus* (King et al. 2005; Faurby et al. 2010) by including collections from 12 previously untested spawning aggregates identified here by ***bolded*** and *italicized* text

DNA, 1× PCR buffer (10 mM Tris-HCl, pH 8.3, 50 mM KCl), 2 mM $MgCl_2$, 0.15 % Tween 20, 0.25 mM dNTPs, 0.5 μM forward and reverse primer, and 0.1 U *Taq* DNA polymerase (Promega, Madison, WI, USA) in a total volume of 10–20 μl.

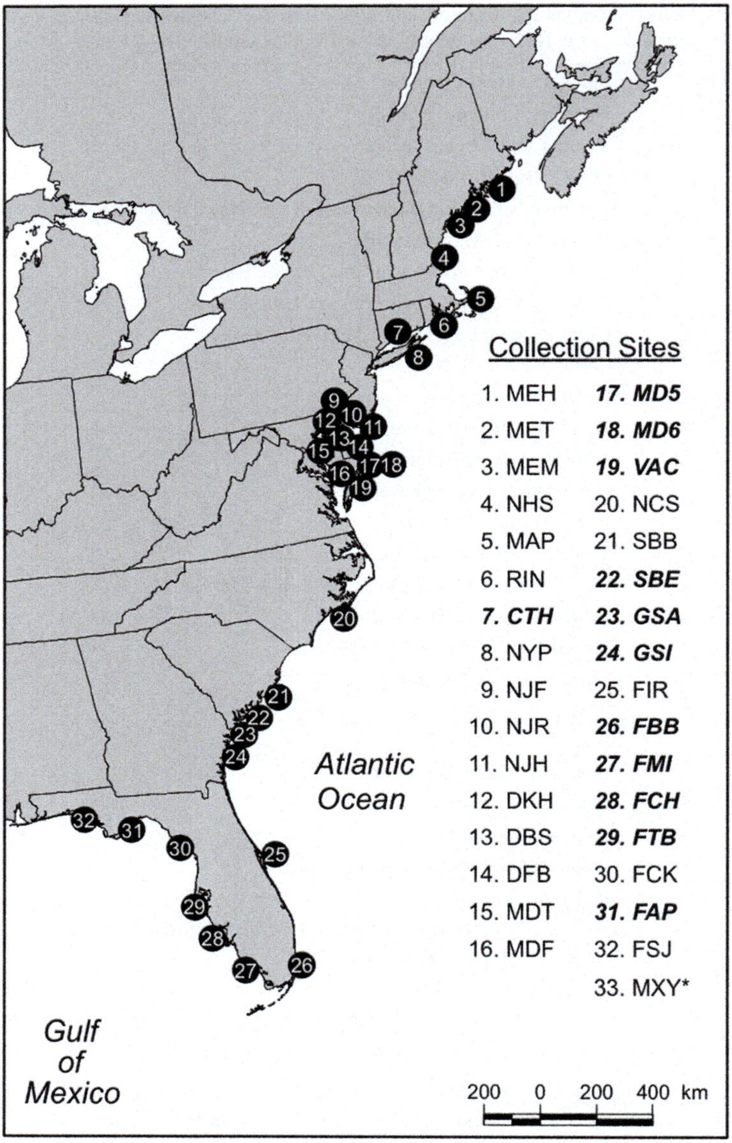

Fig. 4.1 Map showing general locations for 32 of 33 collection sites of *Limulus polyphemus* along the Atlantic and Gulf coasts of the United States. Collection names are provided in Table 4.1. The 12 new collections are identified by *bolded* and *italicized* text. The collection from a site on Mexico's Yucatan Peninsula is not depicted

Amplifications were carried out on either a PTC-200 or PTC-225 Thermal Cycler (MJ Research) using the following procedure: initial denaturing at 94 °C for 2 min; 35 cycles of 94 °C for 40 s, 58 °C for 40 s, 72 °C for 1 min; and a final extension at

72 °C for 5 min. Fragment electrophoresis and scoring were performed according to protocols described by King et al. (2005).

4.2.2 Statistical Analyses

4.2.2.1 Basic Population Genetic Analyses

Genetic diversity of all collections was quantified using GenAlEx (Peakall and Smouse 2006, 2012) to calculate allelic frequencies, number of alleles per locus (N_A), effective number of alleles (A_E), observed heterozygosity (H_O), unbiased expected heterozygosity (uH_E), and the average (across loci) inbreeding coefficient (F_{IS}). Observed genotype frequencies were tested for consistency with Hardy-Weinberg and linkage equilibrium expectations using randomization tests implemented by GENEPOP 4.3 (Raymond and Rousset 1995). The Hardy-Weinberg test used the Markov chain randomization test of Guo and Thompson (1992) to estimate exact two-tailed *p*-values for each locus in each sample. Global tests combined these results over loci and sampling locations using Fisher's method (Sokal and Rohlf 1994). Linkage disequilibrium tests used the randomization method of Raymond and Rousset (1995) for all pairs of loci. Sequential Bonferroni adjustments (Rice 1989) were used to determine statistical significance for these and all other multiple tests.

4.2.2.2 Family Structure Analysis

Unidentified family structure can be problematic for detection of hidden population structure using Bayesian clustering programs like STRUCTURE, as collections dominated by one or a few families can lead to the false interpretation of genetic differentiation or an entire population being out of Hardy-Weinberg equilibrium (Ramilo and Wang 2012). Therefore, to determine whether our collections consisted of a small number of families, we analyzed each collection for the presence of full-sibling families using the program COLONY v2.0 (Wang and Santure 2009). Settings for COLONY analyses included the assumption of male and female polygamy, no genotyping error information, no inbreeding, long run length with the full likelihood analysis method, high likelihood precision, no allele frequency updates, and no sibship prior for Bayesian analysis. Samples were analyzed as offspring without separation into candidate male and female genotypes and progeny, as these data were not available. Although the inference of family relationships is weakened in this situation with no sex, age, relationship information, and the assumption of polygamy for both sexes, COLONY is considered to be more accurate than pairwise estimates of relationships (Wang and Santure 2009). As a representation of the genetic effects of breeding structure, family size, and previous inbreeding, an estimate of the effective number of breeders, also known as the effective population size, N_e, was estimated by sibship assignment using the COLONY full likelihood method assuming random mating.

4.2.2.3 Estimates of Population Differentiation (F'_{ST})

We calculated F'_{ST} in GenAlEx using the scaling approach of Meirmans (2006), which provides a true measure of allelic differentiation between populations where $F'_{ST}=1$ when populations share no alleles. Significance of all pairwise F'_{ST} comparisons was assessed through 9,999 permutations. We estimated the effective number of migrants ($N_e m$) among collections using the private alleles method of Barton and Slatkin (1986) implemented in GENEPOP, instead of the F_{ST} -based estimate. These estimates of $N_e m$ are intended to approximate the relative magnitude of historical gene flow among collections.

4.2.2.4 Analysis of Molecular Variance

To describe differentiation among various regional assemblages of populations, we used a hierarchical AMOVA (Excoffier et al. 1992) implemented in GenAlEx to partition the genetic variance among collections relative to the total variance (F_{RT}), as well as among collections within embayments (F'_{ST}). Within regional groupings, the collections were further divided into various groups of collections to investigate whether there was significant partitioning of genetic variance among spawning assemblages. Significance of all hierarchical AMOVA analyses was assessed through 9,999 permutations.

4.2.2.5 Evolutionary Relationships Among Populations

Evolutionary relationships among the *L. polyphemus* collections were visualized by analysis of the pair-wise genetic distance matrices calculated using the Cavalli-Sforza and Edwards (1967) chord distance in BIOSYS (Swofford and Selander 1981). Principal coordinate analyses were used to graphically compare the pair-wise chord distances without imposing the appearance of a bifurcating evolutionary history (ordinated with PAlaeontological STatistics ver. 2.17c, PAST; Hammer et al. 2001).

The Bayesian clustering program STRUCTURE ver 2.3.1 (Pritchard et al. 2000) was utilized to determine the number of populations (or clusters, K) present among the collections sampled. Unlike F'_{ST} estimates of population structure, STRUCTURE does not rely on a priori grouping of populations for inference. Because a single STRUCTURE analysis on a set of populations may only reveal the uppermost level of population structure (Evanno et al. 2005), we performed a hierarchical STRUCTURE analysis similar to that employed by King et al. (2006). In the initial phase, $K=1$ to $K=33$ clusters were considered for all collections pooled together using 100,000 iterations discarded as burn-in followed by 200,000 iterations, and ten independent runs for each K using the admixture model and un-correlated allele frequencies. Subsequent analysis of each cluster tested $K=1$ to $K=C+3$ (the number of collections [C] included in the subset plus three) using the admixture and correlated allele frequencies model with the same number of burn-in and iterations. Sample location was incorporated as prior information, as populations separated by extreme distances presumably do not exchange migrants. The number of clusters for each analysis was determined using the ΔK method of Evanno et al. (2005) performed in the program Structure Harvester (Earl and vonHoldt 2011).

Isolation-by-distance as a mechanism to explain phylogeographic structuring was examined by correlating the matrices of genetic (D_C) and geographic distances (kilometers) between each pair-wise comparison. Geographic distance was estimated as the shortest ocean distance between collection sites. The statistical significance of the correlation between genetic and geographic distance matrices was assessed with a Mantel randomization test performed by the MXCOMP routine in NTSYS-PC 2.10 (Rohlf 2000).

4.2.2.6 Assignment Testing

Maximum likelihood assignment tests (after Paetkau et al. 1995) were used to determine the likelihood of each individual's multilocus genotype being found in the collection from which it was sampled (without replacement) using the program GeneClass II (Piry et al. 2004). In the event of null frequencies, a constant likelihood of 0.01 was assumed. Based on the results of population structure analyses (see below), we also assessed assignment success to various region of origin scenarios.

4.3 Results

4.3.1 Basic Population Genetic Parameters

Genotype data were collected at 13 microsatellite DNA loci for 1,684 *L. polyphemus* sampled from 33 locations encompassing the range of the species. A high degree of genetic diversity was detected; 251 alleles were observed across the 13 loci ranging from 11 at *Lpo*A37 and LpoA315 to 65 at *Lpo*D6. The mean number of alleles per locus was lowest at the extremes of the species' range (3.0, MEH and 5.5, MXY) and greatest in the collections from the Gulf coast of the U.S. (e.g., 16.0, FTB). Heterozygosity was lowest in animals collected at the extremes of the range (Table 4.2) and relatively uniform (average 76 %) for the remainder of the collections. Estimates of individual pair-wise genetic distances, using the proportion of shared alleles, indicated that levels of genetic diversity observed among the 13

Table 4.2 Summary genetic data for 33 Atlantic horseshoe crabs *Limulus polyphemus* collections from throughout their native range in the coastal USA and Mexico

Collection	N	N_a	A_E	uH_E	H_O	HWE	F_{IS}	Private alleles	SE private alleles	N_e	Lower CI_{95}	Upper CI_{95}	NFSF
MEH	47	3.00	1.93	0.42	0.40	0.05	0.06	0.00	0.00	22	14	42	3
MET	45	6.85	4.01	0.72	0.72	0.00	0.01	0.00	0.00	56	36	86	0
MEM	48	8.23	3.71	0.71	0.70	0.50	0.01	0.00	0.00	62	42	96	1
NHS	48	7.92	4.11	0.73	0.75	0.83	–0.04	0.00	0.00	50	33	79	0
MAP	48	8.92	4.77	0.77	0.75	***<0.001***	0.03	0.08	0.08	60	41	92	0
RIN	48	9.38	4.82	0.76	0.76	0.72	0.02	0.00	0.00	65	44	99	0
CTH	48	9.23	4.51	0.75	0.71	***<0.001***	0.05	0.15	0.10	56	38	87	0
NYP	48	9.31	4.90	0.76	0.72	0.55	0.04	0.00	0.00	64	43	101	1
NJF	48	9.15	4.94	0.77	0.73	0.24	0.05	0.00	0.00	69	46	105	0
NJR	48	9.23	4.70	0.76	0.74	0.17	0.03	0.00	0.00	62	41	96	0
NJH	49	9.15	4.53	0.75	0.72	0.35	0.04	0.00	0.00	71	49	107	1
DKH	36	8.46	4.40	0.75	0.72	0.05	0.04	0.00	0.00	68	42	121	1
DBS	31	8.46	4.67	0.76	0.74	0.22	0.03	0.08	0.08	53	33	90	0
DFB	47	9.31	4.82	0.75	0.71	0.08	0.05	0.00	0.00	66	45	102	2
MDT	30	7.77	4.24	0.72	0.70	0.15	0.01	0.00	0.00	40	24	70	1
MDF	29	8.31	4.66	0.75	0.76	0.06	0.00	0.00	0.00	54	34	94	0
MD5	48	9.46	4.97	0.76	0.73	0.20	0.04	0.00	0.00	65	44	101	0
MD6	48	8.62	4.78	0.76	0.74	0.01	0.06	0.00	0.00	65	44	100	2
VAC	48	9.15	4.94	0.77	0.74	0.01	0.05	0.00	0.00	54	37	87	0
NCS	55	10.00	5.16	0.78	0.80	0.86	-0.03	0.00	0.00	78	55	116	0
SBB	53	10.31	5.15	0.75	0.72	***<0.001***	0.07	0.08	0.08	82	56	124	1
SBE	48	9.62	5.13	0.75	0.73	0.17	–0.01	0.00	0.00	55	37	87	0

GSA	48	9.77	5.06	0.75	0.73	0.46	0.02	0.00	0.00	68	46	105	0
GSI	32	9.23	5.07	0.75	0.74	0.02	0.02	0.08	0.08	57	36	97	0
FIR	46	8.46	4.51	0.74	0.71	0.49	0.04	0.08	0.08	70	47	107	0
FBB	20	9.31	6.34	0.81	0.80	0.75	0.02	0.08	0.08	63	37	138	0
FMI	81	13.54	6.87	0.79	0.78	0.52	0.02	0.00	0.00	106	77	146	1
FCH	51	13.15	7.42	0.83	0.82	0.25	0.00	0.08	0.08	82	56	124	0
FTB	141	16.00	7.78	0.82	0.78	0.00	0.05	1.00	0.59	187	126	232	4
FCK	132	14.62	7.32	0.80	0.78	0.04	0.04	0.23	0.12	145	110	189	0
FAP	92	13.46	7.15	0.81	0.79	***<0.001***	0.01	0.08	0.08	112	82	152	0
FSJ	23	10.00	6.18	0.80	0.78	0.13	0.00	0.00	0.00	92	51	254	0
MXY	20	5.46	3.01	0.47	0.50	1.00	−0.07	0.08	0.08	38	21	83	0

Data include the sample size (N), mean number of alleles observed (*Na*), effective number of alleles (A_E), unbiased expected heterozygosity (uH_E), observed heterozygosity (H_O), the multi-locus *P*-value of conformance to Hardy-Weinberg equilibrium (HWE) determined using Fisher's method output by GENEPOP, the inbreeding coefficient (F_{IS}) output by GENEPOP averaged over loci, the mean and standard error of private alleles per population, the effective population size (with lower and upper 95 % confidence intervals) determined by COLONY, and the number of full-sibling families (NFSF) identified by COLONY. Values for the diversity measures are averaged over the 13 microsatellite DNA loci. Bold italicized values are those that remained significant after sequential Bonferroni correction microsatellite loci were sufficient to produce unique multilocus genotypes (i.e., genetic distances > zero) for all animals surveyed.

Randomization tests showed that genotypes for most collections and most loci surveyed in this study were consistent with Hardy-Weinberg expectations. When *p*-values were combined over loci and analyzed for significance using Fisher's method, four collections deviated from HWE expectations (MAP, CTH, FAP, and SBB). Minimal linkage disequilibrium was observed as only 11 of 2,808 (0.4 %) comparisons among loci by population were found to be significant after correction for multiple tests (overall $\alpha = 0.00002$, $p \leq 0.0004$). In each instance, disequilibria involved different pairs of loci from different collections suggesting that the disequilibrium was likely a result of sampling error, null alleles, inbreeding, year-class mixing, population mixing, or a combination of the three rather than physical linkage among loci.

4.3.2 Demographics and Family Structure

American geneticist Sewall Wright (1931) defined effective population size (N_e) as "the number of breeding individuals in an idealized population that would show the same amount of dispersion of allele frequencies under random genetic drift or the same amount of inbreeding as the population under consideration." The effective population size is usually less than the census population size (*N*) as not every individual within a population spawns with equal success. In general, the genetic variability levels in this study paralleled the estimates of effective population size, N_e, which ranged from lows of 22 (MEH) and 38 (MXY), to a high of 187 (FTB). The average N_e over all collections was 71. COLONY analyses indicated that no single collection was dominated by a small number of large families, with the largest number of inferred full-sib relationships occurring in the MEH and FTB collections ($n = 3$ and $n = 4$ full-sibling families respectively). Estimates of the inbreeding coefficient F_{IS} also were low, ranging from −0.07 to 0.07.

4.3.3 Levels of Population Differentiation

The fixation index, F'_{ST}, is a measure of differentiation between populations. Pair-wise estimates of the fixation index, F'_{ST}, (Table 4.3; above diagonal) ranged from <0.01 (multiple comparisons within the Mid-Atlantic region from MAP to NCS) to 0.91 between the two collections from the extremes of the range (MEH and MXY). Of the 528 tests of significance in pair-wise F'_{ST} values, 431 (82 %) were significantly greater than zero ($p < 0.0005$; data not shown) indicating the presence of considerable population structure throughout the species' range. Inter-regional comparisons between collections comprising the five groups identified previously by King et al. (2005) were all statistically significant (Table 4.4).

Table 4.3 Pairwise matrix of F'_{ST} values (above diagonal) for collections of Atlantic horseshoe crabs, *Limulus polyphemus* (negatives converted to zero). Pairwise chord distances (Cavalli-Sforza and Edwards 1967) are provided below the diagonal

	MEH	MET	MEM	NHS	MAP	RIN	CTH	NYP	NJF	NJR	NJH	DKH	DBS	DFB	MDT	MDF	MD5	MD6
MEH	0.000	0.479	0.511	0.520	0.494	0.516	0.545	0.508	0.484	0.499	0.503	0.491	0.500	0.549	0.583	0.606	0.528	0.533
MET	0.506	0.000	0.021	0.113	0.079	0.076	0.126	0.086	0.077	0.091	0.084	0.062	0.086	0.148	0.146	0.164	0.097	0.123
MEM	0.512	0.184	0.000	0.090	0.109	0.132	0.176	0.155	0.136	0.163	0.150	0.108	0.122	0.216	0.189	0.219	0.156	0.182
NHS	0.521	0.227	0.221	0.000	0.087	0.096	0.132	0.133	0.113	0.130	0.088	0.117	0.075	0.180	0.161	0.146	0.133	0.134
MAP	0.516	0.243	0.251	0.259	0.000	0.005	0.013	0.016	0.000	0.013	0.017	0.015	0.003	0.087	0.052	0.037	0.026	0.028
RIN	0.527	0.251	0.268	0.253	0.196	0.000	0.009	0.011	0.002	0.013	0.018	0.020	0.003	0.081	0.059	0.043	0.000	0.017
CTH	0.543	0.266	0.281	0.270	0.192	0.172	0.000	0.012	0.005	0.017	0.014	0.033	0.010	0.096	0.026	0.016	0.018	0.016
NYP	0.533	0.263	0.277	0.269	0.204	0.193	0.189	0.000	0.000	0.000	0.000	0.000	0.002	0.056	0.036	0.010	0.007	0.018
NJF	0.519	0.246	0.260	0.255	0.170	0.168	0.166	0.194	0.000	0.000	0.000	0.000	0.000	0.062	0.045	0.016	0.008	0.018
NJR	0.520	0.241	0.266	0.266	0.178	0.182	0.182	0.168	0.164	0.000	0.000	0.000	0.003	0.070	0.045	0.018	0.011	0.025
NJH	0.520	0.228	0.259	0.236	0.173	0.173	0.178	0.181	0.170	0.173	0.000	0.009	0.000	0.068	0.038	0.012	0.017	0.031
DKH	0.516	0.243	0.248	0.247	0.204	0.203	0.197	0.188	0.184	0.191	0.193	0.000	0.011	0.091	0.042	0.044	0.011	0.029
DBS	0.527	0.273	0.287	0.273	0.214	0.205	0.212	0.218	0.198	0.207	0.206	0.217	0.000	0.060	0.032	0.013	0.012	0.026
DFB	0.546	0.275	0.308	0.291	0.217	0.210	0.227	0.213	0.217	0.215	0.200	0.242	0.227	0.000	0.149	0.092	0.069	0.095
MDT	0.555	0.290	0.312	0.307	0.226	0.243	0.232	0.236	0.233	0.224	0.227	0.246	0.249	0.283	0.000	0.000	0.047	0.069
MDF	0.567	0.314	0.326	0.306	0.233	0.221	0.217	0.219	0.206	0.212	0.222	0.241	0.236	0.247	0.207	0.000	0.042	0.048
MD5	0.525	0.257	0.285	0.273	0.190	0.171	0.179	0.194	0.178	0.174	0.170	0.209	0.217	0.207	0.225	0.225	0.000	0.010
MD6	0.524	0.245	0.265	0.253	0.180	0.176	0.184	0.177	0.190	0.167	0.174	0.183	0.210	0.202	0.238	0.236	0.181	0.000
VAC	0.545	0.245	0.260	0.261	0.185	0.203	0.201	0.224	0.190	0.196	0.189	0.210	0.216	0.236	0.253	0.247	0.207	0.208
NCS	0.544	0.259	0.271	0.263	0.173	0.190	0.189	0.174	0.175	0.184	0.175	0.202	0.214	0.213	0.232	0.236	0.183	0.169
SBB	0.560	0.310	0.319	0.304	0.235	0.224	0.228	0.226	0.224	0.235	0.218	0.237	0.261	0.266	0.267	0.246	0.218	0.239
SBE	0.578	0.302	0.318	0.299	0.251	0.233	0.227	0.229	0.232	0.243	0.223	0.249	0.262	0.267	0.272	0.248	0.231	0.234
GSA	0.574	0.303	0.323	0.310	0.247	0.238	0.232	0.237	0.224	0.239	0.232	0.253	0.268	0.263	0.280	0.252	0.231	0.244
GSI	0.554	0.314	0.324	0.309	0.256	0.256	0.251	0.243	0.246	0.257	0.253	0.257	0.263	0.285	0.280	0.264	0.255	0.255
FIR	0.600	0.479	0.467	0.477	0.423	0.393	0.413	0.408	0.412	0.411	0.419	0.422	0.413	0.428	0.450	0.417	0.400	0.412
FBB	0.631	0.486	0.459	0.461	0.431	0.416	0.430	0.435	0.437	0.437	0.438	0.433	0.448	0.445	0.476	0.450	0.425	0.431
FMI	0.636	0.484	0.461	0.459	0.432	0.411	0.421	0.422	0.425	0.432	0.430	0.437	0.440	0.432	0.475	0.449	0.412	0.431
FCH	0.645	0.490	0.468	0.465	0.437	0.425	0.427	0.431	0.436	0.435	0.434	0.438	0.446	0.436	0.474	0.451	0.425	0.435
FTB	0.638	0.481	0.455	0.451	0.431	0.415	0.421	0.420	0.425	0.426	0.426	0.432	0.436	0.433	0.469	0.440	0.417	0.430
FCK	0.659	0.489	0.464	0.463	0.427	0.417	0.424	0.421	0.427	0.431	0.428	0.440	0.439	0.434	0.471	0.443	0.422	0.435
FAP	0.666	0.493	0.469	0.464	0.437	0.426	0.432	0.430	0.437	0.441	0.435	0.448	0.444	0.442	0.479	0.453	0.434	0.446
FSJ	0.675	0.504	0.484	0.477	0.434	0.420	0.419	0.432	0.427	0.438	0.432	0.446	0.442	0.437	0.479	0.447	0.434	0.441
MXY	0.791	0.679	0.683	0.676	0.664	0.656	0.667	0.684	0.666	0.676	0.665	0.677	0.662	0.647	0.689	0.673	0.656	0.676

	VAC	NCS	SBB	SBE	GSA	GSI	FIR	FBB	FMI	FCH	FTB	FCK	FAP	FSJ	MXY
MEH	0.582	0.533	0.542	0.595	0.566	0.527	0.617	0.676	0.682	0.710	0.683	0.708	0.710	0.744	0.913
MET	0.150	0.109	0.197	0.190	0.172	0.140	0.389	0.349	0.405	0.405	0.385	0.404	0.386	0.412	0.746
MEM	0.195	0.156	0.246	0.239	0.240	0.185	0.406	0.344	0.406	0.405	0.391	0.395	0.375	0.414	0.741
NHS	0.183	0.107	0.188	0.187	0.191	0.133	0.409	0.364	0.435	0.421	0.412	0.410	0.393	0.403	0.758
MAP	0.067	0.002	0.084	0.085	0.070	0.044	0.321	0.294	0.354	0.357	0.338	0.339	0.331	0.315	0.763
RIN	0.054	0.012	0.080	0.057	0.073	0.052	0.297	0.268	0.326	0.333	0.307	0.316	0.305	0.308	0.739
CTH	0.065	0.017	0.075	0.048	0.062	0.065	0.331	0.342	0.378	0.377	0.357	0.368	0.352	0.326	0.771
NYP	0.092	0.000	0.058	0.062	0.050	0.049	0.304	0.318	0.343	0.352	0.322	0.334	0.330	0.316	0.776
NJF	0.076	0.000	0.070	0.079	0.059	0.047	0.319	0.301	0.349	0.355	0.335	0.346	0.342	0.317	0.765
NJR	0.095	0.001	0.069	0.087	0.065	0.069	0.331	0.315	0.358	0.357	0.327	0.338	0.341	0.318	0.770
NJH	0.095	0.004	0.063	0.061	0.054	0.062	0.327	0.327	0.378	0.369	0.349	0.361	0.348	0.332	0.773
DKH	0.092	0.019	0.088	0.107	0.099	0.093	0.337	0.310	0.362	0.359	0.343	0.361	0.351	0.349	0.783
DBS	0.077	0.004	0.079	0.081	0.075	0.053	0.306	0.300	0.358	0.359	0.347	0.356	0.336	0.343	0.755
DFB	0.147	0.077	0.150	0.138	0.128	0.127	0.366	0.351	0.377	0.370	0.371	0.369	0.367	0.345	0.714
MDT	0.112	0.043	0.108	0.092	0.097	0.122	0.365	0.397	0.446	0.437	0.425	0.430	0.412	0.403	0.783
MDF	0.109	0.019	0.060	0.060	0.050	0.065	0.336	0.364	0.414	0.413	0.395	0.396	0.392	0.365	0.771
MD5	0.033	0.016	0.081	0.058	0.070	0.070	0.306	0.278	0.307	0.298	0.293	0.311	0.301	0.319	0.753
MD6	0.037	0.025	0.110	0.083	0.097	0.090	0.314	0.323	0.370	0.378	0.354	0.368	0.365	0.352	0.784
VAC	0.000	0.075	0.148	0.100	0.123	0.134	0.363	0.325	0.352	0.345	0.339	0.356	0.337	0.358	0.776
NCS	0.193	0.000	0.045	0.057	0.038	0.034	0.311	0.282	0.339	0.332	0.318	0.315	0.311	0.273	0.751
SBB	0.245	0.204	0.000	0.019	0.003	0.006	0.313	0.329	0.389	0.392	0.380	0.372	0.366	0.317	0.764
SBE	0.257	0.207	0.158	0.000	0.018	0.038	0.310	0.329	0.366	0.363	0.351	0.346	0.329	0.291	0.726
GSA	0.248	0.202	0.180	0.170	0.000	0.008	0.341	0.332	0.388	0.393	0.378	0.373	0.366	0.315	0.770
GSI	0.280	0.232	0.206	0.213	0.229	0.000	0.298	0.304	0.368	0.378	0.357	0.347	0.344	0.303	0.740
FIR	0.425	0.411	0.388	0.391	0.401	0.409	0.000	0.223	0.339	0.361	0.351	0.349	0.360	0.332	0.687
FBB	0.422	0.420	0.419	0.421	0.417	0.439	0.381	0.000	0.065	0.083	0.080	0.104	0.098	0.162	0.663
FMI	0.424	0.409	0.421	0.416	0.415	0.439	0.426	0.276	0.000	0.009	0.011	0.045	0.049	0.152	0.691
FCH	0.427	0.415	0.421	0.425	0.420	0.443	0.433	0.290	0.166	0.000	0.018	0.053	0.037	0.169	0.692
FTB	0.420	0.407	0.417	0.414	0.417	0.434	0.428	0.280	0.151	0.161	0.000	0.030	0.040	0.142	0.695
FCK	0.418	0.409	0.416	0.408	0.412	0.432	0.426	0.309	0.197	0.203	0.183	0.000	0.015	0.059	0.649
FAP	0.427	0.415	0.420	0.412	0.420	0.439	0.444	0.305	0.205	0.204	0.184	0.154	0.000	0.092	0.663
FSJ	0.426	0.410	0.415	0.397	0.410	0.432	0.432	0.369	0.323	0.330	0.314	0.258	0.272	0.000	0.630
MXY	0.655	0.651	0.656	0.647	0.661	0.658	0.643	0.633	0.621	0.619	0.625	0.593	0.612	0.578	0.000

Table 4.4 Pairwise matrix of F_{ST} values (below diagonal) for collections of Atlantic horseshoe crabs, *Limulus polyphemus* (negative values were converted to zero)

	MEH	MET	MEM	NHS	MAP	RIN	CTH	NYP	NJF	NJR	NJH	DKH	DBS	DFB	MDT	MDF	MD5	MD6
MEH	0.000	0.001	0.001	0.001	0.001	0.001	0.001	0.001	0.001	0.001	0.001	0.001	0.001	0.001	0.001	0.001	0.001	0.001
MET	0.206	0.000	0.022	0.001	0.001	0.001	0.001	0.001	0.001	0.001	0.001	0.001	0.001	0.001	0.001	0.001	0.001	0.001
MEM	0.222	0.006	0.000	0.001	0.001	0.001	0.001	0.001	0.001	0.001	0.001	0.001	0.001	0.001	0.001	0.001	0.001	0.001
NHS	0.220	0.031	0.025	0.000	0.001	0.001	0.001	0.001	0.001	0.001	0.001	0.001	0.001	0.001	0.001	0.001	0.001	0.001
MAP	0.198	0.020	0.028	0.021	0.000	0.269	0.094	0.066	0.461	0.089	0.064	0.095	0.371	0.001	0.001	0.013	0.013	0.007
RIN	0.208	0.019	0.034	0.024	0.001	0.000	0.163	0.116	0.357	0.089	0.048	0.049	0.368	0.001	0.001	0.011	0.464	0.074
CTH	0.222	0.032	0.046	0.033	0.003	0.002	0.000	0.111	0.267	0.052	0.064	0.007	0.206	0.001	0.024	0.107	0.038	0.063
NYP	0.208	0.022	0.041	0.034	0.004	0.003	0.003	0.000	0.454	0.455	0.454	0.451	0.414	0.001	0.011	0.193	0.181	0.045
NJF	0.196	0.020	0.035	0.028	0.000	0.001	0.001	0.000	0.000	0.469	0.435	0.435	0.474	0.001	0.001	0.091	0.199	0.050
NJR	0.204	0.023	0.043	0.033	0.003	0.003	0.004	0.000	0.000	0.000	0.459	0.435	0.371	0.001	0.003	0.083	0.138	0.010
NJH	0.207	0.022	0.040	0.023	0.004	0.004	0.003	0.000	0.000	0.000	0.000	0.180	0.460	0.001	0.006	0.140	0.053	0.006
DKH	0.206	0.016	0.029	0.030	0.003	0.005	0.008	0.000	0.000	0.000	0.002	0.000	0.182	0.001	0.008	0.005	0.163	0.019
DBS	0.211	0.022	0.032	0.019	0.001	0.001	0.002	0.000	0.000	0.001	0.000	0.003	0.000	0.002	0.039	0.166	0.145	0.038
DFB	0.224	0.038	0.057	0.045	0.020	0.019	0.023	0.013	0.015	0.017	0.016	0.022	0.014	0.000	0.001	0.001	0.001	0.001
MDT	0.256	0.040	0.053	0.043	0.013	0.015	0.006	0.009	0.011	0.011	0.010	0.011	0.008	0.037	0.000	0.474	0.004	0.001
MDF	0.262	0.044	0.060	0.038	0.009	0.010	0.004	0.002	0.004	0.004	0.003	0.011	0.003	0.022	0.000	0.000	0.006	0.006
MD5	0.209	0.024	0.039	0.032	0.006	0.000	0.004	0.002	0.002	0.002	0.004	0.003	0.003	0.015	0.011	0.010	0.000	0.132
MD6	0.214	0.031	0.047	0.033	0.006	0.004	0.004	0.004	0.004	0.006	0.007	0.007	0.006	0.022	0.017	0.011	0.002	0.000
VAC	0.225	0.036	0.047	0.043	0.014	0.012	0.014	0.020	0.016	0.021	0.021	0.020	0.016	0.032	0.026	0.024	0.007	0.008
NCS	0.210	0.027	0.040	0.026	0.000	0.003	0.004	0.000	0.000	0.000	0.001	0.004	0.001	0.018	0.010	0.005	0.004	0.006
SBB	0.223	0.053	0.067	0.049	0.020	0.019	0.018	0.014	0.017	0.017	0.016	0.022	0.019	0.037	0.028	0.015	0.019	0.026
SBE	0.243	0.049	0.063	0.048	0.020	0.013	0.011	0.015	0.018	0.021	0.015	0.026	0.019	0.033	0.023	0.015	0.013	0.019
GSA	0.233	0.045	0.064	0.049	0.016	0.017	0.015	0.012	0.014	0.016	0.013	0.024	0.018	0.031	0.025	0.012	0.016	0.023
GSI	0.225	0.037	0.050	0.035	0.010	0.012	0.016	0.012	0.011	0.017	0.015	0.023	0.013	0.031	0.032	0.016	0.016	0.021
FIR	0.258	0.104	0.111	0.108	0.077	0.072	0.081	0.076	0.078	0.082	0.082	0.084	0.076	0.090	0.096	0.086	0.072	0.075
FBB	0.282	0.083	0.084	0.084	0.061	0.056	0.073	0.069	0.063	0.068	0.072	0.067	0.064	0.075	0.091	0.080	0.056	0.067
FMI	0.247	0.093	0.095	0.097	0.072	0.067	0.079	0.072	0.072	0.075	0.081	0.076	0.074	0.079	0.099	0.089	0.061	0.075
FCH	0.259	0.088	0.090	0.090	0.068	0.064	0.074	0.070	0.069	0.071	0.075	0.071	0.070	0.073	0.092	0.083	0.055	0.072
FTB	0.234	0.086	0.089	0.090	0.067	0.062	0.073	0.067	0.068	0.067	0.073	0.071	0.070	0.076	0.092	0.083	0.057	0.071
FCK	0.246	0.091	0.091	0.091	0.068	0.064	0.076	0.070	0.071	0.070	0.076	0.075	0.073	0.076	0.095	0.084	0.061	0.074
FAP	0.252	0.087	0.086	0.087	0.066	0.062	0.072	0.069	0.069	0.071	0.073	0.073	0.069	0.075	0.090	0.083	0.059	0.073
FSJ	0.311	0.100	0.103	0.096	0.067	0.066	0.071	0.070	0.068	0.070	0.075	0.077	0.075	0.076	0.095	0.083	0.066	0.075
MXY	0.511	0.287	0.288	0.286	0.270	0.263	0.277	0.281	0.272	0.278	0.282	0.289	0.279	0.257	0.304	0.294	0.262	0.277

	VAC	NCS	SBB	SBE	GSA	GSI	FIR	FBB	FMI	FCH	FTB	FCK	FAP	FSJ	MXY
MEH	0.001	0.001	0.001	0.001	0.001	0.001	0.001	0.001	0.001	0.001	0.001	0.001	0.001	0.001	0.001
MET	0.001	0.001	0.001	0.001	0.001	0.001	0.001	0.001	0.001	0.001	0.001	0.001	0.001	0.001	0.001
MEM	0.001	0.001	0.001	0.001	0.001	0.001	0.001	0.001	0.001	0.001	0.001	0.001	0.001	0.001	0.001
NHS	0.001	0.001	0.001	0.001	0.001	0.001	0.001	0.001	0.001	0.001	0.001	0.001	0.001	0.001	0.001
MAP	0.001	0.380	0.001	0.001	0.001	0.003	0.001	0.001	0.001	0.001	0.001	0.001	0.001	0.001	0.001
RIN	0.001	0.096	0.001	0.001	0.001	0.003	0.001	0.001	0.001	0.001	0.001	0.001	0.001	0.001	0.001
CTH	0.001	0.043	0.001	0.002	0.001	0.001	0.001	0.001	0.001	0.001	0.001	0.001	0.001	0.001	0.001
NYP	0.001	0.452	0.001	0.001	0.001	0.001	0.001	0.001	0.001	0.001	0.001	0.001	0.001	0.001	0.001
NJF	0.001	0.472	0.001	0.001	0.001	0.002	0.001	0.001	0.001	0.001	0.001	0.001	0.001	0.001	0.001
NJR	0.001	0.386	0.002	0.001	0.001	0.001	0.001	0.001	0.001	0.001	0.001	0.001	0.001	0.001	0.001
NJH	0.001	0.276	0.001	0.001	0.001	0.001	0.001	0.001	0.001	0.001	0.001	0.001	0.001	0.001	0.001
DKH	0.001	0.036	0.001	0.001	0.001	0.001	0.001	0.001	0.001	0.001	0.001	0.001	0.001	0.001	0.001
DBS	0.001	0.329	0.001	0.001	0.001	0.002	0.001	0.001	0.001	0.001	0.001	0.001	0.001	0.001	0.001
DFB	0.001	0.001	0.001	0.001	0.001	0.001	0.001	0.001	0.001	0.001	0.001	0.001	0.001	0.001	0.001
MDT	0.001	0.002	0.001	0.001	0.001	0.001	0.001	0.001	0.001	0.001	0.001	0.001	0.001	0.001	0.001
MDF	0.001	0.064	0.001	0.001	0.001	0.001	0.001	0.001	0.001	0.001	0.001	0.001	0.001	0.001	0.001
MD5	0.006	0.052	0.001	0.001	0.001	0.001	0.001	0.001	0.001	0.001	0.001	0.001	0.001	0.001	0.001
MD6	0.004	0.014	0.001	0.001	0.001	0.001	0.001	0.001	0.001	0.001	0.001	0.001	0.001	0.001	0.001
VAC	0.000	0.001	0.001	0.001	0.001	0.001	0.001	0.001	0.001	0.001	0.001	0.001	0.001	0.001	0.001
NCS	0.016	0.000	0.001	0.001	0.002	0.008	0.001	0.001	0.001	0.001	0.001	0.001	0.001	0.001	0.001
SBB	0.033	0.011	0.000	0.024	0.314	0.239	0.001	0.001	0.001	0.001	0.001	0.001	0.001	0.001	0.001
SBE	0.022	0.013	0.005	0.000	0.047	0.003	0.001	0.001	0.001	0.001	0.001	0.001	0.001	0.001	0.001
GSA	0.027	0.009	0.001	0.004	0.000	0.213	0.001	0.001	0.001	0.001	0.001	0.001	0.001	0.001	0.001
GSI	0.030	0.008	0.002	0.009	0.002	0.000	0.001	0.001	0.001	0.001	0.001	0.001	0.001	0.001	0.001
FIR	0.082	0.074	0.080	0.077	0.086	0.075	0.000	0.001	0.001	0.001	0.001	0.001	0.001	0.001	0.001
FBB	0.062	0.058	0.074	0.071	0.073	0.066	0.050	0.000	0.004	0.002	0.001	0.001	0.001	0.001	0.001
FMI	0.067	0.069	0.084	0.077	0.083	0.078	0.074	0.012	0.000	0.138	0.024	0.001	0.001	0.001	0.001
FCH	0.061	0.063	0.081	0.072	0.079	0.076	0.075	0.014	0.001	0.000	0.015	0.001	0.001	0.001	0.001
FTB	0.064	0.063	0.081	0.072	0.079	0.074	0.075	0.015	0.002	0.003	0.000	0.001	0.001	0.001	0.001
FCK	0.067	0.063	0.080	0.072	0.079	0.073	0.075	0.019	0.008	0.009	0.005	0.000	0.005	0.001	0.001
FAP	0.063	0.062	0.078	0.068	0.077	0.072	0.077	0.018	0.009	0.006	0.007	0.003	0.000	0.001	0.001
FSJ	0.071	0.058	0.073	0.064	0.071	0.068	0.076	0.031	0.029	0.030	0.027	0.011	0.017	0.000	0.001
MXY	0.262	0.262	0.281	0.262	0.281	0.278	0.255	0.237	0.222	0.219	0.216	0.203	0.209	0.227	0.000

Probability of value being greater than zero is provided above the diagonal

Most of the non-significant F'_{ST} values were observed among intra-regional comparisons. Of the 97 non-significant values, 72 were observed in comparisons among collections from the Mid-Atlantic region from Massachusetts to Maryland. Both the MEH and MXY collections from the extremes of the distribution were highly differentiated from all other collections with pair-wise F'_{ST} values averaging 0.57 and 0.74, respectively.

4.3.4 *Evolutionary Relationships Among Populations*

4.3.4.1 Genetic Distance

Pair-wise genetic distance (D_C) values between all pairs of collections (Table 4.3, below diagonal) were consistent with F'_{ST} estimates as the greatest distances were observed between the collections representing the extremes of the range (0.791 between MEH and MXY) and lowest among populations within regions (0.184 between MEM and MET; 0.164 between NJR and NJF; 0.158 between SBB and SBE; 0.151 between FMI and FTB). Initial principal coordinates analyses (PCoA) of genetic distances showed that the geographically extreme collections surveyed in this study (MEH and MXY) exhibited large genetic distances that distorted the overall structure, and these data were eliminated from further graphical comparisons. In a PCoA ordination of the 31 remaining collections, seven distinct clusters (regional groupings) of collections were apparent (Fig. 4.2): (1) MET, MEM, and NHS (hereafter referred to as Gulf of Maine); (2) MAP to NCS (Mid-Atlantic); (3) SBB to GSI (Southeast Atlantic); (4) FIR (Florida East); (5) FBB (Florida South); (6) FMI to FAP (Florida Gulf of Mexico); and (7) FSJ. At the broadest scale, clustering corresponded to the geographical distribution of the collections, with greater distances separating the southernmost regional collections from the northern collections; at this scale there appeared little correspondence to geographic distribution within the groupings. Upon closer inspection of the collections within each grouping by subsequent PCoA, additional zones of genetic discontinuity were visible along the coasts of the Atlantic Ocean (Fig. 4.3) and Gulf of Mexico (Fig. 4.4) coasts. Along the Atlantic coast (Fig. 4.3), a moderate to high degree of relatedness (or recent gene flow; Table 4.5) was inferred among collections within the Gulf of Maine, from Cape Cod, Massachusetts (MAP) to the outer banks of North Carolina (NCS), and among the collections from South Carolina and Georgia. Within the Mid-Atlantic grouping, a slightly higher degree of relatedness was observed among collections within Delaware Bay than among those with immediate access to the Atlantic coast. In contrast, the two collections from the Chesapeake Bay (MDT, MDF) were moderately differentiated from the other Mid-Atlantic collections. As with the range-wide PCoA, the greatest zones of genetic discontinuity within the study area occurred to the north (GSI) and south (FBB) of the Indian River, FL (FIR) collection (Figs. 4.3 and 4.4).

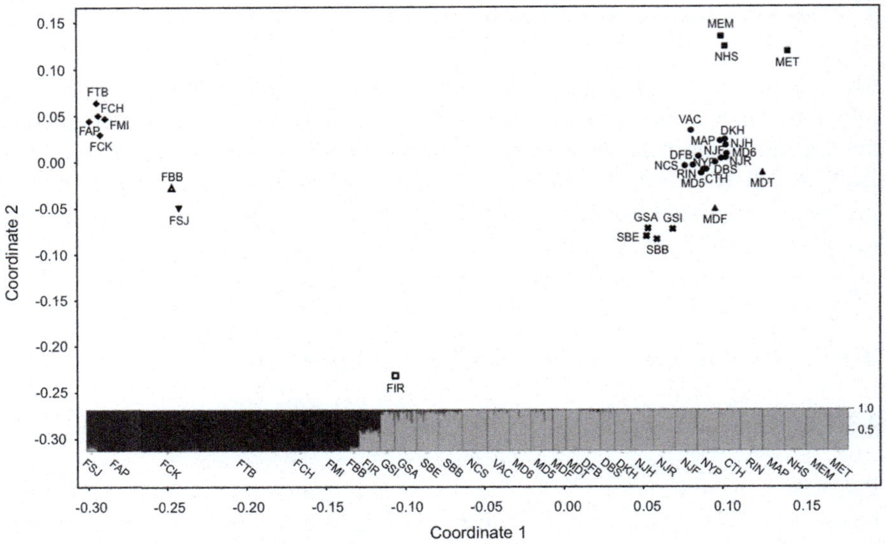

Fig. 4.2 Combined graphical representation of principal coordinates (scatter plot of pairwise chord distance, Cavalli-Sforza and Edwards 1967) and STRUCTURE (histogram) analyses among 1,617 *Limulus polyphemus* sampled from 31 locations along the Atlantic and Gulf coasts of the United States (collections from Hog Bay, Franklin, ME and Mexico's Yucatan Peninsula were eliminated as outliers) surveyed at 13 microsatellite DNA loci. For the STRUCTURE histograms, each individual is represented by a *single vertical bar*, broken into $K=2$ colored segments, the length of which is proportional to the membership fraction in each of the K clusters. *Black lines* partition the collections

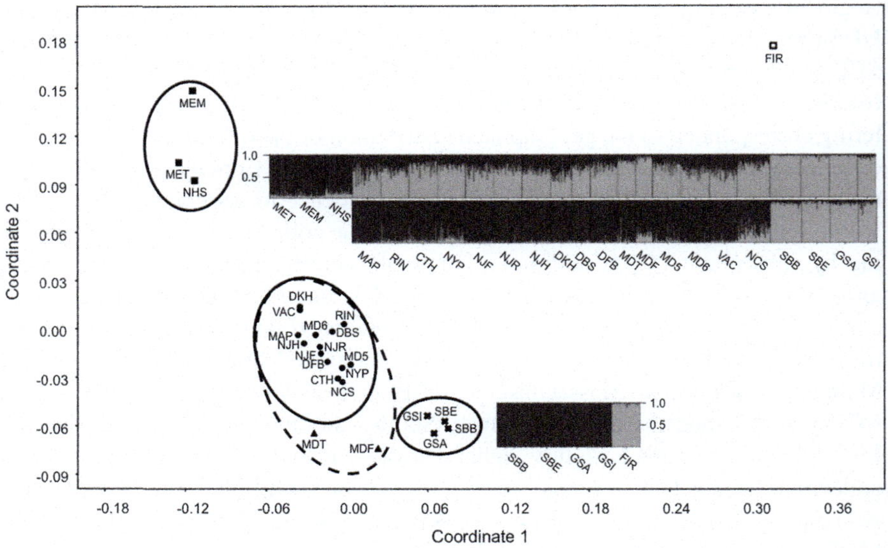

Fig. 4.3 Combined graphical representation of principal coordinates (scatter plot of pairwise chord distance, Cavalli-Sforza and Edwards 1967) and STRUCTURE (histogram) analyses among *Limulus polyphemus* sampled from 24 locations along the Atlantic coast of the United States (collections from Hog Bay, Franklin, ME and Biscayne Bay, FL eliminated) surveyed at 13 microsatellite DNA loci. Ellipses enclose groups of similar collections. For the STRUCTURE histograms, each individual is represented by a *single vertical bar*, broken into K colored segments, the length of which is proportional to the membership fraction in each of the K clusters. *Black lines* partition the collections

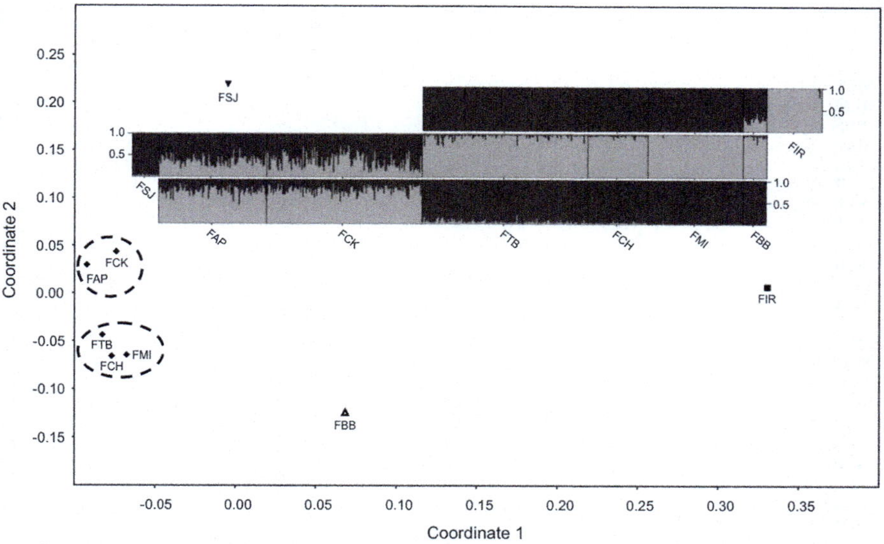

Fig. 4.4 Combined graphical representation of principal coordinates (scatter plot of pairwise chord distance, Cavalli-Sforza and Edwards 1967) and STRUCTURE (histogram) analyses among *Limulus polyphemus* sampled from eight locations along the Florida Atlantic and Gulf coasts of the United States surveyed at 13 microsatellite DNA loci. *Ellipses* enclose groups of similar collections and *boxes* highlight single collections. For the STRUCTURE histograms, each individual is represented by a *single vertical bar*, broken into *K* colored segments, the length of which is proportional to the membership fraction in each of the *K* clusters. *Black lines* partition the collections

In addition to the deep zone of discontinuity that exists to the north and south of Indian River, FL, a regional PCoA of the Florida coast identified shallow but significant differentiation along the central portion of the Gulf coast that was not detectable in the range-wide analysis (Fig. 4.4). This brings to four the number of genetically discontinuous zones detected within this state: (1) FIR (Florida East) and FBB (Florida South); (2) FBB and FMI (Florida Southwest; (3) Florida Southwest and FCK (Florida West); and 4) Florida West and the collection from St. Joseph Bay (FSJ).

4.3.4.2 Population Structure

Population structure can be inferred from the results of the individual-based STRUCTURE analyses among 33 *L. polyphemus* collections, which were congruent with the PCoA (Figs. 4.2, 4.3, and 4.4). The number of inferred clusters (*K*) determined by STRUCTURE for the initial (uppermost hierarchical level) analysis was two, corresponding to the Atlantic coast collections north of Indian River, Florida (FIR) and collections south of FIR around to St. Joseph Bay, Florida the western most collection of the Gulf of Mexico included in the analysis (Fig. 4.2). Due to the

Table 4.5 Pair-wise estimates of the effective number of migrants per generation, N_em, among 31 collections of Atlantic horseshoe crabs *Limulus polyphemus* sampled from throughout their native range in the coastal USA and Mexico

	MEH	MET	MEM	NHS	MAP	RIN	CTH	NYP	NJF	NJR	NJH	DKH	DBS	DFB	MDT	MDF	MD5	MD6
MET	0.5																	
MEM	0.6	9.9																
NHS	0.5	6.6	6.5															
MAP	0.8	4.7	4.9	4.8														
RIN	0.8	4.3	3.5	3.5	6.1													
CTH	0.7	5.0	6.3	4.6	6.0	7.7												
NYP	0.8	5.8	6.4	4.7	6.7	7.2	7.5											
NJF	0.8	4.2	5.7	4.0	6.0	6.7	8.6	6.9										
NJR	0.8	7.3	7.0	4.6	8.5	7.9	7.5	7.8	9.5									
NJH	0.8	9.1	6.3	5.1	8.2	8.2	8.9	8.1	8.5	9.1								
DKH	0.7	4.8	7.2	5.6	6.9	8.5	9.7	8.2	8.2	7.5	10.7							
DBS	0.8	4.1	4.4	3.9	5.4	9.1	7.1	6.8	5.8	6.5	7.3	5.4						
DFB	0.8	3.2	2.4	2.6	8.4	4.2	7.0	7.7	3.5	7.2	8.3	7.9	4.7					
MDT	0.7	3.1	3.4	4.4	6.8	6.4	5.3	5.7	5.6	6.3	8.0	4.5	4.3	3.5				
MDF	0.6	2.5	3.2	3.1	4.2	5.6	6.0	6.1	5.9	5.8	5.7	4.6	4.4	3.0	4.0			
MD5	0.9	5.4	5.2	5.4	5.8	10.4	10.8	7.8	8.5	12.2	9.9	8.9	7.8	4.6	5.5	5.4		
MD6	0.8	4.7	3.9	2.9	8.0	8.1	7.4	7.4	6.1	9.3	8.6	8.6	8.1	7.6	3.8	3.8	6.9	
VAC	0.8	4.7	6.5	4.5	9.2	7.0	7.5	6.9	7.4	10.2	10.6	8.3	5.3	5.9	5.4	5.2	7.9	5.8
NCS	0.7	5.5	6.1	3.9	10.0	7.3	7.8	8.8	8.3	10.5	10.5	7.3	7.5	8.4	6.5	5.2	9.0	9.9
SBB	0.7	3.8	4.0	3.8	6.6	5.5	7.5	7.1	6.6	5.8	7.5	8.4	5.9	3.5	6.5	6.4	8.9	5.2
SBE	0.6	3.4	4.7	3.5	3.8	5.6	6.6	5.6	6.4	6.0	4.8	7.1	6.2	3.6	3.3	3.8	7.1	7.4
GSA	0.7	3.2	3.8	3.3	4.3	4.4	6.6	5.8	5.2	5.4	6.1	5.2	4.0	4.0	4.6	4.2	7.9	4.8
GSI	0.8	2.9	3.6	3.6	4.6	5.6	5.2	5.8	4.4	5.1	6.5	5.2	5.0	3.7	3.4	4.4	6.9	5.8
FIR	0.6	0.9	1.0	0.9	1.8	2.3	1.6	1.9	1.4	1.9	1.5	1.8	1.9	1.7	1.6	1.5	1.9	1.6
FBB	0.5	0.9	1.2	1.1	1.0	1.5	1.4	1.5	1.0	1.3	1.4	1.6	1.2	1.1	1.4	1.3	1.7	1.2
FMI	0.7	1.2	1.8	1.8	1.7	2.1	2.2	2.2	1.4	1.7	1.9	2.3	1.6	1.8	1.7	2.0	2.1	2.0
FCH	0.8	1.4	1.8	1.9	1.6	1.9	2.0	2.2	1.4	1.6	1.9	2.5	1.8	1.9	1.8	2.1	2.2	2.1
FTB	0.5	0.9	1.3	1.2	1.1	1.4	1.4	1.4	1.0	1.2	1.3	1.4	1.1	1.1	1.2	1.3	1.4	1.2
FCK	0.6	1.0	1.2	1.3	1.4	1.7	1.9	1.8	1.3	1.4	1.9	1.7	1.3	1.6	1.5	1.7	1.7	1.6
FAP	0.5	1.1	1.3	1.5	1.3	1.5	1.5	1.7	1.1	1.3	1.8	1.7	1.4	1.5	1.4	1.5	1.4	1.5
FSJ	0.4	0.8	1.1	1.1	1.3	1.5	1.5	1.5	1.1	1.5	1.7	1.7	1.3	1.4	1.5	1.5	1.3	1.5
MXY	0.1	0.2	0.3	0.3	0.4	0.4	0.4	0.4	0.4	0.4	0.4	0.5	0.5	0.4	0.4	0.4	0.3	0.3

	VAC	NCS	SBB	SBE	GSA	GSI	FIR	FBB	FMI	FCH	FTB	FCK	FAP	FSJ
MET														
MEM														
NHS														
MAP														
RIN														
CTH														
NYP														
NJF														
NJR														
NJH														
DKH														
DBS														
DFB														
MDT														
MDF														
MD5														
MD6														
VAC														
NCS	9.5													
SBB	6.0	7.0												
SBE	4.5	6.3	12.0											
GSA	6.0	6.8	9.9	9.3										
GSI	3.7	4.7	7.2	6.6	5.7									
FIR	1.6	1.7	2.1	1.6	1.4	2.0								
FBB	1.2	1.3	1.5	1.2	1.1	1.4	1.4							
FMI	1.7	2.1	1.9	2.1	2.0	1.9	1.8	5.3						
FCH	1.7	2.0	2.0	2.0	1.9	2.0	2.2	5.6	9.4					
FTB	1.2	1.3	1.6	1.4	1.2	1.3	1.5	3.9	13.0	11.1				
FCK	1.5	1.7	2.0	1.9	1.5	1.5	1.7	3.6	7.9	7.4	9.3			
FAP	1.3	1.5	1.8	1.6	1.5	1.4	1.5	3.1	8.6	6.2	11.1	8.4		
FSJ	1.2	1.8	2.4	1.9	1.9	1.7	2.4	2.4	4.3	3.8	3.4	8.6	4.1	
MXY	0.5	0.4	0.4	0.4	0.4	0.5	0.4	0.8	0.9	1.0	0.6	0.7	0.8	1.0

apparent complex migration patterns (zones of genetic discontinuity) existing among the two regions, a sequential method of inferring the number of clusters was employed to identify within-cluster structure. Sequential STRUCTURE analysis of the Atlantic coast collections (MEH to GSI) indicated $K=2$ delineating the Gulf of Maine collections from the Mid-Atlantic and Southeast Atlantic (SBB to GSI) collections (Fig. 4.3). Upon removal of the Gulf of Maine collections, analysis of the remaining collections again identified $K=2$ clusters discriminating between the Mid-Atlantic and Southeast Atlantic collections. Including the highly differentiated Florida East (FIR) collection this brings to four the number of distinct clusters of *L. polyphemus* populations identified along the Atlantic coast from Maine to central Florida (Figs. 4.3, 4.4, and 4.5). A relatively high degree of relatedness was observed among the *L. polyphemus* collections grouped within the identified clusters.

Multiple sequential STRUCTURE analyses of all eight Florida collections (FIR to FSJ) were consistent with the patterns observed for genetic distance in the PCoA ordination (Fig. 4.4). Three analyses were performed with each resulting in $K=2$ clusters. The Indian River (FIR) and St. Joseph Bay (FSJ) collections were identified as two distinct clusters highly differentiated from adjacent populations and from each other. These analyses resulted in a total of four clusters (excluding FIR) bringing the number of total clusters to eight. However, the level of differentiation among the Florida South (FBB), Florida Southwest (FMI to FTB), and Florida West (FCK to FAP) clusters did not appear as great as that observed between the St. Joseph Bay, FL (FSJ) collection and the other Gulf of Mexico collections or as seen among the Atlantic coast clusters.

4.3.4.3 Patterns of Genetic Variation: A Test for Isolation by Distance

A Mantel matrix regression test identified a statistically significant correlation ($r=0.74$; $t=10.3$; $p<0.0001$) between pair-wise matrices of ocean (km) and genetic (chord) distances among 31 *L. polyphemus* collections (excluding the MEH and MXY collections; Fig. 4.6). This graphical representation of the matrix comparison illustrates the deep genetic discontinuity that exists along the Florida Atlantic coast around the Indian River (FIR) collection. All pair-wise comparisons with the FIR collection (denoted by ■) exhibited large genetic distances (>0.35) relative to geographic distance. Most other pair-wise comparisons (denoted by ○) exhibited an isolation-by-distance pattern in which the comparisons with collections south and west of FIR were separated by a distributional gap that appears to correspond with a zone of deep genetic discontinuity.

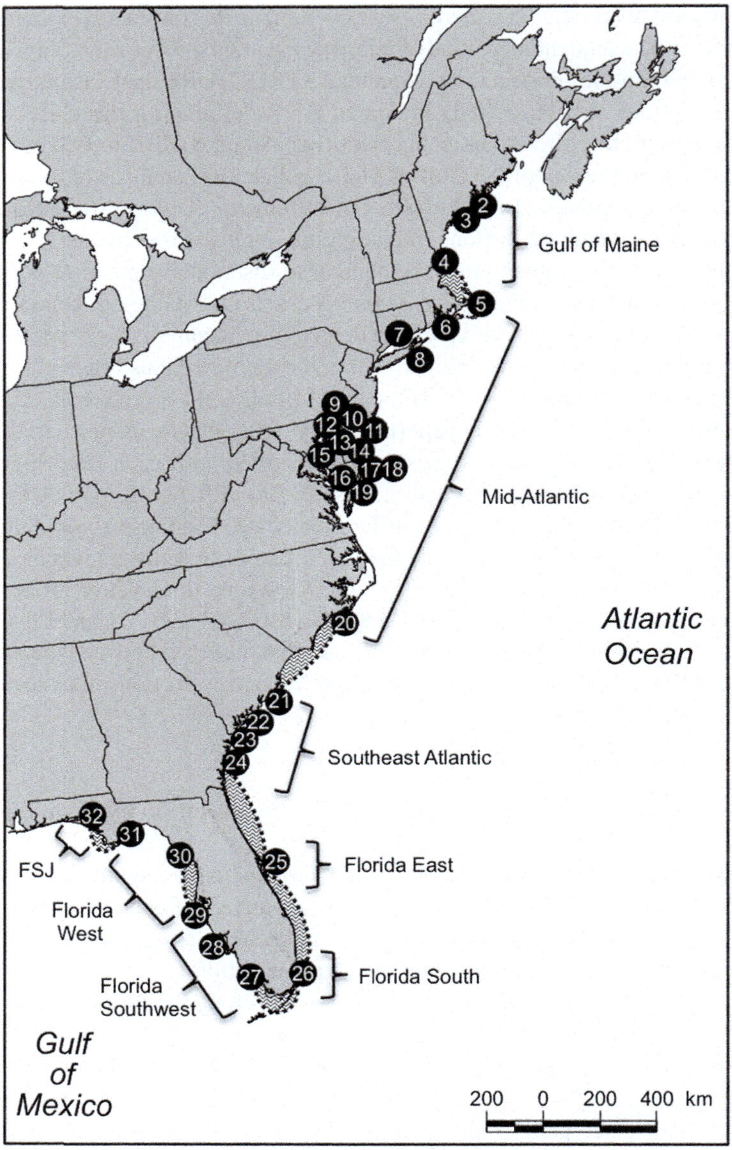

Fig. 4.5 Map showing locations of 32 of 33 collection sites of *Limulus polyphemus* along the Atlantic and Gulf coasts of the United States with *hashed* regions (▒) demarking zones of genetic discontinuity and areas in need of additional sampling. Collections within each delineated area could be considered management units. The collections representing the extremes of the species' range from Hog Bay, Maine and Mexico's Yucatan Peninsula (not depicted) are not shown

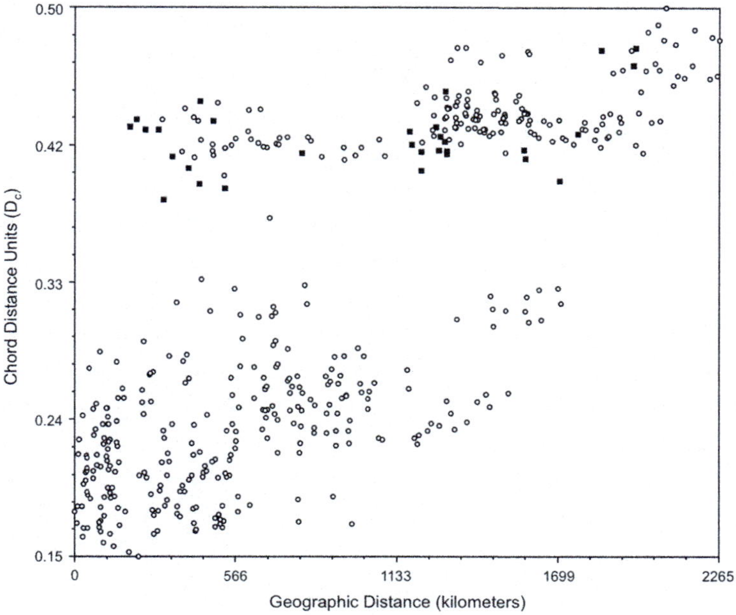

Fig. 4.6 Scatterplot illustrating the significant correlation between geographic distance (coastal distance measured in kilometers) and genetic distance (chord distance; Cavalli-Sforza and Edwards 1967) for 31 *Limulus polyphemus* collections (Hog Bay, Maine, and the Yucatan Peninsula were omitted; Mantel test $r=0.74$; $t=10.3$; $P<0.0001$). Pairwise comparisons with the highly differentiated Indian River, Florida (FIR) collection are denoted by the filled square (■)

4.3.4.4 Assignment Testing

Maximum likelihood assignment tests also revealed structuring of genotypic frequencies in *L. polyphemus*. Individual animals were correctly assigned to collection of origin 25.1 % of the time, on average, across collections, and all assignments to collection of origin were higher than expected by chance ($p<0.001$). When the *L. polyphemus* collections were pooled into eight regional groups including the MEH and MXY collections, 1,399/1,684 (83.1 %) of the specimens were correctly classified to region of origin (Table 4.6). Of the 285 misclassified animals, 263 (92.2 %) were assigned incorrectly to the adjacent regional grouping. The MEH and MXY collections exhibited 100 % correct assignment, and no animals were incorrectly assigned to either collection. Removal of the MEH and MXY collections from this analysis reduced the correct assignment rate slightly to 82.4 %. Attempts to take into consideration shallower levels of differentiation resulted in lower assignment success (Table 4.6). Splitting the Florida Gulf coast collections FMI to FSJ into FMI-FTB and FCK-FSJ groupings resulted in correct assignment of 75.2 % of *L. polyphemus*. At a finer level of resolution the FCK to FSJ collections were split into FCK-FAP and FSJ, which resulted in correct assignment of 73.2 %.

Table 4.6 Results of maximum likelihood assignment tests for all *Limulus polyphemus* collections using multilocus genotypes derived from 13 microsatellite DNA markers

Region	N	MEH	Gulf of Maine	Mid-Atlantic	SE	FL-east	FL-south	FL-gulf	MXY	% Correct assign
MEH	47	**47**	0	0	0	0	0	0	0	**100.0**
Gulf of Maine	141	0	**114**	23	4	0	0	0	0	**80.9**
Mid-Atlantic	709	0	66	**534**	103	4	1	1	0	**75.3**
Southeast	181	0	5	34	**140**	0	1	1	0	**77.4**
FL-east	46	0	0	1	0	**44**	0	1	0	**95.7**
FL-south	20	0	0	0	0	0	**9**	11	0	**45.0**
FL-gulf	520	0	0	0	3	0	26	**491**	0	**94.4**
MXY	20	0	0	0	0	0	0	0	**20**	**100.0**
Eight Regions: (MEH) (MET-NHS) (MAP-NCS) (SBB-GSI) (FIR) (FBB) (FMI-FSJ) (MXY) – 83.1 %										
Six Regions: (MET-NHS) (MAP-NCS) (SBB-GSI) (FIR) (FBB) (FMI-FSJ) – 82.4 %										
Seven Regions: (MET-NHS) (MAP-NCS) (SBB-GSI) (FIR) (FBB) (FMI-FTB) (FCK-FSJ) – 75.2 %										
Eight Regions: (MET-NHS) (MAP-NCS) (SBB-GSI) (FIR) (FBB) (FMI-FTB) (FCK-FAP) (FSJ) – 73.2 %										

Values along the diagonal are the number of correct assignments to that regional grouping. Regions consist of eight proposed groupings delineated by zones of genetic discontinuity observed in analyses of phylogeographic structuring: Hog Bay, Maine (MEH); Gulf of Maine (MET-NHS); Mid-Atlantic (MAP-NCS); Southeast (SBB-GSI); Florida-East (FIR); Florida-South (FBB); Florida-Gulf (FMI-FSJ); and Yucatan Peninsula, Mexico (MXY). Percent correct classification for six, seven, and eight (without extreme collections) region models are also provided

4.3.4.5 Partitioning of Molecular Variance

When *L. polyphemus* collections were grouped into regions according to ordination of genetic distances and clusters inferred from STRUCTURE analysis, quantitative estimates of hierarchical gene diversity indicated significant genetic population structure at every level, the greatest amount owing to variation within collections; a pattern commonly seen in intraspecific comparisons utilizing microsatellite DNA markers. Grouping all collections into eight regions (six as described above and two collections from the extremes of the species' range) revealed that 8 % ($p<0.0001$) of the genetic variation was distributed among regions, 1.0 % ($p<0.0001$) among collections within regions, and 91 % ($p<0.0001$) within collections (Table 4.7A). Elimination of the northernmost and southernmost collections from this analysis reduced the amount of variation among regions to 6 % ($p<0.001$) and increased that

Table 4.7 Analysis of molecular variance among 33 *Limulus polyphemus* collections from (A) Maine to the Yucatán Peninsula pooled into eight regional groupings (including the regional extreme populations MEH and MXY); (B) 31 collections excluding the two regional extremes (Hog Bay, Franklin, Maine and Ria Lagartos, Yucatan Peninsula, Mexico) pooled into six regional groupings; (C) the same collections with the Florida Gulf of Mexico collections split between Tampa Bay (FTB) and Cedar Key (FCK); and (D) the St. Josephs Bay (FSJ) collection split from the other northern Gulf of Mexico collections Each variance component and fixation index (F'_{ST}) was significantly greater than zero at $\alpha=(0.05)$ ($P<0.001$)

Source	*Df*	SS	Variance	% Variance	F'_{ST}
A. Eight Regions: (MEH) (MET-NHS) (MAP-NCS) (SBB-GSI) (FIR) (FBB) (FMI-FSJ) (MXY)					
Among regions	7	1074.6	0.43	8 %	0.33
Among pops. within regions	25	229.7	0.04	1 %	0.04
Within pops	3335	16592.8	4.98	91 %	
Total	3367	17897.1	5.44	100 %	
B. Six Regions: (MET-NHS) (MAP-NCS) (SBB-GSI) (FIR) (FBB) (FMI-FSJ)					
Among regions	5	748.7	0.32	6 %	0.27
Among pops. within regions	25	229.7	0.04	1 %	0.04
Within pops	3203	16219.1	5.06	93 %	
Total	3233	17197.4	5.42	100 %	
C. Seven Regions: (MET-NHS) (MAP-NCS) (SBB-GSI) (FIR) (FBB) (FMI-FTB) (FCK-FSJ)					
Among regions	6	772.9	0.30	6 %	0.26
Among pops. within regions	24	205.4	0.04	<1 %	0.03
Within pops	3203	16219.1	5.06	94 %	
Total	3233	17197.4	5.40	100 %	
D. Eight Regions: (MET-NHS) (MAP-NCS) (SBB-GSI) (FIR) (FBB) (FMI-FTB) (FCK-FAP) (FSJ)					
Among regions	7	784.1	0.30	6 %	0.26
Among pops. within regions	23	194.3	0.03	<1 %	0.03
Within pops	3203	16219.1	5.06	94 %	
Total	3233	17197.4	5.40	100 %	

within collections to 93 % ($p < 0.001$) (Table 4.7B). Other groupings based on shallower levels of differentiation (Table 4.7C, D) failed to increase the amount of variation among regions. The low level of variation observed among collections within the regions (1.0 %) supports the appropriateness of these groupings in all models.

4.4 Discussion

The pattern of genetic variation observed in this study is consistent with that identified previously in surveys of morphological and genetic variation (Shuster 1979; Riska 1981; King et al. 2005), suggesting a series of discontinuities across the species' range that could reflect regional adaptive significance and/or vicariant geographic events (e.g., a change in sea level giving rise to the Florida peninsula). Regional groupings (Fig. 4.5) that were consistent across analyses warrant management unit status based on the presence of statistically significant allele frequency heterogeneity, allocation of genetic diversity (Table 4.6), and a high percentage of correct classification to region of origin (Table 4.7). There appears to be substantial gene flow between each population and its nearest neighbors, with some notable exceptions (discussed below). Moreover, genetic diversity was sufficiently high at these nDNA markers that each individual surveyed possessed a unique multilocus genotype. This allowed assessment of family structure and estimates of effective population sizes.

American horseshoe crabs, *L. polyphemus*, often are considered to be evolutionarily static due to their morphological similarity to mid-Mesozoic taxa and have been referred to as phylogenetic relics (Selander et al. 1970). However, the presence of considerable variability and geographic differentiation in both morphology (Shuster 1979; Riska 1981) and genetic diversity illustrate that this species is far from evolutionarily static. Genetic distances estimated in the present study suggest the presence of eight or more regional management units of *L. polyphemus* within the United States, at a minimum: Hog Bay, Maine, Gulf of Maine, Mid-Atlantic, Southeast Atlantic, Florida East, Florida South, Florida Southwest, Florida West, St. Joseph Bay, Florida. This regional genetic structuring conforms generally to patterns of geographic variation in morphometric characters reported by Shuster (1979) and the findings of Riska (1981). Furthermore, the molecular genetic characteristics lend support to the suggestion of Shuster (1979) that there may be physiological races within the species.

Genetic distance analyses indicate clustering at the regional level (Figs. 4.2, 4.3, and 4.4) that conforms to the estimates of gene flow between each population and its nearest neighbors within groupings. Notable exceptions are the Hog Bay, Maine (MEH), Indian River, Florida (FIR), St. Joseph Bay, Florida (FSJ), and Yucatan Peninsula Mexico (MXY) collections (Table 4.5). The MEH collection is both the northernmost sampling location and a body of water that is almost completely isolated from the Gulf of Maine by strong currents that form a retention zone, trapping *L. polyphemus* in the bay and effectively eliminating migration/gene flow in or out of the bay. If it were to be shown that movement of *L. polyphemus* into Hog Bay from other bays does occur, then one would have to invoke some strong selective force to prevent successful reproduction of immigrants. A more plausible explanation is that geographic isolation combined with small population (and founder) size and inbreeding has afforded stochastic processes (such as random genetic drift) the opportunity to dramatically reduce the level of genetic diversity and heterozygosity in MEH.

Along the Florida coasts, both Indian River Lagoon (FIR) and St. Joseph Bay (SJB) are physically isolated populations that have recently undergone major population declines. Indian River Lagoon is a 156-mile estuary that lies behind barrier islands along the east coast of Florida. Exchange of lagoon water with the Atlantic Ocean is limited to one tidal inlet at the north end, and four tidal inlets in the southern part of the estuary act to geographically isolate FIR *L. polyphemus* populations to the north and south. In addition, the environment of the Indian River Lagoon is quite different from that of other coastal areas in the southeast: salinity is highly variable and depends on distance from the inlets and the magnitude of freshwater inputs (e.g. groundwater, rivers) (Ehlinger and Tankersley 2004), and most of the lagoon is microtidal (tidal fluctuations occur only near the inlets; Ehlinger and Tankersley 2009). American horseshoe crabs in the Indian River Lagoon are also thought to be in decline (Provancha 1997) due to disease (Scheidt and Lowers 2001), entrapment in power plant intakes (Ehlinger and Tankersley 2007) and shoreline development. Genetic differences between Indian River horseshoe crabs and populations to the north or south are likely due to their physical isolation, recent population declines and to the unique environment of the lagoon to which they are adapted.

St. Joseph Bay is a nearly enclosed, 24 km long by 10 km wide bay that is open to the Gulf of Mexico only at its north end. In 1999 bait harvesters removed nearly 100,000 adult horseshoe crabs from this bay (Gerhart 2007). Moreover, the FSJ collection consisted of the smallest sample size along the Florida Gulf coast. Therefore, the genetic differences between FSJ and other Gulf coast populations may be due to sampling error, geographic isolation and/or demographic stochasticity associated with bottleneck effects.

Interestingly, reduced genetic variation and high levels of differentiation also were observed in the collection from the southern extreme of the range (MXY; Yucatan Peninsula, Republic of Mexico). *Limulus polyphemus* has a disjunct distribution in the Gulf of Mexico. Horseshoe crabs are found in the Florida Keys and along the west coast of Florida to the Mississippi and Louisiana barrier islands (Fulford and Haehn 2012), but are not generally found elsewhere in the Gulf of Mexico until one reaches the west (Campeche) and northern (Yucatan) coasts of the Yucatan Peninsula (Shuster 1979; Gomez-Aguirre 1993; Zaldivar-Rae et al. 2009) where they nest on beaches that are not unlike those found in Florida. The MXY collection was from a large population that occupies an extensive estuarine and coastal area along the northern coast of the Yucatan. Geographic isolation from other *L. polyphemus* populations combined with the absence of larval migration between proximal bay systems (reported by Botton and Loveland 2003), affords the possibility that founder effect, inbreeding, and genetic drift likely have played important roles in this population's genetic diversity. Indeed, the level of genetic differentiation between southeastern U.S. and Yucatan Peninsula *L. polyphemus* combined with the observations of discrete populations based on morphometric differentiation (Shuster 1979) may warrant further investigations regarding taxonomic revision.

Several shallow and moderate zones of genetic discontinuity are inferred from all analyses conducted, separating the: (1) Gulf of Maine (MET-NHS) from Mid-Atlantic (MAP-NCS); (2) Mid-Atlantic from Southeast Atlantic (SBB-GSI); (3) Southeast Atlantic from Florida East (FIR); (4) Florida East from Florida South (FBB); (5) Florida South from Florida Southwest (FMI-FTB); (6) Florida Southwest from Florida West (FCK-FAP); (7) Florida West from St. Joseph Bay, Florida (FSJ); and (8) St. Joseph Bay from Yucatan Peninsula, Mexico (MXY). Moderate (or narrower) zones of genetic discontinuity are evident between: (a) the Gulf of Maine and Mid-Atlantic collections; (b) Mid-Atlantic and Southeast collections; and (c) Atlantic and Gulf of Mexico Florida collections. This phylogeographic pattern implies that there are at least eight demographically distinct assemblages across the species' range that are relevant to conservation considerations. In addition to support for recognition of these zones of discontinuity, the observation of genetic differentiation suggested low levels of gene exchange between collections on either side of these genetic discontinuities. The presence of demographically distinct and evolutionary significant lineages delineated by zones of genetic discontinuity is consistent with the findings of researchers assessing behavioral patterns and large differences observed by Graham et al. (2009) in body size within spawning aggregates sampled along the Northeastern Atlantic coast. Integration of the information from the nuclear genome with previously identified variation in allozymes (Selander et al. 1970) and mitochondrial DNA (Saunders et al. 1986; Pierce et al. 2000), and with ecological data should prove essential to developing an ecologically and evolutionarily sound management strategy.

The correlation between genetic distance and geographic distance among collections supports isolation-by-distance as a mechanism underlying the population structure of *L. polyphemus* along the Atlantic coast of North America (Fig. 4.6). Even within the continuum of isolation-by-distance, and providing for gene flow between neighboring collections, there still are regions that are discrete with respect to each other, as indicated by clustering of specific populations using the PCoA and STRUCTURE results (Figs. 4.2, 4.3, and 4.4). For example, the NH and ME collections (minus MEH) form a discrete Gulf of Maine population unit, the southern New England collections (MAP, from the Atlantic side of Cape Cod, MA, and RIN, Rhode Island) group with the collections from CT, NY, NJ, DE, MD, and NC to form a large mid-Atlantic cluster, the SC and GA collections cluster, and all but one Florida collections (FIR) form a cluster. In addition, the collection from the north coast of Mexico's Yucatan Peninsula is more differentiated from the Florida collections than the Florida collections are from the more northerly populations. This population is, however, still within the bounds of the isolation-by-distance model developed for the U.S. populations.

The percentage of genetic variance among populations (7 %) and/or regions (6–8 %) bodes well for using these molecular data for stock identification and mixed-stock analysis. Furthermore, the high probability (83.1 %) of correctly assigning individuals to the management unit from which they were collected indicated by multilocus assignment tests confirms that these data will be useful for identifying the source of *L. polyphemus* that are intercepted as by-catch in commercial fisheries. Of the 285 crabs mis-assigned to management units other than their collection of origin (16.9 % error rate), only 22 were mis-assigned to management units that were not adjacent to their true source (1.3 % of misclassifications). This finding adds further support to the isolation-by-distance model for explaining the observed patterns of genetic variation.

Pair-wise estimates of F'_{ST}, hierarchical gene diversity, and PCoA results suggest relative homogeneity among collections within each of the regions (proposed management units) defined above. Most pair-wise F'_{ST} values that were not significantly different from zero were between collections within the same proposed management unit. Hierarchical gene diversity analysis shows only 1 % of the total genetic variation being found among collections within proposed management units, as opposed to 6 % of genetic variation being found among these regional units (that figure increases to 8 % when the MEH and MXY collections are included; Table 4.4).

Given the large latitudinal range (19–45° North) encompassed by the extant distribution of *L. polyphemus*, it is instructive to investigate the extent to which Pleistocene ice ages and the associated palaeoclimatic oscillations may have influenced observed patterns of allelic diversity. The zone of discontinuity observed between *L. polyphemus* collections surveyed north and south of Cape Cod, MA (~41.7° north latitude) corresponds roughly to the median latitude of maximum Pleistocene glacial advance in North America and has been observed to be a breakpoint between nucleotide diversity and latitudinal distribution of freshwater and anadromous fish species (Bernatchez and Wilson 1998), with nucleotide diversity decreasing with increasing latitude. However, *L. polyphemus*, an estuarine species that exhibits anadromous tendencies (Shuster 1979), showed no general trend of reduced allelic diversity with increasing latitude in the present study. We do not assert that recent glacial and interglacial periods have not impacted *L. polyphemus* in this portion of the species' range, only that any effects appear to have been ameliorated by natural and anthropogenic gene flow, sufficient available habitat, and/or demographic robustness sufficient to have prevented the loss of allelic diversity through stochastic processes. Hence, the zone of discontinuity that exists in this area may simply reflect that Cape Cod is a zoogeographic barrier that influences oceanic and near-shore currents in such a manner that ultimately results in restricted gene flow among *L. polyphemus* populations.

Although a modicum of caution should be exercised when drawing phylogeographic inferences from microsatellite DNA variation (due to the potential for homoplasy in this class of marker), patterns observed in the present study are consistent with the well-documented intraspecific phylogeographic relationships observed among a host of marine taxa in the southeastern U.S. zoogeographic province (Avise 1992), including previous findings for *L. polyphemus* (Selander et al. 1970; Saunders et al. 1986). Microsatellite DNA variation in *L. polyphemus* reveals clear genetic separation of Atlantic and Gulf of Mexico collections, suggesting that the Florida Peninsula has served as a barrier to gene exchange. However, as stated above, physical isolation of the FIR and FBB collections from other Florida populations cannot be excluded as a cause for the observed levels of differentiation in this area. Additional sampling along the Atlantic coastline to south Florida will be required to better understand the population structure of this region (Fig. 4.5).

4.5 Management Implications

Management units (MUs) are defined as populations that are demographically independent of one another (Allendorf and Luikart 2007), meaning that their population dynamics depend mostly on local birth and death rates, and not on genetically effective migration from other spawning assemblages. Identification of MUs – similar in concept to "stocks" widely referred to in fisheries management – is useful for short-term management, such as delineating fishing areas, setting harvest rates, and monitoring population status. Offering an operational definition, Moritz (1994) suggested that MUs are populations that have substantially divergent allele frequencies at many loci. One possible limitation of this approach, however, is that allele frequency differentiation cannot be interpreted directly as evidence for demographic independence (Allendorf and Luikart 2007). A related issue is the difficulty in determining whether migration from nearby spawning assemblages would be sufficient to reestablish an MU should it become overharvested or extinct. Palsbøll et al. (2007) proposed that the identification of MUs from population genetic data be based upon the amount of genetic divergence at which populations become demographically independent; wherein MU status would be assigned when the observed estimate of genetic divergence is significantly greater than a pre-defined threshold value. To illustrate the application of the MU concept, Ramstad et al. (2004) analyzed approximately 100 sockeye salmon from 11 spawning sites throughout the Lake Clark drainage of the Bristol Bay system in Alaska at 11 microsatellite DNA loci to determine whether the spawning assemblages were demographically isolated. The effective population size, N_e, for each of the Lake Clark spawning sites was ~1,000. Using the criterion of at least 10 % exchange (Hastings 1993), groups spawning at these sites would be demographically isolated if they exchanged fewer than about 100 adults, which corresponds to genetic differentiation (F_{ST}) of 0.0025 under a classical Wright-Fisher island model of migration-drift equilibrium. Therefore, one might conclude that these spawning sites constitute separate MUs if their genetic divergence, F_{ST}, exceeds 0.0025. The overall value of F'_{ST} among these sites excluding one outlier was ~0.007 (95 % CI of 0.004–0.010), which was greater than the threshold of 0.0025. The authors concluded that these 11 spawning aggregations were demographically isolated and should be considered separate MUs.

Considering the current data for American horseshoe crabs, it seems clear that there exist multiple MUs, corresponding to the proposed regional management areas. In addition to the demographically discrete lineages identified for American horseshoe crabs, a series of metapopulations and other individual collections delineated within each discrete lineage may be considered distinct management/recovery units for future management planning purposes. Furthermore, metapopulations may exist in the Northern Atlantic (Gulf of Maine and New Hampshire collections), the Mid-Atlantic region (with some substructure within), the upper Chesapeake Bay collections (MDT, MDF), the Southeast Atlantic (SBB, SBE, GSA, and GSI), southwest Florida Gulf of Mexico (FMI, FCH, FTB), and the northwest Florida Gulf of Mexico (FCK, FAP). Within areas bounded by zones of genetic discontinuity, there appears to be substantial gene flow between each population and its nearest neighbors; the presence of these metapopulations appears to bode well for the demographic viability of horseshoe crabs in some regions.

Given the restriction in female-mediated gene flow observed by King et al. (2005) and that of Pierce et al. (2000), and the absence of larval migration reported by Botton and Loveland (2003), the appreciable amount of gene flow we have documented along North America's Atlantic coast is most likely to be the result of effective migration by sub-adult or adult males. These findings suggest that management efforts may best be targeted at local populations rather than at a regional level, as an absence of an effective number of females may result in local functional extinctions. In the most extreme cases, it may become necessary to institute harvest limitations on female *L. polyphemus* for populations known to have low census sizes. The patterns of genetic differentiation observed in this study also suggest that relocations from adjacent embayments could serve as a source for future restoration efforts.

Acknowledgements The U.S. Geological Survey (State Partnership and Quick Response programs), the Atlantic States Marine Fisheries Commission, and the AAAS Women's International Science Collaboration Program funded this research. The following individuals graciously provided *L. polyphemus* tissue samples: S. Schaller, C. McBane, K. Tuxbury, C. Grahn, B. Battelle, M-J. James-Pirri, T. O'Connell, J. Mattei, R. Chapman, G. Ehlinger, L. Barton, T. Summers, C. Callahan, C. Morrison, D.R. Smith, K. Voges, A. Garcia, Z. Johnson, R. Johnson, K. Hovatter, V. Robbins, and P. Pooler. Assistance with the Yucatan field collection was provided by G. V. Rios, Centro Regional de Investigación Pesquera, Yucalpetén, Mexico. We thank the State of Florida Fish and Wildlife Conservation Commission and Ryan Gandy and Michael Tringali (St. Petersburg Office) for their kind help in providing genotype data on the newest Florida collections. Laboratory assistance was provided by B. Eleby, R. Long, R. Johnson, and M. Meadows. We thank J. Young for assistance in generating the collection site map. Funding for EH's participation in this work was provided in part by the Virginia Agricultural Experiment Station and the Hatch Program of the National Institute of Food and Agriculture, U.S. Department of Agriculture. Use of trade, product, or firm names does not imply endorsement by the U.S. Government.

References

Allendorf FW, Luikart G (2007) Conservation and the genetics of populations. Blackwell Publishing, Malden

ASMFC (Atlantic States Marine Fisheries Commission) (1998) Interstate fishery management plan for horseshoe crab. ASMFC, Fishery Management report no. 32, Washington, DC

Avise JC (1992) Molecular population structure and the biogeographic history of a regional fauna: a case history with lessons for conservation biology. Oikos 63:62–76

Baker AJ, Gonzalez PM, Piersma T et al (2004) Rapid population decline in red knots: fitness consequences of decreased refueling rates and late arrival in Delaware Bay. Proc R Soc Lond B Biol Sci 271:875–882

Barton NH, Slatkin M (1986) A quasi-equilibrium theory of the distribution of rare alleles in a subdivided population. Heredity 56:409–415

Bernatchez L, Wilson CC (1998) Comparative phylogeography of Nearctic and Palearctic fishes. Mol Ecol 7:431–452

Botton ML, Loveland RE (2003) Abundance and dispersal potential of horseshoe crab (*Limulus polyphemus*) larvae in the Delaware estuary. Estuaries 26:1472–1479

Brockmann HJ, Colson T, Potts W (1994) Sperm competition in horseshoe crabs (*Limulus polyphemus*). Behav Ecol Sociobiol 35:153–160

Brockmann HJ, Nguyen C, Potts W (2000) Paternity in horseshoe crabs when spawning in multiple male groups. Anim Behav 60:837–849

Cavalli-Sforza LL, Edwards AWF (1967) Phylogenetic analysis: models and estimation procedures. Evolution 21:550–570

Cornuet J-M, Piry S, Luikart G et al (1999) New methods employing multilocus genotypes to select or exclude populations as origins of individuals. Genetics 153:1989–2000

Earl DA, vonHoldt BM (2012) STRUCTURE HARVESTER: a website and program for visualizing STRUCTURE output and implementing the Evanno method. Conserv Genet Resour 4(2):359–361

Ehlinger GS, Tankersley RA (2004) Survival and development of horseshoe crab (*Limulus polyphemus*) embryos and larvae in hypersaline conditions. Biol Bull 206:87–94

Ehlinger GS, Tankersley RA (2007) Reproductive ecology of the American horseshoe crab *Limulus polyphemus* in the Indian River Lagoon: an overview. Fla Sci 70:449–463

Ehlinger GS, Tankersley RA (2009) Ecology of horseshoe crabs in microtidal lagoons. In: Tanacredi JT, Botton M, Smith DR (eds) Biology and conservation of horseshoe crabs. Springer, New York, pp 149–162

Ellegren H (2004) Microsatellites: simple sequences with complex evolution. Nat Rev Genet 5:435–455

Evanno G, Regnaut S, Goudet J (2005) Detecting the number of clusters of individuals using the software STRUCTURE: a simulation study. Mol Ecol 14:2611–2620

Excoffier L, Smouse PE, Quattro JM (1992) Analysis of molecular variance inferred from metric distances among DNA haplotypes: application to human mitochondrial DNA restriction data. Genetics 131:479–491

Faurby S, King TL, Obst M et al (2010) Population dynamics of American horseshoe crabs–historic climatic events and recent demographic pressures. Mol Ecol 19:3088–3100

Fulford RS, Haehn RA (2012) An evaluation of Mississippi barrier islands as spawning and nesting habitat for the American horseshoe crab, *Limulus polyphemus*, with implications for island restoration. Gulf Caribb Res 24:51–62

Gerhart SD (2007) A review of the biology and management of horseshoe crabs with emphasis on Florida populations. Fish and Wildlife Research Institute, St. Petersburg, http://research.myfwc.com. Accessed Jan 2015

Gomez-Aguirre S (1993) Cacerolita de Mar (*Limulus polyphemus* L.) en la Peninsula de Yucatan. In: Salazar-Vallejo SI, Gonzalez NE (eds) Biodiversidad Marina ye Costera de Mexico. Com. Nal. Biodiversidad y CIQRO, Mexico, D.F., pp 650–659

Goudet J, Perrin N, Waser P (2002) Tests for sex-biased dispersal using bi-parentally inherited genetic markers. Mol Ecol 11:1103–1114

Graham LJ, Botton ML, Hata D et al (2009) Prosomal-width-to-weight relationships in American horseshoe crabs (*Limulus polyphemus*): examining conversion factors used to estimate landings. Fish Bull 107:235–243

Guo SW, Thompson EA (1992) Performing the exact test of Hardy-Weinberg proportions for multiple alleles. Biometrics 48:361–372

Hallerman EM (2003) Population genetics: principles and applications for fisheries scientists. American Fisheries Society, Bethesda

Hammer Ø, Harper DAT, Ryan PD (2001) Past: paleontological statistics software package for education and data analysis. Palaeontol Electron 4(1):9pp, art. 4, 178kb. http://palaeoelectronica.org/2001_1/past/issue1_01.htm

Hastings A (1993) Complex interactions between dispersal and dynamics–lessons from coupled logistic equations. Ecology 74:1362–1372

Höss M, Pääbo S (1993) DNA extraction from Pleistocene bones by a silica-based purification method. Nucleic Acids Res 21:3913–3914

King TL, Eackles MS, Spidle AP et al (2005) Regional differentiation and sex-biased dispersal among populations of the horseshoe crab *Limulus polyphemus*. Trans Am Fish Soc 134:441–465

King TL, Switzer JF, Morrison CL et al (2006) Comprehensive genetic analyses reveal evolutionary distinction of a mouse (*Zapus hudsonius preblei*) proposed for delisting from the US Endangered Species Act. Mol Ecol 15:4331–4359

Levin J, Bang FB (1968) Clottable protein in Limulus: its localization and kinetics of its coagulation by endotoxin. Thromb Diath Haemorrh 19(1):186–197

Meirmans PG (2006) Using the AMOVA framework to estimate a standardized genetic differentiation measure. Evolution 60:2399–2402

Moritz C (1994) Defining 'evolutionary significant units' for conservation. Trends Ecol Evol 9:373–375

Paetkau D, Calvert W, Stirling I et al (1995) Microsatellite analysis of population structure in Canadian polar bears. Mol Ecol 4:347–354

Palsbøll PJ, Berube M, Allendorf FW (2007) Identification of management units using population genetic data. Trends Ecol Evol 22:11–16

Peakall ROD, Smouse PE (2006) GENALEX 6: genetic analysis in Excel. Population genetic software for teaching and research. Mol Ecol Notes 6:288–295

Peakall R, Smouse PE (2012) GenAlEx 6.5: genetic analysis in Excel. Population genetic software for teaching and research—an update. Bioinformatics 28:2537–2539

Pierce JP, Tan G, Gaffney PM (2000) Delaware Bay and Chesapeake Bay populations of the horseshoe crab (*Limulus polyphemus*) are genetically distinct. Estuaries 23:690–698

Piry S, Alapetite A, Cornuet J-M et al (2004) GeneClass2: a software for genetic assignment and first-generation migrant detection. J Hered 95:536–539

Pritchard JK, Stephens M, Donnelly P (2000) Inference of population structure using multilocus genotype data. Genetics 155:945–959

Provancha J (1997) Annual report for sea turtle netting in Mosquito Lagoon. NMFS Permit #942, FL Permit #114

Ramilo ST, Wang J (2012) The effect of close relatives on unsupervised Bayesian clustering algorithms in population genetic structure analysis. Mol Ecol Resour 12:873–884

Ramstad KM, Woody CA, Sage GK et al (2004) Founding events influence genetic population structure of sockeye salmon (*Oncorhynchus nerka*) in Lake Clark, Alaska. Mol Ecol 13:277–290

Raymond M, Rousset F (1995) GENEPOP (version 1.2): population genetics software for exact tests and ecumenicism. J Hered 86:248–249

Rice WR (1989) Analyzing tables of statistical tests. Evolution 43:223–225

Riska B (1981) Morphological variation in the horseshoe crab *Limulus polyphemus*. Evolution 35:647–658

Rohlf FJ (2000) NTSYS-PC: numerical taxonomy and multivariate analysis systems, Version 2.10. Exeter Software, Setauket

Ryder O (1986) Species conservation and systematics: the dilemma of subspecies. Trends Ecol Evol 1:9–10

Saunders NC, Kessler LG, Avise JC (1986) Genetic variation and geographic differentiation in mitochondrial DNA of the horseshoe crab, *Limulus polyphemus*. Genetics 112:613–627

Scheidt D, Lowers R (2001) Using an aerial survey to document the extent of a horseshoe crab die-off in Florida. In: American Fisheries Society annual meeting, Baltimore, MD

Selander RK, Yang SY, Lewontin RC et al (1970) Genetic variation in the horseshoe crab (*Limulus polyphemus*), a phylogenetic "relic". Evolution 24:402–414

Shuster CN Jr (1979) Distribution of the American horseshoe "crab", *Limulus polyphemus* (L.). In: Cohen E (ed) Biomedical applications of the horseshoe crab (Limulidae). Alan R. Liss, Inc., New York, pp 3–26

Shuster CS, Barlow RB, Brockmann HJ (eds) (2003) The American horseshoe crab. Harvard University Press, Cambridge, MA

Sokal RR, Rohlf FJ (1994) Biometry: the principles and practice of statistics in biological research, 3rd edn. W.H. Freeman & Co., New York

Swofford DL, Selander RB (1981) BIOSYS-1: A FORTRAN program for the comprehensive analysis of electrophoretic data in population genetics and systematics. J Hered 72:281–283

Walls EA, Berkson J, Smith SA (2002) The horseshoe crab, *Limulus Polyphemus*: 200 million years of existence, 100 years of study. Rev Fish Sci 10:39–73

Wang J, Santure AW (2009) Parentage and sibship inference from multilocus genotype data under polygamy. Genetics 181:1579–1594

Wright S (1931) Evolution in Mendelian populations. Genetics 16:97–159

Zaldivar-Rae J, Sapien-Silva RE, Rosales-Raya M et al (2009) American horseshoe crabs, *Limulus polyphemus*, in Mexico: open possibilities. In: Tanacredi JT, Botton M, Smith DR (eds) Biology and conservation of horseshoe crabs. Springer, New York, pp 97–113

Chapter 5
Florida, Horseshoe Crabs: Populations, Genetics and the Marine-Life Harvest

H. Jane Brockmann, Tiffany Black, and Tim L. King

Abstract American horseshoe crabs (*Limulus polyphemus*) breed in every coastal county of Florida, U.S.A., and yet the data available on the status of this species in Florida are limited. The east and west coasts of Florida have genetically distinct populations of horseshoe crabs, but it is not known if there is further genetic structuring. To fill these gaps, we compile data from a number of sources: genetic analysis of tissue samples from 13 sites, breeding activity from an on-line public database, population estimates from multi-year trawl and seine surveys, and spawning activity from a long-term study. Neighbor-joining and STRUCTURE analyses support at least five genetically distinct management units within both the Atlantic and Gulf coast populations of Florida horseshoe crabs. We find that in some parts of Florida, horseshoe crabs breed throughout the year, but in other parts there are distinct breeding seasons. Florida populations are generally smaller than those in more northern states, particularly along the east coast, but there are few indications of decline except in the Indian River area. Finally, we compile data on the harvest of horseshoe crabs in Florida and discuss the effect of the marine-life fishery, which has not been tracked in other states.

Keywords *Limulus polyphemus* • Population genetic structure • Breeding survey • Trawl survey • Management

---

H.J. Brockmann (✉)
Department of Biology, University of Florida, Gainesville, FL 32611, USA
e-mail: HJB@ufl.edu

T. Black
Department of Marine Fisheries Biology, Florida Fish and Wildlife Conservation Commission, Senator George Kirkpatrick Marine Laboratory, Cedar Key, FL 32625, USA

T.L. King
U.S. Geological Survey, Biological Resources Division, Leetown Science Center, Aquatic Ecology Branch, 11649 Leetown Road, Kearneysville, WV 25430, USA

R.H. Carmichael et al. (eds.), *Changing Global Perspectives on Horseshoe Crab Biology, Conservation and Management*, DOI 10.1007/978-3-319-19542-1_5

5.1 Introduction

American horseshoe crabs, *Limulus polyphemus*, are widely distributed along the Atlantic coast of the U.S. from Maine to Florida and in the Gulf of Mexico from Florida to Louisiana and the Yucatan Peninsula (Shuster 1982; Anderson and Shuster 2003). Because of their ecological and commercial importance, they are managed throughout most of this broad range (Shuster et al. 2003). Although it is all one species, both genetic and phenotypic studies reveal distinct populations of *L. polyphemus*. A survey of allozyme variation suggested that Atlantic and Gulf of Mexico populations were genetically differentiated (Selander et al. 1970), and a subsequent study of mitochondrial DNA (mtDNA) identified a major genetic discontinuity between northern and southern populations of horseshoe crabs in the vicinity of Cape Canaveral (Saunders et al. 1986). More recently using microsatellite markers, King et al. (2005) confirmed the phylogenetic break between Atlantic and Gulf populations and between South Carolina and Florida populations. Not only are Florida horseshoe crabs genetically distinct from populations to the north, but they are physically (Shuster 1979; Riska 1981) and behaviorally different: they are smaller (Brockmann and Smith 2009); they show differences in the timing (Penn and Brockmann 1994; Sekiguchi and Shuster 2009) and location of spawning (Badgerow and Sydlik 1989; Brockmann 2003), differences in the environmental cues that affect spawning (Rudloe 1985; Barlow et al. 1986; Brockmann 2003; Ehlinger et al. 2003; Fulford and Haehn 2012), differences in sex ratios and satellite behavior (Brockmann and Smith 2009; Mattei et al. 2010), and differences in the timing of juvenile molting (Rudloe 1981). Taken together these studies demonstrate that the horseshoe crabs of Florida differ from the better-studied populations to the north. Here we pull together for the first time the available data on horseshoe crab populations from different parts of Florida, evaluate what is known about the status of these populations and make recommendations on their management.

In Florida, the Atlantic States Marine Fisheries Commission (ASMFC) Horseshoe Crab Fisheries Management Plan (FMP) went into effect in March 2000 (Gerhart 2007). There are three categories of commercial use for horseshoe crabs in Florida: as bait in the eel and whelk fisheries, as marine-life specimens for the aquaria trade, research and education, and as specimens for biomedical use (to extract their blood). All three uses are regulated by the Florida Fish and Wildlife Conservation Commission (FWCC). A biomedical license is required for collecting horseshoe crabs for bleeding, but currently there is no active biomedical fishery in Florida. A saltwater products license is required for harvesting horseshoe crabs for bait and it sets a limit on the number that can be collected each day: 25 animals are allowed per day per person. When this license has a marine-life endorsement, the limit is 100 per person per day. Historically, horseshoe crabs used for eel bait were harvested by trawl, but in 2000 new regulations limited their harvest to hand or gig collection only, which reduced bait collection considerably (Gerhart 2007). The marine-life fishery collects mainly juvenile horseshoe crabs for the aquaria trade. In some areas they are collected in such numbers (18,000–32,000 collected per year in Florida over the past 10 years) that populations may be affected (Gerhart 2007).

The FMP sets a limit on the total number of adult horseshoe crabs that can be collected for bait at the number that were collected annually by each state prior to the FMP going into effect. In the case of Florida, the total number of horseshoe crabs that can be harvested annually for bait on the Atlantic coast is 9,455. There is no overall limit on the total number of horseshoe crabs that can be collected in the Florida marine-life fishery. Gulf coast fisheries are governed by the Gulf States Marine Fisheries Commission (GSMFC), which does not regulate horseshoe crabs.

To describe and evaluate the status of horseshoe crabs in Florida, we use data from a number of different sources. (a) Previous research has shown that the horseshoe crabs from the east and west coasts of Florida are genetically distinct (King et al. 2005), and the present study adds to the previous data with tissue samples from three new sites on the Atlantic coast and four new sites along the Gulf coast of Florida. These new analyses allow us to determine whether horseshoe crab populations on each coast are further subdivided or are part of one interbreeding population. (b) The State of Florida Fish and Wildlife Conservation Commission (FWCC) conducts an on-line survey of horseshoe crab observations made by the public. We use these reports to determine the distribution and seasonality of horseshoe crabs in different regions of Florida. (c) Data on the numbers of horseshoe crabs captured in monthly FWCC Fisheries-Independent Monitoring (FIM) trawl and (d) seine sampling surveys allow comparisons of populations in different regions of Florida. These data also allow us to evaluate whether the populations are stable or changing across years. (e) We present population data from the only long-term survey of horseshoe crabs in Florida (Brockmann and Johnson 2011). The present study adds two additional years of data that allow us to evaluate the amount of variation in the numbers of breeding animals, sex ratios and sizes of individuals across 22 years in a population that has not been harvested. (f) Those with licenses to harvest horseshoe crabs in Florida are required to submit reports to the FWCC on the numbers of horseshoe crabs collected (i.e. landings). These reports allow us to determine whether harvests are increasing or decreasing in different regions of Florida. Taken together, these six sources of information allow us to evaluate the status of horseshoe crabs in different regions of Florida. We also provide recommendations for the management of Florida horseshoe crabs.

5.2 Materials and Methods

5.2.1 Regions of Florida

Since horseshoe crabs from the east (Atlantic) and west (Gulf) coasts of Florida are known to be genetically distinct (King et al. 2005), we present data from the two coasts separately. Further, because of the long distances involved, we divide the state into four regions: north and south of the Pinellas-Pasco County line along the Gulf coast and north and south of the Indian River-Brevard County line along the East coast (Fig. 5.1; in each case approximately halfway down the coast of

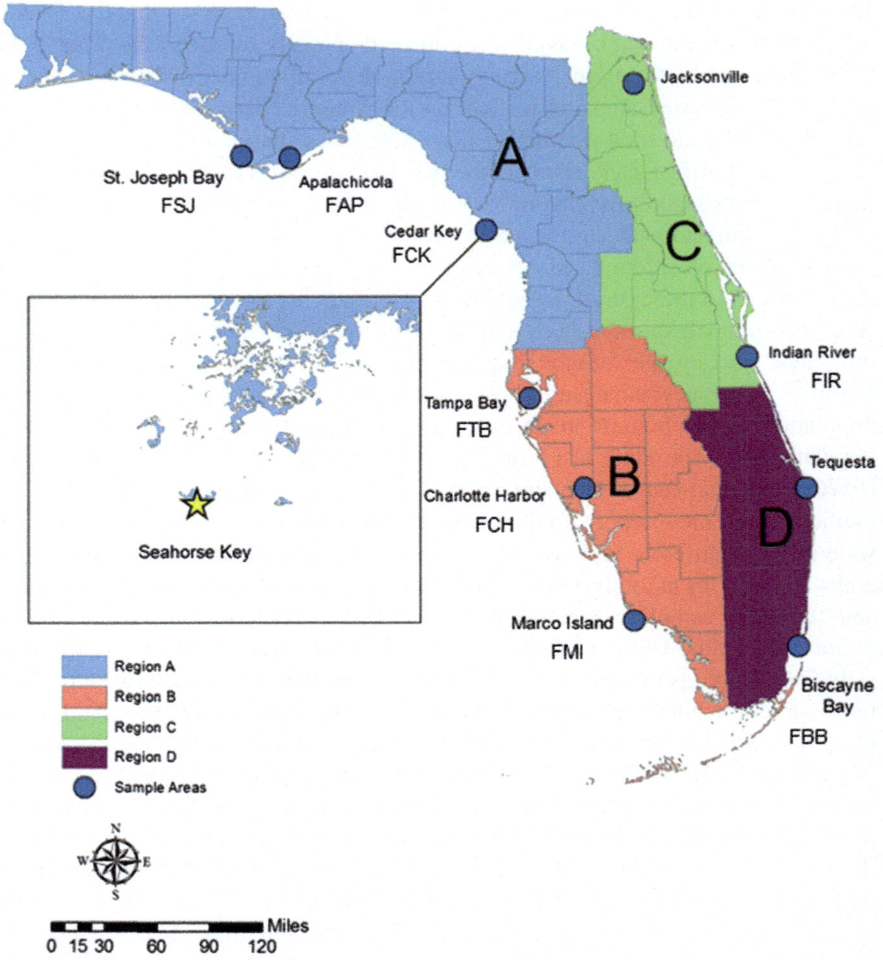

Fig. 5.1 Map of Florida showing the regional designations (**A**, **B**, **C**, **D**), the trawl survey locations (Apalachicola Bay, Cedar Key, Tampa Bay, Charlotte Harbor, Indian River, Jacksonville), the seine survey locations (Apalachicola Bay, Cedar Key, Charlotte Harbor, Tequesta, Indian River, Jacksonville) and the sites where genetic samples were collected (FSJ, FAP, FCK, FTB, FCH, FMI, FBB, FIR)

peninsular Florida, roughly separating genetically distinct northeast from southeast populations and southwest from northwest Florida populations).

5.2.2 *Population Genetic Analysis*

Tissue samples were obtained from seven new sites around Florida (Fig. 5.1) and compared with samples from the three Florida sites used in King et al. (2005). The new Florida tissue samples in the present analysis were collected by FWCC staff during trawl surveys. The DNA extraction and isolation, PCR (polymerase chain reaction) and fragment analysis procedures and the statistical analyses were the same as those used in previous studies (King and Eackles 2004; King et al. 2005, 2015).

Genetic distances between each pair of collections were summarized with genetic distance matrices calculated using the Cavalli-Sforza and Edwards (1967) chord distance in BIOSYS (Swofford and Selander 1981). The evolutionary relationships among the *L. polyphemus* collections were visualized by constructing a Neighbor-Joining tree (Saitou and Nei 1987). The strength of support for each node in the phylogenetic tree was tested by bootstrapping over loci using *njbpop* (J.M. Cornuet, INRA, France). MEGA5 (Tamura et al. 2011) software was used to visualize the tree.

The Bayesian clustering program STRUCTURE ver 2.3.1 (Pritchard et al. 2000) was utilized to determine the number of populations (or clusters, *K*) present among the collections sampled. Because a single STRUCTURE analysis on a set of populations may only reveal the uppermost level of population structure (Evanno et al. 2005), we performed a hierarchical STRUCTURE analysis similar to that employed by King et al. (2006). In the initial phase, $K=1$ to $K=13$ clusters were considered for all collections pooled together using 100,000 iterations discarded as burn-in followed by 200,000 iterations, and ten independent runs for each *K* using the admixture model and un-correlated allele frequencies. Subsequent analysis of each cluster tested $K=1$ to $K=C+3$ (the number of collections [*C*] included in the subset plus three) using the admixture and correlated allele frequencies model with the same number of burn-in and iterations. Sample location was incorporated as prior information, since populations separated by extreme distances presumably do not exchange migrants. The number of clusters for each analysis was determined using the ΔK method of Evanno et al. (2005) performed in the program Structure Harvester (Earl and von Holdt 2011).

5.2.3 On-line Surveys

The FWCC collects reports from the public on horseshoe crab sightings in Florida (see http://myfwc.com/research/saltwater/crustaceans-marine-arthropods/horseshoe-crabs/). The survey was initiated in 2002 but the reporting method was changed in 2010 to reflect the observed numbers more precisely and for this reason we do not use the earlier data. Here we use data collected on this website between 2010 and 2013 to provide the range of dates for horseshoe crab breeding activity in Florida. Of course, the number of voluntary reports from the public differs in different parts of Florida depending on the numbers of horseshoe crabs present, the numbers of observers present and how well the public is educated about reporting horseshoe crab sightings to the FWCC website. For this reason we give the number of reports and the total number of horseshoe crab sightings during each month for each region in Florida during the 4-year period. Because of the extreme variability in the data, we report the median number of horseshoe crabs sighted from all reports for each month in each region.

5.2.4 Trawl and Seine Surveys

Scientists from the Florida Fish and Wildlife Research Institute (which is part of FWCC) collect data each month from trawl and seine surveys as part of their Fisheries-Independent Monitoring (FIM) program. Both trawl and seine surveys are taken at five locations (Apalachicola Bay, Cedar Key, Charlotte Harbor, Indian River and Jacksonville) and one additional trawl survey (Tampa Bay) and one additional seine (Tequesta) survey is taken (Fig. 5.1). All trawl surveys are conducted according to stratified-random sampling protocols using a 6.1 m otter trawl (straight tow with 3.1 mm liner and tickler chain) at depths of 0–43 m (mean depth 3.07 ± 0.01 m). The seine surveys utilize a 183 m center-bag seine (net set from a skiff in a rectangular pattern on shore in water depth <0.5 m). The sampling protocols are designed to estimate finfish abundance and population trends, which means that horseshoe crabs are bycatch in these trawl and seine surveys, but their numbers have been recorded consistently over the years. FIM data have been collected at most sites since 1996, but only since 2001 at the Jacksonville site. The number of trawls varies from 1 month, year or site to the next, so it is important to include total effort when comparing sites and months. Here we compile data from FIM trawling and seining surveys using the total catch (for each year) over the total effort (i.e. yearly catch/total effort). For the trawl surveys, effort is a function of the time (usually 10 min) the trawl was towed and the number of trawls conducted that year. For seine surveys effort is simply a function of the number of seining surveys conducted that year (which was uniform at a site). We use 13 years to calculate recent trends since this is the approximate generation time for horseshoe crabs (Sweka et al. 2007). It is worth noting, however, that generation time is not known for Florida horseshoe crabs and may be shorter than that calculated for Delaware Bay since the animals are smaller and active throughout the year and therefore may develop more quickly.

5.2.5 Seahorse Key Spawning Surveys and Measurements on Horseshoe Crabs

We conducted spawning surveys of horseshoe crabs across 22 years (1992–2014) along a 1 km stretch of beach at Seahorse Key (SK; 29.096 °N, 83.065 °W), which is part of the Cedar Keys National Wildlife Refuge in northern Florida (Fig. 5.1; Brockmann and Johnson 2011). We observed horseshoe crabs during each high tide (two per day) from 2 days before the new or full moon to 5 days after during the spring from late February to May. Horseshoe crabs also nest at SK during the fall (August-October) (Brockmann et al., unpublished manuscript) but there are many fewer at this time and the fall data were collected for only a few years, so they are not included here. During each high tide, from 15 min before to 15 min after the time of the predicted maximum high tide at SK, we walked along the shoreline and counted all horseshoe crabs that were on the beach or swimming within 1 m of the shoreline. We identified whether the individual was female or male and if male whether he was paired with a female (i.e. an attached pair in amplexus) or unpaired. Unpaired males may join spawning pairs, and when they do, previous research has shown that they fertilize many of the eggs that females lay (Brockmann et al. 1994, 2000). Since we do not understand why it is that some high tides do not have spawning horseshoe crabs, we consider here only those high tides on which at least one pair spawned.

During 10 of the years that the spawning survey was conducted at SK (1992–1997, 2008–2010, 2014), we also measured the size (the width of the prosoma at the widest point) and evaluated the surface condition of a sample of male and female horseshoe crabs. Horseshoe crabs have a terminal molt (Koons 1883; Shuster and Sekiguchi 2003; Smith et al. 2010), so the size they attain at maturity is the size they remain as adults. We evaluated whether males and females measured in different years were different sizes using an ANOVA with the Holm-Sidak all pairwise multiple comparison procedure (Systat Software Inc 2011). As the adult horseshoe crab ages, the condition of its surface deteriorates (Penn and Brockmann 1995; Brockmann 2002): they become darker in color, scratched and pitted, the posterior opisthosomal spines become worn, and the lateral eyes become soft (e.g. from chitinoclastic bacteria) or covered with epibionts (Grant 2001; Duffy et al. 2006). We measured the proportion of individuals each year that were in perfect condition, i.e. newly molted (individuals with a light-colored prosoma, no pits, scratches, scars or worn spines and both eyes in good condition). This is a rough estimate of the proportion of individuals that were recently recruited into the population.

5.2.6 Harvest Data

Wholesale dealers and retailers who collect horseshoe crabs must have a license, and this license requires them to submit "trip tickets" that report all saltwater products taken on each collecting trip. The trip tickets are submitted to the FWCC where they are entered into the Marine Fisheries Information System (Gerhart 2007). Since 1997 the bait and marine life landings have been recorded separately. We use these data to determine the pattern of harvesting and harvest intensity in different regions of Florida.

5.3 Results and Discussion

5.3.1 Population Genetics of Florida Horseshoe Crabs

Patterns of genotypic variation in the nuclear DNA at the individual and population scales suggest the presence of multiple zones of genetic discontinuity around Florida (Figs. 5.2 and 5.3). Upon inspection of the neighbor-joining phenogram depicting structure within the chord distance matrix among 13 *Limulus polyphemus*

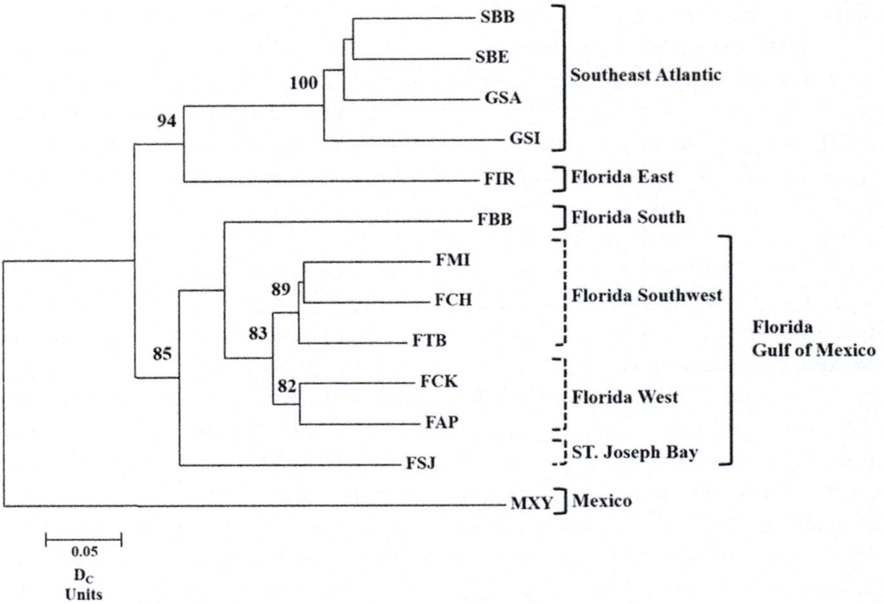

Fig. 5.2 Neighbor-joining phenogram depicting the underlying structure within the pairwise chord distance (D_c) matrix (Cavalli-Sforza and Edwards 1967) among seven *Limulus polyphemus* collections sampled from the Atlantic and Gulf coasts of Florida. Brackets group collections into suggested management units. *Dashed lines* indicate shallow but significant differentiation observed along the Florida Gulf coast. See Fig. 5.1 for the locations of the Florida sites and the regions in which they are found. *SBB* Bulls Bay, South Carolina (N=53), *SBE* Beaufort, South Carolina (N=48), *GSA* Savannah, Georgia (48), *GSI* Sapelo Island, Georgia (32), *FIR* Indian River, Florida (Atlantic coast) (47), *FBB* Biscayne Bay (20), *FMI* Tiger Tail Beach, Marco Island, Florida (Gulf coast) (81), *FCH* Charlotte Harbor, Florida (51), *FTB* Tampa Bay, Florida (141), *FCK* Seahorse Key, Cedar Keys NWR, Florida (132), *FAP* Alligator Point, Apalachicola Bay, Florida (92), *FSJ* St. Joseph Bay, Florida (23)

collections sampled from the Atlantic and Gulf coasts of Florida, seven regional groupings ("clades") were delineated by six zones of genetic discontinuity. The groupings are (Fig. 5.2): (1) SBB to GSI (Southeast Atlantic); (2) FIR (Florida East); (3) FBB (Florida South); (4) FMI to FTB (Florida Southwest); (5) FCK to FAP (Florida West); (6) FSJ (St. Joseph Bay, Florida); and (7) MXY (Yucatan Peninsula, Mexico). At every scale, clustering corresponded to the geographical distribution of the collections, with greater distances separating the southernmost regional collections from the Southeast Atlantic collections. Along the Atlantic coast, Fig. 5.3 suggests a moderate to high degree of relatedness (or recent gene flow) among the Southeast Atlantic collections. The greatest zones of genetic discontinuity within the study area occurred to the north (GSI) and south (FBB) of the Indian River, FL (FIR) collection. In addition to the deep zone of discontinuity that exists to the north and south of Indian River, FL, a regional perspective of the Florida coast identified shallow but significant differentiation along the central portion of the Gulf coast.

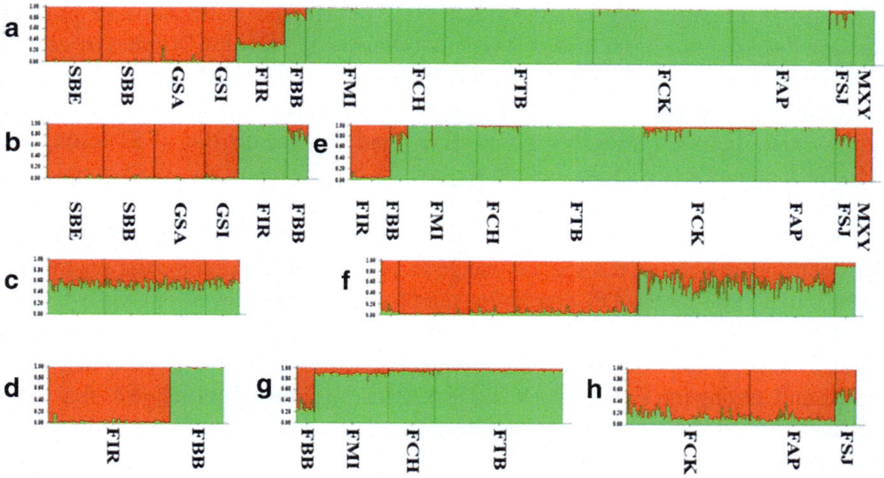

Fig. 5.3 Graphical representation of STRUCTURE analyses among 13 *Limulus polyphemus* collections sampled from the southeast Atlantic and Gulf coasts and genotyped at 13 microsatellite DNA loci. For the STRUCTURE histograms, each individual is represented by a single vertical bar, broken into $K=2$ colored segments, the length of which is proportional to the membership fraction in each of the K clusters. Black lines partition the collections. (**a**–**h**) represent comparisons among different groups of populations. See Fig. 5.2 for site abbreviations

Results from the individual-based STRUCTURE analyses among the 13 *L. polyphemus* collections (Fig. 5.3) were congruent with the neighbor-joining tree (Fig. 5.2). The number of inferred clusters (K) determined by STRUCTURE for the initial (uppermost hierarchical level) analysis was two, corresponding to the Atlantic coast collections north (SBB-GSI) of Indian River, Florida (FIR) and south of FIR around to St. Joseph Bay, Florida, the western most Florida collection (Fig. 5.3a). Due to the apparent complex migration patterns (zones of genetic discontinuity) existing among the two regions, a sequential method of inferring the number of clusters was employed to identify within-cluster structure. Sequential STRUCTURE analysis of the Atlantic Southeast collections (SBB to FBB) indicated $K=2$ delineating the Southeast Atlantic (SBB to GSI) collections from the two Florida Atlantic coast collections (FIR and FBB) (Fig. 5.3b). Analysis of the Southeast Atlantic grouping suggested that the four collections were panmictic as $K=1$ (Fig. 5.3c). Consistent with the neighbor-joining clustering, STRUCTURE indicated that the two Florida Atlantic coast collections were genetically distinct ($K=2$; Fig. 5.3d). Multiple sequential STRUCTURE analyses of all eight Florida collections (FIR to FSJ) and Mexico (MXY) were consistent with the patterns observed for genetic distance in the neighbor-joining tree. Three analyses were performed with each resulting in $K=2$ clusters. The Indian River (FIR) and Yucatan Peninsula (MXY) collections were identified as two distinct clusters highly differentiated from adjacent populations and from each other (Fig. 5.3e). However, the level of differentiation among the Florida South (FBB), Florida Southwest (FMI to FTB), and Florida West (FCK to FAP) clusters did not appear as great as that observed between the St. Joseph Bay,

Florida (FSJ) collection and the other Gulf of Mexico collections or as seen among the Atlantic coast clusters (Fig. 5.3f). Narrower ("shallow") zones of genetic discontinuity were evident between the Tampa Bay (FTB) and Cedar Key (FCK) collections on the Gulf coast (Fig. 5.3g). An additional zone of discontinuity apparently exists between Alligator Point (FAP) and St. Joseph Bay (FSJ) along the Florida panhandle (Fig. 5.3h; both in Region A), but the relatively small sample size from St. Joseph Bay suggests the use of caution at this time.

This phylogeographic pattern implies that there are at least five demographically distinct lineages across the species' range in Florida. In addition to support for recognition of these zones of discontinuity, our data suggest low levels of gene exchange between collections on either side of the larger zones of genetic discontinuity (Fig. 5.3f, h). Metapopulations may exist among the Florida Southwest (FMI, FCH, FTB) and among the Florida West (FCK, FAP) collections. This suggests that management efforts in Florida might best be targeted at a finer scale. These findings also provide justification for release of horseshoe crabs sampled by scientific or commercial interests near the original collection site. The low levels of gene exchange also suggest that juvenile horseshoe crabs have limited dispersal (Botton and Loveland 2003), which may make them particularly vulnerable to local extirpation.

5.3.2 *On-line Surveys*

The public has reported seeing horseshoe crabs in every coastal county of Florida over the past 4 years (736 reports in 2010–2013; Table 5.1). Far more horseshoe crabs were reported from the Gulf coast of Florida (Region A and B) than from the Atlantic coast (Region C and D) (Table 5.1). Further, on both coasts, far more horseshoe crabs were reported from southern (Regions B and D) than from northern counties (Regions A and C) (Table 5.1).

Horseshoe crabs have been reported spawning in every month of the year in Florida. Most areas show a marked seasonality in breeding with the largest numbers spawning in March, April and May (Table 5.1). Along the Gulf coast there appears to be a second peak of breeding in the fall from August to October. Most reports from the Atlantic coast (Regions C and D) are from February to June or July with no obvious increase in the fall (although breeding animals are occasionally reported in the fall). In the southern part of the state (Regions B and D), horseshoe crabs have been observed breeding throughout the year, but in the northern part of the state (Regions A and C) no animals were observed in December and very few in January.

There are far more reports of horseshoe crab sightings to the on-line survey from the Gulf coast than from the Atlantic coast. To some extent this difference reflects the much longer coastline and the presence of more people at a time when horseshoe crabs are breeding. The difference may also reflect more habitats for horseshoe crabs along the Gulf as compared with the Atlantic coastlines. Horseshoe crabs nest most commonly on beaches (of varying substrate from fine sand to shell) along shorelines

Table 5.1 Seasonal variation in the number of horseshoe crabs reported to the Florida FWCC public, on-line database

	Region A			Region B			Region C			Region D		
2010–2013	(Northern Gulf coast)			(Southern Gulf coast)			(Northern Atlantic coast)			(Southern Atlantic coast)		
Month	Count	Reports	Median	Count	Reports	Median	Count	Reports	Median	Count	Reports	Median
January	0	0	0	37	4	8	2	1	2	7	2	4
February	148	8	20	231	8	4	24	1	24	54	3	2
March	4,025	71	20	1,073	45	4	122	15	2	62	6	4
April	3,801	68	16	19,170	106	9	656	31	4	148	23	3
May	1,976	29	4	3,099	98	6	135	18	4	2,287	28	4
June	198	17	3	212	21	4	32	7	1	47	8	5
July	2,585	14	2	109	15	5	2	2	1	106	5	2
August	39	4	4	1,514	14	11	0	0	0	3	2	2
September	116	3	2	532	15	12	2	1	2	0	0	0
October	815	13	8	550	13	8	35	1	35	3	2	2
November	4	1	2	11	4	3	5	3	2	6	2	3
December	0	0	0	120	3	30	0	0	0	2	1	2
Total	13,707	228	10	26,658	346	6	1,015	80	3	2,725	82	5

See Fig. 5.1 for regions. Region A includes counties along the northern west (Gulf) coast of Florida (Bay, Citrus, Dixie, Escambia, Franklin, Gulf, Hernando, Levy, Pasco, Taylor and Wakulla counties). Region B includes counties along the southern Gulf coast (Charlotte, Collier, Hillsborough, Lee, Manatee, Monroe, Pinellas, Sarasota). Region C includes counties along the northern east (Atlantic) coast of Florida (Brevard, Duval, Flagler, Nassau, St. Johns, Volusia) and Region D includes counties along the southern Atlantic coast of Florida (Broward, Indian River, Martin, Miami Dade, Palm Beach and St. Lucie)

Count the total number of horseshoe crabs reported in all reports for that month and region, *Reports* the total number of reports filed on the on-line database for that month and region, *Median* the number of horseshoe crabs observed on the median report filed for that month and region across the 4 years (and the total is the median for the entire region across all months) protected from high waves (Botton et al. 1988; Smith 2002), which are a habitat that is far more common along the Gulf than along the Atlantic coast of Florida.

5.3.3 Trawl and Seine Surveys

The FIM trawl surveys collected very few horseshoe crabs, often just a few animals per 100 trawling runs (Table 5.2). For this reason, it is impossible to determine trends across years at some sites (Apalachicola, Cedar Key and Indian River; Fig. 5.4). Where the numbers are sufficient to consider trends, horseshoe crab populations appear to be stable or declining on the west coast (Tampa Bay and Charlotte Harbor) and stable on the east coast (Jacksonville) (Fig. 5.4). More crabs were caught by seining than by trawling, although the numbers were still very low at some sites (Apalachicola, Jacksonville and Tequesta). The FIM seine surveys suggest that there was a rise in numbers after 2000 followed by a decline that is particularly apparent since 2006 especially along the Gulf coast (Fig. 5.5).

Trawl surveys were conducted in every month of the year and horseshoe crabs were caught throughout the year along the southern Gulf coast (Tampa Bay and Charlotte Harbor) and northeast Atlantic coast (Jacksonville), but were largely absent from trawl samples along the northern Gulf coast after June (Table 5.2). In all areas the largest numbers of animals were collected in the spring (January to May). With the much larger samples from the seine surveys, marked seasonality is apparent along the northern Gulf coast with the largest numbers in March and April and a second smaller peak in September and October (Table 5.3). Horseshoe crabs were collected by seine in the Indian River throughout the year with a spring peak.

Both the FIM trawl surveys and the seine surveys suggest that populations of horseshoe crabs are higher on the west coast of Florida than on the east coast and higher at the more southern sites than at the northern sites (Table 5.2 and 5.3, Figs. 5.4 and 5.5). Although there are some overall similarities between the numbers of horseshoe crabs collected by trawl and seine surveys (e.g. Charlotte Harbor has the highest populations in both surveys), the two methodologies show quite different results at some sites such as Indian River where the trawl data suggest very low numbers and the seine data suggest much larger numbers (in comparison with other sites). Trawl and seine surveys at Cedar Key and the spawning beach surveys at nearby SK (Brockmann and Johnson 2011) agree in showing strong seasonality to horseshoe crab breeding with large spring and smaller fall breeding activity peaks. Our results suggest that a mixture of protocols taken throughout the year may be the best way to estimate horseshoe crab populations.

Although the FIM trawl data are useful in the sense that they have been taken in a consistent manner at the same sites over a long time period, this sampling method captures very few horseshoe crabs (Table 5.2) and the variance is very high. Multispecies, finfish trawl surveys have been criticized as a way of estimating horseshoe crab populations (ASMFC 1999; Hata and Berkson 2004), and therefore conclusions based on these data should be considered tentative.

Table 5.2 Seasonal variation in the number of horseshoe crabs collected in FIM trawl surveys at different sites in Florida

	Region A		Region B		Region C	
Month	Apalachicola	Cedar Key	Tampa Bay	Charlotte Harbor	Indian River	Jacksonville
January	3	17	44	22	1	12
February	5	14	58	28	1	16
March	2	5	50	15	0	10
April	1	4	37	6	0	11
May	0	2	17	2	0	10
June	0	0	4	12	0	2
July	1	0	21	7	1	3
August	0	1	8	4	0	3
September	0	0	7	9	0	14
October	0	0	14	2	0	2
November	0	2	8	8	0	13
December	0	6	26	25	0	11
Totals	12	51	294	140	3	107
Trawls/month[a]	288–377	209–275	422–502	427–476	114–138	599–740
Years surveyed	1999–2013	1997–2013	1996–2013	1996–2013	1996–2013	2001–2013

See Fig. 5.1 for the regions and locations of sites

[a]The number of trawls conducted varied with site, month and year. Trawls/month is the range in the number of trawls conducted for each month across all years at this site

5.3.4 *Seahorse Key Spawning Survey*

There was a large increase in the numbers of horseshoe crabs spawning at SK in 2004 as compared with the earlier years of the survey, 1992–2000 (Brockmann and Johnson 2011) (one-way ANOVA, $F=8.263$, $P<0.001$, $df=12, 627$). This large increase continued through 2010, but in 2014 the numbers were again down; i.e. 2014 had the lowest number of any year since 2000 (but 2014 was still higher than 2000; Fig. 5.6a).

Our long-term study at SK reveals considerable year-to-year variation in the numbers of horseshoe crabs observed spawning (Fig. 5.6a). The variation is in part due to differences in high tide heights at the time that we conducted the surveys. All surveys were taken on spring tides from February to May, but in some years those surveys coincided with weather conditions such as a storm or cold front and north wind, which makes spawning less likely (Brockmann and Johnson 2011). However, the large and sustained increase in the numbers spawning at SK from 2004 to 2010 seems likely to reflect an actual increase in population size. Interestingly, the trawl and seine samples from nearby Cedar Key show a similar increase from 2002 to 2006 (Figs. 5.4 and 5.5). Possible explanations for the increase in numbers include increased development in nearby mainland areas and sea level rise causing erosion of nearby islands, which makes SK particularly attractive for nesting (SK is an

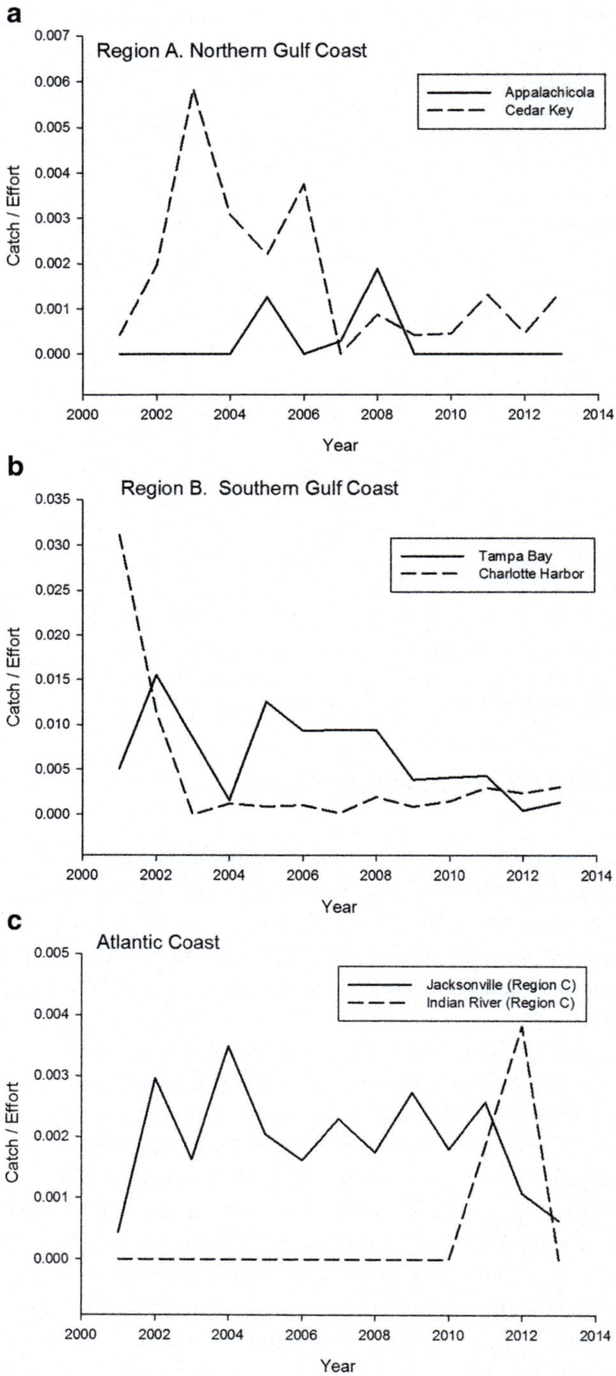

Fig. 5.4 The number of horseshoe crabs captured during trawl surveys at six different sites in Florida. Catch/effort refers to the number caught during all trawls in that year divided by the total trawling effort that year at that site

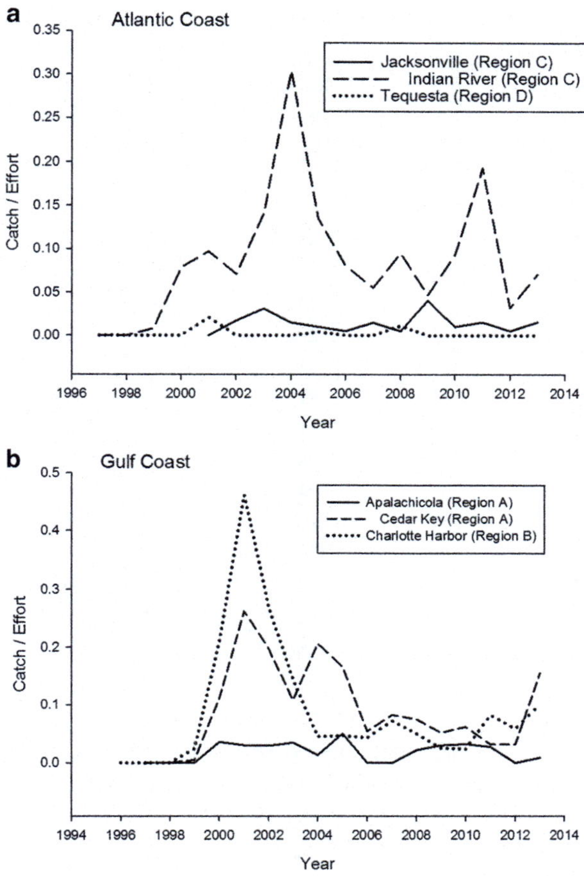

Fig. 5.5 The number of horseshoe crabs captured during seine surveys at six different sites in Florida. Catch/effort refers to the number caught during all seine surveys in that year divided by the total number of seining runs (= effort) that year at that site

ancient sand dune and while it has eroded considerably in the last 30 years, the beaches remain among the best in the area). This explanation does not account for the increase in the numbers from the Cedar Key trawl sampling. Another possibility is that a ban on net fishing in 1995 might have reduced the numbers of horseshoe crabs killed or injured in the entangling nets (Brockmann and Johnson 2011). At the same time, a clam farming industry was initiated in the Cedar Key area (Colson and Sturmer 2000), which may have resulted in an increased food supply for juvenile horseshoe crabs after 2000.

Operational sex ratios (proportion male) remained male biased (56–65 % male) throughout the years (Fig. 5.6b) with 1994, 2000 and 2004 being significantly higher (more male biased) than the most recent years of the study (2009, 2010, 2014; one-way ANOVA, $F = 5.133$, $P < 0.001$, $df = 11, 559$). Overall SK is considerably less male biased than was found at Mashes Sands, Apalachee Bay (78 % male;

Table 5.3 Seasonal variation in the number of horseshoe crabs collected in FIM seine surveys at different sites in Florida

	Region A		Region B	Region C		Region D
Month	Apalachicola	Cedar Key	Charlotte Harbor	Jacksonville	Indian River	Tequesta
January	0	2	5	4	28	2
February	2	9	12	1	55	0
March	30	46	35	5	82	4
April	24	104	43	3	72	6
May	9	38	78	0	36	0
June	1	10	29	9	20	0
July	2	12	11	0	11	0
August	1	4	51	0	14	0
September	13	41	31	3	17	0
October	2	34	28	2	27	0
November	3	15	24	5	42	2
December	0	2	25	6	23	0
Total	87	317	372	38	427	14
Seines/month[a]	304–432	257–281	325–362	198–228	339–439	508–568
No. years	2000–2013	1997–2013	1996–2013	2001–2013	1997–2013	1997–2013

[a]Seines/month is the range in the number of seining surveys conducted for each month across all years at this site

Rudloe 1980). Exceptionally male-biased sex ratios occur in Delaware Bay and other areas where there is preferential harvesting of females, but little or no harvest has occurred in Florida.

5.3.5 *Seahorse Key Size and Condition of Female and Male Horseshoe Crabs*

Females measured during the early years of the study (1992–1997) were slightly larger than those measured in the later years (2008–2014; one-way ANOVA, $F=5.688$, $P<0.001$, $df=9$, 3934). This pattern was not found in males: only the males measured in 1996 were slightly smaller than those measured in 1992 and 1995 (one-way ANOVA, $F=3.743$, $P<0.001$, $df=9$, 6648).

There is considerable year-to-year variation in the proportion of the population that is newly molted (females range from 3 to 18 %; mean 7.9±1.4 %; Fig. 5.6c). The proportion of males that are newly molted is more variable (1–26 %; mean 11.8±2.8 %), which is probably related to the numbers of satellite versus attached males that we collected each year (satellites are older; Brockmann and Penn 1992). While there were significant differences among years in the numbers of animals recruited to the population (i.e. the numbers in perfect condition; Chi Square=97.5, $df=9$, $P<0.001$), there were no differences between the early and later years of the study (females: early years 8.3 % are newly molted, later years 6.6 %,

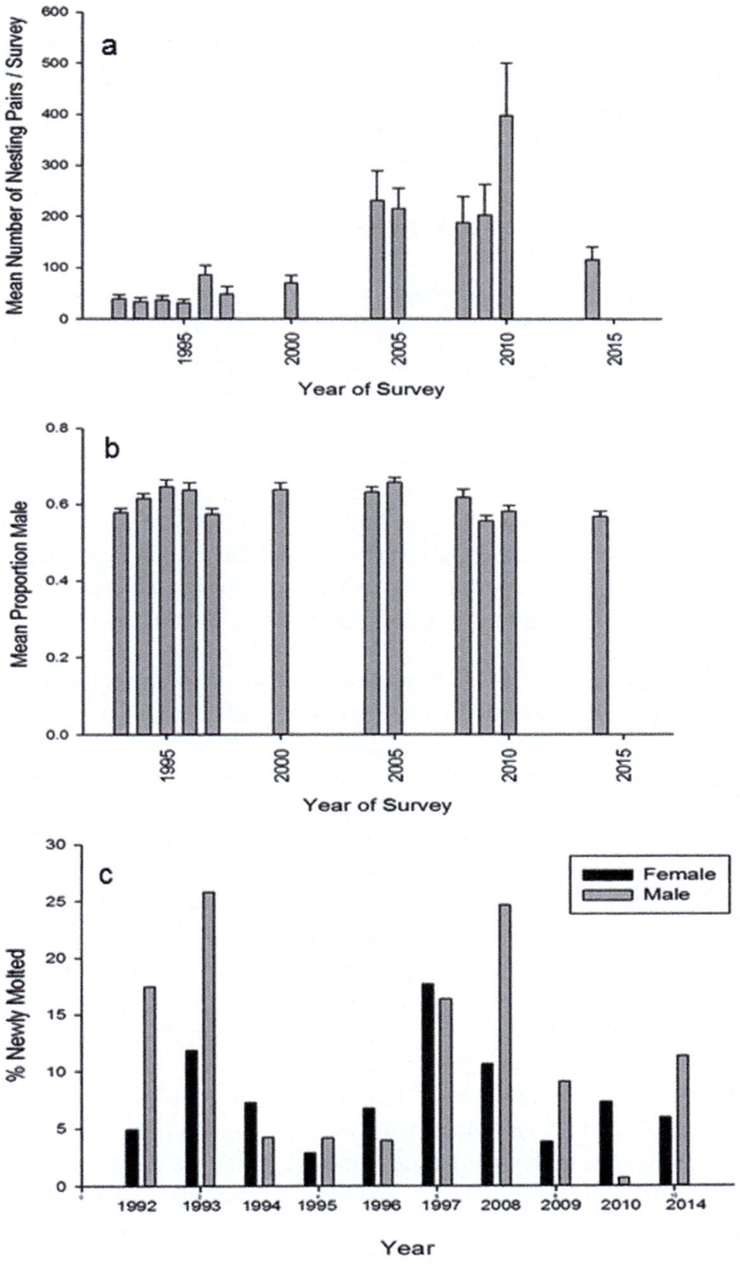

Fig. 5.6 Results of Seahorse Key spawning surveys, 1992–2014. (**a**) The mean number of horseshoe crabs sighted across all surveys in that year, (**b**) The mean proportion of horseshoe crabs that are male across all surveys in that year, (**c**) The mean proportion of animals that are newly molted across all surveys in that year

Chi Square=2.668, $P=0.102$, $df=1$; males: early 11.4 %, later 12.6 %, Chi Square=1.544, $P=0.214$). Nothing is known about the factors that affect size variation or recruitment in horseshoe crab populations, although certainly the availability of food as juveniles is a likely factor.

5.3.6 Bait and Marine-Life Harvest

With the exception of 1999 and 2000, only small numbers of horseshoe crabs have been collected for bait in Florida (Table 5.4; Fig. 5.7a). The exceptional harvest in 1999 took place on the Gulf coast when 110,000 were removed from St. Joseph Bay (Region A on Fig. 5.1; Wallace 1999). This exceptional harvest was associated with new regulations in other states when Florida did not yet have regulations in place (Gerhart 2007). Since regulations have been in place, nearly all of the bait harvest (used mostly as eel bait) has been along the southern Atlantic coast (Region D on Fig. 5.1), and it remains well below the limit set by the FMP of 9,455. This assumes of course, that those harvesting horseshoe crabs for bait are, in fact, reporting their landings. Enforcement of reporting requirements may be lax in Florida due to a very long coastline, the large numbers of species requiring licenses, and the few FWCC enforcement officers.

Since 2000 Florida landings in the marine-life fishery have varied from 18,593 in 2012 to 55,460 in 2002 (Fig. 5.7b). Landings have declined since 2007 but the number of trips has not, which means that the CPUE (catch per unit effort, or horseshoe crabs per trip) has declined over the last 4 years (Table 5.4). Along the Atlantic coast, nearly all of the marine-life harvest has been in the southern region (Region D on Fig. 5.1) where up to 12,385 specimens have been collected annually since 2000. The number of trips, landings and CPUE (catch per unit effort) declined sharply after 2007 (from 78 to 199 crabs/trip in 2003–2007 to 26–71 crabs/trip since 2007; Table 5.4). Along the west coast, the pattern is somewhat different. In Region A the numbers taken in the marine-life fishery have increased steadily, particularly since 2007 when the number of trips expanded sharply; over the past 4 years, however, there has been a decline in the CPUE (Table 5.4). In Region B the number of trips increased markedly after 2000 and rose steadily over the next 5 years; in the past 4 years, the number of trips has increased, and the CPUE has declined, so that the total marine-life harvest is down from the peak period (2001–2005).

There are patterns in the marine-life harvest data that are cause for concern. The marine life harvest is large (Table 5.4) and rivals the numbers taken for bait in some northern states (ASMFC 2009). On the Atlantic coast, the number of animals collected, the number of trips and the CPUE are all down since 2005; on the Gulf coast, the number of animals being collected has increased as has the number of trips, but the CPUE has decreased since 2009. A decrease in the number of trips can be explained by a downturn in the economy (or decreased interest in *L. polyphemus* in the aquaria trade), a decline in the availability of habitat for collecting, or reduced reporting or enforcement. However, assuming that the data are accurately

Table 5.4 Bait and marine life landings of horseshoe crabs in Florida from 1990 to 2012

	Atlantic coast landings[a]					Gulf coast landings[a]									Total Florida landings		
	Region D[b]					Region A				Region B							
Year	Bait	Marine life no.	Marine life trips	Marine life CPUE	Total no[b]	Bait	Marine life no.	Marine life trips	Marine life CPUE	Marine life no.	Marine life trips	Marine life CPUE	Total no.		Bait	Marine life	Total
1990		7,317	58	126.2	7,317		302	5	60.4	3,784	73	51.8	4,086		0	11,403	11,403
1991		5,543	59	93.9	5,543		32	3	10.7	5,232	116	45.1	5,264		0	10,807	10,807
1992		2,951	40	73.8	2,951					6,583	109	60.4	6,583		0	9,534	9,534
1993		41,970	60	699.5	41,970		5	3	1.7	11,276	149	75.7	11,281		0	53,251	53,251
1994		7,585	76	99.8	7,585		112	4	28.0	11,712	171	68.5	11,824		0	19,409	19,409
1995		7,697	67	114.9	7,697		2,719	18	151.1	6,708	101	66.4	9,427		0	17,124	17,124
1996		9,438	62	152.2	9,438		2,522	21	120.1	5,357	87	61.6	7,879		0	17,317	17,317
1997		9,455	65	145.5	9,455		1,481	14	105.8	53,302	69	772.5	54,783		0	64,238	64,238
1998	200	6,691	44	152.1	6,891	0	4,701	66	71.2	14,326	77	186.1	19,027		200	25,718	25,918
1999	19,446	8,063	56	144.0	27,509	110,879	3,203	46	69.6	7,973	86	92.7	122,055		130,325	19,239	149,564
2000	10,462	8,720	64	136.3	19,182	620	2,824	52	54.3	8,296	118	70.3	11,740		11,082	19,840	30,922
2001	0	8,588	71	121.0	8,588	0	1,598	38	42.1	18,239	110	165.8	19,837		0	28,425	28,425
2002	200	12,385	73	169.7	12,585	0	5,713	50	114.3	37,362	125	298.9	43,075		200	55,460	55,660
2003	1,628	9,223	69	133.7	10,851	0	5,011	50	100.2	11,829	183	64.6	16,840		1,628	26,063	27,691
2004	0	11,568	58	199.4	11,568	75	4,112	53	77.6	17,152	166	103.3	21,339		75	32,832	32,907
2005	0	4,135	42	98.5	4,135	0	6,804	66	103.1	10,909	157	69.5	17,713		0	21,848	21,848
2006	469	2,174	28	77.6	2,643	0	9,319	60	155.3	7,947	97	81.9	17,266		469	19,440	19,909
2007	186	2,518	27	93.3	2,704	0	19,450	124	156.9	8,319	97	85.8	27,769		186	30,287	30,473
2008	50	1,011	15	67.4	1,061	0	21,418	157	136.4	6,703	79	84.8	28,121		50	29,132	29,182
2009	0	658	25	26.3	658	0	18,896	197	95.9	7,938	106	74.9	26,834		0	27,492	27,492

(continued)

Table 5.4 (continued)

Year	Atlantic coast landings[a]					Gulf coast landings[a]									Total Florida landings		
	Region D[b]					Region A				Region B							
	Bait	Marine life no.	Marine life trips	Marine life CPUE	Total no[b]	Bait	Marine life no.	Marine life trips	Marine life CPUE	Marine life no.	Marine life trips	Marine life CPUE	Total no.		Bait	Marine life	Total
2010	993	892	22	40.5	1,885	0	16,478	192	85.8	6,214	100	62.1	22,692		993	23,584	24,577
2011	0	1,222	21	58.2	1,222	0	11,633	156	74.6	6,906	104	66.4	18,539		0	19,761	19,761
2012	0	1,155	26	44.4	1,155	0	11,403	140	81.5	6,035	100	60.4	17,438		0	18,593	18,593
2013	0	4,560	64	71.3	4,560	0	13,501	179	75.4	14,215	116	122.5	27,716		0	27,716	27,716

[a]Landings refers to the number of individuals reported on trip tickets required by the FWC

[b]All Atlantic coast landings were in Region D with the exception of 2006 when there were bait landings in Region C: 194 landings in 2 trips for a CPUE of 97

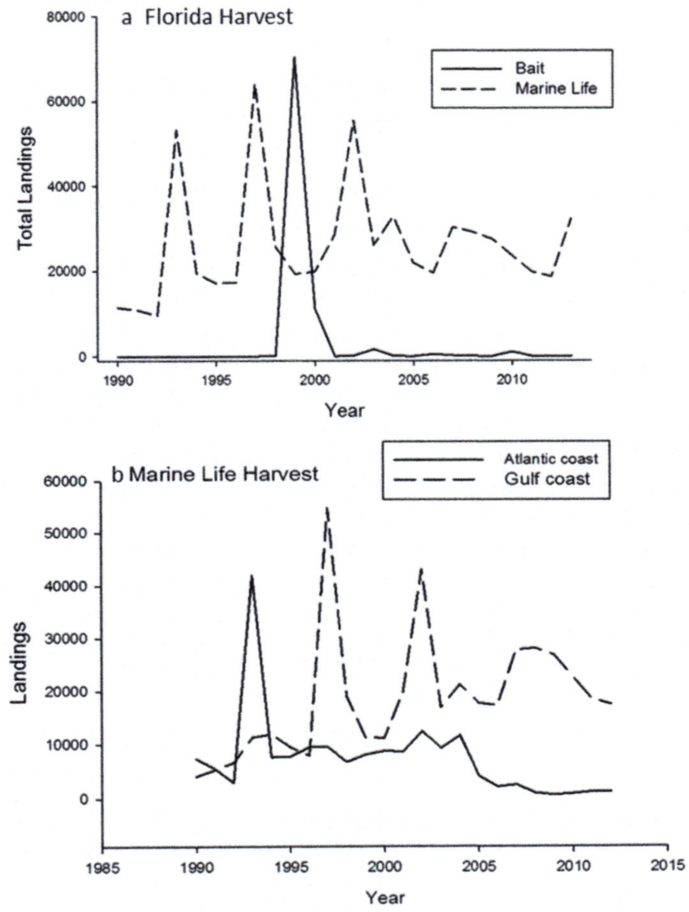

Fig. 5.7 The number of horseshoe crabs harvested in Florida. (**a**) A comparison between the bait and marine life harvests. The 1999 bait harvest was truncated to preserve the scaling–it was actually 130,325, (**b**) A comparison between the marine life harvests on the Atlantic and Gulf coasts

reported, a decrease in the CPUE means that there is a decrease in the number of horseshoe crabs available to marine-life harvesters. This decline in CPUE is most noticeable on the Atlantic coast but is also apparent since 2009 along the Gulf coast (both Regions A and B, Table 5.4).

Juvenile horseshoe crabs are popular aquarium pets and are readily obtained from both retail and on-line saltwater aquaria dealers. Small individuals (2.5–5 cm) sell for US$9–18 and large individuals (5–10 cm) for US$20–23. The internet sites that sell horseshoe crabs identify the source of the animals only as "Atlantic Ocean" which means that animals originally collected in Florida might be released into genetically very different populations.

5.4 General Discussion

5.4.1 *Are Florida Populations of Horseshoe Crabs Stable, Increasing or Decreasing?*

Taken together, the data suggest that some populations of Florida horseshoe crabs are declining: some trawl surveys from the last 5 years have counted substantially fewer horseshoe crabs than in previous years; Seahorse Key spawning surveys suggest a recent decline; and the CPUE in the marine-life harvest data shows a decline in recent years. The one exception is at the Jacksonville trawl site where the numbers have been stable. While most of the data point in one direction, there is considerable uncertainty associated with these data. The trawl sampling methods were not designed for capturing horseshoe crabs, the numbers are very low and the variance high; the Seahorse Key spawning survey is missing a number of crucial years; and the declining CPUE in the harvest data assumes consistent collection and accurate reporting over many years. More reliable data would include long-term and consistent spawning surveys and/or mark-recapture studies (Smith et al. 2002b, 2009), trawl sampling that is designed for horseshoe crabs (Hata and Berkson 2004), and consistent and reliable reporting of harvest data from all localities. Adding to the concern is that Florida horseshoe crabs appear to be fragmented into genetically distinct populations with little interchange between them, which means that local declines could result in local extinctions.

5.4.2 *What Might Explain a Decline?*

Any widespread decline in horseshoe crabs is undoubtedly the result of many factors. Causes that have been identified in other areas include (a) commercial harvest (Davis et al. 2006; Grady and Valiela 2006), (b) habitat loss (Botton et al. 1994; Jackson and Nordstrom 2009) and (c) pollution and water quality issues (Strobel and Brenowitz 1981; Wallace 1999; Botton and Itow 2009). Water quality issues may be particularly important in Florida where red tides are common in nearshore communities particularly in southwest Florida where young horseshoe crabs are one of the affected species (Galtsoff 1949). (d) Horseshoe crabs were the most abundant invertebrate bycatch species in shrimp trawls in Florida (Steele et al. 2002), and even though there are new regulations that require bycatch reduction devices on all shrimp nets, significant numbers of horseshoe crabs are still being caught in shrimp nets (Steele et al. 2002). Even when bycatch is returned to the sea, many individuals are injured (D. Smith, unpublished data).

Sandy beaches that are protected from surf are the most common nesting habitat for horseshoe crabs, and a lack of beaches with adjacent nursery grounds for juveniles (intertidal mud and sand flats) may limit horseshoe crab populations (Rudloe 1982; Botton et al. 1988; Thompson 1999; ASMFC 2009). Beach erosion and human development are concerns along the entire Atlantic coastline (Botton 2001; Jackson and Nordstrom 2009) but particularly in Florida where beach development covers most coastlines (McLachlan et al. 2013). Where sandy beaches remain, beach driving can result in the death of nesting animals (personal observation). Erosion control through hardening the shoreline with bulkheads, groins, rip rap and other structures can reduce the beach and offshore nursery grounds and trap adults as they come ashore (Jackson and Nordstrom 2009). Beach replenishment, on the other hand, can enhance habitat for both adults and juveniles if managed properly (Jackson et al. 2007). Other shoreline modifications, such as power plant intakes, can be particularly deadly to horseshoe crabs (Ehlinger and Tankersley 2007) that naturally follow water currents (Rudloe and Herrnkind 1976, 1980).

Low population densities and declining numbers of horseshoe crabs in the Indian River have been a concern for many years (Ehlinger and Tankersley 2007). In the late 1990s efforts to net sea turtles in Mosquito Lagoon resulted in many fewer horseshoe crabs being caught per unit effort than in 1979, and there has been a shift in the predominant species of sea turtles from loggerhead turtles, which feed heavily on horseshoe crabs, to green sea turtles, which feed largely on plants (Provancha et al. 2006). An enormous increase in the human population density adjacent to the Indian River Lagoon, with its associated loss of habitat and reduced water quality, are likely factors contributing to the decline (Ehlinger and Tankersley 2007). Disease is another possibility: in 1999 an estimated 100,000 horseshoe crabs died in the southern portion of Mosquito Lagoon from unknown causes when no other species were affected (Scheidt and Lowers 2001). Another problem was revealed by a 1-year study at two power plants on the Indian River which recorded a total of 39,097 horseshoe crabs trapped on the intake screens at the Florida Power and Light Cape Canaveral Plant and 53,121 at the Orlando Utilities Commission Indian River Plant (Ehlinger and Bush 2001; Ehlinger and Tankersley 2007). A previous study conducted in 1975 estimated 69,662 at the Canaveral Plant, and 104,000 horseshoe crabs were trapped annually at the intakes of the Indian River plant. This level of mortality alone could easily account for a decline in the Indian River population (Ehlinger and Tankersley 2007).

Commercial harvest is known to deplete horseshoe crab populations along the mid-Atlantic and New England coasts (Widener and Barlow 1999; Carmichael et al. 2003; Rutecki et al. 2004; Schaller et al. 2005; Smith et al. 2009; McGowan et al. 2011). In general, Florida populations are much smaller than those in other areas and therefore more vulnerable to overharvest. The large number of animals taken from St. Joseph Bay in 1999 resulted in Rudloe and others pointing out that there were not enough animals to sustain such heavy harvesting (Wallace 1999; Gerhart 2007), which prompted the Florida Marine Fisheries Commission (now FWC Division of Marine Fisheries Management) to close the fishery temporarily. Today bait harvest of adults is not a threat. However, the marine-life fishery is another matter.

The ASMFC Management Board has expressed concern over the relatively large number of horseshoe crabs landed by the marine-life fishery in Florida (Gerhart 2007). Marine-life harvesters are taking animals from areas with relatively low population densities and harvesting large numbers of juveniles. About half of the reported marine-life landings are from the Florida Keys which have relatively low numbers of horseshoe crabs and where suitable spawning habitat is rare (Gerhart 2007). Many are also taken from areas where populations are likely declining (Table 5.4). Horseshoe crabs have an unusual life history: they require 9–10 years to mature into an adult when breeding begins (Shuster and Sekiguchi 2003). Modeling studies have shown that survival during the early life stages is more important to population growth than adult survival and therefore management actions that reduce mortality in the early juvenile stages will have a greater effect on horseshoe crab populations than adjusting adult harvest levels (Sweka et al. 2007). This suggests that the marine-life harvest of juveniles may be particularly harmful to horseshoe crab populations (Gerhart 2007) and should be limited. Little is known about the requirements, life history or behavior of juvenile horseshoe crabs in Florida (Rudloe 1981; Ehlinger and Tankersley 2003, 2004), so additional studies that assess and monitor spawning, early life stages, nursery habitats and life-history patterns would be particularly valuable in sustaining horseshoe crab populations in Florida.

5.5 Conclusions and Recommendations

Despite the fact that horseshoe crabs are found in every coastal county of Florida and every month of the year, little is known about their status anywhere in Florida. It seems likely that the numbers of horseshoe crabs are declining in some areas, although the available data are poor. We make two types of suggestions: (a) improve data collection on horseshoe crabs in Florida and (b) reduce the marine-life harvest and other sources of mortality on juveniles.

1. There is no overall limit on the number of horseshoe crabs that can be harvested in the marine-life fishery. Given the importance of juvenile survival for population recruitment and evidence of declining numbers in some areas, limits on the harvest of juveniles should be established at least for those areas where a decline is suspected.
2. Since there is some evidence that horseshoe crabs are declining in Florida, reducing the number of animals that can be collected in the marine-life harvest would be appropriate. Horseshoe crabs are easy to rear, and it is particularly easy to rear them to the early juvenile stages (French 1979; Brown and Clapper 1981; Carmichael et al. 2009; Schreibman and Zarnoch 2009; Carmichael and Brush 2012). This means that the animals used by the aquaria trade could come from captive rearing.
3. Harvest and other regulations on horseshoe crabs should be applied equally on the Gulf and Atlantic coasts of Florida. The distance between the two coasts is short, and any limitations on one coast would be met with increased harvest on the other coast. With the high price for horseshoe crabs in some areas (Kreamer and Michels 2009), it seems likely that Florida's horseshoe crab populations will be exploited in the future.

4. Since horseshoe crabs on the two coasts are from genetically distinct populations, animals collected on the east coast should not be released on the west coast or vice versa.
5. Genetic analysis suggests that there may be population structure along the Gulf and Atlantic coasts, with northern and southern populations on each coast being genetically different. If these results are corroborated with additional sampling, this suggests that management efforts in Florida should be targeted at a finer scale.
6. Somewhere north of Indian River on the Atlantic coast, there is a genetic discontinuity and similarly, there appears to be a genetic break somewhere between Tampa Bay and Cedar Key. In both cases we do not know whether there is an abrupt break, or whether the two populations hybridize or remain apart in an area of overlap. These are crucial questions to defining the population structure of Florida *L. polyphemus* and should be investigated.
7. The FMP requires states to identify potential horseshoe crab spawning and nursery habitat and categorize and prioritize critical habitats. States are also asked to consider land acquisition of key spawning and nursery sites and to pursue restriction of all-terrain vehicle and water craft activity at spawning sites (ASMFC 1999). The FWCC identified horseshoe crab spawning sites along Florida shorelines in 2001 (Anne L. McMillen-Jackson, personal communication), but this study should be updated, and the next steps have not been taken; i.e. categorizing and prioritizing critical habitats and taking steps to conserve horseshoe crabs in Florida (Wallace 1999; Gerhart 2007).
8. Long-term population data are needed from multiple sites in Florida similar to the spawning surveys that are used to evaluate populations in Delaware Bay (Smith et al. 2002a). Spawning surveys have been conducted at Mashes Sands in Wakulla County in 1976–1978 (Rudloe 1980) and at Seahorse Key near Cedar Key in Levy County, Florida from 1990 to 2014 (Brockmann and Johnson 2011; this study) and on the Atlantic coast at 21 sites in Indian River Lagoon in 1998–2000 (Ehlinger et al. 2003). Shorter term surveys have been carried out in a number of other locations in Florida. These are prime areas for long-term and/or follow-up studies. Mark and recapture studies would be particularly valuable for estimating population size (Smith et al. 2006; Beekey and Mattei 2009).
9. FIM trawl surveys for finfish do not provide reliable horseshoe crab abundance information (ASMFC 2009). A protocol for benthic trawl surveys designed for horseshoe crabs has been developed (Hata and Berkson 2003, 2004) and should be instituted at various sites in Florida; we suggest Cedar Key, Charlotte Harbor and/or Tampa Bay and Indian River. Seining surveys capture more horseshoe crabs but have very high variance in numbers collected in some areas. Seine surveys should be used in preference to finfish trawl surveys for estimating relative abundance.
10. Tracking studies indicate that some populations of horseshoe crabs remain in a localized area (Rudloe 1980; James-Pirri et al. 2005; James-Pirri 2010; Schaller et al. 2010; Watson and Chabot 2010) whereas others move widely throughout their lives (Brousseau et al. 2004; Swan 2005; Smith et al. 2010). Tracking of

individuals would provide valuable information on the vagility of Florida horseshoe crab populations.

11. The larval and early juvenile stages of horseshoe crabs are crucial to population recruitment yet little is known about the conditions that favor their development (Rudloe 1979a, b, 1981; Ehlinger et al. 2003; Ehlinger and Tankersley 2003, 2004, 2006). What proportion of young survives their first year and are there differences among populations? What are the sources of mortality? What are the correlates of success and do these differ between populations? How often do they molt and what is the life-history of Florida horseshoe crabs? Does the life-history differ among populations within Florida?
12. Identifying particularly important nursery areas for juvenile development and targeting those areas for preservation and no harvesting could be a valuable way to enhance population growth and to allow the Florida horseshoe crab marine-life fishery to be sustainable. There is good evidence that such set asides are effective: a marine sanctuary at the mouth of the Delaware Bay has been credited with increasing horseshoe crab populations in the bay (Jackson and Nordstrom 2009).

Acknowledgments We want to thank the staff from the State of Florida Fish and Wildlife Conservation Commission for their kind help in providing data to us: Ryan Gandy (FWC St. Petersburg) compiled and helped with the analysis of the FWC data from public sightings of horseshoe crabs from around Florida; Mandy Tyler-Jedlund and Bob McMichael compiled FIM data on horseshoe crabs collected by the Fish and Wildlife Research Institute (FWC, St. Petersburg, Florida); and Mike Tringali provided valuable assistance in collecting and preparing tissue samples for genetic analysis. We also thank the graduate students and field assistants who helped collect the Seahorse Key data and the UF Marine Laboratory at Seahorse Key and the Lower Suwannee National Wildlife Refuge for permission to conduct research at Seahorse Key. We also thank the reviewers and editors for their helpful comments on the manuscript. Use of trade, product, or firm names does not imply endorsement by the U.S. Government.

References

Anderson LI, Shuster CN (2003) Throughout geologic time: where have they lived? In: Shuster CN Jr, Barlow RB Jr, Brockmann HJ (eds) The American horseshoe crab. Harvard University Press, Cambridge, MA, pp 189–223

ASMFC (1999) Horseshoe crab stock assessment report for peer review. Atlantic States Marine Fisheries Commission, Washington, DC

ASMFC (2009) Horseshoe crab stock assessment report. Atlantic States Marine Fisheries Commission, Washington, DC

Badgerow JP, Sydlik MA (1989) Nest site selection in horseshoe crabs (*Limulus polyphemus*) using a Cape Cod Beach. Am Zool 29(4):36A

Barlow RB Jr, Powers MK, Howard H et al (1986) Migration of *Limulus* for mating: relation to lunar phase, tide height, and sunlight. Biol Bull 171:310–329

Beekey MA, Mattei JH (2009) What long term mark/recapture studies reveal about horseshoe crab population dynamics in Long Island Sound. In: Biennial Long Island Sound research conference 2008, Connecticut College, Groton

Botton ML (2001) The conservation of horseshoe crabs: what can we learn from the Japanese experience? In: Tanacredi JT (ed) *Limulus* in the limelight. Kluwer Academic, New York, pp 41–52

Botton ML, Itow T (2009) The effects of water quality on horseshoe crab embryos and larvae. In: Tanacredi JT, Botton ML, Smith DR (eds) Biology and conservation of horseshoe crabs. Springer, New York, pp 439–454

Botton ML, Loveland RE (2003) Abundance and dispersal potential of horseshoe crab (*Limulus polyphemus*) larvae in the Delaware estuary. Estuaries 26(6):1472–1479

Botton ML, Loveland RE, Jacobsen TR (1988) Beach erosion and geochemical factors: influence on spawning success of horseshoe crabs (*Limulus polyphemus*) in Delaware Bay. Mar Biol 99:325–332

Botton ML, Loveland RE, Jacobsen TR (1994) Site selection by migratory shorebirds in Delaware Bay, and its relationship to beach characteristics and abundance of horseshoe crab (*Limulus polyphemus*) eggs. Auk 111(3):605–616

Brockmann HJ (2002) An experimental approach to altering mating tactics in male horseshoe crabs (*Limulus polyphemus*). Behav Ecol 13:232–238

Brockmann HJ (2003) Nesting behavior: a shoreline phenomenon. In: Shuster CN Jr, Barlow RB Jr, Brockmann HJ (eds) The American horseshoe crab. Harvard University Press, Cambridge, MA, pp 33–49

Brockmann HJ, Johnson SL (2011) A long-term study of spawning activity in a Florida Gulf coast population of horseshoe crabs (*Limulus polyphemus*). Estuar Coasts 34:1049–1067

Brockmann HJ, Penn D (1992) Male mating tactics in the horseshoe crab, *Limulus polyphemus*. Anim Behav 44:653–665

Brockmann HJ, Smith MD (2009) Reproductive competition and sexual selection in horseshoe crabs. In: Tanacredi JT, Botton M, Smith DR (eds) Biology and conservation of horseshoe crabs. Springer, New York, pp 199–221

Brockmann HJ, Colson T, Potts W (1994) Sperm competition in horseshoe crabs (*Limulus polyphemus*). Behav Ecol Sociobiol 35:153–160

Brockmann HJ, Nguyen C, Potts W (2000) Paternity in horseshoe crabs when spawning in multiple-male groups. Anim Behav 60:837–849

Brockmann HJ, Johnson SL, Sasson DA et al (Unpublished manuscript) Seasonal variation in horseshoe crab reproduction (*Limulus polyphemus*)

Brousseau LJ, Sclafani M, Smith DR et al (2004) Acoustic-tracking and radio-tracking of horseshoe crabs to assess spawning behavior and subtidal habitat use in Delaware Bay. N Am J Fish Manag 24(4):1376–1384

Brown GG, Clapper DL (1981) Procedures for maintaining adults, collecting gametes, and culturing embryos and juveniles of the horseshoe crab, *Limulus polyphemus* L. In: Hinegardner R, Atz J, Fay R, Fingerman M, Josephson R, Meinkoth N (eds) Laboratory animal management, marine invertebrates. National Academy Press, Washington, DC, pp 268–290

Carmichael RH, Brush E (2012) Three decades of horseshoe crab rearing: a review of conditions for captive growth and survival. Rev Aquac 4:32–43

Carmichael RH, Rutecki D, Valiela I (2003) Abundance and population structure of the Atlantic horseshoe crab *Limulus polyphemus* in Pleasant Bay, Cape Cod. Mar Ecol Prog Ser 246:225–239

Carmichael R, Gaines E, Sheller Z et al (2009) Diet composition of juvenile horseshoe crabs: implications for growth and survival of natural and cultured stocks. In: Tanacredi JT, Botton ML, Smith DR (eds) Biology and conservation of horseshoe crabs. Springer, New York, pp 521–534

Cavalli-Sforza LL, Edwards AWF (1967) Phylogenetic analysis: models and estimation procedures. Evolution 21:550–570

Colson S, Sturmer LN (2000) One shining moment known as clamelot: the Cedar Key story. J Shellfish Res 19(1):477–480

Davis ML, Berkson J, Kelly M (2006) A production modeling approach to the assessment of the horseshoe crab (*Limulus polyphemus*) population in Delaware Bay. Fish Bull 104(2):215–225

Duffy EE, Penn DJ, Botton ML et al (2006) Eye and clasper damage influence male mating tactics in the horseshoe crab, *Limulus polyphemus*. J Ethol 24(1):67–74

Earl DA, von Holdt BM (2011) STRUCTURE HARVESTER: a website and program for visualizing STRUCTURE output and implementing the Evanno method. Conserv Genet Resour 4:359–361

Ehlinger GS, Bush M (2001) Identification of cause and extent of horseshoe crab population declines at Canaveral National Seashore. National Park Service, Canaveral National Seashore, Titusville

Ehlinger GS, Tankersley RA (2003) Larval hatching in the horseshoe crab, *Limulus polyphemus*: facilitation by environmental cues. J Exp Mar Biol Ecol 292(2):199–212

Ehlinger GS, Tankersley RA (2004) Survival and development of horseshoe crab (*Limulus polyphemus*) embryos and larvae in hypersaline conditions. Biol Bull 206(2):87–94

Ehlinger GS, Tankersley RA (2006) Endogenous rhythms and entrainment cues of larval activity in the horseshoe crab *Limulus polyphemus*. J Exp Mar Biol Ecol 337(2):205–214

Ehlinger GS, Tankersley RA (2007) Reproductive ecology of the American horseshoe crab *Limulus polyphemus* in the Indian River Lagoon: an overview. Fla Sci 70(4):449–463

Ehlinger GS, Tankersley RA, Bush MB (2003) Spatial and temporal patterns of spawning and larval hatching by the horseshoe crab, *Limulus polyphemus*, in a microtidal coastal lagoon. Estuaries 26(3):631–640

Evanno G, Regnaut S, Goudet J (2005) Detecting the number of clusters of individuals using the software STRUCTURE: a simulation study. Mol Ecol 14:2611–2620

French KA (1979) Laboratory culture of embryonic and juvenile *Limulus*. In: Cohen E (ed) Biomedical applications of the horseshoe crab (Limulidae). Alan R. Liss, Inc., New York, pp 61–71

Fulford RS, Haehn RA (2012) An evaluation of Mississippi barrier islands as spawning and nesting habitat for the American horseshoe crab, *Limulus polyphemus*, with implications for island restoration. Gulf Caribb Res 24:51–62

Galtsoff PS (1949) The mystery of the red tide. Sci Mon 68(2):108–117

Gerhart SD (2007) A review of the biology and management of horseshoe crabs with emphasis on Florida populations. Fish and Wildlife Research Institute, St. Petersburg. http://research.myfwc.com. Accessed 1 June 2014

Grady SP, Valiela I (2006) Stage-structured matrix modeling and suggestions for management of Atlantic horseshoe crab, *Limulus polyphemus*, populations on Cape Cod, Massachusetts. Estuar Coasts 29(4):685–698

Grant D (2001) Living on *Limulus*. In: Tanacredi JT (ed) *Limulus* in the limelight. Kluwer Academic, New York, pp 135–146

Hata D, Berkson J (2003) Abundance of horseshoe crabs (*Limulus polyphemus*) in the Delaware Bay area. Fish Bull 101(4):933–938

Hata D, Berkson J (2004) Factors affecting horseshoe crab *Limulus polyphemus* trawl survey design. Trans Am Fish Soc 133(2):292–299

Jackson NL, Nordstrom KF (2009) Strategies to conserve and enhance sandy barrier habitat for horseshoe crabs (*Limulus polyphemus*) on developed shorelines in Delaware Bay, United States. In: Tanacredi JT, Botton ML, Smith DR (eds) Biology and conservation of horseshoe crabs. Springer, New York, pp 399–416

Jackson NL, Smith DR, Tiyarattanachai R et al (2007) Evaluation of a small beach nourishment project to enhance habitat suitability for horseshoe crabs. Geomorphology 89(1–2):172–185

James-Pirri MJ (2010) Seasonal movement of the American horseshoe crab *Limulus polyphemus* in a semi-enclosed bay on Cape Cod, Massachusetts (USA) as determined by acoustic telemetry. Curr Zool 56(5):575–586

James-Pirri MJ, Tuxbury K, Marino S et al (2005) Spawning densities, egg densities, size structure, and movement patterns of spawning horseshoe crabs, *Limulus polyphemus*, within four coastal embayments on Cape Cod, Massachusetts. Estuaries 28(2):296–313

King TL, Eackles MS (2004) Microsatellite DNA markers for the study of horseshoe crab (*Limulus polyphemus*) population structure. Mol Ecol Notes 4(3):394–396

King TL, Eackles MS, Spidle AP et al (2005) Regional differentiation and sex-biased dispersal among populations of the horseshoe crab *Limulus polyphemus*. Trans Am Fish Soc 134(2):441–465

King TL, Switzer JF, Morrison CL, Eackles MS, Young CC, Lubinski BA, Cryan P (2006) Comprehensive genetic analyses reveal evolutionary distinction of a mouse (*Zapus hudsonius preblei*) proposed for delisting from the US Endangered Species Act. Mol Ecol 15:4331–4359

King TL, Eackles MS, Aunins AW, Brockmann HJ, Hallerman EM, Brown BB (2015) Conservation genetics of the American horseshoe crab (*Limulus polyphemus*): allelic diversity, zones of genetic discontinuity, and regional differentiation. In: Carmichael RH, Botton ML, Shin PKS, Cheung SG (eds) Changing global perspectives on horseshoe crab biology, conservation and management. Springer, Cham pp 65–96

Koons BF (1883) Sexual characters of *Limulus*. Am Nat 17:1297–1299

Kreamer G, Michels SE (2009) History of horseshoe crab harvest in Delaware Bay. In: Tanacredi JT, Botton ML, Smith DR (eds) Biology and conservation of horseshoe crabs. Springer, New York, pp 299–313

Mattei JH, Beekey MA, Rudman A et al (2010) Reproductive behavior in horseshoe crabs: does density matter? Curr Zool 56(5):634–642

McGowan CP, Smith DR, Sweka JA et al (2011) Multispecies modeling for adaptive management of horseshoe crabs and red knots in the Delaware Bay. Nat Resour Model 24(1):117–156

McLachlan A, Defeo O, Jaramillo E et al (2013) Sandy beach conservation and recreation: guidelines for optimising management strategies for multi-purpose use. Ocean Coast Manag 71:256–268

Penn D, Brockmann HJ (1994) Nest-site selection in the horseshoe crab, *Limulus polyphemus*. Biol Bull 187:373–384

Penn D, Brockmann HJ (1995) Age-biased stranding and righting in horseshoe crabs, *Limulus polyphemus*. Anim Behav 49:1531–1539

Pritchard JK, Stephens M, Donnelly P (2000) Inference of population structure using multilocus genotype data. Genetics 155:945–959

Provancha JA, Lowers R, Mota M et al (2006) Trials and tribulations of tracking sea turtles in Mosquito Lagoon–trends in abundance and results from the passive acoustic monitoring network. In: Frick M, Panagopoulou A, Rees AF, Williams K (eds) 26th Annual symposium on sea turtle biology and conservation, Island of Crete, Greece, 2006. International Sea Turtle Society, pp 315–316

Riska B (1981) Morphological variation in the horseshoe crab *Limulus polyphemus*. Evolution 35(4):647–658

Rudloe A (1979a) Locomotor and light responses of larvae of the horseshoe crab, *Limulus polyphemus* (L.). Biol Bull 157:494–505

Rudloe AE (1979b) Release of larval horseshoe crabs, *Limulus polyphemus* from nesting in the field. Am Zool 19(3):989–989

Rudloe A (1980) The breeding behavior and patterns of movement of horseshoe crabs, *Limulus polyphemus*, in the vicinity of breeding beaches in Apalachee Bay, Florida. Estuaries 3(3):177–183

Rudloe A (1981) Aspects of the biology of juvenile horseshoe crabs, *Limulus polyphemus*. Bull Mar Sci 31(1):125–133

Rudloe A (1982) Man's influence as an environmental threat to Limulus. In: Bonaventura J, Bonaventura C, Tesh S (eds) Physiology and biology of horseshoe crabs: studies on normal and environmentally stressed animals, vol 81. Alan R. Liss, Inc., New York, pp 297–300

Rudloe A (1985) Variation in the expression of lunar and tidal behavioral rhythms in the horseshoe crab, *Limulus polyphemus*. Bull Mar Sci 36(2):388–395

Rudloe A, Herrnkind WF (1976) Orientation of *Limulus polyphemus* in the vicinity of breeding beaches. Mar Behav Physiol 4:75–89

Rudloe AE, Herrnkind WF (1980) Orientation by horseshoe crabs, *Limulus polyphemus*, in a wave tank. Mar Behav Physiol 7:199–211

Rutecki D, Carmichael RH, Valiela I (2004) Magnitude of harvest of Atlantic horseshoe crabs, *Limulus polyphemus*, in Pleasant Bay, Massachusetts. Estuaries 27(2):179–187

Saitou N, Nei M (1987) The neighbor-joining method: a new method for reconstructing phylogenetic trees. Mol Biol Evol 4:406–425

Saunders NC, Kessler LG, Avise JC (1986) Genetic variation and geographic differentiation in mitochondrial DNA of the horseshoe crab, *Limulus polyphemus*. Genetics 112:613–627

Schaller S, Thayer P, Hanson S et al (2005) Maine horseshoe crab (*Limulus polyphemus*) spawning surveys, 2004. Maine Department of Marine Resources, Buxton

Schaller SY, Chabot CC, Watson WH (2010) Seasonal movements of American horseshoe crabs *Limulus polyphemus* in the Great Bay estuary, New Hampshire (USA). Curr Zool 56(5): 587–598

Scheidt D, Lowers R (2001) Using an aerial survey to document the extent of a horseshoe crab die-off in Florida. In: American Fisheries Society Annual Meeting, Baltimore

Schreibman MP, Zarnoch CB (2009) Aquaculture methods and early growth of juvenile horseshoe crabs (*Limulus polyphemus*). In: Tanacredi JT, Botton ML, Smith DR (eds) Biology and conservation of horseshoe crabs. Springer, New York, pp 501–512

Sekiguchi K, Shuster CN (2009) Limits on the global distribution of horseshoe crabs (Limulacea): lessons learned from two lifetimes of observations: Asia and America. In: Tanacredi JT, Botton ML, Smith DR (eds) Biology and conservation of horseshoe crabs. Springer, New York, pp 5–24

Selander RK, Yang SY, Lewontin RC et al (1970) Genetic variation in the horseshoe crab (*Limulus polyphemus*), a phylogenetic "relic". Evolution 24:402–414

Shuster CN Jr (1979) Distribution of the American horseshoe "crab", *Limulus polyphemus* (L.). In: Cohen E (ed) Biomedical applications of the horseshoe crab (Limulidae). Alan R. Liss, Inc., New York, pp 3–26

Shuster CN Jr (1982) A pictorial review of the natural history and ecology of the horseshoe crab *Limulus polyphemus*, with reference to other Limulidae. In: Bonaventura J, Bonaventura C, Tesh S (eds) Physiology and biology of horseshoe crabs: studies on normal and environmentally stressed animals, vol 81. Alan R Liss, Inc., New York, pp 1–52

Shuster CN Jr, Sekiguchi K (2003) Growing up takes about ten years and eighteen stages. In: Shuster CN, Barlow RB, Brockmann HJ (eds) The American horseshoe crab. Harvard University Press, Cambridge, MA, pp 103–132

Shuster CN Jr, Botton ML, Loveland RE (2003) Horseshoe crab conservation: a coast-wide management plan. In: Shuster CN, Barlow RB, Brockmann HJ (eds) The American horseshoe crab. Harvard University Press, Cambridge, MA, pp 358–377

Smith DR (2002) Horseshoe crab spawning activity in Delaware Bay: a preliminary report on 2001 and a comparison from 1999 to 2001. Atlantic States Marine Fisheries Commission, Washington, DC

Smith DR, Pooler PS, Loveland RE et al (2002a) Horseshoe crab (*Limulus polyphemus*) reproductive activity on Delaware Bay beaches: interactions with beach characteristics. J Coast Res 18:730–740

Smith DR, Pooler PS, Swan BL et al (2002b) Spatial and temporal distribution of horseshoe crab (*Limulus polyphemus*) spawning in Delaware Bay: implications for monitoring. Estuaries 25:115–125

Smith DR, Millard MJ, Eyler S (2006) Abundance of adult horseshoe crabs (*Limulus polylphemus*) in Delaware Bay estimated from a bay-wide mark-recapture study. Fish Bull 104(3):456–464

Smith DR, Millard MJ, Carmichael RH (2009) Comparative status and assessment of *Limulus polyphemus* with emphasis on the New England and Delaware Bay populations. In: Tanacredi JT, Botton ML, Smith DR (eds) Biology and conservation of horseshoe crabs. Springer, New York, pp 361–386

Smith DR, Brousseau DJ, Mandt MT et al (2010) Age and sex specific timing, frequency, and spatial distribution of horseshoe crab spawning in Delaware Bay: insights from a large-scale radio telemetry array. Curr Zool 56(5):563–574

Steele P, Bert TM, Johnston KH et al (2002) Efficiency of bycatch reduction devices in small otter trawls used in the Florida shrimp fishery. Fish Bull 100(2):338–350

Strobel CJ, Brenowitz AH (1981) Effects of Bunker C oil on juvenile horseshoe crabs (*Limulus polyphemus*). Estuaries 4:157–159

Swan BL (2005) Migrations of adult horseshoe crabs, *Limulus polyphemus*, in the Middle Atlantic Bight: a 17-year tagging study. Estuaries 28(1):28–40

Sweka JA, Smith D, Millard MJ (2007) An age-structured population model for horseshoe crabs in the Delaware Bay area to assess harvest and egg availability for shorebirds. Estuar Coasts 30:277–286

Swofford DL, Selander RB (1981) BIOSYS-1: A FORTRAN program for the comprehensive analysis of electrophoretic data in population genetics and systematics. J Hered 72:281–283

Systat Software Inc (2011) SigmaPlot for Windows. 12.0 edn

Tamura K, Peterson D, Peterson N et al (2011) MEGA5: molecular evolutionary genetics analysis using maximum likelihood, evolutionary distance, and maximum parsimony methods. Mol Biol Evol 28:2731–2739

Thompson M (1999) Assessments of the population biology and critical habitat for the horseshoe crab, *Limulus polyphemus* in South Carolina. M.S. thesis, Medical University of South Carolina

Wallace GE (1999) Potential effects of a horseshoe crab fishery on migratory shorebirds in the state of Florida. Bureau of Wildlife Diversity Conservation. Florida Fish and Wildlife Conservation Commission, Panama City

Watson WH, Chabot CC (2010) High resolution tracking of adult horseshoe crabs *Limulus polyphemus* in a New Hampshire estuary using fixed array ultrasonic telemetry. Curr Zool 56(5):599–610

Widener JW, Barlow R (1999) Decline of a horseshoe crab population on Cape Cod. Biol Bull 197:300–302

Chapter 6
Genetic Features of *Tachypleus tridentatus* in Japan and an Alien (Non-native) Population Founded at Ise-Mikawa Bay, Chubu Region, Central Japan

Shin Nishida, Kenji Kuroyanagi, and Hiroko Koike

Abstract The tri-spine horseshoe crab *Tachypleus tridentatus*, known as a living fossil and endangered, is distributed from coastal Southeast Asia to western Japan. The Japanese population represents the northernmost range for this species. To examine the genetic structure and diversity of the Japanese population, over 600 samples from ten localities were analyzed using mitochondrial DNA AT-rich region sequences as a maternal inheritance marker and over 250 samples using nucleic compound microsatellite loci as biparental markers. Genetic subdivisions were detected in Japanese populations using both markers. Two genetically distinct groups, the western and the eastern group, were detected and six tentative local management units were defined. Locality and genetic units should be taken into account for conservation of Japanese populations.

Furthermore, we report on an alien (non-native) population founded in Japan. Sixty-two individuals including 13 adults were captured in 12 years (from 2001) from Ise-Mikawa Bay, which is out of range of their natural habitat. Genetic analysis and other information indicated that the origin of these individuals was mainland China. This indicates the possibility that even an endangered species can become an alien and invasive one.

Keywords *Tachypleus tridentatus* • Population genetic structure • mtDNA • Microsatellite DNA • Alien (non-native) species/population • Japan

---

S. Nishida (✉)
Biology, Science Education, Faculty of Education and Culture, University of Miyazaki,
1-1 Gakuen-Kibanadai-Nishi, Miyazaki, Miyazaki 889-2192, Japan
e-mail: shin.nsd@nifty.com

K. Kuroyanagi
Minamichita Beach Land Aquarium, 428-1 Okuda, Mihama,
Chita-Gun, Aichi 470-3233, Japan

H. Koike
The Kyushu University Museum, 6-10-1 Hakozaki, Higashi-Ku,
Fukuoka, Fukuoka 812-8581, Japan

R.H. Carmichael et al. (eds.), *Changing Global Perspectives on Horseshoe Crab Biology, Conservation and Management*, DOI 10.1007/978-3-319-19542-1_6

6.1 Genetic Features of Japanese Populations of *Tachypleus tridentatus*

6.1.1 Distribution and Habitats of T. tridentatus *in Japan*

The Japanese population represents the northernmost range for *Tachypleus tridentatus*. Habitats in Japan are intermittently distributed along the coastal region of Northern Kyushu (from Omura Bay) and the Seto Inland Sea. Current habitats with recognized spawning sites (Fig. 6.1) are: Omura-Bay, Kujukushima-Sasebo, and Iki Island (Nagasaki Pref.); Imari Bay (Saga Pref.); Kafuri Bay, Hakata Bay, Tsuyazaki tidal flat, Sone tidal flat, and Kanda (Fukuoka Pref.); Nakatsu, Usa and Morie Bay (Oita Pref.); Chidori Beach, Yamaguchi Bay, and Hirao Bay (Yamaguchi Pref.); Toyo Region (Ehime Pref.); Hiroshima Bay and Takehara (Hiroshima Pref.); and Kasaoka (Okayama Pref.).

The population size of each habitat is decreasing rapidly and drastically due to loss of tidal flats and spawning beaches and the deterioration of coastal waters (reviewed in Sekiguchi 1988; Itow 1993). Sea sand mining to supply material for concrete is probably one of the factors contributing to the destruction of habitats of

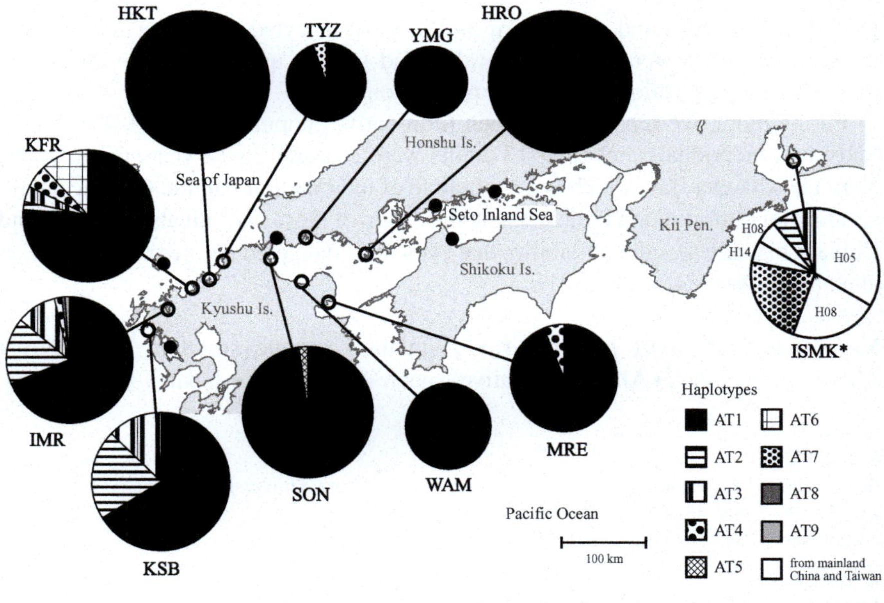

Fig. 6.1 Geographic distribution of mtDNA AT-rich region haplotypes of Japanese *Tachypleus tridentatus*. The size of the *circle* is scaled relative to the sample size. *Closed circles* show other breeding sites of this species. *KSB* Kujukushima-Sasebo Bay, *IMR* Imari Bay, *KFR* Kafuri Bay, *HKT* Hakata Bay, *TYZ* Tsuyazaki tidal flat, *SON* Sone tidal flat, *YMG* Yamaguchi Bay, *WAM* Wama tidal flat, *MRE* Morie Bay, *HRO* Hirao Bay. *ISMK** Ise-Mikawa Bay is an alien (non-native) population in Japan

adult individuals. At a few sites, there is conflict with fishermen because horseshoe crabs sometimes break fish nets when caught. Over 80 % of habitats disappeared in the Seto Inland Sea from 1930 to 1994 (Sekiguchi and Shuster 2009). In particular, many habitats in the eastern Seto Inland Sea area have been destroyed in the last 60 years. *T. tridentatus* is included in the Red List and the Red Data Book of Japan as critically endangered (CR + EN) (Ministry of the Environment, Japan 2000, 2006). Some local laws for the protection of this species have been established in response to these threats. For example, the spawning site at Oehama Beach, Kasaoka, Okayama Pref. is a protected habitat as a national, natural monument in Japan, listed in 1928. However, this area was drained by the large-scale land reclamation of Kasaoka Bay from 1969, and the Konoshima Channel was added to this protected area (Sekiguchi 1988). Spawning sites in the Toyo Region in Ehime Pref. and Imari City in Saga Pref. are also protected as natural monuments by their respective local governments. Due to the decline in population size and habitat available for spawning, it is important to establish management units for this species. Here we describe genetic analyses that allow us to define the genetic structure of horseshoe crab populations in Japan.

6.1.2 *Ecology and Behavior of* **T. tridentatus** *from Field and Breeding Surveys*

In Japan, *T. tridentatus* generally inhabit a cove or bay that is protected from surf and becomes intertidal mud flats during low tide (Sekiguchi 1988). In an 11-year laboratory study, Harada (2003) found that *T. tridentatus* take 11 years and 14 molts to grow from the larval stage to sexual maturity.

The behaviors of *T. tridentatus* indicate that they stay in a relatively limited area near their natal beach. They lay their eggs in relatively small sandy beaches near the high tide line. Larvae of *T. tridentatus* swim very little compared with those of *Limulus polyphemus* and usually stay in their nest after hatching and pass the winter there (Sekiguchi 1988). Observations of the behavior of larvae at Imari Bay (Sakemi 1997) suggest that during a spring tide event, most of the larvae move planktonically during the night at high tide from the sandy beach where they hatched towards the tidal flat with the ebbing tide, exhibiting phototaxis (Sakemi 1997). The juvenile stage (prosomal width up to 7 cm) is spent in the intertidal mud flat near the hatching beach, and locomotive activity is relatively low (50 m/month) (Kawahara 1982). Larger juvenile and sub-adult individuals may gradually move to deeper water depending on their growth stage. Adult individuals winter offshore in deep water (20 m deep or more) and move to shallow water near sandy beaches for spawning when the seawater temperature increases in the late spring (Nishii 1975; Sekiguchi 1988). This permanence should lead to genetic differences between spawning sites (Nishida and Koike 2009).

6.1.3 Genetic Subdivision of Japanese Populations

6.1.3.1 Analysis of Mitochondrial AT-Rich Region

Molecular genetic analysis is a powerful tool for investigating genetic structure, diversity, and differentiation within a population through its history. These types of information are needed to identify management units for conservation efforts of this endangered species.

Sugawara et al. (1988) was the first phylogeographic study for *T. tridentatus*. They analyzed mitochondrial (mt) DNA using restriction fragment length polymorphisms (RFLP) between the Hakata Bay, Kasaoka Bay (located in the Seto Inland Sea in Okayama Pref.), and Amoy (China) populations. They failed to detect any genetic differentiation among them. This result seemed to indicate a close relationship between the Japanese and Chinese populations, although the sample size was small. The resolution of RFLP is relatively low, so another high-resolution genetic marker is needed to estimate genetic relationships between local populations within a relatively small geographic range.

The mtDNA AT-rich region (control region) is a highly variable, noncoding region that is useful for phylogeographic studies and population genetics surveys as a maternal inheritance genetic marker. We analyzed over 600 individuals from ten localities (Nishida and Koike 2009, 2011, Table 6.1): Kujukushima-Sasebo

Table 6.1 List of ten local populations and samples analyzed in this study

Local population/ Samples	Hemolymph	Egg	Muscle	Total	mtDNA	MS
Kujukushima-Sasebo Bay (KSB)	127			127	127	38
Imari Bay (IMR)	99	4	4	107	107	26
Kafuri Bay (KFR)	23	21	7	51	51	25
Hakata Bay (HKT)	132	4	2	138	138	32
Tsuyazaki tidal flat (TYZ)	11		8	19	19	16
Sone tidal flat (SON)	29	19	3	51	51	25
Wama Bay (WAM)	7			7	7	7
Morie Bay (MRE)	14	20		34	34	8
Yamaguchi Bay (YMG)	17		8	25	22	25
Hirao Bay (HRO)	37			61	61	61
Total	496	68	32	620	617	263

mtDNA number analysed using mtDNA from Nishida and Koike (2009, 2011), *MS* number analysed using microsatellites from Nishida and Koike (2011)

Bay (n=127) in Nagasaki Pref.; Imari Bay (n=107) in Saga Pref.; Kafuri Bay (n=51), Hakata Bay (n=138), Tsuyazaki tidal flat (n=19), and Sone tidal flat (n=51) in Fukuoka Pref.; Wama tidal flat (n=7) and Morie Bay (n =34) in Oita Pref.; and Yamaguchi Bay (n=22) and Hirao Bay (n=61) in Yamaguchi Pref. All sites have spawning beaches except Wama tidal flat. Almost all of the samples were hemolymph (86 %) from individuals of various ages, and individuals were released at the same place they were caught after sampling. The other samples were eggs and muscle from dead specimens. The analyzed region sequences were AT-rich mtDNA with 395/396 bp and were amplified using a primer set from Nishida and Koike (2009).

Nine haplotypes (AT01 to AT09) were found in Japanese waters (Nishida and Koike 2011; GenBank accession Nos. AB353283 to AB353289, LC002283 and LC002284; Table 6.2 and Fig. 6.1). Thirty-six haplotypes had been reported and deposited on GenBank from Japan (AT series: Nishida and Koike 2009, 2011), the Taiwan Strait (H series: Yang et al. 2007), and mainland China (HC series: Weng et al. 2013). We constructed a haplotype network from these data by the median-joining method (Bandelt et al. 1999, Fig. 6.2). All haplotypes were closely related

Table 6.2 Distribution of the mtDNA AT-rich region haplotypes and haplotype diversity (h_{mt}) for ten local populations in Japan

Local populations/ haplotypes	AT1	AT2	AT3	AT4	AT5	AT6	AT7	AT8	AT9	Total	h_{mt}
Kujukushima-Sasebo Bay (KSB)	84	27	16							127	0.51
Imari Bay (IMR)	74	18	11	2				1	1	107	0.49
Kafuri Bay (KFR)	39	1	2	3		6				51	0.40
Hakata Bay (HKT)	138									138	0.00
Tsuyazaki tidal flat (TYZ)	18						1			19	0.11
Sone tidal flat (SON)	50				1					51	0.04
Wama Bay (WAM)	7									7	0.00
Morie Bay (MRE)	32			2						34	0.11
Yamaguchi Bay (YMG)	22									22	0.00
Hirao Bay (HRO)	61									61	0.00
Whole	525	46	29	7	1	6	1	1	1	617	0.27

and connected by a single substitution. AT02, AT03, and AT07 were shared between Japan and Taiwan-China populations. The remaining six haplotypes including the major haplotype AT01 were endemic in Japan, showing that Japanese populations are genetically independent from populations in Taiwan and China, even though the populations are genetically close. AT02 and AT03 were posited to be intermediate between haplotypes from Taiwan-China (white circles in Fig. 6.2) and from Japan

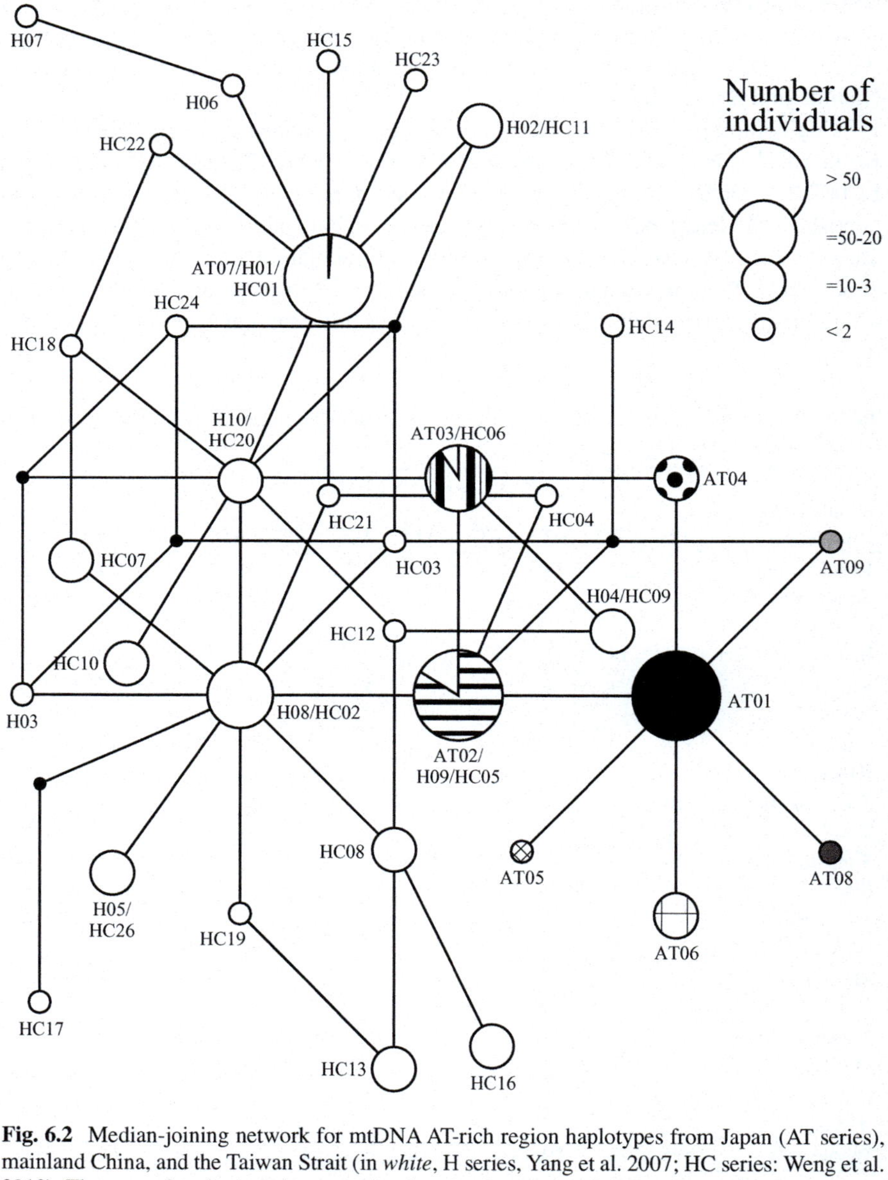

Fig. 6.2 Median-joining network for mtDNA AT-rich region haplotypes from Japan (AT series), mainland China, and the Taiwan Strait (in *white*, H series, Yang et al. 2007; HC series: Weng et al. 2013). The area of each *circle* is proportional to the frequency of the haplotype. Each *line* represents a mutational step

(other circles), and AT01 was posited to have arisen from AT02, indicating that these two haplotypes would be the ancestral haplotypes for the Japanese populations.

Figure 6.2 and Table 6.2 show the distributions and frequencies of haplotypes in Japan. AT01 was the dominant haplotype, observed in all regions with high frequency (greater than 0.65). AT02 and AT03 were only detected in the west of the area analyzed and defined the western group (Kujukushima-Sasebo Bay, Imari Bay, and Kafuri Bay). AT06 was found in Kafuri Bay with relatively high frequency (0.12). Only Imari Bay had AT08 and AT09. Four individuals with the AT04 haplotype were from either Kafuri Bay or Morie Bay, which are geographically separate. The eastern group (Hakata Bay, Tsuyazaki tidal flat, Sone tidal flat, Wama Bay, Morie Bay, Yamaguchi Bay, and Hirao Bay) was defined as almost monomorphic for AT01. AT05 and AT07 were detected only in the Sone tidal flat and Tsuyazaki tidal flat, respectively.

The degree of genetic differentiation was measured using F_{ST} and Φ_{ST} (Table 6.3). Under an infinite-allele mutation model, F_{ST} values ranged from zero, where all subpopulations have equal allele frequencies, to one, where all the subpopulations are fixed for different alleles. The parameter Φ_{ST} is analogous to F_{ST}, but incorporates genetic distances among alleles (for a review, see Allendorf and Luikart 2007). Typically, an F_{ST} value above 0.15 is considered an indication of significant differentiation among fragments (Frankham et al. 2010).

Table 6.3 Pairwise F_{ST} values (lower matrix) and Φ_{ST} (upper matrix) from mtDNA analysis of ten local populations in Japan

	1	2	3	4	5	6	7	8	9	10
1 Kujukushima-Sasebo Bay (KSB)		0.00	0.12*	0.28*	0.10	0.19*	0.11	0.16*	0.17*	0.22*
2 Imari Bay (IMR)	0.00		0.08*	0.23*	0.06	0.15*	0.06	0.11*	0.12	0.17*
3 Kafuri Bay (KFR)	0.06*	0.04		0.16*	0.01	0.06*	−0.02	0.03	0.04	0.09*
4 Hakata Bay (HKT)	0.26*	0.24*	0.25*		0.16	0.02	0.00	0.12	0.00	0.00
5 Tsuyazaki tidal flat (TYZ)	0.11*	0.09	0.05	0.16		0.03	−0.07	0.00	0.01	0.07
6 Sone tidal flat (SON)	0.17*	0.14*	0.11*	0.02	0.00		−0.08	0.00	−0.02	0.00
7 Wama Bay (WAM)	0.10	0.07	0.02	0.00	−0.07	−0.08		−0.05	0.00	0.00
8 Morie Bay (MRE)	0.13	0.10*	0.05	0.12	−0.01	0.01	−0.05		0.01	0.06
9 Yamaguchi Bay (YMG)	0.15*	0.12*	0.09	0.00	−0.01	−0.02	0.00	0.01		0.00
10 Hirao Bay (HRO)	0.20*	0.17*	0.16*	0.00	0.07	0.00	0.00	0.06	0.00	

*$P < 0.05$, with Bonferroni correction

High F_{ST} and Φ_{ST} values of 0.18 and 0.19 between western (Kujukushima-Sasebo Bay to Kafuri Bay) and eastern groups (Hakata Bay to Hirao Bay) were detected, respectively, and were significant ($P<0.01$), supporting these two genetic groups as distinct. The Kafuri Bay population in the western group, located at the border region between the western and eastern groups, was significantly different from its neighboring sites of Imari Bay ($\Phi_{ST}=0.08$) and Hakata Bay ($F_{ST}=0.25$; $\Phi_{ST}=0.16$). Only the Kafuri Bay population had AT06, suggesting that this is an independent population, and the western group was divided two genetic units. There were no significant differences among sites in the eastern group, because this group was almost monomorphic for AT01.

6.1.3.2 Microsatellite DNA Analyses

Microsatellite loci are often applied in population genetics as a nuclear, biparental genetic marker with high resolution. Microsatellites (MS) are short tandem repeats of DNA sequences and have a motif of one to six bases that can differ in repeat number between individuals. We applied this genetic marker to re-evaluate the genetic structure of the Japanese populations and to try to detect genetic differences in the eastern genetic group, which were almost completely monomorphic according to mtDNA.

Over 250 individuals from ten localities were analyzed (Nishida and Koike 2009, 2011; Tables 6.1 and 6.4): Kujukushima-Sasebo Bay ($n=38$) in Nagasaki Pref.; Imari Bay ($n=26$) in Saga Pref.; Kafuri Bay ($n=25$), Hakata Bay ($n=32$), Tsuyazaki tidal flat ($n=16$), and Sone tidal flat ($n=25$) in Fukuoka Pref.; Wama tidal flat ($n=7$) and Morie Bay ($n=8$) in Oita Pref.; and Yamaguchi Bay ($n=25$) and Hirao Bay

Table 6.4 Summary of results of microsatellite analysis from eight loci at ten local sampling sites

Local populations/ haplotypes	N	N_a	H_O	H_E
Kujukushima-Sasebo Bay (KSB)	38	4.00	0.58	0.57
Imari Bay (IMR)	26	3.30	0.50	0.57
Kafuri Bay (KFR)	25	3.10	0.53	0.54
Hakata Bay (HKT)	32	3.00	0.52	0.53
Tsuyazaki tidal flat (TYZ)	16	2.80	0.54	0.51
Sone tidal flat (SON)	25	2.80	0.45	0.53
Wama Bay (WAM)	7	2.50	0.41	0.45
Morie Bay (MRE)	8	2.60	0.45	0.52
Yamaguchi Bay (YMG)	25	2.80	0.49	0.50
Hirao Bay (HRO)	61	2.80	0.42	0.42
Total/Average	263	2.97	0.49	0.51

These data from Nishida and Koike (2011)
N numbers of individuals analyzed, N_a average of number of alleles, H_O observed heterozygosity, H_E expected heterozygosity

($n=61$) in Yamaguchi Pref. Eight compound microsatellite loci were used for the analysis and two to nine alleles were detected for each locus (Nishida and Koike 2010). The average number of alleles for each site was 2.5 (Wama) to 4.0 (Kujukushima-Sasebo).

Figure 6.3 shows the results of STRUCTURE analysis (Pritchard et al. 2000), which is used to investigate population structure and assignment of individuals with a Bayesian-based algorithm incorporating geographic information. The number of population clusters (K; putative populations) inferred from this analysis was $K=4$. Each population cluster had clines, although each site had multiple clusters and did not form genetically monophyletic populations. For example, the white colored putative population cluster dominated the Kujukushima-Sasebo Bay and the Imari Bay populations of the western group defined by mtDNA data. This distinguished the western and eastern groups in the same division as the mtDNA data did. However, contrary to the mtDNA data, it seems that the Kafuri Bay population belonged to the eastern group. Further, many individuals from Hirao Bay have relatively unique genetic compositions, leading to its difference from other populations.

Values from pairwise genetic differentiation by MS analysis (Table 6.5: F_{ST} and R_{ST}; R_{ST} is analogous to $F_{ST,}$ but is under a stepwise mutation model) had higher resolution than mtDNA in detecting genetic differentiations within each genetic group. In the western group, the Kujukushima-Sasebo Bay and the Imari Bay populations were genetically distinct. The mtDNA analysis failed to detect this population difference. Furthermore, the Hirao Bay population, the most eastern sampling site in this analysis, was different from other neighboring populations with high F_{ST} and R_{ST} values. MS analysis thus revealed the existence of population genetic structures within the eastern group.

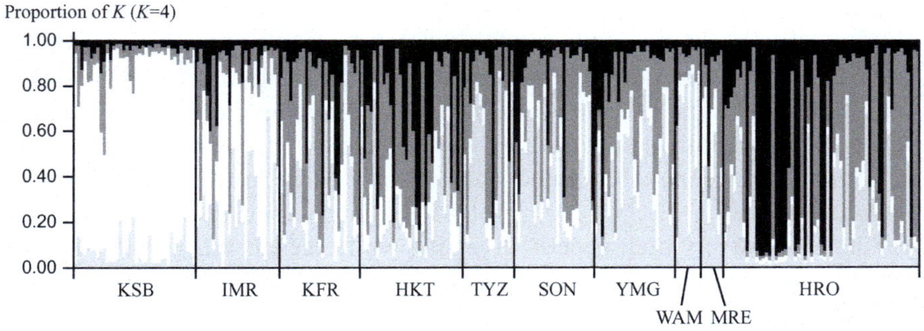

Fig. 6.3 Results of STRUCTURE analyses of Japanese populations with $K=4$ (Modified from Nishida and Koike 2011). Each individual is represented as a *vertical line* partitioned into population clusters (K). Abbreviations of site names are the same as in Fig. 6.1

Table 6.5 Pairwise R_{ST} values (lower matrix) and F_{ST} values (upper matrix) from MS analysis between ten local populations in Japan

	1	2	3	4	5	6	7	8	9	12
1 Kujukushima-Sasebo Bay (KSB)		0.05*	0.10*	0.10*	0.09*	0.08*	0.09*	0.09*	0.10*	0.19*
2 Imari Bay (IMR)	0.06*		0.03*	0.04*	0.04*	0.01	0.03	0.03	0.04*	0.10*
3 Kafuri Bay (KFR)	0.23*	0.09*		0.00	0.00	0.02	0.05	0.05	0.03	0.04*
4 Hakata Bay (HKT)	0.33*	0.18*	0.03		0.01	0.03	0.06	0.06*	0.04*	0.03*
5 Tsuyazaki tidal flat (TYZ)	0.25*	0.12*	–0.01	0.02		0.02	0.07*	0.02	0.03	0.05*
6 Sone tidal flat (SON)	0.16*	0.06*	0.02	0.10*	0.04		0.02	0.02	–0.01	0.05*
7 Wama Bay (WAM)	0.07	0.03	0.06	0.17*	0.10	0.01		0.02	0.03	0.12*
8 Morie Bay (MRE)	0.16*	0.10*	0.04	0.14*	0.04	0.04	0.04		0.03	0.15*
9 Yamaguchi Bay (YMG)	0.18*	0.08*	0.01	0.10*	0.02	0.00	0.04	0.01		0.07*
10 Hirao Bay (HRO)	0.38*	0.21*	0.03	0.01	0.04	0.13*	0.22*	0.22*	0.14*	

*$P < 0.05$, with Bonferroni correction

6.1.4 Level of Genetic Diversity of Japanese Populations

Typical value ranges of haplotype genetic diversity indicate that high levels correspond to over 0.8, middle levels are 0.8–0.5, and low levels are under 0.5. The haplotype diversities (h_{mt}) of mtDNA for each sampling site are shown in Table 6.2. Haplotype diversity was 0.27 for the overall Japanese population. Values from sampling sites in the western group (h_{mt}=0.40–0.51, overall; h_{mt}=0.49) were remarkably higher than in the eastern group (h_{mt}=0.00–0.11, overall; h_{mt}=0.06). The western group maintained ancestral polymorphisms leading to relatively high genetic diversity, while the eastern group may have been strongly influenced by the founder effect, thereby exhibiting less genetic diversity (Nishida and Koike 2009).

The observed and expected heterozygosities from microsatellite DNA analysis also showed low- to middle-range values (Table 6.4). Expected heterozygosities from the western group (0.54–0.57, average of 0.56) were slightly higher than those from the eastern group (0.42–0.53, average of 0.49). The Hirao Bay population had the smallest value with 0.42.

The mtDNA haplotype diversities of populations on mainland China ranged from 0.54 to 0.94 (Yang et al. 2009; Weng et al. 2013), averaging 0.79, showing high genetic diversity and remarkably higher values than those of Japanese populations. Similarly, all expected heterozygosity values from mainland China were higher than Japan, with $H_E=0.60–0.88$, and an average of 0.65 (Li et al. 2009; Weng et al. 2013). The Japanese populations of *T. tridentatus* are the farthest north, located at the edge of the natural habitat. Our genetic analysis indicates that Japanese populations are relatively young, with ancestral populations likely having migrated from the south during the middle to late Quaternary period with glacial and warmer interglacial events. During the glacial events when the sea level was low, the East China Sea was dry, and the western Kyushu was connected to the southern Asian continent. This climate change allowed this species to migrate into Japan around the coastal line (Nishida and Koike 2009). Levels of genetic diversity in Japanese populations were low, likely affected by the founder effect throughout their population history.

6.1.5 Genetic Groups and Management Units

Populations within species may sometimes be sufficiently differentiated in terms of adaptive characteristics or genetic composition to require separate, specialized management (Frankham et al. 2010).

Genetic subdivisions were identified in Japanese populations of *T. tridentatus* occupying a small geographic area using analysis of both the mtDNA AT-rich region and nucleic microsatellite DNA markers. The distribution of mtDNA haplotypes clearly showed two main genetic groups: the western and the eastern group. The western group was defined by haplotypes AT02 and AT03. In contrast, the eastern group was almost monomorphic for AT01. AT02 and AT03 were ancestral haplotypes for the Japanese populations.

The geographic boundary between these two groups is the Itoshima Peninsula located between Kafuri and Hakata Bay. The mtDNA analysis indicated that the Kafuri Bay population was significantly different genetically from neighboring populations in Imari and Hakata Bay; Kafuri Bay only had haplotype AT06. However, microsatellite analysis failed to detect genetic differences between the Kafuri and Hakata Bay populations. Contrary to the mtDNA analysis, it appeared that the Kafuri Bay belonged to the eastern group. This conflict in results could be explained by incomplete sorting of microsatellite alleles or male mediated gene flow, as has been found for *Limulus polyphemus* (King et al. 2005). Probably, Kafuri and Hakata Bay connected around 6,500 years ago, when the sea level was lower than today, allowing some individuals to move to Hakata Bay from Kafuri Bay through a sea channel on Itoshima Peninsula. These individuals went to form the eastern group. Consequently, the Kafuri Bay population should be treated as an independent management unit.

The microsatellite analysis was able to detect other genetic differentiations. In the western group, the Kujukushima-Sasebo and the Imari Bay populations were slightly different from each other. Further, individuals of the Hirao Bay population in the eastern group showed unique genetic features with high F_{ST} and R_{ST} values when compared with neighboring sites, although it is difficult to find geographic barriers to which to attribute this differentiation. This finding also indicates that the far eastern habitat in the Seto Inland Sea exhibited differentiations, but we could not perform an analysis for them (e.g. Toyo Region, Hiroshima, and Kasaoka) because population size was very small and it was difficult to collect enough native individuals for analysis.

Considering the distribution of habitats and genetic information, we propose two core management groups and six local management units under each core unit. Figure 6.4 shows a summary schematic for the genetic structures of Japanese populations and tentative management units. The western and eastern groups are the most basic management units. In the western unit, each local population (the Kujukushima-Sasebo, the Imari Bay, and the Kafuri Bay populations) should be independently managed. Hakata Bay and the Tsuyazaki tidal flat were combined into one unit due to their geographic proximity and lack of genetic differentiation. The Sone tidal flat, the Wama tidal flat, and Yamaguchi Bay are located in the same sea area (Suo-Nada, the western part of the Seto Inland Sea), and there is no genetic difference compared with the Morie Bay population. Therefore, considering the potential for migration of individuals within this unit, it is better considered as one management unit. The remaining management unit is the Hirao Bay population. Microsatellite analysis strongly suggested limited gene flow between this population and neighboring sites. Human activities may lead to and accelerate this genetic differentiation, so management activities ought to pay particular attention to this population.

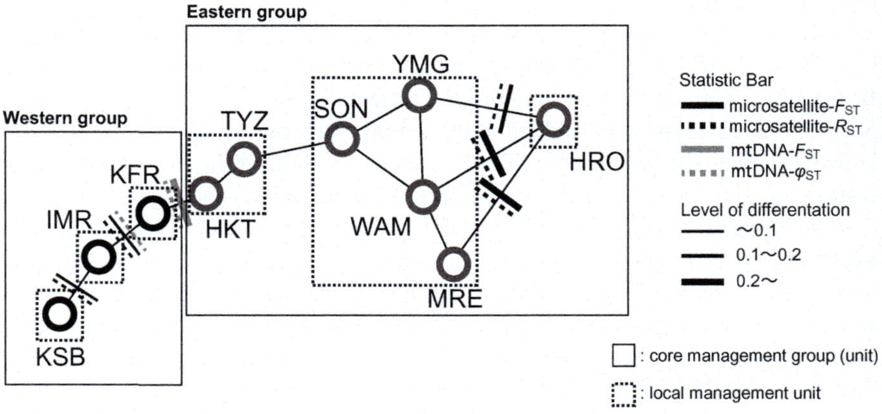

Fig. 6.4 Schematic of population genetic structures and tentative management units in Japan. Abbreviations of site names are the same as in Fig. 6.1

6.2 Report on an Alien (Non-native) Population Founded in Japan

6.2.1 Alien Horseshoe Crab Population Founded in Japan

Alien populations and species are those that have been transported beyond their natural range and have established a new habitat in a new area. In many cases, they strongly affect the original ecosystem and many living native species and populations. Here, we report on an incident of an alien horseshoe crab population that was found in Japan.

The Ise-Mikawa Bay is separated into two bays, Ise Bay (1,738 m^2) and Mikawa Bay (602 m^2). The mouth of the Mikawa Bay connecting to the Ise Bay is located along the Pacific coast of the Chubu Region in central Japan. This is a shallow and semi-closed bay with an average depth of 19.5 m in Ise Bay and 9.2 m in Mikawa Bay, and a maximum depth of 38 m. Both the Ise and Mikawa Bays are listed on the list of 500 important wetlands in Japan (Ministry of the Environment, Japan 2001). As mentioned in the previous section, the distribution range of *T. tridentatus* in Japan is from the Seto Inland Sea to Kujukushima-Sasebo waters. As such, the Ise-Mikawa Bay is completely outside the normal range. There has been no scientific evidence of their existence at this location until now.

The first incident occurred in August 2001, when one horseshoe crab was captured by a trawling net. The fisherman brought it to a local aquarium, Minamichita Beach Land. Aquarium staff and researchers identified the individual as *T. tridentatus* by its morphology. This was the first recorded capture of *T. tridentatus* in the Ise-Mikawa Bay outside their normal range in Japan. This sub-adult individual had a prosomal carapace width of 14.2 cm and was estimated to have an instar number of 12 and to be around 10 years old (Harada 2009). The next year, in 2002, captured individuals increased drastically, reaching 15 animals with an average carapace width of 15.2 cm (Table 6.6). After 2002, horseshoe crabs have been consistently found in this area, although the numbers of individuals captured have been decreasing in recent years. Average carapace width significantly increased in 2004 (*t*-test, $P<0.05$) compared with previous years (2001–2003). The size increased significantly from 2005 to 2007 (*t*-test, $P<0.05$), with the average prosomal width increasing from ~15 cm to over 25 cm during this time. Furthermore, adult individuals were first found in 2006. According to the growth rate and size of *T. tridentatus*, a prosomal width of 15 cm indicates instar numbers of 12–13, and 25 cm indicates near-final instar numbers of 14–16 (Harada 2009). These data clearly show that alien individuals in Ise-Mikawa Bay are growing successfully into adults from juveniles. In 2008, a pair of individuals, female and male, was captured; the pair later laid eggs at an aquarium, where these eggs were hatched. Table 6.6 shows the number of captured individuals and their size distribution from 2001 to 2013. All individuals were

Table 6.6 Number and size distribution of individuals found yearly in Ise-Mikawa Bay

	Carapace width (cm)																									Sex		Carapace width (cm)	
	−11	−12	−13	−14	−15	−16	−17	−18	−19	−20	−21	−22	−23	−24	−25	−26	−27	−28	−29	−30	−31	−32	−33	U	Total	♂	♀	Ave.	SD
2001					1				1																2			16.4	3.1
2002	1		2	2	2	3	1	2	2																15			15.2	2.3
2003			1	2		1	1	1	1	2	1													1	11			16.8	2.8
2004							3	1		3	1		1			1									10			19.5	3.0
2005							1			1	1	3												1	7			20.0	1.9
2006						1													**1**			**1**		1	4	1	1	25.0	8.4
2007																		**1**	1						2	1		27.7	0.8
2008														**1**	**1**	**2**			**2**				**1**		7	4	3	26.9	2.9
2009														**1**				**1**							2	2		25.4	3.2
2010																									0				
2011																									0				
2012																		1							1			27.5	
2013																					**1**				1		1	30.8	
Total	1	0	3	4	3	5	6	4	4	6	3	3	1	2	1	3	0	3	4	0	1	1	1	3	62	8	5		

Bold and underline adult individual

bycatch from fishing nets (including trawling, set nets, and gill nets) from April to November, with numbers peaking in the summer season (June to August). The numbers of captured individuals from this area reached 62 in 12 years, including 13 adults of both sexes. Thus, the Ise-Mikawa Bay appears to have hosted an alien population of *T. tridentatus* for over 10 years.

6.2.2 Genetic Features of the Alien Population in Ise-Mikawa Bay

What is the origin of this population in Ise-Mikawa Bay? Molecular genetic analysis was applied to answer this question. Eighteen individuals from 2002 to 2009 were analyzed and six haplotypes were detected (Fig. 6.1). Six individuals had H05 (HC26), four had H08 (HC02), four had H01 (AT07), and the remaining individuals were either H09 (AT02; HC05), HC06 (AT03), HC08, or HC14. According to our analysis above, H05, H08, HC08, and HC14 were not distributed in Japan, and H01 (AT07), H09 (AT02; HC05), and HC06 (AT03) were shared haplotypes of Japan and Taiwan-China populations (Nishida and Koike 2009; Yang et al. 2009; Weng et al. 2013). Therefore, all of the haplotypes from this population originated from Taiwan-China populations. In addition, the haplotype components and frequency were completely different from other native populations in Japan. These genetic data strongly suggest that the Ise-Mikawa Bay population is not native and not derived from other native populations in Japan, instead originating from a population in Taiwan or China.

6.2.3 History of the Alien Ise-Mikawa Bay **Population and Implications**

How and why were these individuals introduced into the Ise-Mikawa Bay? Aquarium staff collected information about this incident and received an important testimony. In 1996, a man who was an importer of crabs imported horseshoe crabs from Pintan, Fujian, China, located on the continental side of the Taiwan Strait, to sell to Japanese customers. (The intended customers remain unclear, but were perhaps aquaria or pet shops). The number of individuals imported was approximately 1,000; he sent 200 of these individuals to an acquaintance in Okinawa and kept the remaining 800 individuals at Katana Port, Minami-Chita, Aichi Pref. in Mikawa Bay using a net basket. According to his testimony, 200 of the horseshoe crabs kept in the net basket accidentally escaped to sea because of a broken net. The body length of the individuals at the time of their escape was around 10 cm with an estimated prosomal width of approximately 5 cm. This would mean that their estimated instar would be seven to eight. Five years later, the first individual was caught, which was estimated to be about 14–15 in instar number. This growth rate is similar to reference data (Harada 2003). In conclusion, taking into account our molecular genetic data, growth rate, and this man's testimony, the alien population in Ise-Mikawa Bay was most likely artificially introduced from Pintan, China in 1996 with an initial population size of at least 200. Although we could not obtain information about the remaining 600 individuals in the net basket, we suspect that the number of escaped individuals exceeded 200, considering the relatively large number that have been captured since.

The fact that horseshoe crabs can survive outside their normal habitat range is surprising and provides us with a lesson for the conservation of this species as well as for natural habitats of all other organisms. Even an endangered species is able to become an alien invasive population. Potentially, any introduced individual affects and disturbs natural ecosystems. If horseshoe crabs increase in non-native habitats, the composition and population dynamics of benthic species, especially including human-edible bivalves, may be influenced. Benthic species, as lower trophic level organisms in ecosystem pyramids are an important food source for other higher-level organisms. Conservation managers have an obligation to remove the alien horseshoe crabs from non-native habitats to maintain the original ecosystem; however, horseshoe crabs are an endangered and protected animal. That a female-male pair was caught that possessed reproductive ability is cause for great concern, because it indicates that horseshoe crabs will likely continue to increase in Ise-Mikawa Bay. The aquarium Minamichita Beach Land and others have announced this situation to local fisheries and citizens, and are collecting horseshoe crabs to keep them in the aquarium in an effort to curtail their invasive effects on nature in the years to come.

6.3 Implications for Conservation of Japanese Populations

Genetic subdivisions were detected with a maternal mtDNA marker and biparental nucleic DNA markers. Two main genetic groups and six local management units were defined. Each local population has a corresponding population history that shows potentially unique adaptations to each local habitat. Release programs and events for larva and sub-adults bred from eggs are carried out in some habitats by local governments, NGOs, or other groups for recovery of the population. In addition, some habitats need to introduce individuals from other habitats to maintain existence of this species, although recovery of environmental conditions is needed first to ensure their survival. In any case, the locality and genetic units for the origins of eggs and individuals should be taken into account during the planning of such breeding and rerelease programs, because there are genetically distinct populations in Japan. In terms of diversity, this northernmost population is relatively low. Therefore, fragmentation of habitats by human activities would likely lead to a rapid loss of genetic diversity and promptly increase the risk of extinction.

The Japanese Society for the Preservation of the Horseshoe Crab (*Nippon Kabutogani wo Mamoru Kai*) was established in 1978 for the conservation of this species in Japan, and has branch offices in several localities and prefectures. This society, some local NGOs, and other groups are actively engaging local people and are conducting conservation efforts such as field surveys, beach clean-ups, social education, and more. However, these conservation activities and strategies have lacked genetic information on Japanese populations. The environments surrounding the horseshoe crab in Japan are still under severe threat. We have to accelerate conservation actions to protect horseshoe crabs while taking their regional genetic characteristics.

Acknowledgements We thank the following institutions and individuals for their support and collection of samples: Hasegawa, S., director of Minamichita Beach Land aquarium, and Emeritus Professor Dr. Niwa, M., Fukuoka branch of *Nihon Kabutogani wo Mamoru Kai* (Japanese Society for the Preservation of the Horseshoe Crab); the Kujukushima Pearlsea Resort Umikirara Aquarium, the Marine World Umino-Nakamichi, Kyushu Environmental Evaluation Association, Environmental Bureau of Fukuoka City, Federation of Fisheries Cooperative Association of Kafuri, and Imari High School; and Kawakami, Y., Hisano, C., Hayashi, O., Takahashi, S., Hayakawa, O., Sakemi, R., Sugimoto, S., Aizu, M., Nishihara, S., Shiokawa, N., Harada, N., Ono, G., Teshima, T., Maeda, K., Wada, T., Okamura, T., Kotoh, S., Ohira, Y., Michiyama, A., Kido, Y., Shibata, K., Mansyo, M., and Hamada, M. This work was supported in part by the Global COE Program (Center of excellence for Asian conservation ecology as a basis of human-nature mutualism), MEXT, Japan, and JSPS KAKENHI Grant Number 23770090.

References

Allendorf AW, Luikart G (2007) Conservation and the genetics of populations. Blackwell Publishing, Oxford

Bandelt HJ, Forster P, Rohl A (1999) Median-joining networks for inferring intraspecific phylogenies. Mol Biol Evol 16:37–48

Frankham R, Ballou JD, Briscoe DA (2010) Introduction to conservation genetics, 2nd edn. Cambridge University Press, Cambridge

Harada N (2003) Long-term breeding of *Tachypleus tridentatus*, egg to adult. Kabutogani 23:16–17 (In Japanese)

Harada N (2009) Growth rate for larvae of horseshoe crab (*Tachypleus tridentatus*) from a field survey. Kabutoganis 29:33–34 (In Japanese)

Itow T (1993) Crisis in the Seto Inland Sea: the decimation of the horseshoe crab. EMECS Newsl 3:10–11

Kawahara D (1982) Investigations on ecology of horseshoe crab larvae. Aquabiology (*Kaiyou to Seibutsu*) 4:380–382. (In Japanese)

King TL, Eackles MS, Spidle AP et al (2005) Regional differentiation and sex biased dispersal among populations of the horseshoe crab *Limulus polyphemus*. Trans Am Fish Soc 134:441–465

Li Q-Z, Li Q, Liu J et al (2009) Isolation and characterization of microsatellite loci in the horseshoe crab (*Tachypleus tridentatus*). Conserv Genet 10:1879–1881

Ministry of the Environment, Japan (2000) Red list of threatened wildlife of Japan: arthropods. http://www.biodic.go.jp/J-IBIS.html. Accessed 10 Sept 2014. (In Japanese)

Ministry of the Environment, Japan (2001) 500 important wetlands in Japan, MOE, Tokyo. http://www.sizenken.biodic.go.jp/wetland/. Accessed 10 Sept 2014

Ministry of the Environment, Japan (2006) Threatened wildlife of Japan: red data book, vol 7, 2nd edn. Arthropods, Japan Wildlife Research Center, Tokyo

Nishida S, Koike H (2009) Genetic structure of Japanese populations of *Tachypleus tridentatus* by mtDNA AT-rich region sequence analysis. In: Tanacredi JT, Botton ML, Smith DR (eds) Biology and conservation of horseshoe crabs. Springer, New York, pp 183–196

Nishida S, Koike H (2010) Isolation and characterization of compound microsatellite loci in the tri-spine horseshoe crab (*Tachypleus tridentatus*). Conserv Genet Resour 2:31–33

Nishida S, Koike H (2011) Genetic population structure of *Tachypleus tridentatus* in Japan and attempt to comparative phylogenetic study with parasitic planarian species on Asian horseshoe crabs. In: Abstracts of the international workshop on the science and conservation of Asian Horseshoe Crabs, Hong Kong, China, 13–16 June 2011

Nishii H (ed) (1975) *Kabutogani-Jiten* (Encyclopedia of horseshoe crabs), Add. Rev. ed. Private Publishing, Kasaoka. (In Japanese)

Pritchard JK, Stephens M, Donnelly P (2000) Inference of population structure using multilocus genotype data. Genetics 155:945–959

Sakemi R (1997) Observations of larvae behavior in horseshoe crabs at Imari Bay. Kabutogani 17:30–31 (In Japanese)

Sekiguchi K (1988) Biology of horseshoe crabs. Science House Co., Tokyo

Sekiguchi K, Shuster CN Jr (2009) Limits on global distribution of horseshoe crabs (Limulacea): lessons learned from two lifetimes of observations: Asia and America. In: Tanacredi JT, Botton ML, Smith DR (eds) Biology and conservation of horseshoe crabs. Springer, New York, pp 5–24

Sugawara K, Yonekawa H, Tagashima Y et al (1988) X-4 mitochondrial DNA polymorphisms. In: Sekiguchi K (ed) Biology of horseshoe crabs. Science House Co., Tokyo, pp 375–382

Weng ZH, Xie YJ, Xiao ZQ et al (2013) Microsatellite and mitochondrial DNA analysis of the genetic structure of Chinese horseshoe crab (*Tachypleus tridentatus*) in southeast China coast. Afr J Biotechnol 12:2088–2099

Yang M-C, Chen CA, Hsieh H-L et al (2007) Population subdivision of the tri-spine horseshoe crab, *Tachypleus tridentatus*, in Taiwan Strait. Zool Sci 24:219–224

Yang M-C, Chen C-P, Hsieh H-L et al (2009) Phylogeography, demographic history, and reserves network of horseshoe crab, *Tachypleus tridentatus*, in the South and East China seaboards. In: Tanacredi JT, Botton ML, Smith DR (eds) Biology and conservation of horseshoe crabs. Springer, New York, pp 163–181

Part II
Ecology and Population Dynamics

Chapter 7
Preliminary Home Range Study of Juvenile Chinese Horseshoe Crabs, *Tachypleus tridentatus* (Xiphosura), Using Passive Tracking Methods

Billy K.Y. Kwan, Paul K.S. Shin, and Siu Gin Cheung

Abstract Three passive tracking methods, using colored plastic tag (CPT), passive integrated transponder (PIT) tag, and CPT plus PIT were compared to track the movements of juvenile Chinese horseshoe crabs, *Tachypleus tridentatus*, in summer 2013 on a mudflat in Hong Kong. The study site was re-visited within 2 months after the release of the tagged individuals on ten occasions during low tides. The cumulative recovery rates of the tagged juveniles ranged from 70 to 82 %, with the number of position fixes where each individual was recovered during the repeated field visits varying from one to seven. Based on the adaptive kernel density estimation approach, the mean home range size (indicated by 95 % utilization distribution) tracked by different methods varied between 269 and 462 m^2. While the effects of tracking methods on home range size of the juveniles were statistically indistinguishable, PIT tags could be used as a longer-term tracking device for juvenile horseshoe crabs because the tag injected into animal bodies did not cause noticeable effects on survival, growth and molting. Within the body size range of 7th–10th instar juveniles (prosomal width 31–59 mm) used in the present study, no significant correlation was found between body size and home range area. From the analysis of utilization distribution, the juveniles were noted to exhibit residential and nomadic movement patterns, which might be associated with the distribution of food patches on the mudflat. The restricted movements of juveniles in this study suggest potential for high susceptibility to localized habitat disturbance.

Keywords Juvenile *Tachypleus tridentatus* • Home range area • Passive integrated transponder tag • Minimum convex polygon • Kernel density estimation

---

B.K.Y. Kwan
Department of Biology and Chemistry, City University of Hong Kong, Hong Kong, Hong Kong

P.K.S. Shin • S.G. Cheung (✉)
Department of Biology and Chemistry, City University of Hong Kong, Kowloon, Hong Kong

State Key Laboratory in Marine Pollution, City University of Hong Kong, Kowloon, Hong Kong
e-mail: bhsgche@cityu.edu.hk

R.H. Carmichael et al. (eds.), *Changing Global Perspectives on Horseshoe Crab Biology, Conservation and Management*, DOI 10.1007/978-3-319-19542-1_7

7.1 Introduction

Determining the movement pattern and space utilization of an animal is fundamental to the understanding of its biology and ecology (Scopel et al. 2009). The concept of home range was first defined by Burt (1943) as "area traversed by an individual in its normal activities of food gathering, mating and caring for its young". In fact, most animals share the common characteristic to restrict their movements to specific home ranges (White and Garrot 1990), and these small-scale movement steps are the resultant patterns of their interactions with the environment, including food availability (Davies 1978), habitat requirements (Murai et al. 1982), population density (Mackay and Underwood 1977), territoriality (Wada 1993), predation pressure and reproductive activities (Boydston et al. 2003). Many studies also demonstrated that sex (Secor and Piccoli 1996), body size (Shepherd and Clarkson 2001) and life cycle (McCormick and Makey 1997) determine the extent of home range of an animal.

Though home range behavior is an essential element of the biology of an animal, few studies have estimated the home range of marine invertebrates (Pittman and McAlpine 2003). A recent review by Laver and Kelly (2008) summarized 141 home range studies for all groups of animals published in 2004–2006, in which only 1 % of the published papers reported the home range data of arthropods, not to mention marine arthropods. Though updates of home range studies published in 2007–2014 are lacking, our preliminary search using the Institute for Scientific Information Web of Science with the search word "home range" in title similarly found that about 1 % of the relevant literature (out of 520 studies) was related to marine invertebrates. Even among these marine invertebrate studies, most were focused on species with commercial importance, especially lobsters and swimming crabs, in terms of their temporal and spatial dynamics (Kelly 2001), dispersal pattern (Bowlby et al. 2007) and mass migration behavior (Herrnkind and McLean 1971) with the final goals of using such home range data for stock enhancement, habitat improvement and sustainable fisheries resources management. Albeit limited, other studies investigated the biotic and abiotic factors contributing to the homing behaviors and home range sizes of fiddler crabs (Murai et al. 1982; Wada 1993), limpets (Mackay and Underwood 1977) and hermit crabs (Hazlett 1983) in relation to their territorial or feeding patterns. Data on home range and habitat use of marine animals have increasing implications for determining the extent of marine reserves or protected areas to rebuild declining populations due to over harvest and habitat degradation, such as for lobsters (Moland et al. 2011) and fishes (Kramer and Chapman 1999; Lowe et al. 2003).

Home range data would be useful to guide management of the Chinese horseshoe crab (*Tachypleus tridentatus*), an ancient marine arthropod for which declining populations have been reported in many places such as China (Liao et al. 2001), Japan (Itow 1993), Hong Kong (Shin et al. 2009) and Taiwan (Chen et al. 2004). The exploitation of Chinese horseshoe crabs is generally associated with human consumption and biomedical use, despite habitat degradation and loss (Hsieh and Chen 2009; Hong 2011). Marine reserves for Chinese horseshoe crabs have been established in Kinmen, Taiwan (Chen et al. 2004) and coastal regions of southeastern China including Zhejiang, Fujian, Guangdong Provinces and Guangxi Zhuang Autonomous Region (Hong 2011) to provide no-take marine zones for protecting adult reproductive stock and sustainable larvae recruitment. Information on the movement patterns and space use of juvenile horseshoe crabs in their habitats, however, is currently poor and may preclude effective assessment of the designated reserve areas used by these animals.

Previous capture-recapture studies on adult American horseshoe crabs, *Limulus polyphemus*, have demonstrated that these animals can have a restricted movement range close to their original tagging location and tend to repeatedly return to the same beach to spawn in the same or subsequent seasons (Rudloe 1980; James-Pirri et al. 2005). More recent investigations using telemetry technologies also showed that year-round movements of some American horseshoe crab populations are likely to remain within the same estuary or bay (Watson et al. 2009; James-Pirri 2010; Schaller et al. 2010). Daily fine-scale movements were observed both in the field (Watson et al. 2009) and in the laboratory (Chabot et al. 2004, 2007), with clear evidences that adult *L. polyphemus* expressed both daily and tidal rhythms of locomotion governed by factors such as light (Watson et al. 2008), changes in water depth (Watson et al. 2008) and water temperature (Chabot and Watson 2010). For juvenile horseshoe crabs, information on their home ranges and movement patterns on their nursery shores is virtually unknown. A field observation by Rudloe (1981) indicated that juvenile *L. polyphemus* would migrate seawards with increasing body size and eventually move to shallow waters where they became adults, generally taking 10–15 years (approximately 16–18 molts) to reach maturity (Sekiguchi 1988). A more recent study indicated that similar to the adults, juvenile horseshoe crabs can express both daily and tidal patterns of locomotion, which are driven by endogenous clocks (Dubofsky et al. 2013).

Coincidental with the advances in tracking technologies, the concept of home range and associated models have been proposed (reviewed by Worton 1987; Larkin and Halkin 1994; Laver and Kelly 2008). Despite the area traveled within a specific time, which can be estimated by the minimum convex polygon (MCP) home range estimation (Mohr 1947); Worton (1987) pointed out that most animals do not utilize their whole home range at the same intensity; certain areas are used more heavily than others. This concept prompted the recent development in home range estimators, in which kernel density estimation (KDE) has become an influential tool to estimate utilization distribution (UD) of an animal (Worton 1989). While MCP and KDE are the two most common home range estimators in the literature, many ecological and statistical studies have made suggestions to optimize their calculations, especially in the selection of the smoothing parameter in KDE (Worton 1995; Horne and Garton 2006). For tracking methods, fixed array ultrasonic and radio-telemetry technologies have been widely applied in home range studies in turtles (Luschi et al. 2003), fishes (Pine et al. 2003), birds (Nichols and Kaiser 1999) and mammals (Harris et al. 1990), in which the equipment sends out signals that can be detected by a receiver to record the position of the tagged animal. Apart from the above active tracking devices, simple plastic and passive integrated transponder (PIT) tags can also be deployed, especially for obtaining preliminary or coarse-scale information on home range and habitat use of the tagged animals (Moland et al. 2011). In the PIT method, a scanner is required to reveal the unique identifier of the tag inserted inside the body of the tracked animal (Cucherousset et al. 2009).

In this study, we estimated the home range dynamics for 7th–10th instar juvenile Chinese horseshoe crabs on an intertidal mudflat within a 2-month survey period, using three passive tracking methods: colored plastic tag (CPT), passive integrated transponder (PIT) tag, and CPT plus PIT. The aim of this research was to assess the applicability of these methodologies, especially the PIT tracking system, in home range studies to enhance data collection on the movement patterns and space utilization of juvenile horseshoe crabs in the field.

7.2 Materials and Methods

7.2.1 Study Site

San Tau mudflat (22°17′N, 113°55′E) is situated on the northern coast of Lantau Island (Fig. 7.1), under direct influence from the Pearl River. The shore is opposite to largely reclaimed Hong Kong International Airport, separated by a narrow (<1 km) waterway. The site was designated as a Site of Special Scientific Interest in 1994 for its ecologically important seagrass *Halophila ovalis* and *Zostera japonica* beds, occupying a narrow belt of intertidal mudflat at about 1.3–1.5 m above Chart Datum (Lee 1997; Chiu and Morton 2003). The shore is largely sheltered from strong tidal actions, with about 3.6 ha of sandy-mudflat exposed at low tides (Li 2008), supporting the growth of dwarf mangroves *Kandelia obovata*, *Aegiceras corniculatum*, *Bruguiera gymnorrhiza* and *Avicennia marina*, and habitats for scattered oyster *Saccostrea cucullata* beds, fiddler crabs, and juvenile horseshoe crabs.

7.2.2 Tagging Procedures

A total of 86 7th–10th instar juvenile Chinese horseshoe crabs, *Tachypleus tridentatus*, (prosomal width 44.3 ± 6.9 mm) were collected at the study site on 19 June 2013. The prosomal carapace of juveniles was individually labeled with a waterproof red-colored plastic tag (CPT; size: 0.25 cm^2; Dymo LetraTag®; USA) with a unique number using superglue. The CPT has been successfully deployed in a previous study (Morton and Lee 2011), and our preliminary laboratory studies showed that the tag can remain on the individual for at least 3 months (or until the next molt) without noticeable effects on survival and growth.

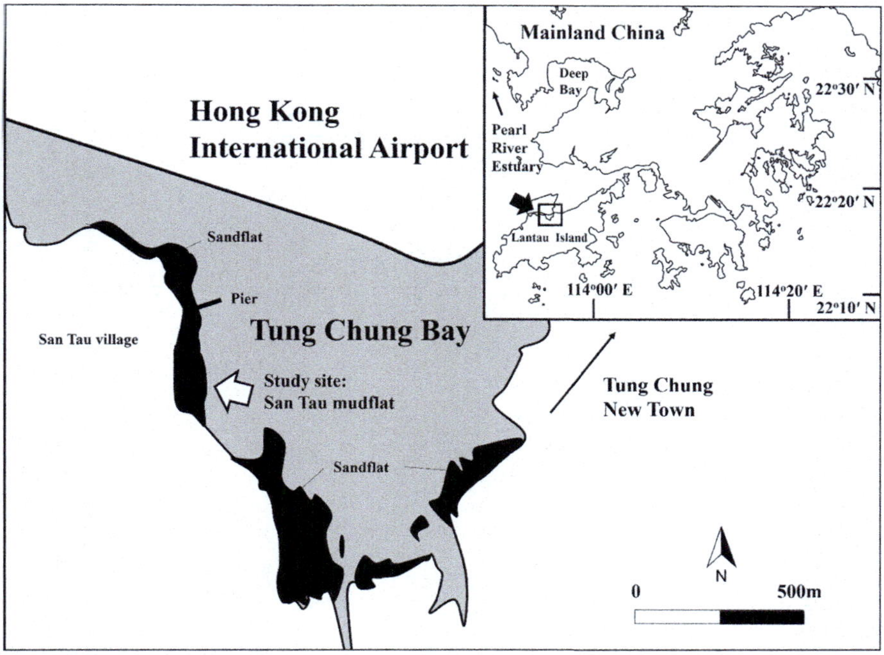

Fig. 7.1 A map of Hong Kong showing location of the study site, San Tau mudflat

To explore a longer-term tagging method on juvenile horseshoe crabs, a passive integrated transponder tag (PIT; length: 9 mm, 0.076 mg in air, 134.2 kHz, Biomark®; USA) was inserted into the ventral subfrontal area of 30 juveniles (prosomal width 42.2 ± 5.0 mm) collected at the same site using a syringe-style implanter (model: MK10, Biomark®; USA) equipped with a sterile 6 gauge 2-in. replacement needle (model: N206, Biomark®; USA). The ventral subfrontal area of each individual was wiped with 70 % ethanol before and after the tag injection to avoid sepsis. While a needle scar was left after the PIT tag injection, the effects of the tag on survival, growth, and molting were not affected (Liao 2011). Field studies conducted in 2010 and 2012 also demonstrated that all recovered juveniles had positive growth rates and molted at least once after release (Table 7.1).

Another 34 juveniles (prosomal width 42.1 ± 4.5 mm) were collected and labeled with red CPT plus PIT tags. All these 7th–10th instar juveniles were released on the same day within 1 m^2 around the location point at 22°17′18.0″ N, 113°55′20.3″ E. This release location was chosen because many individuals were spotted in that area during routine population surveys (unpublished data) and it was not far away (<5 m) from seagrass beds, dwarf mangroves and a freshwater stream, where the juveniles could choose their preferred homing locations.

Table 7.1 Effects of PIT tagging on the growth rate (mm/day), specific growth rate in terms of prosomal width (SGR_{PW}, %/day), and number molting (out of 16 individuals) of juvenile *T. tridentatus* at nursery sites in Hong Kong

Release date	PW at release (mm)	Instar stage at release	Number of days elapsed between release and recovery (d)	PW at recovery (mm)	Instar stage at recovery	Growth rate (mm/day)	SGR_{PW} (%/d)	Number of molting
17-9-2010	44.36	8	260	84.13	10	0.16	0.24	2
17-9-2010	42.67	8	260	78.69	10	0.16	0.24	2
17-9-2010	37.55	7	260	68.61	9	0.15	0.26	2
17-9-2010	45.00	8	243	76.74	10	0.13	0.22	2
17-9-2010	51.89	8	243	74.76	10	0.09	0.15	2
17-9-2010	42.42	8	243	60.87	9	0.08	0.15	1
17-9-2010	44.77	8	243	63.19	9	0.08	0.14	1
17-9-2010	57.48	9	243	80.43	10	0.09	0.14	1
17-9-2010	39.86	7	77	50.00	8	0.13	0.29	1
14-7-2012	36.30	7	58	47.10	8	0.19	0.45	1
14-7-2012	33.00	6	58	42.00	7	0.16	0.42	1
14-7-2012	38.55	7	58	53.10	8	0.25	0.55	1
14-7-2012	28.30	6	87	38.40	7	0.12	0.35	1
14-7-2012	44.90	8	87	62.00	9	0.20	0.37	1
14-7-2012	46.80	8	101	61.70	9	0.15	0.27	1
14-7-2012	36.80	7	101	44.80	8	0.08	0.19	1

Growth rate was calculated as (PW at recovery − PW at release)/days elapsed between release and recovery, whereas SGR_{PW} was determined as $\{[\ln(\text{PW at recovery}) - \ln(\text{PW at release})] / \text{days elapsed between release and recovery}\} \times 100$ (Hu et al. 2013)

7.2.3 Tracking Procedures

The study site was re-visited on ten occasions during summer low tides (22 June–22 August 2013), within 2 months following release. This sampling period was chosen because juvenile horseshoe crabs have been reported to actively feed on the sandy mudflats in summer (June to August) during low tides after hibernation in the winter and early spring (Chiu and Morton 2004; Hu et al. 2009). A laboratory study on emerging behavior of juvenile *T. tridentatus* also indicated that no juvenile emerged from the sediment under simulated tidal cycles when water temperature was below 20 °C, whereas 23 % and 5 % of the juveniles emerged during simulated low and high tides respectively at 25–30 °C (Lee and Morton 2009). Hence, it was assumed in this study that we could recover the highest number of tagged juveniles during low tides in summer and that undetected crabs likely stayed in the same location but buried deeper in the sediment.

For each field visit, the same three trained researchers recovered tagged juveniles for 90 min in the pre-defined study area (approximately 1.4 ha) where juvenile horseshoe crabs were spotted according to observations of previous population surveys (BKY Kwan personal observations). Other areas within the study site are covered with gravel and coarse sand, which are not suitable nursery habitat for juvenile horseshoe crabs (Chiu and Morton 2003). The first field visit was 3 days after the release date (22 June 2013) on the premise that the juveniles had adapted to the field conditions and resumed normal daily activities. To verify this assumption, the location data of juveniles with CPT collected 3 days and 7 days after the release were compared for home range estimation by different models. Unpaired t-test showed that there was no significant difference ($p > 0.05$) in these two sets of data for the estimation of home range area.

For those juveniles that were recaptured with a needle scar, they were scanned through a portable tag reader connected to a waterproof racket antenna (Model 2001 F-ISO; Destron Fearing™; USA) to obtain the unique identifier of PIT tagged individuals. The maximum detection distance of the racket antenna used in this study ranged from 10 to 15 cm in the air, therefore only tagged juveniles that emerged from the sediment could be detected. The location where a tagged juvenile was recovered during each field visit was considered as an independent position fix and recorded using a hand-held Global Positioning System device (GPSmap 76S; Garmin; USA). Human disturbance (not associated with this study) to the tagged juveniles was considered negligible because of the remoteness of the study site, which is about 3 km from the nearest residential area and not publically accessible. The fishermen from San Tau village may collect clams and oysters from the study site, but generally at much lower intertidal shore (about 0–0.7 m above Chart Datum) where the density of juvenile horseshoe crabs is generally low (Li 2008).

7.2.4 *Home Range Calculations*

Home range size was estimated from the locations of tagged individuals recovered with at least four positional fixes, using the minimum convex polygon (MCP, Mohr 1947) home range estimation and adaptive kernel density estimation (KDE) of the utilization distribution (UD, Worton 1989). For the most accurate home range determination, at least 30 possible position fixes of each recovered individual are recommended (Seaman et al. 1999). Although relatively limited, the data set used in this study provides a preliminary estimation of home range for the juveniles at nursery sites. The MCP method is the calculation of an enclosed convex polygon area by connecting all the location data points (Mohr 1947), while the KDE is the calculation of probability density that generates a UD of an animal within a given area (Worton 1989). In this study, home range area was defined as the smallest area containing 95 % and 50 % of the estimated utilization distribution (UD95 and UD50) (Scopel et al. 2009; Moland et al. 2011). To determine the degree of smoothing along the perimeter of each home range area, the KDE method requires the selection of a smoothing parameter. Likelihood Cross-Validation (CVh) was used to choose the smoothing parameter for UD95, which was reported to perform better than Least Squares Cross-Validation (LSCVh) in terms of surface fit and variability, especially in this case where the sample size was less than 50 (Horne and Garton 2006). The smoothing parameter for each individual was determined using the Animal Space Use software program (Beta version 1.3, Horne and Garton 2009). The location data in latitude and longitude were converted to Universal Transverse Mercator (UTM) coordinate system with easting and northing coordinates before smoothing parameter and home range calculations using the CALHOME software package (version 1.0, Kie et al. 1996).

7.2.5 *Statistical Analysis*

Data were first examined for normality and homogeneity of variance using Shapiro-Wilk and Levene's tests, respectively. If the above tests failed, data were subjected to transformation. One-way analysis of variance (ANOVA) was used to test the differences of recovery rates of released individuals and home ranges among three different tracking methods. Two-way ANOVA (Tracking method [fixed] × Home range model [fixed]) was employed to assess home range sizes of the juveniles. In the case that a significant difference was detected for these factors but not for their interaction, Tukey post-hoc tests were conducted for multiple comparisons, if necessary. Pearson product-moment correlation was applied to determine if changes in home range areas coincided with differences in prosomal width of the experimental juveniles. All statistical analyses were conducted using SPSS software version 16.0 (SPSS, Inc., Chicago, IL, USA).

7.3 Results

Preliminary field studies conducted in 2010 and 2012 to investigate the effects of PIT tagging on the growth and survival of juveniles showed that of the 16 individuals recovered, their prosomal width growth rate ranged from 0.08 to 0.25 mm/day, and the corresponding specific growth rate ranged from 0.14 to 0.55 %/day (Table 7.1). While most recovered individuals molted one time, five tagged individuals experienced two molting events from 8th to 10th instar (except one individual from 7th to 9th instar) after 243–260 days following release.

During the 2-month field visits, 69.8 % CPT, 76.7 % PIT and 82.4 % CPT plus PIT tagged juveniles were successfully recovered at least once (Table 7.2). The mean recovery rates for tracking methods CPT, PIT and CPT plus PIT were 12.7 %, 19.3 % and 22.1 % per visit, with no significant differences among tagging methods (ANOVA, $F=2.552$, $p>0.05$). The positional fixes for each individual recorded during repeated field visits varied from 1 to 7, with a mean of 2.3 (Fig. 7.2).

The resulting home range sizes were estimated from those tagged individuals with at least four positional fixes using MCP home range estimation and KDE of the utilization distribution. The overall home range area estimated by UD95 using different tracking methods ranged from 11 to 1,126 m^2, with a mean range of 269–462 m^2 (Table 7.2). The overall UD50 home range area, which represented the core area of activity, had a mean range of 37–73 m^2. Similar to UD95, MCP home range area for tracked juveniles was from 9 to 1,161 m^2, with the mean MCP area range of 131–354 m^2 (Table 7.2). One-way ANOVA concluded that the tracking methods, including CPT ($n=9$), PIT ($n=6$) and CPT plus PIT ($n=9$), produced statistically similar estimates ($p>0.05$) of the home ranges of juveniles in terms of UD95, UD50 and MCP (Table 7.2). The tracking method, home range model and their interaction also had no significant effect on the home range size of the juveniles (ANOVA, Tracking method: $F=2.23$, $p>0.05$, Model: $F=0.62$, $p>0.05$, Tracking method × Model: $F=0.06$, $p>0.05$) (Fig. 7.3).

Estimated home range areas showed the greatest overlap near the release point (Fig. 7.4). By plotting the utilization distribution areas of each tagged individual using KDE (UD95 and UD50), two general movement patterns were identified (Fig. 7.5), including residential (i.e., an individual was recaptured several times at

Table 7.2 Recovery rates and home range sizes (mean ± SEM) for three tracking methods of juvenile *T. tridentatus* during 22 June to 22 August 2013

Tracking method	Recovery rate (%/visit)	Cumulative recovery rate (%)	Area within UD95 (m^2)	Area within UD50 (m^2)	Area within MCP (m^2)
CPT	12.7 ± 2.4	69.8	462.0 ± 147.9	73.4 ± 17.5	354.1 ± 139.7
PIT	19.3 ± 3.6	76.7	295.1 ± 156.4	36.6 ± 11.5	325.7 ± 104.4
CPT + PIT	22.1 ± 4.9	82.4	269.1 ± 91.0	45.0 ± 16.6	131.1 ± 60.9

Data on recovery rates were transformed using arcsine square root function prior to ANOVA. No statistical differences were detected among tracking methods ($p>0.05$)

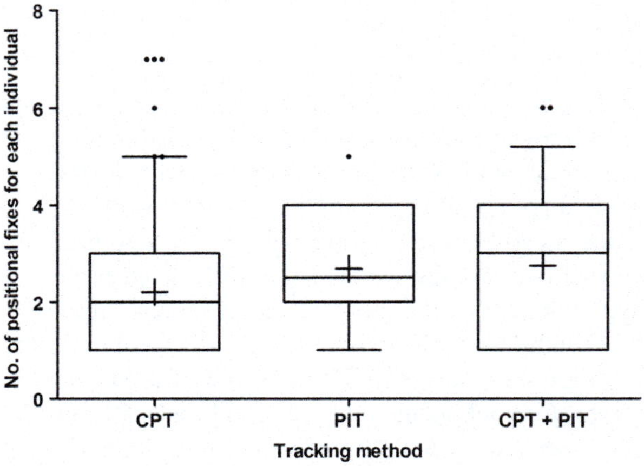

Fig. 7.2 *Box* and *whisker* plots of the number of position fixes recovered from different tracking methods during the ten site visits after the release of the tagged juveniles. The *box* displays 25–75 % and *whisker* 10–90 % data range. The *line* in the *box* represents the median value, while the cross symbol (+) indicates the mean value. Outliers are represented by a *dot* (•) beyond the plot limit. No statistical differences were detected among the three tracking methods (one-way ANOVA, $p > 0.05$)

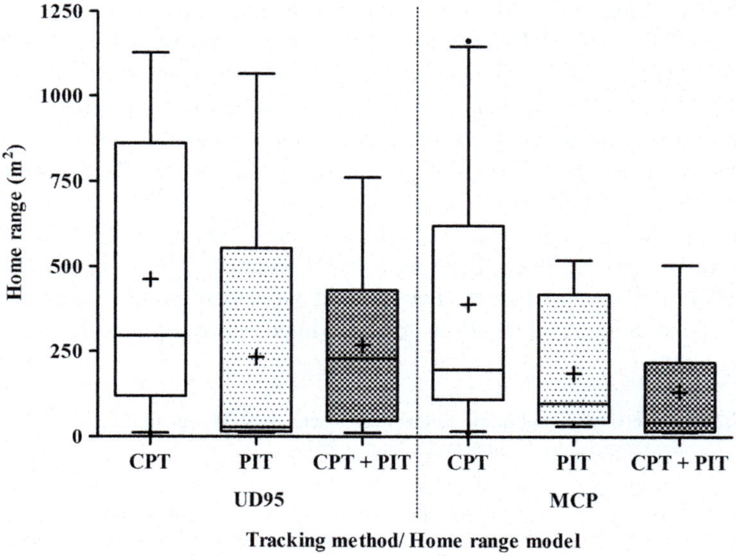

Fig. 7.3 *Box* and *whisker* plots (refer to Fig. 7.2 for explanation of symbols) of the home range sizes of juvenile *T. tridentatus* estimated from 95 % utilization distribution (UD95) and minimum convex polygon (MCP) models for different tracking methods. No statistical differences were detected among tracking methods and between home range models (two-way ANOVA, $p > 0.05$)

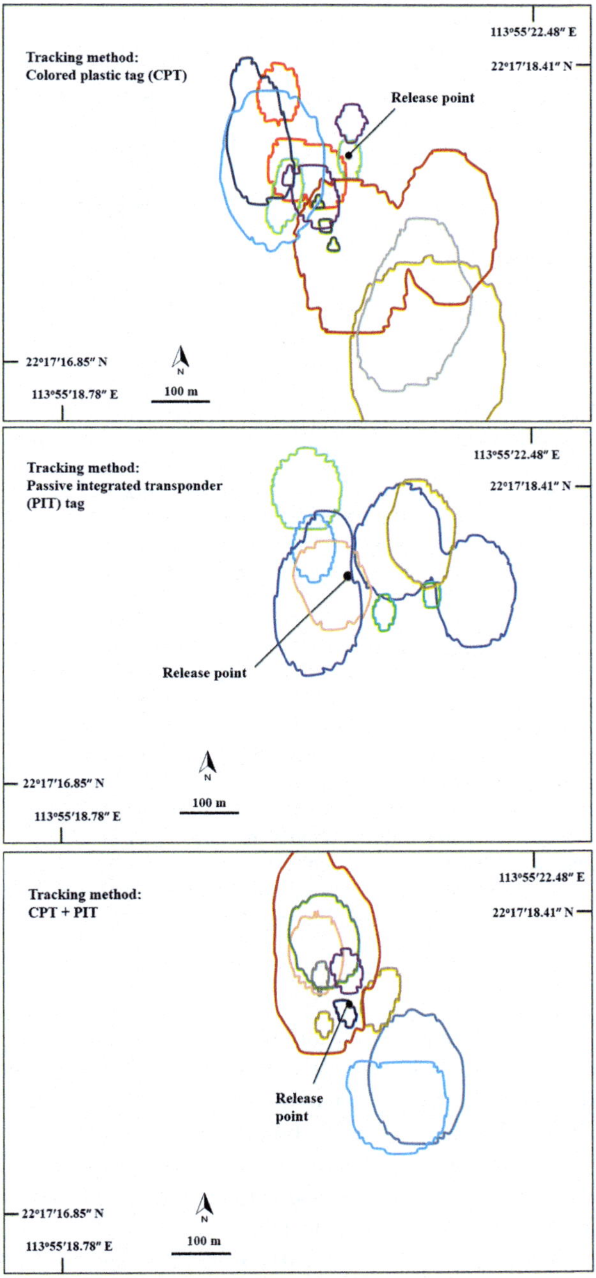

Fig. 7.4 UD95 home ranges (*solid circle lines*) exhibited by tagged individuals of juvenile *T. tridentatus* from different tracking methods at the study site during the observation period from 22 June to 22 August 2013. The *black dot* represents the release point of these tagged juveniles on 19 June 2013

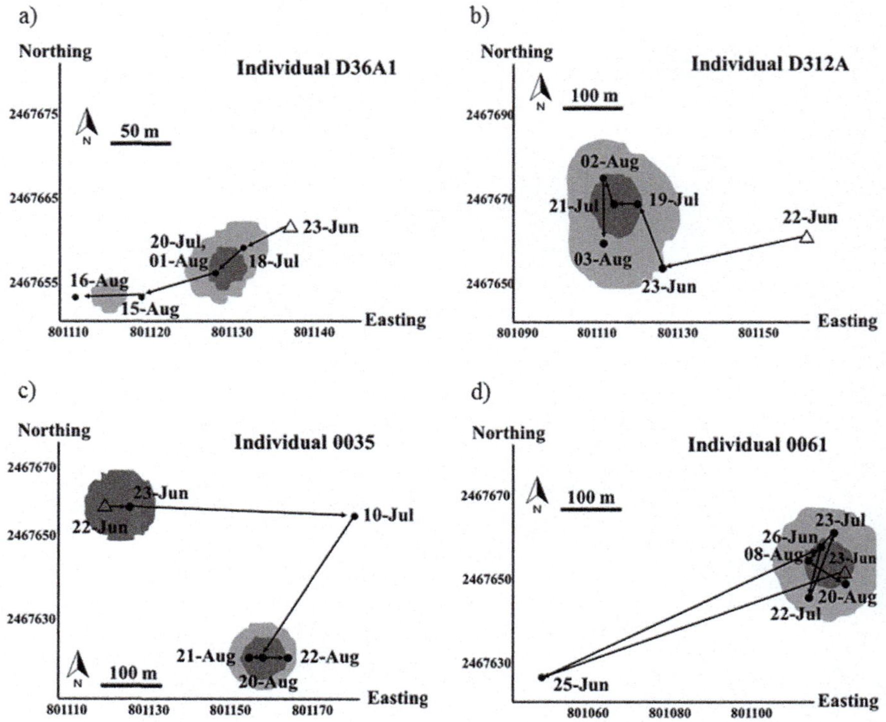

Fig. 7.5 Movement patterns (**a-d**) exhibited by four tagged individuals over the entire observation period from 22 June to 22 August 2013, which were identified by plotting the utilization distribution areas using kernel density estimation, UD95 (in *light grey*) and UD50 (in *dark grey*). The first location where the tagged juvenile was recovered is indicated by a *triangle* (△), while subsequent locations where the same juvenile was detected (with recovery date) are indicated by a *dot* (•). The *arrows* show the movement directions of the tagged juveniles. The locations are defined by the Universal Transverse Mercator (UTM) coordinate system

the same location) and nomadic (i.e., an individual was repeatedly recaptured at a new location during the study period) movements that subsequently established one, two, or more UD areas (Fig. 7.5a–c), as well as returning to the established UD area once the individual had left the UD boundary (Fig. 7.5d).

Within the size range (prosomal width 30.6–58.6 mm) of juveniles studied, there was no correlation between prosomal width and home range area estimated by different models in CPT (Pearson's correlation, df = 8, $r_{UD95} = 0.17$; $r_{MCP} = 0.15$, $p > 0.05$), PIT (df = 5, $r_{UD95} = 0.51$; $r_{MCP} = 0.55$, $p > 0.05$) and CPT plus PIT (df = 8, $r_{UD95} = 0.33$; $r_{MCP} = -0.22$, $p > 0.05$) individuals.

7.4 Discussion

This study is the first to document the use of passive tracking methods in the study of home range of juvenile Chinese horseshoe crabs in the field. The present findings showed no statistical difference in the recovery rates and home range size estimates from CPT, PIT, and CPT plus PIT juveniles, indicating the suitability of using these tags for tracking purposes. In particular, PIT tags recently have been applied in a wide range of animal species, including crustaceans (Pengilly and Watson 1994), fishes (Roussel et al. 2000), amphibians (Cucherousset et al. 2009), reptiles (Jemison et al. 1995; McDonald and Dutton 1996), birds (Elbin and Burger 1994), and mammals (Ball et al. 1991). Since the injected PIT tag is not known to cause the tracked juveniles any noticeable effects on survival, growth, or molting (Table 7.1), the use of PIT tags may allow for longer-term (>3 months) observations of animal movements especially in the case of home range determination, which generally requires a minimum of 30 positional fixes for calculations (Seaman et al. 1999). The PIT tracking system may also increase the efficacy to recover tagged juvenile horseshoe crabs by screening and detecting them on mudflat surfaces or even shallowly buried (<3 cm) in sediment (Lee and Morton 2009). Such portable PIT-antennae have been successfully used to locate tagged amphibians by moving the device above the water surface (Cucherousset et al. 2009) or to detect tagged fish below an ice-covered stream (Linnansaari et al. 2007).

The present 2-month tracking study only provides preliminary data on the home range sizes of juvenile horseshoe crabs. A longer monitoring period would be required to obtain more position fixes of the tagged juveniles and derive a more accurate estimate of space use. Some 30 %, 23 %, and 18 % of CPT, PIT, and CPT plus PIT juveniles respectively were unsuccessfully recovered after initial release (Table 7.2). Assuming that no juveniles had left the survey area, a relatively greater number of CPT horseshoe crabs were not detected during the 2-month study possibly as a result of molting. Some molts with CPT were observed on the mudflat during the study period. In rare cases observed in the laboratory, PIT tags were found to exit the body of juvenile horseshoe crabs during molting. Hence, tracking of an individual using both CPT and PIT may guarantee a higher recovery rate, in which the probability of tag loss (and loss of associated data) could be minimized.

The movement distance and patterns of an animal are the outcome of various interacting factors. The acquisition of benefits is balanced with the costs of such movement, associated with energetic expenditure, foraging time and predation risk (Hazlett 1983). During the 2-month study period, the overlapped home range areas near the release point (Fig. 7.4) might indicate restricted movements and the absence of territorial behavior in these juveniles. Such confined home ranges are further confirmed by the median values of home range sizes estimated from different tracking groups (Fig. 7.3), in which the medians were always lower than the mean values. The range of lower quartile data (10–25 %, bottom whisker) was also smaller than that of the upper quartile (75–90 %, top whisker). The high-frequency positions with short movement distance near the release point might reflect the availability of ample food sources near the release point and a lesser requirement for foraging farther away for food, and/or the high specific location fidelity similar to other intertidal invertebrates as discussed above. In addition, as there are oyster beds on the east and south of the study site, the oyster clusters (>5 per m^2) may act as obstacles that limited the foraging or dispersal activities and, in turn, the estimated home range size of the juveniles. A more detailed examination of the fine-scale movement patterns in Fig. 7.5 demonstrated that the juveniles continuously changed their location points with increasing observation time, which could be related to the search for food sources (or other normal drivers of dispersal through time) on the mudflat.

The present findings showed no correlation between size (31–59 mm prosomal width) of juvenile horseshoe crabs and their home ranges (UD95 and MCP). While it is generally accepted that terrestrial animals exhibit larger home range areas with increasing body size to acquire more food sources (e.g., McLoughlin and Ferguson 2000), there is no consensus on the relationship among locomotion activity, home range area, and body size in aquatic invertebrates. The available data for marine decapods suggests a poor correlation between home range scale and body size, such as in American lobsters (Scopel et al. 2009), European lobsters (Moland et al. 2011) and Caribbean spiny lobsters (Bertelsen and Hornbeck 2009). However, some studies report an inverse relationship between body size and range, indicating that smaller lobsters may move farther than larger sized animals (Kelly and MacDiarmid 2003; Linnane et al. 2005). Similarly, for intertidal invertebrates, some crabs can exhibit significant size-activity area correlations while others do not (Hazlett 1983; Wada 1993). For horseshoe crabs, in view of the relatively greater distance (4–34 km) traveled by tagged adult American horseshoe crabs (Shuster 1982; Thompson 1998), such body size and home range relationships remain to be further tested with a wider size range of juveniles or adults. In addition, a more robust home range determination of juvenile horseshoe crabs and the underlying reasons affecting their movements and space use, including food availability, population density, life cycle, and shoreline or other habitat characteristics such as sediment parameters, are worthwhile for future investigations.

Acknowledgements The work described in this paper was funded by the Ocean Park Conservation Foundation, Hong Kong. We thank S.M. Cheung for his advice on home range calculations and K.C. Siu and Virginia K.Y. Un for their assistance in the field. We also appreciate the constructive comments provided by two reviewers, which helped improve the content of this manuscript.

References

Ball DJ, Argentieri G, Krause R et al (1991) Evaluation of a microchip implant system used for animal identification in rats. Lab Anim Sci 41:185–186

Bertelsen RD, Hornbeck J (2009) Using acoustic tagging to determine adult spiny lobster (*Panulirus argus*) movement patterns in the Western Sambo Ecological Reserve (Florida, United States). NZ J Mar Freshwat 43:35–46

Bowlby HD, Hanson J, Hutchings JA (2007) Resident and dispersal behavior among individuals within a population of American lobster *Homarus americanus*. Mar Ecol Prog Ser 331:207–218

Boydston EE, Kapheim KM, Szykman M et al (2003) Individual variation in space use by female spotted hyenas. J Mammal 84:1006–1018

Burt WH (1943) Territoriality and home range concepts as applied to mammals. J Mammal 24:346–352

Chabot CC, Watson WH (2010) Circatidal rhythms of locomotion in the American horseshoe crab *Limulus polyphemus*: underlying mechanisms and cues that influence them. Curr Zool 56:499–517

Chabot CC, Kent J, Watson WH (2004) Circatidal and circadian rhythms of locomotion in *Limulus polyphemus*. Biol Bull 207:72–75

Chabot CC, Betournay SH, Braley NR et al (2007) Endogenous rhythms of locomotion in the American horseshoe crab, *Limulus polyphemus*. J Exp Mar Biol Ecol 345:79–89

Chen CP, Yeh HY, Lin PF (2004) Conservation of the horseshoe crab at Kinmen, Taiwan: strategies and practices. Biodivers Conserv 13:1889–1904

Chiu HM, Morton B (2003) The sediment and hydrographic characteristics of three horseshoe crab nursery beaches in Hong Kong. J Ocean Univ Qingdao 2:35–43

Chiu HM, Morton B (2004) The behaviour of juvenile horseshoe crabs, *Tachypleus tridentatus* (Xiphosura), on a nursery beach at Shui Hau Wan, Hong Kong. Hydrobiologia 523:29–35

Cucherousset J, Marty P, Pelozuelo L et al (2009) Portable PIT detector as a new tool for non-disruptively locating individually tagged amphibians in the field: a case study with Pyrenean brook salamanders (*Calotriton asper*). Wildl Res 35:780–787

Davies NB (1978) Territorial defence in the speckled wood butterfly (*Pararge aegeria*), the resident always wins. Anim Behav 26:138–147

Dubofsky EA, Simpson SD, Chabot CC et al (2013) Patterns of activity expressed by juvenile horseshoe crabs. Biol Bull 225:42–49

Elbin SB, Burger J (1994) In my experience: implantable microchips for individual identification in wild and captive populations. Wildl Soc B 22:677–683

Harris S, Cresswell WJ, Forde PG et al (1990) Home-range analysis using radio-tracking data–a review of problems and techniques particularly as applied to the study of mammals. Mamm Rev 20:97–123

Hazlett BA (1983) Daily movement in the hermit crabs *Clibanarius tricolor* and *Calcinus tibicen*. J Crustac Biol 3:223–234

Herrnkind WF, McLean R (1971) Field studies of homing, mass emigration and orientation in the spiny lobster, *Panulirus argus*. Ann NY Acad Sci 188:359–377

Hong S (2011) Feeding habits. In: Hong S (ed) Biology of horseshoe crabs, *Tachypleus tridentatus*. Xiamen University Press, Xiamen, pp 79–81

Horne JS, Garton EO (2006) Likelihood cross-validation versus least squares cross-validation for choosing the smoothing parameter in kernel home-range analysis. J Wildl Manag 70:641–648

Horne JS, Garton EO (2009) Animal space use 1.3. College of Natural Resources, University of Idaho, Idaho, USA

Hsieh HL, Chen CP (2009) Conservation program for the Asian horseshoe crab *Tachypleus tridentatus* in Taiwan: characterizing the microhabitat of nursery grounds and restoring spawning grounds. In: Tanacredi JT, Botton ML, Smith DR (eds) Biology and conservation of horseshoe crabs. Springer, New York, pp 417–438

Hu M, Wang Y, Chen Y et al (2009) Summer distribution and abundance of juvenile Chinese horseshoe crabs *Tachypleus tridentatus* along an intertidal zone in southern China. Aquat Biol 7:107–112

Hu M, Wang Y, Cheung SG et al (2013) Comparison of different frozen natural foods on survival and growth of juvenile Chinese horseshoe crab *Tachypleus tridentatus* (Leach, 1819): implications on laboratory culture. Aquacult Res 44:567–573

Itow T (1993) Crisis in the Seto Inland Sea: the decimation of the horseshoe crab. EMECS Newsl 3:10–11

James-Pirri MJ (2010) Seasonal movement of the American horseshoe crab *Limulus polyphemus* in a semi-enclosed bay on Cape Cod, Massachusetts (USA) as determined by acoustic telemetry. Curr Zool 56:575–586

James-Pirri MJ, Tuxbury K, Marino S et al (2005) Spawning densities, egg densities, size structure, and movement patterns of spawning horseshoe crabs, *Limulus polyphemus*, within four coastal embayments on Cape Cod, Massachusetts. Estuaries 28:296–313

Jemison SC, Bishop LA, May PG et al (1995) The impact of PIT-tags on growth and movement of the rattlesnake, *Sistrurus miliarus*. J Herpetol 29:129–132

Kelly S (2001) Temporal variation in the movement of the spiny lobster *Jasus edwardsii*. Mar Freshw Res 52:323–331

Kelly S, MacDiarmid AB (2003) Movement patterns of mature spiny lobsters, *Jasus edwardsii*, from a marine reserve. NZ Mar Freshw 37:149–158

Kie JG, Baldwin JA, Evans CJ (1996) CALHOME: a program for estimating animal home ranges. Wildl Soc B 24:342–344

Kramer DL, Chapman MR (1999) Implications of fish home range size and relocation for marine reserve function. Environ Biol Fish 55:65–79

Larkin RP, Halkin D (1994) A review of software packages for estimating animal home ranges. Wildl Soc B 22:274–287

Laver PN, Kelly MJ (2008) A critical review of home range studies. J Wildl Manag 72:290–298

Lee SY (1997) Annual cycle of biomass of a threatened population of the intertidal seagrass *Zostera japonica* in Hong Kong. Mar Biol 129:183–193

Lee CN, Morton B (2009) Emergence behavior of juvenile *Tachypleus tridentatus* under simulated tidal conditions in the laboratory and at two different sediment temperatures. In: Tanacredi JT, Botton ML, Smith DR (eds) Biology and conservation of horseshoe crabs. Springer, New York, pp 275–283

Li HY (2008) The conservation of horseshoe crabs in Hong Kong. MPhil thesis, City University of Hong Kong

Liao SH (2011) Tri-spine horseshoe crab culture and restock. In: Zhang ZW (ed) Epic of horseshoe crabs. National Museum of Marine Biology and Aquarium, Republic of China, pp 45–48. (in Chinese)

Liao YY, Hong SG, Li XM (2001) A survey on the horseshoe crabs in the north of South China Sea. Acta Zool Sin 47:108–111 (Chinese with English Abstract)

Linnane A, Dimmlich W, Ward T (2005) Movement patterns of the southern rock lobster, *Jasus edwardsii*, off South Australia. NZ J Mar Freshwat Res 39:335–346

Linnansaari T, Roussel JM, Cunjak RA et al (2007) Efficacy and accuracy of portable PIT-antennae when locating fish in ice-covered streams. Hydrobiologia 582:281–287

Lowe CG, Topping DT, Cartamil DP et al (2003) Movement patterns, home range, and habitat utilization of adult kelp bass *Paralabrax clathratus* in a temperate no-take marine reserve. Mar Ecol Prog Ser 256:205–216

Luschi P, Hays GC, Papi F (2003) A review of long-distance movements by marine turtles, and the possible role of ocean currents. Oikos 103:293–302

Mackay DA, Underwood AJ (1977) Experimental studies on homing in the intertidal patellid limpet *Cellana tramoserica* (Sowerby). Oecologia 30:215–237

McCormick MI, Makey LJ (1997) Post-settlement transition in coral reef fishes: overlooked complexity in niche shifts. Mar Ecol Prog Ser 153:247–257

McDonald DL, Dutton PH (1996) Use of PIT tags and photoidentification to revise remigration estimates of leatherback turtles (*Dermochelys coriacea*) nesting in St. Croix, U.S. Virgin Islands, 1979–1995. Chelonian Conserv Biol 2:148–152

McLoughlin PD, Ferguson SH (2000) A hierarchical pattern of limiting factors helps explain variation in home range size. Ecoscience 7:123–130

Mohr CO (1947) Table of equivalent populations of North American small mammals. Am Midl Nat 37:223–249

Moland E, Olsen EM, Andvord K et al (2011) Home range of European lobster (*Homarus gammarus*) in a marine reserve: implications for future reserve design. Can B Fish Aquat Sci 68:1197–1210

Morton B, Lee CN (2011) Spatial and temporal distributions of juvenile horseshoe crabs (Arthropoda: Chelicerate) approaching extirpation along the northwestern shoreline of the New Territories of Hong Kong SAR, China. J Nat Hist 45:227–251

Murai M, Goshima S, Nakasone Y (1982) Some behavioral characteristics related to food supply and soil texture of burrowing habitats observed on *Uca vocans vocans* and *U. lactea perplexa*. Mar Biol 66:191–197

Nichols JD, Kaiser A (1999) Quantitative studies of bird movement: a methodological review. Bird Study 46:S289–S298

Pengilly D, Watson LJ (1994) Automated detection of internally injected tags in red king crabs at crab processing facilities. Fish Res 19:293–300

Pine WE, Pollock KH, Hightower JE et al (2003) A review of tagging methods for estimating fish population size and components of mortality. Fisheries 28:10–23

Pittman SJ, McAlpine CA (2003) Movements of marine fish and decapod crustaceans: process, theory and application. Adv Mar Biol 44:205–294

Roussel JM, Haro A, Cunjak RA (2000) Field test of a new method for tracking small fishes in shallow rivers using passive integrated transponder (PIT) technology. Can B Fish Aquat Sci 57:1326–1329

Rudloe A (1980) The breeding behavior and patterns of movement of horseshoe crabs, *Limulus polyphemus*, in the vicinity of breeding beaches in Apalachee Bay, Florida. Estuaries 3:177–183

Rudloe A (1981) Aspect of the biology of juvenile horseshoe crabs, *Limulus polyphemus*. B Mar Sci 31:125–133

Schaller SY, Chabot CC, Watson WH III (2010) Seasonal movements of American horseshoe crabs *Limulus polyphemus* in the Great Bay Estuary, New Hampshire (USA). Curr Zool 56:587–598

Scopel DA, Golet WJ, Watson WH III (2009) Home range dynamics of the American lobster, *Homarus americanus*. Mar Freshw Behav Physiol 42:63–80

Seaman DE, Millspaugh JJ, Kernohan BJ et al (1999) Effects of sample size on kernel home range estimates. J Wildl Manag 63:739–747

Secor DH, Piccoli PM (1996) Age-and sex-dependent migrations of striped bass in the Hudson River as determined by chemical microanalysis of otoliths. Estuaries 19:778–793

Sekiguchi K (1988) Biology of horseshoe crabs. Science House Co. Ltd., Tokyo

Shepherd SA, Clarkson PS (2001) Diet, feeding behaviour, activity and predation of the temperate blue-throated wrasse, *Notolabrus tetricus*. Mar Freshw Res 52:311–322

Shin PKS, Li HY, Cheung SG (2009) Horseshoe crabs in Hong Kong: human exploitation and current population status. In: Tanacredi JT, Botton ML, Smith DR (eds) Biology and conservation of horseshoe crabs. Springer, New York, pp 347–360

Shuster CN Jr (1982) A pictorial review of the natural history and ecology of the horseshoe crab *Limulus polyphemus*, with reference to other Limulidae. In: Bonaventura J, Bonaventura C, Tesh S (eds) Physiology and biology of horseshoe crabs: studies on normal and environmentally stressed animals-progress in clinical and biological research. Alan R. Liss Inc., New York, pp 1–52

Thompson M (1998) Assessments of the population biology and critical habitat for the horseshoe crab, *Limulus polyphemus*, in the South Atlantic Bight. MPhil thesis, University of Charleston

Wada K (1993) Territorial behavior, and sizes of home range and territory, in relation to sex and body size in *Ilyoplax pusilla* (Crustacea: Brachyura: Ocypodidae). Mar Biol 115:47–52

Watson WH, Bedford L, Chabot CC (2008) Dissociation between circadian rhythms of visual sensitivity and circatidal rhythms of locomotion in the American horseshoe crab *Limulus polyphemus*. Biol Bull 215:46–56

Watson WH III, Schaller SY, Chabot CC (2009) The relationship between small-and large-scale movements of horseshoe crabs in the Great Bay estuary and *Limulus* behavior in the laboratory. In: Tanacredi JT, Botton ML, Smith DR (eds) Biology and conservation of horseshoe crabs. Springer, New York, pp 131–147

White GW, Garrot RA (1990) Analysis of wildlife radiotracking data. Academic, San Diego

Worton BJ (1987) A review of models of home range for animal movement. Ecol Model 38:277–298

Worton BJ (1989) Kernel methods for estimating the utilization distribution in home-range studies. Ecology 70:164–168

Worton BJ (1995) Using Monte Carlo simulation to evaluate kernel–based home range estimators. J Wildl Manag 59:794–800

Chapter 8
Population Structure and Growth of Juvenile Horseshoe Crabs *Tachypleus tridentatus* and *Carcinoscorpius rotundicauda* (Xiphosura) in Southern China

Menghong Hu, Billy K.Y. Kwan, Youji Wang, Siu Gin Cheung, and Paul K.S. Shin

Abstract The abundance, population structure and growth of juvenile Asian horseshoe crabs *Tachypleus tridentatus* and *Carcinoscorpius rotundicauda* are rarely reported in the literature. This study investigated the density and growth allometry of these sympatric species on the intertidal areas of three spawning and nursery beaches, Xi Bei Ling (XBL), Jin Hai Wan (JHW) and Xi Chang (XC) along the coast of Beibu Gulf, southern China. While *T. tridentatus* and *C. rotundicauda* coexisted on JHW and XC beaches, only *T. tridentatus* was found at XBL. The density of *T. tridentatus* (0.9–3.2 individuals/100 m^2) was significantly higher than that of *C. rotundicauda* (0.1–1.7 individuals/100 m^2) at JHW, but densities were similar at XC. Juvenile *T. tridentatus* ranged in size (prosomal width) from 14 to 89 mm and were estimated to represent instars 4–12 and ages 2–8 years based on previous studies. *C. rotundicauda* juveniles were 16–75 mm in prosomal width and represented instars 5–15. The Hiatt growth model on the relationship between pre- and post-molt size data showed similar growth increments in juvenile *T. tridentatus* and *C. rotundicauda*. Growth rates for *T. tridentatus* varied from 4.4 to 19.7 mm/year,

---

Co-first authors: Menghong Hu and Billy K.Y. Kwan

M. Hu • Y. Wang
Department of Biology and Chemistry, City University of Hong Kong, Hong Kong, Hong Kong

College of Fisheries and Life Science, Shanghai Ocean University, Shanghai, China

B.K.Y. Kwan
Department of Biology and Chemistry, City University of Hong Kong, Hong Kong, Hong Kong

S.G. Cheung • P.K.S. Shin (✉)
Department of Biology and Chemistry, City University of Hong Kong, Kowloon, Hong Kong

State Key Laboratory in Marine Pollution, City University of Hong Kong, Kowloon, Hong Kong
e-mail: bhpshin@cityu.edu.hk

R.H. Carmichael et al. (eds.), *Changing Global Perspectives on Horseshoe Crab Biology, Conservation and Management*, DOI 10.1007/978-3-319-19542-1_8

depending on instar stage and site. Growth rates were not determined for *C. rotundicauda* due to lack of data. These data establish a baseline for density and growth allometry of juvenile *T. tridentatus* and *C. rotundicauda* horseshoe crabs in Asian waters and will support future monitoring and assessment of juvenile populations and habitat resources.

Keywords Population density • Growth allometry • Baseline • China

8.1 Introduction

Horseshoe crabs are evolutionarily important as one of the oldest marine living fossils, whose origin can be dated back to 445 million years ago (Rudkin and Young 2009). They play key roles in coastal ecosystems as a predator in the benthic food web, and their eggs serve as a major protein source for migratory shorebirds (Botton et al. 2003). A significant decline in populations of horseshoe crabs has been noted in many Asian habitats, including China (Liao et al. 2001), Hong Kong (Shin et al. 2009), Japan (Itow 1993), and Taiwan (Chen et al. 2004), due to exploitation for human consumption and biomedical use, along with habitat loss (Hsieh and Chen 2009; Hong 2011). Baseline information on the status of juvenile Asian horseshoe crabs, which are important to recruitment and future survival, in their natural spawning and nursery habitats is rarely reported (Carmichael et al. 2003).

In the life history of horseshoe crabs, adults generally spawn near the high-tide zone, while juveniles inhabit the adjacent intertidal areas and gradually migrate to the deeper subtidal zone prior to maturation (Rudloe 1979; Sekiguchi 1988; Chen et al. 2004). The restricted dispersal ability of juvenile horseshoe crabs thus leads to substantial population subdivisions on small geographic scales (Pierce et al. 2000; Botton and Loveland 2003; King et al. 2005). To manage the harvest of horseshoe crab populations, baseline data on the abundance, size, growth rate and mortality rate of juvenile horseshoe crabs on their spawning/nursery grounds are essential for estimating annual recruitment and predicting future populations in the wild. Such data can also be used as an indicator of the suitability of the nursery habitats since larval development is dependent on the environmental conditions of the shores where they live (Chen et al. 2004).

Growth of juvenile horseshoe crabs is similar to crustaceans, with the hard exoskeleton being periodically replaced in successive molts to allow for an increase in body size (Shuster 1982; Sekiguchi 1988; Carmichael et al. 2003). However, defining age classes as well as estimating growth within horseshoe crab populations can be difficult. At present, growth of horseshoe crabs is estimated based on the size increments between molts and the duration of the inter-molt period in laboratory-rearing studies (Sekiguchi 1988; Zadeh et al. 2009). In field studies, juvenile growth can be estimated directly from the sampling of cohorts or inferred indirectly from the analysis of the size-frequency distributions. By comparing the mean prosomal width of each instar stage using both direct and indirect cohort measurements,

Carmichael et al. (2003) demonstrated that the growth of juvenile American horseshoe crabs *Limulus polyphemus* in Pleasant Bay, USA was significantly faster than that of adults, while the growth rate for male and female adults was statistically indistinguishable. Using similar methodology for growth estimation of the Asian horseshoe crab species in their natural habitats, however, has not been reported.

The present study aimed to determine the abundance, population structure and growth of juvenile horseshoe crabs *T. tridentatus* and *C. rotundicauda* on different spawning and nursery beaches along Beibu Gulf, southern China. At present, Beibu Gulf is regarded as one of the most important habitats of *T. tridentatus* in Chinese waters (Liao and Li 2001). It is also noted that on some shores along Beibu Gulf, both *T. tridentatus* and *C. rotundicauda* juveniles coexist (Liao et al. 2001). Results will establish a current baseline for density and growth allometry of these two sympatric species in Asian waters that will be useful in understanding current and future distribution patterns at natural spawning and nursery habitats.

8.2 Materials and Methods

8.2.1 Study Beaches

The study was conducted within three nursery beaches along Beibu Gulf, Behai Municipality, Guangxi Province in southern China (Fig. 8.1). Many rivers flow into Beibu Gulf, including Fengjiajiang and Dafengjiang Rivers, forming an extensive estuarine ecosystem (Zhao 1999). Based on the preliminary survey data from earlier studies by the Guangxi Institute of Fisheries and Hu et al. (2009), three study sites were chosen: Xi Bei Ling (XBL), Jin Hai Wan (JHW) and Xi Chang (XC), which are fringed by mangroves with expanse of intertidal zone during low tides and can be considered as representative nursery beaches for juvenile horseshoe crabs. JHW is included as part of the Beihai Mangrove Reserve, while XBL and XC are fishing villages (Hu 2011). Although the Chinese horseshoe crab is listed as a Class II protected species by China and a key protected species in Guangxi Province, there is an estimated harvest of one million Chinese horseshoe crab adult pairs per year along Beibu Gulf shores for consumption and biomedical use due to illegal fishing and collection (Li and Hu 2011). The loss of breeding and nursery habitats resulting from urbanization along the coastline of Beihai Municipality may also contribute to the decline in horseshoe crab populations.

8.2.2 Juvenile Horseshoe Crab Measurements

Survey data were collected from May to June 2009, during daytime at low tides, when juveniles are known to emerge from the sand to feed (Lee and Morton 2009). To determine the abundance of juvenile horseshoe crabs on each beach, we used six

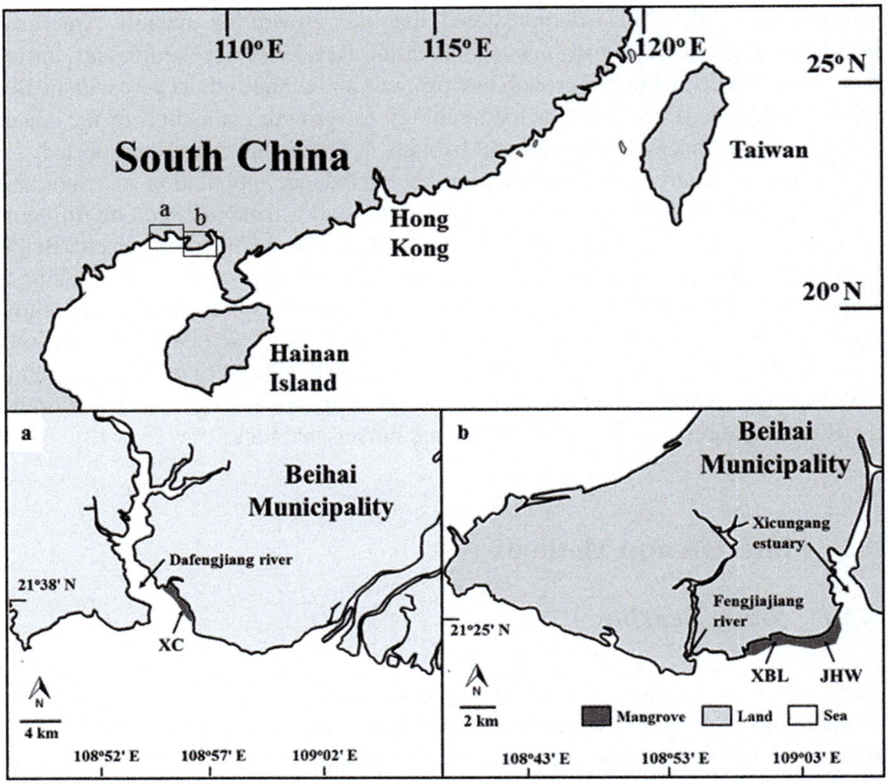

Fig. 8.1 Location of the study beaches in Beibu Gulf, southern China

horizontal transects (12 m width) set equally apart from tidal heights of 0.6–1.6 m above chart datum (i.e., the lowest astronomical tide as defined by the British Hydrographic Office). The length of transects was similar to the width of each beach (Table 8.1). To avoid disturbing the fringe of mangrove area and any unhatched horseshoe crab eggs, no attempts were made to sample the earliest instar (1st–3rd) stage juveniles that were expected above the 1.6 m tidal height, closest to nesting sites at the three nursery beaches. The top 5 cm of the sediment along each transect was searched using hand-held trowels by a team of six persons. This sampling depth was chosen based on laboratory observations that most juveniles burrow to a depth of less than 3 cm (Lee and Morton 2009). All found juveniles were collected from each transect. *T. tridentatus* was identified by the presence of three distinct immovable spines above the insertion of the telson (Mikkelsen 1988; Chiu and Morton 1999; Morton and Lee 2003), whereas *C. rotundicauda* was characterized by a telson that is relatively circular in cross section and smooth on the dorsal side (Mikkelsen 1988; Chiu and Morton 2003). The prosomal width of the juvenile specimens was also measured using a vernier caliper to the nearest 0.1 mm. For each study beach, the lowest shoreline transect (0.6 m above chart datum) was

Table 8.1 Mean density (± SE) of juvenile horseshoe crabs on three study beaches in southern China

	Shore width (m^2)	Total individuals	Total sampling area (100 m^2)	Mean density (ind/100 m^2)	CV (%)
T. tridentatus					
XBL	1,350	1,742	972	1.79 ± 0.16	9.13
JHW	1,200	2,753	864	3.19 ± 0.28	8.74
XC	700	441	504	0.88 ± 0.16	17.77
C. rotundicauda					
JHW	1,200	121	864	0.14 ± 0.02	13.77
XC	700	872	504	1.73 ± 0.14	8.18

Mean density is the average of population density (total individuals along each transect divided by sampling area) for each transect. *CV* coefficient of variation (standard error divided by mean × 100). Total sampling area = shore width × transect width × transect number
XBL Xi Bei Ling, *JHW* Jin Hai Wan, *XC* Xi Chang

sampled first, followed by the next higher shoreline transect, and so on. Each beach was sampled within 1 day to minimize the chance of recounting the same individuals.

8.2.3 Cohort Analysis

For each beach collection, prosomal width data from each juvenile horseshoe crab species were grouped into size classes of 1 mm intervals and a size-frequency histogram was plotted. Individual size cohorts were separated using the modal progression analysis software FiSAT II (FAO-ICLARM Fish Stock Assessment Tool, http://www.fao.org/fishery/topic/16072/en). FiSAT II applies the maximum likelihood concept of separating the normal distribution components of size-frequency samples, which allows for an accurate demarcation of the component cohorts from the composite polymodal population size-frequency distribution. The instar stage for juvenile cohort was defined using the mean prosomal width data from Seikiguchi (1988) and Yeh (1999) for *T. tridentatus* and from Sekiguchi (1988) and Zadeh et al. (2009) for *C. rotundicauda*.

8.2.4 Growth Analysis

Growth of each horseshoe crab species was evaluated by recording the size increment of each cohort collected sequentially, based on the data of cohort demarcation. Data from Sekiguchi (1988) on the mean duration of each instar for each species were applied to estimate the mean instar stages of the aggregated cohorts.

The relationship between post- and pre-molt size was used to compare the difference between the two horseshoe crab species, according to the Hiatt growth model (Hiatt 1948; Kurata 1962). The model describes growth as a linear equation: $L_{t+1} = aL_t + b$, where L_t is the pre-molt prosomal width at instar stage *t*, L_{t+1} is the post-molt prosomal width at instar stage *t +1*; *a* is the slope; and *b* is the y-intercept of the regression line. The slopes of the growth regression of the two species at the study beaches were compared using analysis of covariance (ANCOVA). All calculations on growth analysis were conducted by the software SPSS 16.0 (SPSS, Inc., Chicago, IL, USA). Differences were considered significant at $p<0.05$.

The estimated mean age of juvenile *T. tridentatus* was also determined by referring to the data in Sekiguchi (1988). Juvenile *T. tridentatus* growth rates were ascertained by comparing the prosomal width of instars with the estimated mean age of instars. Owing to the lack of information with respect to the estimated mean age of juvenile *C. rotundicauda*, growth rates of *C. rotundicauda* at different study beaches were not estimated in the present study.

8.2.5 Statistical Analysis

Data were first examined for normality and homogeneity of variance using Shapiro-Wilk and Levene's tests, respectively (Zar 1996). If the above tests failed, data were subjected to transformation. One-way analysis of variance (ANOVA) was used to test the differences between population densities for juvenile *T. tridentatus* on different study beaches. In the case that a significant difference was detected, Tukey post-hoc test was conducted for multiple comparisons. An unpaired t-test was used to compare the differences in population density between the two species at a particular beach or population density of juvenile *C. rotundicauda* between two different beaches. All statistical analyses were also conducted using SPSS 16.0.

8.3 Results

8.3.1 Abundance

While *T. tridentatus* and *C. rotundicauda* coexisted at JHW and XC, only *T. tridentatus* was found at XBL. The juvenile *T. tridentatus* densities among the three study beaches ranged from 0.88 to 3.19 individuals/100 m^2. Juvenile *C. rotundicauda* had a lower density of 0.14–1.73 individuals/100 m^2 (Table 8.1). The density of juvenile *T. tridentatus* at JHW was significantly higher than that at XC (ANOVA: $F=5.29$, $p=0.02$). For juvenile *C. rotundicauda*, their densities at JHW and XC were statistically indistinguishable (Unpaired t-test: $t=-2.03$, $p=0.07$). The density of juvenile *T. tridentatus* was significantly higher than that of juvenile *C. rotundicauda* at JHW (Unpaired t-test: $t=3.21$, $p<0.01$), whereas densities of the two species at XC were not statistically similar (Unpaired t-test: $t=-1.66$, $p=0.13$).

8.3.2 Population Structure

Among the three nursery beaches, juvenile *T. tridentatus* ranged in size (prosomal width) from 14.0 to 88.8 mm, whereas *C. rotundicauda* was smaller, from 15.5 to 75.3 mm (Table 8.2). Based on the size-frequency histograms and the data from Sekiguchi (1988), nine consecutive 4th–12th instar stages of juvenile *T. tridentatus* were identified at XBL, JHW and XC (Fig. 8.2). The population of *C. rotundicauda* in the JHW intertidal area was too small (N = 121) to define cohorts. Therefore, Fig. 8.3 only depicts the polymodal population structure based on the size-frequency histogram of juvenile *C. rotundicauda* collected at XC, in which 11 consecutive 5th–15th instar stages were identified.

The mean prosomal width of successive juvenile *T. tridentatus* cohorts showed a possible site difference (Table 8.2). Information on growth rates at various instar stages for *C. rotundicauda* was limited, and hence no comparison could be made between the present and previous studies (Table 8.2). Juveniles *T. tridentatus* at the 7th instar stage were in the majority at XBL, and at both JHW and XC juveniles *T. tridentatus* at 8th instar stage were the most abundant (Fig. 8.2). For juvenile *C. rotundicauda*, the 8th instar stage was most abundant at XC (Fig. 8.3).

Table 8.2 Instar stage and mean prosomal width (± SE) for juvenile cohorts defined for the three study beaches and in previous studies

	Mean prosomal width (mm)							
	T. tridentatus					*C. rotundicauda*		
Instar stage	Sekiguchi (1988)	Yeh (1999)	XBL	JHW	XC	Sekiguchi (1988)	Zadeh et al. (2009)	XC
1	6.0	5.8				5.0	3.7	
2	7.8	8.7				7.6	5.0	
3	10.7	11.8				10.2	7.0	
4	13.3	15.3	14.1 ± 0.5	14.0 ± 1.2	15.5 ± 1.0	12.4	9.4	
5	17.5	21.2	18.8 ± 1.0	19.7 ± 1.0	18.7 ± 0.6		12.6	15.5 ± 0.9
6	22.2		22.7 ± 1.2	23.8 ± 1.0	23.0 ± 0.5		17.1	20.8 ± 0.9
7	28.0		28.9 ± 1.5	30.5 ± 0.9	30.1 ± 0.5		23.4	27.2 ± 1.5
8	37.3		37.4 ± 1.7	39.6 ± 1.2	39.7 ± 1.0			30.9 ± 0.6
9	42.7		43.2 ± 1.3	45.0 ± 0.7	45.0 ± 0.9			35.9 ± 0.9
10	48.4		51.2 ± 2.0	50.8 ± 0.9	51.0 ± 0.5			40.6 ± 0.8
11	70.0		68.6 ± 3.6	70.5 ± 1.0	70.3 ± 0.6			44.5 ± 1.2
12	87.0		87.8 ± 1.4	88.1 ± 1.4	88.8 ± 0.8			50.5 ± 0.8
13	113.0							53.7 ± 0.6
14	147.0							61.8 ± 2.5
15								75.3 ± 3.2

Instar stages were estimated using data from Sekiguchi (1988) and Yeh (1999) for *T. tridentatus*, and Sekiguchi (1988) and Zadeh et al. (2009) for *C. rotundicauda*

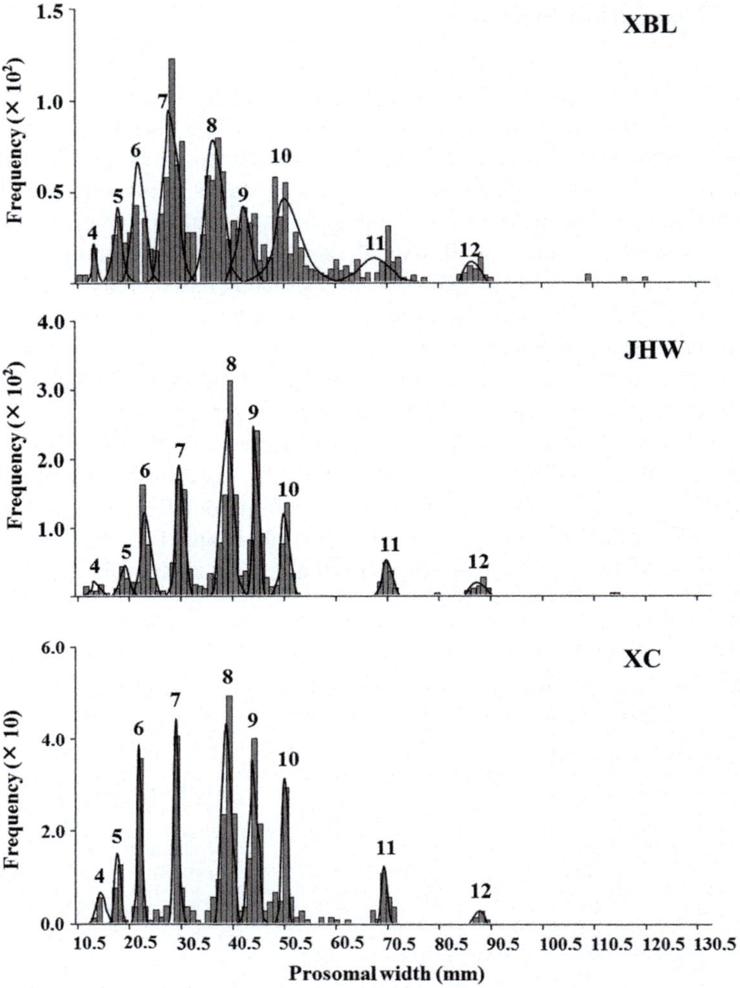

Fig. 8.2 Juvenile *T. tridentatus* cohorts based on size (prosomal width) frequency data on the three study beaches. Instar stage of the juveniles was based on information from Sekiguchi (1988). The numbers represent the identifiable size cohort shown in Table 8.2

8.3.3 Growth

The relationships between mean post- to pre-molt size among juveniles of the two species were compared by Hiatt growth statistics (Table 8.3). At the three study beaches and from data by Sekiguchi (1988), the slopes of growth regression for juvenile *T. tridentatus* were not different (ANCOVA: $F=1.13$, $p=0.36$). A similar growth pattern was observed for juvenile *C. rotundicauda* at XC and in data by Sekiguchi (1988) (ANCOVA: $F=0.09$, $p=0.77$). The slopes of the Hiatt growth

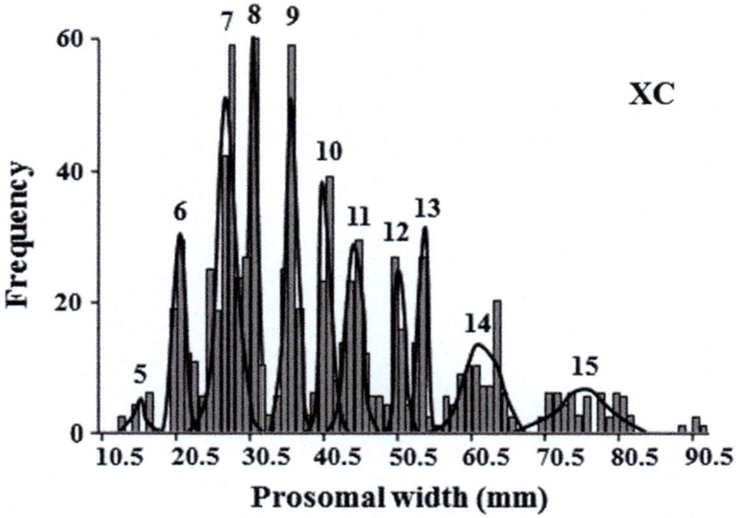

Fig. 8.3 Juvenile *C. rotundicauda* cohorts based on size (prosomal width) frequency data on XC nursery beach. Instar stage of the juveniles was based on information from Sekiguchi (1988). The numbers represent the identifiable size cohort shown in Table 8.2

Table 8.3 Relationship between mean post-molt (L_{t+1}) and pre-molt (L_t) prosomal widths of juvenile horseshoe crabs (Hiatt growth model) sampled for this study and in a previous study (Sekiguchi 1988)

	Equation	R^2
T. tridentatus		
XBL	$L_{t+1} = 1.33\ L_t - 1.86$	0.95
JHW	$L_{t+1} = 1.24L_t + 0.31$	0.97
XC	$L_{t+1} = 1.27L_t - 0.86$	0.98
Sekiguchi (1988)	$L_{t+1} = 1.12L_t + 2.30$	0.99
C. rotundicauda		
XC	$L_{t+1} = 1.11L_t + 1.87$	0.98
Sekiguchi (1988)	$L_{t+1} = 0.92\ L_t + 3.05$	0.99

equation for these two species were also statistically indistinguishable (ANCOVA: F=6.78, p=0.13).

The growth rates and ages of juvenile *T. tridentatus* at the three study beaches were estimated based on the aggregated data and comparison to data from Sekiguchi (1988) (Table 8.4). The relationships between the mean prosomal width and age of the juveniles are depicted in Fig. 8.4, with the slope of each line showing the approximate growth rate. For juvenile *T. tridentatus*, the growth rates were generally low, ranging 4.42–19.70 mm/year (Table 8.4). Growth rates of juvenile *T. tridentatus* were not different among the three beaches or from data by Sekiguchi (1988) (ANCOVA: F=0.07, p=0.98).

Table 8.4 Growth rates for estimated instars of juvenile *T. tridentatus* collected from three study sites in southern China (XBL, JHW, XC) compared to laboratory data from individuals of know age by Sekiguchi (1988)

Instar	Age (year) (Sekiguchi 1988)	Growth rate (prosomal width, mm/year)			
		Sekiguchi (1988)	XBL	JHW	XC
1	0.5	2.25	—	—	—
2	1.3	9.66	—	—	—
3	1.6	13.00	—	—	—
4	1.8	10.50	11.75	14.25	8.00
5	2.2	7.83	6.50	6.83	7.17
6	2.8	7.25	8.25	8.38	8.88
7	3.6	10.33	9.11	10.11	10.67
8	4.5	4.50	4.67	4.50	4.42
9	5.7	6.33	9.00	6.44	6.67
10	6.6	21.60	17.4	19.70	19.30
11	7.6	17.00	19.2	17.60	18.50

Data not collected = —

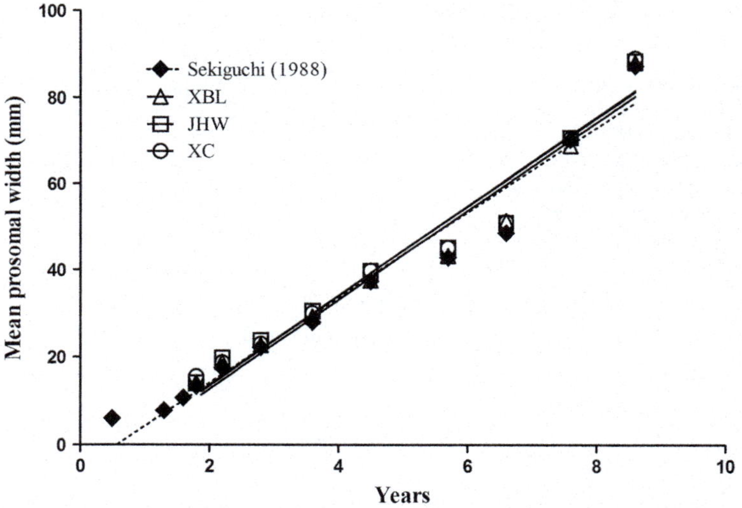

Fig. 8.4 Growth rates of juvenile *T. tridentatus*, measured as increase in mean prosomal width through time. Juvenile growth rates were determined based on size to age relationships in Table 8.4. Model II regression was used to derive equations for the best-fitted lines describing the relationships for juveniles in data from Sekiguchi (1988): $y=9.79x-5.58$, $R^2=0.96$, $F_{reg}=246.20$; juveniles at XBL: $y=10.22x-7.84$, $R^2=0.96$, $F_{reg}=151.30$; juveniles at JHW: $y=10.23x-6.89$, $R^2=0.96$, $F_{reg}=155.50$; juveniles at XC: $y=10.30x-7.22$, $R^2=0.96$, $F_{reg}=151.10$. All F_{reg} are significant to $p<0.001$

8.4 Discussion

Inter-beach differences in the abundance of juvenile horseshoe crabs were noted in this study. The higher density of juvenile *T. tridentatus* at JHW might be attributed to the designation of JHW as part of the Beihai Mangrove Reserve, which can restrict public access and offer protection for the spawning stock of horseshoe crabs from disturbance and harvest. The other two study sites XC and XBL are villages, in which traditional fishing activities are still intensive (Hu 2011). The harvesting of breeding adults at these sites could affect the annual larval recruitment, given that an estimated one million pairs of Chinese horseshoe crabs along Beibu Gulf shores may be collected annually without proper stock management measures (Li and Hu 2011). The loss of breeding and nursery habitats due to urbanization and tourism activities along the coastline in Beihai Municipality might also limit the density of horseshoe crabs at all of the study sites.

In this study, consecutive instar stages of juvenile horseshoe crabs were identified that allowed estimation of growth rates. According to the instar stages and age relationships reported by Sekiguchi (1988), *T. tridentatus* juveniles may take 4 years to grow from the 1st to 7th instar. After the 7th instar stage, Sekiguchi (1988) estimated in a more conservative way that the inter-molt time would be about 1 year until the juveniles become sexually mature. *T. tridentatus* at the three study beaches were recorded to the 12th instar stage, in which their estimated age was about 8 years old. Hence, these juveniles could molt four to five more times before reaching maturity, which is estimated to occur after 13–14 years and at a prosomal width of 237 mm for male and 265 mm for female (Chiu and Morton 2001). If the inter-molt period were more than 1 year, then growth may be slower than estimated here, and the juveniles could have a longer life expectancy for each size class, becoming sexually mature even later. Juvenile *C. rontundicauda* in this study were identified to 15th instars at XC. *C. rotundicauda* is hypothesized to have a growth rate similar to *T. tridentatus* (Sekiguchi 1988), a concept supported by the similar growth increments found between species in this study. Laboratory data for the growth stages of *C. rotundicauda*, however, are unavailable for comparison to this study because of a shorter rearing period in the laboratory (Sekiguchi 1988). From field observations, Chiu and Morton (2001) concluded that juvenile *C. rotundicauda* molted once in the first year and three times in the second year. These data along with the similar growth increment data between species suggest that the estimated age of juvenile *C. rotundicauda* at XC could be over 10 years old. In view of such generally slow growth to maturity, once a horseshoe crab population is locally extirpated, it will take a long period of time for recovery (ASMFC 1998).

Ideally, growth analyses should follow individuals from hatching to maturity in a natural habitat. This approach, however, is impractical in many cases due to the difficulty in consistently locating and continuously monitoring juveniles in the field. Instead, horseshoe crab growth primarily has been evaluated based on size increments between molts and the duration of the inter-molt period in laboratory-reared juveniles. Sekiguchi et al. (1988) conducted an 8-year study on the post-embryonic development of *L. polyphemus* and *T. tridentatus* by rearing the animals under laboratory conditions, from artificially fertilized eggs until death of the last individual, which was at the 10th instar stage. Based on this information they made the projection that *T. tridentatus* males become mature in the 13th year and females in the 14th year, after undergoing 15 and 16 molts, respectively and suggested the size-age relationships for different horseshoe crab species (Sekiguchi et al. 1988). However, it might be erroneous to age horseshoe crabs simply from the size-age relationship as species-specific instar duration often varies between laboratory and field studies (Kawahara 1982; Sekiguchi 1988). Many factors, such as hatching asynchrony (Ehlinger and Tankersley 2004), environmental variables (Jegla and Costlow 1982; Shuster 1982; Yeh 1999) and potentially different sex-specific growth rates (Tomaschko et al. 1997) could possibly explain such deviations. Consequently, the estimation of instar stages and age of juvenile horseshoe crabs on the study beaches in this study is only tentative. Accurate estimation can only be made based on long-term tagging experiments (Morton and Lee 2003) and more temporally detailed direct field observations to complement laboratory studies.

Hiatt (1948) proposed that the relationship between post- and pre-molt size of crustaceans could be generally fitted with a linear growth model. Kurata (1962) demonstrated that the length of the inter-molt period increased either directly as the cube of the body length or logarithmically with body length. Such linear or log-linear relationship between post- and pre-molt size (Hiatt diagram) has frequently been used to estimate the average molt increment in size and the number of molts necessary to reach a given size (e.g., Diesel and Horst 1995; Comeau et al. 1998). The present data suggested that the relationship between the post- and pre-molt size of juvenile horseshoe crabs can be generally fitted with the Hiatt growth model, and this finding was consistent with the previous study of *L. polyphemus* (Carmichael et al. 2003). The prosomal width increment of horseshoe crabs might be influenced by the environmental and biotic differences of their habitats (Kawahara 1982; Carmichael et al. 2003; Lee and Morton 2005). However, in the present study, the average prosomal width of each species did not show significant differences among study beaches or compared to data reported by Sekiguchi (1988). Growth rates of juvenile *T. tridentatus* also were not different among sites. These findings may reflect an overall similarity of environmental conditions among the three study beaches along the southern coast of Beihai Municipality, separated by Dafengjiang and Fengjiajiang Rivers (see Fig. 8.1) and/or a phenotypic plasticity of juveniles within the range of environmental variation among sites.

The results in this study document the density, population structure and dynamics of two juvenile Asian horseshoe crab species at three representative spawning and nursery sites in southern China. The collection of these data provides essential baseline information needed to support future monitoring and assessment of juvenile populations and habitat resources. To be better assess the horseshoe crab population, further information regarding juvenile survival and mortality, sex ratio and spawning stock biomass of horseshoe crabs on these shores is required.

Acknowledgements We would like to thank Ms. Q.Z. Li and staff from the Guangxi Institute of Fisheries (Guangxi Province, China) for their assistance in the field. M. Hu and Y. Wang were supported by Postgraduate Studentships from the City University of Hong Kong, and B.K.Y. Kwan was partially funded by Ocean Park Conservation Foundation for his Postgraduate Studentship from the City University of Hong Kong. Constructive comments from the reviewers are much appreciated.

References

Atlantic States Marine Fisheries Commission (ASMFC) (1998) Interstate fishery management plan for horseshoe crab. Fish Manag Rep 32:1–39

Botton ML, Loveland RE (2003) Abundance and dispersal potential of horseshoe crab (*Limulus polyphemus*) larvae in the Delaware estuary. Estuaries 26:1472–1479

Botton ML, Shuster CN Jr, Keinath JA (2003) Horseshoe crabs in a food web: who eats whom. In: Shuster CN Jr, Barlow RB, Brockmann HJ (eds) The American horseshoe crab. Harvard University Press, Cambridge, MA, pp 133–153

Carmichael RH, Rutecki D, Valiela I (2003) Abundance and population structure of the Atlantic horseshoe crab *Limulus polyphemus* in Pleasant Bay, Cape Cod. Mar Ecol Prog Ser 246:225–239

Chen CP, Yeh HY, Lin PF (2004) Conservation of the horseshoe crab at Kinmen, Taiwan: strategies and practices. Biodivers Conserv 13:1889–1904

Chiu HMC, Morton B (1999) The distribution of horseshoe crabs (*Tachypleus tridentatus* and *Carcinoscorpius rotundicauda*) in Hong Kong. Asian Mar Biol 16:185–194

Chiu HMC, Morton B (2001) Growth and allometry of two horseshoe crab species, *Tachypleus tridentatus* and *Carcinoscorpius rotundicauda* (Xiphosura), in Hong Kong. Asian Mar Biol 18:129–141

Chiu HMC, Morton B (2003) The morphological differentiation of two horseshoe crab species, *Tachypleus tridentatus* and *Carcinoscorpius rotundicauda* (Xiphosura), in Hong Kong with a regional Asian comparison. J Nat Hist 37:2369–2382

Comeau M, Conan GY, Maynou F et al (1998) Growth, spatial distribution, and abundance of benthic stages of the snow crab (*Chionoecetes opilio*) in Bonne Bay, Newfoundland, Canada. Can J Fish Aquat Sci 55:262–279

Diesel R, Horst D (1995) Breeding in a snail shell: ecology and biology of the Jamaican Montane crab *Sesarma jarvisi* (Decapoda: Grapsidae). J Crustac Biol 15:179–195

Ehlinger GS, Tankersley RA (2004) Survival and development of horseshoe crab (*Limulus polyphemus*) embryos and larvae in hypersaline conditions. Biol Bull 206:87–94

Hiatt RW (1948) The biology of the lined shore crab, *Pachygraspus crassipes* Randall. Pac Sci 2:135–213

Hong S (2011) Biology of horseshoe crabs, *Tachypleus tridentatus*. Xiamen University Press, Xiamen

Hsieh HL, Chen CP (2009) Conservation program for the Asian horseshoe crab *Tachypleus tridentatus* in Taiwan: characterizing the microhabitat of nursery grounds and restoring spawning grounds. In: Tanacredi JT, Botton ML, Smith DR (eds) Biology and conservation of horseshoe crabs. Springer, New York, pp 417–438

Hu M (2011) Aspects of ecology and biology of two sympatric horseshoe crab species (*Tachypleus tridentatus* and *Carcinoscorpius rotundicauda*). PhD thesis, City University of Hong Kong

Hu M, Wang Y, Chen Y et al (2009) Summer distribution and abundance of juvenile Chinese horseshoe crabs *Tachypleus tridentatus* along an intertidal zone in southern China. Aquat Biol 7:107–112

Itow T (1993) Crisis in the Seto Inland Sea: the decimation of the horseshoe crab. EMECS Newsl 3:10–11

Jegla TC, Costlow JD (1982) Temperature and salinity effect on developmental and early posthatch stages of *Limulus*. In: Bonaventura J, Bonaventura C, Tesh S (eds) Physiology and biology of horseshoe crabs. Alan R Liss, New York, pp 103–113

Kawahara D (1982) Investigation on ecology of horseshoe crab larvae. Aquabiology 4:380–382

King TL, Eackles MS, Spidle AP et al (2005) Regional differentiation and sex-biased dispersal among populations of the horseshoe crab *Limulus polyphemus*. Trans Am Fish Soc 134:441–465

Kurata H (1962) Studies on the age and growth of Crustacea. Bull Hokkaido Reg Fish Res Lab 24:1–115

Lee CN, Morton B (2005) Experimentally derived estimates of growth by juvenile *Tachypleus tridentatus* and *Carcinoscorpius rotundicauda* (Xiphosura) from nursery beaches in Hong Kong. J Exp Mar Biol Ecol 318:39–49

Lee CN, Morton B (2009) Emergence behavior of juvenile *Tachypleus tridentatus* under simulated tidal conditions in the laboratory and at two different sediment temperatures. In: Tanacredi JT, Botton ML, Smith DR (eds) Biology and conservation of horseshoe crabs. Springer, New York, pp 275–283

Li QZ, Hu MH (2011) Status of Chinese horseshoe crab population and conservation advance in Beihai, Guangxi. Mar Environ Sci 30:131–134 (Chinese with English Abstract)

Liao YY, Li XM (2001) Present situation of horseshoe crab resources in the sea area of China and tactics of preservation. Resour Sci 23:53–57 (Chinese with English Abstract)

Liao YY, Hong SG, Li XM (2001) A survey on the horseshoe crabs in the north of South China Sea. Acta Zool Sin 47:108–111 (Chinese with English Abstract)

Mikkelsen T (1988) The secret in the blue blood. Science Press, Beijing

Morton B, Lee CNW (2003) The biology and ecology of juvenile horseshoe crabs along the Northwestern coastline of the New Territories, Hong Kong: prospects and recommendations for conservation. Final report to China Light and Power Company Limited, The Swire Institute of Marine Science, The University of Hong Kong, Hong Kong

Pierce JC, Tan G, Gaffney PM (2000) Delaware Bay and Chesapeake Bay populations of the horseshoe crab *Limulus polyphemus* are genetically distinct. Estuaries 23:690–698

Rudkin DM, Young GA (2009) Horseshoe crabs–an ancient ancestry revealed. In: Tanacredi JT, Botton ML, Smith DR (eds) Biology and conservation of horseshoe crabs. Springer, New York, pp 25–44

Rudloe A (1979) Locomotor and light responses of larvae of the horseshoe crab, *Limulus polyphemus*. Biol Bull 157:494–505

Sekiguchi K (1988) Biology of horseshoe crabs. Science House, Tokyo

Sekiguchi K, Seshimo H, Sugita H (1988) Post-embryonic development of the horseshoe crab. Biol Bull 174:337–345

Shin PKS, Li HY, Cheung SG (2009) Horseshoe crabs in Hong Kong: current population status and human exploitation. In: Tanacredi JT, Botton ML, Smith DR (eds) Biology and conservation of horseshoe crabs. Springer, New York, pp 347–360

Shuster CN Jr (1982) A pictorial review of the natural history and ecology of the horseshoe crab, *Limulus polyphemus*, with reference to other Limulidae. In: Bonaventura J, Bonaventura C, Tesh S (eds) Physiology and biology of horseshoe crabs. Alan R Liss, New York, pp 1–52

Tomaschko KH, Wilhelm E, Buckmann D (1997) Growth and reproduction of *Pycnogonum litorale* (Pycnogonida) under laboratory conditions. Mar Biol 129:595–600

Yeh HY (1999) Life cycle, juvenile habitat and conservation strategies of the horseshoe crab (*Tachypleus tridentatus*) in Kinmen. MPhil thesis, National Taiwan University

Zadeh SS, Christianus A, Saad CR et al (2009) Comparisons in prosomal width and body weight among early instar stages of Malaysian horseshoe crabs, *Carcinoscorpius rotundicauda* and *Tachypleus gigas* in the laboratory. In: Tanacredi JT, Botton ML, Smith DR (eds) Biology and conservation of horseshoe crabs. Springer, New York, pp 267–274

Zar J (1996) Biostatistical analysis. Prentice Hall, Upper Saddle River

Zhao HT (1999) The physiognomy and environment of the coasts in South China and islands in South China Sea. Science Press, Beijing

Chapter 9
Distribution and Abundance of Two Sympatric Species of Horseshoe Crabs Along the Odisha Coast, India

Satyaranjan Behera, Basudev Tripathy, K. Sivakumar, B.C. Choudhury, and Punyasloke Bhadury

Abstract Two sympatric species of horseshoe crab *Tachypleus gigas* and *Carcinoscorpius rotundicauda* are found along the coastlines of Odisha, India facing the Bay of Bengal. The distribution and abundance of these two species were surveyed for a period of 1 year from two coastal sites at Ekakula and Balaramgadi of Odisha in link with phases of the moon. At Ekakula, both species were encountered during the study period; however, over 95 % of the horseshoe crabs recorded were dead. While the numbers of horseshoe crabs counted during new and full moon phases were statistically similar for both species, relatively fewer sightings were recorded in the winter period. Combining all live and dead specimens, a higher mean density of 1.34 individuals/m^2 was recorded for *T. gigas* than that of 0.18 individuals/m^2 for *C. rotundicauda*. In addition, a clear habitat preference was noted, in which *T. gigas* was found on sandy beaches whereas *C. rotundicauda* near mangrove fringes. At Balaramgadi, only sightings of dead *T. gigas* were encountered, and abundance was found to be very low, ranging from zero to four specimens per sampling occasion. At both sites, survival

S. Behera (✉)
Odisha Biodiversity Board, Regional Plant Resource Centre Campus, Nayapally, Bhubaneswar 751015, Odisha, India
e-mail: behera.satyaranjan@gmail.com

B. Tripathy
Malacology Division, Prani Vigyan Bhawan, M-Block, Zoological Survey of India, New Alipore, Kolkata 700053, West Bengal, India

K. Sivakumar
Department of Endangered Species Management, Wildlife Institute of India, Dehradun 248001, Uttarakhand, India

B.C. Choudhury
Wildlife Trust of India, Noida 201301, Uttar Pradesh, India

P. Bhadury (✉)
Integrative Taxonomy and Microbial Ecology Research Group, Department of Biological Sciences, Indian Institute of Science Education and Research Kolkata, Mohanpur, 741246 Nadia, West Bengal, India
e-mail: pbhadury@iiserkol.ac.in

R.H. Carmichael et al. (eds.), *Changing Global Perspectives on Horseshoe Crab Biology, Conservation and Management*, DOI 10.1007/978-3-319-19542-1_9

of horseshoe crabs was severely affected because of entanglement of individuals in various types of nets along the shore due to intense fishing activities. Threats that persist to horseshoe crab populations on Odisha coast also included high predation pressure from wild animals and loss of spawning ground due to coastal developments. Establishment of routine monitoring programmes and conservation reserves and training of enforcement agencies to protect horseshoe crab populations in India are proposed.

Keywords *Tachypleus gigas* • *Carcinoscorpius rotundicauda* • Ekakula • Balaramgadi • Bay of Bengal • Odisha • India

9.1 Introduction

Horseshoe crabs (Xiphosura) are an archaic group of marine arachnids with the oldest fossils dating back to at least 445 million years ago (Rudkin and Young 2009). Among the four species of horseshoe crabs that inhabit the coastlines globally, *Limulus polyphemus* (Linnaeus 1758) is native to the eastern shoreline of North America, from Southern Maine to the Yucatan Peninsula, with Delaware Bay as the centre of the population (Shuster 2001). The other three species, *Carcinoscorpius rotundicauda* (Latreille 1802), *Tachypleus tridentatus* (Leach 1819) and *Tachypleus gigas* (Müller 1785) are found along coastlines of Asia. *Tachypleus gigas* and *Carcinoscorpius rotundicauda* are distributed in the Indo-Pacific region from the Bay of Bengal to Indonesia, whereas *T. tridentatus* is recorded from the coast of the Philippines to the south western seas of Japan (Sekiguchi 1988).

Understanding the natural history and ecology of horseshoe crabs is paramount to implement efforts for their conservation (Sekiguchi and Shuster 2009). In India, *T. gigas* and *C. rotundicauda* are known to inhabit the north-eastern coastal stretches along the Bay of Bengal, mostly confined to the coastlines of Andhra Pradesh, Odisha and West Bengal (Chatterji 2005). In particular, the Odisha coast is reportedly the home of the largest population of horseshoe crabs, and the highest concentration is found on Balaramgadi beach near Balasore (Chatterji 1994). In Odisha, the horseshoe crab is known locally as *Ram Laxmania Kankada*, *Laxmanpatia*, *Samudra Bichha* and *Belangkar*. While several studies have been carried out on Indian horseshoe crabs by different workers in the past (e.g. Roonwall 1944; Sekiguchi et al. 1976; Chatterji et al. 1988; Debnath 1991; Mishra 1994), there is a severe dearth of information on their current population data especially from the East Coast.

The status of horseshoe crabs along the Odisha coast has not been documented in detail for more than a decade. Except for some studies on biomedical use of the blue blood of horseshoe crabs and artificial breeding in captivity (Chatterji et al. 1994), there is very little information available on the occurrence of different species, population status, breeding season and various natural and anthropogenic threats along the Odisha coast. The objective of this study was to record the occurrence, abundance and distribution of these two sympatric species of horseshoe crabs from two coastal sites of Odisha.

9.2 Materials and Methods

9.2.1 Study Area

The Odisha coast is bestowed with a variety of coastal habitats, such as tidal creeks, backwaters, brackish water lagoons, estuaries, mangroves, mudflats and salt marshes. Some of the major rivers of Odisha including Subarnarekha, Budhabalanga, Brahmani, Baitarani, Mahanadi, Devi and Rushikulya and several smaller rivers, such as Hansua, Kadua, Kushabhadra and Bahuda, drain into the Bay of Bengal. Two sites, one near the estuary of river Budhabalanga, locally known as Balaramgadi (Chandipur) (20° 42′N; 87° 02′E) and another place called Ekakula (19° 16′N; 84° 53′E), at the mouth of rivers Brahmani and Baitarani were selected for this study (Fig. 9.1). Ekakula is part of the Bhitarkanika National Park, which is rich in mangrove-associated floral and faunal diversity. Balaramgadi is a large intertidal zone facing the Bay of Bengal with predominance of sandy and clay sediment. At low

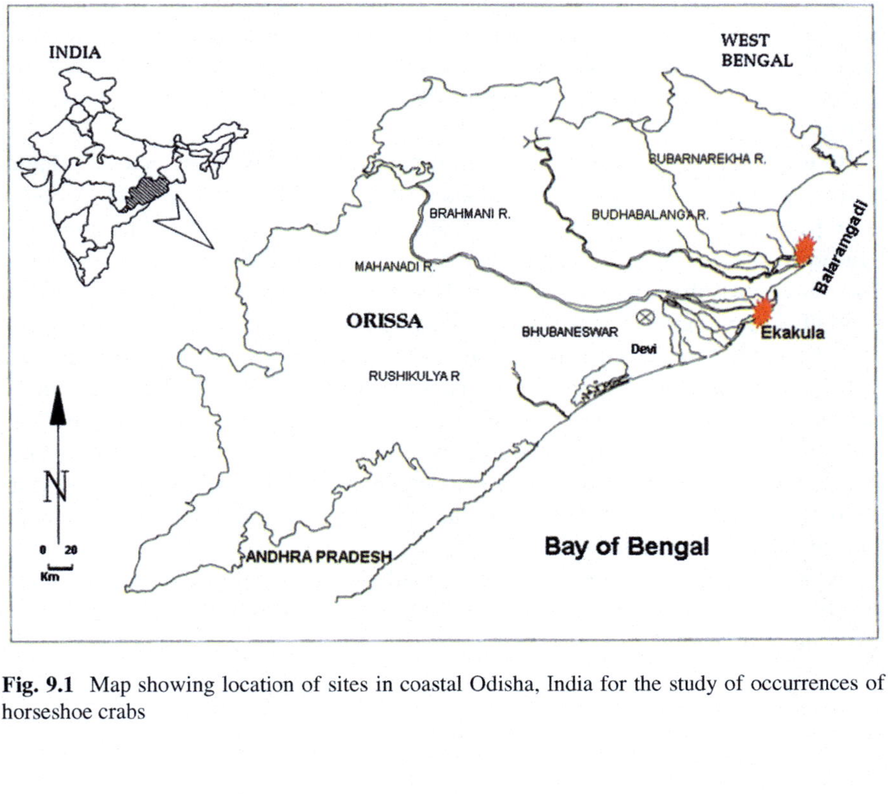

Fig. 9.1 Map showing location of sites in coastal Odisha, India for the study of occurrences of horseshoe crabs

tides the intertidal zone in Balaramgadi can extend up to 3 km. The Ekakula site also has an intertidal zone and is represented by habitats such as mudflats, mangroves and some sandy beaches. During the entire study period, only two species *T. gigas* and *C. rotundicauda* were recorded, and data collection was focused on these two species since the former prefers sandy habitats while the latter mostly live on mangrove mudflats.

9.2.2 Field Sampling and Data Analysis

The study was conducted from April 2009 to March 2010 to assess the population status of *T. gigas* and *C. rotundicauda*. In Balaramgadi, a 1 km × 1 km strip transect was laid in the intertidal zone to count the animals, in view of the very large vast shore area. In Ekakula, a 5 m (width) × 900 m (length) belt transect was adopted parallel to the river mouth. This transect was further divided into three sections, in which the counting of horseshoe crabs was conducted along each 300 m section, based on published protocols (Chalmers and Parker 1989; Cartwright-Taylor et al. 2009, 2011). These three transect sections represented typical habitats of sandy beach towards the river mouth, mixed sand and mud area in the mid-transect section, and muddy sediment at the mangrove fringes. The abundance data were collected fortnightly, during new moon and full moon high tides across these two study sites. Both live and dead horseshoe crabs washed ashore were counted in the transects at an interval of 2–3 days during full moon and new moon days. The dead horseshoe crabs were marked on their carapaces with white and yellow paint and were dragged up the shore in order to avoid repeated counting during subsequent surveys.

Information was further collected from fishermen living in the coastal villages surrounding the study sites to understand more about the population status of these two species of horseshoe crabs. Local people were interviewed regarding horseshoe crab occurrence in their localities. They were also asked about the number of horseshoe crabs that they have seen on the beach, season for such sightings, nesting, mortality, if any, and the peak season and cause of mortality. Other information collected included incidental catch of horseshoe crabs in various types of artisanal fishing gears and the nature of nets used for fishing. Additionally, development activities along the coastline of Odisha such as plantations, industries, land use pattern as well as socio-economic status were taken into account during the study. Visits were undertaken in some of the fishing harbours situated in close proximity to the study sites so as to collect information on incidental catch of horseshoe crabs during fishing operations.

A paired sample t-test (Zar 1984) was conducted to compare the sighting of two horseshoe crab species between new and full moon phases at the same survey site over the study period.

9.3 Results and Discussion

9.3.1 Population Trends

A total of 6,050 live and dead specimens of *T. gigas* were recorded at Ekakula during the present study, whereas 804 individuals of *C. rotundicauda* were counted from the same site. At Ekakula, the number of horseshoe crab sightings during the new (Fig. 9.2) and full moon phase (Fig. 9.3) was statistically similar for *T. gigas* ($t=-0.425$, $p=0.679$) and *C. rotundicauda* ($t=0.734$, $p=0.478$). However, relatively fewer sightings were recorded for both species over the winter months (December to March). During the survey period very few numbers of live horseshoe crabs were encountered, and the dead individuals accounted for 98.8 % and 96.5 % respectively for *T. gigas* and *C. rotundicauda*. Only 75 live *T. gigas* with 29 males and 46 females, and 28 live *C. rotundicauda* with 18 males and 10 females were sampled. At Balaramgadi, no live horseshoe crabs were seen, with only 22 dead specimens of *T. gigas* recorded during the study period (Table 9.1).

9.3.2 Abundance and Population Density

At Ekakula, although the abundance of horseshoe crabs varied in the three transect sections and over time, the mean density of live and dead *T. gigas* and *C. rotundicauda* was 1.34 and 0.18 individuals/m^2, respectively (Table 9.2). During the month of May the highest sighting was observed for both species. According to Chatterji (1994), *T. gigas* prefers to breed on sandy beaches near estuaries whereas *C. rotundicauda* spawns mostly on mangrove mudflats. Being a sandy beach, the Ekakula site

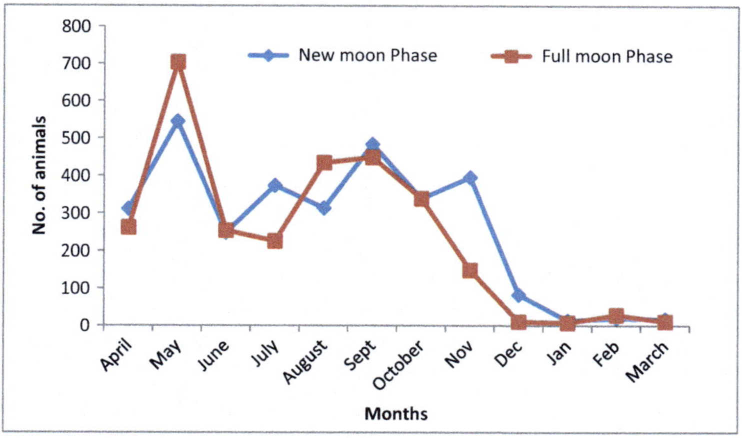

Fig. 9.2 Sightings of *T. gigas* along the transect in new and full moon phases at Ekakula, Odisha coast

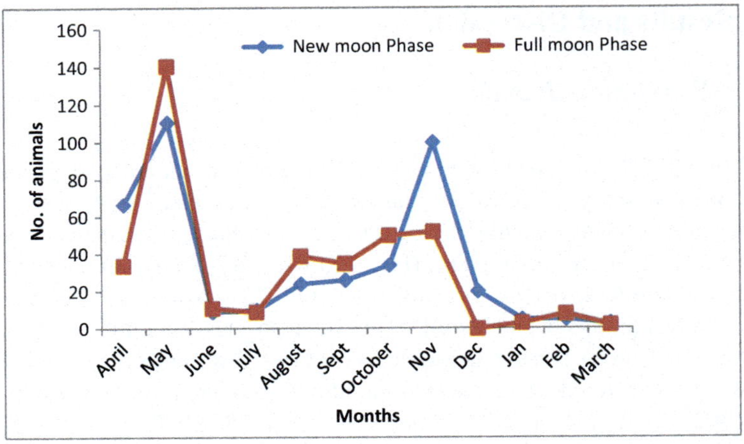

Fig. 9.3 Sightings of *C. rotundicauda* along the transect in new and full moon phases at Ekakula, Odisha coast

Table 9.1 Specimens of dead *T. gigas* encountered in the transect area at Balaramgadi, Odisha coast

Sampling month	New moon	Full moon	Total
April 2009	1	2	3
May	1	0	1
June	NA	NA	0
July	NA	NA	0
August	NA	NA	0
September	NA	NA	0
October	2	0	2
November	1	0	1
December	1	1	2
January 2010	2	3	5
February	2	1	3
March	4	1	5
Total	14	8	22

NA not available

thus attracts more mating pairs of *T. gigas* than *C. rotundicauda*. Moreover, *T. gigas* prefers to live in high saline conditions in close proximity to the mouth of the estuary, whereas *C. rotundicauda* likes to strive in low saline conditions near mangrove mudflats (Tripathy et al. 2013). From the transect data, maximum sightings of live or dead horseshoe crabs belonging to *T. gigas* were found near the sea mouth or sandy habitats, whereas *C. rotundicauda* individuals were found near the mangrove fringe areas where salinity gradient was comparatively lower (Table 9.2). The present observations supported such habitat preference with highest sightings of *T. gigas* individuals during summer, which corresponded with high salinity in the study area.

Table 9.2 Density of live and dead *T. gigas* and *C. rotundicauda* along the transect sections at Ekakula, Odisha coast

Transect	Abundance	Density (individual/m^2)
Tachypleus gigas		
1	2,934	1.96
2	2,913	1.94
3	203	0.14
Total	6,050	1.34
Carcinoscorpius rotundicauda		
1	24	0.02
2	264	0.18
3	516	0.34
Total	804	0.18

1 sandy shore near estuary, *2* mixed sand and mud mid-shore, *3* mudflat near mangrove fringe

Most of the specimens of horseshoe crabs recorded in this study were found to be dead, except for very few live individuals in the field. Many dead horseshoe crabs were entangled in standing nets, locally called ‘Bedha Jala’ and died suspiciously. Incidentally, these nets were put up by local fishermen during new moon and full moon high tide days when the highest incidence of dead horseshoe crab specimens was observed. Additionally, many of the live horseshoe crabs were found to be entangled among the pneumatophores of mangrove patches at Ekakula after the receding tide, and they were predated by wild pigs. This might have increased the incidences of encountering more numbers of dead horseshoe crabs compared to live ones during the field surveys at Ekakula.

At Balaramgadi, relatively fewer horseshoe crabs were noted as compared to Ekakula, and the mortality of horseshoe crabs there was largely due to the use of standing nets locally known as ‘Mala Jaala’ in the intertidal areas (Fig. 9.4). These nets are usually fixed during low tide, and horseshoe crabs entering the intertidal zone during high tide are entangled when the tide ebbs (Tripathy et al. 2007) (Fig. 9.5). As per information of local villagers, there are comparatively higher numbers of horseshoe crabs entangled in these nets during winter months when fishing is at its peak at Balaramgadi.

9.4 Conclusion

Not much is known in terms of predation and other ecological factors that can affect survival of horseshoe crab populations along the coastlines of India, in particular the Odisha coast. In order to develop a viable conservation strategy, there is a need to train local law enforcement agencies, especially forest officials, about horseshoe crabs and establish monitoring protocols that can be applied on a regular basis

Fig. 9.4 Dead horseshoe crab specimens encountered due to intensive fishing activities at Ekakula, Odisha coast

Fig. 9.5 Standing nets along the intertidal areas at Balaramgadi and Ekakula, resulting in entanglement and mortality of horseshoe crabs

across the vulnerable intertidal areas of the Odisha coast. Non-government conservation organizations can also be included in this process. Conservation of horseshoe crabs will become more successful if fishermen living along coastal Odisha are actively involved. To this end, mass awareness programmes should be initiated to explain the need for conservation of horseshoe crab populations locally.

One of the major threats facing the survival of horseshoe crabs along the Odisha coast is the unregulated use of mechanized bottom trawler net, prawn-fishing trammel net, monofilament gill net, as well as beach shore seine and standing nets by the fishermen, resulting in massive entanglements and ultimate mortality of many individuals on the seabed or when horseshoe crabs come to the shore for spawning. This has been clearly observed as part of this study undertaken across two sites in Odisha. Additionally, habitats of horseshoe crabs such as muddy shores along with mangrove vegetation are lost for construction of fishing jetties and other coastal developments, leading to a gradual decline of their breeding habitats. A sustainable coastal development programme should be developed so as to ensure the viability of the horseshoe crab populations. Predation also poses a threat to the populations of horseshoe crabs. For example observations at Ekakula revealed that high predation pressures from wild pigs, jackals and hyenas affect the breeding pairs of horseshoe crabs. Although the Odisha coast is reported to be a major spawning ground for horseshoe crabs, except for the 35 km stretch of the Gahirmatha coast in the Bhitarkanika Wildlife Sanctuary, which is protected, other important nesting sites of horseshoe crabs such as Balaramgadi (Chandipur), Kasaphala, Talshari and Udaypur need special protection, and the government can declare these sites as conservation reserves, especially during the spawning season. Apart from the above, information on the effect of predation and other ecological factors affecting survival of older juveniles of horseshoe crabs is also limited, both from the Indian coastlines and in other parts of the world. These clearly reflect the vital gaps in our knowledge on the biology and ecology of horseshoe crabs, which should be addressed in future studies.

The horseshoe crabs found in India are categorized as "Data Deficient" in the International Union for Conservation of Nature (IUCN) Red List categories of 2009, and both species have been declared as a Schedule-IV species under the Wildlife Protection Act (1972) in 2009 by the Indian Government. The inclusion of both species under Schedule-IV provides an opportunity for protection including conservation of their habitats and prohibition of illegal trading. However, to avoid a further decline in horseshoe crab populations along the Indian coastlines and to safeguard their survival for future generations, a consultative process between the forestry and fishery departments, together with the involvement of local fishermen, needs to be initiated as soon as possible.

References

Cartwright-Taylor L, Lee J, Hsu CC (2009) Population structure and breeding pattern of the mangrove horseshoe crab *Carcinoscorpius rotundicauda* in Singapore. Aquat Biol 8:61–69

Cartwright-Taylor L, Yap VB, Hsu CC et al (2011) Distribution and abundance of horseshoe crabs *Tachypleus gigas* and *Carcinoscorpius rotundicauda* around the main island of Singapore. Aquat Biol 13:127–136

Chalmers N, Parker P (1989) The OU project guide: fieldwork and statistics for ecological projects, 2nd edn. Field Studies Council, Shrewsbury

Chatterji A (1994) The horseshoe crab–a living fossil. A Project Swarajya Publication, Cuttack

Chatterji A (2005) New record of the sympatric distribution of two Asian species of the horseshoe crab. Curr Sci 77:746–747

Chatterji A, Vijaykumar R, Parulekar AH (1988) Growth and morphometric characteristic in the horseshoe crab, *Carcinoscorpius rotundicauda* (Latreiide) from Canning (West Bengal), India. Pak J Sci Ind Res 33:352–355

Chatterji A, Mishra JK, Vijayakumar R et al (1994) Length-weight relationship of the Indian horseshoe crab, *Tachypleus gigas* (Müller). Indian J Fish 41:58–60

Debnath R (1991) Studies on species morphology and sexual dimorphism of Indian horseshoe crabs. Gobios New Rep 10:128–131

Latreille PA (1802) Histoire naturelle, générale et particulière des crustacés et des insectes: ouvrage faisant suite aux oeuvres de Leclerc de Buffon, et Partie du Cours Complet d'Histoire Naturelle Rédigé par C.S. Sonnini. F. Dufart, Paris

Leach WE (1819) Entomostraca, vol 14, Dictionnaire des Sciences Naturelles. Levrault and Schoell, Paris

Linnaeus C (1758) Systema Naturae per regna tria naturae, secundum classes, ordines, genera, species, cum characteribus, differentiis, synonymis, locis. Editio decima, reformata. Laurentius Salvius, Holmiae

Mishra JK (1994) Horseshoe crab, an amazing creature of Odisha Coast. Project Swarajya Publication, Odisha

Müller OF (1785) Entomostraca seu insecta testacea, quae in aquis Daniae et Norvegiae reperit, descripsit et iconibus illustravit Otho Fridericus Müller. F.W. Thiele, Lipsiae/Havniae

Roonwall ML (1944) Some observations on the breeding biology, and on the swelling, weight, water-content and embryonic movements in the developing eggs of the *Tachypleus gigas* (Müller) [Arthropoda, Xiphosura]. Proc Indian Acad Sci 20:115–129

Rudkin DM, Young GA (2009) Horseshoe crabs–an ancient ancestry revisited. In: Tanacredi JT, Botton ML, Smith DR (eds) Biology and conservation of horseshoe crabs. Springer, New York, pp 25–44

Sekiguchi K (1988) Biology of horseshoe crabs. Science House Press, Tokyo

Sekiguchi K, Shuster CN Jr (2009) Limits on the global distribution of horseshoe crabs (Limulacea): lessons learned from two lifetimes of observations: Asia and America. In: Tanacredi JT, Botton ML, Smith DR (eds) Biology and conservation of horseshoe crabs. Springer, New York, pp 5–24

Sekiguchi K, Nakamura K, Sen TK et al (1976) Morphological variation and distribution of a horseshoe crab, *Tachypleus gigas*, from the Bay of Bengal and the Gulf of Siam. Proc Jap Soc Syst Zool 15:24–30

Shuster CN (2001) Two perspectives: horseshoe crabs during 420 million years, worldwide, and the past 150 years in the Delaware Bay area. In: Tanacredi JT (ed) *Limulus* in the limelight. Kluwer Academic/Plenum Publishers, New York, pp 17–40

Tripathy B, John S, Choudhury BC et al (2007) Status report of horseshoe crabs along the East Coast of India. Wildlife Institute of India, Dehradun, pp 2–32

Tripathy B, Sivakumar K, John S et al (2013) Assessment of the population status and threats to the horseshoe crabs along the Northern East Coast of India. In: Venkataraman K, Sivaperuman C, Raghunathan C (eds) Ecology and conservation of tropical marine faunal communities. Springer, Heidelberg, pp 137–146

Wildlife Protection Act (1972) Inclusion of horseshoe crab in the Schedule-IV of the Wildlife (Protection) Act, 1972. The Gazette of India, S.O. 2293(E)

Zar JH (1984) Biostatistical analysis, 2nd edn. Prentice-Hall, Englewood Cliffs

Chapter 10
Studies of Horseshoe Crabs Around Singapore

Lesley Cartwright-Taylor

Abstract Results from students' research on horseshoe crabs are not often readily available to other workers. A review of these studies has shown that during 6 years since the first published studies, the distribution of horseshoe crabs around Singapore's main island has remained unchanged, the density of mangrove horseshoe crabs, *Carcinoscorpius rotundicauda*, at two sites has remained stable, and periods of high and low reproductive activity during the year have been confirmed. The student studies indicated that juvenile *C. rotundicauda* concentrated at the high tide zones of the mudflats along the Strait of Johor, and smaller juveniles were found on the mud surface and in the small streamlets. Adults were found spread more evenly over the mudflats from the high to low tide zones, on average about 3 m apart, and commonly buried or found in the larger streams among the mangroves. Feeding preference studies suggest mangrove horseshoe crabs are not particularly selective feeders. Sex ratios remained approximately 1:1 all year and at all sites, and mean percentage growth during ecdysis was 34 %, with little difference between the smallest crab of 1.1 cm and the largest of 90.1 cm. A mortality study found death rates of mangrove horseshoe crabs higher at a polluted site (19.8 %) than at a less polluted site (7.5 %). Island wide distribution studies showed that while there is a thriving population of mangrove horseshoe crabs around the north-west, by contrast, *Tachypleus gigas*, the coastal horseshoe crab, is still very rare around Singapore, but an encouraging finding was the first reported sighting of *T. gigas* juveniles at one site. These studies, although small and of short duration, are important for the regular monitoring of the current status of horseshoe crab populations around Singapore and offer a major contribution to the understanding of horseshoe crab ecology in the country.

Keywords Singapore • Mangrove • Coastal • Density • Distribution • Mudflats • *Carcinoscorpius rotundicauda* • *Tachypleus gigas* • Demography • Habitat • Diet preferences • Growth • Conservation

L. Cartwright-Taylor (✉)
Nature Society (Singapore), 510 Geylang Road, #02-05, The Sunflower 389466, Singapore
e-mail: cartwrighttaylor@gmail.com

R.H. Carmichael et al. (eds.), *Changing Global Perspectives on Horseshoe Crab Biology, Conservation and Management*, DOI 10.1007/978-3-319-19542-1_10

10.1 Introduction

The Mandai mudflats at Kranji in the north of the main island of Singapore along the Strait of Johor are home to a thriving population of mangrove horseshoe crabs, *Carcinoscorpius rotundicauda*. This area is the largest expanse of mudflats remaining in Singapore and is also an important site for resident shore birds and migratory birds travelling along the East Asian Australasian Flyway.

Fortuitously, the site is accessible, and horseshoe crabs are easy to find. Horseshoe crabs do not run away when approached and they do not sting, bite or pinch, so they make an ideal subject of study for students needing a topic for their degree dissertations. The site offers a valuable educational resource for fieldwork. Most studies are of short duration and are designed so that they can be completed in just a couple of months. So far no student in Singapore has remained involved in horseshoe crab research long enough to continue to a PhD or other studies that would require a longer-term project with more in-depth research, using more sophisticated techniques. In most cases once the study is completed and the student moves on, the dissertation is filed in the library of the university or college without publication in a peer-reviewed journal or other publicly accessible source. Thus, there is a view that there are no or very few data on the biology of horseshoe crabs in Singapore, and knowledge that is based on empirical data rather than anecdotes is fragmentary and not easy to find. In many cases the students have put in much hard work and the studies are thorough and well supervised. Although there is often an element of repetition, these small studies help to confirm earlier findings and provide frequent monitoring of horseshoe crab sites, which can be valuable in determining emerging threats. Hence, if made accessible, the work in these dissertations has potential to make an important contribution to the overall understanding of horseshoe crab ecology in Singapore.

Since the Nature Society of Singapore (NSS) embarked on a research project, born out of a horseshoe crab rescue program, larger studies of longer than a few weeks duration have been undertaken. Unlike students who move on, most of the NSS members have a long-standing involvement in the Society and an abiding interest in nature and wildlife, providing continuity in the NSS work. Once the NSS horseshoe crab program was underway, the NSS soon became able to field a large number of volunteers (students and staff from schools and colleges, NSS members, and the public) to help with field searches and data collection. This effort allowed more ambitious studies to be done and more sophisticated equipment, such as acoustic tracking devices, to be used. Gathering together the data from the short, unpublished works as well as these longer published studies will give a fuller picture of what is known and not known about the status of the two species of horseshoe crabs (*C. rotundicauda* and *T. gigas*) in Singapore. It will then be possible to guide research to fill the gaps in the knowledge of the ecology of these species such that practical conservation measures can be taken to protect them.

All the field studies on the horseshoe crabs in Singapore have used the same methods of search for *C. rotundicauda* and *Tachypleus gigas*, which is occasionally seen around Singapore. At low tide the site is searched visually for surface crabs on the substrate or in small streams and, for buried crabs, by probing the surface with fingers or small sticks. Individuals of all sizes, from juveniles of approximately 0.5 cm in carapace width to adults of approximately 16 cm and crabs in amplexus can be found at any site visit throughout the year. Juveniles are considered to be those less than 8 cm in carapace width, based on the smallest size of males found in amplexus. There is no mass spawning event on a seasonal basis as with the American species of horseshoe crabs, and spawning individuals are very rarely seen. Sex determination in the adults is always based on the structure of the first pair pedipalps. Sexing of juveniles using genital pore structure is not undertaken in the field because of the difficult conditions under which the work has to be done.

10.2 Distribution

10.2.1 NSS Surveys

In March 2009 the NSS team conducted the first island-wide survey to determine if there were populations elsewhere around Singapore (Cartwright-Taylor et al. 2011). Loss of habitat and pollution are listed as the two main threats to horseshoe crabs in Singapore (Davidson et al. 2008), so periodic surveys of possible remaining habitats are important in monitoring any emerging threats to the remaining populations or, indeed, in identifying if individuals are returning to formerly deserted sites. This first survey was conducted at eight sites around the main island including the Mandai mudflats near Kranji, where horseshoe crabs have been well studied (Fig. 10.1).

Western Sites Further along the coast to the west of Kranji and past the Sungei Buloh Wetlands Reserve (SBWR), a known habitat of horseshoe crabs, is Lim Chu

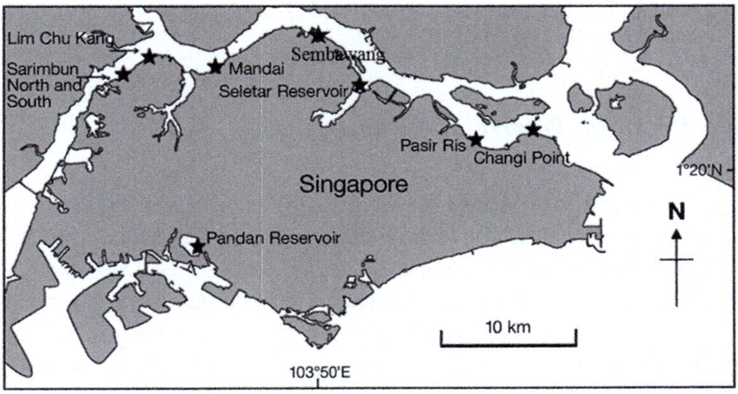

Fig. 10.1 Map of Singapore showing sites where horseshoe crab surveys have been conducted

Kang, a muddy shore with mangrove fringes, where a population of *Carcinoscorpius rotundicauda* was found during the survey. Although this area was known to support a population of *C. rotundicauda*, very little work has been done on the horseshoe crabs at this site. Occasionally *Tachypleus gigas* was also found in the vicinity, and the NSS survey yielded one dead *T. gigas* at this site suggesting some co-habitation with *C. rotundicauda*. The local fishermen also say they occasionally find *T. gigas*. More work at this site would be useful, but the mud here can be very deep, there is a great deal of debris, and moving around the site is not easy.

Further to the west, Sarimbun, a muddy mangrove shore, and the Pandan Reservoir canal to the south were also included in the first survey. *C. rotundicauda* was found at both sites, but neither site is well studied. The mud at the site at Pandan Reservoir, like Lim Chu Kang, is very deep making work here difficult. This site on the south coast is far removed from other known horseshoe crab habitats, all of which are contiguous along the north-west coast, so genetic studies on the Pandan population would be of interest to determine how isolated it has become.

Eastern Sites To the east along the Strait of Johor, no horseshoe crabs were observed at Pasir Ris, a sandy beach with some mud, two live *C. rotundicauda* were found at Lower Seletar, also a mix of sand and mud, and one *C. rotundicauda* was seen at Changi Beach, a true sandy beach more suited to *T. gigas*. The survey also yielded six live *T. gigas* at Changi Beach, all trapped in nets. Clearly there are some *T. gigas* around the shore but where they spawn or forage is unclear. According to elderly fishermen and most Singaporeans in their 50s and 60s, *T. gigas* frequently was seen at all three sites some years ago. However, with coastal development and large areas of land reclaimed from the sea around the south and east of the island, most of the sandy habitats have been destroyed or are now man-made and too steep for horseshoe crabs. Hence, sightings of *T. gigas* are now rare. There is a clear and urgent need for more studies on this species, particularly on spawning sites and timing. Tracking studies, as done on *C. rotundicauda* (Cartwright-Taylor et al. 2012), would help determine how far this species travels and which sites it prefers. Similarly, studies on *T. gigas* molts as a proxy for the live individuals would provide information on population dynamics, molting and growth rates as has been done on *L. polyphemus* in USA (Carmichael et al. 2011; Estes et al. 2015).

10.2.2 Republic Polytechnic Surveys

Many of these same sites around the north of Singapore were again surveyed for horseshoe crabs by Republic Polytechnic students during March and April (Lee 2012) and between September and November of 2012 (Lim 2013). All sites were along the Strait of Johor, with the exception of the Changi site to the east. Sarimbun was replaced with Sembawang because a crocodile had been sighted at Sarimbun and, for safety, access to the shore was denied. Sembawang site is a sandy beach with a substrate suitable for *T. gigas* where a canal opens to the Strait. The other side of the mouth of the canal is an area of inaccessible mangroves that may be suitable for *C. rotundicauda*.

C. rotundicauda was the only species found at Mandai and Lim Chu Kang, while nothing was seen at Changi beach or at Sembawang. However, during the two April surveys, ten and six live juvenile *T. gigas* were found at Sungei Tampines near Pasir Ris. This site is a mix of sand and mud, and although it is also at the mouth of a canal, the location of the search transects was along the Strait. Live *T. gigas* are rarely found and a cluster of juveniles, in particular, may indicate a return of this species to this site, a possibility that needs to be followed up.

10.2.3 National University of Singapore (NUS) Survey

More recently an NUS student (Leng 2013) studied the mangrove horseshoe crab population in the small tidal streams among the mangroves of Sungei Mandai Kechil. This is one of the largest remaining expanses of mangrove forest left in Singapore and is contiguous with and to the east of the Mandai mudflats, yet this area has not been included in previous horseshoe crab surveys or studies largely because of the difficulty of access. The area extends from the outflow of the canal, Sungei Mandai, to the causeway across the Strait to Johor and from the shore inland to the old Malaysia railway line. It is an area that has been untouched by development because the Malaysia railway line running parallel to the shore has protected the mangroves to its seaward side from easy access. However the closure and removal of this line means this area is now more accessible. Although this is good for study of the area, it is of grave concern for the future of the mangroves, as the site is now available for development. The area is not protected by law, and efforts by the NSS in 2009 to include this area in a protected wetland were unsuccessful. The area contains a rich biodiversity of flora and fauna, including locally endangered and rare mangrove species, and the mangroves form an important habitat for mangrove horseshoe crabs and a nursery area for many marine species.

Leng (2013) indicated that there is a healthy population of mangrove horseshoe crabs of all sizes in the tidal streams including adults in amplexus. This finding suggests that horseshoe crabs were breeding in the area, and the tidal streams are an important habitat for all stages of the life cycle.

10.2.4 National Parks Department Survey

The third island-wide survey was conducted by the National Parks Department of Singapore between May and December 2013 using belt transects and visual searches at 18 sites around the main island (Cheo and Lee 2014). *C. rotundicauda* was found at only 6 of the 18 sites where the species has been observed in previous surveys, all along the Strait of Johor on the north coast. *T. gigas* was observed only at one site, Changi Creek Mangrove, close to Changi Beach where *T. gigas* has been seen in previous surveys. With the possible exception of the canal at the Pandan reservoir in

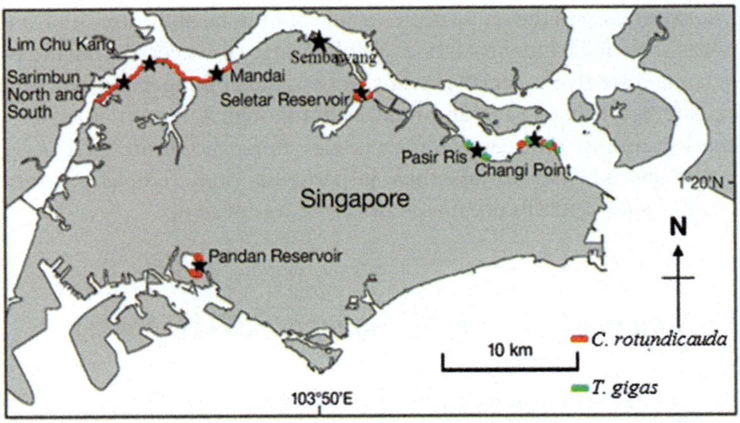

Fig. 10.2 Map of Singapore showing sites where horseshoe crabs have been observed during island-wide surveys

the south west, it seems clear that horseshoe crabs are no longer breeding and can no longer be found around the south coast of the main island, except perhaps for occasional sightings of single individuals washed up on the shore or trapped in nets. Figure 10.2 shows the distribution of horseshoe crabs from all surveys in Singapore.

10.2.5 Summary

C. rotundicauda From the surveys over time described above, it is clear that the situation for the mangrove horseshoe crab has not deteriorated over the years. There are still a few sites where *C. rotundicauda* is thriving, mainly concentrated around the north west of Singapore island. A contiguous population of mangrove horseshoe crabs exists from the causeway to the west past the Mandai mudflats, the SBWR, Lim Chu Kang, Sarimbun and down the west coast, probably along the untouched military reserve site as far as the second link bridge at Tuas across to Malaysia. In addition, a discontinuous population of this species exists along the east of the Strait of Johor. It is not known how much integration there is among all these populations or with sites across the Strait along the Malaysia coast. Genetic comparisons between individuals from all these sites and with crabs from sites across the Strait would be interesting to determine the genetic diversity, if any. Fragmentation of the habitat of the horseshoe crabs means that it is ever more important to protect and preserve the whole area of the north and west where there is a healthy breeding and probably contiguous population.

T. gigas By contrast, *T. gigas*, has been a cause for concern around Singapore for many years, and the situation has not improved. Individual sightings do occur and two adults in amplexus were caught by fishermen during tracking studies of mangrove horseshoe crabs in the Strait of Johor (Cartwright-Taylor et al. 2011) but were not included in the report, which concentrated on *C. rotundicauda*. A tag was attached to the male of this pair and was tracked for 6 days as he travelled a few miles along the Strait to the west. It is not clear if such individuals represent a small breeding population around Singapore or if they wandered across from Malaysia or elsewhere. Encouragement can be drawn from the fact that juvenile *T. gigas* have been seen on two occasions where they were not seen for many previous years. There also is an intermittent presence of *T. gigas* along the Strait particularly towards the extreme east. This species has been rather neglected as it has been so difficult to find, so work concentrating on *T. gigas* is urgently needed to determine why this species is so uncommon when it seems to co-habitat with *C. rotundicauda* in the Strait. Future work should also consider repopulation of suitable sites with individuals from Malaysia before they also disappear.

10.2.6 Density of Horseshoe Crabs

Only two sites in Singapore have populations of *C. rotundicauda* sizeable enough for density studies, Lim Chu Kang and Mandai mudflats at Kranji. Several studies have been conducted at Mandai since 2007, three by students and two by NSS volunteers, using transects of different sizes and shapes. All studies included juveniles and adults, both solitary and in amplexus, of all sizes as these form the normal population on the mudflats. All studies used collections from measured transects on the mudflats, but the size and shape of the transects differed among studies. Some studies used round quadrats but these present more practical difficulties on the mud surface than the more conventional rectangular or square shapes. These sampling differences could account for the lower density findings in some studies. All studies used visual searches for crabs on the surface of the mudflat and probing with fingers or small sticks for buried crabs. Rakes cannot be used on this surface.

The mean density for all studies combined was 0.46 per m^2 with a minimum and maximum of 0.16 per m^2 and 0.90 per m^2, respectively. The consistency of findings among studies and across months suggests there was no apparent seasonal pattern to the densities (Fig. 10.3). It is encouraging to note that the findings of the studies in 2012 closely match those of the earlier in studies in 2007–2009 (Cartwright-Taylor et al. 2011), suggesting that density has changed very little in the intervening years. The site is an important habitat for small juveniles as well as mature and mating adults. Any urbanization of this coastline, therefore, is expected to have a detrimental impact on this thriving population.

At Lim Chu Kang, further along the coast from Mandai, densities of 0.21 and 0.15 per m^2 were recorded by Lim (2013) at two site visits in March 2009. At the same site in Sept and Oct densities of 0.09 and 0 were recorded by Lee (2012). These are only slightly different from densities of 0.012 and 0.05 per m^2 recorded in two transects at the same site by NSS volunteers in March 2009 (unpublished data). However, in all these studies only two small transects were used each time at this site.

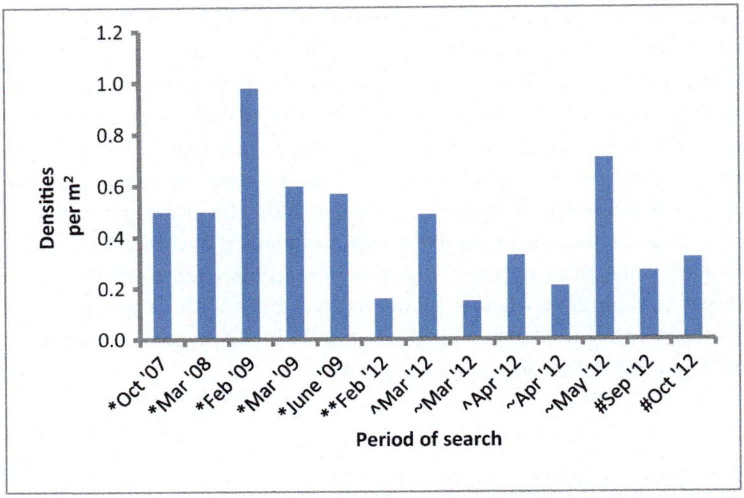

Fig. 10.3 Densities of *Carcinoscorpius rotundicauda* at Mandai mudflat (Compiled from *Cartwright-Taylor et al. 2011; ^Lee 2012; #Lim 2013; **NSS unpublished data; ~Saad 2013)

One recent study (Leng 2013) examined the density of *C. rotundicauda* in two main tidal streams in the mangroves at Sungei Mandai Kechil, further east of the mudflats at Kranji near the causeway to Johor where several streams branch throughout the mangroves. This area has a very different habitat from the open mudflat and comprises stands of mature mangroves cut by tidal streams draining into the Strait. From ten sampling dates between December 2012 and February 2013 at different moon phases, Leng calculated mean densities of adults and juveniles together of 0.037 per m^2 in one stream and 0.068 per m^2 in the other. The range for Stream 1 was 0.011–0.056 per m^2 and for Stream 2 slightly higher at 0.033–0.099 per m^2. It is interesting to note that these densities are an order of magnitude lower than those at the open Mandai mudflats about 200 m away (compare to Fig. 10.3). Indeed the mangrove forest of Sungei Mandai Kechil adjoins the open mudflats so there is unbroken habitat for the horseshoe crabs from the open mudflat site to deep into the mangroves until the causeway. It is not clear why the density should be so much higher just a short distance along the shore, but studies on the salinity of the tidal streams and other environmental parameters may provide an answer. Nevertheless, even these densities are higher than most population studies conducted elsewhere for Asian Pacific horseshoe crabs (Hu et al. 2009; Morton and Lee 2011).

Leng's findings on density showed a trend towards a link with moon phase such that both adult and juvenile densities tended to increase during new and full moons and decrease during half moon phases in both the streams studied. This trend was not confirmed by statistical analysis, possibly because the sample numbers were small, and therefore, require further confirmation. Moon phase and tidal changes are so inextricably linked, however, that it would be difficult to determine which of these may be the driver for any density change. This work certainly is worth repeating on a larger and longer scale and at more sites. These fluctuations in density suggest that there may be movement to and from the tidal streams, possibly to the nearby mudflats, triggered by moon phase, tidal height or changes in salinity in the streams. This study appears to be the first in Singapore to correlate population density with moon phases, a topic that warrants further work. More work is needed on the horseshoe crabs in the mangrove streams and the mudflats as a single entity to determine the link between the two habitats.

Another student study examined, among other things, the spatial and temporal distribution of *C. rotundicauda* on the Mandai mudflats (Ong 2012a). While adults were evenly distributed on the mudflats from the shore to the low water zone, the juveniles were concentrated 1–23 m from the high tide line. Since the eggs are laid at the high tide level it is to be expected that the juveniles would also be found in this region on the mudflats. The exposed mudflat from the high to low spring tide line covers a distance of about 250 m with a maximum tide height of about 2.5 m, a wide expanse where the horseshoe crabs can be found. Thus 23 m from shore is a very small part of the total area exposed at low tide. However, this distribution means that any coastal development that includes the high tide area of the shoreline will have a profound effect on the population because there will be no place for the eggs and young juveniles.

The students also found a trend toward decreasing numbers of both adults and juveniles as the distance from shore increased, with the highest population density between 10 and 50 m from shore. In addition, they observed that from March to December, excluding June to August, the mean distance between a total of 437 *C. rotundicauda* individuals at Mandai was 313 cm (SD 401.78), demonstrating solitary rather than social behavior. This observation supports the view of Shuster et al. (2003) that it is in the nature of horseshoe crabs to avoid one another. Clustering of mangrove horseshoe crabs is not seen on the open mudflats (personal observation).

10.3 Demography and Population Structure

All demography studies have been done on *C. rotundicauda* because *T. gigas* is too rare now for such studies. An NSS study conducted during 1 year starting in 2007 (Cartwright-Taylor et al. 2009) with annual follow up for a further 3 years (Cartwright-Taylor and Hsu 2012) demonstrated a pattern of monthly changes in the population structure of *C. rotundicauda* at the Mandai mudflats.

The proportion of small juveniles (1–3 cm prosomal width) was highest in November 2007 and January 2008, 25 % and 30 % respectively. In June and July only 8 % and 4 % respectively were in these same size classes. By June and July recruitment to the larger size classes (10–12 cm) was approximately 35 %. In captivity it takes at least 369 days for hatching and growth to 1.7 cm prosomal width (Zadeh et al. 2009). At Mandai, juveniles less than 2 cm were not found in June, suggesting that there may be a rest period of low or no breeding activity from May to July each year resulting in none of the smallest sizes in June and July of the following year. The follow-up work (Cartwright-Taylor and Hsu 2012) indicated that this was not just a 1 year phenomenon but that it occurred year-after-year.

These findings are supported by more recent student studies (Ong 2012a; Saad 2013) on this population of *C. rotundicauda* at the same site, during the same time period (Figs. 10.4 and 10.5). Both studies also reported increasing proportions of adults (>7.9 cm carapace width) in the population coupled with decreasing proportions of juveniles (<8 cm carapace width) during visits from March to May 2012. Saad (2013) recorded carapace widths in only three size classes.

Site visits were repeated between September and December 2012 (Ong 2012b) during which time adult horseshoe crabs continued to dominate the population. A total of 507 horseshoe crabs were collected during these 4 months, of which 89.2 % were adult (45 % males, 44.2 % females) and only 10.8 % were juveniles, compared with 51.3 % adults (21.4 % males, 28.9 % females) and 49.7 % juveniles from a total of 532 horseshoe crabs collected between March and May 2012 (Ong 2012a).

One student study (Leng 2013) focused on the population structure of *C. rotundicauda* in the small tidal streams deep among the mangroves at Sungei Mandai Kechil, 200 m along the shore to the east of the Mandai mudflats at Kranji, and a very different habitat from the open mudflats. There, the population structure of horseshoe crabs in two streams in Sungei Kechil from December 2012 to February 2013 (Leng 2013) showed differences in spite of the proximity of the streams to one another; 30 % in one stream were juveniles (<8 cm carapace width) compared with 69 % in the other stream. However, the numbers collected from the first stream were substantially smaller than the second stream, and this may have affected the proportions to some extent. Nevertheless these figures support those of Cartwright-Taylor and Hsu (2012) where higher proportions of juveniles occurred from December to March. These streams, therefore, may be important nursery areas for this population, and coastal urbanization or other destruction of this habitat could be detrimental to the population of *C. rotundicauda* on the nearby mudflats.

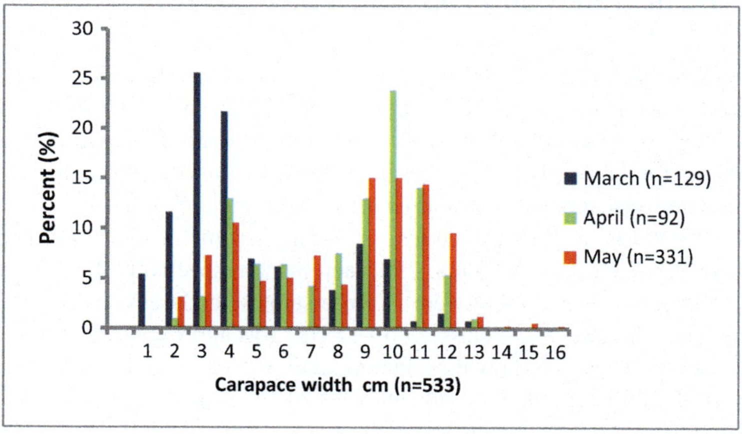

Fig. 10.4 Frequency of size classes (prosomal width) of *Carcinoscorpius rotundicauda* collected from March to May 2012 at Mandai mudflats (From raw data in Ong 2012a)

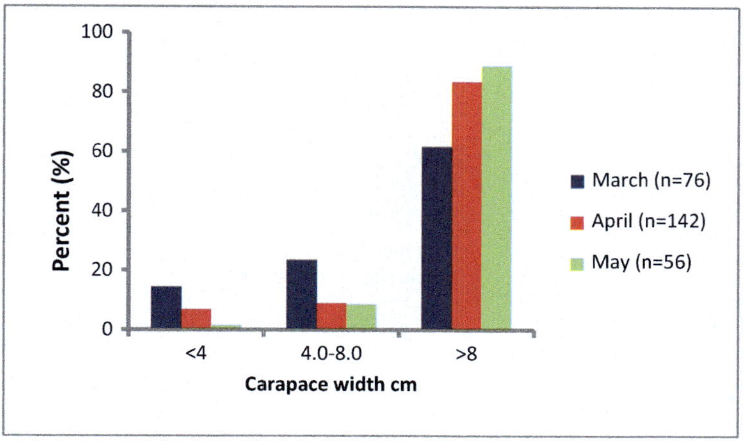

Fig. 10.5 Frequency of size classes (carapace width) of *Carcinoscorpius rotundicauda* collected from March to May 2012 at Mandai mudflats (From raw data in Saad 2013)

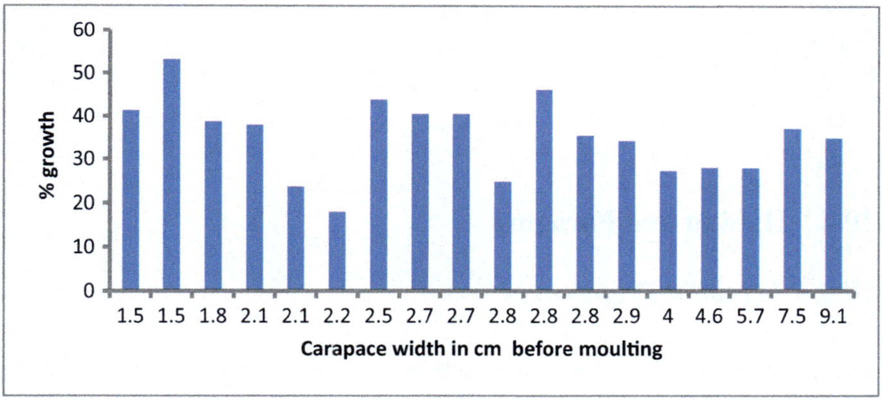

Fig. 10.6 Growth rates during ecdysis of 18 individual *Carcinscorpius rotundicauda* from Mandai mudflats

10.3.1 Growth

From 2007 to 2012 a total of 18 individual *C. rotundicauda* from the wild population at Mandai mudflats were collected while in the process of ecdysis. The prosomal width was measured before and after complete ecdysis, and the percentage growth recorded. Pre-molting widths varied from 1.5 to 9.1 cm, and mean incremental growth was 35.4 % (SD 8.8) (Fig. 10.6). Percentage growth in the largest crab was not very different from that in the smallest. Further studies on molting and growth in the wild population would be interesting to compare with such studies on captive individuals.

Table 10.1 Sex ratios of *Carcinscorpius rotundicauda* at Mandai mudflats from student studies

	Mean (range)	
	March–May	Sept–Dec
Lim (2013)		0.74 (0.60–0.88)
Lee (2012)	0.89 (0.86–0.92)	
Saad (2013)	1.05 (0.4–1.7)	
Ong (2012a, b)	0.74[a]	1.02[a]
Goh (2010)		0.82[b]

[a]Four months combined
[b]Single site visit Sept 2009

10.3.2 Sex Ratios

Sex ratios of *C. rotundicauda* at the Mandai mudflats from all the student studies in 2012 (Table 10.1) support published figures (mean 1.28; SD 0.78) from the same site during a 4-year period prior to 2011 (Cartwright-Taylor and Hsu 2012). The ratios altered little from 1 throughout the year. All studies confirmed that mating activity was not strongly seasonal, as occurs in temperate climates. In addition pairs in amplexus were seen year-round in varying numbers and with no discernible monthly pattern.

10.4 Habitat and Foraging

10.4.1 Habitat Preferences

Two studies have been conducted on sub-habitat preferences and size class among the mangroves that are flooded at high tide. Between 5th February and 3rd March 2004 one study was conducted along the Kranji Nature Trail (KNT) (Hong 2004) a few miles west of the Mandai mudflats. The second study (Leng 2013) was conducted from December 2012 to February 2013 at Sungei Mandai Kechil, 200 m to the east of the Mandai mudflats. Both sites comprise a range of sub-habitats deep within the mangroves, and the studies included streams over mud where water is fast flowing, streamlets over mud that drain into the streams and where water velocity is slower and the bed muddier, small pools, mudflats, and, in the eastern site, sea grass, *Halophila beccarii*, over mud, and sand, a grittier substrate than mud. Mud and sand were differentiated by the texture of sediment. In both studies, adults and juveniles of different size classes showed different preferences for these microhabitats.

Juveniles (<8.6 cm) comprised 70 % of 162 individuals found at KNT (Hong 2004). Later studies on *C. rotundicauda* at Mandai have used 8 cm as the threshold for adults based on sizes of pairs in amplexus, and indeed this may be too high because rarely smaller males in amplexus have been found during recent visits to

Mandai (7.5–7.9 cm; unpublished data). The high percentage of juveniles is not surprising as we now know from the studies on population structure over the years, that February and March are months when the proportion of juveniles to adults is high (as previously described).

In both studies, adults (above 8.0 cm) and the larger juveniles (up to 7.0 cm) were more commonly found in larger streams and mud, while smaller juveniles (down to 1.0 cm) were more numerous in small pools, shallow streamlets, and on the mud surface. Hong (2004) observed no juvenile <6.0 cm carapace width was buried in the mud, but juveniles larger than this size were found buried and were thus absent from streams. She postulated that the small juveniles were not strong enough to bury themselves in mud.

At the Sungei Mandai Kechil site, deep in the mangroves, among juveniles, size was significantly different among habitats (Kruskal Wallis $p=2.33\times10^{-7}$), indicating that juveniles of different sizes preferred different habitats (Leng 2013). Smaller juveniles (10–42.6 mm) tended to be in sea grass habitats, sand (10–40.4 mm) or in small streams (10–60.3 mm), while larger juveniles (12.6–76.7 mm) were found in faster flowing streams. Leng (2013) postulated that this preference was due to the higher oxygen content in streamlets and sea grass that juveniles prefer. Morton and Lee (2011) found higher numbers of *T. tridentatus* juveniles in substrates with high oxygen concentration (8–14 mg/L). Water velocity in streams, however, may wash away the small juveniles resulting in a higher proportion of larger juveniles in streams.

There was no statistically significant difference in the size of adults among habitats (Kruskal Wallis $p=7.97\times10^{-1}$), indicating that adults were randomly distributed among the different habitats. The sex ratios were similar in both studies, 0.9 in KNT (Hong 2004) and 2.0 and 1.5 in the two streams at Sungei Kechil (Leng 2013), suggesting that males and females had similar habitat preferences. Sand appeared to be the least preferred habitat for adult and juvenile mangrove crabs alike.

The study by Hong (2004) was one of the earliest studies on horseshoe crabs in Singapore. In spite of the generally accepted local view that breeding occurred year round, Hong postulated a seasonality of mating that has now been demonstrated as a period of low breeding activity rather than true seasonal breeding. Her recommendation that year-round monitoring is required to determine such seasonality is ongoing by the NSS. She also noted that the numbers of amplexed pairs, albeit only seven, but with no gravid females, might also imply that January to March is not the mating season. Longer studies now suggest that the time of low breeding activity is probably from May to August. These findings highlight the value of longer and more intensive studies to follow up and corroborate smaller and shorter duration studies.

There also is a need for more studies on the length of time *C. rotundicauda* spends in amplexus and how much of this time is spent nesting and spawning in a region with no seasons. Anecdotal evidence from local people and from pairs caught in the wild but kept in captivity by the Singapore Zoo suggests that *C. rotundicauda* remains in amplexus for months rather than weeks, and much of this time does not seem to involve spawning. This pattern of behavior may be different compared with species in temperate climates, where coupling and spawning are seasonal and happen in just a few weeks in the spring, and amplexus out of season is uncommon. The assumption that adults in amplexus indicate a breeding season may not be correct if pairs remain together for extended periods. Studies using tracking devices on pairs of crabs in amplexus would be useful to determine how long they stay together.

Observations of Leng (2013) on microhabitat preference support those of Hong (2004), nearly a decade earlier in a different site but with similar microhabitats, making these studies valuable as a baseline for the population of *C. rotundicauda* in this habitat. Within the mangroves the tidal streams are not homogenous, resulting in many different microhabitats. These studies show that adults and juveniles have different preferences for microhabitats. *C. rotundicauda* selects for different environmental parameters at different stages of the life cycle. These tidal streams among the mangroves may be important for spawning and as nursery areas safe from predators such as large crabs, but spawning surveys are needed to determine when and where spawning occurs as it has rarely been seen and few egg nests have been found.

10.4.2 Habitat Degradation and Mortality

The percentage of dead to live *C. rotundicauda* found in defined areas at two different locations of the Mandai mudflat at Kranji have been compared to determine the influence pollution may have on the survivability of horseshoe crabs (Saad 2013). One location was heavily polluted and nearer the shore while the other location, 10 m further down the beach, was not so polluted. In addition to flotsam and jetsam, the mudflats are polluted with oils and road run off from the many small, contaminated rivulets and canals that drain from the town out to the mudflats. Patches of black mud with an unpleasant smell indicating the presence of sulphides can also be seen along the whole of the intertidal zone, particularly near the shoreline.

From seven site visits, the percentage of dead crabs was consistently higher in the polluted location (19.8 %) than in the less polluted location (7.5 %). This difference may be due to greater pollutant loads, which have been recorded along the Strait (Bayen et al. 2005; Cuong et al. 2005) or to entrapment of carcasses by debris at the more littered site. It would be interesting to pursue the mortality study by continuing with chemical or other tests that could suggest possible causes of the higher mortality rate in the high tide zone. Regular monitoring of organic and inorganic pollutants in the horseshoe crabs' habitat is important, particularly if further development of industry along that coast is planned. However, it is a feature of these student projects that, by their nature, there is little or no follow up or subsequent study to build on what was found.

10.4.3 Epibionts

Ong (2012b) compared the number of the common acorn barnacle, *Semibalanus balanoides*, attached to male and female mangrove horseshoe crabs in amplexus to determine if these epibionts would affect amplexus. Of 157 amplexus pairs 56.7 % of the males (89/157) and 32 % (51/157) of the females had barnacles. The higher rate of attachment was attributed to the larger surface area available for epibionts to attach while the pairs are in ecdysis, which may be several weeks. Epibionts do not seem to affect amplexus formation. In addition, the numbers of barnacles were higher on the larger crabs, but they were also more numerous on males than females of the same size. A positive correlation between horseshoe crab size and the size of the largest barnacles of *Ballanus* species had also been observed by Goh (2010) who also observed that females carried larger barnacles (mean 9.3 mm) than males (mean 8.6 mm).

A similar study 25 years ago (Jeffries et al. 1989) of the pedunculate barnacle, (*Octolasmis warwickii*), on both species of horseshoe crab collected from around Singapore, showed three times greater abundance of this species on the 18 female *T. gigas* than on the 38 males, and all were on the ventral prosoma, while none of 19 *C. rotundicauda* carried this barnacle. Twenty five years ago *T. gigas* was still to be found around the Singapore shores, the highest number (43) being at Tuas on the west coast of the main island, an area that has not been included in recent surveys, but should be considered for future work, particularly on *T. gigas*.

10.4.4 Diet Preferences

In two small studies to determine food preferences in *C. rotundicauda* collected at Mandai (Ong 2012a, b), an individual horseshoe crab was turned on its back, selected food was put onto the mouth area, and the time taken for the food to be consumed was measured. It is not normal for horseshoe crabs to feed this way so the findings need to be interpreted with caution. Red bristle worms (Polychaeta, species not given) were popular with the adults and juveniles alike, while seaweed (species not given but probably *Halophila beccarii*) and the flesh of blood cockles (*Anadara granosa*) were not consumed at all. Only the adults consumed the flesh of white clams (species not given) but took an average of 40 min to do so, possibly because of the size of the food offered (Ong 2012a).

In a follow up study (Ong 2012b) red bristled worms, blood worms (*Chironomidae tetans*), white clams and mussels (*Perna viridis*) were used. In this study 23 horseshoe crabs were offered each of these items as food, two to three times, and the numbers of crabs feeding on the offering were counted. No preferences were observed among the food sources offered (not significant at the 1 % level), and the conclusion was that horseshoe crabs were not particularly selective feeders. Generally horseshoe crabs are considered to be omnivorous scavengers feeding on whatever they encounter, such as bivalves, worms, molluscs and even algae (Goh 2001; Ng and Sivasothi 2002).

Ong (2012a, b), however, was offering various food species to the horseshoe crabs in the laboratory to determine preferences. This approach does not identify what horseshoe crabs would actually eat in their normal habitat. By contrast, Akbar et al. (2012) conducted a study in Malaysia on feeding ecology and food preferences by examining gut contents of *C. rotundicauda*. They collected individuals from two known nesting sites on the east coast of Malaysia in Pahang. Pairs of crabs were collected every month for a year shortly after nesting, and in the laboratory gut contents were identified to group level (bivalves, gastropods, crustaceans, polychaetes and other). Macrobenthic samples were also collected to determine the availability of food sources. The feeding choices of the crabs differed according to the monsoon period, with preference changing from polychaetes to bivalves in periods between monsoons and back to polychaetes during monsoon periods. Overall, molluscs, especially gastropods, were preferred over other macrobenthos groups.

It would be useful to conduct similar studies on gut contents of horseshoe crabs from sites in Singapore. Gut contents, however, provide information only on recently consumed food and food that is hard to digest and remains in the gut. A more fruitful method of determining foraging ranges and food sources may be to compare stable isotope ratios in the horseshoe crabs with food sources in different locations as has been done with *Limulus polyphemus* in USA (Carmichael et al. 2004). Not only would such information help with the captive breeding of *C. rotundicauda* and in determining migration patterns, but it would be particularly interesting to determine if the waste food from under the fish farms in the Strait of Johor forms part of the diet of the resident crabs. The whole area along the north west of Singapore, although small, supports a very high density of horseshoe crabs, so it would be useful to know how much the fish farms contribute to this success, if at all.

10.4.5 Dietary Contaminants

The study by Ong (2012b) included one species that was also included in a study on persistent organic pollutants (POPs; Bayen et al. 2005) in marine samples collected nearby along the coast. The green mussel (*Perna viridis*) is a filter feeder at the top end of the trophic scale (3.1) among the species studied. Since horseshoe crabs feed on this species when offered, it can be assumed that horseshoe crabs may be a trophic level above green mussels, and they are likely to have similar concentrations to the higher trophic level species. The highest concentration of POPs was found in predator species in the higher trophic levels and in detritivores (e.g. rodong shell), indicating a biomagnification in the higher trophic levels (Bayer et al. 2005). Since horseshoe crabs forage on detritus and smaller aquatic species, they may well have POP concentrations similar to those species studied.

10.5 Future Work

Although much research has been published on the biochemistry of the blood of the mangrove horseshoe crab in Singapore, little has been published on the ecology and behavior of this species and even less on the coastal horseshoe crab in Singapore. Hence, any studies, however small, may make a useful contribution to increasing knowledge of these creatures. Although the existing student studies are necessarily small and often of short duration, the findings prove useful and important because they contribute to the continuous monitoring of the horseshoe crab populations around Singapore. This work is particularly important at Kranji where there is a healthy population of mangrove horseshoe crabs. The methodology described in these studies is straightforward, so results through time and among studies can be compared with caveats being considered for experimental differences.

These studies were conducted by students of different institutions, they are not published, and the findings are often not easily available to students or researchers at other institutions, resulting in the risk of duplication. Now that the Nature Society of Singapore has become known for its work on horseshoe crabs through publications, publicity and the NSS journal, some students are sending their dissertations to the NSS horseshoe crab team for comment. This step is very helpful to allow findings to be collated, compared, and used for planning future work, which is required to build on these studies.

The NSS has an outreach program with most of the colleges, so is in a good position to encourage, suggest and guide future work to plug knowledge gaps. With so few habitats in Singapore remaining for either of the horseshoe crabs species, but in particular *T. gigas*, all research findings should be gathered and used to help the efforts of the NSS and other interest groups to preserve what is left of the natural sea shore of Singapore.

The horseshoe crab habitats along the Strait of Johor are important in their own right, but also as support for the protected area of the Sungei Buloh Wetland Reserve (SBWR), the Mandai mudflats, and Sungei Mandai Kechil streams that are vital for the survival of the Reserve. The whole stretch of coastline needs to be considered as an integrated, inter-dependent and inter-connected ecosystem, and there is much ongoing discussion from many interested groups to press for the whole area to be listed as a Ramsar site (The Convention on Wetlands of International Importance). Importantly, the work completed to-date strongly suggests that urbanization or industrial development that destroys any part of this coastline will significantly impair recovery of the remaining horseshoe crab populations and will also risk adversely affecting the SBWR, one of the major wetland systems of the country.

References

Akbar JB, Kamaruzzaman BY, Jalal KCA et al (2012) Feeding ecology and food preferences of *Carcinoscorpius rotundicauda* collected from the Pahang nesting grounds. Sains Malaysiana 41(7):855–861

Bayen S, Wurl O, Karuppiah S et al (2005) Persistent organic pollutants in mangrove food webs in Singapore. Chemosphere 61:303–313

Carmichael RH, Rutecki D, Annett B et al (2004) Position of horseshoe crabs in estuarine food webs: N and C stable isotopic study of foraging ranges and diet composition. J Exp Mar Biol Ecol 299:231–253

Carmichael RH, Estes MG, Brady AJ (2011) Molts reveal life-history patterns of American horseshoe crabs in fringe habitats. International workshop on the science and conservation of Asian horseshoe crabs, Hong Kong, 13–16 June 2011

Cartwright-Taylor L, Hsu CC (2012) Follow-up study on population structure and breeding pattern of the mangrove horseshoe crab *Carcinoscorpius rotundicauda* in Singapore. Aquat Biol 14:217–222

Cartwright-Taylor L, Lee J, Hsu CC (2009) Population structure and breeding pattern of the mangrove horseshoe crab *Carcinoscorpius rotundicauda* in Singapore. Aquat Biol 8:61–69

Cartwright-Taylor L, Bing YP, Hsu CC et al (2011) Distribution and abundance of horseshoe crabs *Tachypleus gigas* and *Carcinoscorpius rotundicauda* around the main island of Singapore. Aquat Biol 13:127–136

Cartwright-Taylor L, Ng NN, Goh TY (2012) Tracked mangrove horseshoe crab *Carcinoscorpius rotundicauda* remain resident in a tropical estuary. Aquat Biol 17:235–245

Cheo PR, Lee ISH (2014) Population distribution of the horseshoe crabs *Carcinoscorpius rotundicauda* and *Tachypleus gigas* in Singapore. Poster presentation at biodiversity research symposium 2014, Singapore, 24 May 2014

Cuong DT, Bayen S, Wurl O et al (2005) Heavy metal contamination in mangrove habitats of Singapore. Mar Pollut Bull 50:1713–1744

Davidson GWH, Ng PKL, Ho HC (eds) (2008) The Singapore Red Data Book: threatened plants and animals of Singapore. The Nature Society (Singapore), Singapore

Estes MG Jr, Carmichael RH, Macdonald PDM et al (2015) Molts reveal life-history patterns for juvenile American horseshoe crabs in fringe habitats. In: Carmichael RH, Botton M, Shin PKS, Cheung SG (eds) Changing global perspectives on horseshoe crab biology, conservation and management. Springer, Cham, pp 255–278

Goh L (2001) Survivors of time. Wetlands 8(2). https://www.sbwr.org.sg/wetlands/text/01-8-2-1.htm. Accessed 29 Jan 2014

Goh M (2010) Population studies and estimation of age and moulting frequency of *Carcinoscorpius rotundicauda*. Research paper Natural Sciences and Science Education Academic Group, NIE, Singapore

Hong RF (2004) Population and distribution of horseshoe crab, *Carcinoscorpius rotundicauda* at the Kranji Nature Trail estuaries, Western Johor Straits. Dissertation, National University of Singapore

Hu MH, Wang Y, Chen Y et al (2009) Summer distribution and abundance of juvenile Chinese horseshoe crabs *Tachypleus tridentatus* along an intertidal zone in southern China. Aquat Biol 7:107–112

Jeffries WB, Voris HK, Yang CM (1989) Observations on the incidence of the pedunculate barnacle, *Octolasmis warwickii*, (Gray, 1825) on horseshoe crabs (Xiphisura) in the seas adjacent to Singapore. Raffles Bull Zool 37(1&2):58–62

Lee SKI (2012) Baseline survey and analysis of the intertidal distribution of horseshoe crabs in Singapore. Dissertation, Republic Polytechnic, Singapore

Leng JMG (2013) Population and distribution of mangrove horseshoe crab (*Carcinoscorpius rotundicauda*) in the tidal streams of Mandai Kechil 3 stream system of Sungei Mandai Kechil. Dissertation, National University of Singapore

Lim SM (2013) Baseline survey & analysis of the intertidal distribution of horseshoe crabs in Singapore. Dissertation, Republic Polytechnic, Singapore

Morton B, Lee CN (2011) Spatial and temporal distribution of juvenile horseshoe crab (Arthropoda: Chelicerata) approaching extirpation along the northwestern shoreline of the New Territories of Hong Kong SAR, China. J Nat Hist 45:227–251

Ng PKL, Sivasothi N (2002) A guide to the mangroves of Singapore: I (the ecosystem and plant diversity), 2nd edn, BP Guide to Nature series. Singapore Science Centre, Singapore

Ong HY (2012a) Foraging patterns and spatial and temporal distribution of horseshoe crabs *Carcinoscorpius rotundicauda*. SAS-Sem2-2012-0096 Dissertation, Republic Polytechnic, Singapore

Ong HY (2012b) Foraging patterns and spatial and temporal distribution of horseshoe crabs *Carcinoscorpius rotundicauda*. SAS-Sem1-2012-0104 Dissertation, Republic Polytechnic, Singapore

Saad MSBM (2013) Monitoring the demographics of horseshoe crabs. Dissertation, Republic Polytechnic, Singapore

Shuster CN, Barlow RB, Brockmann HJ (2003) The American horseshoe crab. Harvard University Press, Cambridge, MA

Zadeh S, Christianus A, Saad CR et al (2009) Comparisons in prosomal width and body weight among early instar stages of Malaysian horseshoe crabs, *Carcinoscorpius rotundicauda* and *Tachypleus gigas*, in the laboratory. In: Tanacredi JT, Botton ML, Smith DR (eds) Biology and conservation of horseshoe crabs. Springer, New York, pp 267–274

Chapter 11
The Population Size and Movement of Coastal Horseshoe Crab, *Tachypleus gigas* (Müller) on the East Coast of Peninsular Malaysia

Mohamad Faridah, Noraznawati Ismail, Amirrudin Bin Ahmad, Azwarfarid Manca, Muhammad Zul Fayyadh Azizo Rahman, Muhammad Farhan Saiful Bahri, Muhd Fawwaz Afham Mohd Sofa, Izzatul Huda Abdul Ghaffar, Amirul Asyraf Alia'm, Nik Hafiz Abdullah, and Mohd Mustakim Mohd Kasturi

Abstract Horseshoe crabs have recently gained popularity as new exotic delicacies in Malaysia, especially in Johor. It has been reported and claimed by some fishermen that horseshoe crabs, particularly females, are being transported in bulk to Thailand. This claim has sparked interest to identify population size and study the behaviour of horseshoe crabs along the east coast of Peninsular Malaysia in an attempt to determine the effect of this demand on the existing population. A study was carried out from December 2011 to March 2012 to determine the coastal horseshoe crab *Tachypleus gigas* population size in Chendor and Cherating, Pahang, about 10 km apart on the east coast of Peninsular Malaysia, using a Capture-Mark-Recapture method. The sampling was carried out three times at both sites. The population size was calculated using Bailey's Triple Catch formula with the assumption of an open population. The horseshoe crabs' 'traveling distance' was also determined by attaching numbered, white button tags, each bearing the researcher's phone number, to a total of 99 horseshoe crab individuals (61 in Chendor, 38 in Cherating). The total estimated populations were 7,140 in Chendor and 9,900 in

---

This article is dedicated to our MSc student, Izzatul Huda, who died of cancer at the very young age of just 26 on October 20th, 2014.

M. Faridah (✉) • A.B. Ahmad • A. Manca (✉) • M.Z.F.A. Rahman • M.F.S. Bahri
M.F.A. Mohd Sofa
School of Marine and Environmental Sciences, Horseshoe Crab Research Group (HCRG), Universiti Malaysia Terengganu, 21030 Kuala, Terengganu, Malaysia
e-mail: mfaridah@umt.edu.my; azwarfarid@ymail.com

N. Ismail • A.A. Alia'm
Institute of Marine Biotechnology, Horseshoe Crab Research Group (HCRG), Universiti Malaysia Terengganu, 21030 Kuala, Terengganu, Malaysia

I.H.A. Ghaffar • N.H. Abdullah • M.M. Mohd Kasturi
School of Fundamental Sciences, Horseshoe Crab Research Group (HCRG), Universiti Malaysia Terengganu, 21030 Kuala, Terengganu, Malaysia

R.H. Carmichael et al. (eds.), *Changing Global Perspectives on Horseshoe Crab Biology, Conservation and Management*, DOI 10.1007/978-3-319-19542-1_11

Cherating. The two populations were found to be interchangeable as the tagged animals from Chendor were recaptured in Cherating and vice versa, thus confirming emigration and immigration. About 25 % of tagged animals were recaptured in various places (at Chendor, Cherating and also outside both locations) as reported by fishermen. Animals tagged in Cherating were caught in a few locations in Pahang, including Tanjung Lumpur which is more than 50 km to the south. Most recaptures (40 %) were from different locations remote from the site where the horseshoe crabs were tagged and released. This study is on-going and notifications are still expected regarding the future location of the tagged animals.

Keywords *Tachypleus* • Malaysia • Tagging • Population size • Movement pattern

11.1 Introduction

The coastal horseshoe crab *Tachypleus gigas* is one of the three Asian horseshoe crab species of sub-family Tachypleinae. This creature, along with three other species, tri-spine horseshoe crab *Tachypleus tridentatus*, mangrove horseshoe crab *Carcinoscorpius rotundicauda* and the American horseshoe crab *Limulus polyphemus* (sub-family: Limulinae), are believed to have lived and crawled in this world's Precambrian seas more than 400 million years ago, surviving all five massive extinction periods with little morphological change from their ancestor (Chatterji 1994; Mattei and Beekey 2008). Three horseshoe crab species can be found in the waters of Malaysia (Fig. 11.1). The coastal horseshoe crab *T. gigas* and mangrove

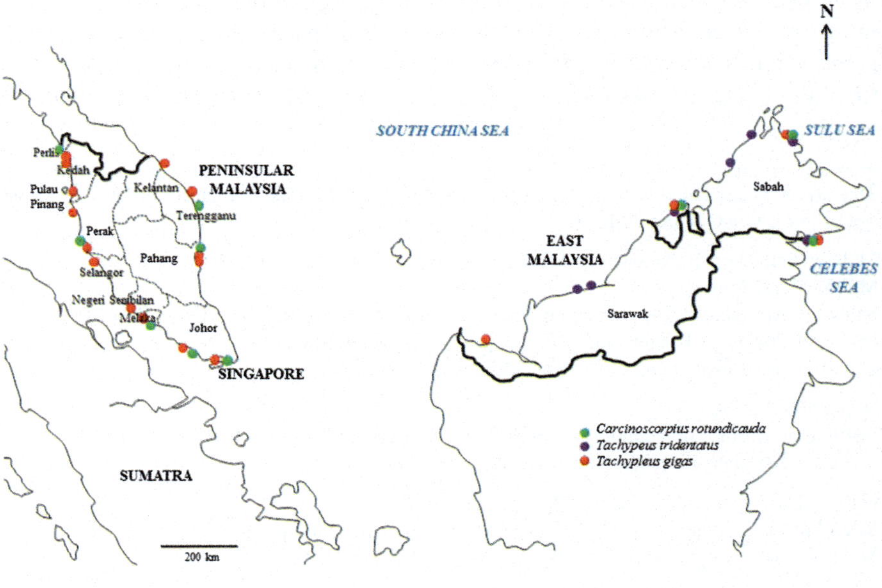

Fig. 11.1 The distribution of all three Asian horseshoe crab species in Malaysia (Hajeb et al. 2009; Srijaya et al. 2010; Ismail and Sarijan 2011; Akbar et al. 2012a; Tan et al. 2012; Zaleha et al. 2012; Rozihan et al. 2013; Chatterji and Pati 2014; Robert et al. 2014, and personal observations)

horseshoe crab *C. rotundicauda* have been reported from both the east and west coasts of Peninsular Malaysia (Kassim et al. 2008; Srijaya et al. 2010; Ismail and Sarijan 2011; Tan et al. 2012; Zaleha et al. 2012), whereas *T. tridentatus* can only be found in Malaysian Borneo.

Like its counterpart in India (Mishra 2009), horseshoe crabs in Malaysia are not subjected to exploitation for biomedical, fertilizer, or bait purposes. However, recently *T. gigas* has gained popularity as a new exotic delicacy in Malaysia, especially in the state of Johor. We believe this harvest is now the greatest threat to horseshoe crabs, particularly *T. gigas* in Malaysia. Export to Thailand has also been reported (Christianus and Saad 2009), and some fishermen claim that horseshoe crabs, especially females, are being transported in bulk. Some have claimed that, at certain times, tons of animals are transported to Thailand for consumption.

Interviews with local fishermen on the east coast of Peninsular Malaysia reveal that they have been selling horseshoe crabs to middle men for more than a decade now, and they have noticed a declining catch ever since. However, reports on the size of the population of horseshoe crabs in Malaysia are lacking. Many studies on this animal in Malaysia focused more on the biochemical and other ecological aspects (Akbar et al. 2010, 2011, 2012a, b, 2013; Zaleha et al. 2010, 2012; Ismail and Sarijan 2011; Tan et al. 2011, 2012; Ismail et al. 2012; Rozihan and Ismail 2012; Rozihan et al. 2013). To fill that gap, this study was conducted to provide an estimate of the population size of *T. gigas* at two sites and to determine the movements of horseshoe crabs in east coast waters. We hope to define the current status and improve understanding of this animal on the east coast of Peninsular Malaysia to guide sustainable management and conservation efforts, especially in Malaysia.

11.2 Materials and Methods

11.2.1 Study Sites

This study was carried out in two different locations; Chendor (N 04° 10.642′; E 103° 25.261) and Cherating (N 04° 07.309; E 103° 23.155), which are located at the northern part of Pahang on the east coast of Peninsular Malaysia (Fig. 11.2). Both study sites are sandy beaches (Fig. 11.3), popular among tourists with many motels and resorts in the vicinity. These locations were selected based on reports of horseshoe crab sightings by locals. This study was carried out from December 2011 to March 2012 with three sampling sessions at each site.

11.2.2 Estimation of Population Size (Capture-Mark-Recapture Methods)

Horseshoe crabs were caught using gill nets with 12.7 cm mesh size and 5 m (width) × 200 m (length), a type of net commonly used by local fishermen to catch crabs within the study area. The nets were deployed with the help of hired fishermen

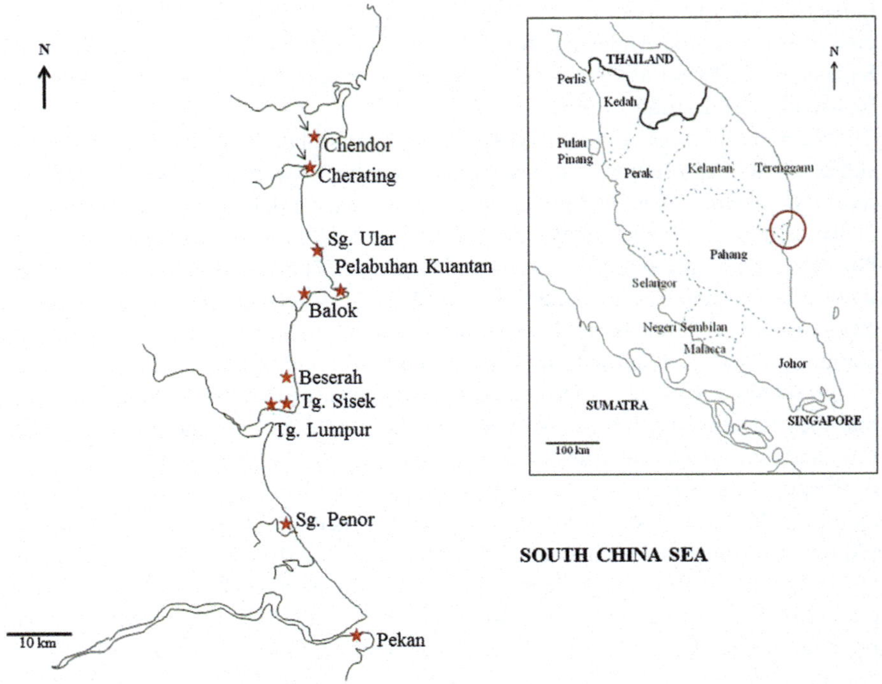

Fig. 11.2 The locations of sampling sites are indicated by arrows, Cherating and Chendor. The (*star*) mark shows the location of recapture from tagged animals. *Inset* is the map of Peninsular Malaysia showing the tagging and recaptures locations

2–4 km offshore in waters approximately 30 m deep, and checked once 24 h after deployment. All tangled horseshoe crabs were removed from the net and placed in the boat for marking. Each individual horseshoe crab was marked on the right side of the prosoma with a unique coloured paint according to the capture time (i.e. yellow for first time capture, red for second and green for third time capture). All marked horseshoe crabs were then immediately released at the capture point. The next recapture was done at least 48 h after the release, allowing sufficient time for the marked individuals to mingle and distribute themselves among the unmarked population. The same marking protocols were applied for second and third sampling sessions. All the data were recorded and the population size (P) was estimated using Bailey's Triple-Catch formula as follows:

$$\mathrm{P} = \mathrm{N} + \mathrm{Va} \tag{11.1}$$

$$\mathrm{N} = a_2 (n_2 + 1) r_{31} / (r_{21} + 1)(r_{32} + 1) \tag{11.2}$$

$$\mathrm{Var} = \mathrm{N}^2 - a_2^2 (n_2 + 1)(n_2 + 2)(r_{31} - 1) r_{31} / (r_{21} + 1)(r_{21} + 2)(r_{32} + 1)(r_{32} + 2) \tag{11.3}$$

Fig. 11.3 The beaches of the study sites, Cherating (*above*) and Chendor (*below*)

Where P is the estimated population size, a_2 is the number of newly marked individuals released from second catch, n_2 is the number caught in the second sample, r_{31} is the number of individuals caught in the third catch that were marked in first catch, r_{21} is the number of individuals caught in the second catch that were marked in the first catch, and r_{32} is the number of individuals caught in the third catch that were marked in the second catch.

11.2.3 Determination of Horseshoe Crab Movement: White Button Tagging

The horseshoe crabs collected as described in the previous section were also marked with simple white button tags to determine the horseshoe crab movement after being released at the capture point. The button tags were 44 mm in diameter, each bearing a unique ID number (numbers 001–100), the researcher's address and phone number, and instructions for reporting and releasing the tagged animals. Any epibionts present on the dorsal surface of the prosoma were removed and a small hole was drilled into the left side of the prosoma (latero-posterior point) using a special awl with a stopper on it, and the tag was secured (Fig. 11.4). The location, time and date of tagging and the sex of the tagged animals were recorded. Any recapture information obtained from the callers (i.e. time, location of recapture, condition of the horseshoe crab) was recorded on a Microsoft Excel worksheet. Everyone who called to report the capture of tagged animals was rewarded with a mobile phone top-up (a gift card for cell phone minutes) and was given brief information on the tagged horseshoe crabs they captured.

The recapture rate of tagged animals was determined as below:

$$\text{Recapture rate, \%} = \left(\text{number of individual recaptured} / \text{total tagged individuals}\right) \times 100 \tag{11.4}$$

Fig. 11.4 The placement and position of the white button tag on the horseshoe crab

11.3 Results

11.3.1 Population Size of Horseshoe Crab

It is estimated that the total population of *T. gigas* was 7,140 and 9,900 in Chendor and Cherating, respectively (Table 11.1). This is the first estimation of horseshoe crab population size in Malaysia, therefore no comparison could be made with other data. Observations should be repeated at these sites during other months of the year to identify any seasonal peak in numbers of this species on the east coast of Peninsular Malaysia.

11.3.2 Tagged Horseshoe Crabs

A total of 99 individuals were tagged, including 60 (male n=56, female n=4) tagged at Chendor and 39 (male n=19, female n=20) at Cherating, respectively. Five groups of tagged horseshoe crabs were released with two groups in Chendor and three in Cherating (Table 11.2). In total, 29 individuals were recaptured out of the 99 released, giving a recapture rate of 29.3 %. Of these, 12 in Chendor (20.0 %) and 17 (43.6 %) in Cherating were recaptured. Some individuals were recaptured more than once; 13 recaptures from 12 animals at Chendor and 24 recaptures from 17 animals at Cherating.

11.3.3 The Movements of Tagged Horseshoe Crabs

The location of recaptured horseshoe crabs was used to determine their movements. From the total recaptured animals, only 15 (41 %) were recaptures in the same tagging location, whereas 22 (60 %) were reported at different locations. Table 11.3 shows the details of the number of recaptures for the animals tagged at Chendor and Cherating. Five individuals were recaptured twice, two individuals were recaptured three times, and one individual (tag 070, male) was recaptured four times until end of 2013. Most of the tagged horseshoe crabs travelled out of the tagging location as most recaptures were reported in different locations, i.e. Sungai Ular, Pelabuhan

Table 11.1 The number of horseshoe crabs captured based on the components of Bailey's triple catch formula

Site	a_2	n_2	r_{21}	r_{31}	r_{32}	N	Var	P
Chendor	12	13	1	1	0	84	7056	7140
Cherating	18	21	3	1	0	99	9801	9900

Description of ***a***, ***n***, ***r*** are listed in Sect. 11.2.2

Table 11.2 Number of horseshoe crabs tagged and recapture rates (%) according to groups released in Chendor and Cherating, Pahang

Location of tagging	Date	Tag number	No. of individual tagged	Recapture %
Chendor	Dec 2011–Jan 2012	001–060	60	12 (20.0 %)
Cherating	Jan 2012–Mar 2012	061–100 (071 excluded)	39	17 (43.6 %)

Kuantan, Balok, Beserah, Tanjung Sisek, Tanjung Lumpur, and Sungai Penor. Some of these places are shown in Fig. 11.2.

Female horseshoe crabs were likely to remain in the same area for at least a short time, as they were recaptured in the same location a week after being released. In contrast, tagged males were more likely to travel elsewhere. One male (tag 031) that was tagged in Cherating was recaptured on the following day in Chendor, which is about 10 km away. Another male (tag 067) from Cherating was recaptured after 4 days, also in Chendor (Table 11.4).

One female (tag 063) and four males were recovered more than once. All four males were recaptured outside their tagging locations. The female was recaptured within the same location (Cherating) even after 5 weeks of being tagged and released. The most frequently recaptured individual (tag 070) was a male tagged in Cherating that was recaptured four times. On the first to third recapture (weeks 19, 20 and 32 after tagging), this male was recaptured in Sungai Ular, which is located approximately 10 km south of the tagging location, and it was again recaptured on week 73 in Balok a further 20 km south of the previous recapture location (Fig. 11.5).

11.4 Discussion

11.4.1 Population Size of Horseshoe Crab

To our knowledge, there are no previous reports on the estimated number of horseshoe crabs in Malaysian waters. Our estimations of the population sizes of *T. gigas* in Chendor of 7,140 and Cherating of 9,900 are therefore the first estimates reported for our area. These estimates are much lower compared with the estimated populations of *L. polyphemus* on the U.S. Atlantic coast (Carmichael et al. 2003; Hata and Berkson 2003), such as in Delaware Bay where 20 million horseshoe crabs are estimated in the spawning population alone (Smith et al. 2006). Our study was based on non-spawning horseshoe crabs in deeper waters; we expect the spawning population in Malaysia could be even lower because animals spawning at any given time typically represent a portion of the total population of adults.

Unfortunately, there is only anecdotal evidence that *T. gigas* was once abundant, with no historical data on horseshoe crab population size in Malaysia to compare

Table 11.3 Recapture number and location of tagged horseshoe crab

Tagging location	Sex		First recapture		Second recapture		Third recapture		Fourth recapture	
	M	F	Same location	Different location	Same location	Different location	Same location	Different location	Same location	Different location
Chendor	11	1	4	8	1	0	0	0	0	0
Cherating	6	11	9	8	1	3	0	2	0	1
Total	17	12	13	16	2	3	0	2	0	1

M male, *F* female

Table 11.4 Time elapsed between recovery and location of recovery of tagged horseshoe crab with two or more times of recovery

TL	LR	Recovery																								
		D1	D4	D7	W1	W4	W5	W6	W8	W9	W10	W16	W17	W18	W19	W20	W25	W27	W29	W32	W38	W43	W54	W73	W86	W88
Chendor	Chendor				[a]					[a](2)							[a]		M[b] #031							
	Cherating	M[b] #031						[a]			[a]			[a]	[a]											
	Sg. Ular																	[a]				[a]				
	Sg. Penor									[a]																
Cherating	Chendor		[a]																							
	Cherating		F[b] #063 (4)[a]	[a]			[a]	F[b] #063					[a]												[a]	
	Sg. Ular								[a]						M[d] #070	M[d] #070				M[d] #070						
	Pel. Kuantan							M[b] #094																		
	Balok					M[b] #094				M[c] #100														M[d] #070		
	Berserah																									[a]
	Tg. Sisek																[a]									
	Tg. Lumpur																				M[c] #100					
	Pekan																						M[c] #100			

TL tagging location, *LR* location of recovery, *D* day, *W* week, *F* female, *M* male, (*x*) number of individual, #*number* tag number for crabs with multiple recapture

[a]Recovery once

[b]Recovery twice

[c]Recovery three times

[d]Recovery four times

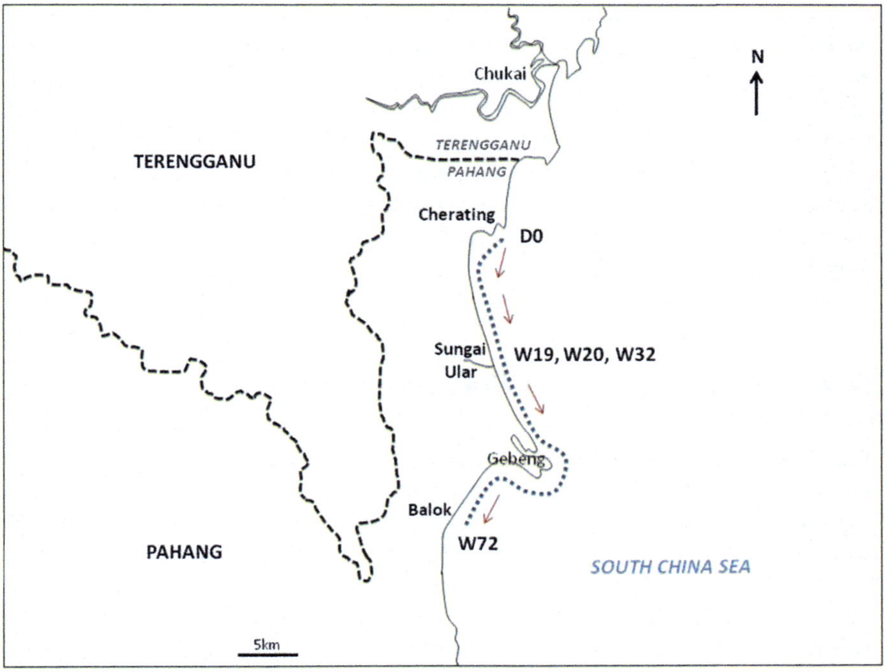

Fig. 11.5 Movements of a male horseshoe crab (tag #070) that was recaptured four times, starting from the tagging day (D0) in Cherating and heading south where he was recaptured three times in weeks 19 (W19) 20 (W20), and 32 (W32) in Sungai Ular before being recaptured a fourth time in week 72 (W72) at Balok, approximately 30 km from the tagging location. *Arrows* show the direction of travel

with our data. A study by Cartwright-Taylor et al. (2011) reported a very small population of *T. gigas* in Singapore, whereas the other horseshoe crab species, *C. rotundicauda*, was found at one site to be much more abundant (0.57–0.98 individual/m^2). However, these numbers were also from a population that included both adults and juveniles congregating on mud flats rather than an adult population in deeper waters. The higher number of horseshoe crabs in Cherating might indicate a larger population compared with Chendor. However, the results from our tagging study revealed that the same individuals of *T. gigas* forage in both areas. While this estimation could represent the number of horseshoe crabs in the study area, these values should be used with caution.

Information gathered from the locals, primarily fishermen, suggests that both Chendor and Cherating might act as nursery and foraging areas for *T. gigas*. The animals can be seen coming to shallow waters in certain months of the year, especially during the peaks in May to June (Akbar et al. 2010). In our study, we also found more males to females in Chendor (14:1). This male biased sex-ratio might indicate the impact of a selective horseshoe crab trade that has preferentially harvested females in that area for more than a decade. The sex ratio in Cherating, however, was 1:1, which is similar to the operational sex ratios previously reported for spawning populations of *T. tridentatus* (Botton et al. 1996) and other Asian horseshoe crab species with low population density (Mattei et al. 2010). Thus, we speculate that the animals caught in Cherating might be the ones preparing to spawn. There is a possibility that Cherating is the spawning ground, whereas these animals could forage elsewhere including Chendor. Future studies may benefit if surveys were conducted year-round and included spawning populations to capture natural seasonal variation in abundance (Kassim et al. 2008; Zaleha et al. 2010) and be consistent with previous study.

11.4.2 The Movements of Horseshoe Crabs

The tagged horseshoe crabs in Chendor and Cherating were found to travel from the tagging areas up to 30 km southward, intermingling with other horseshoe crabs along Sungai Ular, Pelabuhan Kuantan, Balok, Beserah, Tanjung Sisek, and down to Tanjung Lumpur and Sungai Penor (Fig. 11.2). As we recaptured many Chendor-tagged *T. gigas* in Cherating, and vice versa, it is safe to say that both Chendor and Cherating might share the same population with a large home range. This travel distance is known to be possible because another species of horseshoe crab, *L. polyphemus*, showed a large home range of 64.1 ± 20.6 ha (Moore and Perrin 2007).

Most recaptures came from Cherating suggesting that this area has a large number of *T. gigas* in deep waters, no matter where they were tagged (Table 11.4). This recapture rate is also in agreement with the larger population size calculated for Cherating compared with Chendor. However, we believe another reason for this higher number of recaptures might be due to the higher number of fishermen in Cherating and, therefore, a higher chance of netting tagged animals. Chendor is more remote and has a far smaller self-built jetty that houses just a few fishing boats compared with Cherating (Fig. 11.6).

According to those who reported the tagged horseshoe crabs, *T. gigas* was found in the open seas, close to shore coming to nest, in estuaries, or caught in nets on river banks. Some were caught while in amplexus and some were singles, indicating that the tag did not interfere with the animals' activities. The callers were mostly fishermen (some were briefed about this tagging project), beach visitors, and joggers who were curious about the tag carried on the side of the animal.

Generally, *T. gigas* is not confined to a small area in the waters of Peninsular Malaysia. The males were found to travel out of the tagging locations while females tended to remain within the same tagging area. A single male tagged in Chendor that was recaptured three times in three different locations travelling further southward within a period of 54 weeks, provides a clue about the ability of this animal to travel long distances. It suggests that they might travel elsewhere along the east coast, but may escape recapture because some of the beaches are inaccessible. It also indicates that the tag we used is tightly secure and safe for the animals.

Fig. 11.6 Jetty house built by fishermen in the study site, Chendor (*above*) and Cherating (*below*)

11.4.3 Possible Threats to the Horseshoe Crab Population

Apart from being captured and sold as seafood, during this study period, we also observed some other possible threats to *T. gigas* populations on the east coast of Peninsular Malaysia. Most of the direct threats came from the fishermen themselves. Some throw the males out of the boat while they are fishing at sea. Horseshoe crabs are described as a problem for nets because they are very hard to remove once they are entangled. To prevent recurrence, some fishermen's solution is to throw the trapped horseshoe crabs far up the beach leaving them to die from desiccation and heat (Fig. 11.7). This is done when they bring the net (together with the tangled horseshoe crab) and clean it on the beach.

Fig. 11.7 The stranded male horseshoe crabs left to die at the beach (*top*). A bundle of cleaned fishermen's net can be seen in the background (*bottom*)

11.5 Conclusion

The population size of *T. gigas* on the east coast of Peninsular Malaysia is very small, and believed to be shrinking. As the animals were found to travel some distance in waters of this area, conservation efforts involving many localities need to be coordinated. As the local fishermen are most directly involved with the horseshoe crabs, educating them about the animals and the importance of sustainable management should be prioritized in these conservation efforts. Inclusion of this species into the Malaysian Wildlife Conservation Act 2010 [Act 716] could give some legal protection that may be able to curb over harvesting of this species for local markets. Local government agencies such as the Department of Wildlife and Nature Parks (DWNP) could be engaged with the Department of Fisheries (DoF) in providing meaningful enforcement to control the export of this animal abroad. Although DWNP has been successful in controlling illegal trafficking of many protected animals on land, more enforcement is needed with cooperation from various parties to successfully manage this limited resource.

Acknowledgements We thank the government of Malaysia and Universiti Malaysia Terengganu for funding this research. We are grateful to Prof. Jennifer Mattei from Sacred Heart University for her technical support in this study.

References

Akbar JB, Jalal KCA, Kamaruzzaman BY et al (2010) Mechanism in the clot formation of horseshoe crab blood during bacterial endotoxins invasion. J Appl Sci 10(17):1930–1936

Akbar JB, Kamaruzzaman BY, Jalal KCA et al (2011) Hydrology of the horseshoe crab nesting grounds at Pahang coast, Malaysia. Orient J Chem 27(4):1475–1483

Akbar JB, Sihin S, Kamaruzzaman BY et al (2012a) Sediment profiling of the nesting grounds of horseshoe crabs at east peninsular Malaysia. Int J Biol 4(2):159–165

Akbar JB, Kamaruzzaman BY, Jalal KCA et al (2012b) Feeding ecology and food preferences of *Carcinoscorpius rotundicauda* collected from the Pahang nesting grounds. Sains Malaysiana 41(7):855–861

Akbar JB, Jalal KCA, Kamaruzzaman BY (2013) Macrobenthic diversity in horseshoe crab nesting ground-Balok station, Pahang, Malaysia. Orient J Chem 29(4):1311–1318

Botton ML, Shuster CN Jr, Sekiguchi K (1996) Amplexus and mating behaviour in the Japanese horseshoe crab, *Tachypleus tridentatus*. Zool Sci 13:151–159

Carmichael RH, Rutecki D, Valiela I (2003) Abundance and population structure of the Atlantic horseshoe crab, *Limulus polyphemus*, in Pleasant Bay, Cape Cod. Mar Ecol Prog Ser 246:225–239

Cartwright-Taylor L, Bing YV, Chi HC et al (2011) Distribution and abundance of horseshoe crabs *Tacypleus gigas* and *Carcinoscorpius rotundicauda* around the main island of Singapore. Aquat Biol 13:127–136

Chatterji A (1994) The horseshoe crab–a living fossil. Project Swarajya Publication, Orissa

Chatterji A, Pati S (2014) Allometric relationship in the adult population of the Malaysian horseshoe crab (*Tachypleus tridentatus*; Leach). Int J Res 1(11):1378–1385

Christianus A, Saad CR (2009) Traditional uses of horseshoe crabs in Malaysia and Thailand. In: Tanacredi JT, Botton ML, Smith DR (eds) Biology and conservation of horseshoe crabs. Springer, New York, p 616

Hajeb P, Christianus A, Ismail A et al (2009) Heavy metal concentration in horseshoe crab (*Carcinoscorpius rotundicauda* and *Tachypleus gigas*) eggs from Malaysian coastline. In: Tanacredi JT, Botton ML, Smith DR (eds) Biology and conservation of horseshoe crabs. Springer, New York, pp 455–464

Hata D, Berkson J (2003) Abundance of horseshoe crabs (*Limulus polyphemus*) in the Delaware Bay area. Fish Bull 101:933–938

Ismail N, Sarijan S (2011) Phylogenetic inference from 18S rRNA gene sequences of horseshoe crabs, *Tachypleus gigas* between Tanjung Dawai, Kedah & Cherating, Pahang, Peninsular Malaysia. Int J Biol Vet Agric Food Eng 5(8):66–69

Ismail N, Jolly JJ, Dzulkiply SK et al (2012) Allometric variations of horseshoe crab (*Tachypleus gigas*) populations collected from Chendor and Cherating, Pahang, peninsular Malaysia. J Sustain Sci Manag 7(2):164–169

Kassim Z, Shahuddin H, Shaharom F et al (2008) Abundance of three species of the horseshoe crab along the coast of Malaysia. J Bombay Nat Hist Soc 105:209–211

Mattei JH, Beekey MA (2008) The horseshoe crab conundrum: can we harvest and conserve? Wrack Lines 8(1):2–7

Mattei JH, Beekey MA, Rudman A et al (2010) Reproductive behaviour in horseshoe crabs: does density matter? Curr Zool 56(5):634–642

Mishra JK (2009) Horseshoe cabs, their eco-biological status along the northeast coast of India and the necessity for ecological conservation. In: Tanacredi JT, Botton ML, Smith DR (eds) Biology and conservation of horseshoe crabs. Springer, New York, pp 89–96

Moore S, Perrin S (2007) Seasonal movement and resource-use patterns of resident horseshoe crab (*Limulus polyphemus*) populations in a Maine, USA estuary. Estuar Coasts 30(6):1016–1026

Robert R, Muhammad Ali SH, Amelia-Ng PF (2014) Demographics of horseshoe crab population in Kota Kinabalu, Sabah, Malaysia with emphasis on *Carcinoscorpius rotundicauda* and some aspects of its mating behavior. Pertanika J Trop Agric Sci 37(3):375–388

Rozihan M, Ismail E (2012) Impact of Malaysia continental drift on the genetic diversity of horseshoe crab inferred through mtDNA sequence analysis. Int J Biol 4(1):104–110

Rozihan M, Ismail E, Akbar JB et al (2013) Genetic diversity of horseshoe crab (*Tachypleus gigas*) in Malaysia revealed using microsatellite markers. Asian J Anim Vet Adv 8(1):63–72

Smith DR, Millard MJ, Eyler S (2006) Abundance of adult horseshoe crabs (*Limulus polyphemus*) in Delaware Bay estimated from a bay-wide-recapture study. Fish Bull 104(3):456–464

Srijaya TC, Pradeep PJ, Mithun S et al (2010) A new record on the morphometric variations in the populations of horseshoe crab (*Carcinoscorpius rotundicauda* Latreille) obtained from two different ecological habitats of Peninsular Malaysia. Our Nat 8:204–211

Tan AN, Christianus A, Abdul Satar MK (2011) Epibiont infestation on horseshoe crab *Tachypleusgigas* (Müller) at pantai Balok in peninsular Malaysia. Our Nat 9:9–15

Tan AN, Christianus A, Shakibazadeh S et al (2012) Horseshoe crab, *Tachypleus gigas* (Müller, 1785) spawning population at Balok beach, Kuantan, Pahang, Malaysia. Pak J Biol Sci 15(13):610–620

Zaleha K, Kamaruzzan BY, Akbar JB et al (2010) Cd, Cu and Pb concentration levels in horseshoe crab nesting grounds of Pahang coast, Malaysia. J Biol Sci 10(8):790–794

Zaleha K, Akbar JB, Erni Artika H et al (2012) Spawning and nesting behaviour of *Tachypleus gigas* along the east coast of peninsular Malaysia. Int J Biol 4(2):102–111

Chapter 12
Habitat Inventory Trend Analysis of *Limulus polyphemus* Populations on Long Island, U.S.A.: From the Tip of Brooklyn to the Tip of Montauk, 2003–2014

John T. Tanacredi and Sixto Portilla

Abstract Considerable concern regarding the abundance of the American horseshoe crab (HSC) *Limulus polyphemus*, along the east coasts of New Jersey and Delaware prompted past moratoriums on collecting HSC for bait in New Jersey. The parallel population decline in migratory shorebirds such as Red knots *Calidris canatus*, Ruddy turnstones *Arenaria interpres* and others that seasonally feed on the copious quantities of HSC eggs laid along this shoreline resulted in reduced HSC collection permits to numbers considered sustainable. In New York State's Marine District, which is mostly comprised of the Long Island coastline, there is no reliable or routine habitat inventory network existing for determining HSC populations or habitat. Shorebird data, which have been collected by Audubon Chapters, the National Park Service and the US Fish and Wildlife Service, as well as academia, have hinted at declining HSC populations. However, due to the lack of a formal and extensive or reliable inventory network, assessing changing trends in HSC population levels is unattainable or mostly inaccurate. Molloy College's Long Island HSC Network provides survey forms and a website to (1) collect data on Long Island sites which support HSC; (2) count HSC for as reliable an estimate of the HSC population as practical; (3) sex and age individual HSC at each site; and most importantly (4) establish a network that can be repeated annually to detect precipitous changes in HSC population numbers, distributions, and habitat. Data collection for HSC will aid in protecting the HSC population as well as bird species number which require HSC eggs as food during significant migratory periods.

J.T. Tanacredi (✉) • S. Portilla
Center for Environmental Research and Coastal Oceans Monitoring (CERCOM), Molloy College, Rockville Center, NY 11570, USA
e-mail: jtanacredi@molloy.edu

R.H. Carmichael et al. (eds.), *Changing Global Perspectives on Horseshoe Crab Biology, Conservation and Management*, DOI 10.1007/978-3-319-19542-1_12

Results for the last 11 years for 68 sites monitored annually reveal (1) considerably reduced numbers of HSC than "remembered in the past"; (2) sites along the Long Island coastline now believed to support HSC have been found to have few to modest numbers of HSC; (3) a preliminary number projection of *Limulus* on Long Island at approximately 15,000 adult breeding individuals on beaches surveyed; (4) of the 68 beaches monitored since 2003, an increase of 8.2 % had been observed for the number of beaches exhibiting "no breeding activity"; and (5) there has been observed an average of 1 % per yr decline in the total number of HSC observed.

Keywords Captive breeding • Long-term monitoring • Habitat suitability • Inventory • Monitoring re-nesting

12.1 Introduction

At the turn of the new millennium in 2001 the senior author's book "*Limulus* in the Limelight" (Tanacredi 2001) portended to be a "silent siren" for conservation biologists to make all efforts to protect and conserve a group of organisms that have existed on Earth for over 445 million years, surviving mass extinction events. This last 14-year period is even more significant to our conservation efforts as biodiversity vanishes globally. We noted then that, "some of the most basic biological information about (this) species still eludes us in describing horseshoe crab biology." This statement rings true today for all of the four species of horseshoe crabs, as attempts by a recently formed team of scientists and conservationists, who are working to have them listed on the IUCN's (International Union of Conservation for Nature) Red Listing. Hopefully this may provide some added protection against the ultimate demise of the species from overharvesting and/or habitat loss. This effort has been a daunting task, as little information exists on the three Asian species, all considered to be at ordinarily low population levels (Tanacredi et al. 2009), as well as an absence of detailed inventory of numbers and habitats of the North American species. Anecdotal reports of HSC along Long Island provided the impetus to establish an inventory of habitat supporting breeding horseshoe crabs, as Long Island's HSC populations have historically been plundered for fertilizer, bait, and for bleeding.

Long Island, the largest island in the United States, provides over 1,000 miles (ca. 1,600 km) of coastline with estuarine ecosystems supporting breeding horseshoe crabs dating back to before formal records were required by fisherman to gain a permit to harvest them. The largest continuous acreage of tidal wetlands and greatest diversity of habitat in New York State (NYS) is encompassed by the south shore of Long Island (Pataki and Daniels 2001), providing a major nursery ground for juvenile fish and shellfish maturation (Roman et al. 2000). Today NYS Department of Environmental Conservation (DEC) issues permits to collect a total of 135,000 horseshoe crabs each year for bait, bleeding and education/scientific research. It is the only state along the Atlantic Coast of the US that has not placed a total moratorium on harvesting HSC during their critical breeding times. The commercial take of *Limulus* for conch (*Busycon carica*) and eel (*Anguilla rostrata*) fisheries and for pharmaceutical use, creates a lucrative industry for local fishermen and for LAL (*Limulus* amoebocyte lysate) production in the US.

During the peak breeding periods along the Atlantic coastline of high tide/full moon or new moon, thousands of spawning adults gather along this tidal coastal fringe, accessible for harvesting which can "clear a beach" of all horseshoe crabs in a matter of hours. Those crabs permitted by NYS DEC to be harvested for "bleeding" are taken to Massachusetts where one company bleeds up to 135,000 animals and releases them, unfortunately to local waters in Cape Cod, not back to NYS waters as required by the NYS DEC permit. Many of those animals are "re-harvested" for bait and sold back to NYS fishermen (anonymous reference), at an average cost of US $5/crab, thus making it extremely economical to all involved.

There has been no formal, consistent or reliable, horseshoe crab habitat inventory established on Long Island. An annual survey since 2011 of eight beaches, supported by NYS DEC, by Cornell Cooperative Extension, has concentrated on characterization of the breeding horseshoe crabs on those beaches. Their annual total numbers of horseshoe crabs per site have been included as sites for the Molloy College Long Island Horseshoe Crab (LI HSC) inventory. The Molloy College inventory is identifying breeding habitats, defined by observations of breeding pairs and eggs in sand. The majority of 68 breeding sites (Fig. 12.1) have been visited at least one time each season. If a site is visited more than once in a season, the average of the count is used as "site–use" by breeding horseshoe crabs. All 68 sites have been subject to illegal harvesting, which can have untold impact on the sites productivity. There are sites with added protection against poaching such as Jamaica Bay, in New York City, where the National Park Service has a total ban restriction on harvesting crabs for bait, due to this habitat site being within the boundary of a US National Park Unit, the Gateway National Recreation Area (Tanacredi 2001).

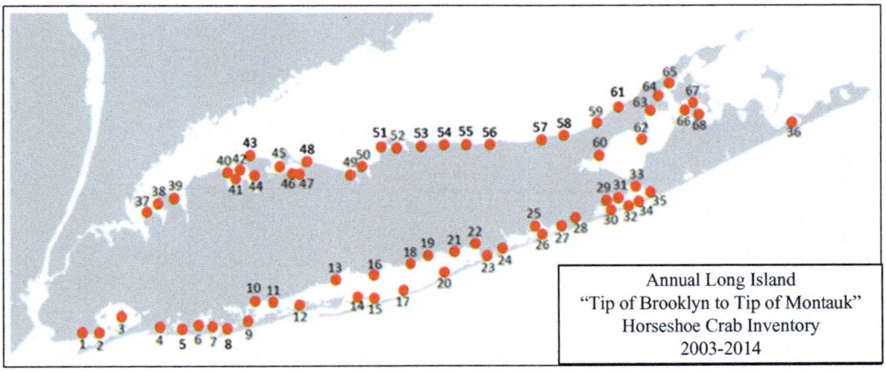

Fig. 12.1 Map of the Long Island Horseshoe Crab Inventory with 68 sites denoted by *red circles*

It has been documented that such beach inventories of spawning horseshoe crabs, with similar water quality conditions, sediment quality and limited access for harvesting, reveal inconsistencies in population numbers over time (Fig. 12.2). Geographic Information System maps of LI HSC populations establish that successive generations annually use the same beaches for breeding and that HSC exhibit considerable site fidelity (Berkson and Shuster 1999).

12.2 Materials and Methods

Annually commencing in 2003, volunteer "Beach Captains" after a 3-h orientation session, visit each of the 68 beaches identified in Fig. 12.1. The survey location was recorded by GPS (global positioning system) for repeated sampling. Over a period of 1–2 h, and a minimum of 50 yards (ca. 46 m), each beach is walked and all HSC counted. Animals are counted, sexed and any "unique observations" recorded on a standard form (Table 12.1). Additional beaches to a grand total of 101 sites mentioned may have been counted; however, only 68 beaches were monitored since 2003, the first year of the inventory because they were all "originally" identified as sites where there were horseshoe crabs historically in "large" numbers.

12.3 Results

Results of the 68 sites monitored annually reveal (1) considerable reduction of HSC population than "remembered in the past"; (2) sites along the Long Island coastline now believed to support HSC have been found to have few to modest numbers of HSC; (3) a very preliminary number projection of *Limulus* on Long Island at an average of 5,000+ adult breeding individuals on the beaches surveyed; (4) of the 68 beaches monitored since 2003, an increase of 8.2 % had been observed for the number of beaches exhibiting "no breeding activity" (Fig. 12.2); and (5) there has been observed an average of 1 % per year decline in the total number of HSC observed (Fig. 12.3).

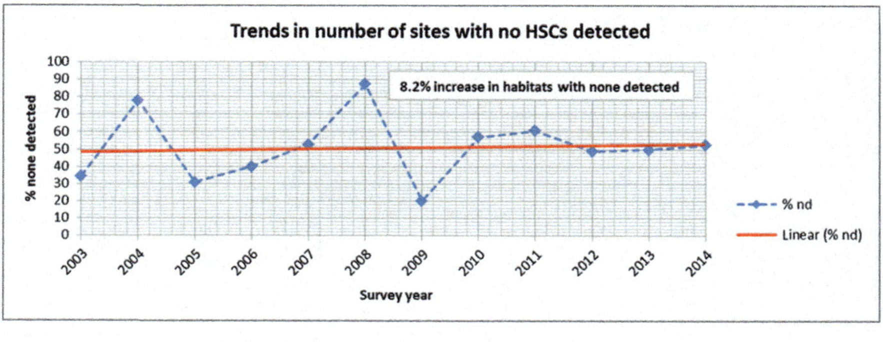

Fig. 12.2 Trend in number of spawning habitat sites with no horseshoe crabs detected

Table 12.1 Long Island Horseshoe Crab Inventory data submission form

Molloy College Long Island Horseshoe Crab Network*

Date: ___________ Year: ___________ Team # _______ (Registered Teams only)

Location: ____

____

____

GPS coordinates: ________.______N ________.______W

Tide: High ☐ Tide ☐ Tide ☐

Observations:

Number of Horseshoe Crabs at site ___________

Number of Males ______ (see back for description) Number of Females ______ (see back for description)

Prosoma Measurements (see back for description)

One measurement per box

Encrusting Organisms: ____

____

____

____

____

____

Recorded by ________________________ Telephone # ____________________

* Call Dr. John Tanacredi at (516) 323-3591 for more information

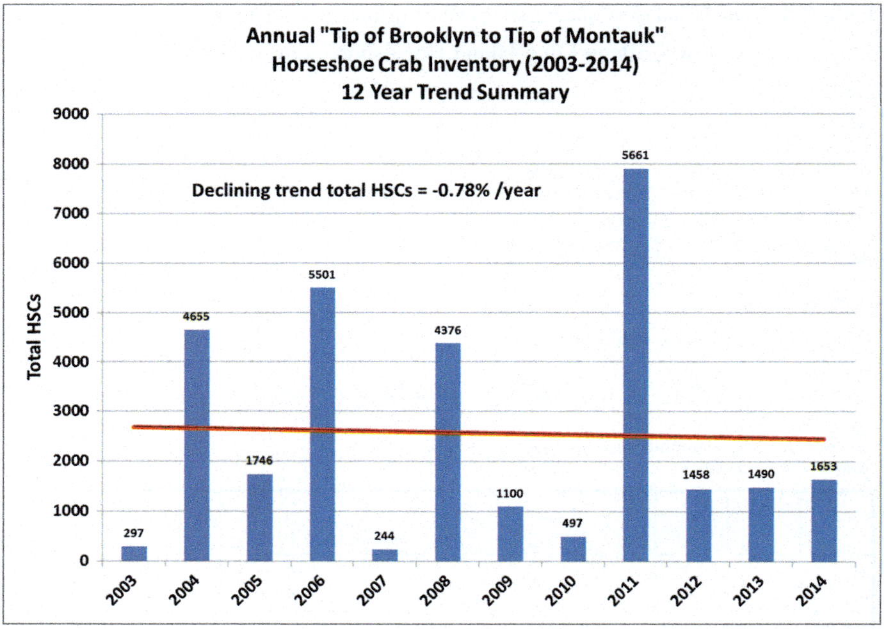

Fig. 12.3 Long Island Horseshoe Crab Inventory from 2013 to 2014

12.4 Discussion and Conclusions

Urbanization of sprawl on Long Island has increased coastal development requiring greater volumes of wastewater and surface runoff to empty into estuarine habitat resulting in decreasing water quality for natural ecosystem functioning. Comprehensive long-term monitoring programs are rarely found to be able to track all the variables that can affect sustainable populations in natural systems (Botton 2001). Moratoriums have been suggested on harvesting HSC, yet they are usually for trial periods too short to adequately investigate all symptoms contributing to potential extinction events (Sekigchi 1988). Observed declines that are not precipitous can always lull our efforts into complacency, with offers of experimenting with "reductions in harvests," "synthetic baits and pharmaceuticals" and "increased conservation efforts" to stave off the chronic reductions observed each year. Even with a fair debate of the actual numbers of HSC on Long Island, there is a clear continuous downward trend that should not be taken lightly. Considerable progress in preserving these animals by restricting their harvesting for bait and recommending alternative baits for these niche fisheries are proposed. In the pharmaceutical procedures in harvesting for blood, a single mandatory stipulation that all HSC collected for bleeding must be returned to the waters where they were collected with immediate notification to the regulatory agency responsible to observe this action as it occurs so as to assure local population maintenance. If added protection can be established with protection of "breeding beaches" or coastlines from any further habitat loss by development or habitat degradation, all the necessary ingredients for HSC protection could be applied on a global scale (Shuster 1981).

On a human scale, HSC (specifically *Limulus polyphemus*) have been of exceptional value to our health as a model for vision research (which has resulted in a Nobel Prize); used in the biopharmaceutical industry as well as in immunological products; and, in ecological terms critical in the preservation of migratory avian species with synchronous migrating patterns over eons of time. HSC will continue to provide important insights as long as humankind ensures the conservation of this rare living fossil. The future appears foreboding (Cohen and Bang 1979; Brockman 1996; Botton et al. 1998). The current status and conservation needs are *not* definitively known for all extant species. There are only four species globally (Sekiguchi and Shuster 2009). At present the IUCN lists three of these species as "data deficient". There is a need for a global assessment of the ongoing research and conservation needs for HSC including such aspects as "by-catch" impacts, prime habitat loss, shoreline development pollution concerns, pharmaceutical harvesting, food consumption and use for bait in eel and conch fisheries (Botton 2001).

Horseshoe crabs are relicts of an ancient class of Arthropods, most members of which have been extinct for hundreds of millions of years. The eggs of these "living fossils" (Eldredge and Stanley 1984) laid each spring on intertidal shores from Maine to the Yucatan in Mexico, with their reproductive epicenter located in Delaware Bay, provide the vital nourishment for hundreds of thousands of migrating shorebirds. This annual spectacle of HSC spawning and bird feeding is critical to the survival, or, most certainly the individual species that maintain their sustenance, to get from South America to breeding grounds in the Arctic. Adult HSC are also food for Loggerhead sea turtles (*Caretta caretta*) and their eggs are food for a variety of juvenile fishes (see http://www.horseshoecrab.org/nh/eco.html).

Limulus amoebocyte lysate is today a required test on most injectable drugs. LAL is a product from the blue blood of HSC (copper based) and is a rapid test for endotoxemia, saving countless lives in hospital stays. Even National Aeronautics and Space Administration (NASA) had used LAL to determine more precisely the level of contamination from microbes, whether dead or living, that may cause "life-detection instrument systems" on the Mars missions to be fooled in our search for extra-terrestrial life forms. Exploration of the HSC genetics is also critically important to determine future local population pressures.

References

Berkson J, Shuster CN Jr (1999) The horseshoe crab: the battle for a true multiple use resource. Fisheries 24(11):6–10

Botton M (2001) The conservation of horseshoe crabs: what can we learn from the Japanese experience? In: Tanacredi JT (ed) *Limulus* in the limelight. Academic/Plenum Publishers, New York, pp 41–51

Botton M, Johnson K, Helleby L (1998) The effects of copper and zinc on embryos and larvae of the horseshoe crab, *Limulus polyphemus*. Arch Environ Con Tox 35:25–32

Brockmann HJ (1996) Satellite male groups in horseshoe crabs, *Limulus polyphemus*. Ethology 102:1–21
Cohen E, Bang FB (eds) (1979) Biomedical applications of the horseshoe crab (Limulidae): proceedings of symposium held at Marine Biological Laboratories, Woods Hole, Massachusetts, October 1978
Eldredge N, Stanley S (eds) (1984) Living fossils. Springer, New York
Pataki G, Daniels R (2001) South Shore estuary reserve final master plan. Long Island South Shore Estuary Reserve Council. NYS, DOS, Albany
Roman C, Jaworski N, Short F et al (2000) Estuaries of the Northeastern United States: habitat and land use signatures. Estuaries 23:743–764
Sekiguchi K (1988) Biology of horseshoe crabs. Science House Press, Tokyo
Sekiguchi K, Shuster CN Jr (2009) Limits on the global distribution of horseshoe crabs (Limulacea): lessons learned from two lifetimes of observations: Asia and America. In: Tanacredi JT, Botton ML, Smith DR (eds) Biology and conservation of horseshoe crabs. Springer, New York, pp 5–24
Shuster CN Jr (1981) A pictorial review of the natural history and ecology of the horseshoe crab *Limulus polphemus*, with reference to other Limulidae. In: Bonaventura J, Bonaventura C (eds) Physiology and biology of horseshoe crabs: studies on normal and environmentally stressed animals. Alan R. Liss, New York
Tanacredi J (2001) Horseshoe crabs imperiled? The fate of a species 350 million years in the making. In: Tanacredi J (ed) *Limulus* in the limelight. Springer, New York, pp 1–7
Tanacredi JT, Botton ML, Smith DR (eds) (2009) Biology and conservation of horseshoe crabs. Springer, New York, p 659

Chapter 13
The Life History Cycle of *Limulus polyphemus* in the Great Bay Estuary, New Hampshire U.S.A.

Helen Cheng, Christopher C. Chabot, and Winsor H. Watson III

Abstract The overall goal of this chapter is to provide an overview of the life history cycle of the American horseshoe crabs (*Limulus polyphemus*) that reside in the Great Bay Estuary, New Hampshire, U.S.A. Great Bay horseshoe crabs generally spawn during high tides in the spring. Based on recent work and studies, spawning appeared to be triggered by increases in water temperature, and animals seemed to prefer to spawn in the warmest sections of the estuary. However, in contrast to horseshoe crabs in some other areas of the U.S.A., peaks of spawning activity did not necessarily correspond with the new and full moons, or with the highest tides, and similar numbers of animals were observed spawning during day and night high tides. Once the eggs hatch, it is hypothesized that their planktonic larvae are likely transported to the upper regions of the estuary where they settle on the expansive mudflats that characterize most of the Great Bay Estuary. At ~9 years (about the 17th instar stage), males appear to reach sexual maturity, while it appears that females molt one more time before reaching sexual maturity. This difference, along with a tendency for males to approach mating beaches more often than females, may contribute to a sex ratio that is skewed towards males at most spawning beaches in the estuary.

Keywords *Limulus polyphemus* • Great Bay Estuary • Spawning • Juveniles • Mating • Thermal preferences • Estuary • Sex ratios • Tidal rhythms

13.1 Introduction

In this overview, our goal is to bring together the results of published and more recent unpublished studies to provide an overview of the life history cycle of the horseshoe crabs that reside in the Great Bay Estuary, New Hampshire, U.S.A.

H. Cheng • W.H. Watson III (✉)
Department of Biological Sciences, University of New Hampshire, Durham, NH 03824, USA
e-mail: win@unh.edu

C.C. Chabot
Department of Biological Sciences, Plymouth State University, Plymouth, NH 03264, USA

R.H. Carmichael et al. (eds.), *Changing Global Perspectives on Horseshoe Crab Biology, Conservation and Management*, DOI 10.1007/978-3-319-19542-1_13

This estuary is located at the northern end of the range of horseshoe crabs and, perhaps as a result, there are both similarities and differences in the behavior and life history of this subpopulation of *Limulus* compared to other populations that have been extensively studied in other areas. We will begin this overview in the early spring, when overwintering adults begin to migrate to intertidal areas to spawn, and we will end with the juvenile horseshoe crabs that grow up in the estuary and reach sexual maturity after ~17 molts (Sekiguchi et al. 1988).

13.2 The Great Bay Estuary of New Hampshire, U.S.A.

The Great Bay Estuary is a complex, semi-enclosed, embayment near the New Hampshire-Maine border. It is separated into three different areas: the Piscataqua River, Little Bay, and Great Bay (Fig. 13.1). The rivers that flow into the Great Bay Estuary drain a watershed that extends more than 1,600 km^2, and this convergence of land and water shapes the features and uses of the ecosystem (Short 1992). There are five dominant aquatic habitats: eelgrass meadows, mudflats, salt marsh, channel bottom, and rocky intertidal. Additionally, the large quantities of water that move in and out of the estuary create some of the strongest tidal currents in North America

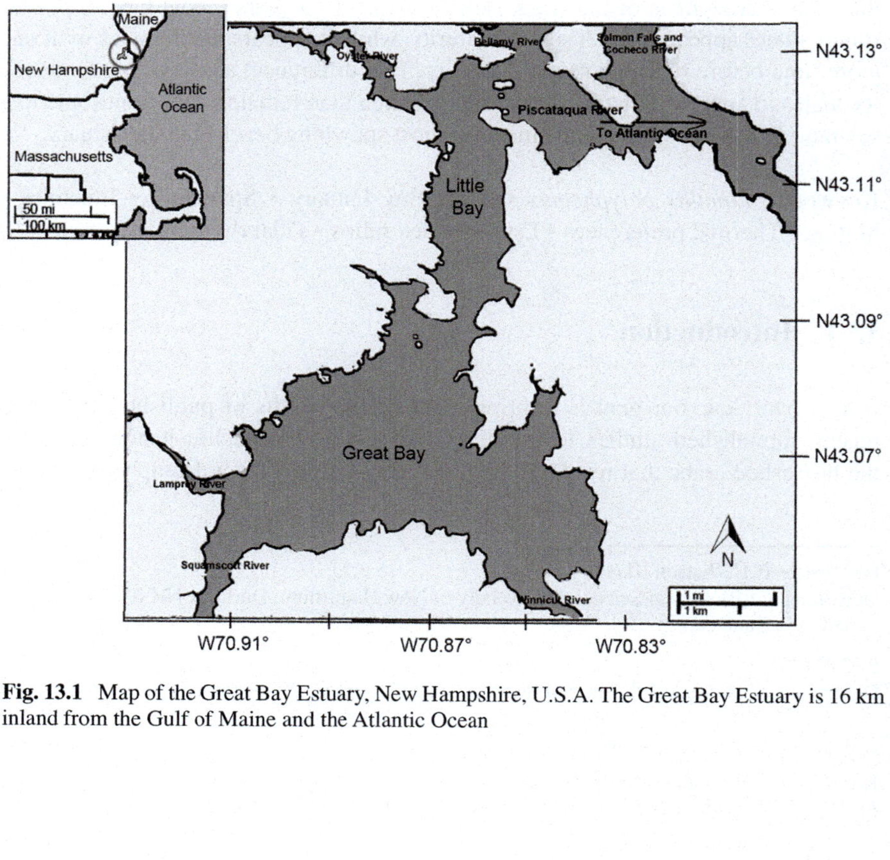

Fig. 13.1 Map of the Great Bay Estuary, New Hampshire, U.S.A. The Great Bay Estuary is 16 km inland from the Gulf of Maine and the Atlantic Ocean

(Short 1992). This tidal exchange structures the Great Bay Estuary ecosystem by affecting water quality, habitat extent, and species distributions.

The Great Bay Estuary appears to support a large population of horseshoe crabs (Watson et al. 2009; National Marine Fisheries Service 2010; Schaller et al. 2010; Watson and Chabot 2010). Using tag-recapture methods and the simplified Lincoln-Peterson Index calculation (Krebs 1998), preliminary population estimates in 2012 and 2013 indicated that there were approximately 122,430 horseshoe crabs that spawned along the shorelines of the Great Bay Estuary in the spring (Cheng, unpublished data).

13.3 Spawning Activity

Horseshoe crab spawning in the estuary generally occurs during the months of May and June, when both mature males and females approach spawning beaches. More than 95 % of the females approaching spawning beaches have a male attached (amplexed) to them (Cheng 2014), and interestingly, since we often observe male: female pairs while diving throughout the year, pair formation is not unique to the mating season. Beach approaches by single males are also common; however, amplexed pairs outnumber single males by ~3:1 (2.86 pairs to each single male).

13.3.1 Large-Scale Movements to Spawning Beaches

In the Great Bay Estuary, horseshoe crabs tend to overwinter in deep areas that are 2–5 km from where they spawn (Schaller et al. 2010; Watson and Chabot 2010). In contrast to reports from the mid-Atlantic States such as New York, New Jersey, Maryland, Delaware and Virginia (Shuster and Botton 1985; Botton and Ropes 1987; Walls et al. 2002), horseshoe crabs in northern bays and estuaries, such as the Great Bay Estuary, Taunton Bay, Maine and Pleasant Bay, Massachusetts, do not appear to move into coastal waters or offshore at any time of the year (Moore and Perrin 2007; James-Pirri 2010; Schaller et al. 2010). Based on high resolution tracking of animals equipped with ultrasonic transmitters, horseshoe crabs are very sedentary during the coldest months of the year (Schaller et al. 2010; Watson and Chabot 2010). Then, in the spring, as estuarine waters warm and exceed ~8–10 °C, they become active and migrate further up into the estuary where it tends to be warmer and the salinity is often lower. These data are consistent with similar telemetry studies in Taunton Bay and Pleasant Bay (Moore and Perrin 2007; James-Pirri 2010) and with laboratory investigations showing the close relationship between activity rhythms and water temperature (Watson et al. 2009; Chabot and Watson 2010; Chabot et al. 2011).

While this pattern of moving up estuary, or into bays, appears to be consistent at different locations, it is not clear what cues these animals use to find their spawning beaches. While chemical, tactile, and visual cues likely aid their ability to locate each other (Barlow 1983; Barlow et al. 1984, 2001; Powers and Barlow 1985; Barlow and Powers 2003; Schwab and Brockmann 2007; Saunders et al. 2010) and suitable beaches for laying their eggs (Botton et al. 1988), these same cues are probably not useful for guiding them long distances to the general area where they spawn because odors would be rapidly diluted and dispersed in most locations; tactile cues only are effective in their immediate vicinity; and visual cues are only useful for distances of a few meters, especially in an estuary. Therefore, it is more likely that horseshoe crabs are detecting physical cues and gradients of things such as salinity and/or temperature. James-Pirri (2010) found that female horseshoe crabs were located in the shallow upper-regions of Pleasant Bay in the spring, moved towards the deeper portions of the bay in the fall, and then moved up to the upper regions again the following spring. Schaller et al. (2010) also found a similar trend in seasonal movements in the Great Bay Estuary and further reported that when temperatures exceeded 10–11 °C in the spring, horseshoe crabs started to become active and moved up-estuary. Then, when water temperatures became colder in the fall, the horseshoe crabs moved to deeper areas with more stable conditions where they spent the winter. Moreover, we find many spawning horseshoe crabs in the warmest parts of the estuary and up into the rivers that empty into the estuary, and our preliminary laboratory studies, described later in this overview, also suggest that horseshoe crabs might seek the warmest regions of bays and estuaries to spawn.

13.3.2 *Timing of Mating Activity*

While there is some evidence from certain study locations that horseshoe crabs tend to mate during the days that coincide with the new and/or full moons (Rudloe 1980; Cohen and Brockmann 1983; Barlow et al. 1986; Smith et al. 2002; Brockmann 2003; Table 13.1), in our study location there was no clear bias towards increased

Table 13.1 Comparison of major factors found to influence spawning activity among locations, including the Great Bay Estuary (Cheng 2014)

	Apalachee key (Florida)[a]	Seahorse key (Florida)[b]	Delaware Bay (New Jersey, Delaware)[c]	Mashnee Dike (Massachusetts)[d,e]	Great Bay Estuary (New Hampshire)
Lunar cycle	More spawning animals during full moon than new moon	More spawning animals during full moon than new moon	Only first half of spawning season, more spawning animals during full and new moon	Spawning occurred throughout May and first week of June, regardless of moon phase[d] More spawning animals during full and new moon[e]	Spawning occurred throughout May and June, regardless of moon phase

(continued)

Table 13.1 (continued)

	Apalachee key (Florida)[a]	Seahorse key (Florida)[b]	Delaware Bay (New Jersey, Delaware)[c]	Mashnee Dike (Massachusetts)[d,e]	Great Bay Estuary (New Hampshire)
Highest high tides	Spawning only occurred on spring tides	Spawning only occurred on spring tides; more observed during high tides of full moon than high tides of new moon	Poor indicator alone of spawning activity	More spawning animals observed during spring tides, though significant numbers spawn on neap tides (lowest high tides)[e]	No relationship of spawning animals to highest high tides (spring tides); spawning occurred on spring tides and neap tides (lowest high tides)
	[f][Mean high tide height during full and new moon ~1.1 m; ↑ ~ 23.2 % from neap high tides (quarter moons)]	[f][Mean high tide height during full and new moon ~1.1 m; ↑ ~ 20.3 % from neap high tides (quarter moons)]	[f][Mean high tide height during full and new moon ~1.6 m; ↑ ~10.4 % from neap high tides (quarter moons)]	[f][Mean high tide height during full and new moon ~1.2 m; ↑ ~7.28 % from neap high tides (quarter moons)]	[f][Mean high tide height during full and new moon ~2.1 m; ↑~ 6.80 % from neap high tides (quarter moons)]
Day versus night	More spawning animals observed at night	More spawning animals observed at day than at night		Spawning only observed at night[4]	In 2012, more spawning animals observed at day; in 2013, no difference between day and night
	[Day tides height higher than night]	[Day tides higher than night by ~ ≤0.2 m]		Spawning observed day and night (dependent on highest tide of a given day)	[Night tides higher than day by ~0.5 m during spring tides, ~0.1 m during neap tides]
				[Daily tide difference depends on lunar phase, varies during spring tides by ~1 m][e]	

[a]Rudloe (1980)
[b]Cohen and Brockmann (1983)
[c]Smith et al. (2002)
[d]Cavanaugh (1975)
[e]Barlow et al. (1986)
[f]Tide heights were gathered from May 2014 NOAA Tide Predictions

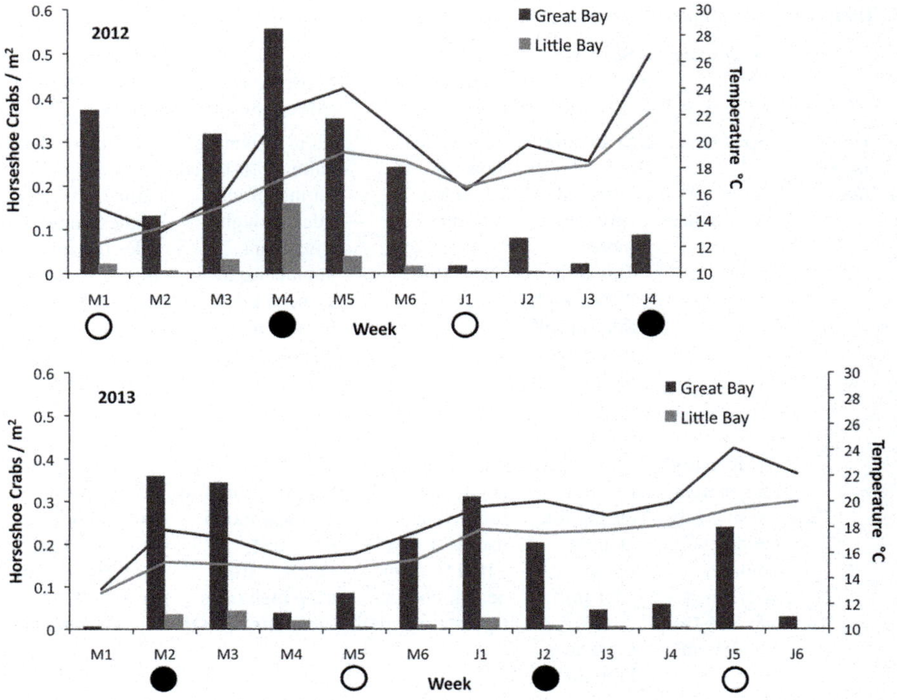

Fig. 13.2 The relationship between spawning horseshoe crab density (number of animals per square meter) and mean temperature per 5-day interval for May (*M*) and June (*J*) of 2012 and 2013. Five-day intervals when the full moon (*open circle*) and new moon (*darkened circle*) occurred are indicated

spawning at these times or during the highest high tides (Fig. 13.2; Table 13.1; Cheng 2014). There was no significant difference in the densities of spawning horseshoe crabs during the days around the new moon, full moon, and equivalent time periods outside of the moon cycles for both 2012 and 2013 combined ($p=0.430$; Kruskal-Wallis test). This could be due to the characteristics of the tidal system and spawning beaches in the Great Bay Estuary. For example, certain well-studied Florida beaches (such as north and west of Apalachee Key and Seahorse Key; Table 13.1) have minimal tides throughout most of the month, except during the full and new moons. Therefore, if horseshoe crabs use water depth as a cue for the occurrence of high tides (and this appears to be the case; Watson et al. 2008), and thus for aggregating on spawning beaches, these might be the only times of the month when that cue might be above their detection threshold. In contrast, tide heights outside the full and new moons (neap tides) in the Great Bay Estuary are always large enough to entrain horseshoe crabs to the tidal rhythm (Chabot and Watson 2010), and the small increases in tide height that occur around the new and full moons (6.8 % increase from neap tide to spring tides; Table 13.1) might not be significant enough to cause a change in their behavior.

Another difference between the spawning beaches surrounding the Great Bay Estuary and those in some other areas along the East Coast is that the Great Bay beaches are generally shallow and not very long. Therefore, during a large high tide,

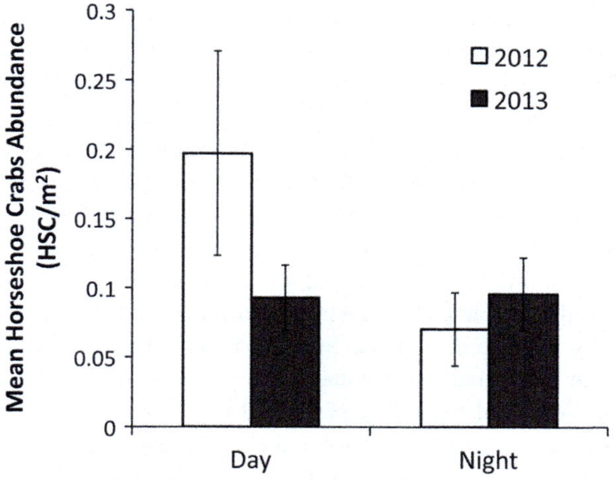

Fig. 13.3 Mean spawning horseshoe crab densities during day and night high tides in 2012 ($n=12$ survey dates) and 2013 ($n=39$ survey dates). *Bars* represent standard error of the means

the water can completely submerge a spawning beach, leaving little room for mating where the water meets the beach. This might actually lead to fewer animals mating during these higher tides and thus might, in part, explain why we did not find a close relationship between tide height, or a lunar cycle, and the number of mating horseshoe crabs (Fig. 13.2; Cheng 2014).

In contrast to a number of previous reports from studies conducted in other parts of the geographic range of *Limulus*, there was no clear tendency for horseshoe crabs to spawn during the night high tides in comparison to daytime high tides (Table 13.1; Fig. 13.3; Cheng 2014). In fact, in 2012 there were significantly more crabs spawning during the day ($p=0.016$, Wilcoxon matched-pairs signed-ranks test, $n=12$ survey dates; Fig. 13.3). Powers et al. (1991), Chabot et al. (2007), and Watson et al. (2008) proposed that horseshoe crabs evolved the ability to increase their visual sensitivity at night so they could effectively mate during either, or both, day and night high tides, which they appear to do in many locations. However, even though horseshoe crabs might be able to see well at night, due to the turbidity of many estuaries and the low amount of contrast and color difference between horseshoe crabs and the substrate, it might be somewhat more advantageous for them to mate during the day when they can probably see each other and other objects better (Barlow 1983; Barlow et al. 1984, 2001; Krutky et al. 2000).

13.3.3 Temperature

In the Great Bay Estuary both the timing and location of horseshoe crab mating are closely correlated with water temperature. For example, while spawning typically commences each year in May, in 2012, as the result of a warm winter and an unusual warm spell in March and April, horseshoe crabs moved towards spawning areas sooner and spawned about 2 weeks earlier than usual (Fig. 13.2; Cheng 2014). In contrast, in 2013 spawning activity started at the more typical time in mid-May, when water temperatures began to warm and surpassed 11–13 °C (Fig. 13.2), and then continued into June. This is consistent with reports that the greatest spawning activity generally occurs at this time at other locations within their geographic range (Shuster 1979, 1982; Shuster and Botton 1985; Barlow et al. 1986).

The first large surge in spawning activity in the Great Bay Estuary is usually followed by two to three more peaks, which are separated by 2–8 days, and these peaks are generally associated with rising water temperatures. Thus, spawning may be modulated by the rate of increase in water temperature, rather than just higher temperatures. An additional factor causing the periodic peaks in spawning activity might be a physiological refractory period during which female horseshoe crabs need to recharge their ability to produce and extrude mature eggs. Thus, even in the absence of higher tides associated with the full and new moons, or peaks in water temperature, there will be oscillations in the number of females spawning throughout the season.

The greatest numbers of spawning animals were always observed at the beaches that were located furthest up into the estuary, where the water temperature was also significantly warmer (Fig. 13.4; these beaches do not appear to differ in any obvious way from other beaches in the estuary; Cheng 2014). This observation is consistent with recent behavioral assay laboratory studies demonstrating that horseshoe crabs significantly spend more time on the warmer side than the colder side of a Y-maze ($p=0.023$; unpaired t-test, Fig. 13.5; Cheng 2014). Thus, at least in the Great Bay Estuary, seasonal and daily changes in water temperature appear to be the primary drivers of mating activity and strongly influence both when and where animals mate.

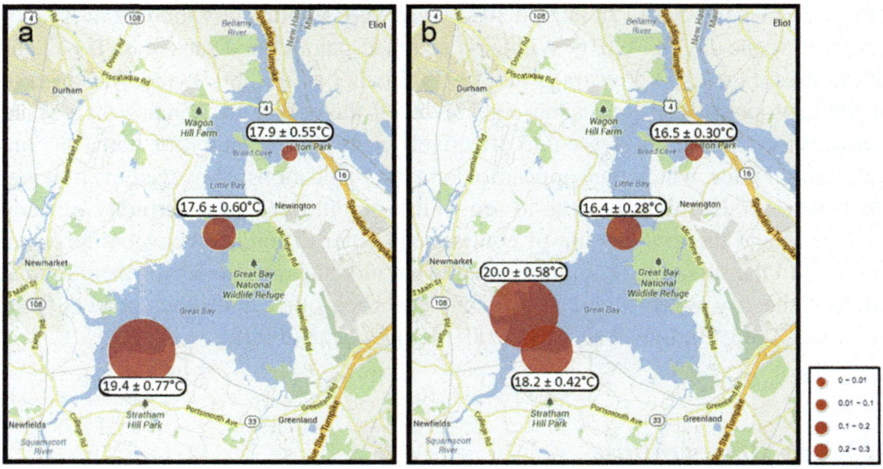

Fig. 13.4 Graduated bubble map of spawning horseshoe crab density (horseshoe crabs/m^2) at each survey site, along with the mean temperatures at each site in 2012 (**a**) and 2013 (**b**)

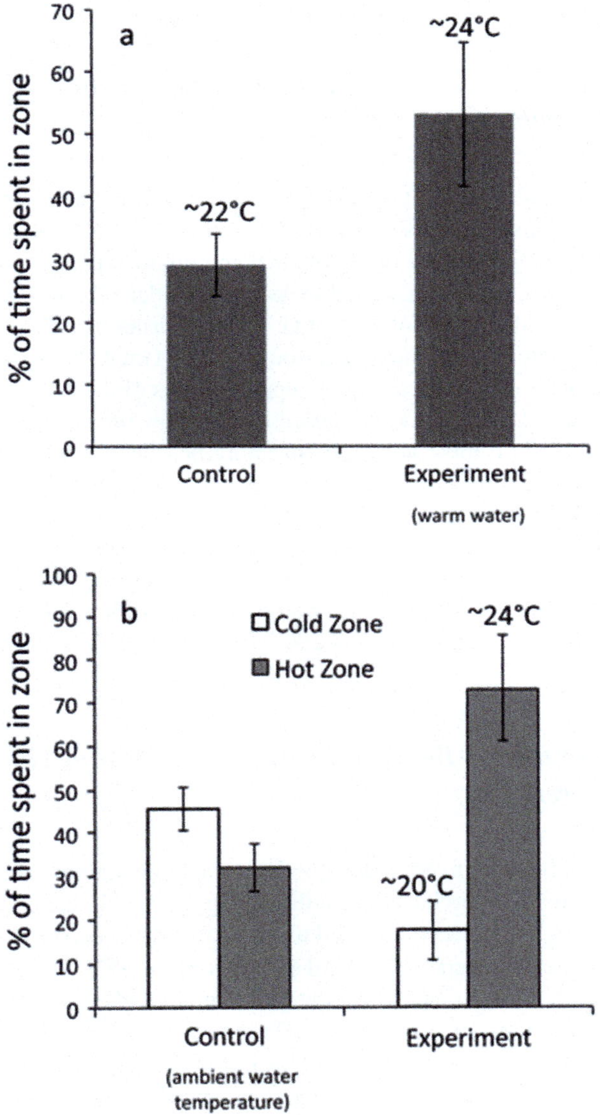

Fig. 13.5 The mean percent time spent by horseshoe crabs in the experimental zone (warm water) during control and experiment phases of Y-maze behavioral studies. Each experiment included a 1-h control period and a 1-h experimental period. In the first experiment (**a**), animals were given a choice between ambient water and warmer water ($n = 6$), while in the second experiment (**b**), animals were given a choice between ambient and ambient (Control), and then cold and warm water (Experiment; $n = 7$). *Bars* represent the standard error of the means

13.4 Growing Up in the Great Bay Estuary

13.4.1 Juvenile Horseshoe Crabs and the Location of Nursery Habitats

The extensive mudflats of the Great Bay Estuary appear to be important horseshoe crab feeding habitats for adults and juveniles (Lee 2010). These areas are generally characterized by an abundance of meiofaunal and infaunal prey and the absence of predators, which are usually restricted to deeper, subtidal habitats (Gunderson et al. 1990; Morrison et al. 2002; Holsman et al. 2006). During SCUBA surveys carried out at times of high tides on mudflats throughout the Great Bay Estuary, we found the most juvenile horseshoe crabs in the upper regions of the estuary, in Great Bay proper, and in areas that were not exposed to air at low tide. Juveniles were found both adjacent to adult horseshoe crab spawning beaches, as well as in areas that were at least 0.5–2.5 km from a known breeding beach (Fig. 13.6; Cheng 2014). Interestingly, there were no juvenile horseshoe crabs found during any of the surveys in Little Bay, despite surveying adjacent to documented spawning beaches. This may be due to the way larvae are carried by tidal currents, and perhaps the way that larvae move up and down in the water column (Rudloe 1979; Shuster 1982; Sekiguchi 1988; Botton and Loveland 2003; Ehlinger and Tankersley 2006).

13.4.2 Interactions Between Larval Behavior and Tidal Currents

Many estuarine, benthic invertebrates produce planktonic larvae that have evolved strategies to increase their chances of either being retained in the estuary or carried out of the estuary. This is accomplished using tidal-influenced vertical migration or selective tidal-stream transport (Forward and Tankersley 2001; Forward et al. 2003). After hatching, horseshoe crabs become planktonic (Rudloe 1979; Shuster 1982; Sekiguchi 1988; Botton and Loveland 2003) and often migrate vertically within the water column (Ehlinger and Tankersley 2006). However, it is unclear how long they stay planktonic, or if they time their excursions into the water column in a manner that would influence their transport within the estuary.

Tidal currents vary throughout the Great Bay Estuary, with fast moving currents ranging from 0.75 to 1.5 m/s in some areas of Little Bay to currents less than 0.5 m/s in the vast expanses of Great Bay (Short 1992). Drifter studies, using devices designed to mimic the passive drift of larvae, have been conducted in the Great Bay Estuary to investigate American lobster (*Homarus americanus*) larval transport (Goldstein 2012) and in all cases the drifters were retained within the estuary, regardless of whether they were released on a flood or ebb tide. Moreover, those released in Little Bay ended up in Great Bay ("upstream"). Recently, we repeated some of these drifter studies, releasing them adjacent to selected mating beaches in Little Bay and Great Bay, and most of them, as predicted from Goldstein's (2012)

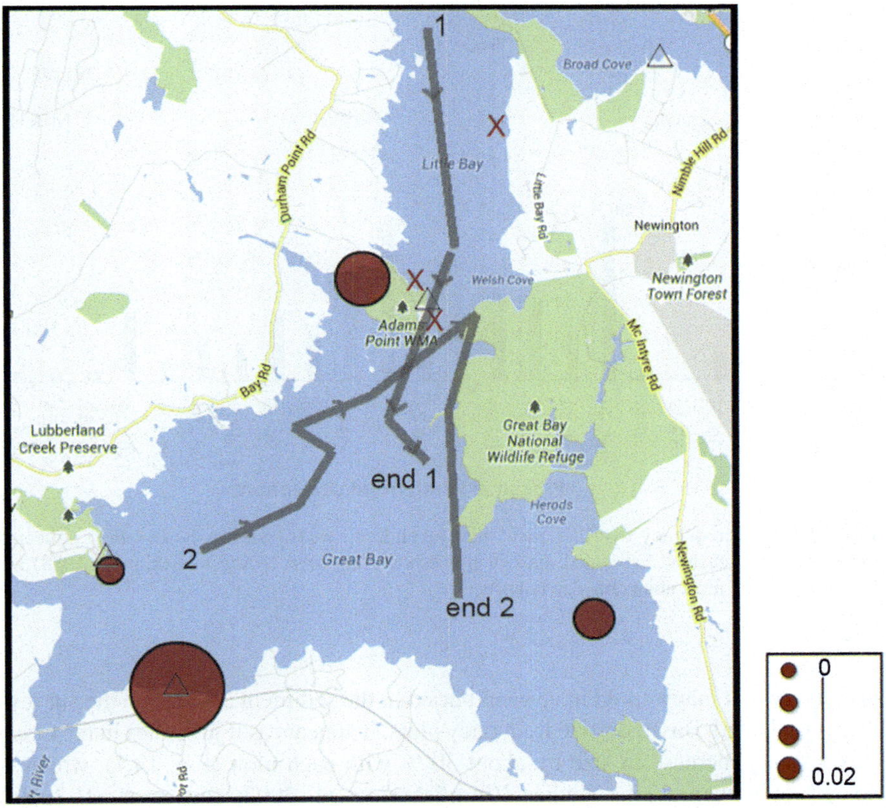

Fig. 13.6 Graduated bubble map of juvenile horseshoe crab densities (horseshoe crabs/m^2) in Great Bay and Little Bay, based on dive surveys in 2012 and 2013 combined. Adult spawning survey beaches (Cheng 2014) are indicated by Δ's (also, refer to Fig. 13.4). X's indicate that no juveniles were found during surveys in these areas. The *gray lines* show the paths taken by two drifters that were released in Little Bay (*1*) and Great Bay (*2*) in 2014. Note that there is a tendency for the drifters to move up into Great Bay

work, were transported up into Great Bay (Fig. 13.6; Watson et al. unpublished data). Thus, it would appear that planktonic larvae are rarely exported out of the system and may, in fact, use these currents to increase their retention in the estuary. Specifically, horseshoe crab larvae that are hatched in spawning grounds in Little Bay are likely transported to areas in Great Bay where the current speeds are reduced in magnitude, and then they settle on appropriate soft bottom sediments.

13.4.3 Growth and Maturation

The majority of the juvenile horseshoe crabs found SCUBA diving were 45–85 mm in prosomal width (Fig. 13.7). Smaller juveniles were either in different areas or not captured because they were hard to see due to their size and cryptic coloration. It is

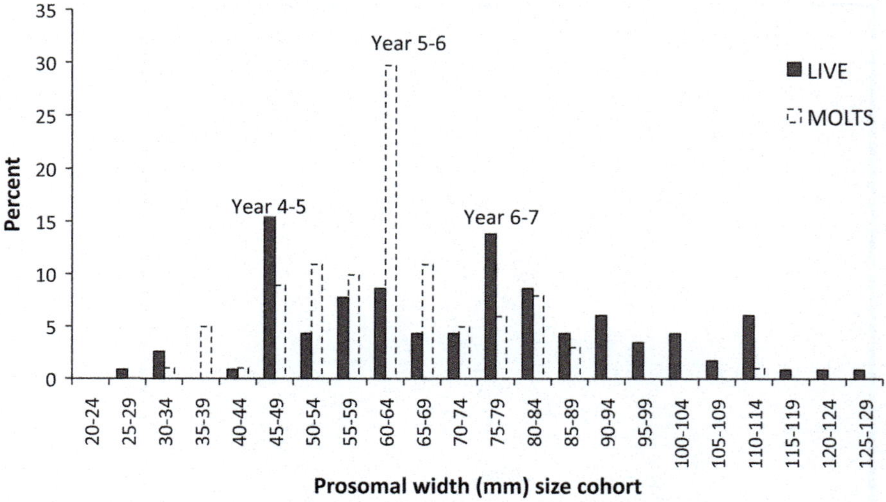

Fig. 13.7 Size-frequency distribution of live juveniles ($n = 116$) and molts ($n = 101$) collected diving and on beaches. Estimated ages of horseshoe crabs (According to Sekiguchi 1988) are indicated for the most abundant size cohorts

also likely that many could have been buried in the sediment at the time the surveys were conducted. Based on size frequency plots, it appears as if juveniles in the Great Bay Estuary increase in size by about 30 % with each molt (Fig. 13.7), which is consistent with previous studies (Carmichael et al. 2003; Burton et al. 2009). Carmichael et al. (2003) measured the growth of juvenile horseshoe crabs in Pleasant Bay, Massachusetts and also found that horseshoe crabs grew ~25–35 % larger with each molt. Based on the sizes of juveniles collected in the Great Bay Estuary, these horseshoe crabs were likely 1–8 years old (according to Table I from Sekiguchi et al. 1988 and Figure VI-44 from Sekiguchi 1988).

During shoreline surveys on nearby spawning beaches, the most abundant small males were 120 mm in prosomal width, and the most abundant small females were 150 mm (Fig. 13.8; Cheng 2014). Thus, we rarely saw horseshoe crabs that were 100–115 mm in either our dive surveys or spawning surveys. It is likely that this "missing" cohort of juveniles were beginning to move into areas that were deeper than where the smaller juveniles were found and yet were not sexually mature, so they were also not observed on the mating beaches.

After undergoing approximately 17 molts, in the span of 7–11 years, horseshoe crabs reach their terminal molt and stop growing (Shuster 1950; Walls et al. 2002). Given the sizes of the smallest males and females observed spawning in the Great Bay Estuary (Fig. 13.8), it appears as if males reach sexual maturity one molt sooner than females. This was also been observed in laboratory studies (Sekiguchi 1988; Sekiguchi et al. 1988). They suggest that males reach sexual maturity at year 9, at

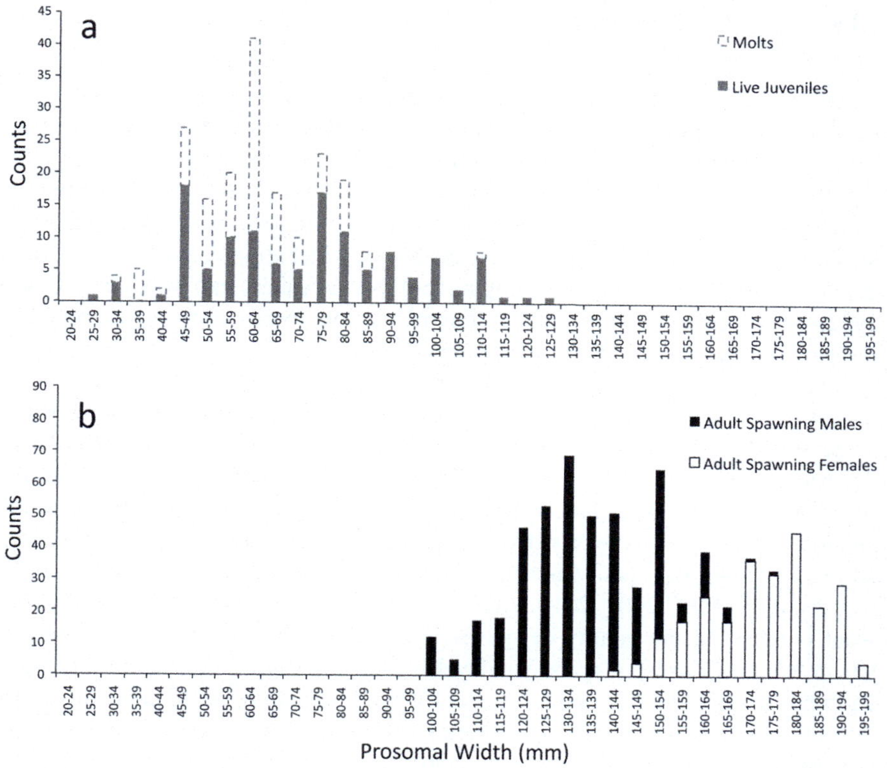

Fig. 13.8 A comparison of the size frequency distribution of (**a**) molts and live juvenile horseshoe crabs ($n = 225$), and (**b**) adult spawning horseshoe crabs ($n = 668$)

the 17th instar stage, and females reach sexual maturity at year 10, at the 18th instar stage. If males reach sexual maturity one molt sooner than females, this might explain both their size differences and, in part, the male-skewed sex ratios observed on mating beaches, both in New Hampshire (1.4 males to 1 female; Cheng 2014) and elsewhere (Rudloe 1980; Cohen and Brockmann 1983; Smith et al. 2002; Carmichael et al. 2003; James-Pirri et al. 2005). For example, in Florida, Rudloe (1980) and Cohen and Brockmann (1983) observed sex ratios of four males to one female (4:1), and in Delaware Bay, sex ratios have been reported to reach up to 5:1 (Shuster and Botton 1985; Smith et al. 2002). In some areas the sex ratio is even more skewed, but this might be the result of harvesting for the larger females by the biomedical and bait industries (James-Pirri et al. 2005). Data gathered from Pleasant Bay, a location where horseshoe crabs historically are harvested in large numbers for biomedical purposes (Rutecki et al. 2004; James-Pirri et al. 2005), showed nine males for every one female (James-Pirri et al. 2005).

13.5 Final Comments

The results from these collective studies on the American horseshoe crabs that reside in the Great Bay Estuary shed new light on this northern population. The Great Bay Estuary is characterized by large seasonal fluctuations in temperature and salinity, typical of high latitudinal temperate estuaries (Watson et al. 2009). Maximum temperatures occur during mid-summer through the fall. In addition, tidal cycles cause temperature and salinity fluctuations, with ebb tide water temperatures being warmer than flood tides, and flood tides bringing more saline waters into the estuary (Short 1992). All of these environmental factors play a role in the activities of horseshoe crabs in this location, and thus the behaviors and distributions observed in this study may be a result of local adaptations to the Great Bay Estuary.

The overall numbers of adult horseshoe crabs that return to spawn year after year along the shorelines, embayments, and estuaries of the U.S. Atlantic Coast are decreasing, or have decreased, in a number of regions. Though there are populations, such as those in Delaware Bay, that have shown annual fluctuations (Smith and Michels 2006) and are reported to be stable or increasing in recent years, populations in New York and Massachusetts show continuing declines and have not recovered (ASMFC 2009, 2010, 2013). Additional pressure from coastal development is reducing the quality of spawning beaches and possible nursery habitats where juvenile horseshoe crabs can develop. The Great Bay Estuary could serve as an important baseline for comparisons of horseshoe crab population health across the entire U.S. Atlantic coast since there has been little to no reported commercial harvest for horseshoe crabs in recent years (J. Carloni 2013 personal communication) and no history of biomedical company activity (State of New Hampshire 2001, 2002). Continued monitoring and assessments of the Great Bay horseshoe crab population might be useful in determining whether shifts in sex ratios, distribution, abundance, growth rates and other characteristics are the result of climate change or harvesting.

Acknowledgements The authors thank the many research assistants, volunteers, and talented SCUBA divers, who have assisted with field surveys and behavioral studies and experiments. Thanks also to David Shay for his daily maintenance of the facilities at Jackson Estuarine Laboratory. This work was supported by grants from the University of New Hampshire School of Marine Science and Ocean Engineering, New Hampshire Sea Grant, and NSF (NSF-IOS 0920342) to CCC and WHW III.

References

ASMFC (Atlantic States Marine Fisheries Commission) (2009) Terms of reference and advisory report to the horseshoe crab stock assessment peer review. Stock Assessment Report No. 09-02. Atlantic States Marine Fisheries Commission, Washington, DC

ASMFC (Atlantic States Marine Fisheries Commission) (2010) Stock assessment overview: horseshoe crab. Atlantic States Marine Fisheries Commission, Washington, DC

ASMFC (Atlantic States Marine Fisheries Commission) (2013) Horseshoe crab stock assessment update. Atlantic States Marine Fisheries Commission, Washington, DC
Barlow RB Jr (1983) Circadian rhythms in the *Limulus* visual system. J Neurosci 3:856–870
Barlow RB Jr, Powers MK (2003) Seeing at night and finding mates: the role of vision. In: Shuster CN Jr, Barlow RB Jr, Brockmann HJ (eds) The American horseshoe crab. Harvard University Press, Cambridge, MA, pp 83–102
Barlow RB Jr, Powers MK, Kass L et al (1984) Vision in *Limulus* mating behavior during the day and at night. Biol Bull 167:522
Barlow RB Jr, Powers MK, Howard H et al (1986) Migration of *Limulus* for mating: relation to lunar phase, tide height, and sunlight. Biol Bull 171:310–329
Barlow RB Jr, Hitt JM, Dodge FA (2001) *Limulus* vision in the marine environment. Biol Bull 200(2):169–176
Botton ML, Loveland RE (2003) Abundance and dispersal potential of horseshoe crab (*Limulus polyphemus*) larvae in the Delaware Estuary. Estuaries 26(6):1472–1479
Botton ML, Ropes JW (1987) Populations of horseshoe crabs, *Limulus polyphemus*, on the northwestern Atlantic continental shelf. Fish Bull 85(4):805–812
Botton ML, Loveland RE, Jacobsen TR (1988) Beach erosion and geochemical factors: influence on spawning success of horseshoe crabs (*Limulus polyphemus*) in Delaware Bay. Mar Biol 99:325–332
Brockmann HJ (2003) Nesting behavior, a shoreline phenomenon. In: Shuster CN Jr, Barlow RB Jr, Brockmann HJ (eds) The American horseshoe crab. Harvard University Press, Cambridge, MA, pp 33–49
Burton WH, Kelley FS, Franks EA (2009) Distribution of juvenile horseshoe crabs in subtidal habitats of Delaware Bay using a suction-dredge sampling device. In: Tanacredi JT, Botton ML, Smith DR (eds) Biology and conservation of horseshoe crabs. Springer, New York, pp 285–293
Carmichael RH, Rutecki D, Valiela I (2003) Abundance and population structure of the Atlantic horseshoe crab *Limulus polyphemus* in Pleasant Bay, Cape Cod. Mar Ecol Prog Ser 246:225–239
Cavanaugh CM (1975) Observations on mating behavior in *Limulus polyphemus*. Biol Bull 149:419–453
Chabot CC, Watson WH III (2010) Circatidal rhythms of locomotion in the American horseshoe crab, *Limulus polyphemus*: underlying mechanisms and cues that influence them. Curr Zool 56(5):499–517
Chabot CC, Betournay SH, Braley NR et al (2007) Endogenous rhythms of locomotion in the American horseshoe crab, *Limulus polyphemus*. J Exp Mar Biol Ecol 345(2):79–89
Chabot CC, Yelle JF, O'Donnell CB et al (2011) The effects of water pressure, temperature, and current cycles on circatidal rhythms expressed by the American horseshoe crab, *Limulus polyphemus*. Mar Freshwater Behav Physiol 44(1):43–60
Cheng H (2014) The environmental influences on American horseshoe crab (*Limulus polyphemus*) behavior and distribution in the Great Bay Estuary, New Hampshire, U.S.A. M.S. thesis, University of New Hampshire
Cohen JA, Brockmann HJ (1983) Breeding activity and mate selection in the horseshoe crab, *Limulus polyphemus*. Bull Mar Sci 33:274–281
Ehlinger GS, Tankersley RA (2006) Endogenous rhythms and entrainment cues of larval activity in the horseshoe crab, *Limulus polyphemus*. J Exp Mar Biol Ecol 337(2):205–214
Forward RB Jr, Tankersley RA (2001) Selective tidal-stream transport of marine animals. Oceanogr Mar Biol 39:305–353
Forward RB Jr, Tankersley RA, Welch JM (2003) Selective tidal-stream transport of the blue crab *Callinectes sapidus*, an overview. Bull Mar Sci 72:347–365
Goldstein JS (2012) The impact of seasonal movements by ovigerous American lobsters (*Homarus americanus*) on egg development and larval release. Dissertation, University of New Hampshire
Gunderson DR, Armstrong DA, Shi YB et al (1990) Patterns of estuarine use by juvenile English sole (*Parophrys vetulus*) and Dungeness crab (*Cancer magister*). Estuaries 13(1):59–71

Holsman KK, McDonald P, Armstrong DA (2006) Intertidal migration and habitat use by subadult Dungeness crab *Cancer magister* in a NE Pacific estuary. Mar Ecol Prog Ser 308:183–195

James-Pirri MJ (2010) Seasonal movement of the American horseshoe crab *Limulus polyphemus* in a semi-enclosed bay on Cape Cod, Massachusetts (USA) as determined by acoustic telemetry. Curr Zool 56(5):575–586

James-Pirri MJ, Tuxbury K, Marino S et al (2005) Spawning densities, egg densities, size structure, and movement patterns of spawning horseshoe crabs, *Limulus polyphemus*, within four coastal embayments on Cape Cod, Massachusetts. Estuaries 28(2):296–313

Krebs CJ (1998) Estimating abundance: mark-recapture. In: Krebs CJ (ed) Ecological methodology, 2nd edn. Benjamin Cummings, Menlo Park, pp 19–69

Krutky MA, Atherton JL, Smith S et al (2000) Do the properties of underwater lighting influence the visually guided behavior of *Limulus*? Biol Bull 199:178–179

Lee WJ (2010) Intensive use of intertidal mudflats by foraging adult American horseshoe crabs *Limulus polyphemus* in the Great Bay Estuary, New Hampshire. Curr Zool 56:611–617

Moore S, Perrin S (2007) Seasonal movement and resource-use patterns of resident horseshoe crab (*Limulus polyphemus*) populations in a Maine, USA Estuary. Estuar Coast 30(6):1016–1026

Morrison M, Francis M, Hartill B et al (2002) Diurnal and tidal abundance changes in the fish fauna of a temperate tidal mudflat. Estuar Coast Shelf Sci 54:793–807

National Marine Fisheries Service Management Division (2010) Programs improving management of ASMFC managed species in New Hampshire: annual report. New Hampshire Fish and Game Department, Durham

Powers MK, Barlow RB Jr (1985) Behavioral correlates of circadian rhythms in the *Limulus* visual system. Biol Bull 169(3):578–591

Powers MK, Barlow RB Jr, Kass L (1991) Visual performance of horseshoe crabs day and night. Vis Neurosci 7:179–189

Rudloe A (1979) Locomotor and responses of larvae of the horseshoe crab, *Limulus polyphemus*. Biol Bull 157:494–505

Rudloe A (1980) The breeding behavior and patterns of movement of horseshoe crabs *Limulus polyphemus* in the vicinity of breeding beaches in Apalachee Bay, Florida. Estuaries 3:177–183

Rutecki D, Carmichael RH, Valiela I (2004) Magnitude of harvest of Atlantic horseshoe crabs, *Limulus polyphemus*, in Pleasant Bay, Massachusetts. Estuaries 27(2):179–187

Saunders KM, Brockmann HJ, Watson WH III et al (2010) Male horseshoe crabs *Limulus polyphemus* use multiple sensory cues to locate mates. Curr Zool 56:611–617

Schaller SY, Chabot CC, Watson WH III (2010) Seasonal movements of American horseshoe crabs *Limulus polyphemus* in the Great Bay Estuary, New Hampshire (USA). Curr Zool 56:587–598

Schwab RL, Brockmann HJ (2007) The role of visual and chemical cues in the mating decisions of satellite male horseshoe crabs, *Limulus polyphemus*. Anim Behav 74:837–846

Sekiguchi K (1988) Biology of horseshoe crabs. Science House, Tokyo

Sekiguchi K, Seshimo H, Sugita H (1988) Post-embryonic development of the horseshoe crab. Biol Bull 174:337–345

Short FT (ed) (1992) The ecology of the Great Bay Estuary, New Hampshire and Maine: an estuarine profile and bibliography. NOAA – Coastal Ocean Program Publication, Durham

Shuster CN Jr (1950) Observation on the natural history of the American horseshoe crab, *Limulus polyphemus*. Woods Hole Oceanogr Inst Contrib 564:18–23

Shuster CN Jr (1979) Distribution of the American horseshoe crab, *Limulus polyphemus*. In: Cohen E (ed) Biomedical applications of the horseshoe crab (Limulidae). Alan R. Liss, New York, pp 3–26

Shuster CN Jr (1982) A pictorial review of the natural history and ecology of the horseshoe crab *Limulus polyphemus*, with reference to other Limulidae. In: Bonaventura J, Bonaventura C, Tesh S (eds) Physiology and biology of horseshoe crabs. Alan R. Liss, New York, pp 1–52

Shuster CN Jr, Botton ML (1985) A contribution to the population biology of horseshoe crabs, *Limulus polyphemus* (L.), in Delaware Bay. Estuaries 8:363–372

Smith DR, Michels SF (2006) Seeing the elephant: importance of spatial and temporal coverage in a large-scale volunteer-based program to monitor horseshoe crabs. Fisheries 31(10):485–491

Smith DR, Pooler PS, Swan BL et al (2002) Spatial and temporal distribution of horseshoe crab (*Limulus polyphemus*) spawning in Delaware Bay: implications for monitoring. Estuaries 25:115–125

State of New Hampshire (2001) Horseshoe crab fisheries and management program annual report for 2000. New Hampshire Fish and Game, Concord

State of New Hampshire (2002) Horseshoe crab fisheries and management program annual report for 2001. New Hampshire Fish and Game, Concord

Walls EA, Berkson J, Smith SA (2002) The horseshoe crab, *Limulus polyphemus*: 200 million years of existence, 100 years of study. Rev Fish Sci 10:39–73

Watson WH III, Chabot CC (2010) High resolution tracking of adult horseshoe crabs (*Limulus polyphemus*) in a New Hampshire estuary using fixed array ultrasonic telemetry. Curr Zool 56(5):599–610

Watson WH III, Bedford L, Chabot CC (2008) Dissociation between circadian rhythms of visual sensitivity and circatidal rhythms of locomotion in the horseshoe crab *Limulus polyphemus*. Biol Bull 215:46–56

Watson WH III, Schaller SY, Chabot CC (2009) The relationship between small- and large-scale movements of horseshoe crabs in the Great Bay Estuary and *Limulus* behavior in the laboratory. In: Tanacredi JT, Botton ML, Smith DR (eds) Biology and conservation of horseshoe crabs. Springer, New York, pp 131–148

Chapter 14
Molts Reveal Life-History Patterns for Juvenile American Horseshoe Crabs in Fringe Habitats

Maurice G. Estes Jr., Ruth H. Carmichael, Peter D.M. Macdonald, Angela J. Brady, and John McFadyen

Abstract In many regions of the world, horseshoe crab populations are sparse or difficult to study. We investigated the utility of molted shells, which may be more accessible than live animals, to provide essential data on population structure for juvenile horseshoe crabs. We collected and compared horseshoe crab (*Limulus polyphemus*) molts from two representative habitats near the edge of horseshoe crabs' known range in the U.S., Taunton Bay, Maine (2005, 2006) and Petit Bois Island, Mississippi (2008, 2009, 2011). Molts were collected and measured every 2 weeks and characterized in terms of abundance, prosomal width, and sex. The greatest number of molts was found during full moon periods, corresponding to known spawning periods and following seasonal temperature increases at both sites. Molt prosomal width ranged from 3 to 169 mm at Petit Bois Island and 19–165 mm at Taunton Bay. Molt size frequency distributions and estimated growth increments were similar between sites and to size classes known for living juvenile *L. polyphemus* found elsewhere. Male to female sex ratios of molts ranged from 1.0 to 2.3, similar to sex ratios measured among live juveniles elsewhere and among live

M.G. Estes Jr. (✉)
Atmospheric Science, University of Alabama in Huntsville,
320 Sparkman Drive, Huntsville, AL 35805, USA

Biological and Environmental Sciences, Alabama A&M University,
Normal, AL 35810, USA
e-mail: maury.estes@nsstc.uah.edu

R.H. Carmichael
Dauphin Island Sea Lab, University of South Alabama, Dauphin Island, AL 36528, USA

Department of Marine Sciences, University of South Alabama, Mobile, AL 36688, USA

P.D.M. Macdonald
Mathematics and Statistics, McMaster University, Hamilton, ON L8S 4K1, USA

A.J. Brady
Marine Sciences, University of Maine-Machias, Machias, ME 04654, USA

J. McFadyen
Biology, Dauphin Island Sea Lab, Dauphin Island, AL 36528, USA

R.H. Carmichael et al. (eds.), *Changing Global Perspectives on Horseshoe Crab Biology, Conservation and Management*, DOI 10.1007/978-3-319-19542-1_14

subadult and adult horseshoe crabs previously measured at the study sites. Molt numbers and size distributions on Petit Bois Island following the Deepwater Horizon oil spill confirmed survival of young crabs through the oil spill period. The overall similarity in molt frequency patterns and size distributions between distinct habitat types suggest some endogenous plasticity in horseshoe crab molting and subsequent growth patterns, despite regional differences in environmental attributes. Hence, data from molted shells have potential as an alternative to live animal studies by at least providing data on timing and rate of growth under different environmental conditions. These data will be particularly beneficial in habitats where live horseshoe crab populations are difficult to study or there are few background data to support live animal studies.

Keywords Molt • Habitat • Prosomal • Cohort • Environmental • Juvenile • Abundance

14.1 Introduction

Data on population dynamics and status of horseshoe crab populations are needed to inform conservation and management plans. The American horseshoe crab (*Limulus polyphemus*) ranges from Maine in the U.S. to the Yucatan Peninsula of Mexico, and west to Texas in the northern Gulf of Mexico (nGOM) (Sekiguchi and Shuster 2009; Zaldivar-Rae et al. 2009). Three other species of horseshoe crabs (*Tachypleus gigas*, *Tachypleus tridentatus*, *Carcinoscorpius rotundicauda*) inhabit Indo-Pacific waters (Sekiguchi and Shuster 2009; Shin et al. 2009). Apex horseshoe crab populations in the mid-Atlantic U.S. are the most abundant and well-studied (Walls et al. 2002; Smith et al. 2009), while the sparser populations in Asia and at the extreme boundaries of their range (fringe habitat areas) in the U.S. are relatively poorly investigated. Concerns about low or declining horseshoe crab abundance have raised interest in collecting baseline data among populations world-wide to assess and manage previously unstudied populations as well as to understand ecological links between apex and fringe populations (Berkson et al. 2009).

While less abundant, horseshoe crab populations in fringe habitats in the U.S. and populations in the Indo-Pacific may be increasingly important as harvest pressures, climate change and urbanization continue to affect horseshoe crab populations in ways that are not fully understood (Moore and Perrin 2007; Kreamer and Michels 2009; Shuster and Sekiguchi 2009). Horseshoe crabs are used in the U.S. as bait in the eel and whelk fisheries and worldwide to produce amebocyte lysate, a biomedical product that detects endotoxins pathogenic to humans (Berkson and Shuster 1999; Levin et al. 2003; Novitsky 2009). The U.S. populations on the mid-Atlantic coast appear to have stabilized with implementation of fishery management practices, but habitat loss remains a concern, and the status of horseshoe crabs in fringe areas, where they are not commercially harvested (e.g. the northern U.S. Atlantic and nGOM coasts), is not as well known. In the Indo-Pacific, harvest and habitat loss have threatened horseshoe crabs to near extinction in some areas.

However, Asian horseshoe crabs are listed by the International Union for the Conservation of Nature (IUCN) as Data Deficient (World Conservation Monitoring Centre 1996) and in need of more study. The American horseshoe crab is included in the Mexican list of species "in danger of extinction" (Baillie and Groombridge 1996; Berkson et al. 2009; Zaldivar-Rae et al. 2009). Hence, more data on abundance and habitat stressors such as changes in climate or habitat size and quality for sparse populations are needed to understand species-wide resiliency to pressures that threaten abundance and alter distributions (Foley 1994).

Traditionally, horseshoe crab population data have been collected by trawl or spawning surveys, which may be less effective for determining the status of sparser populations and typically leave out juveniles. Both trawl and spawning surveys rely on higher numbers of larger sized animals that are easy to find and sample, even in turbid waters (Botton et al. 2003; Carmichael et al. 2003; Ehlinger et al. 2003; Brockmann and Smith 2009; Shin et al. 2009). Trawl surveys may miss small or buried animals, and spawning surveys are limited to adult horseshoe crabs that cluster on the beach and may be biased toward detection of males that aggregate to maximize reproductive potential (Carmichael et al. 2003; Smith et al. 2009; Brockmann and Johnson 2011). Direct observation and counting of live animals in intertidal and subtidal waters is less common (Carmichael et al. 2004) and can better capture overall population structure, but relies on clear and calm waters or large shallow intertidal nursery areas that are not typical to all habitats. Furthermore, juveniles and subadults, particularly 5–100 mm prosomal width, can be hard to sample by any of these survey methods due to their small size, camouflaged appearance, and rapid swimming ability compared to larger horseshoe crabs (Carmichael personal observations, Carmichael et al. 2004; Burton et al. 2009). In fringe habitat areas of the U.S. such as northern Maine and the nGOM, turbidity, tidal range, temperature extremes, and low numbers make even spawning surveys problematic, and whole population surveys are difficult using established methods (Moore and Perrin 2007; Fulford and Haehn 2012). Indo-Pacific population surveys present similar difficulties (Chen et al. 2004; Mishra 2009; Morton and Lee 2011; Cartwright-Taylor 2015). Alternative approaches, therefore, are needed to describe these hard to study horseshoe crab populations.

One option to potentially resolve sampling difficulties is to combine traditional survey methods with molt sampling and use molts as a proxy for live juvenile and subadult (large animals near maturity) animals. In most areas where crabs live and spawn, molted shells wash ashore (often intact) and may reflect the number and size of young actively growing (molting) animals recently living in nearby waters. In this study, we tested the use of molt characteristics as an alternative method to obtain data on juvenile and subadult population dynamics for two sparse horseshoe crab populations in fringe habitats (at the edge of the know range) in the U.S. for which no such data currently exist. We hypothesize that molts could serve as a proxy for live juveniles to provide information on population structure, molt timing, and relative growth rates for these comparatively small and hard to study populations of horseshoe crabs. This information could be applied to evaluate regional-scale differences in these life-history traits or local response to pulse environmental perturbations (such as the Deepwater Horizon oil spill or severe storm events). To do this, we compared data on the number, size (prosomal width), and sex ratio of molts from American horseshoe crabs in Taunton Bay, Maine on the northeast Atlantic coast with data from molts at Petit Bois Island, Mississippi on the nGOM coast. This approach has potential to contribute important information on population structure for any juvenile horseshoe crab population, but may be most important in areas where crabs are low in abundance or environmental attributes make them hard to locate and sample.

14.2 Methods

14.2.1 Study Sites

The habitats sampled in this study represent the extreme range of known habitats used by horseshoe crabs in U.S. waters, from the quiet, muddy and marsh-fringed intertidal areas of northern Maine to the dynamic, sandy and relatively narrow, sloping beaches of barrier islands in the nGOM.

Taunton Bay, Maine (**ME**) Taunton Bay is the northernmost known horseshoe crab spawning habitat in the U.S. (Fig. 14.1; Moore and Perrin 2007; Schaller and Dorsey 2011). Taunton Bay discharges into Frenchman Bay, which drains into the Gulf of Maine. From Frenchman Bay inland, the bay meanders with large portions enclosed from extreme marine tidal and wind effects. The intertidal area is typically comprised of cobbled mud flats surrounded by salt marsh and experiences regular tidal fluctuation of ~2 m (FTB 1991). The area has a typical annual monthly mean air temperature range of −6 to 20 °C (TWC 2014). The 1,400 ha Taunton Bay estuary is situated within a rural area in which spawning sites (and breeding adult horseshoe crab populations) are known and have been monitored since 2001 (Moore and Perrin 2007; Schaller and Dorsey 2011). There are no known studies or data available on juvenile or subadult horseshoe crab populations in Taunton Bay, and the site has no known history of horseshoe crab harvest. Genetic analyses and preliminary telemetry data suggest the Taunton Bay horseshoe crab population is distinct and may be locally isolated (King et al. 2005; Moore and Perrin 2007).

Petit Bois Island, **Mississippi** (**MS**) Petit Bois Island is an uninhabited barrier island off the coast of Mississippi (Fig. 14.1), which separates Mississippi Sound from the Gulf of Mexico. The island is about 0.7 km wide, 10 km in length and has an average tidal range of ~0.5 m. Some horseshoe crab spawning sites have been identified and described on the northern (Mississippi Sound) side of Petit Bois Island and other nearby barrier islands in the nGOM (Carmichael personal observations, Fulford and Haehn 2012). The portion of the island that was surveyed for this study supports horseshoe crab spawning and is characterized by narrow, gently sloping, medium-grained sandy beaches and includes island lagoon and pond complexes (Boone 1973; Turner et al. 1987; Fulford and Haehn 2012). The island typically experiences annual monthly mean air temperatures of 11–29 °C (TWC 2014).

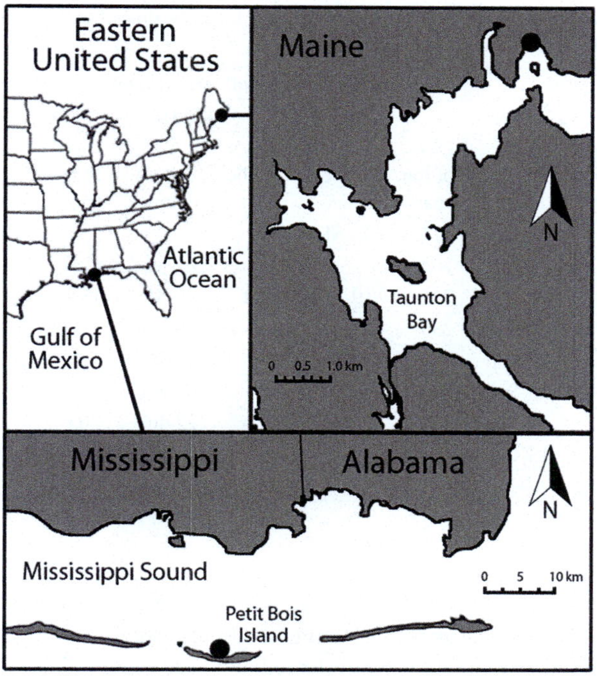

Fig. 14.1 Study sites in Taunton Bay, Maine (44.5868 °N, 68.2368 °W) on the northern U.S. Atlantic coast and on Petit Bois Island, Mississippi (30.2017 °N, 88.4637 °W) in the northern Gulf of Mexico

Petit Bois Island is part of the Gulf Islands National Seashore managed by the National Park Service, designated as a Gulf Islands Wilderness Area, and horseshoe crab harvest of adults or juveniles is not permitted. Few data are available on juvenile or subadult horseshoe crabs in the nGOM (Shuster 1955; Rudloe 1981), and no data exist from Petit Bois Island. No genetic analyses and limited spawning surveys have been conducted in the region (Fulford and Haehn 2012).

14.2.2 Field Surveys

Molt samples were collected by hand and processed in the lab for an ~3.0 m wide by 0.5 km long transect running parallel to the waterline. At macrotidal Taunton Bay, the area was sampled at low tide (because the site was typically submerged at high tide) and included intertidal flat adjacent to marsh. At microtidal Petit Bois Island, the area sampled included intertidal beach from the high water line seaward to ~1.0 m below the water line (subtidal) and was sampled within 2 h of high tide due to attempts to also sample live animals that are typically found at higher water

Table 14.1 Dates and corresponding days of the year for horseshoe crab molt surveys in Taunton Bay, Maine and on Petit Bois Island, Mississippi

Taunton Bay, Maine				Petit Bois Island, Mississippi					
2005		2006		2008		2009		2011	
Survey date	Day of the year	Survey date	Day of the year	Survey date	Day of the year	Survey date	Day of the year	Survey date	Day of the year
—	—	23-Jun	177	26-Mar	86	—	—	—	—
—	—	21-Jul	202	—	—	01-May	121	—	—
—	—	04-Aug	216	19-May	140	26-May	146	12-May	132
17-Aug	229	18-Aug	230	—	—	06-Jun	157	01-Jun	152
31 Aug	243	01-Sep	244	19-Jun	171	19-Jun	170	15-Jun	166
14-Sep	257	15-Sep	258	—	—	03-Jul	184	29-Jun	180
28-Sep	271	29-Sep	272	—	—	22-Jul	203	13-Jul	194
12-Oct	285	13-Oct	286	—	—	05-Aug	217	—	—
26-Oct	299	—	—	21-Aug	234	24-Aug	236	—	—

— no sampling

at the site (Carmichael personal observations). The transect at each site was sampled every 2 weeks corresponding to full and new moons, except for the 2008 Petit Bois Island survey, for which data were collected at 4-week intervals (Table 14.1). This sampling scheme was chosen because juvenile horseshoe crabs have been found to molt with lunar periodicity in the nGOM (Rudloe 1981) and are known to hatch and molt at roughly 2–4 week intervals (Carmichael et al. 2003) following periods of spawning that also often occur relative to moon phases (Smith et al. 2002). Starting and ending GPS (global positioning system) points for the survey transects were confirmed with a handheld Garmin GPS and visible landmarks, and field notes were taken on weather conditions that might affect the number of molts on the beach.

Molts were collected from 17 August through 26 October 2005 and 23 June through 13 October 2006 in Taunton Bay, Maine. Molts were collected from 26 March through 21 August 2008 and from 1 May through 24 August 2009 on Petit Bois Island, Mississippi (Table 14.1). Due to high numbers of small sized molts (<16 mm) in the wrack line in summer 2009, beginning with the 22 July 2009 survey, all wrack (comprised largely of dried seagrass) was harvested from transects, dried and sorted to collect any trapped molts. Because the dataset for Petit Bois Island, Mississippi included only four possible sampling dates in 2008, we opted to continue sampling during an additional season for this site. The Deepwater Horizon MC252 oil spill (DWHOS), however, prevented access to Petit Bois Island during spring and summer 2010, and samples were collected at the next possible opportunity, from 12 May through 13 July 2011. This sampling scheme serendipitously provided the opportunity to make pre- and post-spill comparisons of molt number and size at this site.

14.2.3 Measuring Molt Size and Sex

Molts were measured for prosomal width (PW) to the nearest 0.1 mm using vernier calipers for animals <16 mm and to the nearest 1.0 mm for larger sized molts. Sex determinations were made by identifying genital slits or pores under the first set of book gills in combination with height of the prosomal arch and mating scars, if present (Shuster 1982; Sekiguchi 1988; Carmichael et al. 2003). From these data, molts were categorized as male, female, or unknown.

14.2.4 Environmental Data

To relate molt timing, number and size to differences in temperature between regions, data on air temperature for the location and period of study were obtained from National Oceanic and Atmospheric Administration's (NOAA) National Climatic Data Center (NCDC) (http://www.ncdc.noaa.gov/). Air temperature was used instead of water temperature because sources for in situ water temperature data were not continuously available at both survey sites during study. Relevant in situ measurements cannot be made retrospectively, and remotely sensed sea surface temperature data are not reliable for all types of waterbodies, particularly small and irregularly shaped embayments, such as Taunton Bay, that would result in unusable mixed pixels (land and water in same pixel) for determining water temperature. While air temperatures are more variable than water temperatures during the spring to summer season, they are only 1–2 °C higher (Cayan 1980; Jaswal et al. 2012), making them sufficiently similar for the purpose of this study (to test regional-scale differences in molt patterns relative to temperature change). Air temperature, therefore, was considered a reasonable alternative in the absence of site-specific water temperature (McCloy and Dolan 1973; Kresback et al. 2012). Within the NCDC, global surface conditions for air temperature were provided by the Automated Surface Observing System (ASOS), which serves as a primary climatological observing network in the U.S. Data from Hancock County Bar Harbor, Maine were applied to the Taunton Bay site, and Grand Bay, Mississippi station data were collected for the Petit Bois Island site. Based on lunar calendars, full and new moon periods were assigned days of the year and separated into biweekly periods of equivalent days (Gore 1996) for comparison of air temperature patterns among survey periods and between sites.

14.2.5 Data Analysis

Although molts of all sizes were collected, only molts ≥40 mm PW were found at all the sites and in each year. Hence, molts <40 mm PW were not included in statistical comparisons between sites and through time. Descriptive statistics for each site by year, linear regression analyses of the number and size (prosomal width) of molts collected through time, and analyses of size frequency histograms were performed using Microsoft Excel 2007 and 2010. IBM SPSS Statistics Version 21 was used for Pearson Chi Square, Analysis of Variance (ANOVA), and Pearson's *R* correlation tests.

The size-frequency histograms of prosomal width for Petit Bois 2008, 2009, 2011 and Taunton Bay 2005, 2006 samples were analyzed as mixture distributions using the grouped-data maximum likelihood method of Macdonald and Pitcher (1979) as implemented by the mixdist package (Macdonald and Du 2012) in the R 3.1.1 environment (R Core Team 2014). The sample sizes were not large, so constraints on the parameters were necessary to avoid over-parameterization. In all cases except for Petit Bois 2011, the assumptions of equally-spaced means (i.e. constant growth increments) and constant coefficient of variation (i.e. the ratio of standard deviation to mean was assumed to be the same for each group found within a histogram) appeared to be justified; the goodness-of-fit test was acceptable, and the fit was not significantly improved by relaxing the constraints. The Petit Bois 2011 histogram was fitted without the constraint of equally-spaced means. While the goodness-of-fit test did not accept the fit for 2009 and 2011 data, at least visually and qualitatively, the fits were consistent with fits for other years. The mixed distribution fit to each histogram was plotted in R and is provided in Fig. 14.2.

ANOVA followed by Tukey's Honest Significance Difference post-hoc test was used to detect differences in mean molt sizes among years and between sites. Pearson's *R* correlation was used to compare the size frequency distributions of molts between years.

Chi Square analysis was used to define the independence of cumulative frequency data for sex ratios between sampling years at each site. A chi-squared probability of ≤ 0.05 was used to determine if the paired observations (male and female frequency) by year were independent of each other. A significance value of $P < 0.05$ was applied to all analyses.

14.3 Results

Nearly twice as many juvenile molts were found, representing a wider range of sizes, at the study site in the nGOM (Petit Bois Island, Mississippi) compared to the site in northern New England (Taunton Bay, Maine) (Table 14.2). The greatest number of molts was collected in 2009 at the Petit Bois site primarily due to a larger number of molts <10 mm in PW. These small sized molts were also found on Petit Bois Island following the DWHOS in 2011, but were not found earlier (in 2008) at Petit Bois or at any time in Taunton Bay, Maine (Table 14.2).

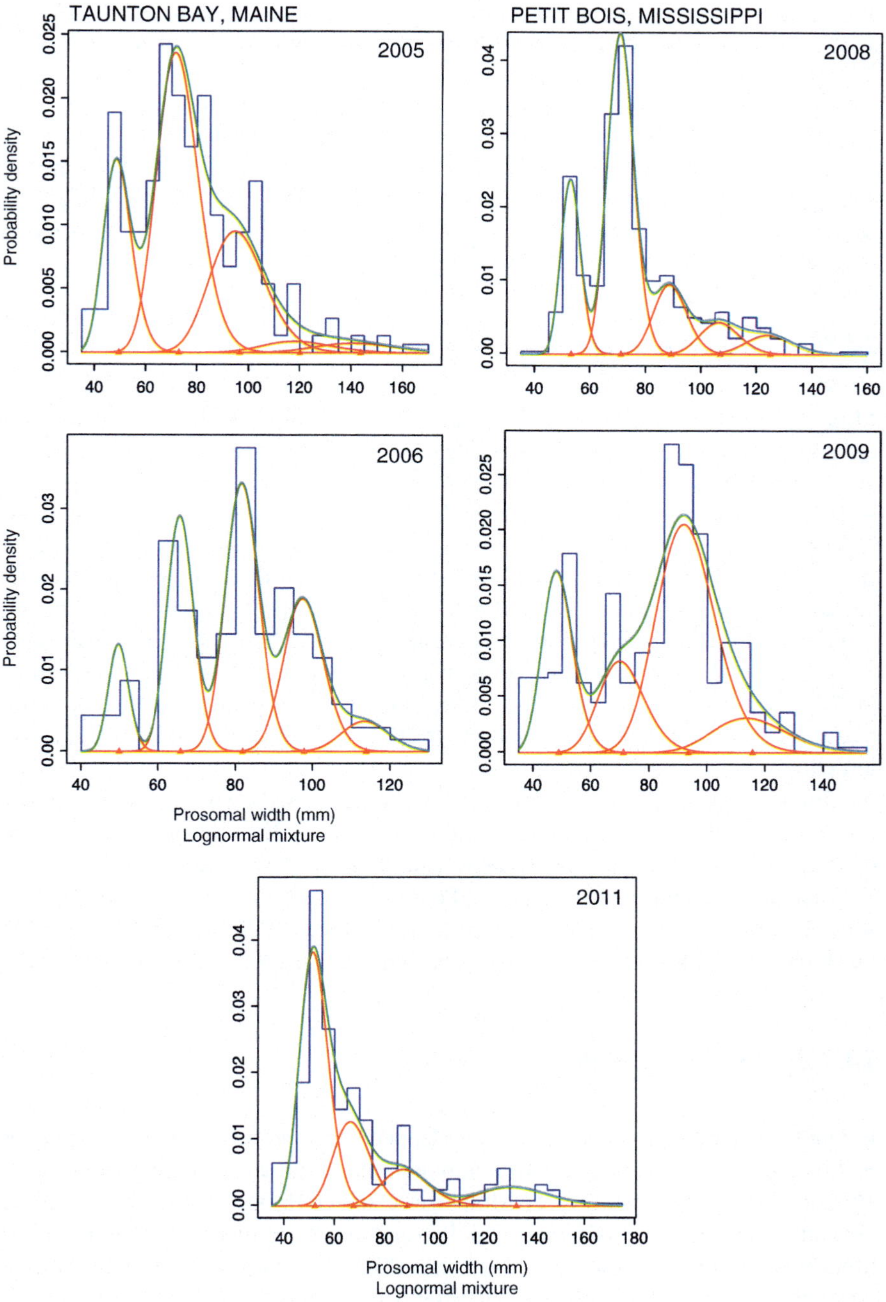

Fig. 14.2 Modeled mixed distributions fit to each size frequency histogram (shown in Figs. 14.4 and 14.6) for horseshoe crab molts collected in Taunton Bay, Maine (2005, 2006) and Petit Bois Island, Mississippi (2008, 2009, 2011)

Table 14.2 Total number and size (including the range and mean ± standard error of prosomal widths) of horseshoe crab molts sampled from sites in Taunton Bay, Maine and Petit Bois Island, Mississippi during 5 years

	Taunton Bay, Maine		Petit Bois Island, Mississippi		
	2005	2006	2008	2009	2011
N	153	84	287	506	270
Range (mm)	19.8–165	18.7–124	21–152	3.0–150	4.9–169
Mean (≥40 mm)	77.7 ± 1.9	80.6 ± 2.1	75.8 ± 1.2	82.9 ± 1.6	69.2 ± 1.7

14.3.1 Molt Number

Molt number changed through time such that peak molt abundance occurred at different times at each site, but within each site patterns were the same for males and females (Fig. 14.3). The overall number of molts decreased from late spring to early fall for each sampling period at both sites, but this pattern was only significant at Taunton Bay in 2005 for males and females and at Petit Bois for males in 2009 (Fig. 14.3). Parabolic relationships were found between molt number and date for all sample types in 2006 at Taunton Bay; however, the relationships were only significant for males. Peaks in molt abundance occurred from day of the year 226 to 256 (August–September) at Taunton Bay and from day 138 to 155 (May–June) at the Petit Bois Island site. A similar pattern of molt number through time was observed at the Petit Bois Island site in 2011, following the DWHOS (Fig. 14.4, left panel).

Molt surveys in Taunton Bay began after known peak spawning periods reported for the region (Schaller et al. 2005; Schaller and Dorsey 2011, black bars, Fig. 14.3) or in apex habitats of the mid-Atlantic (Smith et al. 2002; grey bars, Fig. 14.3). Highest molt abundance occurred at the Petit Bois Island site during periods reported for peak spawning on the Mississippi barrier islands (Fulford and Haehn 2012; black bars, Fig. 14.3) and in apex habitats (Smith et al. 2002; gray bars, Fig. 14.3).

14.3.2 Prosomal Width

For molts ≥40 mm, prosomal widths overlapped between sites, with mean prosomal widths ranging from 69 ± 2 to 83 ± 2 mm and maximum sizes ranging from 124 to 169 mm across sites (Table 14.2). Molt sizes changed through time at each site, but did not differ between males and females, and smaller molts were found close to times of known peak spawning at both sites (Figs. 14.4 and 14.5). At Taunton Bay, mean prosomal width of molts did not differ between years, but increased during the

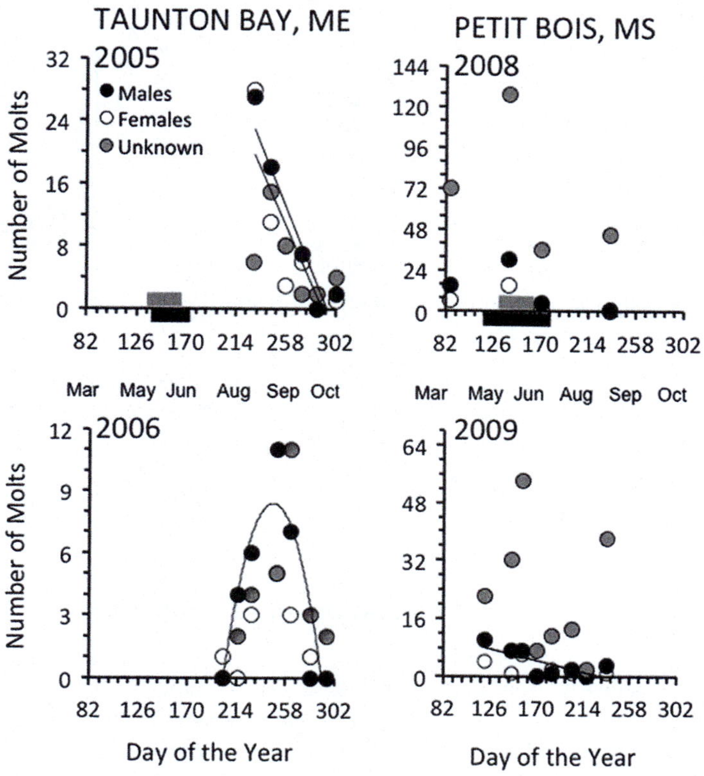

Fig. 14.3 Number of molts collected through time at Taunton Bay, Maine (2005, 2006) and on Petit Bois Island, MS (2008, 2009). 2005: Males: $y = -0.36x + 105.13$, $R^2 = 0.86$, $F_5 = 24.92$, $P = 0.01$; Females: $y = -0.33x + 94.10$, $R^2 = 0.70$, $F_5 = 9.10$, $P = 0.04$; 2008: Males; $y = -0.004x^2 + 2.13x - 254.56$, $R^2 = 0.80$, $P = 0.04$; 2009: Males: $y = -0.73 + 16.76x$, $R^2 = 0.55$, $F_7 = 7.22$, $P = 0.04$. *Black bars* indicate peak spawning periods previously reported at Taunton Bay and Petit Bois Island habitats (Data from Schaller et al. 2005; Schaller and Dorsey 2011; Fulford and Haehn 2012). *Grey bars* indicate peak spawning periods reported for apex habitats in the mid-Atlantic (Data from Smith et al. 2002)

survey period (summer-fall) in 2005 (Fig. 14.5, left panels). At Petit Bois Island, molt prosomal width differed among all survey years, with greatest mean prosomal width measured among molts in 2009 and smallest (smaller than in other years at any site) in 2011 following the DWHOS (Table 14.2; ANOVA: $F_4 = 2.57$, $P < 0.001$; Tukey's HSD: $P < 0.02$ for all significant comparisons). Molt size decreased during the survey period (spring-late summer) on Petit Bois Island for all samples in 2009 (Fig. 14.5, right panels) and for female molts in 2011 (Fig. 14.4, middle panel).

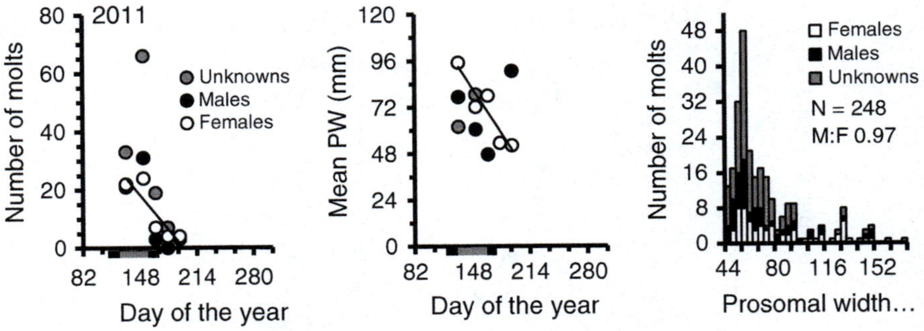

Fig. 14.4 Number of molts and prosomal width compared to sampling day of the year and size frequency distributions of horseshoe crabs collected at Petit Bois Island, Mississippi following the DWHOS in 2011. Number of molts: Females: $y=0.36x+71.94$, $R^2=0.77$, $F_5=10.04$, $P=0.05$. Prosomal width: Females: $y=0.69x+184.66$, $R^2=0.87$, $F_5=19.86$, $P=0.02$. *Black bars* indicate peak spawning periods documented at Taunton Bay and Petit Bois Island, and *grey bars* show peak spawning periods documented in mid-Atlantic apex habitats (Delaware, New Jersey)

14.3.3 Size Frequency Distribution and Cohort Analyses

Size frequency distributions of juvenile horseshoe crab molts were generally similar among years and between sites (Fig. 14.6). At Taunton Bay, molt size frequency distributions were correlated between years ($R=0.75$, $P=0.001$) and with molts at Petit Bois Island in 2008 (compared to Taunton Bay 2005: $R=0.71$, $P=0.001$; 2006: $R=0.53$, $P=0.001$) (Fig. 14.6, right panels). In these years, five cohorts were detected ranging in mean prosomal width from 49 to 144 mm (Tables 14.3 and 14.4; Fig. 14.2). Molt size frequency distributions at Petit Bois Island in 2009 were weakly correlated to two other sampling periods, Taunton Bay 2006 ($R=0.49$, $P=0.003$) and Petit Bois Island 2008 ($R=0.35$, $P=0.04$). In 2011, overall molt size distributions were correlated to Taunton Bay in 2005 ($R=0.39$, $P=0.02$) and Petit Bios Island in 2008 ($R=0.56$, $P<0.001$), likely due to the presence of cohorts >125 mm in all 3 years (Table 14.4 and Figs. 14.4 and 14.6). In 2009 and 2011, four cohorts were detected ranging in mean prosomal width from 49 to 133 mm (Table 14.4 and Figs. 14.2 and 14.6). These modeled cohorts visually and quantitatively fit distributions in previous years but did not significantly fit the data for these 2 years (Table 14.3). Each juvenile cohort was composed of crabs collected during any of our sampling times, at which they had achieved a certain size. We interpret these cohorts as representing individuals of the same age at the time of molting (as has been previously described for live juvenile horseshoe crabs using this approach; Carmichael et al. 2003; Smith et al. 2009; Table 14.4).

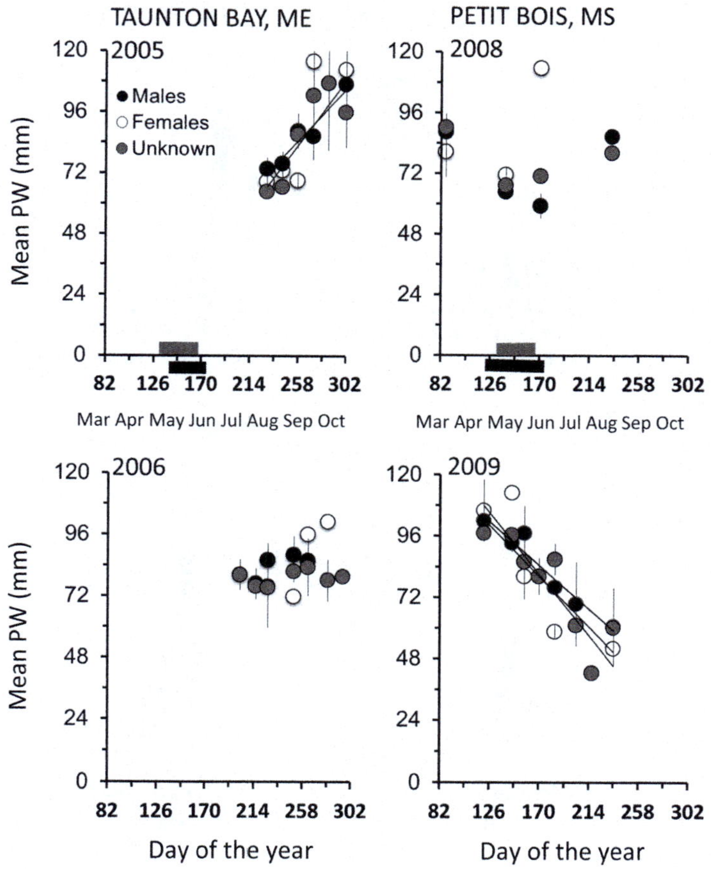

Fig. 14.5 Mean prosomal width ± SE of molts collected through time at Taunton Bay, Maine (2005, 2006) and on Petit Bois Island, Mississippi (2008, 2009). 2005: Males: $y=0.46x-34.33$, $R^2=0.93$, $F_4=37.70$, $P=0.01$; Unknowns: $y=0.58x-66.68$, $R^2=0.73$, $F_5=10.74$, $P=0.03$ and males, females and unknowns for 2009 at Petit Bois (Males: $y=-0.39x+150.88$, $R^2=0.95$, $F_5=69.21$, $P=0.00$; Females: $y=-0.55x+174.31$, $R^2=0.77$, $F_4=10.01$, $P=0.05$; Unknowns: $y=-0.45x+157.05$, $R^2=0.77$, $F_7=20.51$, $P=0.003$). *Black bars* indicate peak spawning periods in Taunton Bay or Petit Bois Island, and *grey bars* indicate peak spawning periods in mid-Atlantic apex habitats (Delaware, New Jersey)

14.3.4 Sex Ratios

The sex of juvenile molts was determined in 72 % of samples collected from Taunton Bay in 2005 and 56 % in 2006, and for ~26 %, 20 %, and 52 % of molts sampled in 2008, 2009, and 2011 respectively at Petit Bois Island. Sex ratios among recovered molts were somewhat male dominated, ranging from 1.3 to 2.3 during

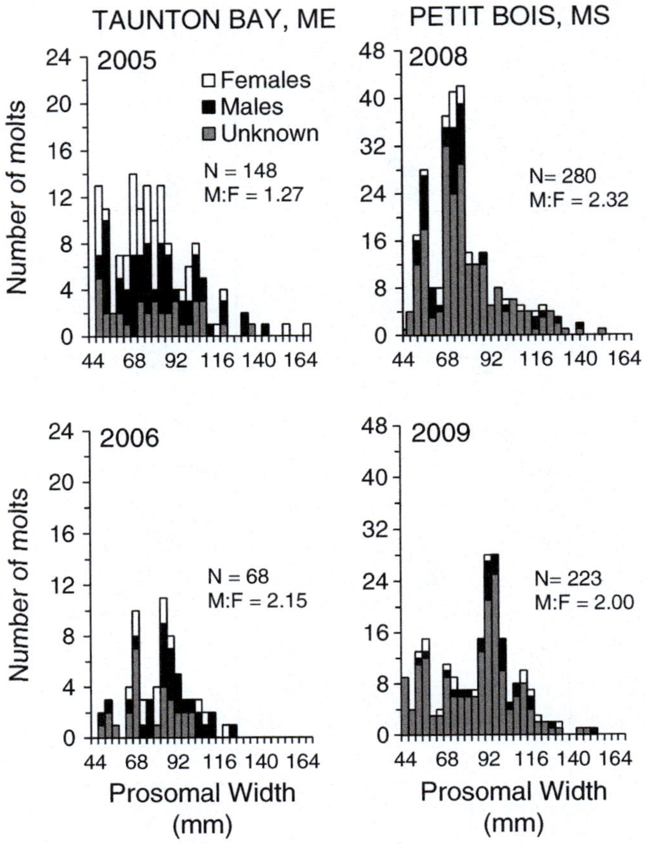

Fig. 14.6 Size frequency distribution of horseshoe crab molts collected in Taunton Bay, ME (2005, 2006) and Petit Bois Island, Mississippi (2008, 2009), separated by sex. "Unknown" indicates molts for which sex could not be determined

Table 14.3 Statistics for mixed distribution model fits to prosomal widths of molts collected at Taunton Bay, Maine and Petit Bois Island, Mississippi

	Taunton Bay, Maine		Petit Bois Island, Mississippi		
	2005	2006	2008	2009	2011
df	17	8	15	15	17
χ^2	19.88	6.56	17.12	42.19	48.91
p	0.28	0.58	0.31	<0.001	<0.001

Outputs correspond to distributions plotted in Figs. 14.2, 14.4 and 14.6

Table 14.4 Mean ± standard deviation of prosomal width (mm) for molt cohorts from Taunton Bay, Maine and Petit Bois Island, Mississippi during 5 years, determined by mixed distribution analyses (μ ± σ outputs)

Taunton Bay, Maine		Petit Bois Island, Mississippi			Smith et al. (2009)	Carmichael et al. (2003)	Sekiguchi (1988)
2005	2006	2008	2009	2011			
49±6	50±3	53±4	49±5	52±6	—	49±2 (instar 10)	50 (instar 12)
73±8	66±3	71±5	71±8	68±7	—	63±3 (instar 11)	64 (instar 13)
96±11	82±4	89±6	94±10	89±10	91–96 (age 7)	91±3 (instar 13)	81 (instar 14)
—	98±5	107±7	—	—	104 (age 8)	103±4 (instar 14)	—
120±13	114±6	—	116±13	—	113–116 (age 8)	116±4 (instar 15)	—
144±16	—	125±8	—	133±14	136–145 (age 9)	134±3 (instar 16)	—

Data are compared to cohort sizes reported for live animals in previous studies, including estimated age in years (Smith et al. 2009) or instar stages (Shuster 1955, 1982; Sekiguchi 1988; Carmichael et al. 2003)

— no cohort identified at this mean size

most sampling periods, with the exception of the 2011 Petit Bois Island survey, when the sex ratio was ~1.0 (Figs. 14.4 and 14.6, insets). Male to female sex ratios did not differ between consecutive years at Taunton Bay ($\chi^2=1.918$, $P=0.166$) or Petit Bois Island ($\chi^2=0.132$, $P=0.716$), but were significantly lower in 2011 at Petit Bois Island compared to other sampling years except 2005 (2011 compared to 2006: $\chi^2=4.37$, $P=0.04$; 2008: $\chi^2=7.93$, $P<0.01$; 2009: $\chi^2=4.03$, $P=0.045$).

14.3.5 Environmental Influences

At each site, annual and seasonal air temperature patterns were similar between years but differed in magnitude between sites (Fig. 14.7). Peak molt number was consistently associated with the period of peak air temperatures and typically during the 2-week period including a full moon at both sites. At Taunton Bay, most molts were found at temperatures of 14–15 °C, just after the summer temperature peak (19.6±0.5 in 2005 and 20.6±0.4 in 2006) and during the late summer when mean air temperatures were decreasing (Fig. 14.7, top panels). At Petit Bois Island, most molts were found at temperatures of 20–22 °C, during the spring and early summer while mean air temperatures were increasing toward the annual summer peak (29.7±0.4 in 2008, 30.3±0.4 in 2009, and 32.4±0.4 in 2011) (Fig. 14.7, middle and bottom panels).

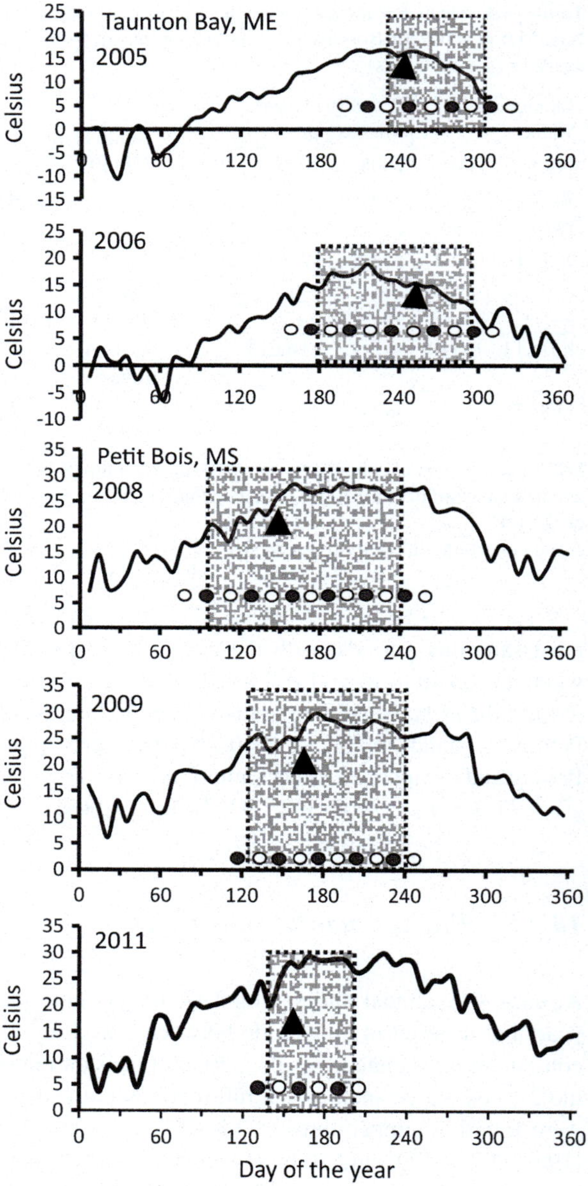

Fig. 14.7 Mean weekly air temperatures and lunar phases for the weeks in each year when horseshoe crab molts were collected from sites in Taunton Bay, Maine and on Petit Bois Island, Mississippi. *Triangles* show peak molting periods during the survey period for each site. Temperature data were not available for mid-November through December 2005 for the Taunton Bay site

14.4 Discussion

Molts provided information to assess aspects of juvenile life history in two environmentally distinct fringe habitats where these data were not previously available for molts or live juvenile horseshoe crabs. Applying demographic data from molts enabled regional-scale comparisons of molt timing, size distributions and sex ratios through time and following a major environmental perturbation, the Deepwater Horizon oil spill. Our results suggest that, if applied with care, juvenile molts could be useful as a proxy for understanding changes in live juvenile populations in several ways.

14.4.1 Molt Number

The observation that higher numbers of molts occurred during or just following known periods of peak spawning and with some apparent lunar periodicity is consistent with previous study (Rudloe 1981; Carmichael et al. 2003). Larval horseshoe crabs are known to emerge from nests ~2 weeks following spawning, with molting and growth into subsequent instar stages every 2–4 weeks, resulting in a roughly lunar periodicity to growth and molting patterns of early instars (Carmichael et al. 2003). This pattern would not necessarily be expected among older juveniles (>1 year) like those sampled in this study because the intermolt period becomes progressively longer as horseshoe crabs grow (Sekiguchi 1988). Molts at or near 40 mm typically represent instars 9–11 (years 3–4) among juveniles reared in culture and live animals in New England and on the Gulf of Mexico in Florida (Rudloe 1981; Sekiguchi 1988; Carmichael et al. 2003). Nonetheless, previous studies of juvenile horseshoe crabs along the Florida nGOM coast found peaks in the number of molts among these older juveniles, each occurring on a full moon during December through March (Rudloe 1981). In this study, the parabolic pattern in molt abundance at Taunton Bay in 2006 defined a peak molting period in summer that was similar to the patterns in molt abundance previously described by Rudloe (1981) in winter. The significant decline in numbers of molts from late spring through summer during the sampling period at Taunton Bay in 2005 and Petit Bois Island also were consistent with the descending aspect of a lunar periodic molt peak, which happened to correlate with known peak periods of spawning in the area. These findings suggest a common lunar periodicity in molt occurrence among regions. The number of summertime molts found during this study, however, was considerably higher than found off the Florida nGOM coast where molt peaks occurred in winter (Rudloe 1981). These patterns in molt timing could reflect a common periodicity but difference in seasonal molt frequency among locations (at least during some years) that require additional study.

14.4.2 Molt Size

Comparisons of molt numbers within different size classes between successive years were useful to follow growth cohorts at each site. The size classes of molts sampled at both sites corresponded to known size classes for *L. polyphemus* cohorts defined elsewhere (Sekiguchi 1988; Botton et al. 2003; Carmichael et al. 2003;

Smith et al. 2009). For example, cohorts in the 49–50 mm prosomal width range have been shown to represent instars 10–12 (year 4+); sizes near 81–91 mm are typically instars 13–14 (year 7+); and molt sizes >100 mm may correspond to instars 14+ (years 8+) (Carmichael et al. 2003; Smith et al. 2009). These data represent the first attempt at quantitatively defining cohorts and estimating size-at-age relationships for horseshoe crabs in the northcentral Gulf of Mexico. More data from smaller size classes will help clarify relationships between instar stage and year of life for horseshoe crabs at both sites.

Shifts in prosomal width distributions through time (cohort progression) have been used to estimate growth rates in live juvenile *L. polyphemus* (Carmichael et al. 2003). If we apply this concept to Taunton Bay molt frequency distributions, which showed peaks in the 49 mm and 73 mm ranges in 2005 that shifted to 66 mm and 82–98 mm ranges in 2006, we estimate ~20 mm of annual growth (roughly 0.05 mm/day) between cohorts at these sizes. A very similar pattern was found for Petit Bois Island for cohorts near 53 mm and 71 mm in 2008 that shifted to 71 mm and 94 mm in 2009, also suggesting a possible growth increment of ~20 mm per year. This growth estimate based on annual change in molt size distribution is consistent with growth rates estimated from live juveniles on Cape Cod, Massachusetts and in culture (Sekiguchi 1988; Carmichael et al. 2003; Table 14.4). These detectable progressions in molt size in consecutive years indicate that molt data have potential to be used similarly to data from live animals to track growth and possibly mortality patterns of juveniles through time. These data also suggest remarkably consistent size classes and growth patterns among environmentally diverse habitats, from the continental climate of Taunton Bay to temperate Cape Cod and subtropical Petit Bois Island.

14.4.3 Molt Sex Ratio

Molts provided data useful to assess population structure relative to sex ratio, depending on the condition of recovered molts. Due to poor condition in some cases, we were not able to determine sex of molts as often or as confidently as with live animals (e.g. Carmichael et al. 2003). Although it was not possible to directly compare the sex ratio of molts with sex ratios of live juveniles during this study (because live juveniles were not found), the sex ratios we found among juvenile molts were consistent with data from live adult animals previously found at each site (M:F = 2.1 for Petit Bois Island, Estes unpublished data; and M:F = 1.7–2.7 for Taunton Bay, Schaller et al. 2005) and among live horseshoe crabs in the same size ranges elsewhere (Carmichael et al. 2003). These observations corroborate the potential utility of molts to reflect sex ratios in local populations of live animals.

Adult operational sex ratios (M:F) for *L. polyphemus* typically range from 1.1 to 5.8 and vary widely within and among populations (Carmichael et al. 2003; Brockmann and Smith 2009). Males typically spend more time near the water line during spawning season (Brockmann and Johnson 2011), potentially biasing sex ratios determined during spawning surveys compared to those determined during whole population surveys (Rudloe 1981; Smith et al. 2002; Carmichael et al. 2003). This reproductive strategy could bias molt sex ratios toward males if subadult males congregating near the beach were also molting, so that their molts were more likely to wash ashore and be in better condition. Rudloe (1981) and others have found that older juveniles live farther from shore, which may reduce the likelihood that larger molts reach the shoreline. Further research is needed to determine the effect of molt size on the probability of a molt to reach the shoreline. Overall, molt sex ratios measured during this study were relatively low and more consistent with sex ratios measured in whole population studies rather than spawning surveys, suggesting that molt sex ratios did not suffer from significant bias and may be a good proxy for estimating population level sex ratios.

14.4.4 Deepwater Horizon Oil Spill

In 2011, after the DWHOS, horseshoe crab molts collected on Petit Bois Island were on average significantly smaller, and sex ratios were closer to 1:1 than measured in other years. This finding is consistent with potentially greater spawning and recruitment of young crabs just prior to the DWHOS. Because horseshoe crabs analyzed for this study (>40 mm), are estimated to represent instars 9–10 at ages 4 years or older, crabs measured on Petit Bios Island in 2011 were likely spawned in 2006 or earlier and survived through the DWHOS to molt in 2011. Although 2011 field sampling started later and ended earlier than in 2008 and 2009, sampling overlapped for 2 months (five surveys) from mid-May to mid-July when peak molt abundance was found at the Petit Bois site during all surveys, making comparisons possible for this key period. Continued field surveys will be needed to confirm longer-term effects on *L. polyphemus* in the nGOM due to the DWHOS. Importantly, molt data provided an approach to effectively assess some population level effects of this catastrophic environmental perturbation on these long-lived animals in the absence of sufficient pre- or post-spill data on population dynamics for live animals in this understudied region.

14.4.5 Environmental Attributes

Despite very different environmental conditions between sites, juvenile and subadult horseshoe crabs at Taunton Bay (near subarctic) and Petit Bois Island (subtropical) showed similar mean prosomal widths, apparent cohort distributions and male to female sex ratios. This finding is somewhat surprising given that mean and peak seasonal temperatures, which differed considerably among these sites, are thought to affect horseshoe crab growth and subsequent molting (Rudloe 1981; Sekiguchi 1988; Carmichael and Brush 2012). Accordingly, previously reported spawning periods at Taunton Bay begin about 1 week later than spawning at apex habitats (Schaller et al. 2005; Schaller and Dorsey 2011), and at Petit Bois Island spawning begins about 2 weeks earlier and extends about 1 week longer than at apex habitats (Smith et al. 2002; Fulford and Haehn 2012). Rather than molting in response to absolute temperatures, our data suggest that crabs responded to temperature changes relative to the ambient temperatures to which they were acclimated. Peak molting periods occurred 2–4 weeks after a temperature increase of several days duration at both sites and more often during full than new moon phases. An inverse relationship between the two sites existed between survey date and air temperature due to the timing of sampling relative to ambient seasonal air temperatures. In Taunton Bay, initial molt surveys occurred near the air temperature peak for the survey period and declined during the remainder of the survey season. Conversely, Petit Bois Island surveys began in the spring when air temperatures steadily increased for most of the survey season and began to decline at the time of the last survey. Overall, the similarity in molt frequency patterns and sizes among sites despite absolute differences in temperature suggest plasticity in horseshoe crab molting and subsequent growth patterns; factors independent of absolute temperature may be more important to mediate the timing of molting patterns in wild horseshoe crabs across different habitat types. Molts from animals in fringe habitats such as Taunton Bay and Petit Bois Island, in turn, may have the greatest potential to provide insights regarding response to environmental perturbations because fringe populations must be adaptable to environmental extremes and will be among the first to be affected by environmental change (Ehlinger et al. 2003).

Environmental conditions, including currents, tides and storms, may affect the distribution of molts within a study area, potentially limiting the application of molt data to assess local populations. If molt distributions are affected differently by physical processes among years or across sites, it could affect molt number or size distributions through time and raise questions as to whether molts found in a given location originated elsewhere. For the periods of study at Taunton Bay and Petit Bois Island, no severe storms (hurricanes, tropical storms or depressions that are most likely to perturb these systems and affect molt transport) made landfall at or near enough to the survey sites to likely influence molt distribution among sites or years (U.S. Department of Commerce 2014 with searchable archive for historical storm tracks and reports for 2005, 2006, 2008, 2009, 2011). During 2011 tropical storm Lee made landfall in the northcentral Gulf of Mexico in early September, after surveys had ended (Brown 2011). Furthermore, Taunton Bay is a protected embayment that is more than 30 km from the discharge of the Bay into the Gulf of Maine (Fig. 14.1). Due to distance and the irregular Maine coastline, the Taunton Bay site is not affected by near shore oceanic currents. While the daily tidal range of ~3 m could move molts on and off the shoreline and potentially degrade their quality, the distance from the sampling site to open waters significantly reduces the effect of tidal flushing (Arter 2007). While it is still possible some molts may wash away from the site due to tidal flushing, it is highly unlikely that molts from other areas would be deposited here. Hence, it is highly unlikely that molts found at Taunton Bay originated elsewhere. While Petit Bois Island is more exposed to open waters, the survey site was located on the leeside of the island, protected from onshore ocean winds and near shore currents (Fig. 14.1). Microtidal ranges of less than 1 m at Petit Bois Island (Seim et al. 1987) also reduce the likelihood of substantial molt redistribution among adjacent barrier islands. It is likely, therefore, that molts collected during this study represent live animals from the area where they were collected and that molts were affected by relatively comparable physical processes among survey years. Prior to using molts to evaluate juvenile populations for an area, environmental influences should be considered, and application may be best suited to comparisons of populations across spatial scales that are larger than the range over which molts are likely to be redistributed by local hydrodynamic processes.

14.5 Conclusions

Molts provided an inexpensive and relatively easily acquired dataset on juvenile horseshoe crab life-history patterns in areas where live animals are rarely observed or difficult to study. We found similarity in molt timing and size frequency distributions between sites despite regional differences in temperature and timing of spawning. These similarities suggest endogenous plasticity in horseshoe crab molting and growth patterns across different habitat types. Molts may be less useful for making inferences about some types of population parameters such as abundance because molt presence may be affected more than live animal presence by physical processes, and molts should not be expected to reflect total population size. Molts also may not be useful to compare population dynamics on fine spatial scales that are smaller than the distance across which molts travel as they wash ashore. Our data suggest, however, that if collected diligently, molt data can be used to identify time periods of growth, define possible cohorts and growth rates, and assess some response to environmental perturbations.

Given the value of data from juveniles to understand growth, survival and recruitment to adult stages and the difficulty studying wild juveniles, an alternative sampling approach is highly beneficial to complement live animal studies. The need for an alternative sampling approach is evidenced by the very few studies that have sampled wild juvenile *Limulus* populations (the best studied species) in the past 60 years (Shuster 1955; Rudloe 1981; Botton et al. 2003; Carmichael et al. 2003; Smith et al. 2009). Analysis of molt characteristics will be particularly beneficial in fringe habitats like Maine and the nGOM or sites in Asia, where horseshoe crab populations are sparse, there are few or no background data, or sites are difficult to access due to water depth, turbidity, or temperature and weather extremes. In locations where live juveniles are discovered, molt data should be directly calibrated to living populations to provide the most reliable proxy possible for live animal data. Consequently, combining molt and live animal data may allow longer-term whole population data records to be maintained to support further research and the development of comprehensive conservation plans.

Acknowledgements Work in Taunton Bay, Maine was funded in part by a Maine Sea Grant development award to R. Carmichael at the University of Maine Machias. Work on Petit Bois Island was funded by the Dauphin Island Sea Lab and partially supported by funding from Dr. X. Chen at Alabama A&M University. Additional support was provided by S. Carmichael, N. Jillette, K. Rehn, A. Clapp, M. Buckley and K. Rose, who provided field assistance in Maine; C. Pabody, N. Taylor, C. Kovacs, M. Gilroy who assisted at the Petit Bois Island site and provided editorial assistance; E. Hieb who edited and formatted the final manuscript; USRA scientist Dr. Mohammad Al-Hamdan who assisted with GIS and graphics; and Dr. Yong Wang at Alabama A&M University who consulted on statistical analyses. The authors thank the National Park Service for their permission to survey Petit Bois Island in the Gulf Islands National Seashore.

References

Arter BS (2007) Taunton Bay mudflat management: a case study in mudflat fisheries management using ecosystem-based principles. http://www.seagrant.umaine.edu/hosting/TB/07TBplan.pdf. Accessed 15 Sept 2014

Baillie J, Groombridge B (1996) 1996 IUCN red list of threatened animals. International Union for Conservation of Nature, Gland

Berkson J, Shuster CN Jr (1999) The horseshoe crab: the battle for a true multiple-use resource. Fisheries 24:6–10

Berkson J, Chen C, Mishra J et al (2009) A discussion of horseshoe crab management in five countries: Taiwan, India, China, United States, and Mexico. In: Tanacredi JT, Botton ML, Smith DR (eds) Biology and conservation of horseshoe crabs. Springer, New York, pp 465–475

Boone PA (1973) Depositional systems of the Alabama, Mississippi, and western Florida coastal zone. Trans Gulf Coast Assoc Geol Soc 23:266–277

Botton ML, Loveland RE, Tiwari A (2003) Distribution, abundance, and survivorship of young-of-the-year in a commercially exploited population of horseshoe crabs *Limulus polyphemus*. Mar Ecol Prog Ser 265:175–184

Brockmann HJ, Johnson SL (2011) A long-term study of horseshoe crabs in Florida. Estuar Coast 34:1049–1067

Brockmann HJ, Smith MD (2009) Reproductive competition and sexual selection in horseshoe crabs. In: Tanacredi JT, Botton ML, Smith DR (eds) Biology and conservation of horseshoe crabs. Springer, New York, pp 199–221

Brown DP (2011) Tropical cyclone report: tropical storm Lee, AL132011, National Hurricane Center. http://www.nhc.noaa.gov/data/tcr/AL132011_Lee.pdf. Accessed June 2014

Burton WH, Kelley FS, Franks EA (2009) Distribution of juvenile horseshoe crabs in subtidal habitats of Delaware Bay using a suction-dredge sampling device. In: Tanacredi JT, Botton ML, Smith DR (eds) Biology and conservation of horseshoe crabs. Springer, New York, pp 285–292

Carmichael RH, Brush E (2012) Three decades of horseshoe crab rearing: a review of conditions for captive growth and survival. Rev Aquac 4:32–43

Carmichael R, Rutecki D, Valiela I (2003) Abundance and population structure of the Atlantic horseshoe crab *Limulus polyphemus* in Pleasant Bay, Cape Cod. Mar Ecol Prog Ser 246:225–239

Carmichael RH, Rutecki D, Annett B et al (2004) Position of horseshoe crabs in estuarine food webs: N and C stable isotopic study of foraging ranges and diet composition. J Exp Mar Biol Ecol 299:231–253

Cartwright-Taylor L (2015) Studies of horseshoe crabs around Singapore. In: Carmichael RH, Botton M, Shin PKS, Cheung SG (eds) Changing global perspectives on horseshoe crab biology, conservation and management. Springer, Cham, pp 193–211

Cayan DR (1980) Large-scale relationships between sea surface temperature and surface air temperature. Mon Weather Rev 108:1293–1301

Chen CP, Yeh HY, Lin PF (2004) Conservation of the horseshoe crab at Kinmen, Taiwan: strategies and practices. Biodivers Conserv 13:1889–1904
Ehlinger G, Tankersley RA, Bush MB (2003) Spatial and temporal patterns of spawning and larval hatching by the horseshoe crab, *Limulus polyphemus*, in a microtidal coastal lagoon. Estuaries 26:631–640
Foley P (1994) Predicting extinction times from environmental stochasticity and carrying capacity. Conserv Biol 8:124–137
FTB (Friends of Taunton Bay) (1991) Inventory of the Taunton Bay region, 2nd edn. Friends of Taunton Bay, Hancock
Fulford R, Haehn RA (2012) An evaluation of Mississippi barrier islands as spawning and nesting habitat for the American horseshoe crab, *Limulus polyphemus*, with implications for island restoration. Gulf Caribb Res 24:51–62
Gore PJW (1996) Phases of the moon. http://facstaff.gpc.edu/~pgore/astronomy/astr101/moonphas.htm. Accessed 14 Sept 2014
Jaswal AK, Singh V, Bhambak SR (2012) Relationship between sea surface temperature and surface air temperature over Arabian Sea, Bay of Bengal and Indian Ocean. J Indian Geophys Union 16:41–53
King TL, Eackles MS, Spidle AP et al (2005) Regional differentiation and sex-biased dispersal among populations of the horseshoe crab *Limulus polyphemus*. Trans Am Fish Soc 134:441–465
Kreamer G, Michels S (2009) History of horseshoe crab harvest on Delaware Bay. In: Tanacredi JT, Botton ML, Smith DR (eds) Biology and conservation of horseshoe crabs. Springer, New York, pp 299–312
Kresback KM, Tromble EM, Reid DG et al (2012) Evaluation of Baroclinic ADCIRC using a process-oriented test along a slope. Estuar Coast Model. doi:10.1061/9780784412411.00005
Levin J, Hochstein HD, Novitsky TJ (2003) Clotting cells and *Limulus* amebocyte lysate: an amazing analytical tool. In: Shuster CN Jr, Barlow RB, Brockman HJ (eds) The American horseshoe crab. Harvard University Press, Cambridge, MA, pp 310–334
Macdonald PDM, Du J (2012) Mixdist: finite mixture distribution models. R package version 0.5-4. http://CRAN.R-project.org/package=mixdist. Accessed Oct 2014
Macdonald PDM, Pitcher TJ (1979) Age-groups from size-frequency data: a versatile and efficient method of analyzing distribution mixtures. J Fish Res Board Can 36:987–1001
McCloy JM, Dolan R (1973) Water-air temperature relationship along coastal North Carolina, U.S.A. Geogr Ann A 55:117–121
Mishra JK (2009) Horseshoe crabs, their eco-biological status along the Northeast coast of India and the necessity for ecological conservation. In: Tanacredi JT, Botton ML, Smith DR (eds) Biology and conservation of horseshoe crabs. Springer, New York, pp 89–95
Moore S, Perrin S (2007) Seasonal movement and resource-use patterns of resident horseshoe crab (*Limulus polyphemus*) populations in a Maine, USA estuary. Estuar Coast 30:1016–1026
Morton B, Lee CN (2011) Spatial and temporal distributions of juvenile horseshoe crabs (Arthropoda: Chelicerata) approaching extirpation along the northwestern shoreline of the New Territories of Hong Kong SAR, China. J Nat Hist 45:227–251
Novitsky TJ (2009) Biomedical applications of *Limulus* amebocyte lysate. In: Tanacredi JT, Botton ML, Smith DR (eds) Biology and conservation of horseshoe crabs. Springer, New York, pp 315–326
R Core Team (2014) R: a language and environment for statistical computing. R Foundation for Statistical Computing, Vienna. http://www.R-project.org/. Accessed Oct 2014
Rudloe A (1981) Aspects of the biology of juvenile horseshoe crabs, *Limulus polyphemus*. Bull Mar Sci 3:125–133
Schaller S, Dorsey F (2011) Taunton Bay horseshoe crab tagging study: the 10 year report. Friends Taunton Bay Newsl 40:1–3
Schaller S, Thayer P, Hanson S et al (2005) Maine horseshoe crab (*Limulus polyphemus*) spawning surveys, 2004. Report to Maine Department of Marine Resources, West Boothbay Harbor
Seim HE, Kjerfve B, Sneed JE (1987) Tides of Mississippi Sound and the adjacent continental shelf. Estuar Coast Shelf Sci 25:143–156

Sekiguchi K (1988) Biology of horseshoe crabs. Science House, Tokyo
Sekiguchi K, Shuster CN Jr (2009) Limits on the global distribution of horseshoe crabs (Limulacea): lessons learned from two lifetimes of observations: Asia and America. In: Tanacredi JT, Botton ML, Smith DR (eds) Biology and conservation of horseshoe crabs. Springer, New York, pp 5–22
Shin PKS, Li H, Cheung SG (2009) Horseshoe crabs in Hong Kong: current population status and human exploitation. In: Tanacredi JT, Botton ML, Smith DR (eds) Biology and conservation of horseshoe crabs. Springer, New York, pp 347–359
Shuster Jr CN (1955) On morphometric and serological relationships within the Limulidae, with particular reference to *Limulus polyphemus* (L.). PhD thesis, New York University
Shuster CN Jr (1982) A pictorial review of the natural history and ecology of the horseshoe crab, *Limulus polyphemus*, with reference to other Limulidae. In: Bonaventura J, Bonaventura C, Tesh S (eds) Physiology and biology of horseshoe crabs. Alan R. Liss, New York, pp 1–52
Shuster CN Jr, Sekiguchi K (2009) Basic habitat requirements of the extant species of horseshoe crabs (Limulacea). In: Tanacredi JT, Botton ML, Smith DR (eds) Biology and conservation of horseshoe crabs. Springer, New York, pp 115–127
Smith DR, Pooler PS, Swan BL et al (2002) Spatial and temporal distribution of horseshoe crab (*Limulus polyphemus*) spawning in Delaware Bay: implications for monitoring. Estuaries 25:115–125
Smith DR, Millard MJ, Carmichael RH (2009) Comparative status and assessment of *Limulus polyphemus* with emphasis on the New England and Delaware Bay populations. In: Tanacredi JT, Botton ML, Smith DR (eds) Biology and conservation of horseshoe crabs. Springer, New York, pp 361–381
Turner RE, Schroeder WW, Wiseman WJ Jr (1987) The role of stratification in the deoxygenation of Mobile Bay and adjacent shelf bottom waters. Estuaries 10:13–19
TWC (2014) The weather channel. http://www.weather.com/weather/today/l/USME0149:1:US (Franklin, Maine); http://www.weather.com/weather/today/l/USAL0146:1:US (Dauphin Island, AL). Accessed 26 Jun 2015
U.S. Department of Commerce (2014) National Weather Service North Atlantic hurricane tracking chart. http://www.nhc.noaa.gov/data/#tracks_all. Accessed 14 Sept 2014
Walls EA, Berkson J, Smith SA (2002) The horseshoe crab, *Limulus polyphemus*: 200 million years of existence, 100 years of study. Rev Fish Sci 10:39–73
World Conservation Monitoring Centre (1996) The IUCN red list of threatened species. Version 2014.2. www.iucnredlist.org. Accessed 15 Oct 2014
Zaldivar-Rae J, Sapien-Silva RE, Rosales-Raya M et al (2009) American horseshoe crabs, *Limulus polyphemus*, in Mexico: open possibilities. In: Tanacredi JT, Botton ML, Smith DR (eds) Biology and conservation of horseshoe crabs. Springer, New York, pp 97–109

Part III
Biology and Physiology

Chapter 15
Preliminary Investigation of Effects of Seawater pH on Early Development of *Limulus polyphemus*

John T. Tanacredi and Sixto Portilla

Abstract Estuarine organisms are exposed to periodic strong fluctuations in seawater pH driven by natural O_2/CO_2 biogeochemical cycles. The average pH of ocean surface waters has decreased by about 0.1 units from 8.2 to 8.1, since the beginning of the industrial revolution. This phenomenon, dubbed ocean acidification, may be exacerbated into the future by increasing atmospheric CO_2 from fossil fuel combustion and associated land use changes. The direct biological effect of ocean acidification is not well understood and may vary among organisms. In a preliminary investigation to detect the sensitivity of *Limulus polyphemus* to pH changes, sets of 100 fertilized eggs were exposed to five isolated systems with pH values ranging from 6.0 to 8.0 for 4 weeks. This preliminary investigation chronicled the effects of pH on larval and/or juvenile carapace length and the development to the post-larval juvenile stage over time. Trends in carapace length (without telson) for larva and juveniles (with telson), measured by scanning electron microscopy (SEM), were positively correlated to pH within the experimental range. The long-term consequences of ocean acidification on early stage horseshoe crab development are unknown, and this work prompts further investigation.

Keywords Estuaries • Ocean acidification • pH • Juvenile development • Carapace dimensions

15.1 Introduction

Considerable attention has been given to the influence of atmospheric carbon dioxide levels on seawater chemistry. Influencing factors range from the photosynthesis as a CO_2 sink to CO_2 as an important carbonate ion source utilized in shell and

---

J.T. Tanacredi (✉) • S. Portilla
Center for Environmental Research and Coastal Oceans Monitoring (CERCOM), Molloy College, Rockville Center, NY 11570, USA
e-mail: jtanacredi@molloy.edu

R.H. Carmichael et al. (eds.), *Changing Global Perspectives on Horseshoe Crab Biology, Conservation and Management*, DOI 10.1007/978-3-319-19542-1_15

skeleton formation in biological resources (Sabine et al. 2004). NOAA's Pacific Marine Environmental Laboratory Ocean Acidification Home Page notes, "pH of the ocean surface waters have already decreased from an average of 8.21 SU to 8.10 SU since the beginnings of the industrial revolution" (CONNDEP 2011). The Intergovernmental Panel on Climate Change (IPCC) has predicted a decrease of an additional 0.3 SU by 2100 (Caldeira and Wickett 2005). Time series measurements and other field data have documented the decrease in ocean pH and other related changes in seawater chemistry (Dore et al. 2009).

Although the horseshoe crab (HSC) has survived five global mass extinctions events, one must not fall into complacency that worldwide changes in pH cannot impact their future. The early life history of *Limulus polyphemus* includes a short period of high vulnerability when they are non-motile and rely on the suitability of spawning location. HSC spawning site selection is based on a number of criteria, like beach length and wave energy (Smith et al. 2002) and levels of hydrogen sulfide (Botton and Ropes 1988). The pH of sediment within the tidal range, however, within which the developing eggs are nestled, is the result of dynamic biological processes and geological characteristics that gravid females may not consider for site selection.

Routine water quality monitoring during the HSC spawning season in waters around Long Island New York (Fig. 15.1) by the Center for Environmental Research and Coastal Oceans Monitoring at Molloy College reveal a 9-year drop in average pH (Fig. 15.2). This trend, coupled with concern for the anecdotal decadal decrease in horseshoe crab populations in the same region, provided the impetus for this preliminary investigation to determine if low pH has an impact on the early life stages of *Limulus polyphemus*.

The protective exoskeleton of the horseshoe crab is composed of chitin, whose synthesis and structural integrity if affected by acidic environments could jeopar-

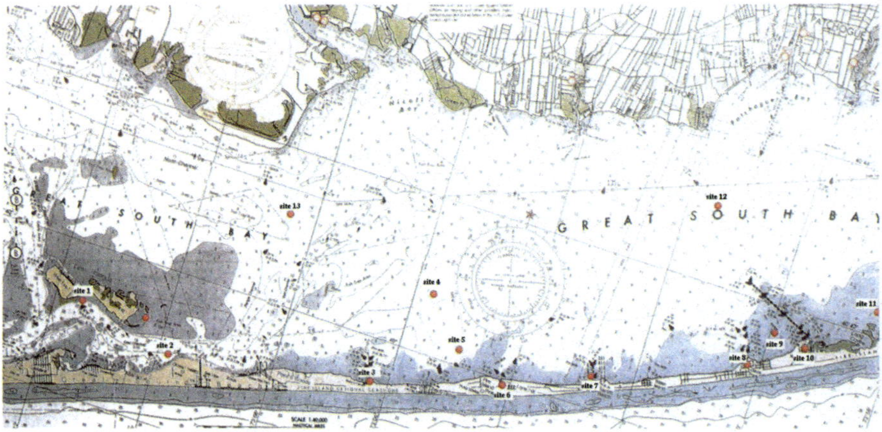

Fig. 15.1 Chart of Great South Bay, New York. Includes sampling sites (*red dots*)

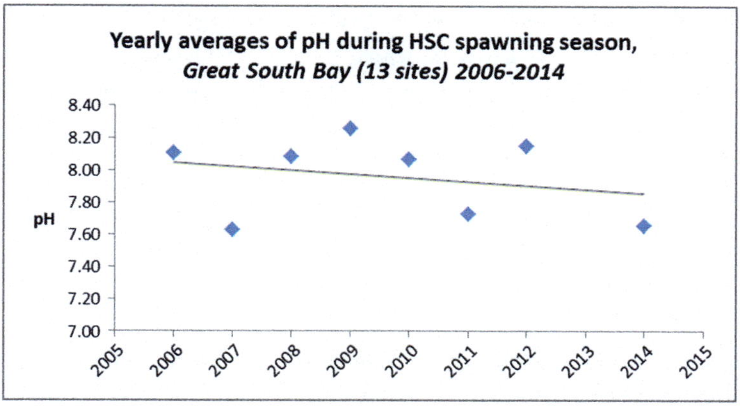

Fig. 15.2 Yearly averages of pH during the horseshoe crab spawning season from late May through August in Great South Bay, New York. Values are calculated from an average of 10 sample dates per season

dize the integrity of the HSC egg sac and ultimately its maturation. Rojas et al. (2001) observed optimum hatching conditions for the chitinous ephippia of the *Moina micrura*, a small bodied cladoceran, fall within the pH range from 5.0 to 9.0 units. Outside this range hatch rate drops precipitously (Rojas et al. 2001). In their study of early stage amphibian development, Horne and Dunson (1995) reported high sensitivity of *Ambystoma jeffersonianum* to low pH environments; pH <4.5 yielding high embryonic mortality. They also cited levels of sensitivity to pH are species specific as other larval amphibians in the same environment were observed to suffer no developmental complications at pH< 4.5 (Horne and Dunson 1995).

The goal of this preliminary investigation is to identify measureable effects on the hatch rate of freshly spawned HSC eggs and alterations in the chitinous larval and juvenile carapaces of individuals subjected to exposure to a select range of pH. The range of pH tested in this investigation conforms to levels currently found in local estuarine waters around Long Island, New York.

15.2 Methods

Five 4 L vessels were established containing filtered seawater at 5 pH levels selected for this investigation; 6.0, 6.5, 7.0, 7.5 and 8.0 units. pH levels were measured by an Orion Star A121 pH meter from ThermoFisher Scientific, Pittsburg, PA and calibrated approximately once each week. Sodium hydroxide was added to raise pH; hydrochloric acid and/or CO_2 enriched aeration with a microbubble diffuser were employed to lower pH in order to obtain the target pH levels ± 0.1 units. Temperature of water was manipulated at 25 °C which is within the typical estuarine range during HSC spawning season. The vessels were set upon thermally controlled stir plates with stir bars spinning at the lowest rate in order to achieve good mixing.

Table 15.1 Mean carapace dimensions at various pH values (28 day growth period)

		Larvae (no telson)		Juveniles (w/telson)	
pH	No. of carapaces[a]	Width (mm)	Length (mm)	Width (mm)	Length (mm)
6.0	9	3.69	3.74	0	0
6.5	9	3.68	3.74	0	0
7.0	9	3.76	3.83	0	0
7.5	13	3.69	3.82	5.44	6.40
8.0	16	3.84	3.94	5.16	6.08

[a]Total Carapaces = 56

One hundred eggs, freshly spawned and fertilized, were placed in each of the five pH and temperature controlled vessels. The eggs were spawned at the Center for Environmental Research and Coastal Oceans Monitoring (CERCOM) at Molloy College. After 28 days all hatched horseshoe crabs were removed from the systems for analysis and placed in sealed vials of 95 % ethanol and stored at −20 °C.

Sample larval and juvenile HSCs were then analyzed by a scanning electron microscopy (SEM) (Hitachi) (a) to observe any superficial carapace irregularities that might occur at the different pH levels, and (b) to measure carapace widths and lengths (with and without telson) for both larva and juveniles carried out before and after critical point drying. Table 15.1 exhibits the results of 56 total carapaces after 28 days.

15.3 Results

Measurements of larval carapaces were made by integrated scale of the Hitachi scanning electron microscope (SEM). Lengths and widths were taken from tip to tip as seen in images below (Fig. 15.3a, b). All five systems produced larva which were measured by SEM. Trends in average lengths (Fig. 15.4a) and average widths (Fig. 15.4b) display a positive correlation within the experimental pH range.

No juveniles survived for measurement in systems with pH 6.0, 6.5 nor 7.0 during this trial, indicating low pH may have a retarding effect on the progression from larva to 2nd instar juvenile stage. The required timeline for this stage of development within the lower pH range may have exceeded the time constraints imposed by this preliminary analysis. Survivorship at pH 7.5 and 8.0 may reflect survivorship to juvenile stage as observed in situ pH ranges.

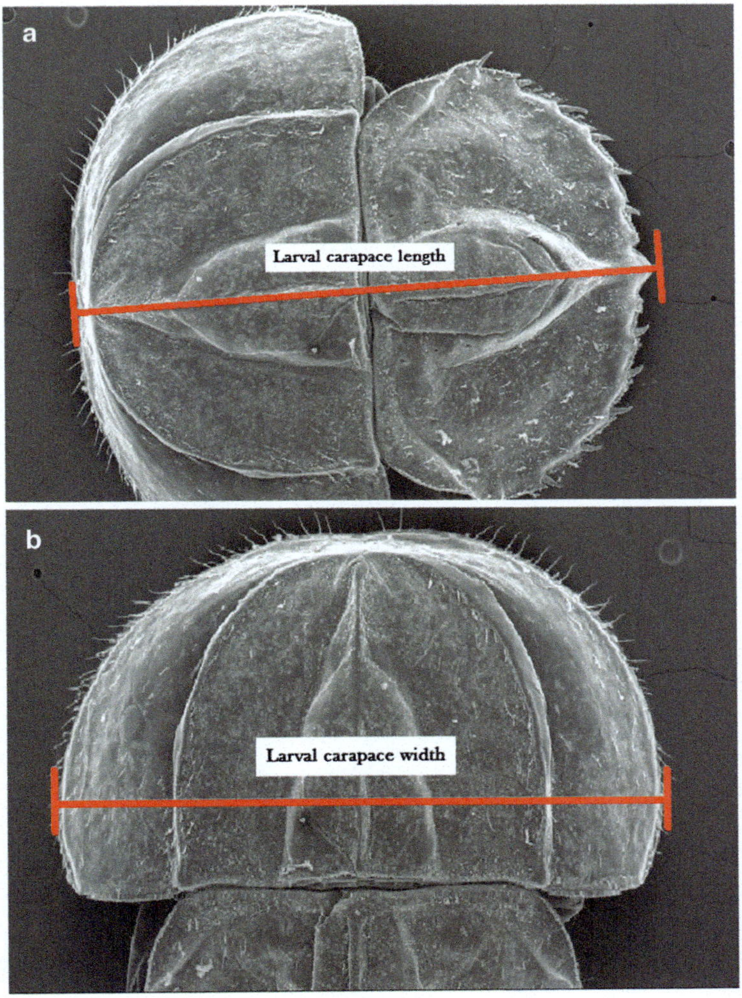

Fig. 15.3 Locations of larval (**a**) length and (**b**) width measurements, observed under SEM

15.4 Discussion and Conclusion

Results provide a preliminary indication that pH has distinct impacts on particular aspects of early stages of HSC development, namely carapace length and width, and retarding developmental advancement.

Data for water column pH is readily available in many natural HSC spawning habitats. However, data from estuarine shoreline sediments, where HSC eggs develop, is not typically acquired in routine water quality monitoring. The results shown here suggest sediment pH or sediment pore-water should be included among other routine parameters (salinity (S%), dissolved oxygen (DO), nutrients and

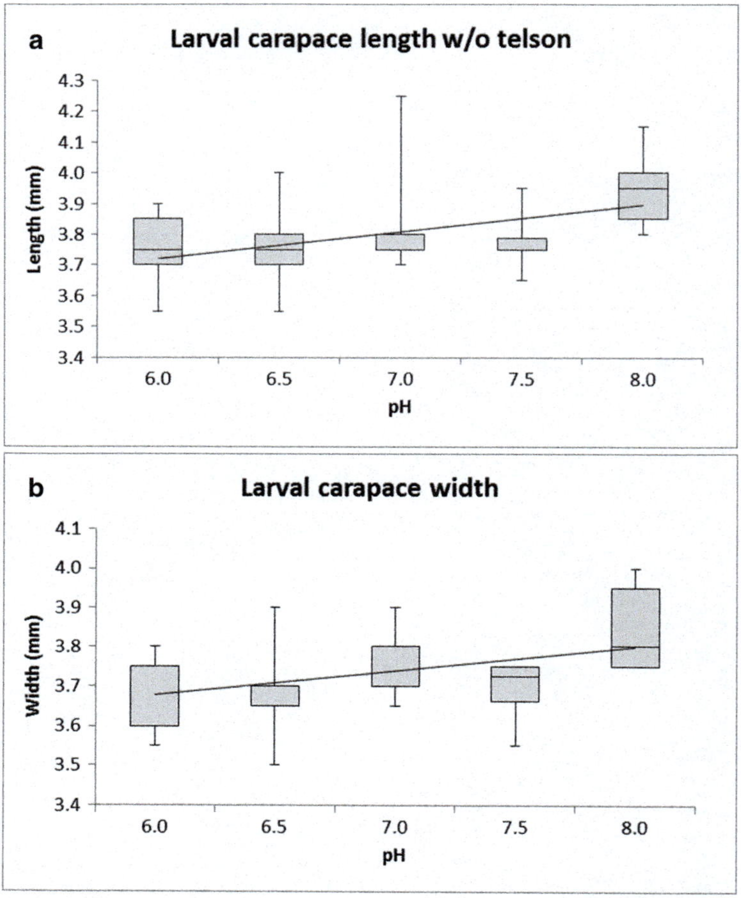

Fig. 15.4 Box plots of larval carapace measurements in millimeters (*mm*). Trends in average measurements reveal positive correlation to pH within experimental range for (**a**) carapace lengths without telson, and (**b**) carapace widths

dissolved inorganic carbon) in HSC spawning habitat assessment (National Research Council 2010).

Since the mid-1970s acid precipitation to the global oceans has been well documented, ranging from 3.7 to 5.7 in Hawaii, 4.5–5.0 in the Indian Ocean, and averaging about 5.6 in the South Pacific near Samoa (Kerr 1981). The entire marine food web dynamic may be impacted at an unprecedented pace and was identified in the Monaco Declaration in 2009 which encouraged world leaders to take ocean acidification as seriously as climate change impacts in general (Anon 2009).

Horseshoe crabs have an unusual reproduction system unlike that of any other arthropod; all four species of HSC have external fertilization and broadcast spawning (Sekiguchi 1988). The intensity of reproductive competition in

L. polyphemus has been suggested as a contributing factor to female nesting densities (Brockmann and Smith 2009). Changes in ocean acidification rates may impact horseshoe crabs globally (Tanacredi 2001).

Acknowledgments This project was supported by Molloy College, CERCOM field station, the BCES Department of Earth and Environmental Sciences degree program. A. Lundstrom and R. Vesia collected data and conducted SEM photographs as part of an undergraduate degree program for visiting students to Molloy College, CERCOM Internship program. Scanning Electron Microscopy was provided by Dr. Alex Smirnov and Dr. Chris Boyko of the Department of Earth and Marine Sciences at Dowling College, Oakdale, NY.

References

Anon (2009) Acid bath. Science Illustrated, Nov/Dec, pp 68–75

Botton ML, Ropes JW (1988) Populations of horseshoe crabs, *Limulus polyphemus*, on the northeast Atlantic continental shelf. Fish Bull 85:805–812

Brockmann HJ, Smith MD (2009) Reproductive competition and sexual selection in horseshoe crabs. In: Tanacredi JT, Botton ML, Smith DR (eds) Biology and conservation of horseshoe crabs. Springer, New York, pp 199–221

Caldeira K, Wickett ME (2005) Ocean model predictions of chemistry changes from carbon dioxide emissions to atmosphere and oceans. J Geophys Res 110:C09S04. doi:10.1029/2004JC002671

CONNDEP Energy and Environment Report (2011) Long island sound annual monitoring report. Connecticut Department of Environmental Protection

Dore JE, Lukas R, Sadler DW (2009) Physical and biogeochemical modulation of ocean acidification in the central North Pacific. Proc Natl Acad Sci U S A 106(30):12235–12240

Horne MT, Dunson WA (1995) Effects of low pH, metals and water hardness on larval amphibians. Arch Environ Contam Toxicol 29:500–505

Kerr RA (1981) Is all acid rain polluted? Science 212(4498):1014

National Research Council (2010) Ocean acidification: a national strategy to meet the challenges of a changing ocean. The National Academy Press, Washington, DC

Rojas NET, Mains MA, Rocha O (2001) The effects of abiotic factors on the hatching of *Moina micrura* Kurz, 1874 (Crustacea: Cladocera) ephippial eggs. Braz J Biol 61(3):371–376

Sabine CL, Feely RA, Gruber N et al (2004) The oceanic sink for anthopogenic CO_2. Science 305(5682):367–371

Sekiguchi K (1988) Ecology. In: Sekiguchi K (ed) Biology of horseshoe crabs. Science House, Tokyo, pp 50–68

Smith DR, Pooler PS, Loveland RE et al (2002) Horseshoe crab (*Limulus polyphemus*) reproductive activity on Delaware Bay beaches: interactions with beach characteristics. J Coast Res 18(4):739–740

Tanacredi JT (2001) *Limulus* in the limelight: the fate of a species 350 million years in the making, and in peril. Springer, New York

Chapter 16
Effects of Sediment Type and Tank Shape on Horseshoe Crab (*Limulus polyphemus*) Growth and Survival in Culture

Elizabeth E. Hieb, Jacy D. Baggett, Allen M. Aven, and Ruth H. Carmichael

Abstract We measured the effects of two conditions, sediment type (natural or clean sand) and tank shape (rectangular or kreisel style), on growth, survival, and movement patterns of juvenile American horseshoe crabs (*Limulus polyphemus*). During the 9-week experiment, horseshoe crabs reared in natural sediment and/or in rectangular tanks had higher growth rates than crabs reared in kreisel style tanks with clean sand. Horseshoe crabs reared in natural sediment also had lower % mortality compared to crabs in clean sand. Hence, by the conclusion of the experiment, horseshoe crabs reared in kreisel style tanks with clean sand showed no growth and 100 % mortality. To determine whether sediment type affected growth and survival by affecting nutrition, we measured N and C stable isotope ratios in horseshoe crab tissues compared to organic matter in sediments and gel diet fed to crabs during the study. Sediment-derived organic matter made a significant contribution to the diet of crabs reared in natural sediments. Enriched N stable isotope ratios, along with the lack of growth and poor survival, suggested that crabs reared in clean sand experienced starvation. The movement patterns of juvenile horseshoe crabs within each tank also varied with sediment and tank type, suggesting links between these attributes and settlement or foraging behavior. Natural sediment may contribute to horseshoe crab success by providing an additional food source to juvenile crabs, and tank shape may have a secondary effect on growth, survival, and movements related to hydrodynamics that merits further study.

E.E. Hieb
University Programs, Dauphin Island Sea Lab, 101 Bienville Blvd, Dauphin Island, AL 36528, USA

J.D. Baggett
Murphy High School, 100 South Carlen Street, Mobile, AL 36606, USA

A.M. Aven • R.H. Carmichael (✉)
Dauphin Island Sea Lab, University of South Alabama, Dauphin Island, AL 36528, USA

Department of Marine Sciences, University of South Alabama, Mobile, AL 36688, USA
e-mail: rcarmichael@disl.org

R.H. Carmichael et al. (eds.), *Changing Global Perspectives on Horseshoe Crab Biology, Conservation and Management*, DOI 10.1007/978-3-319-19542-1_16

Keywords Aquaculture • Diet composition • Stable isotope • Larval growth

16.1 Introduction

Despite increasing interest in horseshoe crab culture in recent years, there are few consistently collected and detailed data available on best practices for captive rearing (Carmichael et al. 2009; Carmichael and Brush 2012). In particular, data are needed to better understand how horseshoe crab growth and survival are linked to environmental conditions and diet composition in culture (Chen et al. 2010; Carmichael and Brush 2012). Type of substrate and holding tank style are two variables that could affect the chemical and physical environment for horseshoe crabs in ways that affect growth and survival, but these effects are rarely directly studied.

Substrate type may affect food availability because natural or "living" sediments can provide an important source of nutrition to juvenile horseshoe crabs. Stable isotope analysis has shown that through the seventh instar, juvenile American horseshoe crabs in the wild may rely almost entirely on a diet supported by particulate organic matter (POM) (Gaines et al. 2002; Carmichael et al. 2009). At early instar stages most available prey species are too large for juveniles to eat, and they instead feed largely on benthic and suspended organic matter (Gaines et al. 2002; Carmichael et al. 2009). Increased growth has also been reported among juvenile Asian horseshoe crabs maintained in tanks with sediments compared to those reared without sediments (Hong et al. 2009). As adults, horseshoe crabs also are known to assimilate organic matter derived from sediments (Carmichael et al. 2004). These data suggest that organic particles derived from natural sediments may be an important component of diet that can benefit culture efforts.

Similarly, tank type may affect horseshoe crab growth and survival by affecting water movement and subsequent agitation of juvenile crabs. An increased level of dissolved oxygen associated with water agitation led to increased growth rates (hatching rates) and survival of Asian horseshoe crab, *Tachypleus tridentatus*, eggs in culture (Chen et al. 2010; Chen 2011). Agitation may further affect growth rates and movement patterns of horseshoe crabs by mimicking the agitating effects of changing tides and, in turn, facilitating hatching or pelagic dispersal (Ehlinger and Tankersley 2003, 2006). Agitation that retains juvenile horseshoe crabs in the water column, however, could negatively affect growth and survival by reducing access to sediment-derived food sources and, among wild horseshoe crabs, could increase mortality by predation, which is high during pelagic dispersal (Botton et al. 2003). Therefore, while agitation may enhance development, it may also impact growth and survival by limiting the animals' ability to settle in the benthos to forage or for shelter.

The aim of the present study was to measure the effects of two conditions, sediment type (natural or clean sand) and tank shape (traditional rectangular or round kreisel style), on growth, survival, and movement patterns of juvenile horseshoe crabs (*Limulus polyphemus*, instars 1–4). Tank shape was used as a proxy for differences in water movement, with kreisel style tanks having greater potential flow velocity, vertical mixing, and agitation compared to traditional rectangular tanks (Oca et al. 2004). We used N and C stable isotope ratios in horseshoe crab tissues and in available food sources to determine whether sediment or tank type affected growth and survival by affecting nutrition. We hypothesized that natural sediments and round tanks would increase growth and survival by providing an additional food source and enhancing development of young crabs, but that enhanced growth may be mediated by reduced time for foraging due to greater agitation and potential for re-suspension of juvenile crabs in round tanks.

16.2 Materials and Methods

16.2.1 Experimental Design

Horseshoe crab larvae were hatched from eggs of local horseshoe crabs (collected from Mississippi Sound along the Mississippi-Alabama coast in the northern Gulf of Mexico) that spawned in recirculating tanks at the Dauphin Island Sea Lab's public Estuarium. First and second instar crabs (3–5 mm prosomal width) were transferred to plastic tanks that were either round (kreisel style) or rectangular and lined with either clean or natural sediment, resulting in four treatment types: round tank with natural sediment, round tank with clean sediment, rectangular tank with natural sediment, and rectangular tank with clean sediment. Each treatment was replicated three times for a total of 12 tanks (Fig. 16.1, top).

Round tanks were constructed using six plastic Cambro 6.6 L buckets, cut at the side and laid sideways to create a kreisel style tank (Fig. 16.1, bottom). Rectangular tanks consisted of 7.3 L plastic bins (Fig. 16.1, bottom). Air stones measuring $9.5 \times 2.0 \times 1.7$ cm were placed on the left side of each tank, affixed near the mid to upper water level in round tanks using 3M Marine Adhesive Sealant Fast Cure 5200 and placed at the bottom of each of the six rectangular tanks to provide similar dissolved oxygen concentrations but facilitate differences in water movement and resulting agitation within each tank type (Oca et al. 2004; Fig. 16.2). Dye tests using 0.02 g/ml carmine alum lake (Carolina Biological Supply) in tank water confirmed that water generally moved from the top to the bottom of rectangular tanks (Fig. 16.2, left middle), showing relatively consistent settlement without substantial turbulent mixing (Fig. 16.2, bottom). In contrast, round tanks were characterized by a rotating flow cell (Fig. 16.2, right middle) that remained turbulent throughout the water column and showed greater mixing (more homogenization of dye in the water column without settlement; Fig. 16.2, bottom). Hence, round tanks were considered to represent greater water mixing and agitation compared to rectangular tanks in which water flow was more consistent with sedimentation.

Each tank was lined with 450 g of either clean or natural sand. Clean sand was collected from the backshore of a local beach on Dauphin Island, Alabama, USA

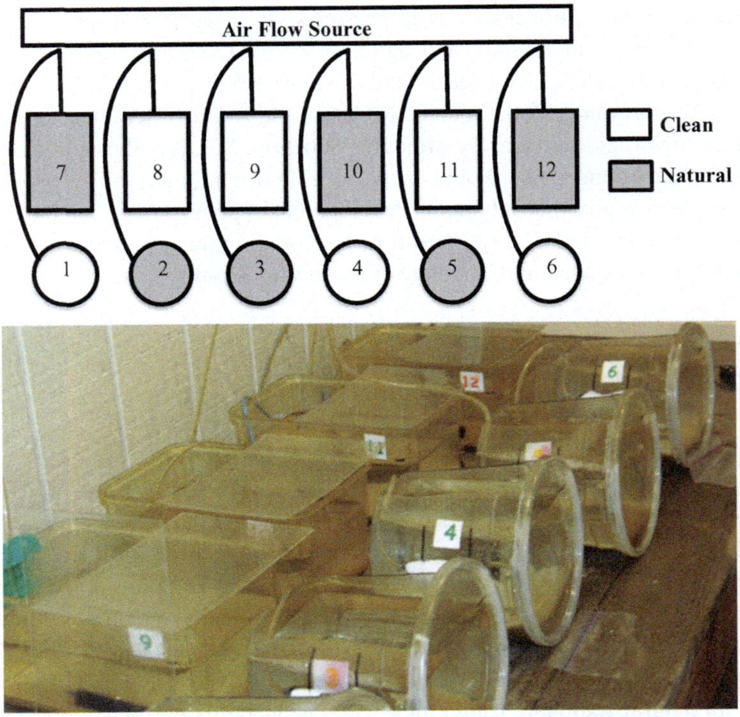

Fig. 16.1 Experimental schematic, showing randomly selected locations of each treatment type (*top*), including round vs. rectangular tanks and clean vs. natural sediments (*bottom*)

and thoroughly washed with de-ionized water to remove particulate organics. Natural sand was collected from the intertidal zone of the same beach and used without washing. Macrofaunal organisms were removed by hand from all sediments. Tanks were filled with 5 L of 200 mm pre-filtered water stock from Mississippi Sound that was adjusted to a salinity of 24 using artificial sea salts (Instant Ocean).

Each treatment tank was stocked with 30 juvenile horseshoe crabs of two size classes, including 3 mm (n = 20) and 5 mm (n = 10). Two different size classes were used to capture natural variation in effects on different size classes within a single cohort, including the first instar that does not forage (and might be expected to primarily respond to physical dynamics) and a foraging stage. To account for potential variation in tank conditions, we measured salinity, dissolved oxygen, and temperature in each tank approximately three times per week, using a YSI 85 hand-held meter. Initial water conditions were measured before adding horseshoe crabs to tanks. Horseshoe crabs were fed in tanks weekly, with each tank receiving 2.5 g of Herbivore Aquatic Gel Diet (Mazuri 5ZJ3), a sufficient quantity for survival in juvenile crabs (Carmichael et al. 2009). This diet was chosen as a commercially available control diet, intended to be similar in composition to particulate organic matter

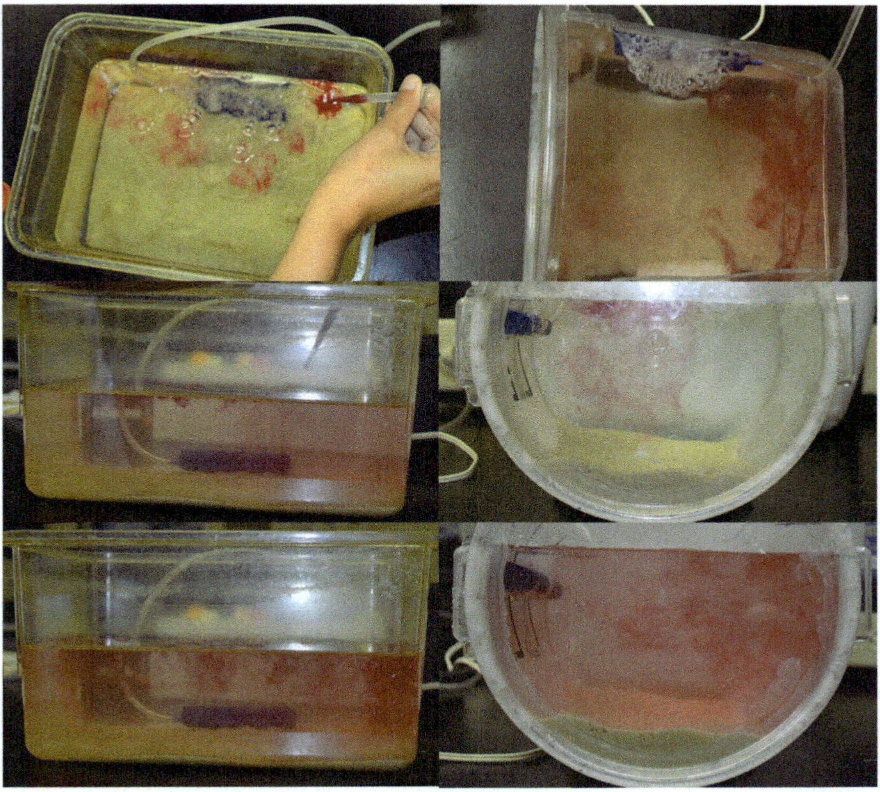

Fig. 16.2 Differences in hydrodynamic patterns in rectangular (*left*) compared to round (*right*) aquaculture tanks in which horseshoe crabs were reared during the study

found in natural sediments, which substantially contributes to horseshoe crab diet during the instar stages studied (Carmichael et al. 2009). Water and sediment were replaced weekly following collection of data.

16.2.2 Sample Collection

To determine the effect of sediment and tank type on horseshoe crabs, we measured growth, survival, and locations of animals in each tank at weekly intervals for approximately 9 weeks (54 days). Growth was measured in terms of change in prosomal width, which was measured to the nearest 1.0 mm using vernier calipers. Survival was measured by counting the number of dead horseshoe crabs in each tank through time. Data on animal movements were recorded as the number of planktonic crabs, number visible on the sediment surface, and number buried in the sediment.

16.2.3 Stable Isotope Sample Preparation and Analysis

Stable isotope (SI) samples were collected from horseshoe crab molts, sediments, and gel diet. Samples were dried at 60 °C, ground using a mortar and pestle, and sent to the University of California-Davis Stable Isotope Facility to determine $\delta^{15}N$ and $\delta^{13}C$ values by isotope ratio mass spectrometry (IRMS). Samples were analyzed by continuous flow-IRMS (CF-IRMS; 20-20 mass spectrometer, PDZ Europa). $\delta^{15}N$ and $\delta^{13}C$ values in crab molts were compared to those in sediments and gel diet based on expected mean fractionation for N (2–4‰) and C (±1‰) for a single trophic shift from food source to consumer (Peterson and Fry 1987; Carmichael et al. 2004).

16.2.4 Statistical Analysis

Growth rates of crabs in each treatment were determined as the slope of the best-fit line through a significant regression of prosomal width against sampling date using R software version 2.15.1 (R Foundation for Statistical Computing, Vienna, Austria). Growth rates were compared using analysis of covariance (ANCOVA), following a test for homogeneity of slopes (Rohlf and Sokal 1995) using StatView 5.1 (SAS Institute, Cary, North Carolina, USA). Analyses of variance (ANOVA) with post hoc tests (Tukey's HSD) were performed using R software to determine effects of sediment type and tank shape on survival and stable isotope ratios in horseshoe crab molts. If the interaction between the sediment type and tank shape was significant, we performed separate one-way ANOVA's for each variable.

16.3 Results and Discussion

16.3.1 Horseshoe Crab Growth

Horseshoe crab growth varied among treatments (Fig. 16.3). Most notably, horseshoe crabs in round (kreisel style) tanks with clean sand showed no growth during the 54-day study (crabs did not molt past their initial 3 or 5 mm size). In contrast, horseshoe crabs in rectangular tanks with clean sand grew at a rate of 0.04 ± 0.01 mm/day, and crabs in tanks with natural sand grew at a rate of 0.06 ± 0.01 mm/day. Only horseshoe crabs in rectangular tanks with natural sand molted to the 9 mm stage (around day 40; Fig. 16.3). While both tank shape and sediment type had a significant effect on horseshoe crab growth (GLM, tank shape: $F_1 = 6.94$, $P < 0.01$; sediment type: $F_1 = 23.8$, $P < 0.001$), growth rates were not significantly different among treatments in which horseshoe crabs grew during the study (ANCOVA: test for homogeneity of slopes: $F_{1,2} = 1.71$, $P = 0.20$; intercepts: $F_{1,2} = 1.96$, $P = 0.16$).

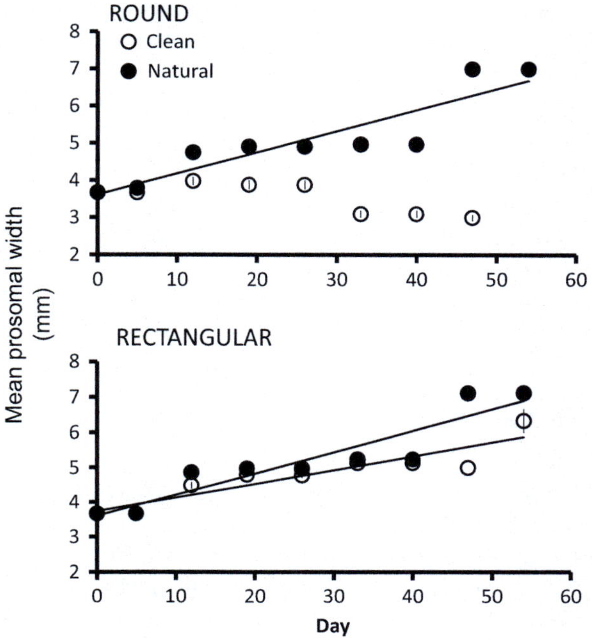

Fig. 16.3 Growth in terms of mean (± standard error) prosomal width of juvenile horseshoe crabs during the 54-day experimental period, for each treatment (tank shape and sediment type). There are no data points for clean, round tanks on day 54 due to 100 % mortality. Rectangular, natural: $y = 0.06x + 3.61$, $R^2 = 0.04$, $F_{reg1,7} = 43.82$; rectangular, clean: $y = 0.04x + 3.74$, $R^2 = 0.84$, $F_{reg1,7} = 35.90$; round, natural: $y = 0.06x + 3.61$, $R^2 = 0.82$, $F_{reg1,7} = 31.96$; round, clean: no growth. All F_{reg} are significant to $P < 0.001$. Where no error *bars* are visible, error was smaller than the symbol

Growth rates of horseshoe crabs during this experiment were comparable to those reported in other culture studies (Carmichael and Brush 2012) and lower than reported for juvenile American horseshoe crabs in the wild (Carmichael et al. 2003). While there are few studies that similarly tested sediment type and tank shape, Hong et al. (2009) also found different rates of growth for Asian horseshoe crabs, *T. tridentatus*, reared on different substrates. Other studies with similar temperature and salinity conditions to our treatments found that horseshoe crabs molted to the 5 mm stage near day 12 and the 9 mm (fourth instar) stage near day 40, similar to horseshoe crabs in our study (Jegla and Costlow 1982; Sekiguchi et al. 1988; Fig. 16.3). There is only one study that has reported growth rates for juvenile *L. polyphemus* at instars 1–4 (3–9 mm) in the wild, making it difficult to evaluate the extent to which natural sediments used in culture may mimic conditions in the natural environment. Carmichael et al. (2003) reported an average growth rate of 0.14 ± 0.07 mm/day between the first and fourth instars for juvenile horseshoe crabs on Massachusetts beaches, compared to an average growth rate of 0.05 ± 0.01 mm/day for the same instar stages of crabs in this study. This finding is not entirely unexpected, however, since slower or comparable growth rates have been reported for other species of horseshoe crabs reared in culture compared to wild stocks (Sekiguchi et al. 1988; Chen et al. 2010) due to variation in environmental factors, including reduced food supply in culture (Chen et al. 2010).

Table 16.1 Percent mortality of horseshoe crabs per replicate tank for each treatment, including natural and clean sediments in round or rectangular tank types

	% Mortality			
	Replicate tank			
Treatment type	1	2	3	Mean
Natural				
Round	60	63	97	73±12
Rectangular	10	27	30	22±6
Clean				
Round	100	100	100	100±0
Rectangular	83	87	100	90±5

16.3.2 Horseshoe Crab Mortality

Horseshoe crab mortality ranged from 10 to 100 % during the experiment and differed by tank shape and sediment type (Table 16.1; GLM, tank shape: $F_1=11.87$, $P<0.001$; sediment type: $F_1=10.57$, $P<0.01$). Mortality was highest in round tanks with clean sand, in which all animals died by the end of the experimental period (Tukey's HSD, for all comparisons: $P<0.01$). Mortality was lowest in rectangular tanks with natural sediment, which showed 22 % mortality on average. Overall, animals reared in clean sediments always showed greater than 80 % mortality, while animals in rectangular tanks fared slightly better with mortality of 83 % or more (Table 16.1). These findings suggest sediment type had a greater effect on mortality than tank shape, which had a secondary but also significant effect.

Juvenile *L. polyphemus* in natural environments reportedly show 98–99 % mortality by the fourth instar (9 mm), largely attributed to predation (Botton et al. 2003; Carmichael et al. 2003). In the absence of predation in culture, horseshoe crabs typically show 87–100 % mortality for these same instar stages (Carmichael and Brush 2012), for reasons that are not entirely clear. On average, juveniles in our experiment showed comparable mortality rates to these previous reports, except in natural sediments, where mortality was among the lowest reported for any culture study on *L. polyphemus* (Carmichael and Brush 2012). Our results, coupled with the results of previous study, suggest that the physical presence of sediment substrate alone is not as important as natural (unwashed) sediments to improve survival in culture.

16.3.3 Animal Movement

In all treatments, planktonic horseshoe crabs were only seen during the first 12 days of the experiment, when animals were still in larval stages. Juvenile horseshoe crabs showed the most movement in round tanks with clean sand, with ~50 % of the animals at the sediment surface and 50 % buried within the sediment throughout the study (Fig. 16.4, top left). In round tanks with natural sand and rectangular tanks with clean sand, the majority of horseshoe crabs (~90 %) remained buried in the

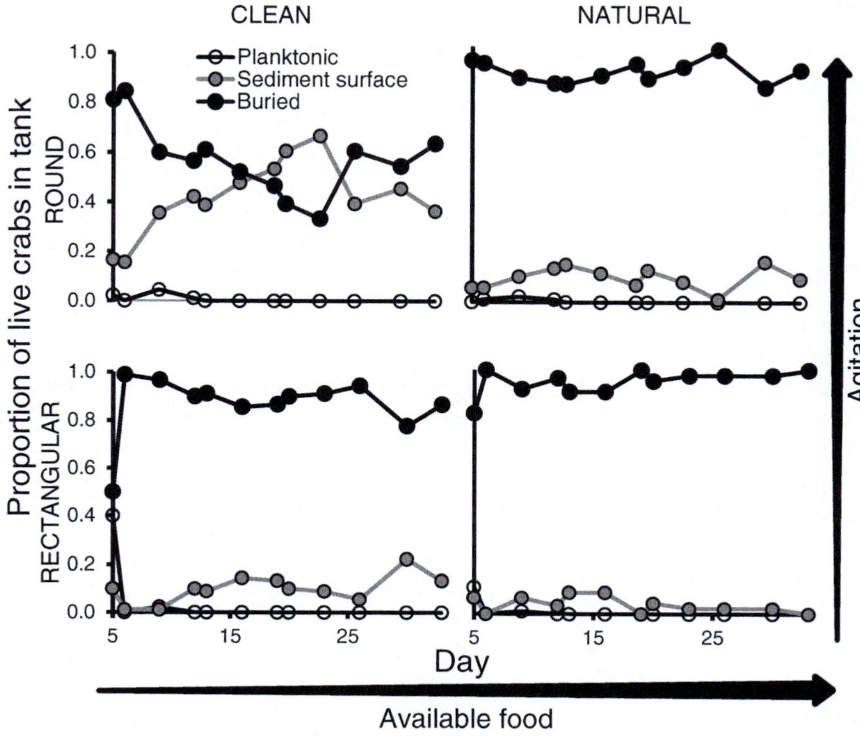

Fig. 16.4 Location of horseshoe crabs (planktonic, visible on sediment surface, buried) in treatment tanks through time during the experiment. *Black arrows* show hypothesized interaction among treatments, assuming round tanks have greater agitation and natural sand provides an additional food source

sediment (Fig. 16.4). Horseshoe crabs showed the least movement in rectangular tanks with natural sand, where ~96 % remained buried in the sediment (Fig. 16.4, bottom right).

The relatively greater movement out of sediments by horseshoe crabs in round tanks with clean sand may be driven by a greater need to forage in tanks without natural sand, the greater agitation associated with water movement that did not promote sedimentation in round tanks, or a combination of these factors. Burrowing to forage is expected for juvenile crabs beyond the first instar stage, when young crabs develop a gut and begin feeding (Gaines et al. 2002; Botton et al. 2003). Accordingly, crabs during this study shifted from planktonic to primarily benthic locations at the molt to the 2nd instar stage near day 10 in all treatments. At subsequent instar stages, natural sediments may have provided a food source to crabs that reduced the need for active foraging at the sediment surface in tanks with natural sand (Carmichael et al. 2004, 2009). Natural sediment has been suggested to promote growth and survival of horseshoe crabs by providing components essential to build chitin, which comprises the horseshoe crab carapace (Carmichael et al. 2004, 2009). Potential food supplement alone, however, did not account for all buried crabs since those in clean sediments also remained buried when they were reared in rectangular

Table 16.2 Summary of tank conditions for each horseshoe crab culture treatment

Treatment type	Dissolved oxygen (mg/L)	Salinity (‰)	Temperature (°C)
Natural			
Round	5.05 ± 0.13	25.76 ± 0.18	21.43 ± 0.08
Rectangular	4.89 ± 0.14	26.17 ± 0.21	21.35 ± 0.08
Clean			
Round	4.95 ± 0.15	25.55 ± 0.17	21.62 ± 0.08
Rectangular	4.80 ± 0.15	26.19 ± 0.18	21.19 ± 0.07

tanks with less water column agitation. This finding emphasizes the potential for differences in tank shape (and associated differences in agitation) also to affect horseshoe crab location within tanks (Fig. 16.2). Wild juvenile *L. polyphemus* are not typically found in highly dynamic areas with scoured sediments (Botton et al. 2003; Carmichael et al. 2003), supporting the idea that local hydrodynamics affect juvenile distributions. Hence, it seems likely that movement patterns in and out of the sediments by horseshoe crabs during this study were driven by the combination of sediment type and tank shape.

16.3.4 Tank Conditions

Mean values for dissolved oxygen (DO) ranged from 4.80 to 5.05 mg/L in each treatment type and did not differ among treatments (Table 16.2). Mean values for salinity and temperature in this study ranged from 25.55–26.19 to 21.19–21.62 °C, respectively, and differed by tank shape (Table 16.2; GLM, salinity: $F_1 = 0.68$, $P < 0.01$; temperature: $F_1 = 10.37$, $P < 0.01$). Other studies have reported increased DO associated with increased water flow as a factor potentially affecting culture success (Chen et al. 2010; Chen 2011), but DO was not a factor in culture success in this study, as there were no differences in DO concentrations associated with tank shape or sediment type.

Although statistically significant, the differences in salinity and temperature with tank shape were small (<1 for salinity and <1 °C for temperature), and all values were within the range suitable for growth and survival of juvenile American horseshoe crabs. In a review of horseshoe crab culture experiments, maximum growth and survival rates of juvenile American horseshoe crabs were observed at salinity ranging from 25 to 35 and temperature ranging from 19 to 34 °C (Carmichael and Brush 2012), consistent with conditions in all treatments during this study. Furthermore, in previous studies, American horseshoe crabs showed no differences in growth rates or survival when reared at salinity or temperatures that differed by as much as 5–10 units (Jegla and Costlow 1982; Sekiguchi et al. 1988). American horseshoe crabs in the wild appear to tolerate an even broader range of salinities (5–34) among estuarine habitats they occupy along the coastal U.S. (Ehlinger and Tankersley 2004). Hence, conditions in all tanks during this study fell well within the normal range for maximum growth and survival and did not likely account for differences in growth, survival, or movements of horseshoe crabs among treatments.

16.3.5 Stable Isotopes

Stable isotope values in horseshoe crabs ranged from 10 to 12‰ for N and −18 to −20‰ for C and differed among treatments (Fig. 16.5, top). $\delta^{15}N$ ratios in horseshoe crabs differed by tank shape and sediment type (GLM, tank shape: $F_1 = 8.73$, $P < 0.01$; sediment type: $F_1 = 12.67$, $P < 0.01$), such that the $\delta^{15}N$ values in horseshoe crabs reared in clean sand were significantly heavier than those in crabs in rectangular tanks with natural sand (Tukey's HSD, for all comparisons: $P \leq 0.03$). $\delta^{13}C$ ratios varied only by sediment type. $\delta^{13}C$ values in horseshoe crabs in tanks with clean sand were significantly lighter than those in horseshoe crabs in tanks with natural sand (GLM, $F_1 = 50.53$, $P < 0.001$).

Mean SI ratios in natural sand were 7.91 ± 0.25‰ for $\delta^{15}N$ and −17.63 ± 0.04‰ for $\delta^{13}C$ (Fig. 16.5, bottom). Mean $\delta^{15}N$ values in horseshoe crabs from tanks with natural sand were 2.85 ± 0.35‰ heavier for $\delta^{15}N$ and 0.49 ± 0.27‰ heavier for $\delta^{13}C$ values compared to natural sand itself. These findings are consistent with expected fractionation from food source to consumer and indicate that natural sand provided a significant food source for horseshoe crabs in these tanks. Clean sand did not

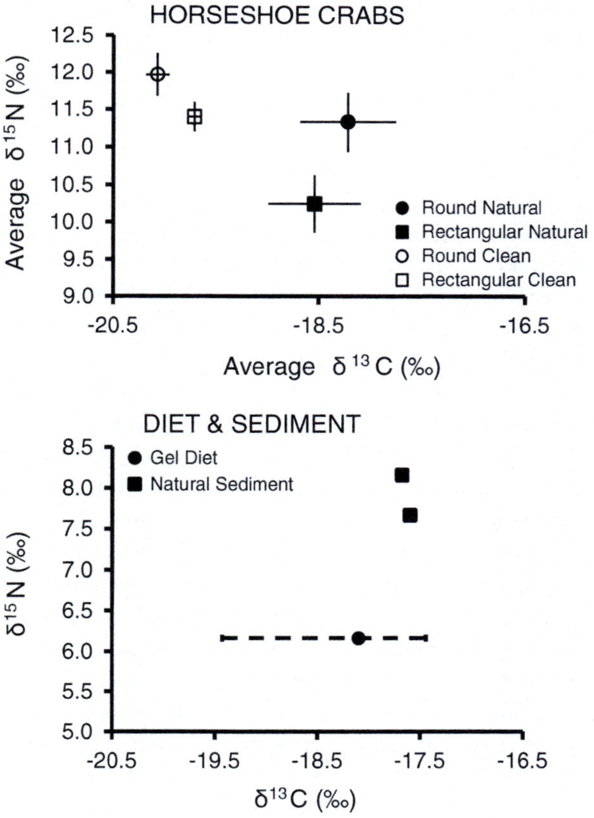

Fig. 16.5 N and C stable isotope ratios in horseshoe crab molts from each treatment (*top*) and in the artificial gel diet and natural sediments provided to horseshoe crabs (*bottom*). Dashed line shows the range of values in gel diet during the study

contain sufficient organic material for SI determination, confirming that it was not a potentially significant food source during the study.

Mean SI ratios in the gel diet fed to horseshoe crabs during the study were 6.16 ± 0.15‰ for $\delta^{15}N$ and -18.10 ± 0.39‰ for $\delta^{13}C$ (Fig. 16.5, bottom). On average, SI ratios in horseshoe crabs from tanks with natural sand were 4.60 ± 0.28‰ heavier than gel diet samples for $\delta^{15}N$ and 0.02 ± 0.47‰ heavier for $\delta^{13}C$. For horseshoe crabs in tanks with clean sand, average isotope ratios were 5.54 ± 0.24‰ heavier than gel diet samples for $\delta^{15}N$ and 1.80 ± 0.40‰ heavier for $\delta^{13}C$. These results suggest that horseshoe crabs in all treatments consumed gel diet, but values for animals from clean sand were somewhat higher than expected for a single trophic shift (usually 2–4‰) from food source to consumer.

Elevated SI values (greater than expected for a single trophic shift) can result from catabolic processes that occur in tissues during starvation (Hobson et al. 1993; Gannes et al. 1997; Olive et al. 2003). Enriched N SI ratios in the absence of growth, as seen among horseshoe crabs in round tanks with clean sand during this study (Figs. 16.3 and 16.5) is likely indicative of starvation conditions. Horseshoe crabs in tanks with clean sand, therefore, were not receiving sufficient food supply from gel diet alone. These findings corroborate the importance of natural sand to supplement the diet of horseshoe crabs in culture and the potential to affect growth, survival, and movement patterns in culture, where diet is regulated and may be limited (Chen et al. 2010).

16.4 Conclusions

Conditions including sediment quality, water circulation, and diet composition have been reported as important to determine success of juvenile horseshoe crabs in culture (Hong et al. 2009; Chen et al. 2010; Chen 2011; Carmichael and Brush 2012). In terms of growth and survival, during the 9-week experimental period, we found that juvenile horseshoe crabs reared in rectangular tanks with natural sand fared ~3 times better than horseshoe crabs in round tanks with natural sand and ~13 times better than horseshoe crabs in any tanks with clean sand. Movement patterns and SI ratios among horseshoe crabs in this study further demonstrated that natural sediment affected horseshoe crab growth and survival by affecting both burrowing behavior and available food supply. Greater agitation magnified these effects such that horseshoe crabs in round (kreisel style) tanks with clean sand had no growth, showed signs of starvation, and experienced 100 % mortality by the end of the study. In contrast, the lower agitation and greater food supply in rectangular tanks with natural sand allowed animals to settle in the sediment and commence foraging behavior, resulting in relatively rapid growth and ~80 % survival. Based on the metrics used in this study, sediment type appeared to have a somewhat greater effect on horseshoe crab success compared to tank shape, which had a secondary, but still significant effect on growth and survival. The combination of these two factors had the greatest effect on all measured metrics. Refining parameters for future horseshoe crab culture efforts will benefit from defining the interaction of multiple factors that affect growth and survival of these known environmental generalists.

Acknowledgments Funding for this project was provided by the Dauphin Island Sea Lab and the University of South Alabama. We thank the Dauphin Island Sea Lab's public Estuarium for providing juvenile horseshoe crabs for this experiment. We also thank Courtney Nelson, Jessica Delo, and Heather Patterson for assistance with data collection and stable isotope sample preparation.

References

Botton ML, Loveland RE, Tiwari A (2003) Distribution, abundance, and survivorship of young-of-the-year in a commercially exploited population of horseshoe crabs *Limulus polyphemus*. Mar Ecol Prog Ser 265:175–184

Carmichael RH, Brush E (2012) Three decades of horseshoe crab rearing: a review of conditions for captive growth and survival. Rev Aquac 4:32–43

Carmichael RH, Rutecki D, Valiela I (2003) Abundance and population structure of the Atlantic horseshoe crab *Limulus polyphemus* in Pleasant Bay, Cape Cod. Mar Ecol Prog Ser 246:228–239

Carmichael RH, Rutecki D, Annett B et al (2004) Position of horseshoe crabs in estuarine food webs: N and C stable isotopic study of foraging ranges and diet composition. J Exp Mar Biol Ecol 299:231–253

Carmichael RH, Gaines E, Sheller Z et al (2009) Diet composition of juvenile horseshoe crabs: implications for growth and survival of natural and cultured stocks. In: Tanacredi JT, Botton ML, Smith DR (eds) Biology and conservation of horseshoe crabs. Springer, New York, pp 521–534

Chen Y, Lau CW, Cheung SG et al (2010) Enhanced growth of juvenile *Tachypleus tridentatus* (Chelicerata: Xiphosura) in the laboratory: a step towards population restocking for conservation of the species. Aquat Biol 11:37–46

Chen CP (2011) The challenges and opportunities for horseshoe crab conservation in Taiwan. In: Oral presentation, International symposium on the science and conservation of Asian Horseshoe crabs, Hong Kong, China 13–16 June 2011. http://www.cityu.edu.hk/bch/iwscahc2011/Download/PT-1%20%20Chang-Po%20Chen.pdf. Accessed 2 Aug 2015

Ehlinger GS, Tankersley RA (2003) Larval hatching in the horseshoe crab, *Limulus polyphemus*: facilitation by environmental cues. J Exp Mar Biol Ecol 292:199–212

Ehlinger GS, Tankersley RA (2004) Survival and development of horseshoe crab (*Limulus polyphemus*) embryos and larvae in hypersaline conditions. Biol Bull 206:87–94

Ehlinger GS, Tankersley RA (2006) Endogenous rhythms and entrainment cues of larval activity in the horseshoe crab *Limulus polyphemus*. J Exp Mar Biol Ecol 337:205–214

Gaines EF, Carmichael RH, Grady SP et al (2002) Stable isotopic evidence for changing nutritional sources of juvenile horseshoe crabs. Biol Bull 203:228–230

Gannes LZ, O'Brien DM, del Rio CM (1997) Stable isotopes in animal ecology: assumptions, caveats, and a call for more laboratory experiments. Ecology 78:1271–1276

Hobson KA, Alisauskas RT, Clark RG (1993) Stable-nitrogen isotope enrichment in avian tissues due to fasting and nutritional stress: implications for isotopic analyses of diet. Condor 95:388–394

Hong S, Zhang X, Zhao Y et al (2009) Effect of sediment type on growth and survival of juvenile horseshoe crabs (*Trachypleus tridentatus*). In: Tanacredi JT, Botton ML, Smith DR (eds) Biology and conservation of horseshoe crabs. Springer, New York, pp 521–534

Jegla TC, Costlow JD (1982) Temperature and salinity effects on developmental and early posthatch stages of *Limulus*. In: Bonaventura J, Bonventura C, Tesh S (eds) Physiology and biology of horseshoe crabs: studies on normal and environmentally stressed animals. Alan R. Liss, New York, pp 103–113

Oca J, Masaló I, Reig L (2004) Comparative analysis of flow patterns in aquaculture rectangular tanks with different water inlet characteristics. Aquac Eng 31:221–236

Olive PJW, Pinnegar JK, Polunin NVC et al (2003) Isotope trophic step fractionation: a dynamic equilibrium model. J Anim Ecol 72:608–617

Peterson BJ, Fry B (1987) Stable isotopes in ecosystem studies. Ann Rev Ecol Syst 18:293–320

Rohlf FJ, Sokal RR (1995) Statistical tables. Macmillan, New York

Sekiguchi K, Seshimo H, Sugita H (1988) Post-embryonic development of the horseshoe crab. Biol Bull 174:337–345

Chapter 17
Bacteria and Fungi Identified on Horseshoe Crabs, *Tachypleus gigas* and *Carcinoscorpius rotundicauda* in the Laboratory

M.I.M. Faizul, H.T. Eng, A. Christianus, and Y.M. Abdel-Hadi

Abstract In Malaysia, reduction of horseshoe crab sightings in their natural habitat has prompted studies on the ecology of this invertebrate and potential threats to its survival. In this study, eggs and newly hatched *Tachypleus gigas*, collected from a natural spawning ground in Banting, Selangor (Malaysia), were conditioned and cultured in the laboratory to test the potential of fungal or bacterial infections to occur and impair development and growth. During the culture period, evidence of bacterial and fungal infestation was observed. Identified species included bacteria (*Shewanella putrefaciens*, *Bacillus cereus*, *Corynebacterium* sp. and *Enterococcus faecalis*) and fungi (*Aspergillus* sp., *Aspergillus niger*, *Peniclillium* sp., *Gliocladium* sp.). Eggs infected with bacteria or fungi usually turned reddish, grey or black and ceased to develop. Larvae became passive and coated with black spore and in some cases died soon after molting into the larval stage. The species with greatest potential to affect the growth and survival of horseshoe crab eggs and larvae sampled in this study were *S. putrefaciens* (a Gram-negative bacterium) and *A. niger* (a fungi). The suite of species identified during this study were different from those previously identified on horsehose crabs in culture, suggesting possible species, location, and/or condition-specific variation in communities of infectious microbes brought in from the natural environment and propagated in culture. Infected horseshoe crabs, therefore, may require different management or treatment strategies and additional study.

M.I.M. Faizul
Institute of Bioscience, Universiti Putra Malaysia, 43400 Serdang, Selangor, Malaysia

H.T. Eng • Y.M. Abdel-Hadi
Department of Aquaculture, Faculty of Agriculture, Universiti Putra Malaysia, 43400 Serdang, Selangor, Malaysia

A. Christianus (✉)
Institute of Bioscience, Universiti Putra Malaysia, 43400 Serdang, Selangor, Malaysia

Department of Aquaculture, Faculty of Agriculture, Universiti Putra Malaysia, 43400 Serdang, Selangor, Malaysia
e-mail: anniechr@yahoo.com

R.H. Carmichael et al. (eds.), *Changing Global Perspectives on Horseshoe Crab Biology, Conservation and Management*, DOI 10.1007/978-3-319-19542-1_17

Keywords Bacteria • Eggs • Fungus • Infection • Larvae

17.1 Introduction

Laboratory cultured horseshoe crabs may have less exposure to microbes than horseshoe crabs in the natural environment, nevertheless they are still susceptible to accidentally introduced pathogens (Smith and Berkson 2005; Faizul et al. 2012). Disease agents such as fungi, viruses, bacteria and parasites can affect the development and growth of eggs, larvae and juvenile horseshoe crabs and survival of adults (Nolan and Smith 2009; Smith et al. 2011; Faizul et al. 2012). These disease agents are commonly found in natural environments such as water, soil, animals and plants. Some of these biological agents are not detectable by naked eye and have the ability to reproduce rapidly. It was the aim of this study to identify the microbes found on *Tachypleus gigas* and *Carcinoscorpius rotundicauda*, particularly those suspected in causing negative effects on the development, growth, and survival of horseshoe crabs reared under culture conditions.

17.2 Materials and Methods

Two hundred eggs and larvae of *Tachypleus gigas* and 20 adults of *Carcinoscorpius rotundicauda* were collected from a natural spawning ground at Banting, Selangor, Malaysia and conditioned in 25 ppt of seawater at ambient temperature (26–29 °C) in the laboratory. Experimental studies were carried out at the Marine Laboratory at MTDC (Malaysian Technology Development Corporation), Putra Science Park, Universiti Putra Malaysia, Serdang, Selangor, Malaysia. These eggs and larvae were then held at 30 ppt in rearing trays with daily water changes. Adult *C. rotundicauda* were maintained in a circular fiberglass tank (500 L capacity). The larvae of *T. gigas* were fed *Artemia* nauplii (*ad libitum*) and adult *C. rotundicauda* with shrimp (once a day). Water parameters (temperature, dissolved oxygen (DO), pH, and ammonia) were measured using a YSI DO meter, pH meter, and ammonia API test kit, respectively, once a week. Samplings of 30 eggs and larvae and microscopic observation (under dissecting and compound microscope at 40, 100, 200 and 400× magnification) were carried out weekly. Eggs were incubated until 80 % of the eggs from each tray hatched. *Tachypleus gigas* larvae were observed for a period of 3 months.

To collect microbial contaminants, fungal infected eggs and larvae were swabbed with cotton on the infected areas and plated for culture on triptic soy agar (TSA) + 1.5 % salt media. Resulting bacterial and fungal growth on the agar plate was photographed. Subsequently, these cultures were sent for identification at the Bacteriology Laboratory, Faculty of Veterinary, Universiti Putra Malaysia, Serdang, Selangor.

17.3 Results and Discussion

17.3.1 Bacterial Infection on Tachypleus gigas *Eggs and Larvae*

Species of bacteria found on the eggs and larvae of *T. gigas* were identified as *Shewanella putrefaciens*, *Enteroccus faecalis*, *Bacillus cereus* and *Corynebacterium* sp. The most commonly found species was *Shewanella putrefaciens*, a Gram-negative rod shaped bacteria. Infected eggs usually turned light pink or red (Fig. 17.1). Some of these infected eggs (Fig. 17.2a) continued to develop; however, larvae (Fig. 17.2b) died immediately after hatching.

Isolated *Shewanella putrefaciens* grew within 8 h after inoculation onto the agar media. During egg incubation, *S. putrefaciens* was detected as a visible white film layer covering the water and eggs, with a distinct odor. *S. putrefaciens* is a saprophytic bacterium and commonly found in all types of water samples such as fresh-

Fig. 17.1 Bacterial infected horseshoe crab (*T. gigas*) eggs that have turned *pink* or *red*

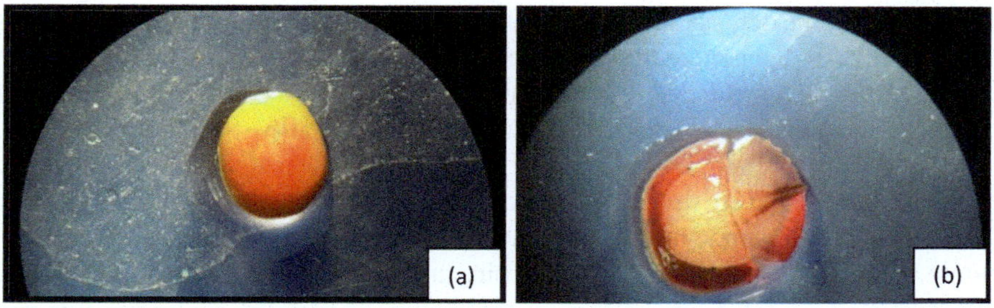

Fig. 17.2 (**a**) Eggs and (**b**) larvae of *T. gigas* infected with *Shewanella putrefaciens*

water, brackish water, marine water, mud water, stagnant water, lake water, river water and sewage water (Chen et al. 1997; Wang et al. 2004). It also is found on frozen poultry carcasses and in petroleum brines, cream, oil emulsions, ground beef, natural gas, milk (Von Graevenitz 1985) and spoiled seafood (The Environment Agency 2012). Clinical signs of skin lesions and septicemia have been found in the proximal intestine of infected perch (Goldschmidt-Clermont et al. 2008). In pufferfish (*Lagocephalus lunaris*), infection has been found in the internal organs and on the skin (Auawithoothij and Noomhorm 2012). European seabass usually displayed characteristics of lethargy, loss of appetite, pale gills, exopthalmus, ulcers over the dorsal part body, haemorrhages (on fins, body, operculum, liver), splenic enlargement, liver and kidney pale and gastro-intestinal tract (Korun et al. 2009). Without water change and proper care, therefore, horseshoe crab eggs may become infected and have potential to contaminate other eggs in the same or nearby incubation trays.

Gram-positive bacteria (Silva et al. 2000), *Enteroccus faecalis*, *Corynebacterium* sp., and *Bacillus cereus* were found on *T. gigas* eggs. These species are different from those previously reported on the American horseshoe crab, *Limulus polyphemus*, including the genera *Vibrio*, *Oscillatoria*, *Pseudomonas*, *Flavobacterium*, *Leucothrix* and *Pasteurella* (Smith et al. 2011). In this study, *E. faecalis* was found on the eggs and larvae of *T. gigas*. Four species from this genus, *E. casseliflavus*, *E. faecium*, *E. mundtii* and *E. raffinosus* have been found in the intestine of farmed fish in Thailand (Petersen and Dalsgaard 2003), occurring in the kidney, liver, spleen and heart of rainbow trout and turbot (Carson et al. 1993; Toranzo et al. 1995). *Bacillus cereus* has been identified as an agent responsible for two types of food-borne diseases (Kenneth 2008). The resulting illnesses are characterized by nausea, vomiting and abdominal cramps or diarrhea with incubation periods from 1 to 16 h, depending on type (Kenneth 2008). *Corynebacterium* sp., a coryneform bacteria (Hindmarch et al. 1990), has been documented to cause exophthalmia or bulging eye in striped bass and salmon (Austin and Austin 1999). *Bacillus cereus* var. *toyoi*, however, is considered a beneficial bacterium and used as a probiotic product for animals and humans (Sanders 2003).

In this study, *E. faecalis* and *B. cereus* were found on the dorsal and ventral part of the prosoma of *T. gigas* larvae. The infection by these bacteria may cause stress on horseshoe crabs by impairing movement of the animal and slowing the blood clotting reaction, and in some cases infection may lead to mortality (Bang 1956). Pathogenic microbes spread in many ways to attack the host and simultaneously avoid its immune defense (Jiang et al. 2007). Bacterial infections in horseshoe crabs can colonize and penetrate the carapace, eyes and gill surfaces, affecting an entire system, involving the deeper tissues of the organs, gills and circulatory sinuses and resulting in extensive tissue necrosis that can cause mortality (Smith et al. 2011).

Hemolymph factors are responsible for the natural innate immune system in horseshoe crabs and are divided into two types: hemocyte and hemoplasma (Kouno et al. 2008). The hemocytes of horseshoe crabs contain various cysteine-rich cationic peptides (polyphemusin, tachyplesin, tachystatin, tachycitin and big defensin), which have inhibitory effects on the growth of bacteria (Gram-positive and Gram-negative bacteria) and fungi (Silva et al. 2000). Blood cells from *T. tridentatus* contain tachycitin (antibacterial substance) to react against both Gram-positive and Gram-negative bacteria. Fresh hemolymph of *T. gigas* exhibited antibacterial activity against *Escherichia coli*, *B. cereus*, and *Vibrio parahaemolyticus*, but demonstrated only partial antibacterial activity against *B. subtillis* and *Sterptococcus aureus* (Shakiba et al. 2011). Hence, horseshoe crabs may become infected despite their innate immune defenses, particularly under stressed conditions or if the immune system is otherwise compromised.

17.3.2 Fungal Infection on Tachypleus gigas *Eggs and Larvae*

Fungus was found to affect the development of *T. gigas* eggs and larvae. Fungus attached on larvae when the spore separated in the water (Fig. 17.3). Fungal infection found on the eggs and larvae of *T. gigas* were *Aspergillus* sp., *Aspergillus niger*, *Penicillium* sp. and *Gliocladium* sp. The fungus *Gliocladium sp.* was found only on eggs of *C. rotundicauda*. Eggs infected by *A. niger* will turn to black, and the eggs cease to develop (Fig. 17.4).

Aspergillus niger is not only found on eggs of *T. gigas*, but also commonly found in the environment where crabs live. On natural spawning grounds, *T. gigas* eggs can be found in highly organic sands at the mangrove areas where *A. niger* is in

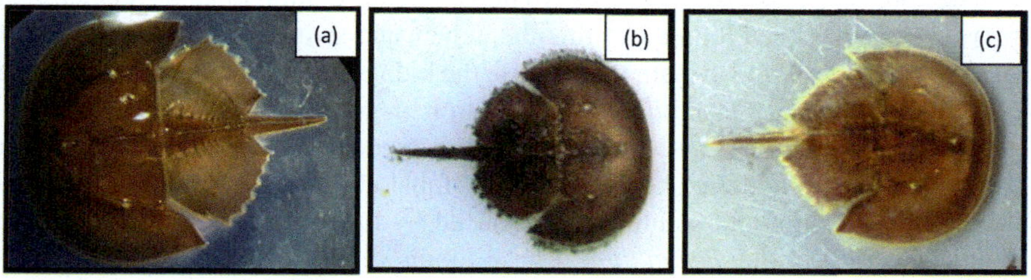

Fig. 17.3 Comparison between horseshoe crabs *T. gigas* that are not infected (**a**) and those that are infected (**b**) and (**c**) by *Aspergilus niger* in culture

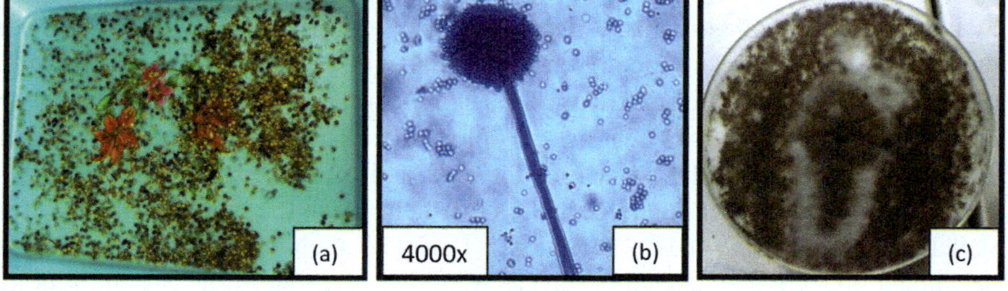

Fig. 17.4 (**a**) *Tachypleus gigas* eggs infected with *Aspergillus niger* during incubation, (**b**) morphology of *A. niger*, (**c**) *A. niger* on agar plate

abundance. *Tachypleus gigas* eggs can tolerate temperatures (28–31.9 °C; Christianus and Saad 2010) in the range suitable for *Aspergillus niger* (6–47 °C). *Aspergillus niger* also is capable of growing at a very wide pH range, from 1.4 to 9.8 (Gautam et al. 2011). The growth ability in various temperature and pH ranges as well as the abundant amount of conidiospores allow the continuous spread of this fungus in horseshoe crab habitats.

The genus *Aspergillus* with 260 species has been studied for several centuries (Samson and Varga 2009). *Aspergillus niger* is a filamentous ascomycete having the ability of fast growth, black color, and having broad pH tolerance (Pitt and Hocking 1997; Perrone et al. 2007; Perfect et al. 2009). *Aspergillus* spp. in carp species infect a variety of body parts from the head and eyes to abdomen and fins, including symptoms such as eroded scales, hemorrhaging along the body surface, body distension, and tissue erosion (Iqbal et al. 2012). *Aspergillus niger* may also produce certain mycotoxins having nephrogenic, heptocarciinogenic, and immunological properties (Gautam et al. 2011). Mycotoxins such as ochratoxin are important food-safety agents (Sweeney and Dobson 1998). *Aspergillus niger* is most commonly used by citric acid and pectinase producers and serves as a model fungal fermentation process (Li and King 1963; Rombout and Pilnik 1978; Kerns et al. 1987; Baker 2006). If inhaled in sufficient quantity (Baker 2006; Gautam et al. 2011), *A. niger* also can infect the lungs of humans (aspergillosis), but the U.S. Food and Drug Administration (USFDA) considers the fungus as generally recognized as safe (GRAS) (Gautam et al. 2011).

The fungus *Penicillium sp.* was also found on eggs of *C. rotundicauda. Penicillium corylophilum* has been found infecting the kidney and swim bladder of red snapper (*Lutjanus campechanus*) where it can cause the kidney and swim bladder to enlarge (Blaylock et al. 2001). *Penicillium* spp. infection on goldfish head, pelvic and abdomen of goldfish (*Carassius auratus*) can cause eroded scales, hemorrhaging over body surface and moderate body distension (Iqbal et al. 2012). Fungal secondary metabolites (citrinin and patulin) are produced by fungal strains belonging to the genus *Aspergillus* and *Penicillium.* These toxins contaminate cereals and fruits (Bennett and Klich 2003) and also have been detected in monascus (natural dietary supplements to prevent cardiovascular disease in Asia) fermentation products (Wei et al. 2003; Liu et al. 2005). Kidney is one of the target organs of citrinin in animals, including dogs, rats, and rabbits (Kogika et al. 1993; Kumar et al. 2007; Singh et al. 2007).

In general, fungi are organisms depending on other organisms for their supply of nutrients, known as eukaryotic organisms (Ozdal et al. 2012). Fungi play an important role in the recycling of nutrients in the ecosystem, and most fungi are living on dead organic material or known as saprophytes (Eduard 2007). Conidia (asexual spores) are attached to the surface of the host and release extracellular enzymes including lipases, chitinase and proteases that help break the host's chitinous exoskeleton (Ozdal et al. 2012). Mycotoxin (basis of their toxicity) production is dependent on a number of factors such as water humidity, substrate, strain of mold, temperature, the presence of chemical preservatives and microbial interactions (Chagas et al. 2000; Varga et al. 2005). Fungi are known to attack all life stages of fish including eggs, fry, fingerlings, adult and brood stock (Iqbal et al. 2012). Fungal infections can cause low production in fish culture and low productivity of fry (Kwanprasert et al. 2007). Unfortunately, most treatments using antibiotics for bacterial infections are generally ineffective. Accordingly, infected horseshoe crabs become inactive, anorectic and eventually die from the infection.

17.4 Conclusion

During the culture period, evidence of microbial infestation was observed, including bacteria (*Shewanella putrefaciens*, *Bacillus cereus*, *Corynebacterium* sp. and *Enterococcus faecalis*) and fungi (*Aspergillus* sp., *Aspergillus niger*, *Peniclillium* sp., *Gliocladium* sp.). Eggs infected with bacteria or fungi usually turned reddish, grey or black and ceased to develop. Larvae became passive and coated with black spore, and in some cases died soon after molting into the larval stage. The species with greatest potential to affect the growth and survival of horseshoe crab eggs and larvae sampled in this study were *S. putrefaciens* (a Gram-negative bacterium) and *A. niger* (a fungi). The species identified during this study were different from those previously identified on *Limulus polyphemus* in culture, suggesting possible species, location, and/or condition-specific variation in communities of infectious microbes brought in from the natural environment and propagated in culture. Infected horseshoe crabs, therefore, may require microbe-specific management or treatment strategies and additional study to ensure successful rearing in culture.

Acknowledgment The authors would like to thank to the Government of Malaysia for providing funding through the Research university grant scheme (Grant No: 04-02-12-2087RU) and to Universiti Putra Malaysia for providing the facilities for this project.

References

Auawithoothij W, Noomhorm A (2012) *Shewanella putrefaciens*, a major microbial species related to tetrodotoxin (TTX)-accumulation of puffer fish *Lagocephalus lunaris*. J Appl Microbiol 113(2):459–465

Austin BB, Austin DA (1999) Bacterial fish pathogens: disease of farmed and wild fish. Springer, Dordrecht

Baker SE (2006) *Aspergillus niger* genomics: past, present and into the future. Med Mycol 44:17–21

Bang FB (1956) A bacterial disease of *Limulus polyphemus*. Bull Johns Hopkins Hosp 98:325–351

Bennett JW, Klich M (2003) Mycotoxins. Clin Microbiol Rev 16:497–516

Blaylock RB, Overstreet RM, Klich MA (2001) Mycoses in red snapper (*Lutjanus campechanus*) caused by two deuteromycete fungi (*Penicilium corylophilum* and *Cladosporiun sphaerospernum*). Hydrobiologia 460:221–228

Carson J, Gudkovs N, Austin B (1993) Characteristics of an *Enterococcus*-like bacterium from Australia and South Africa, pathogenic for rainbow trout, *Oncorhynchus mykiss* Walbaum. J Fish Dis 16:381–388

Chagas F, Freire O, Kozakiewicz Z et al (2000) Mycoflora and mycotoxins in Brazilian black pepper, white pepper and Brazil nuts. Mycopathologia 149:13–19

Chen YS, Liu YC, Yen MY et al (1997) Skin and soft-tissue manifestations of *Shewanella putrefaciens* infection. Clin Infect Dis 25:225–229

Christianus A, Saad CR (2010) Some aspects on the habitat of *Tachypleus gigas* in Malaysia. In: Proceedings of the international workshop on the science and conservation of Asian horseshoe crabs, Hong Kong. 6 Nov 2010

Eduard W (2007) Edible insects. The nordic expert group for criteria documentation of health risks from chemicals. http://www.food-info.net/uk/products/insects/intro.htm. Accessed 1 Nov 2009

Faizul MIM, Faizal MM, Christianus A et al (2012) Incubation and hatching of *Tachypleus gigas* (Muller, 1785) eggs in sand and water media. Asian J Anim Vet Adv. doi:10.3923/ajava.2012

Gautam AK, Sharma S, Avasthi S et al (2011) Diversity, pathogenicity and toxicology of *A. niger*: an important spoilage fungi. Res J Microbiol 6:270–280

Goldschmidt-Clermont E, Wahli T, Frey J et al (2008) Identification of bacteria from the normal flora of perch, *Perca fluviatilis* L., and evaluation of their inhibitory towards Aeromonas species. J Fish Dis 31(5):353–359

Hindmarch JM, Magee JT, Hadfield MA et al (1990) A pyrolysis-mass spectrometry study of *Corynebacterium* spp. J Med Microbiol 31:137–149

Iqbal Z, Sheikh U, Mughal R (2012) Fungal infections in some economically important freshwater fishes. Pak Vet J 32:422–426

Jiang N, Tan NS, Ho B et al (2007) Respiratory protein–generated reactive oxygen species as an antimicrobial strategy. Nat Immunol 8:1114–1122

Kenneth T (2008) *Bacillus cereus* food poisoning. http://textbookofbacteriology.net/B.cereus.html. Accessed 17 Nov 2013

Kerns G, Okunev ON, Ananin VM et al (1987) Enhanced formation of b-glucosidase by *Aspergillus niger* VKMF-2092 in fed batch operation with frequency intermittent glucose addition. Acta Biotechnol 7:535–545

Kogika MM, Hagiwara MK, Mirandola RM (1993) Experimental citrinin nephrotoxicosis in dogs: renal function evaluation. Vet Hum Toxicol 35:136–140

Korun J, Akgun-Dar K, Yazici M (2009) Isolaion of *Shewanella putrefaciens* from cultured European sea bass, (*Dicentrarchus labrax*) in Turkey. Rev Med Vet 160:532–536

Kouno T, Fujitani N, Mizuguchi M et al (2008) A novel B-Defensin structure: a potential strategy of big defensin for overcoming resistance by gram-positive bacteria. Biochemistry 47:10611–10619

Kumar M, Dwivedi P, Sharma AK et al (2007) Ochratoxin A and citrinin nephrotoxicity in New Zealand white rabbits: an ultrastructural assessment. Mycopathologia 163:21–30

Kwanprasert P, Hangavant C, Kitancharoen N (2007) Characteristics of *Achyla bisexualis* isolated from eggs of Nile Tilapia (*Oreochromis niloticus* Linn.). KKU Res J 12:195–202

Li L, King KW (1963) Fractionation of beta-glucosidases and related extracellular enzymes from *Aspergillus niger*. Appl Microbiol 11:320–325

Liu BH, Wu TS, Su MC et al (2005) Evaluation of citrinin occurrence and cytotoxicity in Monascus fermentation products. J Agric Food Chem 53:170–175

Nolan MW, Smith SA (2009) Clinical evaluation, common diseases, and veterinary care of the horseshoe crab, *Limulus polyphemus*. In: Tanacredi JT, Botton ML, Smith D (eds) Biology and conservation of horseshoe crabs. Springer, New York, pp 479–499

Ozdal M, Incekara U, Polat A et al (2012) Isolation of filamentous fungi associated with two common edible aquatic insects, *Hydrophilus piceus* and *Dytiscus marginalis*. J Microbiol Biotechnol Food Sci 2:95–105

Perfect JR, Cox GM, Lee JY et al (2009) The impact of culture isolation of *Aspergillus* species: a hospital-based survey of aspergillosis. Clin Infect Dis 33:1824–1833

Perrone G, Susca A, Cozzi G et al (2007) Biodiversity of *Aspergillus* species in some important agricultural products. Stud Mycol 59:53–66

Petersen A, Dalsgaard A (2003) Species composition and antimicrobial resistance genes of *Enterococcus* spp., isolated from integrated and traditional fish farms in Thailand. Environ Microbiol 5:395–402

Pitt JI, Hocking AD (1997) The ecology of fungal food spoilage. In: Pitt JI, Hocking AD (eds) Fungi and food spoilage. Springer, New York, pp 3–9

Rombout FM, Pilnik W (1978) Enzymes in fruit and vegetable juice technology. Process Biochem 8:9–13

Samson RA, Varga J (2009) What is a species in *Aspergillus*? Med Mycol 47:13–20

Sanders ME (2003) Probiotics: considerations for human health. Nutr Rev 61:91–99

Shakiba SZ, Zaki P, Hajeb P et al (2011) Screening for tachycitin alike substance in blood cells of *Carcinoscorpius rotundicauda* and *Tachypleus gigas*. In: International workshop on the science and conservation of Asian horseshoe crabs, Hong Kong. 12–16 June 2011

Silva PI, Daffre S Jr, Bulet P (2000) Isolation and characterization of gomesin, an 18-residue cysteine-rich defense peptide from the spider *Acanthoscurria gomesiana* hemocytes with sequence similarities to horseshoe crab antimicrobial peptides of the tachyplesin family. J Biol Chem 270:33464–33470

Singh ND, Sharma AK, Dwivedi P et al (2007) Citrinin and endosulfan induced maternal toxicity in pregnant Wistar rats: pathomorphological study. J Appl Toxicol 27:589–601

Smith SA, Berkson J (2005) Laboratory culture and maintenance of the horseshoe crab (*Limulus polyphemus*). Lab Anim 34:27–34

Smith SA, Scimeca JM, Mainous ME (2011) Culture and maintenance of selected invertebrate in the laboratory and classroom. ILAR J 52:153–164

Sweeney MJ, Dobson ADW (1998) Mycotoxin production by *Aspergillus*, *Fusarium* and *Penicillium* species. Int J Food Microbiol 43:141–158

The Environment Agency (2012) *Shewanella putrefaciens* group. http://www.environmentagency.gov.uk/static/documents/Research/Shewanella_putrefaciens_group.pdf. Accessed 16 Nov 2012

Toranzo AE, Cutrin JM, Nunez S et al (1995) Antigenic characterization of *Enterococcus* strains pathogenic for turbot and their relationship with other gram-positive bacteria. Dis Aquatic Org 21:187–191

Varga J, Peteri Z, Tabori K et al (2005) Degradation of ochratoxin A and other mycotoxins by *Rhizopus* isolates. Int J Food Protect 99:321–328

Von Graevenitz A (1985) Ecology, clinical significance and antimicrobial susceptibility of infrequently encountered glucose-nonfermenting gram-negative rods. In: Gilardi GL (ed) Nonfermentative gram-negative rods; laboratory identification and clinical aspects. Marcel Dekker, New York, pp 181–192

Wang I-K, Lee M-H, Chen Y-M et al (2004) Polymicrobial bacteremia caused by *Escherichia coli*, *Edwardsiella tarda*, and *Shewanella putrefaciens*. Chang Gung Med J 27(9):701–705

Wei W, Li C, Wang Y et al (2003) Hypolipidemic and anti-atherogenic effects of long-term cholestin (*Monascus purpureus* – fermented rice, red yeast rice) in cholesterol fed rabbits. J Nutr Biochem 14:314–318

Chapter 18
Marine Bacteria Associated with Horseshoe Crabs, *Tachypleus gigas* and *Carcinoscorpius rotundicauda*

Noraznawati Ismail, Mohamad Faridah, Amirrudin Ahmad, Amirul Asyraf Alia'm, Oh Sheng Khai, Muhd Fawwaz Afham Mohd Sofa, and Azwarfarid Manca

Abstract Bacteria associated with marine organisms have received major attention worldwide due to their potential in pharmaceutical industries. A total of 32 bacterial strains were isolated and identified, 22 from *Tachypleus gigas* and 10 from *Carcinoscorpius rotundicauda* from Balok Beach, Malaysia. Overall, Gram-negative bacteria showed dominancy on both species. A combination of biochemical tests, macro morphology, and micro morphology was used to roughly determine genus of bacterial isolates; where results from 16S rDNA sequencing were taken as the end results due to its high reliability. Results from the molecular approach showed only *Pseudoalteromonas* sp. (38.46 %), *Vibrio* sp. (46.15 %), and *Photobacterium* sp. (15.38 %) might exhibit mutualism with horseshoe crabs, *T. gigas* and *C. rotundicauda*. Study on the diversity of bacteria associated with horseshoe crabs can enhance the understanding of these animals and the potential sustainability of producing useful compounds for defense mechanisms.

Keywords Marine bacteria • *Tachypleus gigas* • *Carcinoscorpius rotundicauda* • Bacteria diversity

N. Ismail (✉) • A.A. Alia'm • O.S. Khai
Institute of Marine Biotechnology, Horseshoe Crab Research Group (HCRG), Universiti Malaysia Terengganu, 21030 Kuala, Terengganu, Malaysia
e-mail: noraznawati@umt.edu.my

M. Faridah • A. Ahmad • M.F.A. Mohd Sofa • A. Manca
School of Marine and Environmental Sciences, Horseshoe Crab Research Group (HCRG), Universiti Malaysia Terengganu, 21030 Kuala, Terengganu, Malaysia

R.H. Carmichael et al. (eds.), *Changing Global Perspectives on Horseshoe Crab Biology, Conservation and Management*, DOI 10.1007/978-3-319-19542-1_18

18.1 Introduction

Horseshoe crabs are an ancient, unique group of marine arthropods that have contributed to many clinical research studies, such as the discovery of *Limulus* amoebocyte lysate (LAL), a lipopolysaccharide (endotoxin) detection compound used to test for the presence of Gram-negative bacteria (Manco et al. 2010). The associations of microbes and marine organisms have received major attention in the past few decades due to the high potential for these interactions to yield pharmaceutical and medical products. Bacteria have been proven to associate with many marine organisms such as sponges, soft corals, and mollusks, which synthesize useful bioactive compounds (Armstrong et al. 2001; Lee et al. 2001; Vasanthabharathi and Jayalakshmi 2012). The association of the marine bacteria with the marine host organism is typically mutualistic, which benefits both the host and the bacteria (Lopanik 2014). Currently, information on the association of bacteria with horseshoe crab species, especially in Malaysia, is still very limited where only a few studies have been reported on the eggs, juvenile and adult horseshoe crabs. This study was conducted to identify and characterize marine bacteria associated with horseshoe crabs.

18.2 Methodology

18.2.1 Sample Collection and Isolation of Bacteria

Bacteria samples were derived from the book gills and mouth of horseshoe crabs from Balok Beach, Malaysia. The exterior of the horseshoe crabs were cleaned by using sterile distilled water several times to avoid contamination by bacteria from the environment. A cotton swab was used to take the sample, which was incubated overnight at 37 °C on marine agar to isolate single colonies.

18.2.2 Phenotypic Characterization

Gram-negative and positive bacteria were identified by using heat to fix the smears and a simple staining procedure using crystal violet, iodine and safranin (Black 2008). The isolates were routinely grown on marine agar and incubated at 37 °C for biochemical tests. Catalase existence was determined by the method of Reiner (2010), and an oxidase test was used to detect cytochrome oxidase (Shields and Cathcart 2010).

The production of sulphide, formation of indole, and motility of the bacteria were detected by a Sulfur, Indole, and Motility (SIM) Test (Vasanthabharathi and Jayalakshmi 2012). A Triple-Sugar-Iron (TSI) Test was used by the method of Ateba and Marumo (2014) to differentiate Gram-negative enteric bacilli based on carbohydrate fermentation and the production of hydrogen sulfide.

18.2.3 Extraction of DNA

Total genomic DNA was extracted by using DNeasy Blood and Tissue Kit (Qiagen, USA). Total DNA was isolated from 1 ml of culture after overnight incubation at 37 °C. The total DNA extracts were electrophoresed in an Agarose gel electrophoresis system. The total DNA extracted from the isolates were amplified by 16S rDNA sequence followed by PCR amplification universal primers 16rDNA primer pair 27F (5′-GAGTTTGATCMTGGCTCAG-3′) and eubacteria-specific primer 1492R (5′-TACGGTTACCTTGTTACGACTT-3′) (Weisburg et al. 1991).

The PCR amplification was conducted in 25 μl containing 2.5 μl of 10× *Taq* buffer, 1.5 μl of $MgCl_2$ (25 mM), 0.5 μl of dNTP mix (10 mM each), 0.5 μl of each primer (10 μM), 0.2 μl of 1 U *Taq* DNA polymerase, 1 μl template DNA and deionized water. The PCR cycling program started with one initial denaturation step of 95 °C for 30 s; 35 cycles consisting of denaturation at 95 °C for 30 s, annealing at 55 °C for 30 s, and elongation at 72 °C for 60 s. The final elongation step was extended at 72 °C for 7 min.

The amplification reaction was performed by using Master-Cycler gradient (Eppendorf, Germany). All purified products, using reverse and forward primers, were sent for DNA sequencing (First BASE Laboratories Sdn. Bhd, Selangor, Malaysia). The resulting gene fragments were aligned using BLAST (Basic Local Alignment Search Tool) search (http://www.ncbi.nml.nih.gov/BLAST) to investigate the region of similarity between sequences, including nearest neighbors (Ismail and Sarijan 2011).

18.3 Results and Discussion

18.3.1 Macro Morphology and Micro Morphology

A total of 32 bacterial strains were isolated from horseshoe crabs; 22 were from *T. gigas*, and the remaining 10 were from *C. rotundicauda*. The macro morphology (shape, elevation and color of colonies) of isolated bacteria from both species was recorded. Based on macro morphology, most isolated bacteria were narrowed down to a few genera or species.

Gram-staining of 22 isolated bacteria from *T. gigas* showed dominancy of Gram-negative bacteria, which included 21 strains (95 %), and only 1 Gram-positive strain (5 %) was observed. Three out of ten bacterial isolates (30 %) from *C. rotundicauda* were Gram-positive, while the remaining seven colonies (70 %) were Gram-negative. This result was in agreement with previous studies by Kaiser and Benner (2008) in which the marine environment was dominated by Gram-negative microorganisms, yet Gram-positive microbes were still present. Our study solely depended on bacteria cultured from the horseshoe crabs, *T. gigas* and *C. rotundicauda*.

18.3.2 Identification of Marine Bacteria Isolates

Identification of bacteria was done with reference to Bergey's Manual of Determinative Bacteriology by Holt et al. (1994). The combination of macro morphology, micro morphology, and biochemical tests enabled identification of bacteria to the genus level. Among the 32 isolates, only 13 were sent for sequencing because some isolates belonged to the same genus or species (Tables 18.1 and 18.2). Of the isolates sequenced, nine were from *T. gigas* and four were from *C. rotundicauda*.

Sequencing results showed that the 13 isolates belong to three different genera, which were *Pseudoalteromonas*, *Vibrio*, and *Photobacterium* from the Gram-negative bacteria group. Therefore, with these sequencing results, 22 isolates from *T. gigas* and 10 isolates from *C. rotundicauda* were grouped into families. In *T. gigas* samples, 13 strains (13/22 = 59 %) were in the Pseudoalteromonadaceae family and 9 strains (9/22 = 41 %) were in the Vibrionaceae family.

On the other hand, among the ten *C. rotundicauda* isolates, six strains (6/10 = 60 %) belonged to the Vibrionaceae family, while the remaining four strains (4/10 = 40 %) were in the Pseudoalteromonadaceae family. According to Drancourt et al. (2000), almost 90 % of identifications from the overall performance of 16S rDNA sequence analysis were outstanding, compared to isolates identified by biochemical profile and Gram-staining, which failed to produce an accurate result. In this study, the sequencing results proved that some biochemical tests were less reliable compared to the 16S rDNA sequencing method.

Pseudoalteromonas sp. is Gram-negative bacillus bacteria. They are motile by a single polar flagellum. Species in this genus are reported to have association with marine invertebrates such as sponges (Vasanthabharathi and Jayalakshmi 2012). Members of *Pseudoalteromonas* are known to produce bioactive substances. For example, *P. phenolica* are antibiotic-producing bacteria when they are associated with their hosts (Isnansetyo and Kamei 2003). This genus is also reported to have anti-bacterial, anti-fungal, and anti-fouling activities (Berborn et al. 2011). The dominancy of *Pseudoalteromonas* sp. showed that they might have a strong association with the horseshoe crab. However, no previous studies have reported that bacteria from this genus showed association with horseshoe crabs.

Vibrio sp. is a genus of Gram-negative bacteria with a curved rod shape, facultative anaerobes, motile by polar flagella with sheaths. This genus was reported to be the most commonly associated with crustaceans, demonstrating pathogenic effects, while at the same time producing bioactive compounds (Jayasinghe et al. 2008). According to Bell et al. (1994), *V. parahaemolyticus* associated with marine organisms showed anti-microbial and hemolytic activities. Isolated bacteria from *T. gigas* and *C. rotundicauda* showed the presence of *Vibrio* sp., suggesting a potential source of bioactive compounds.

Photobacterium sp. is Gram-negative bacteria in the Vibrionaceae family. They are morphologically rod, motile by one to three polar flagella, facultative aerobes, and some strains have the ability to emit light. Many species of *Photobacterium*

Table 18.1 Biochemical tests of 22 bacteria isolated from the horseshoe crab, *T. gigas*

No. of identification	Gram reaction	Micro-morphology	Catalase activity	Oxidase activity	TSI			Indole production	Motility	No. of isolates
					Slant	Butt	H_2S			
1	–	Bacillus	+	/	K	K	+	–	/	10
2	–	Staphylococci	+	/	K	A	+	–	/	3
3	–	Staphylococci	+	/	A	A	+	–	/	2
4	–	Coccus	+	/	K	A	+	–	/	2
5	+	Diplococci	+	/	K	K	–	–	/	1
6	–	Coccus	–	/	A	A	+	–	/	1
7	–	Staphylococci	+	/	K	K	+	–	/	2
8	–	Coccobacillus	+	/	K	A	+	–	/	1

TSI triple sugar iron test, *K* alkaline, *A* acid, (+) positive reaction, (–) negative reaction, (/) nil

Table 18.2 Biochemical tests of ten bacteria isolated from the horseshoe crab, *C. rotundicauda*

No. of identification	Gram reaction	Micro-morphology	Catalase activity	Oxidase activity	TSI			Indole production	Motility	No. of isolates
					Slant	Butt	H_2S			
1	+	Coccus	+	/	K	A	+	+	/	1
2	+	Coccus	+	/	K	A	+	–	/	1
3	+	Diplococci	+	/	K	A	+	+	/	1
4	–	Coccobacillus	+	/	K	A	+	–	/	1
5	–	Coccus	+	/	K	A	+	–	/	2
6	–	Bacillus	+	/	K	K	+	–	/	1
7	–	Staphylococci	+	/	K	K	+	–	/	3

TSI triple sugar iron test, *K* alkaline, *A* acid, (+) positive reaction, (–) negative reaction, (/) nil were proven to live in symbiosis with marine organisms (Chimetto et al. 2010). Many species of marine fish form bioluminescent symbioses with *Photobacterium* spp., including *P. kishitanii*, *P. leiognathi*, and *P. mandapamensis* (Urbanczyk et al. 2011). So far, other species of *Photobacterium*, *P fischeri* were first reported and identified to live symbiotically with light organs of luminous fishes (Ruby and Nealson 1976).

18.4 Conclusion

This study demonstrates that *Pseudoalteromonas* sp., *Vibrio* sp., and *Photobacterium* sp. can be associated with horseshoe crabs, *T. gigas* and *C. rotundicauda*. These marine bacteria might exhibit a mutualistic relationship with the horseshoe crab and may produce useful bioactive compounds. Further study on marine bacteria associated with horseshoe crabs are needed to identify these compounds, which might be important for pharmaceutical industries and contribute to the survival of the horseshoe crab.

Acknowledgements We thank Sabah Biodiversity (SaBC), Malaysia and Universiti Malaysia Terengganu (UMT) for funding this research.

References

Armstrong E, Yan L, Boyd KG et al (2001) The symbiotic role of marine microbes on living surfaces. Hydrobilogia 461:37–40

Ateba CN, Marumo BI (2014) Isolation of enterohaemorrhagic *Escherichia coli* O104 strains from raw meat products in the North West Province, South Africa. J Food Nutr Res 2(6):288–293

Bell R, Carmeli S, Sar N (1994) Vibrindole A, a metabolite of the marine bacterium, *Vibrio parahaemolyticus*, isolated from the toxic mucus of the boxfish *Ostacion cubicus*. J Nat Prod 57(11):1587–1590

Berborn N, Ng YY, Kjelleberg S et al (2011) Marine bacteria from Danish coastal waters show antifouling activity against the marine fouling bacterium *Pseudoalteromonas* sp. strain S91 and zoospores of the green alga *Ulva australis* independent of bacteriocidal activity. Appl Environ Microbiol 77(2):8557–8567

Black JG (2008) Microbiology principles and exploration, 7th edn. Wiley, New York

Chimetto LA, Cleenwerck I, Thompson CC et al (2010) *Photobacterium jeanii* sp. nov., isolated from corals and zoanthids. Int J Syst Evol Microbiol 60(12):2843–2848

Drancourt M, Bollet C, Carlioz A et al (2000) 16S ribosomal DNA sequence analysis of a large collection of environmental and clinical unidentifiable bacterial isolates. J Clin Microbiol 38(10):3623–3630

Holt JG, Krieg NR, Sneath PH et al (1994) Bergey's manual of determinative bacteriology. Williams and Wilkins, Baltimore

Ismail N, Sarijan S (2011) Phylogenetic inference from 18S rRNA gene sequences of horseshoe crabs, *Tachypleus gigas* between Tanjung Dawai, Kedah and Cherating Pahang. Int J Biol Vet Agric Food Eng 56(8):947–950

Isnansetyo A, Kamei Y (2003) *Pseudoalteromonas phenolica* sp. nov., a novel marine bacterium that produces phenolic anti-methcilin-resistant *Staphylococcus aureus* substances. Int J Syst Evol Microbiol 53(2):583–588

Jayasinghe CVL, Ahmed SBN, Kariyawasam MGIU (2008) The isolation and identification of *Vibrio* species in marine shrimps of Sri Lanka. J Food Agric 1(1):36–44

Kaiser K, Banner R (2008) Major bacterial contribution to the ocean reservoir of detrital organic carbon and nitrogen. Limnol Oceanogr 53(1):99

Lee YK, Lee JH, Lee HKM (2001) Microbial symbiosis in marine sponges. J Microbiol 39:254–264

Lopanik NB (2014) Chemical defensive symbioses in the marine environment. Funct Ecol 28:328–340

Manco M, Putignani L, Bottazzo GF (2010) Gut microbiota, lipopolysaccharides, and innate immunity in the pathogenesis of obesity and cardiovascular risk. Endocr Rev 31(6):817–844

Reiner K (2010) Catalase test protocol. American Society for Microbiology. http://www.microbelibrary.org/library/laboratory-test/3226-catalase-test-protocol. Accessed 15 June 2014

Ruby EG, Nealson KH (1976) Symbiotic association of *Photobacterium fischeri* with the marine luminous fish *Monocsntris japonica*: a model of symbiosis based on bacterial studies. Biol Bull 151(3):574–586

Shields P, Cathcart L (2010) Oxidase test protocol. American Society for Microbiology. http://www.microbelibrary.org/library/laboratory-test/3229-oxidase-test-protocol. Accessed 18 June 2014

Urbanczyk H, Ast JC, Dunlap PV (2011) Phylogeny, genomic and symbiosis of *Photobacterium*. FEMS Microbiol Rev 35(2):324–342

Vasanthabharathi V, Jayalakshmi S (2012) Bioactive potential of symbiotic bacteria and fungi from marine sponges. Afr J Biotechnol 11(29):7500–7511

Weisburg WG, Barns SM, Pelletier DA et al (1991) 16S ribosomal DNA amplification for phylogenetic study. J Bacteriol 173(2):697–703

Chapter 19
Mating Tactics of the American Horseshoe Crab (*Limulus polyphemus*)

H. Jane Brockmann, Sheri L. Johnson, Matthew D. Smith, and Daniel Sasson

Abstract The reproductive behavior of horseshoe crabs (*Limulus polyphemus*) is easily observed, yet until recently, little was known about their unusual mating habits. Fertilization is external and occurs beneath the female as her eggs are being laid in the sand. Spawning is often synchronized to the highest spring tides of the year, so the time available for breeding is limited. Males have alternative mating tactics, some search and pair with a female offshore and migrate inshore to spawn, while others search onshore for nesting pairs and engage in sperm competition in group spawning. Females also have alternative tactics such that some spawn with a single attached male and others spawn with multiple males, which results in multiple male paternity of their offspring. Both male and female tactics are condition and context-dependent and are affected by breeding density, operational sex ratio, and the interplay between male and female tactics. To better understand horseshoe crab reproductive behavior, we review studies conducted during the past 25 years in one Gulf of Mexico population in northern Florida at Seahorse Key. We discuss the costs, benefits, and tradeoffs of alternative tactics for males and females. We synthesize the recent literature on mating tactics, resolve some conflicting results, and point to the future by identifying the questions that remain.

Keywords *Limulus polyphemus* • Alternative reproductive tactics • Mate choice • Polyandry • Sperm competition • Compatibility • Explosive breeding • Management implications

H.J. Brockmann (✉) • M.D. Smith • D. Sasson
Department of Biology, University of Florida, Gainesville, FL 32611, USA
e-mail: HJB@ufl.edu; madsmith@ufl.edu; dsasson@ufl.edu

S.L. Johnson
Department of Zoology, University of Otago, Box 56, Dunedin 9054, New Zealand
e-mail: sheri.johnson@otago.ac.nz

R.H. Carmichael et al. (eds.), *Changing Global Perspectives on Horseshoe Crab Biology, Conservation and Management*, DOI 10.1007/978-3-319-19542-1_19

19.1 Introduction

Reproduction is the most conspicuous feature of horseshoe crab biology, since it is the only time when horseshoe crabs are seen in large numbers. During their springtime breeding, horseshoe crabs travel to low energy, coastal beaches and estuaries, sometimes in enormous numbers (Teale 1957; Myers 1986; Kerlinger 1998). They arrive with the incoming tide, usually in amplexus, the male clasping the posterior margin of the female's shell with his modified claws. The pair nests at the top of the high tide line, sometimes entirely out of the water, and quite apparent to any beach visitor. Yet, it has only been in the last 25 years that this highly visible mating and nesting behavior has been studied in detail.

The female lays her eggs in beach sediments where the eggs are fertilized externally as they are being laid by aquatic, free-swimming sperm (Levitan 1995; Brockmann 2003b). The fertilized eggs incubate in the sand for several weeks before emerging as larvae. No other arthropod, indeed no other animal, reproduces in this way. In addition to their unique mode of reproduction, *L. polyphemus* spawning is synchronized and temporally constrained. Spawning begins in the spring as the water warms (Smith et al. 2010) and continues for about 2 months (the dates depend on latitude); most females lay all their eggs for the year during just a few, consecutive high tides (i.e. determinate spawners, Leschen et al. 2006). In contrast, males return repeatedly to the spawning area (Brockmann and Penn 1992). For this reason, there are usually many more males on the nesting beach than females, i.e. the operational sex ratio (males: females or OSR) is usually male biased (Rudloe 1980; Sekiguchi 1988; Leschen et al. 2006; Smith et al. 2010; Brockmann and Johnson 2011). In most areas breeding is synchronized with the high tides around the new and full moon, i.e. the highest tides in a month (Rudloe 1985; Barlow et al. 1986; Penn and Brockmann 1994; Brockmann and Johnson 2011). Where mating and nesting occurs at the extreme high tide line, it results in eggs being left in an environment that is conducive to embryonic and larval development (Penn and Brockmann 1994; Brockmann 2003b; Vasquez et al. 2015). Synchronization with spring high tides means that spawning is usually confined to a limited number of hours each month during a 2–3 month breeding season. Such constrained and synchronized spawning has been referred to as "explosive breeding" with males engaging in scramble competition for females (Emlen and Oring 1977; Alcock 2009).

Horseshoe crabs share a number of mating patterns with other explosively breeding species such as frogs, toads (Brockmann 1990; Wells 2007; Roberts and Byrne 2011), and fish (Martin et al. 2004; Byrne and Avise 2009). For example, like male *L. polyphemus*, males of explosively breeding frogs and toads search out females, whereas prolonged breeding anurans sit and wait for females to arrive (Höglund 1989). Among explosively breeding anurans where scramble competition for mates is the rule, there is little female choice or direct male combat (Wells 2007), no size-assortative mating, and no differences in size between amplexed and non-amplexed males, as there is among prolonged breeders (Höglund and Saterberg 1989). Similarly, studies have found only weak or no size-assortative mating in *L. polyphemus* (Pomerat 1933; Cavenaugh 1975; Botton and Loveland 1992; Suggs et al. 2002), no differences in body size (Brockmann and Penn 1992) or clasper size (Botton et al. 1996) between paired and unpaired males, and no evidence of female choice during pairing (Loveland and Botton 1992). Many features of horseshoe crab mating behavior can be understood in light of their external fertilization and the constraints imposed by "explosive" breeding and scramble competition for mates.

Understanding the reproductive patterns and mating behavior of horseshoe crabs is crucial to their management. Horseshoe crabs play an ecologically important role as predators of mollusks and other invertebrates and as food for sea turtles, intertidal fishes, and migratory shorebirds (Botton et al. 2003; Botton 2009). Commercially, horseshoe crabs are used by the biomedical industry and in the whelk (conch) and eel fisheries (Kreamer and Michels 2009). Dramatic declines in horseshoe crab numbers during the 1990s led to their management (ASMFC 1998; Berkson and Shuster 1999; Shuster et al. 2003). Much has been learned in the last 15 years about horseshoe crab reproduction, which is used to inform management decisions. For example, information on the effect of changes in OSR informed the decision to institute a male-only harvest in Delaware Bay in 2008 (ASMFC 2008). Long-term studies at a single location are particularly valuable for understanding annual and seasonal variation in reproduction as well as the behavior and reproductive success of individuals that experience different conditions.

A long-term study on the reproductive behavior of *L. polyphemus* has been conducted at Seahorse Key, FL, an island breeding site along the west coast of northern peninsular Florida (Brockmann and Johnson 2011). At Seahorse Key (SK), horseshoe crabs nest along a 1 km beach where their eggs develop in the sand. Emerging larvae quickly settle just offshore in an extensive mud, grass and sand nursery where the juveniles feed and grow. Since 1990, the Seahorse Key studies have focused on many aspects of horseshoe crab biology, behavior, life-history traits, and population status. Here we pull together what is known about the mating behavior of *L. polyphemus* at SK, resolve some discrepancies and conflicting results, make comparisons with other populations, and point to the future by identifying some of the questions and puzzles that remain.

19.2 Male Mating Tactics

The evolution and maintenance of two reproductive tactics in a population is puzzling. In horseshoe crabs some males clasp on to a female out at sea and arrive on a beach to spawn in amplexus (i.e., the attached tactic), whereas other males come to the breeding area alone (unattached tactic), roam the shoreline, and join spawning pairs or groups (Rudloe 1980; Cohen and Brockmann 1983; Shuster and Botton 1985; Barlow et al. 1986; Loveland and Botton 1992). These unattached males vie for position around the female and engage in sperm competition with other males, (i.e., satellite behavior, defined as an unattached male that is in physical contact with the female, her attached male or other satellites; Fig. 19.1, Brockmann 1990). Populations differ in the sizes of satellite groups (group sizes are correlated with nesting density), but at most *L. polyphemus* breeding sites, males show both attached

Fig. 19.1 Photograph of horseshoe crabs spawning at Seahorse Key, FL. A pair arrives on the beach, the larger female in front (*F*) with her attached male behind. An unattached male (*U*) approaches the beach and the arriving pair. A monandrous female (*M*) is nesting on the beach, well buried in the sand with her attached male (*AM*) visible behind her. A polyandrous female (*P*) is nesting nearby well buried in the sand and her attached male (*AP*) is visible behind her. There are two satellite males on either side of the polyandrous female (S1 and S2) both touching the AP male's carapace. These two males are in the 1F position with the S1 satellite under the front margin of the AP male's prosoma. This is the position that yields the highest paternity (Photograph by S.L. Johnson)

and unattached tactics (Brockmann and Smith 2009). Normally we would expect that if one tactic were even slightly less successful than the other, selection would eliminate that tactic over time (Brockmann 2001). Hence, in order for two tactics to be maintained within a population, each must be more successful than the other under some situations. In some species, alternative tactics are maintained as a genetic polymorphism, but in most species the tactics depend on the individual's phenotype (condition-dependent tactics), such as body size or age, and the circumstances in which they are living (context-dependent tactics), such as population density or sex ratio (Gross 1996; Zamudio and Chan 2008). These studies suggest the possibility that there may be differences among male horseshoe crabs that are using different tactics or there may be circumstances in which males are consistently attached or unattached.

19.2.1 Are Individual Males Consistently Attached or Unattached?

Mark-recapture studies are a useful tool for studying horseshoe crab mating tactics since males repeatedly return to the same nesting beach. These studies show that males are consistent in their attachment status: if a male is unattached on one high tide, he is very likely to be unattached on the next high tide (Brockmann 2002). This consistency was also demonstrated through a series of experimental field manipulations in which we compared the behavior of males using different tactics. Males that came to the breeding beach unattached were compared with males that arrived attached to females but were then detached (Brockmann and Penn 1992). Unattached males were much more likely to return still unattached on the next high tide (49 %, $N=81$) than the attached males that had been detached (11 %; $N=129$, $\chi^2=37$, $df=1$, $P<0.0001$). When detached (i.e. originally attached) and unattached males were put together in a wading pool along with detached females (not with their original mates), the detached males paired much more quickly than the unattached males (36 %, $N=52$, originally attached males reattached and 0 %, $N=55$ unattached males attached within the first 5 min, $\chi^2=7$, $df=1$, $P<0.02$). They were also more likely to stay attached when released at sea. These studies demonstrate that within a breeding season, individual males are fairly consistent in the mating tactics they use.

19.2.2 Is There Anything Different About Attached and Unattached Males?

Attached and unattached males differ in appearance but not size. There are no differences in the body sizes of males using different mating tactics (Botton and Loveland 1992; Loveland and Botton 1992), and even when one attached male replaces another, he is not larger than the original male that was paired with the female (Brockmann and Penn 1992). This finding is surprising given the competitiveness of horseshoe crab mating behavior. However, on average, attached males are lighter in color compared to unattached males, and their eyes and claspers are more likely to be in good condition (Brockmann and Penn 1992; Duffy et al. 2006). Brockmann (2002) investigated the condition and mating status of 252 males that were marked and returned to the spawning beach at least three times, finding that 26 % were attached (75 % or more of the time), 39 % were unattached (75 % or more of the time), and 35 % showed a mixture of behavioral tactics. These animals differed in original attachment status ($\chi^2=43$, $df=2$, $P<0.0001$) and condition: 40 % of the animals in the best condition category were attached and 24 % were unattached, whereas in the lowest condition category 7 % were attached and 61 % were unattached ($\chi^2=26.9$, $df=6$, $P=0.002$). When original attachment status and condition were considered together, only condition affected whether males spent most of their time unattached (ANOVA; original status: $F=0.016$, $df=1$, $P=0.9$, condition: $F=12.8$, $df=1$, $P=0.0004$, original status by condition $F=2.9$, $df=1$, $P=0.09$; Fig. 19.2). These results suggest that male mating tactics are based on male condition (condition-dependent tactics).

The physical condition of the carapace deteriorates over time (Brockmann and Penn 1992; Brockmann 1996) because horseshoe crabs do not molt as adults (Koon 1883; Shuster 1954; Sokoloff 1978; Shuster and Sekiguchi 2003; Smith et al. 2010).

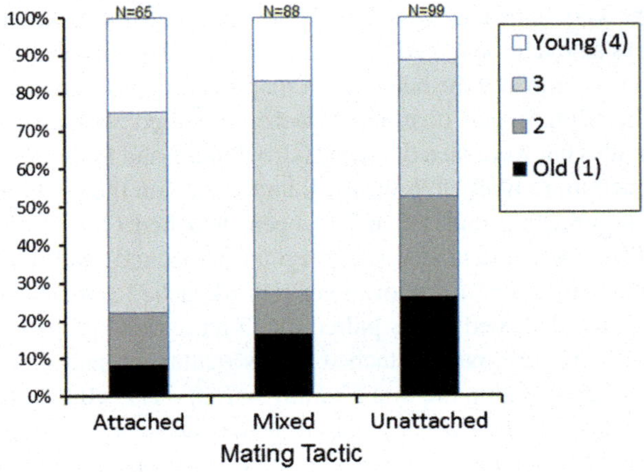

Fig. 19.2 The percent of males using each male mating tactic (attached, mixed, unattached) separated by age (condition) category. The age categories range from young (light color, no pitting or epibionts, mucus present and eyes and spines in perfect condition) to old (deteriorated, dark colored prosoma with pitting, no mucus, and with soft or covered eyes and worn spines). These data are from males (N=252) that were marked, their condition evaluated and released back to sea. They were then re-sighted at least three times on subsequent high tides at the breeding beach to be included in this dataset. Attached refers to males that returned attached on more than 75 % of their visits; mixed refers to those that showed a mixture of tactics on their multiple visits; and unattached were those that returned unattached on more than 75 % of visits (Redrawn from Brockmann 2002)

The prosoma becomes progressively darker; pits and scratches develop; and many fouling organisms invade their shells (Grant 2001). While there is no method to measure absolute adult age in horseshoe crabs, relative age can be determined based on shell condition. For example, the darker color is associated with more pitting caused by chitinoclastic bacteria, algae, or other infections (Leibovitz and Lewbart 2003). Darker animals also tend to have reduced quantities of mucus, which is exuded through pores in the carapace as an anti-fouling and surface immunity agent (Harrington and Armstrong 2000; Harrington et al. 2008). Horseshoe crabs with more deteriorated carapaces also have, on average, larger slipper shells, *Crepidula fornicata* living on them (Botton and Ropes 1988). Since slipper shells attach as larvae, this correlation means that when a crab carapace is in worse condition, the animal has been an adult longer than when the carapace is in better condition (Brockmann 1996, 2002). This age-dependent deterioration and biofouling means that carapace condition can be used as a proxy for relative adult age (Sasson et al. 2012). Moreover, the condition of the carapace correlates with physical performance such as righting behavior (Penn and Brockmann 1995) and mating tactics (Duffy et al. 2006).

The correlation between physical condition and attachment status means that male horseshoe crabs exhibit condition-dependent mating tactics (Brockmann 2001; Brockmann and Taborsky 2008). Our hypothesis (Brockmann 2003a) is that these mating tactics are alternative reproductive strategies in which males that are

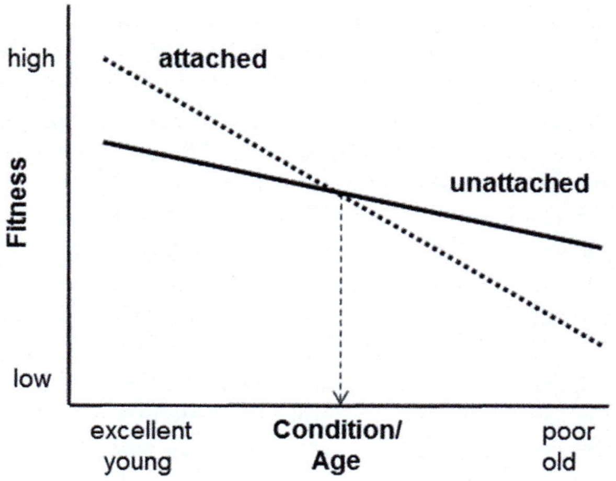

Fig. 19.3 Threshold model for the evolution of alternative mating tactics in male horseshoe crabs. The *arrow* shows the threshold age below which males that show the attached tactic (pairing with females at sea) have higher fitness and above which males that show the unattached tactic (joining spawning pairs on shore) have higher fitness (Redrawn from Brockmann 2002)

younger and in better condition have higher success when adopting the attached tactic, and those that are older and in poorer condition have greater success when adopting the unattached tactic (Fig. 19.3). At a certain threshold age or condition, males maximize fitness by switching from attaching to females out at sea (attached tactic) to coming ashore and looking for mating groups to join (unattached tactic).

While this model is difficult to fully evaluate, there is some supporting evidence from a field experiment in which males (both unattached males and attached males that we detached) were prevented from attaching by placing plastic bags over the claws they use for clasping females (Brockmann 2002). If males are coming ashore as satellites just because they cannot find or hold onto a female (using a making-the-best-of-a-bad-job tactic), then all males with bags should search onshore for pairs to join because none can attach. But, this is not what happened. The experiment showed that males that were in poor condition were much more likely to come ashore as satellites than males that were in good condition (these males apparently stayed offshore searching for females at sea). Both condition and original attachment status (but not size) significantly affected whether males with bags became satellites (logistic regression, $N=225$, $\chi^2=5.55$, $P=0.01$; for condition $\chi^2=10.59$, $P=0.001$; for original status $\chi^2=5.07$, $P=0.02$). These results suggest that the two mating tactics are the product of males making condition-dependent decisions, rather than males making the best-of-a-bad-job.

19.2.3 What Are the Trade-Offs (Costs and Benefits) for Each Male Tactic?

Several tradeoffs involving differences in costs and benefits have been identified for each male mating tactic. During each 1 week spawning cycle (the period of extra high tides around the new and full moon), attached males normally mate with only one female, whereas satellite males may join several mating pairs (Brockmann 2003a). With each clutch of eggs laid, satellite males have similar paternity success compared to attached males (Brockmann et al. 2000). Attached males do not always compete for paternity with satellites because groups do not always form when nesting densities are low (Brockmann 1996), and some nesting pairs do not attract satellite males at all (Johnson and Brockmann 2012). In contrast, satellite males must always engage in sperm competition (Brockmann et al. 1994, 2000). While unattached males appear to mate more often than attached males (Brockmann and Penn 1992), they do not always find a spawning pair and when they do, some older males are unable to get into the best positions around the female (Smith and Brockmann 2014). Lastly, horseshoe crabs are often overturned on the beach, leaving them vulnerable to desiccation and predation (Botton and Loveland 1989). Attached males are less likely to be overturned than satellites and when this happens, younger males are better able to right themselves than older males (Penn and Brockmann 1995). Taken together, attached males appear to have higher success overall than unattached males (Brockmann et al. 2000). If an age or condition threshold for switching tactics has evolved that maximizes fitness, however, then we expect that there should be compensatory costs associated with the attached tactic. One previously unexplored, potential cost to the attached male tactic is nutritional stress caused by a reduced ability to feed while attached to a female (Smith et al. 2013b). During amplexus, the attached male's mouth is dorsal to and covered by the female's telson (Loveland and Botton 1992) and the male cannot bury himself in the substrate to feed in the normal manner by manipulating food into the mouth using the gnathobases that surround the mouth (Botton 1984). If the attached tactic inhibits feeding, then attached males should produce less fecal material than satellite males. In a waste production experiment, satellite males produced 57 % more waste than attached males, and females produced more waste than either type of male (likely due to their larger size) suggesting that attached males were indeed eating less. When gut contents were compared, the gut fullness of satellites was 150 % greater compared to attached males. While lower fecal production and an emptier gut could be due to attached males being less motivated to feed rather than the physical constraint of being attached, a subsequent feeding experiment showed that attached and satellite males ate the same amount of food, supporting the reduced feeding hypothesis. In Florida, males typically remain attached for a 1 week spawning cycle (mean $\pm$ SD length of attachment is 3.7 ± 6.1 days, Brockmann and Penn 1992), but occasionally may stay attached up to 51 days (Brockmann 2003a). In other populations, pairs can remain attached much longer (Shuster 1954); for example, in New England, attached pairs have been observed overwintering together (Barlow et al. 1987; Moore 2004). Overall, these studies suggest that attached males may be fasting for several weeks during the breeding season.

It might seem obvious that fasting is costly but this cost may be offset by the energetic benefits of being attached. Attached males do not spend energy locating spawning pairs as unattached males do, and attached males are 'carried' along by females as they travel to the beach. . Nutritional stress due to a period of fasting (when an animal forgoes feeding in favor of other activities) or starvation (when feeding is prevented due to some extrinsic limitation; McCue 2010) can be inferred from stable isotope values in animal tissues (Hobson et al. 1993; Castillo and Hatch 2007; McCue and Pollock 2008). If an animal is in a negative energetic balance, ^{15}N is preferentially retained, while ^{14}N is excreted, and as a result, $\delta^{15}N$ values increase in tissues over time as the animal feeds on itself (McCue and Pollock 2008). A comparison of $\delta^{15}N$ values in the feces of wild-caught individuals demonstrated that mean $\delta^{15}N$ values for attached males were higher than those for satellite males and females (there was no difference between satellite males and females). A subsequent 4 week starvation experiment demonstrated increased $\delta^{15}N$ values in feces of individuals that were not given food, whereas the values did not change for individuals that were fed. These results support the idea that reduced consumption of food and a period of nutritional stress are costs of the attached tactic. Horseshoe crabs that remain attached longer than those sampled in the Seahorse Key population, may suffer even greater costs of adopting the attached tactic.

Low-energy, alternative phenotypes often evolve as a release from the energetic demands of 'preferred' phenotypes (Taborsky 1998; Widemo 1998; Brockmann and Taborsky 2008; Cummings and Gelineau-Kattner 2009), and in some systems, a male's success for a given tactic may partially depend on his energy reserves (McCauley et al. 2000). We suggest that older males in poorer condition may not be able to afford the cost of reproduction (i.e. periodic fasting) that accompanies being attached to a female during breeding. Consequently, the satellite tactic may allow males to maintain (or regain) a positive energy balance while still obtaining reproductive success. Investigating this hypothesis is the next step to fully understand the evolution and maintenance of alternative tactics in horseshoe crabs.

19.2.4 Sperm Competition and Multiple Paternity

Horseshoe crab sperm remain quiescent until they are in close proximity to the eggs; then chemical cues from the eggs activate the sperm (Brown 1976). Millions of sperm may attach to each egg and undergo acrosome reactions, but only one sperm ultimately succeeds in fertilizing the egg. Sperm competition occurs when the sperm of two or more males compete for the same ova (Parker 1970). As the number of satellite males around a female increases, so does the intensity of sperm competition (Parker 1998). Attached males generally fertilize a majority of the eggs when competing against just one satellite (Brockmann et al. 1994), but attached males are less successful than satellite males when two satellites are present (although there is a lot of variation among males, Brockmann et al. 2000). In larger groups, the paternity share of additional satellite males comes at the expense of the other satellites, and the attached male normally fertilizes at least some of the eggs.

The most important factor affecting fertilization success during sperm competition is the position of the male around the female. Attached males are directly behind the female and thus are close to where eggs are released. Satellite males can take any position around the female, but when satellites first arrive at a pair, they nearly always move into positions over the female's incurrent canals (the channel between the prosoma and opisthosoma, Eldredge 1970), called the 1F positions (Fig. 19.1, Brockmann et al. 2000). When in this position, satellite males have close access to the site of egg release and consequently the highest percentage of satellite paternity comes from males in this position (Brockmann et al. 1994, 2000). Additionally, satellites in this position may work their way under the anterior margin of the attached male's prosoma. This "under" position may result in higher paternity for that satellite male, especially when additional satellite males are present. Once the 1F positions have been occupied, later arriving satellite males must take up other positions around the female, which have reduced paternity (Fig. 19.4).

When sperm from multiple males are in close proximity to the eggs, the outcome of sperm competition likely hinges on the quantity and quality of sperm released (Snook 2005). Sperm quantity is a function of two traits: ejaculate size and sperm concentration (the number of spermatozoa in an ejaculate of a particular size). Augmenting either trait increases the number of spermatozoa that are competing during mating. All else being equal, the male that releases the most sperm should have an advantage over competing males (Parker 1982). We found that attached males have, on average, higher sperm concentrations ($N=102$, mean = 8.5×10^9 sperm/mL, S.E. = 7.2×10^8) than satellite males ($N=44$, mean = 6.5×10^9 sperm/mL, S.E. = 6.0×10^8, $P=0.04$) (Sasson et al. in press).

Age is another factor that affects sperm quantity in horseshoe crabs: younger males have larger ejaculates and more concentrated sperm compared to older males (Sasson et al. 2012). Thus, given the relationship between age and mating tactics, we expect younger attached males to have higher sperm quantity compared to older satellite males. While reproductive tactic is correlated with sperm concentration, we found no difference in ejaculate size between attached ($N=84$) and satellite males

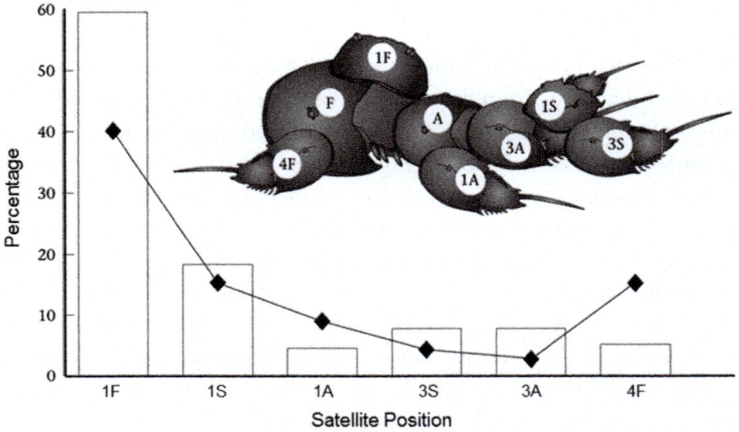

Fig. 19.4 Frequency distribution of satellite male positions and the associated paternity. The *bars* show the frequency with which satellite males were found in different positions around a nesting pair (positions are shown in the picture) and the *line* shows the mean paternity for satellite males when they were in each position (Modified from Brockmann et al. 2000)

(N=37, GLM: $F_{1,114.7}=0.43$, P=0.51) (Sasson et al. in press). One explanation for this apparent contradiction is that attached males may adjust their sperm output when faced with sperm competition, as has been found in some externally fertilizing fish (Candolin and Reynolds 2002; Pilastro et al. 2002). However, we have previously found that ejaculate size remains relatively constant within males across a number of days (Spearman rank correlation, $r_s=0.85$, N=19, P<0.001), suggesting that ejaculate size is not affected by a male's immediate competitive environment (Sasson et al. 2012). While fertilization success in competitive situations can also be affected by sperm quality (e.g., velocity or viability, Snook 2005), neither age nor mating tactic correlates with any measure of sperm quality among horseshoe crabs in the Seahorse Key population (Sasson et al. 2012, in press).

Higher sperm competition risk should select for sperm traits that make males more successful at fertilizing eggs under competitive conditions (Parker 1998; Snook 2005; Alvarez et al. 2014). Populations differ in the risk of sperm competition because of differences in OSR. For example, the Seahorse Key (SK) population of horseshoe crabs averages about 2.1 males per female, while the population in Delaware Bay (DB) has 3.5 males per female (Brockmann and Smith 2009). The higher OSR means that attached males in DB are more likely to face sperm competition than attached males at SK. Accordingly, we found differences in sperm traits between DB and SK populations. The horseshoe crabs in DB had larger ejaculates and more concentrated sperm than those in SK (Sasson and Brockmann in prep). While a number of factors may be involved, sperm competition risk seems to be an important driving force for changes in sperm concentration and a strong contributor to a change in ejaculate size across populations that differ in sperm competition risk (Sasson and Brockmann in prep). Similar patterns have been found across closely related species that differ in sperm competition risk such as frogs and butterflies (Gage 1994; Byrne et al. 2003) and in other externally fertilizing species with alternative reproductive tactics (Fu et al. 2001; Vladic and Jarvi 2001).

19.3 Female Mating Tactics

Group spawning by satellite males can also be viewed as multiple mating by females. Even when the OSR is strongly male biased, some females nest only with their attached males (monandrous females), so their offspring are fathered by one male. At the same time, nearby females may spawn with satellites (polyandrous females), which results in multiple paternity of their offspring (Fig. 19.1, Brockmann 1990; Brockmann et al. 1994, 2000). The frequency of monandry declines as nesting density and OSR increase (Brockmann and Smith 2009), but there are always some monandrous pairs present on the nesting beach. For example, when nesting densities are high at SK, 20–80 % of pairs are monandrous (Fig. 19.5a, Johnson and Brockmann 2012) and in DB, where the OSR is much higher, 5–20 % of pairs are monandrous (Fig. 19.5b, Brockmann 1996). As with males, do females have two alternative mating tactics, or are monandry and polyandry a by-product of male competition for females?

19.3.1 Are Individual Females Consistently Monandrous or Polyandrous?

Two research methods have been used to assess whether females consistently maintain monandrous and polyandrous behavior. (1) Data from mark-recapture studies show that individual monandrous females are more likely to remain monandrous from one nesting to the next than are polyandrous females (Hassler 1999), even when their attached males were removed and they paired again with a new attached male ($\chi^2=21.3$, $df=1$, $P<0.001$; Johnson and Brockmann 2012). (2) Results of field experimental manipulations show that monandrous pairs are more likely to stay monandrous than nearby polyandrous pairs that had all their satellites removed ($\chi^2=96$, $df=1$, $P<0.001$; from data given in Brockmann 1996). These results support the view that individual females are consistently monandrous or polyandrous. Our results (Fig. 19.5) also show that when there are large numbers of unattached males present, all polyandrous females have satellites, whereas when few unattached males are present, some polyandrous females nest without satellites (and thus are recorded as monandrous). Hence, the frequency of monandry is affected by the environment, in this case by nesting density and OSR.

19.3.2 There Are Two Types of Monandrous Females

Horseshoe crabs are an ideal model species in which to study mating tactics because you can experimentally remove and "add" satellite males to females to address questions about the mating system. Satellites are not attracted to monandrous pairs,

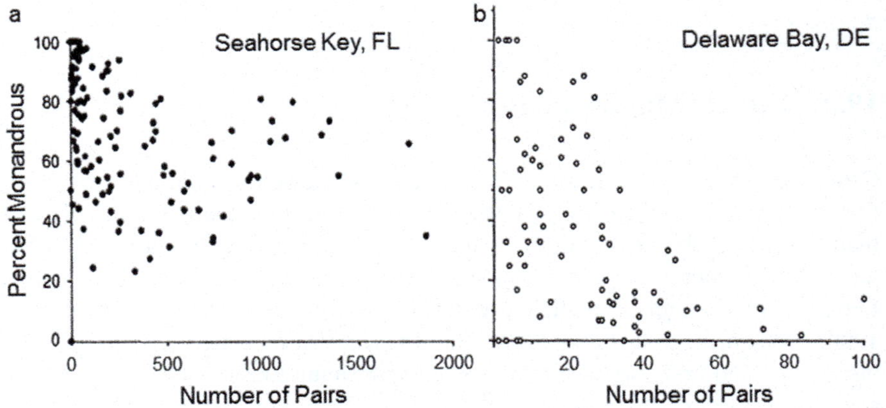

Fig. 19.5 The relationship between the proportion of female horseshoe crabs that were monandrous and the nesting density (number of pairs present on the breeding beach). (**a**) Proportion of monandrous females at SK on 157 tides on which we counted one or more pairs along the 1 km beach (Redrawn from Johnson and Brockmann 2012), (**b**) Proportion of monandrous females at Cape Henlopen, Delaware Bay on 12 high tides from eight 100 m beach sections (Redrawn from Brockmann 1996)

but we could encourage a male to join a monandrous pair by gently guiding him toward the nesting couple (and the control pair was allowed to remain monandrous, Johnson and Brockmann 2010). More than half of the monandrous females (62 %) left the nesting beach within 2 min rather than nest with the added satellite male. We call these exclusively monandrous females "intolerant" (Johnson and Brockmann 2012). Other monandrous females were "tolerant," since they would continue to nest with experimentally added satellite males, although they did not attract (or were not attractive to) satellites naturally on their own.

19.3.3 Are Monandrous and Polyandrous Females Different?

Monandrous females were, on average, slightly smaller (mean prosoma width 13.75 cm ± 0.03 S.E., $N=1{,}059$) and in better condition (younger) than polyandrous females (13.85 cm ± 0.05 SE, $N=507$), and intolerant monandrous females were slightly smaller than tolerant monandrous females (Brockmann 1996; Hassler 1999; Schwab and Brockmann 2007; Johnson and Brockmann 2012). Monandrous females lay fewer eggs on average during each nesting bout than polyandrous females (Table 19.1, Schwab and Brockmann 2007); but there were no differences between monandrous and polyandrous females in the rate of laying eggs, egg size (Johnson and Brockmann 2012), energy content of the eggs (Hassler 1999), fertilization success of the eggs (monandrous: mean = 99.6 %; polyandrous: mean = 99.2 %, $N=29$; Wilcoxon test, $P=0.39$) or any measure of developmental success of the offspring (Johnson and Brockmann 2010, 2012). Although smaller females tend to lay fewer eggs (Leschen et al. 2006), the smaller size of the monandrous females does not fully explain the smaller numbers of eggs that she lays (Johnson and Brockmann 2012), i.e. female mating status explains a significant amount of the variation in egg number (Johnson and Brockmann 2010). While we found no differences between monandrous and polyandrous females in the size or

Table 19.1 Median number of eggs laid in three clusters and the proportion of the clusters that contained eggs for monandrous and polyandrous females (Data from Johnson and Brockmann 2012)

	Median number of eggs laid			Proportion of clusters with eggs (N)	
Year	Mon	Poly	P^a	Mon	Poly
1993	3,300	5,280	0.36	0.88 ($N=36$)	0.88 (N=8)
1994	4,004	4,840	0.38	0.82 (N=22)	1.0 (N=6)
1995	3,696	6,820	0.03	0.86 (N=69)	0.93 (N=28)
2004	3,476	5,192	0.03	1.0 (N=33)	1.0 (N=35)
2005	2,728	3,080	0.76	0.84 (N=50)	0.88 (N=60)
2008	5,500	5,720	0.25	0.95 (N=40)	0.96 (N=55)

[a]Mann-Whitney U Test

condition of their attached males, the attached males of intolerant females were in better condition than the attached males of tolerant females (Johnson and Brockmann 2012).

Differences have been found between the sperm concentration of males attached to monandrous and polyandrous females. The attached males of monandrous females have on average higher sperm concentrations (N=59, mean = 9.6×10^9 sperm/mL, S.E. = 1.2×10^9) than the attached males of polyandrous females (N=43, mean = 6.9×10^9, S.E. = 5.3×10^8, Tukey HSD: P=0.04) or satellite males (N=44, mean = 6.5×10^9 sperm/mL, S.E. = 6.0×10^8, Tukey HSD P=0.01) (Sasson et al. in press). Overall, the evidence suggests that monandry and polyandry are stable tactics (alternative reproductive tactics) that are associated with condition-dependent (e.g. size, age, physical condition) and context-dependent (the particular male that is attached) differences among females, with differences in reproductive investment in egg number (per nesting bout).

19.3.4 What Do Females Gain from Multiple Mating?

Multiple mating is widespread among animals, even when females can acquire all the sperm they need by mating with a single male and even when mating is costly (Jennions and Petrie 2000). So why do females mate with multiple males and, perhaps even more puzzling, why are both patterns maintained in horseshoe crabs? Many explanations for multiple mating have been proposed including increasing the opportunity for a female to mate with a high quality (good genes) or genetically compatible male and increasing the genetic diversity of offspring by mating with multiple males (Zeh and Zeh 2003; Simmons 2005; Slatyer et al. 2012). Alternatively, multiple mating might just be a by-product of male competition (i.e. convenience polyandry, Alcock et al. 1977; Thornhill and Alcock 1983). We explored potential costs and benefits of multiple mating in the field using both observational and manipulative experiments where we monitored nesting time, clutch numbers, and numbers of eggs laid. When females were allowed to choose freely between nesting with or without satellites (i.e., naturally occurring monandrous and polyandrous females), they had equal nesting success and similar duration of nesting, number of eggs, and developmental success.

In an experiment in which satellite males were removed from polyandrous females, we found that spawning with multiple males was costly for these females. Polyandrous females without satellites laid more eggs, largely because they stayed on the beach longer, than polyandrous females nesting with satellites (Johnson and Brockmann 2010). However, to our surprise the developmental success of the eggs from polyandrous females whose satellites had been removed did not differ from the success of polyandrous females with satellites. When we conducted paternity analyses on larvae reared from the satellites-removed nests, we found that 78 % of the clutches had been fertilized by satellites-even when all satellites had been removed 20 min earlier! This finding means that somehow satellite sperm are retained or pulled along by the female as she lays successive clutches of eggs in the sand. When satellites were experimentally added to monandrous females, they laid fewer eggs than monandrous females nesting alone, mainly because "intolerant" females left the beach immediately and hence laid fewer eggs. Overall, these manipulative experiments suggest that mating with multiple males is costly for all females by reducing nesting success, and monandrous females can avoid this costly nesting with satellites by not attracting them in the first place.

19.3.5 Are There Trade-Offs or Benefits for Polyandrous But Not for Monandrous Females?

Artificial fertilization experiments can be conducted easily in horseshoe crabs by obtaining unfertilized eggs and sperm through gentle electro-ovulation and – ejaculation (Brown and Clapper 1981) and *in vitro* fertilization (IVF). These techniques eliminate the male or female pre-mating behavioral component of mate choice and mating success. In one IVF study, we fertilized eggs from polyandrous females with sperm from either their own attached male or a randomly chosen satellite male. The result was lower hatching success with their attached male than with the satellite male. In contrast, eggs from monandrous females that were fertilized by sperm from their attached males had equal success compared to eggs fertilized by satellite males (Fig. 19.6, Johnson and Brockmann 2013). This finding suggests that there was something different about the quality or compatibility of the attached males of

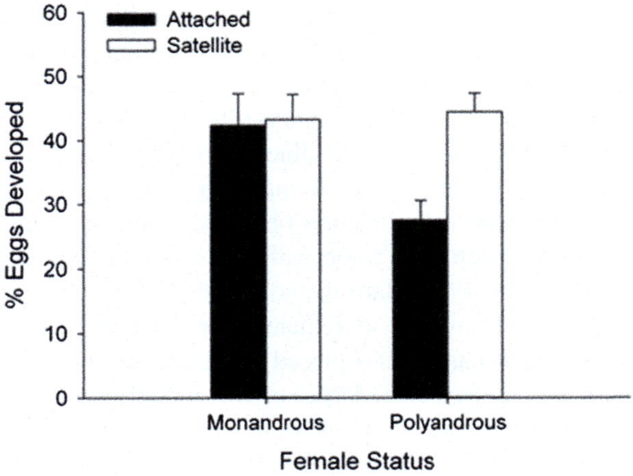

Fig. 19.6 The developmental success (mean ± SE) of eggs from females mated with attached (*black bars*) and satellite (*white bars*) males for monandrous and polyandrous females. Using *in vitro* fertilization methods, a total of 21 females of each type were crossed with their respective attached males and with 2 randomly chosen satellite males (Redrawn from Johnson and Brockmann 2013)

monandrous and polyandrous females, and that there may be indirect (post-mating) benefits to polyandrous females who mate with satellite males.

Both genetic compatibility and good genes (mate quality) may be important factors that polyandrous females derive from mating with multiple males. Further IVF experiments demonstrated that no single male was best for *all* females when it comes to hatching success of their embryos; instead, there was some evidence that compatible male-female crosses result in higher hatching success. However, the rate of development (number of larvae that reach metamorphosis in just 45 days), was affected by both the intrinsic quality of males and male-female compatibility. We also found evidence of female effects, both intrinsic female effects such as female or egg quality and female status effects: polyandrous females tended to have a lower proportion of offspring achieving early metamorphosis compared to monandrous females (4.7 % vs. 8.0 %; when females were crossed with random satellite males). It is reasonable to assume that the heavily armored juveniles are more likely to survive than the non-feeding, unarmed trilobite larvae. Hence, early metamorphosis should result in fitness benefits for horseshoe crabs. Taken together, these results suggest that monandrous and polyandrous females differ in the costs and benefits of multiple mating and that genetic incompatibility between males and females and/or male quality may offset the costs of multiple mating in this system.

19.4 Finding and Attracting Mates

19.4.1 How Males Find Females

The two mating tactics of male horseshoe crabs require different mate-searching behavior (Brockmann 2003a), and the different female mating tactics require different responses to males. Pairing typically occurs offshore exclusively in an aquatic environment, whereas satellite behavior requires locating and identifying spawning pairs that are above or below the high tide line where there may be constant interference from waves. Using cement casts of female horseshoe crabs placed offshore in shallow water, Barlow and his colleagues observed males orienting to females or horseshoe crab shapes when they came within visual range (Barlow et al. 1982, 1986, 1988; Powers et al. 1991; Barlow and Powers 2003). Similarly, unattached males approached cement models of females (Hassler and Brockmann 2001) or pairs (Schwab and Brockmann 2007) placed along the shoreline. In a field choice test, unattached males preferred the larger of two model females that differed by only 2.5 cm (both within the natural range of female sizes; Fig. 19.7). Clearly, then, male horseshoe crabs can locate mates visually. Although horseshoe crab eyes are remarkably sensitive (males can detect horseshoe crab-like objects equally well at night and in the day and when contrast is low, Powers et al. 1991; Passaglia et al. 1997; Herzog et al. 1996), visual cues are not always available because of diseased eyes (Brockmann and Penn 1992; Wasserman and Cheng 1996; Duffy et al. 2006),

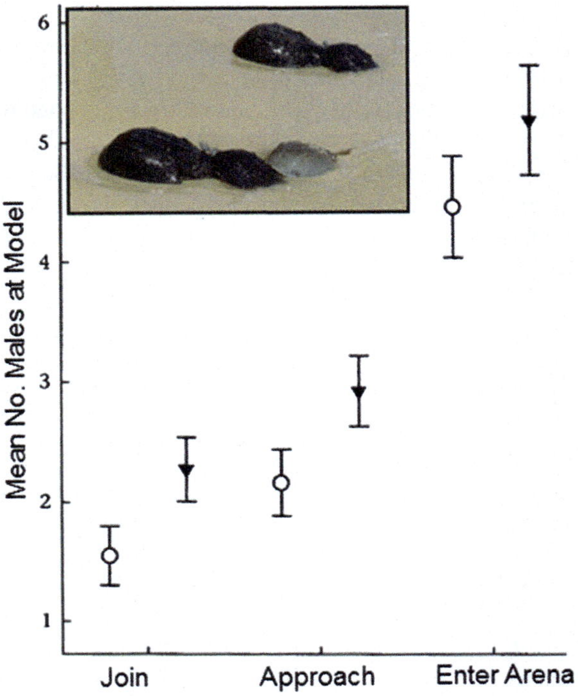

Fig. 19.7 The number of satellite males that were attracted to the large (*dark symbol*) or small (*light symbol*) cement model casts of a pair of horseshoe crabs nesting on the beach. Mean (± SE) number of satellites coming within 1 m of a model (= enter arena), coming within 10 cm of a model (= approach) or touching the model (= join). The inset shows the two cement models used during the tests with a satellite male joining the model in the foreground. The models differed in size by 2.5 cm (Modified from Schwab and Brockmann 2007)

turbid water, or because nesting females are buried in the sand (Schwab and Brockmann 2007). Like many other species (Candolin 2003; Partan and Marler 2005), horseshoe crabs may use multiple cues when searching for mates (Saunders et al. 2010).

Horseshoe crabs are well endowed with chemoreceptors and the neurons for processing chemical information (Barber 1956; Hayes and Barber 1982; Loesel and Heuer 2010), and there is evidence that males use chemical cues to locate mates (Brockmann 2003a). In a field experiment, identical cement horseshoe crab casts (models) were placed over a spot on the beach where a monandrous female had been nesting moments before (the pair was removed), where a polyandrous female had been nesting (and was removed) and a third (control) model was placed over a nearby spot on the sand where no female had been nesting (Hassler and Brockmann 2001). The effect was strong. For up to 10 min after the experiment started, more satellites arrived at the models placed over the polyandrous nesting spots than at the monandrous nesting spots or the controls (Friedman ANOVA: $N=36$, $df=2$; $P=0.01$ at 0 min; $P=0.002$ at 1 min, $P=0.04$ at 4 min, $P=0.04$ at 6 min, and $P=0.004$ at 10 min). In another field experiment, sponges were placed under the two cement models, one sponge was filled with water from underneath a polyandrous female and the other (control) was filled with seawater. Approaching satellites preferred the model with water from the polyandrous female (Hassler and Brockmann 2001; Saunders et al. 2010). We also know that the cues that attract males from a distance are not coming from eggs (which are known to have a sperm attractant, Shoger and Bishop 1967), since satellites approach females as they arrive on the beach well before any eggs are laid. Further, when newly laid eggs are placed under one of the two cement models, satellites are no more attracted to the model with eggs than to the model without, but once in contact with the model, they remain longer with the one that has eggs (Schwab and Brockmann 2007). Nothing is known about the nature of the chemicals that attract males from a distance except that they can remain in the sand for many minutes despite repeated washing by small waves (Hassler and Brockmann 2001).

19.4.2 Do Females Attract Males with Chemical Cues?

Given that males are using chemical cues or semiochemicals to find mates, the question then becomes, who is producing those cues? There are several possibilities. (a) Females might be producing attractants (polyandrous) or deterrents (monandrous) or (b) the attached males might be attracting or deterring satellite males. It is unlikely that selection would favor attached males that produced cues that attracted competitors, but he might produce inadvertent cues that would reveal that he was of low quality. For example, if an attached male had lower sperm volumes, which would render him unable to compete with satellites, then he might be attractive to unattached males if they could detect the sperm. (c) It is also possible that males copy the behavior of other males, i.e. that satellites are attracted to pairs that have already attracted satellites. This behavior would be particularly likely if the presence of a satellite improved the success of subsequent satellites or indicated the presence of a non-competitive attached male or tolerant female. It is difficult to resolve which of these explanations is most likely. In a field experiment, satellites were attracted to a pair nesting in shallow water but hidden from view by an enclosure (the enclosure had many small holes drilled in each side so chemical cues were available) and they were more attracted to a pair than to a lone female or a lone male covered by the enclosures (Saunders et al. 2010). These results suggest that the semiochemicals come from a nesting pair alone, without satellites. But is the female or the attached male producing the mate-attracting semiochemical?

Monandrous females are consistent in not attracting satellites from one nesting bout to the next, even when their attached males are removed and replaced with a new male (Johnson and Brockmann 2012). Polyandrous females also consistently attract satellites, even when their attached male has been replaced. Moreover, some monandrous females do not attract and are intolerant of satellite males and will leave the beach rather than nest as a polyandrous female. These results suggest that the presence of satellites around a female is a property of the female, or an interaction between female and male (e.g. male quality and female mating status) and not due exclusively to the attached male. The female, therefore, may control the amount of sperm competition her attached male faces. Whether the female is monandrous or polyandrous determines the male's reproductive success, as males attached to monandrous females fertilize 100 % of the eggs, while males attached to polyandrous females may have their paternity share drastically reduced. Unless males can detect the reproductive tactic of the female prior to amplexus and before arriving at the beach, it may be impossible for males to predict whether they will face sperm competition. If females become polyandrous when attached to a poor quality or incompatible male, however, then it would be possible for a male to predict his chances of facing sperm competition. Clearly, we need to know much more about pre-copulatory mate choice in this species.

19.5 Conclusions and Future Directions

Research on the reproductive tactics of horseshoe crabs has revealed the importance of studying male and female behavior concurrently. Previous studies failed to consider how female behavior affects males, and as a result, we missed important elements of horseshoe crab reproductive biology.

19.5.1 Mate Choice in Horseshoe Crabs

It is not clear what if any mate choice occurs at the time of amplexus because little is known about the behavior of pairs prior to arriving on the beach. There is ample evidence that attachment is non-random. For example, when males are detached from their mates and then allowed to re-attach to a new female in a wading pool, they pair much more quickly than satellite (unattached) males that have been placed in the same pool (Brockmann and Penn 1992; Duffy et al. 2006). These results demonstrate non-random pairing, which suggests mate choice, but it is not clear whether this non-random pairing is due to motivational differences among males or to choice by the male, the female, or both.

Females are limited in the time available for nesting, and they must have an attached male to spawn (Brockmann 1990). Like other explosive breeders, horseshoe crabs may have few opportunities for active choice prior to mating (amplexus). A lack of pre-copulatory mate choice means that the best way for females to control mate choice may be through their control of polyandry. We have shown that monandrous females that retain their attached males were more likely to remain monandrous than females whose attached males were removed and replaced with new males. This result, together with the finding that polyandrous females have lower developmental success with their attached males than with satellites, suggests that a female's mating tactic is affected by her attached male and is context-dependent. Since chemical cues are used by satellites to identify polyandrous females, it is possible that females are attracting satellite males by producing attractants when an attached male is of low quality or genetically incompatible.

19.5.2 Can Females Discriminate Attached Male Quality or Compatibility?

Mechanisms for detecting underlying male quality (good genes) might include male size or condition, sperm quantity or quality, or parasite load. We have not detected size or condition differences between the attached males of monandrous and polyandrous females, although the attached males of intolerant monandrous females tend to be in better condition (Johnson and Brockmann 2012, 2013). We have detected differences in sperm concentration (Sasson et al. in press), indicating it is possible that females assess male quality based on a male's sperm traits. Parasite load has not been properly investigated, but would be worth studying (Wedekind 1994) since older males have more parasitic flatworms (unpubl. data). It would be advantageous for females to detect genetically compatible males to prevent inbreeding, to increase heterozygosity or compatibility, or to avoid selfish genetic elements or heritable microorganisms (Jennions and Petrie 2000; Tregenza and Wedell 2000; Zeh and Zeh 2003; Chandler and Zamudio 2008). Perhaps the high heterozygosity observed in *L. polyphemus* (King et al. 2005) populations is the result of mate choice for compatible males to avoid inbreeding (Johnson and Brockmann 2013). Studies investigating the relatedness of attached pairs are needed to test this hypothesis. In general, if a female has information about the quality or compatibility of her attached male, we expect that she should attract additional mates when her attached male's sperm are less compatible with her eggs or when the male is of low quality, such as when he has a low sperm concentration.

19.5.3 Multiple Mating: Convenience Polyandry or Female Tactics?

When multiple mating is costly, yet present in a system, as in horseshoe crabs, the usual explanation is sexual conflict, sexual harassment, or convenience polyandry (Thornhill and Alcock 1983; Maklakov and Lubin 2006; Sztatecsny et al. 2006). Our in-depth studies of female mating behavior and reproductive success allow us to reject these hypotheses as the primary explanation for multiple mating in horseshoe crabs. Some females benefit from multiple mating while other females have mechanisms for reducing the effect of coercing males such as leaving the beach or laying fewer eggs, mechanisms that are similar to those observed in other species (Sirot et al. 2003; Hettyey et al. 2009). We find that multiple mating is more likely explained by the interplay between male and female mating tactics. A male may adopt the unattached tactic when he stands to benefit from being a satellite male. A female may attract (or fail to prevent) satellite males when she stands to gain either through improved mate quality or compatibility. However, it seems unlikely that females would need to attract more than a few additional satellite males to ensure that she mates with a high quality or compatible male. Hence, very large group sizes may result from convenience polyandry, where it is more costly for females to resist or stop mating than to allow additional satellites, particularly for polyandrous females.

19.5.4 The Dynamic Interplay Between Male and Female Tactics

Overall, the evidence demonstrates that male tactics are affected by male age (condition) and behavior (what other males are doing and female behavior). Female tactics are the result of male traits and behavior as well as female mating decisions. The different types of females choose different mating tactics, depending on their condition (condition-dependent) and circumstances (context-dependent), when they are free to do so. Female decisions change the opportunities and payoffs for male tactics, just as male tactics affect the reproductive success of females. This interplay between male and female mating tactics and mate choice decisions suggests that we should be studying the interactions within and between the sexes concurrently (Alonzo 2008). Further, the decisions that individuals make are dynamic, affected by OSR, density and other social (e.g. how aggregated is the nesting) and environmental (e.g. wave height that increases the risk of being overturned) factors. Such complexity cannot be evaluated piecemeal, which suggests that modeling may be needed to understand reproductive decisions in horseshoe crabs.

Condition-dependent tactics can be maintained in populations as a result of tradeoffs as has been suggested here: males switch tactics at a threshold age or condition that maximizes fitness. However, condition-dependent tactics may also be maintained by frequency-dependent selection (Brockmann and Taborsky 2008) or through an interaction between frequency and density dependence (Lucas and Howard 2008). When condition-dependence and frequency dependence are involved, the success of alternative tactics is not equal (Calsbeek et al. 2002; Shuster and Wade 2003; Hazel et al. 2004), but there is a unique and stable switch point to which the population returns when perturbed (Repka and Gross 1995; Gross and Repka 1998). The mating tactics of male horseshoe crabs are density dependent, and they are likely to be frequency dependent, since the success of the unattached tactic depends on its frequency in the population; for example, if there were no satellites, then a rare mutant satellite would be highly successful, but as the frequency of satellites increased, their success would decline (Brockmann 2001). Understanding the interacting and dynamic effects involved in the mating tactics of horseshoe crabs may require a dynamic game theoretic modeling approach (Clark and Mangel 2000).

19.5.5 Application to Other Populations: Effects of Density and OSR

Our studies have focused mainly on the *Limulus polyphemus* population at SK, so how might these findings apply elsewhere? Other populations differ from SK in a number of respects, including the timing and seasonality of nesting, and the OSR and density of nesting horseshoe crabs (Brockmann 1996). Genetically distinct populations may also show adaptation to local conditions (King et al. 2005). These factors may affect mating tactics and the intensity of sexual selection (Brockmann and Smith 2009). For example, Mattei et al. (2010) compared the mating behavior of horseshoe crabs from two different areas in Connecticut (CT) that differ from the DB population in spawning density, but still maintain a similar male-biased OSR. They found that the extremely low density of animals in CT resulted in females being almost exclusively monandrous when paired, and many females arrived on the beach without an attached male, which is rare at SK (Brockmann and Johnson 2011). Mattei et al. (2010) suggested that at such low densities horseshoe crabs may have difficulty finding and attracting mates, both when attaching in the water and once they have arrived at the beach. In such low-density environments, the selection pressures acting on reproductive traits may shift. For instance, females may be forced to be monandrous even if they would benefit from mating multiply, potentially lowering their reproductive success. Thus, there may be stronger selection in these populations to choose high quality or compatible mates when pairing. Since sperm competition would be rare for males in these populations, rather than acting on sperm traits, selection should favor a male's ability to find and hold onto females. Furthermore, because at these densities unattached males may have little reproductive success, we predict that the threshold age at which males switch to the unattached tactic would shift to an older age compared to higher density populations. While Mattei et al. (2010) still found many unattached males, they also found no difference in condition between attached and unattached males, suggesting that older males may forgo a switch to the unattached tactic and continue to attach at an older age than males in the SK population. Such a decision may mean that these males are less able to hold onto females and could result in females losing their attached males as they come ashore. This scenario would explain why unattached females are more common in the CT population compared to SK. Overall, these findings suggest that horseshoe crab density, at least at certain levels, may be more important than the OSR in affecting mating behavior and reproductive success. We know little, however, about how pairing occurs at sea other than the cues used, and we do not know whether males change tactics when density or OSR changes.

19.5.6 Implications for Management

Where populations are being depleted to the point that densities are very low, such as parts of New England, Florida's east coast and the northern Gulf of Mexico (Carmichael et al. 2003; James-Pirri et al. 2005; Gerhart 2007; Fulford and Haehn

2012), per capita reproductive success is likely to be affected. Since horseshoe crabs have external fertilization, a female can lay viable eggs only when a male is attached and at very low densities it may be difficult for males and females to find optimal, or any, mates (Mattei et al. 2010). In addition, horseshoe crab spawning may be social in the sense that females may be attracted to and gain from the presence of other females, for example, if many females nesting in close proximity swamp predators (Brockmann and Smith 2009). Low population density would result in negative density-dependent effects on reproduction, which may affect the ability of populations to recover from overharvest, habitat destruction or disease (Mattei et al. 2010).

In 2008 a male-only harvest was instituted in DB as a management strategy to prevent over-harvest of female horseshoe crabs (ASMFC 2008) and more recently an adaptive management approach has been used (McGowan et al. 2011; ASMFC 2012). Adaptive management of horseshoe crabs in DB is designed to allow sustainable harvest while maintaining an adequate supply of horseshoe crab eggs as a food resource for migrating shorebirds (Smith et al. 2013a). Our studies of horseshoe crab mating suggest some additional management considerations. First, since spawning is strongly aggregated, large numbers of females may arrive all at once, each requiring an attached male before laying eggs. Most males do not easily switch from satellite to attached tactics, so the OSR is not a good measure of the males available for pairing. Since younger males are more likely to attach than older males, the male-only harvest should exclude young or already attached males. Second, some females may attract satellite males when the male to which they are paired is of low quality or incompatible. This behavior means that unattached males are not "excess males" but an important part of the mating system of this species. Third, when nesting densities are low, even when the OSR is male biased, some females do not pair and hence do not lay eggs (and at extremely high densities females fail to lay eggs due to interference from other females). Hence, the relationship between nesting density and surface egg density will not be linear. Fourth, satellite males are costly to certain females, and some females will not nest with satellites. Therefore at very high OSRs, spawning would decline due to interference from satellite males. Exceptionally male-biased OSRs may result from selective harvest of females in some populations (Widener and Barlow 1999; James-Pirri et al. 2005), and our findings suggest that nesting at these sites may be particularly costly to females. Finally, there are likely to be population differences in male and female mating behavior, depending on the natural and sexual selection history of the population. For example, if a low OSR is the rule, then we expect different levels of sperm competition than when OSR is often high. Hence, populations are likely to differ in their reproductive response and vulnerability to harvest, habitat loss and disease.

19.5.7 Explosive Breeding in Horseshoe Crabs

Horseshoe crabs are often thought of as explosive breeders with scramble competition among males and, indeed, they seem to share many traits with other explosive breeders (Brockmann 1990). Explosive breeding is thought to allow few opportunities for mate choice, male combat or other sexually selected traits (Wells 1977), yet the OSR is often strongly male biased which usually means strong selection for sexually selected traits (Kvarnemo and Ahnesjö 1996). If some males do not mate under these conditions, then this further intensifies sexual selection (Shuster and Wade 2003). In fact, sexual selection for traits that enable a male to reproduce may be especially intense in explosive breeding scramble competition systems because males only have a brief opportunity to mate each year, which means that many males may be unable to find and secure a mate during that short window of opportunity (Greene and Funk 2009). Under these conditions males would be expected to have particularly well-developed sensory organs, high mobility for locating females, good spatial memory (Andersson and Iwasa 1996) and other traits that make them faster and more effective at pairing (Sztatecsny et al. 2012). Studies on a species of explosively breeding frog, for example, show that although overall size is not associated with mating success, small differences in the structures for holding onto females affect fitness (Greene and Funk 2009). These traits are not associated with male combat, female choice, or the ability of males to take over females once amplexus occurs, but rather with the ability of males to hold onto females until oviposition. We do not know the proportion of male horseshoe crabs that are not breeding, and hence, we do not know the intensity of sexual selection in our system. However, based on studies of anurans, we suspect that mating is not random with respect to characters that affect the ability of males to find and hold onto females.

Explosive breeding also increases opportunities for alternative tactics such as multi-male spawning, clutch piracy and sneaking behavior (Zamudio and Chan 2008). Such behavior may increase female coercion and even injury to females as males vie for position around a female (Sztatecsny et al. 2006). Explosive breeding is a particularly dynamic mating system and many species show differences between years and populations in OSR and the time available for breeding, and hence in the intensity of scramble competition (Olson et al. 1986; Reichard et al. 2007; Alvarez et al. 2014). When population size, the degree of female asynchrony, and the time available for mating competition vary (as they might on a daily to yearly basis), selection will favor an associated change in mating behavior and different mating tactics under different conditions (Mendoza-Cuenca and Macias-Ordonez 2010). Horseshoe crabs have similarly large differences in the amount of time available for breeding and in the intensity of scramble competition from year to year or at different breeding sites. This variation likely results in differences in the frequency of different female and male mating tactics between years and breeding locations.

Acknowledgments The research described in this chapter was supported by the National Science Foundation and the University of Florida Seahorse Key Marine Laboratory. The research was conducted under special use permits from the Cedar Keys National Wildlife Refuge. We especially want to thank the many graduate and undergraduate field and laboratory assistants who have helped us with this research over the years. We also appreciate the reviewers' and editor's comments on our manuscript.

References

Alcock J (2009) Animal behavior, 9th edn. Sinauer Associate, Inc, Sunderland

Alcock J, Eickwort GC, Eickwort KR (1977) The reproductive behavior of *Anthidium maculosum* (Hymenoptera: Megachilidae) and the evolutionary significance of multiple copulations by females. Behav Ecol Sociobiol 2:385–396

Alonzo SH (2008) Conflict between the sexes and alternative reproductive tactics within a sex. In: Oliveira RF, Taborsky M, Brockmann HJ (eds) Alternative reproductive tactics: an integrative approach. Cambridge University Press, Cambridge, pp 435–450

Alvarez D, Viesca L, Nicienza AG (2014) Sperm competitiveness differs between two frog populations with different breeding systems. J Zool 292:202–205

Andersson M, Iwasa Y (1996) Sexual selection. Trends Ecol Evol 11:53–58

ASMFC (1998) Interstate fishery management plan for horseshoe crabs. Atlantic States Marine Fisheries Commission, Washington, DC

ASMFC (2008) Addendum V to the interstate fishery management plan for horseshoe crabs. Atlantic States Marine Fisheries Commission, Washington, DC

ASMFC (2012) Addendum VII to the interstate fishery management plan for horseshoe crabs for public comment. Atlantic States Marine Fisheries Commission, Washington, DC

Barber SB (1956) Chemoreception and proprioception in *Limulus*. J Exp Zool 131(1):51–73

Barlow RB, Powers MK (2003) Seeing at night and finding mates: the role of vision. In: Shuster CN Jr, Barlow RB, Brockmann HJ (eds) The American horseshoe crab. Harvard University Press, Cambridge, MA, pp 83–102

Barlow RB Jr, Ireland LC, Kass L (1982) Vision has a role in *Limulus* mating behaviour. Nature 296(5852):65–66

Barlow RB Jr, Powers MK, Howard H et al (1986) Migration of *Limulus* for mating: relation to lunar phase, tide height, and sunlight. Biol Bull 171:310–329

Barlow RB Jr, Powers MK, Howard H et al (1987) Vision in *Limulus* mating and migration. In: Herrnkind WF, Thistle AB (eds) Signposts in the sea. Florida State University Press, Tallahassee, pp 69–84

Barlow RB Jr, Powers MK, Kass L (1988) Vision and mating behavior in *Limulus*. In: Atema J, Fay RR, Popper AN, Tovolga WN (eds) Sensory biology of aquatic animals. Springer, New York, pp 419–434

Berkson J, Shuster CN (1999) The horseshoe crab: the battle for a true multiple-use resource. Fisheries 24(11):6–10

Botton ML (1984) Diet and food preferences of the adult horseshoe crab *Limulus polyphemus* in Delaware Bay, New Jersey, USA. Mar Biol 81:199–207

Botton ML (2009) The ecological importance of horseshoe crabs in estuarine and coastal communities: a review and speculative summary. In: Tanacredi JT, Botton M, Smith DR (eds) Biology and conservation of horseshoe crabs. Springer, New York, pp 45–63

Botton ML, Loveland RE (1989) Reproductive risk: high mortality associated with spawning by horseshoe crabs (*Limulus polyphemus*) in Delaware Bay, USA. Mar Biol 101(2):143–151

Botton ML, Loveland RE (1992) Body size, morphological constraints, and mated pair formation in four populations of horseshoe crabs (*Limulus polyphemus*) along a geographic cline. Mar Biol 112:409–415

Botton ML, Ropes JW (1988) An indirect method for estimating longevity of the horseshoe crab (*Limulus polyphemus*) based on epifaunal slipper shells (*Crepidula fornicata*). J Shellfish Res 7(3):407–412

Botton ML, Shuster CN, Sekiguchi K et al (1996) Amplexus and mating behavior in the Japanese horseshoe crab, *Tachypleus tridentatus*. Zool Sci 13:151–159

Botton ML, Harrington BA, Tsipoura N et al (2003) Synchronies in migration: shorebirds, horseshoe crabs, and Delaware Bay. In: Shuster CN Jr, Barlow RB, Brockmann HJ (eds) The American horseshoe crab. Harvard University Press, Cambridge, MA, pp 5–32

Brockmann HJ (1990) Mating behavior of horseshoe crabs, *Limulus polyphemus*. Behaviour 114:206–220

Brockmann HJ (1996) Satellite male groups in horseshoe crabs, *Limulus polyphemus*. Ethology 102:1–21

Brockmann HJ (2001) The evolution of alternative strategies and tactics. Adv Study Behav 30:1–51

Brockmann HJ (2002) An experimental approach to altering mating tactics in male horseshoe crabs (*Limulus polyphemus*). Behav Ecol 13:232–238

Brockmann HJ (2003a) Male competition and satellite behavior. In: Shuster CN Jr, Barlow RB, Brockmann HJ (eds) The American horseshoe crab. Harvard University Press, Cambridge, MA, pp 50–82

Brockmann HJ (2003b) Nesting behavior: a shoreline phenomenon. In: Shuster CN Jr, Barlow RB, Brockmann HJ (eds) The American horseshoe crab. Harvard University Press, Cambridge, MA, pp 33–49

Brockmann HJ, Johnson SL (2011) A long-term study of spawning activity in a Florida Gulf coast population of horseshoe crabs (*Limulus polyphemus*). Estuar Coast 34:1049–1067

Brockmann HJ, Penn D (1992) Male mating tactics in the horseshoe crab, *Limulus polyphemus*. Anim Behav 44:653–665

Brockmann HJ, Smith MD (2009) Reproductive competition and sexual selection in horseshoe crabs. In: Tanacredi JT, Botton M, Smith DR (eds) Biology and conservation of horseshoe crabs. Springer, New York, pp 199–221

Brockmann HJ, Taborsky M (2008) Alternative reproductive tactics and the evolution of alternative allocation phenotypes. In: Oliveira RF, Taborsky M, Brockmann HJ (eds) Alternative reproductive tactics: an integrative approach. Cambridge University Press, Cambridge, pp 25–51

Brockmann HJ, Colson T, Potts W (1994) Sperm competition in horseshoe crabs (*Limulus polyphemus*). Behav Ecol Sociobiol 35:153–160

Brockmann HJ, Nguyen C, Potts W (2000) Paternity in horseshoe crabs when spawning in multiple-male groups. Anim Behav 60:837–849

Brown GG (1976) Scanning electron-microscopical and other observations of sperm fertilization reactions in *Limulus polyphemus* L. (Merostomata: Xiphosura). J Cell Sci 22:547–562

Brown GG, Clapper DL (1981) Procedures for maintaining adults, collecting gametes, and culturing embryos and juveniles of the horseshoe crab, *Limulus polyphemus* L. In: Hinegardner R, Atz J, Fay R, Fingerman M, Josephson R, Meinkoth N (eds) Laboratory animal management, marine invertebrates. National Academy Press, Washington, DC, pp 268–290

Byrne RJ, Avise JC (2009) Multiple paternity and extra-group fertilizations in a natural population of California grunion (*Leuresthes tenuis*), a beach-spawning marine fish. Mar Biol 156:1681–1690

Byrne PG, Simmons LW, Roberts JD (2003) Sperm competition and the evolution of gamete morphology in frogs. Proc R Soc Lond B Biol 270:2079–2086

Calsbeek R, Alonzo SH, Zamudio K et al (2002) Sexual selection and alternative mating behaviours generate demographic stochasticity in small populations. Proc R Soc Lond B Biol 269:157–164

Candolin U (2003) The use of multiple cues in mate choice. Biol Rev 78:575–595

Candolin U, Reynolds JD (2002) Adjustments of ejaculation rates in response to risk of sperm competition in a fish, the bitterling (*Rhodeus sericeus*). Proc R Soc Lond B Biol 269:1549–1553

Carmichael RH, Rutecki D, Valiela I (2003) Abundance and population structure of the Atlantic horseshoe crab *Limulus polyphemus* in Pleasant Bay, Cape Cod. Mar Ecol Prog Ser 246:225–239

Castillo LP, Hatch KA (2007) Fasting increases delta N-15 values in the uric acid of *Anolis carolinensis* and *Uta stansburiana* as measured by nondestructive sampling. Rapid Commun Mass Spectrom 21:4125–4128

Cavenaugh CM (1975) Observations on mating behavior in *Limulus polyphemus*. Biol Bull 149:422

Chandler CH, Zamudio KR (2008) Reproductive success by large, closely related males facilitated by sperm storage in an aggregate breeding amphibian. Mol Ecol 17:1564–1576

Clark CW, Mangel M (2000) Dynamic state variable models in ecology: methods and applications. Oxford University Press, Oxford

Cohen JA, Brockmann HJ (1983) Breeding activity and mate selection in the horseshoe crab, *Limulus polyphemus*. Bull Mar Sci 33(2):274–281

Cummings ME, Gelineau-Kattner R (2009) The energetic costs of alternative male reproductive strategies in *Xiphophorus nigrensis*. J Comp Physiol 195:935–946

Duffy EE, Penn DJ, Botton ML et al (2006) Eye and clasper damage influence male mating tactics in the horseshoe crab, *Limulus polyphemus*. J Ethol 24(1):67–74

Eldredge N (1970) Observations on burrowing behavior in *Limulus polyphemus* (Chelicerata, Merostomata), with implications on the functional anatomy of trilobites. Am Mus Novit 2436:1–17

Emlen ST, Oring LW (1977) Ecology, sexual selection, and the evolution of mating systems. Science 197:215–223

Fu P, Neff BD, Gross MR (2001) Tactic-specific success in sperm competition. Proc R Soc B Biol Sci 268:1105–1112

Fulford RS, Haehn RA (2012) An evaluation of Mississippi barrier islands as spawning and nesting habitat for the American horseshoe crab, *Limulus polyphemus*, with implications for island restoration. Gulf Caribb Res 24:51–62

Gage MJG (1994) Associations between body-size, mating pattern, testis size and sperm lengths across butterflies. Proc R Soc B Biol Sci 258:247–254

Gerhart SD (2007) A review of the biology and management of horseshoe crabs with emphasis on Florida populations. Fish and Wildlife Research Institute, St. Petersburg. http://research.myfwc.com. Accessed 1 June 2014

Grant D (2001) Living on *Limulus*. In: Tanacredi JT (ed) *Limulus* in the limelight. Kluwer Academic, New York, pp 135–146

Greene AE, Funk WC (2009) Sexual selection on morphology in an explosive breeding amphibian, the Columbia spotted frog (*Rana luteiventris*). J Herpetol 43:244–251

Gross MR (1996) Alternative reproductive strategies and tactics: diversity within sexes. Trends Ecol Evol 11:92–97

Gross MR, Repka J (1998) Game theory and inheritance in the conditional strategy. In: Dugatkin L, Reeve HK (eds) Game theory and animal behavior. Oxford University Press, Oxford, pp 168–187

Harrington JM, Armstrong PB (2000) Initial characterization of a potential anti-fouling system in the American horseshoe crab, *Limulus polyphemus*. Biol Bull 199:189–190

Harrington JM, Leippe M, Armstrong PB (2008) Epithelial immunity in a marine invertebrate: a cytolytic activity from a cuticular secretion of the American horseshoe crab, *Limulus polyphemus*. Mar Biol 153(6):1165–1171

Hassler CL (1999) Satellite male groups in horseshoe crabs (*Limulus polyphemus*): how and why are males choosing females. M.S. thesis, University of Florida

Hassler C, Brockmann HJ (2001) Evidence for use of chemical cues by male horseshoe crabs when locating nesting females (*Limulus polyphemus*). J Chem Ecol 27:2319–2335
Hayes WF, Barber SB (1982) Peripheral synapses in *Limulus* chemoreceptors. Comp Biochem Physiol 72A(2):287–293
Hazel W, Smock R, Lively CM (2004) The ecological genetics of conditional strategies. Am Nat 163(6):888–900
Herzog ED, Powers MK, Barlow RB (1996) *Limulus* vision in the ocean day and night: effects of image size and contrast. Vis Neurosci 13:31–41
Hettyey A, Baksay S, Vagi B et al (2009) Counterstrategies by female frogs to sexual coercion by heterospecifics. Anim Behav 78:1365–1372
Hobson KA, Alisauskas RT, Clark RG (1993) Stable nitrogen isotope enrichment in avian tissues due to fasting and nutritional stress: implications for isotopic analyses of diet. Condor 95:388–394
Höglund J (1989) Pairing and spawning patterns in the common toad, *Bufo bufo*: the effects of sex ratios and the time available for male-male competition. Anim Behav 38:423–429
Höglund J, Saterberg S (1989) Sexual selection in common toads: correlates with age and body size. J Evol Biol 2:367–372
James-Pirri MJ, Tuxbury K, Marino S et al (2005) Spawning densities, egg densities, size structure, and movement patterns of spawning horseshoe crabs, *Limulus polyphemus*, within four coastal embayments on Cape Cod, Massachusetts. Estuaries 28(2):296–313
Jennions MD, Petrie M (2000) Why do females mate multiply? A review of the genetic benefits. Biol Rev 75:21–64
Johnson SL, Brockmann HJ (2010) Costs of multiple mates: an experimental study in horseshoe crabs. Anim Behav 80:773–782
Johnson SL, Brockmann HJ (2012) Alternative reproductive tactics in female horseshoe crabs. Behav Ecol 23:999–1008
Johnson SL, Brockmann HJ (2013) Parental effects on early development: testing for indirect benefits of polyandry. Behav Ecol 24:1218–1228
Kerlinger P (1998) Showdown at Delaware Bay. Nat Hist 107:56–58
King TL, Eackles MS, Spidle AP et al (2005) Regional differentiation and sex-biased dispersal among populations of the horseshoe crab *Limulus polyphemus*. Trans Am Fish Soc 134(2):441–465
Koons BF (1883) Sexual characters of *Limulus*. Am Nat 17:1297–1299
Kreamer G, Michels SE (2009) History of horseshoe crab harvest in Delaware Bay. In: Tanacredi JT, Botton M, Smith DR (eds) Biology and conservation of horseshoe crabs. Springer, New York, pp 299–313
Kvarnemo C, Ahnesjö I (1996) The dynamics of operational sex ratios and competition for mates. Trends Ecol Evol 11:404–408
Leibovitz L, Lewbart GA (2003) Diseases and symbionts: vulnerability despite tough shells. In: Shuster CN Jr, Barlow RB, Brockmann HJ (eds) The American horseshoe crab. Harvard University Press, Cambridge, MA, pp 245–275
Leschen AS, Grady SP, Valiela I (2006) Fecundity and spawning of the Atlantic horseshoe crab, *Limulus polyphemus*, in Pleasant Bay, Cape Cod, Massachusetts, USA. Mar Ecol Evol Persp 27(1):54–65
Levitan DR (1995) The ecology of fertilization in free-spawning invertebrates. In: McEdward L (ed) Marine invertebrate larvae. CRC Press, New York, pp 123–156
Loesel R, Heuer CM (2010) The mushroom bodies – prominent brain centres of arthropods and annelids with enigmatic evolutionary origin. Acta Zool 91:29–34
Loveland RE, Botton ML (1992) Size dimorphism and the mating system in horseshoe crabs (*Limulus polyphemus* L.). Anim Behav 44:907–916
Lucas JR, Howard RD (2008) Modeling alternative mating tactics as dynamic games. In: Oliveira RF, Taborsky M, Brockmann HJ (eds) Alternative reproductive tactics: an integrative approach. Cambridge University Press, Cambridge, pp 52–82

Maklakov AA, Lubin Y (2006) Indirect genetic benefits of polyandry in a spider with direct costs of mating. Behav Ecol Sociobiol 61:31–38
Martin KLM, van Winkle RC, Drais JE et al (2004) Beach-spawning fishes. Terrestrial eggs and air breathing. Physiol Biochem Zool 77:750–759
Mattei JH, Beekey MA, Rudman A et al (2010) Reproductive behavior in horseshoe crabs: does density matter? Curr Zool 56(5):634–642
McCauley SJ, Bouchard SS, Farina BJ et al (2000) Energetic dynamics and anuran breeding phenology: insights from a dynamic game. Behav Ecol 11:429–436
McCue MD (2010) Starvation physiology: reviewing the different strategies animals use to survive a common challenge. Comp Biochem Physiol A 156:1–18
McCue MD, Pollock ED (2008) Stable isotopes may provide evidence for starvation in reptiles. Rapid Commun Mass Spectrom 22:2307–2314
McGowan CP, Smith DR, Sweka JA et al (2011) Multispecies modeling for adaptive management of horseshoe crabs and red knots in the Delaware Bay. Nat Resour Model 24(1):117–156
Mendoza-Cuenca L, Macias-Ordonez R (2010) Female asynchrony may drive disruptive sexual selection on male mating phenotype in a Heliconius butterfly. Behav Ecol 21:144–152
Moore S (2004) The Taunton Bay assessment. Unpublished manuscript
Myers JP (1986) Sex and gluttony on Delaware Bay. Nat Hist 95:68–76
Olson DH, Blaustein AR, O'Hara RK (1986) Mating pattern variability among western toad (*Bufo boreas*) populations. Oecologia 70:351–356
Parker GA (1970) Sperm competition and its evolutionary consequences in the insects. Biol Rev 45:525–568
Parker GA (1982) Why are there so many tiny sperm – sperm competition and the maintenance of 2 sexes. J Theor Biol 96:281–294
Parker GA (1998) Sperm competition and the evolution of ejaculates: towards a theory base. In: Birkhead TR, Møller AP (eds) Sperm competition and sexual selection. Academic, New York, pp 3–54
Partan SR, Marler P (2005) Issues in the classification of multimodal communication signals. Am Nat 166:231–245
Passaglia CL, McSweeney ME, Stewart KM et al (1997) Visual performance of horseshoe crabs: role of underwater lighting. Biol Bull 193:205–207
Penn D, Brockmann HJ (1994) Nest-site selection in the horseshoe crab, *Limulus polyphemus*. Biol Bull 187:373–384
Penn D, Brockmann HJ (1995) Age-biased stranding and righting in horseshoe crabs, *Limulus polyphemus*. Anim Behav 49:1531–1539
Pilastro A, Scaggiante M, Rasotto MB (2002) Individual adjustment of sperm expenditure accords with sperm competition theory. Proc Natl Acad Sci USA 99:9913–9915
Pomerat CM (1933) Mating in *Limulus polyphemus*. Biol Bull 64(2):243–252
Powers MK, Barlow RB Jr, Kass L (1991) Visual performance of horseshoe crabs day and night. Vis Neurosci 7(3):179–190
Reichard M, Le Comber SC, Smith C (2007) Sneaking from a female perspective. Anim Behav 74:679–688
Repka J, Gross MR (1995) The evolutionarily stable strategy under individual condition and tactic frequency. J Theor Biol 176:27–31
Roberts JD, Byrne PG (2011) Polyandry, sperm competition and the evolution of anuran amphibians. Adv Study Behav 43:1–53
Rudloe A (1980) The breeding behavior and patterns of movement of horseshoe crabs, *Limulus polyphemus*, in the vicinity of breeding beaches in Apalachee Bay, Florida. Estuaries 3(3):177–183
Rudloe A (1985) Variation in the expression of lunar and tidal behavioral rhythms in the horseshoe crab, *Limulus polyphemus*. Bull Mar Sci 36(2):388–395
Sasson DA, Brockmann HJ (in prep) Factors influencing geographical variation in sperm traits of the American horseshoe crab, *Limulus polyphemus*

Sasson DA, Johnson SL, Brockmann HJ (in press) Sperm traits in a system with condition-dependent alternative reproductive tactics. Behav Ecol Sociobiol
Sasson DA, Johnson SL, Brockmann HJ (2012) The role of age on sperm traits in the American horseshoe crab, *Limulus polyphemus*. Anim Behav 84:975–981
Saunders KM, Brockmann HJ, Watson WH et al (2010) Male horseshoe crabs *Limulus polyphemus* use multiple sensory cues to locate mates. Curr Zool 56(5):485–498
Schwab RL, Brockmann HJ (2007) The role of visual and chemical cues in the mating decisions of satellite male horseshoe crabs, *Limulus polyphemus*. Anim Behav 74:837–846
Sekiguchi K (1988) Ecology [of horseshoe crabs]. In: Sekiguchi K (ed) Biology of horseshoe crabs. Science House Co., Ltd., Tokyo, pp 50–68
Shoger RL, Bishop DW (1967) Sperm activation and fertilization in *Limulus polyphemus*. Biol Bull 133:485
Shuster CN Jr (1954) A horseshoe "crab" grows up. Ward's Nat Sci Bull 28(1):1–6
Shuster CN Jr, Botton ML (1985) A contribution to the population biology of horseshoe crabs, *Limulus polyphemus* (L.) in Delaware Bay. Estuaries 8(4):363–572
Shuster CN Jr, Sekiguchi K (2003) Growing up takes about ten years and eighteen stages. In: Shuster CN, Barlow RB, Brockmann HJ (eds) The American horseshoe crab. Harvard University Press, Cambridge, MA, pp 103–132
Shuster SM, Wade MJ (2003) Mating systems and strategies. Princeton University Press, Princeton
Shuster CN Jr, Botton ML, Loveland RE (2003) Horseshoe crab conservation: a coast-wide management plan. In: Shuster CN, Barlow RB, Brockmann HJ (eds) The American horseshoe crab. Harvard University Press, Cambridge, MA, pp 358–377
Simmons LW (2005) The evolution of polyandry: sperm competition, sperm selection, and offspring viability. Ann Rev Ecol Evol Syst 36:125–146
Sirot LK, Brockmann HJ, Marinis C et al (2003) Maintenance of a female-limited polymorphism in *Ischnura ramburi* (Zygoptera: Coenagrionidae): evaluation of hypotheses and new insights. Anim Behav 66:763–775
Slatyer RA, Mautz BS, Backwell PRY et al (2012) Estimating genetic benefits of polyandry from experimental studies: a meta-analysis. Biol Rev 87:1–33
Smith MD, Brockmann HJ (2014) The evolution and maintenance of sexual size dimorphism in horseshoe crabs: an evaluation of six functional hypotheses. Anim Behav 96:127–139
Smith DR, Brousseau DJ, Mandt MT et al (2010) Age and sex specific timing, frequency, and spatial distribution of horseshoe crab spawning in Delaware Bay: insights from a large-scale radio telemetry array. Curr Zool 56(5):563–574
Smith DR, McGowan CP, Daily JP et al (2013a) Evaluating a multispecies adaptive management framework: must uncertainty impede effective decision-making. J Appl Ecol 50:1431–1440
Smith MD, Schrank HE, Brockmann HJ (2013b) Measuring the costs of alternative reproductive tactics in horseshoe crabs, *Limulus polyphemus*. Anim Behav 85:165–173
Snook RR (2005) Sperm in competition: not playing by the numbers. Trends Ecol Evol 20:46–51
Sokoloff A (1978) Observations on populations of the horseshoe crab *Limulus* (=Xiphosura) *polyphemus*. Res Popul Ecol 19:222–236
Suggs DN, Carmichael RH, Grady SP et al (2002) Effects of individual size on pairing in horseshoe crabs. Biol Bull 203:225–227
Sztatecsny M, Jehle R, Burke T et al (2006) Female polyandry under male harassment: the case of the common toad (*Bufo bufo*). J Zool 270:517–522
Sztatecsny M, Preininger D, Freudmann A et al (2012) Don't get the blues: conspicuous nuptial colouration of male moor frogs (*Rana arvalis*) supports visual mate recognition during scramble competition in large breeding aggregations. Behav Ecol Sociobiol 66:1587–1593
Taborsky M (1998) Sperm competition in fish: 'bourgeois' males and parasitic spawning. Trends Ecol Evol 13:222–227
Teale EW (1957) The oldest migration. Nat Hist 66:364–369
Thornhill R, Alcock J (1983) The evolution of insect mating systems. Harvard University Press, Cambridge, MA

Tregenza T, Wedell N (2000) Genetic compatibility, mate choice and patterns of parentage: invited review. Mol Ecol 9:1013–1027

Vasquez MC, Johnson SL, Brockmann HJ et al (2015) Nest site selection minimizes environmental stressor exposure in the American horseshoe crab, *Limulus polyphemus*. J Exp Mar Biol Ecol 463:105–114

Vladic TV, Jarvi T (2001) Sperm quality in the alternative reproductive tactics of Atlantic salmon: the importance of the loaded raffle mechanism. Proc R Soc B Biol Sci 268:2375–2381

Wasserman GS, Cheng Z (1996) Electroretinographic measures of vision in horseshoe crabs with uniform versus variegated carapaces. Biol Signals 5:247–262

Wedekind C (1994) Mate choice and maternal selection for specific parasite resistances before, during and after fertilization. Phil Trans R Soc B 346:303–311

Wells KD (1977) The social behaviour of anuran amphibians. Anim Behav 25:666–693

Wells KD (2007) The ecology and behavior of amphibians. University of Chicago Press, Chicago

Widemo F (1998) Alternative reproductive strategies in the ruff, *Philomachus pugnax*: a mixed ESS? Anim Behav 56:329–336

Widener JW, Barlow R (1999) Decline of a horseshoe crab population on Cape Cod. Biol Bull 197:300–302

Zamudio KR, Chan LM (2008) Alternative reproductive tactics in amphibians. In: Oliveira RF, Taborsky M, Brockmann HJ (eds) Alternative reproductive tactics: an integrative approach. Cambridge University Press, Cambridge, pp 300–331

Zeh JA, Zeh DW (2003) Toward a new sexual selection paradigm: polyandry, conflict and incompatibility. Ethology 109:929–950

Chapter 20
Examination of Large Exuviae with Mating Scars: Do Female American Horseshoe Crabs, *Limulus polyphemus*, Molt After Sexual Maturity?

Ruth H. Carmichael, Elizabeth E. Hieb, Glenn Gauvry, and Carl N. Shuster Jr.

Abstract Determining when a female horseshoe crab reaches sexual maturity, and presumably experiences a terminal molt, is of increasing interest to interpret the dynamics of American horseshoe crab populations for management. Males acquire easily recognizable and biologically determinate sexually dimorphic characteristics at maturity, while the most distinctive traits acquired by females are carapace scars that result from amplexus with another crab, presumably during mating. We found well defined mating scars on the axial ridge of the opisthosoma and clasper depressions on the opisthosomal projections of 35 large-sized (prosomal width 176–257 mm) horseshoe crab molts collected from Delaware Bay. The greatest intensity of scarring per unit area was found on molts >230 mm, a size range that overlaps with known sizes of females at spawning beaches, but is larger than sizes typically reported for subadults. These data represent the first published description and analysis of female horseshoe crab molts with mating scars typical of prolonged amplexus and sufficient to have occurred during spawning. We lack data to definitively prove that the molts examined came from females that engaged in reproductive activity beyond extended amplexus, but prolonged amplexus without reproduction is generally disad-

R.H. Carmichael (✉)
Dauphin Island Sea Lab, University of South Alabama, Dauphin Island, AL 36528, USA

Department of Marine Sciences, University of South Alabama, Mobile, AL 36688, USA
e-mail: rcarmichael@disl.org

E.E. Hieb
University Programs, Dauphin Island Sea Lab, 101 Bienville Blvd, Dauphin Island, AL 36528, USA

G. Gauvry
Ecological Research & Development Group, Inc., 190 Main Street, Little Creek, Dover, DE 19901, USA

C.N. Shuster Jr.
Biological Oceanographer (retired), 3733 North 24th Street, Arlington, VA 22207-5011, USA

R.H. Carmichael et al. (eds.), *Changing Global Perspectives on Horseshoe Crab Biology, Conservation and Management*, DOI 10.1007/978-3-319-19542-1_20

vantageous. Molting after first reproduction, however, may enhance reproductive success through shell repair and increased fecundity. This study describes a quantitative approach that can serve as the basis for additional targeted studies to quantify mating scars on large-sized molts and characterize mating scar formation on live spawning crabs that will be useful to further evaluate molt timing relative to reproduction.

Keywords Prosomal width • Reproduction • Terminal molt • Amplexus • Cohort

20.1 Introduction

Among American horseshoe crabs (*Limulus polyphemus*) the concept of terminal molt has come to be generally accepted as a final molt at sexual maturity during which sexually dimorphic characteristics are acquired and after which individuals begin mating and no longer grow. Among males this molt includes the acquisition of modified claspers for attachment (amplexus) to the female. Among females this final molt is typically associated with development and maturation of eggs (Munson 1898; Shuster 1955; Smith et al. 2009). Historically, the primary evidence for existence of a terminal molt in American horseshoe crabs was the absence of observing any molts with modified claspers (mature male molts). Large-sized female molts have been found, but unlike males, aside from a larger size, sexually mature females have no comparably unambiguous external characteristics to distinguish them from subadults. The occurrence of characteristic mating scars due to the rubbing of the male carapace and claspers on the female opisthosoma are the major distinction, therefore, between large juvenile and adult females. Since it has been postulated that female horseshoe crabs may have one more instar stage than males (Shuster 1982; Sekiguchi 1988), the occurrence of large female exuviae alone is not considered indicative of molting among adults.

Recent evidence, however, supports the idea that some adult horseshoe crabs may continue to molt as adults or at least have an additional molt at adult size before reaching a terminal molt. The presence of statistically defined cohorts (distinct size groupings) among adults and biologically unlikely growth increments required to molt from mean juvenile to adult sizes have been proposed as evidence that crabs may continue to molt as adults in some waters (Carmichael et al. 2003). This idea along with the discovery of four female exuviae with mating scars within a few weeks in Delaware Bay, USA (Fig. 20.1) raised question as to whether (1) some females molt at least once after reaching sexual maturity, or (2) mating scars can be acquired prior to sexual maturity such that they do not accurately define maturity in female horseshoe crabs. These initial observations led to targeted searches for female molts bearing mating scars and application of mathematical models to test potential explanations for the size dimorphism between sexes and determine subsequent relationships to growth, maturity, and the timing of a terminal molt (Smith et al. 2009; Gauvry pers. obs.). Models based on data from thousands of horseshoe crabs collected in Delaware Bay indicated that differential-maturity (an additional molt among females prior to sexual maturity) was the most likely explanation for size dimorphism between

Fig. 20.1 Examples of female American horseshoe crab molts with mating scars. *Top panel*: Two live animals (*left*) showing the color and texture of recently molted females without mating scars, and the exuvia (*right*) from which the lower animal emerged, showing mating scars (*white arrow*) on the opisthosomal arch of the molted shell. *Bottom panels*: Molt from a female (prosomal width 250 mm) with noticeable evidence of amplexus (*enlarged in right panels*), including an axial mating scar with lateral rubbing on the opisthosomal arch (*lower right*) and clasper imprints on the opisthosomal projections

male and female horseshoe crabs (Smith et al. 2009). These data, however, could not account for exuviae bearing mating scars nor exclude the possibility of post-reproductive molting by at least some crabs prior to a terminal molt.

In this study, we conducted the first quantitative analysis of mating scars on female horseshoe crab exuviae to objectively describe the scarring patterns and attempt to determine if the scarring patterns could indicate molt timing relative to sexual maturity. Molts were collected during a 7 year period in the Delaware Bay area, and mating scars were quantified in terms of area and intensity (depth) of scarring using a newly developed approach based on the concepts of visual aging originally described by Shuster (1955). We compared the size (prosomal width) distribution of sampled molts to known size distributions of live spawning females and related the quantitative evidence of amplexus to apparent age class based on size and visual age at the time of molting. In addition to providing initial data to relate molt timing to sexual maturity, which are critical metrics for management, this study describes a quantitative approach that can serve as the basis for future mating scar analyses.

20.2 Methods

20.2.1 Sample Collection

To determine if exuviae from female American horseshoe crabs (*L. polyphemus*) showed evidence of molting after sexual reproduction, we directly collected or otherwise obtained as many molts as possible bearing evidence of mating scars during 2001–2007. Molts were collected from Broadkill Beach, Delaware during October 2001 (n = 1), offshore trawls on the continental shelf in September 2001 (n = 2; Hata and Berkson 2003), and the wrackline on Fowlers Beach, Delaware during September 2002 and in October 2003–2007 (n = 32), following the first nor'easter (macro-scale storm with winds from the northeast along the mid to northern U.S. Atlantic coastline) of the season.

20.2.2 Attributes of Exuviae

Molts were confirmed as female by the presence of female attributes including intact chelate pedipalps and/or genital slits, a low or absent anterior prosomal arch, and large size. Size was determined by measuring prosomal width and interocular distance as independent measures of molt size to the nearest 1.0 mm, using a clear plastic scale. Five exuviae could not be measured due to a partial or damaged prosoma, and these specimens were excluded from prosomal width comparisons. We confirmed that collected molts were within the size range of live adult females by comparing the size distribution of sampled molts to known size distributions of adult females at spawning beaches at the same location for a broad range of time periods, including the sampling period (Table 20.1). We further classified molts as 'young', 'middle aged' or 'old' based on overall carapace wear according to the methods of Shuster (1955).

Table 20.1 Range of prosomal width (PW) among live adult female American horseshoe crabs collected from Delaware Bay, USA compared to molts collected during this study

N	PW (mm)	Year	Source
56	178–258		Shuster (1955, 1979)
100	245–350		Shuster et al. (1961) (unpublished)
14	260–310	1979–1980	Shuster and Botton (1985)
216	258 (0.07)[a]	1986–1989	Botton and Loveland (1992)
9,749	220–330	2003–2005	Smith et al. (2009)
35	176–257	2001–2007	This study (molts)

[a]Data reported as mean PW (coefficient of variation)

20.2.3 *Quantifying Mating Scar Patterns*

To determine whether the patterns of mating scars on each molt could indicate at least one season of amplexus consistent with reproductive activities of adult horseshoe crabs, we characterized mating scar patterns on each molt in terms of area of coverage and intensity of scarring in four regions of the opisthosoma that are known to show wear specific to amplexus (Shuster 1955; Brockmann 2003). These regions included erosion on the axial ridge of the opisthosoma caused by the rubbing of a male prosoma (Regions I and II; Fig. 20.2, left panels) and depressions on the trailing projections of the female opisthosoma from pressure by the male claspers (Regions III and IV; Fig. 20.2).

The area of mating scars was determined by measuring the greatest straight length (anterior to posterior) and width (laterally) of scarring in each region to the nearest 0.1 mm using a clear plastic scale and calipers, depending on the size and shape of the area measured. Intensity of scarring was quantified from least to most intense (deepest scarring) by a rank of 1–3 (examples shown in Fig. 20.3). Scarring with a score of 1 appeared darker than the surrounding exuvia surface with no deeper pitting. A score of 2 indicated the majority of scarring was darker than the surrounding exuvia, with some (a minority) of deeper, lighter colored pitting. A score of 3 indicated the majority of scarring was deeper, revealing lighter colored layers within the shell and substantial pitting. Left, right and/or center sub-regions were measured and scored separately, but treated cumulatively to obtain a total area and total intensity score for each region. These data were then averaged within an individual to make subsequent comparisons among individuals. The extent of scarring in terms of area, intensity, and the combination of these factors (intensity per unit area of scarring) was compared to molt size and visual age to relate amplexus and associated sexual activity with apparent age class at the time of molting.

20.3 Results

During a 7-year period (2001–2007), 35 horseshoe crab molts bearing mating scars were collected. In most cases, the combination of prosomal width and arch structure were typical to adult females (91 %). A smaller number of molts (25 %) also retained

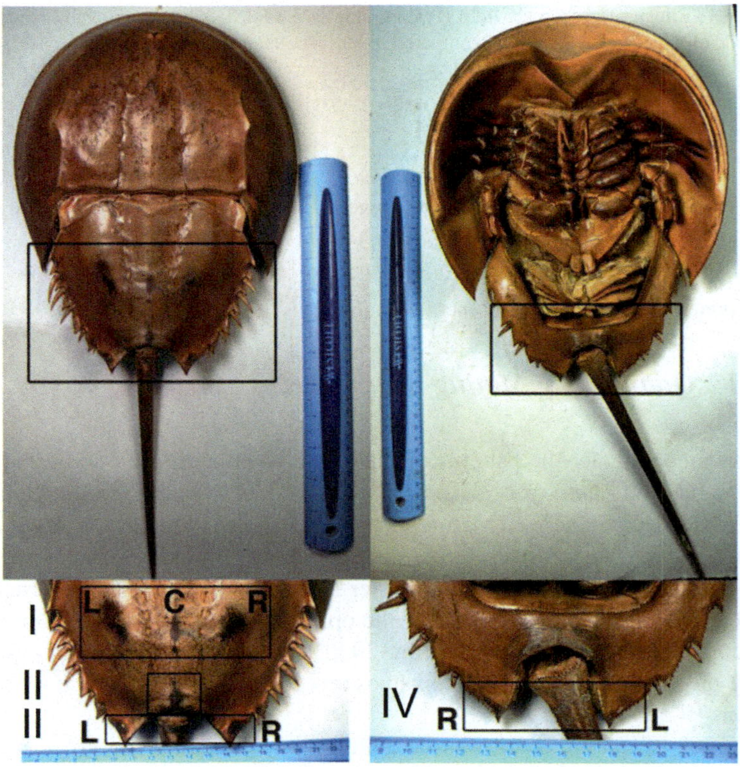

Fig. 20.2 American horseshoe crab molts showing dorsal (*left*) and ventral (*right*) views of mating scars in four regions (I–IV, *enlarged in lower panels*) on the opisthosoma that were measured to determine area and intensity of scarring. *L* left, *R* right, *C* center

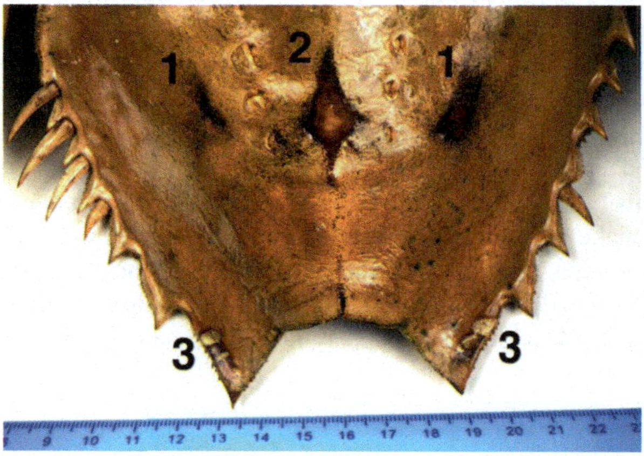

Fig. 20.3 Posterior portion of the opisthosoma, showing examples of the intensity categories used to rank mating scars for this study. *1* low intensity area characterized by dark discoloration, *2* moderate intensity area indicated by appearance of lighter colored deeper shell layers, *3* high intensity area indicated by pitting within lighter areas

intact pedipalps and/or genital operculum to confirm sex, and one molt had indefinite characteristics but was considered to be female due to the presence of mating scars alone. Molts with mating scars averaged 233±3 mm prosomal width (176–257 mm) and 137±2 mm interocular width (107–158 mm) (Fig. 20.4), falling in the range of adult sizes reported for females among some previous studies in Delaware Bay, USA (Table 20.1).

All exuviae had lost the lacquered appearance of a newly molted animal and bore the characteristic signs of amplexus, including erosion on the axial ridge and depressions on the trailing projections of the opisthosoma (e.g.; Figs. 20.1 and 20.2). Based on overall shell wear, the majority of molts were classified as 'young' animals with four apparently 'middle aged' (Fig. 20.4, bottom). Scarring patterns varied among molts, but some regions on the opisthosoma showed greater area or intensity of mating scars than others (Table 20.2). Region I (opisthosomal ridge) had significantly greater mean area of scarring than other regions (ANOVA: $F_{3,136}=31.52$, $P<0.001$; Scheffé test: $P<0.001$ for all comparisons to region I).

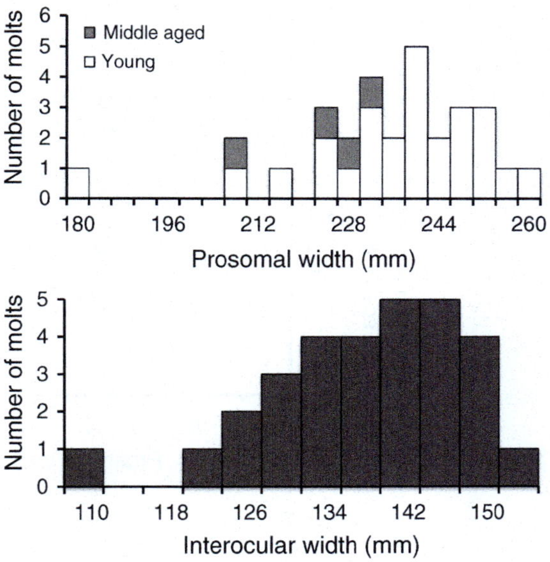

Fig. 20.4 The number of horseshoe crab molts analyzed for this study in terms of prosomal width, including estimated age based on carapace wear (*top*) and interocular distance (*bottom*)

Table 20.2 Mean area and intensity of mate scarring on different regions of the opisthosoma (I–IV, defined in Fig. 20.2) among all sampled horseshoe crab molts

Region	Mean area (cm^2)	Mean intensity
I	11.06±1.64	1.26±0.09
II	2.15±0.36	1.18±0.12
III	2.00±0.25	1.97±0.10
IV	0.63±0.11	1.14±0.11

Mean intensity is based on a rank of 1–3, from lowest to highest intensity (defined in Fig. 20.3)

Region III (dorsal opisthosomal projections) showed significantly higher intensity or depth of scarring compared to other regions ($F_{1,112}=15.60$, $P<0.001$; Scheffé test: $P<0.001$ for all comparisons to region III).

Neither total area nor mean intensity of mating scars was related to horseshoe crab size (Fig. 20.5). The mean area of scarring ranged from 1 to 40 cm^2, with an average value of 16±2 cm^2 per molt. Intensity of scarring ranged from 0.5 to 2.0, with a mean of 1.4±0.1. These two metrics of scarring were weakly but significantly correlated (Pearson: $r=0.43\pm0.01$, $P=0.01$). When area and intensity of scarring were analyzed in tandem as an 'index of scarring' (intensity of scarring per unit area), higher values were found among molts at the mid to large size range (230–256 mm prosomal width) analyzed for this study (Fig. 20.6).

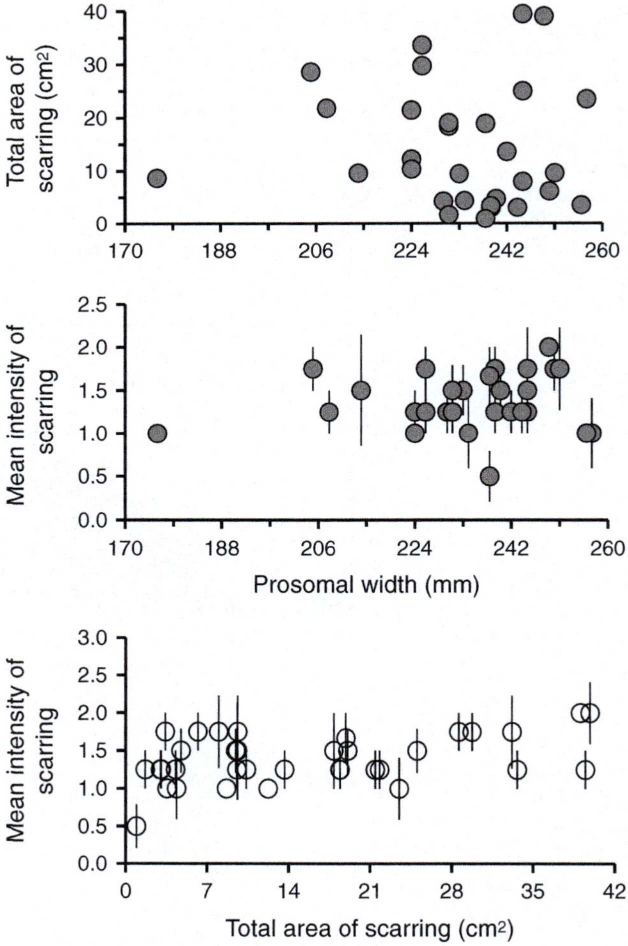

Fig. 20.5 Total area of mating scars from regions I to IV combined (*top*) and mean intensity of mating scars averaged among regions (*middle*) for each individual compared to prosomal width, and the correlation between both metrics of scarring (*bottom*)

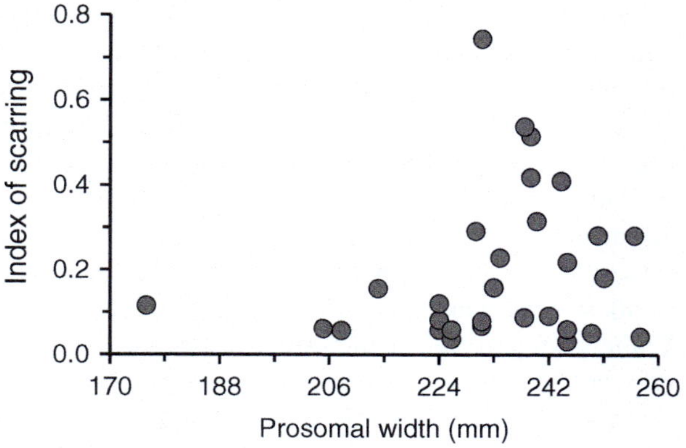

Fig. 20.6 Index of scarring intensity (mean intensity per area of mating scars) relative to prosomal width

20.4 Discussion

Large sized female molts, within the range of prosomal widths common to some spawning females, had mating scars consistent with amplexus and mating activity. The prosomal widths of molts bearing mating scars overlapped with all but the largest sized females previously reported on Delaware Bay spawning beaches, but were larger than sizes typically reported as subadult (<200 mm; Botton and Loveland 1992; Smith et al. 2009). Molts without mating scars have also been found in the size range of molts we examined, confirming this size range for horseshoe crabs prior to reaching maturity (Gauvry pers. obs.). This finding is consistent with the idea that these female molts may represent a range of sizes of sexually mature females who molted at least once after at least one season of amplexus and possibly reproduction. This finding is also consistent with the idea suggested by Carmichael et al. (2003) that prosomal growth increments from the mean juvenile to mean adult sizes among females in Pleasant Bay, Cape Cod, USA were unusually large, in the absence of some molting among horseshoe crabs classified as adults. Using the cohort data from Pleasant Bay horseshoe crabs as an example, if just one additional molt were to occur among females after maturity, the estimated growth increment from mean juvenile size (~150 mm) to the first adult size class that did not overlap with juvenile sizes (~183 mm) would be 18 %. The subsequent molt (potentially a terminal molt) into an adult size would then require a more reasonable increase of 8–56 % (compared to 44–92 % without an additional molt), which is more consistent with demonstrated physiological capabilities of other species that experience a terminal molt (Dawe et al. 1991; Tomaschko et al. 1997; Carmichael et al. 2003). Smith et al. (2009) also suggested that while cohorts within adult size classes could be accounted for by variation in interannual recruitment, adult females may undergo at least one additional molt to achieve larger size. That study, however, determined that this additional molt likely occurred prior to sexual maturity based on apparent maturity (the presence of eggs in females and claspers in males) at later instars in females compared to males. Our data suggest that in at least some adult females this additional molt may occur after sexual maturity.

It is also possible that molts bearing mating scars simply represent subadult females incidentally amplexed by males (Smith et al. 2009). During spawning season males are known to amplex objects or other species that are the appropriate size and shape of an adult female (Barlow et al. 2001; Smith et al. 2009; Carmichael pers. obs.), and this specious amplexus could include subadult females. If molts with mating scars represent subadult females that were not yet sexually mature but accidentally amplexed by males, we would expect relatively low intensity of scarring over a relatively smaller area among molts that overlap in size only with the smallest sized adults and in the range identified as subadults (typically <200 mm). While we found crabs with lower area and intensity of scarring among all sizes, the presence of deeper scarring patterns per unit area and an overall older age characterization among mid to larger sized molts (using our quantitative metrics and the qualitative methods of Shuster 1955) suggests longer-term amplexus among these crabs. There is little evidence of size-related selection for amplexus among males that could otherwise account for this greater level of scarring, particularly among mid to larger sized molts (Cohen and Brockmann 1983; Loveland and Botton 1992; Suggs et al. 2002; Brockmann et al. 2015). Hence, it seems possible that some of these larger sized females bearing mating scars molted after reaching sexual maturity. If true, the terminal molt in these females would occur some time after first reproduction.

Another possibility is that subadult horseshoe crabs do not molt annually and could be incidentally amplexed during multiple years. Rudloe (1978) suggested adult horseshoe crabs molt one time each year or less and Carmichael et al. (2003) assumed annual molt frequency for subadult horseshoe crabs based on molt frequency patterns described among crabs reared in the lab (Sekiguchi 1988; Shuster and Sekiguchi 2003). Actual molt frequency among wild subadult horseshoe crabs is unknown. Given the difference in prosomal width distributions reported for crabs reared in different locations and between wild and lab reared stocks (Shuster 1982; Sekiguchi 1988; Botton and Loveland 2003; Carmichael et al. 2003), molt frequency with age or size likely varies regionally based on environmental attributes, food supply, or endogenous factors that are not yet described and could reasonably occur less often than annually for subadult crabs (Shuster and Sekiguchi 2003). If females remain subadults in the last instar stage before sexual maturity and are incidentally amplexed for multiple years, they could acquire deeper mating scars without reproduction. This explanation, however, does not account for the greater index of scarring among molts larger than 220 mm or why multi-year amplexus would only occur in some subadults and not others (large molts are found without mating scars). Furthermore, prolonged amplexus has been shown to interfere with feeding and movements, potentially affecting metabolic rates, growth, and survival in some species of invertebrates and amphibians (Robinson and Doyle 1985; Roff 1992; Bowcock et al. 2009); whether or not this is the case with *Limulus* is unknown, but nutritional stress has been documented at least among males engaged in longer-term amplexus (Brockmann et al. 2015). Long-term amplexus without reproduction, therefore, can be a costly choice and does not fully explain the data found in this study.

If female horseshoe crabs molt after sexual maturity, it is not clear what proportion of the population they may represent. Sexually mature females that molt would likely represent a minority of the total number of horseshoe crabs that molt in any given year because most molting crabs would be juveniles. They also likely represent a minority of females in the population since most adult females would presumably have already reached a terminal molt. In any given year, many animals that are subadult (males and females) and any sexually mature adult females prior to terminal molt may be molting, resulting in an adult population of mixed apparent age (as found by Carmichael et al. 2003; Smith et al. 2009, and in this study). Through natural attrition with age and growth, we also would expect older, sexually mature females to represent a shrinking portion of the population, and those that may be molting would be a smaller number still. Smith et al. (2009) estimated that molting among adult horseshoe crabs could occur in less than 1 % of the population and would represent a biologically insignificant number. Accordingly, we found relatively few molts with mating scars relative to search effort during this study (35 in 7 years). We additionally searched Gulf of Mexico beaches on Dauphin Island and Petit Bois Island, USA during 2013 and found no molts with mating scars, nor were any such molts noted (although not targeted) in previous survey years at these locations (2008, 2009, 2011).

Even if a relatively small number of horseshoe crabs molt after reaching maturity, it is possible that this number could be biologically meaningful. First, more horseshoe crabs may molt after maturity than known. There is some evidence that older crabs molt in deeper waters or in winter (Rudloe 1978; Shuster and Sekiguchi 2003), and larger and more degraded molts may not persist well, reducing their likelihood to wash ashore and be detected. Second, an additional molt after at least one season (or more) of sexual maturity could be advantageous. Smith et al. (2009) summarized many ecological implications for an additional molt prior to sexual maturity. These same considerations apply as much or more so to a molt after sexual maturity. For example, an additional molt would help females attain their characteristic sexually dimorphic size, which allows carrying more eggs (Shuster and Botton 1985; Leschen et al. 2006). An additional molt would also afford opportunity to repair and regenerate shell structure, which may be damaged by fishing gear or during burrowing and mating, and depending on environmental conditions (Shuster and Sekiguchi 2003). Third, given estimated population sizes of ~500,000 adult horseshoe crabs in Pleasant Bay, Cape Cod and 7–11 million in Delaware Bay, if only 1 % of the population were molting after sexual maturity, this would equate to roughly 5,000 crabs in Pleasant Bay and 70,000–110,000 crabs in Delaware Bay. If these individuals were also increasing their fecundity and potential for survival by increasing size and building unblemished shell, they could also enhance long-term reproductive success and their contribution to subsequent generations. Fourth, it is possible that there is some variation in the synchrony among physiological factors that regulate egg maturation and molting, so that some females have precocious egg development or late molting. Such an occurrence has not been described, but is consistent with the genetic physiological diversity and evolutionary success of this ecological generalist (Sekiguchi and Shuster 2009). For example, females have exuded eggs when out of the typical spawning environment and season such as when found in subtidal or shelf waters and during winter months (Shuster and Carmichael pers. obs.). Finally, since molting (particularly at large sizes) is also associated with increased mortality, it could be advantageous to molt at smaller adult sizes but not to continue molting at largest sizes. This molting strategy is consistent with our observation that no exuviae with mating scars were found to overlap in size with the largest sized living adult females (>260 mm). A similar strategy has been described for other molting species in which the occurrence of a terminal molt has been debated (Dawe et al. 1991).

These data demonstrate a quantitative approach to assess carapace scarring patterns that may be combined with qualitative estimates of age to more accurately describe shell condition and distinguish mating activity from general shell wear that may be environmentally determined. The lack of a relationship between individual scarring metrics and prosomal width is consistent with a single population of molting females within adult size ranges. Because area of scarring likely depends on individual shell shapes of both the female and male in amplexus, greater area of wear could indicate amplexus with multiple males of different shapes and sizes as opposed to a single male for a longer period of time. It is not surprising, therefore, that we found relatively high variation in area of scarring among individuals. In this study, the intensity of scarring per unit area, which may reflect longer-term attachment, was more useful to reveal size-related scarring patterns. These findings suggest that both area and intensity are important to interpret different aspects of mating scar formation, but scarring intensity per unit area may be a better index of scarring relative to the duration of amplexus and potential mating among size classes. Studies to monitor mating scar formation through time on marked wild females or in the laboratory will be useful to quantitatively relate scar formation patterns to specific type and duration of activity, but we cannot further evaluate these possibilities with the available data.

20.5 Conclusion

We provide the first published description and analysis of mating scars on female horseshoe crab molts typical of prolonged amplexus and sufficient to have occurred during spawning, suggesting that at least some females may molt after reaching sexual maturity. While we cannot demonstrate without question that the molts examined came from females that had engaged in reproductive activity beyond extended amplexus, prolonged amplexus without reproduction is generally not a benefit to either sex. There is growing evidence for the occurrence of at least one additional molt among female horseshoe crabs into adult size classes before reaching a terminal molt. We hope this research will stimulate additional targeted studies to help determine whether molting occurs strictly prior to reproduction or if some crabs (by physiological design or variation) may enhance their longevity and reproductive success by molting after first reproduction. Specific studies could include quantifying mating scars on large sized molts among geographic regions and through time as well as research to characterize mating scar formation on live spawning crabs in the field or lab. Through this study we provide a quantitative approach that can serve as the basis for such future assessments.

Acknowledgments We thank Mary Van House, Environmental Education Program Coordinator, Childrens Beach House, Lewes, Delaware, who was one of the first to discover a female molt with mating scars and bring it to the attention of the authors. We also recognize the crew of the F/V *Tony and Ann* out of Ocean City, Maryland, Capt. Jeffrey "Jeff" Eustler, skipper who helped look for and collect specimens; Benjie Lynn Swan, Limuli Laboratories, who provided a base for field operations and specimens for study, Maury Estes, Noel Wingers, and Morgan Waites who searched for large sized molts in northern Gulf of Mexico waters. Funding was provided in part by the Dauphin Island Sea Lab and the Ecological Research and Development Group (ERDG).

References

Barlow RB, Hitt JM, Dodge FA (2001) Limulus vision in the marine environment. Biol Bull 200:169–176

Botton ML, Loveland RE (1992) Body size, morphological constraints, and mated pair formations in four populations of horseshoe crabs (*Limulus polyphemus*) along a geographic cline. Mar Biol 112:409–415

Botton ML, Loveland RE (2003) Abundance and dispersal potential of horseshoe crab (*Limulus polyphemus*) larvae in the Delaware estuary. Estuaries 26:1472–1479

Bowcock H, Brown GP, Shine R (2009) Beastly bondage: the costs of amplexus in cane toads (*Bufo marinus*). Copeia 2009:29–36

Brockmann HJ (2003) Male competition and satellite behavior. In: Shuster CN Jr, Barlow RB, Brockmann HJ (eds) The American horseshoe crab. Harvard University Press, Cambridge, pp 50–82

Brockmann HJ, Johnson SL, Smith MD et al (2015) Mating tactics of the American horseshoe crab (*Limulus polyphemus*). In: Carmichael RH, Botton M, Shin PKS, Cheung SG (eds) Changing global perspectives on horseshoe crab biology, conservation and management. Springer, Cham, pp 321–351

Carmichael RH, Rutecki D, Valiela I (2003) Abundance and population structure of the Atlantic horseshoe crab *Limulus polyphemus* in Pleasant Bay, Cape Cod. Mar Ecol Prog Ser 246:225–239

Cohen JA, Brockmann JH (1983) Breeding activity and mate selection in the horseshoe crab, *Limulus polyphemus*. Bull Mar Sci 33:274–281

Dawe EG, Taylor DM, Hoenig JM et al (1991) A critical look at the idea of terminal molt in male snow crab (*Chionoecetes opilio*). Can J Fish Aquat Sci 48:2266–2275

Hata D, Berkson J (2003) Abundance of horseshoe crabs (*Limulus polyphemus*) in the Delaware Bay area. Fish Bull-NOAA 101:933–938

Leschen AS, Grady SP, Valiela I (2006) Fecundity and spawning of the Atlantic horseshoe crab, *Limulus polyphemus*, in Pleasant Bay, Cape Cod, Massachusetts, USA. Mar Ecol 27:54–65

Loveland RE, Botton ML (1992) Size dimorphism and the mating system in horseshoe crabs, *Limulus polyphemus* L. Anim Behav 44:907–916

Munson JP (1898) The ovarian egg of *Limulus*: a contribution to the problem of the centrosome and yolk-nucleus. J Morphol 15:113–220

Robinson BW, Doyle RW (1985) Trade-off between male reproduction (amplexus) and growth in the amphipod *Gammarus lawrencianus*. Biol Bull 168:482–488

Roff DA (ed) (1992) Evolution of life histories: theory and analysis. Springer, New York
Rudloe A (1978) Some ecologically significant aspects of the behavior of the horseshoe crab, *Limulus polyphemus*. Dissertation, Florida State University, Tallahassee
Sekiguchi K (1988) Biology of horseshoe crabs. Science House, Tokyo
Sekiguchi K, Shuster CN Jr (2009) Limits on the global distribution of horseshoe crabs (Limulacea): lessons learned from two lifetimes of observations: Asia and America. In: Tanacredi JT, Botton ML, Smith DR (eds) Biology and conservation of horseshoe crabs. Springer, New York, pp 5–24
Shuster CN Jr (1955) Morphometric and serological relationships among the Limulidae, with particular reference to *Limulus polyphemus*. Dissertation, New York University, New York
Shuster CN Jr (1979) Distribution of the American horseshoe "crab" *Limulus* (L.). In: Cohen E (ed) Biomedical applications of the horseshoe crab (Limulidae). Alan R. Liss, Inc., New York, pp 3–26
Shuster CN Jr (1982) A pictorial review of the natural history and ecology of the horseshoe crab *Limulus polyphemus*, with reference to other Limulidae. In: Bonaventura J, Bonaventura C, Tesh S (eds) Physiology and biology of horseshoe crabs, studies on normal and environmentally stressed animals. Alan R. Liss, Inc., New York, pp 1–52
Shuster CN Jr, Botton ML (1985) A contribution to the population biology of horseshoe crabs, *Limulus polyphemus* (L.), in Delaware Bay. Estuaries 8:363–372
Shuster CN Jr, Sekiguchi K (2003) Growing up takes about ten years and eighteen stages. In: Shuster CN Jr, Barlow RB, Brockmann HJ (eds) The American horseshoe crab. Harvard University Press, Cambridge, pp 113–115
Smith DR, Mandt MT, MacDonald PDM (2009) Proximate causes of sexual size dimorphism in horseshoe crabs (*Limulus polyphemus*) of the Delaware Bay. J Shellfish Res 28:405–417
Suggs D, Carmichael RH, Grady SP et al (2002) Effects of individual size on pairing in horseshoe crabs. Biol Bull 203:225–227
Tomaschko KH, Wilhelm E, Bückmann D (1997) Growth and reproduction of *Pycnogonum litorale* (Pycnogonida) under laboratory conditions. Mar Biol 129:595–600

Part IV
Conservation and Management

Chapter 21
Emerging Issues in Horseshoe Crab Conservation: A Perspective from the IUCN Species Specialist Group

Mark L. Botton, Ruth H. Carmichael, Paul K.S. Shin, and Siu Gin Cheung

Abstract Horseshoe crabs are valued in the production of the *Limulus* and *Tachypleus* amoebocyte lysate test for bacterial contamination, as bait in several fisheries in the U.S. and Mexico, and for human consumption in Southeast Asia. Increasing commercial exploitation of horseshoe crabs, coupled with losses of essential spawning and nursery habitats, particularly in Southeast Asia, has raised concerns about the efficacy of current fisheries regulations and coastal management practices. By the mid-2000s, a critical mass of scientific research coupled with widespread concerns about population declines, provided momentum for a series of international meetings, leading to the 2012 formation of the Horseshoe Crab Specialist Group within the IUCN (International Union for the Conservation of Nature). The Specialist Group is an advocate for science-based conservation of horseshoe crabs and their habitat, with a priority to update the Red List status for the three Asian horseshoe crab species, which are currently regarded as 'data deficient' based largely on outdated information. A troubling and unexpected new issue in horseshoe crab conservation is the importing of Asian horseshoe crabs to supply the U.S. bait market. This practice is an economic incentive to overharvest Asian stocks that may also introduce diseases and invasive organisms into U.S. waters. Illegal fishing activity in the U.S. and Southeast Asia, driven by the high market value of the crabs, also appears to be increasing. The Specialist Group has taken important actions to support internationally collaborative research and education activities

M.L. Botton (✉)
Department of Natural Sciences, Fordham University,
113 West 60th Street, New York, NY 10023, USA
e-mail: botton@fordham.edu

R.H. Carmichael
Dauphin Island Sea Lab, University of South Alabama, Dauphin Island, AL 36528, USA

Department of Marine Sciences, University of South Alabama, Mobile, AL 36688, USA

P.K.S. Shin • S.G. Cheung
Department of Biology and Chemistry, City University of Hong Kong,
Kowloon, Hong Kong

State Key Laboratory in Marine Pollution, City University of Hong Kong,
Kowloon, Hong Kong

R.H. Carmichael et al. (eds.), *Changing Global Perspectives on Horseshoe Crab Biology, Conservation and Management*, DOI 10.1007/978-3-319-19542-1_21

that will increase awareness of and curtail these emerging threats to horseshoe crab populations before local population extinctions are inexorable.

Keywords IUCN • Red List • Conservation • *Limulus* • *Tachypleus* • *Carcinoscorpius* • Invasive species

21.1 Introduction

The past four decades have seen extraordinary growth in the scientific study and public awareness of the importance of the American horseshoe crab, *Limulus polyphemus*. One major impetus was the discovery and subsequent commercialization of the *Limulus* amoebocyte lysate (LAL) test for bacterial endotoxin (Levin et al. 2003), which turned these animals into an important economic commodity. In the late 1970s and early 1980s, a different group of scientists began to recognize the essential importance of horseshoe crab eggs as a food for migratory shorebirds during their Spring stopover on Delaware Bay, USA, which is now recognized as a key link in their lengthy journey from wintering areas in South America to their breeding sites in the Canadian Arctic (Botton and Harrington 2003; Mizrahi and Peters 2009). The proliferation of popular articles (e.g., Myers 1986) and documentary films about the horseshoe crab-shorebird relationship stimulated ecotourism centered on this spectacle. Yet, some of this wave of enthusiasm was quickly dampened by the exponential increase in an emerging bait industry that used horseshoe crabs in traps for eels and whelks. These competing interests raised questions about whether American horseshoe crab populations could support uses for LAL, ecotourism, and bait (Botton and Ropes 1987; Berkson and Shuster 1999; Shuster et al. 2003), and whether a reduced horseshoe crab population would affect the estuarine food web (Botton and Shuster 2003).

Interest in the biology and conservation of the three Southeast Asian species of horseshoe crabs, *Tachypleus tridentatus*, *T. gigas*, and *Carcinoscorpius rotundicauda*, emerged along a parallel trajectory. The English translation of the late Dr. Koichi Sekiguchi's treatise on horseshoe crab research (Sekiguchi 1988), along with several visits to Delaware Bay in the late 1980s by Sekiguchi and some of his former students, including Tomio Itow, Hiroaki Sugita, Fumio Shishikura, allowed a comparison of perspectives on the plight of American and Southeast Asian horseshoe crabs. Horseshoe crabs are not commercially fished in Japan, but extensive coastline development has decimated most of their former spawning sites (Botton 2001), and water pollution in certain locations negatively impacts the survival of embryos (Botton and Itow 2009). Research trips by veteran horseshoe crab researchers Drs. Carl Shuster and Mark Botton to Japan in 1994 (Botton et al. 1996) and by Dr. Tomio Itow to the US in 1997–1998 (Itow et al. 1998a, b) were critical in establishing a foundation for development and growth of international collaboration that continues today.

Apart from Japan, most research on Asian horseshoe crabs has been conducted in isolation in the region. For example, there has been much information on the biology and ecology of *T. tridentatus* in mainland China as summarized in the life-time work of Hong (2011) and in other literature (e.g., Liao and Li 2001; Liao et al. 2001). Unfortunately, such reports are mainly in Chinese, which make them not easily accessible to a wider readership internationally. In Hong Kong and Taiwan, much work has been conducted since the 2000s to assess population status (Chiu 2003; Shin et al. 2009; Morton and Lee 2011) and examine conservation measures (Chen et al. 2004, 2009). All such findings have pointed to the significant decline in populations of Asian horseshoe crabs due to overharvesting for biomedical and human consumption in mainland China and habitat loss and disturbance in Hong Kong and Taiwan. Concerns about habitat loss and degradation are also apparent in Singapore for their existing population of *C. rotundicauda* (Cartwright-Taylor et al. 2011) and in India for *T. gigas* (Mishra 2009). Regional collaborations on the study and conservation of Asian horseshoe crabs are in progress, with the first establishment of a tripartite alliance for the conservation of horseshoe crabs along the coast of South China Sea among mainland China, Hong Kong, and Taiwan in June 2014.

21.2 Establishing International Cooperation

By the mid-2000s, a critical mass of interest in horseshoe crab research, coupled with growing concerns about their possible overexploitation, particularly in Asian countries, stimulated discussions about the concept of an International Symposium to discuss common goals for conservation. Under the leadership of Dr. John Tanacredi, then at Dowling College in Oakdale, New York, the International Symposium on the Science and Conservation of Horseshoe Crabs was held in June, 2007. More than 100 participants from ten countries attended presentations on the conference theme, "*Science helping to advance conservation; Conservation helping to prioritize science*" (Tanacredi et al. 2009). With a consensus emerging that there were significant data gaps for the three species of horseshoe crabs in Southeast Asia, it was decided that the second International Meeting should focus on their plight.

In 2011, the International Workshop on the Science and Conservation of Asian Horseshoe Crabs was held at the Hong Kong Wetland Park in Hong Kong, China, under the leadership of Drs. Paul Shin and Siu Gin Cheung of the City University of Hong Kong. Although data were available from relatively few places (e.g. Japan, Hong Kong, Taiwan), participants attended from eight countries or areas including new participation from researchers in Malaysia and Singapore. The resulting presentations and discussions at that Hong Kong meeting confirmed that Southeast Asian horseshoe crab populations were broadly threatened by overfishing and habitat loss. Harvest regulations in Southeast Asia were often absent or poorly enforced. There were lively conversations about how the group could best exert influence to raise government awareness of the importance of horseshoe crabs. It was felt that the best possible tool would be through the authoritative and internationally recognized IUCN (International Union for the Conservation of Nature) Red List of Threatened Species.

21.3 A Primer on the IUCN and the Species Survival Commission

Founded in 1948 and currently based in Gland, Switzerland, the IUCN is the world's largest global professional environmental organization. It seeks to provide "a neutral forum for governments, NGOs, scientists, business and local communities to find practical solutions to conservation and development challenges" (http://www.iucn.org/about/). The IUCN is organized into six commissions: Commission on Education and Communication, Commission on Environmental, Economic and Social Policy, World Commission on Environmental Law, Commission on Ecosystem Management, World Commission on Protected Areas, and the Species Survival Commission (SSC). There are more than 130 Species Specialist Groups within the SSC, consisting of experts on diverse groups of flora and fauna from around the world (http://www.iucn.org/about/work/programmes/species/who_we_are/about_the_species_survival_commission_/).

One of the principal roles of the SSC is the collation of data on species' status, population trends and habitat needs into the IUCN Red List of Threatened Species. The Red List "…is the world's most comprehensive information source on the global conservation status of animal, fungi and plant species and their links to livelihoods. Far more than a list of species and their status, the IUCN Red List is a powerful tool to inform and catalyze action for biodiversity conservation and policy change, critical to protecting the natural resources we need to survive…" (http://www.iucn.org/about/work/programmes/species/our_work/the_iucn_red_list/).

Once data are assembled by the leadership of a Species Specialist Group, a recommendation can be made to assign the species into categories of relative conservation concern (Fig. 21.1). If a determination is made that the data are inadequate to properly assess the species' status, the species is classified as 'data deficient.' If the data are adequate, then there is a range of possibilities from 'least concern' to 'extinct' or 'extinct in the wild', meaning that the only living individuals are found in captivity. Within this range, there are three categories of threatened species, referred to as 'critically endangered', 'endangered', or 'vulnerable.' Below that level of threat are species that are 'near threatened.' The IUCN provides comprehensive guidelines for these criteria to be used to determine the appropriate conservation status of a species (IUCN 2001; IUCN Standards and Petitions Subcommittee 2014).

According to IUCN criteria, the three horseshoe crab species in Southeast Asia are currently listed as 'data deficient,' while the American horseshoe crab is presently a species of 'least concern' (Fig. 21.1). At the 2011 Hong Kong workshop, participants agreed that an important step in horseshoe crab conservation would be the establishment of a Specialist Group with the goal to update the IUCN Red List status of Asian horseshoe crabs. It was especially important that the data deficient status for Southeast Asian horseshoe crabs be reconciled with new information on their declining numbers. Although placement of a species on the IUCN Red List is not a guarantee of protection, it does inform policy makers of the situation and may

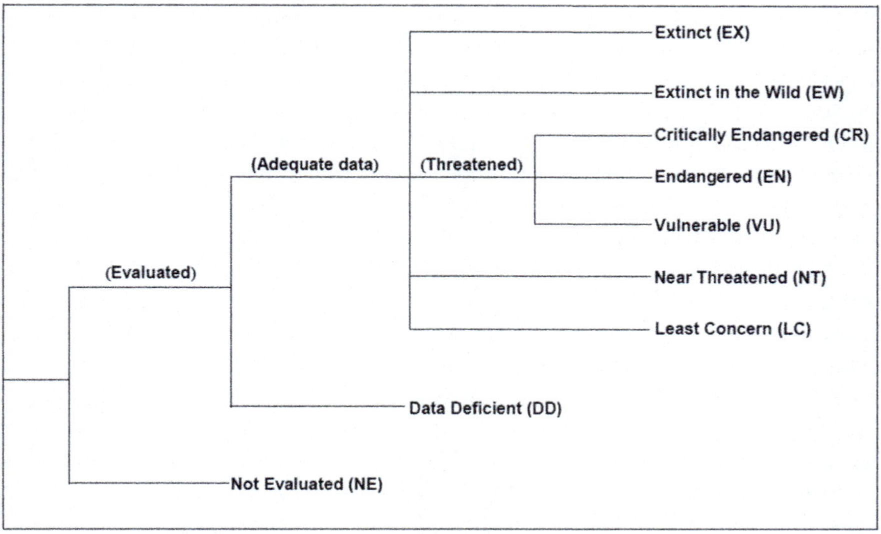

Fig. 21.1 The IUCN Red List Criteria (From IUCN 2001). The three species of horseshoe crabs in Southeast Asia (*Tachypleus tridentatus*, *T. gigas, and Carcinoscorpius rotundicauda*) are currently regarded as 'data deficient' and the American horseshoe crab (*Limulus polyphemus*) is a species of 'least concern'

give greater credibility to conservation efforts at the national or regional level. Thus, a core group was formed, and they prepared an application to the SSC to form the Horseshoe Crab Specialist Group. The proposal was submitted to the IUCN in May 2012 and formally approved in June 2012. Based on the geographic distribution of the four species of horseshoe crabs, a joint Co-Chair structure was formed, with Dr. Mark Botton (USA) and Dr. Paul Shin (Hong Kong) serving as the North American and Southeast Asian Co-Chairs, respectively.

21.4 Priorities of the Horseshoe Crab Species Specialist Group

The Horseshoe Crab Specialist Group is an advocate for science-based conservation of horseshoe crabs and their habitat. To advance this conservation-focused mission, four working groups are established within the Horseshoe Crab Specialist Group, drawing upon the expertise of scientists from throughout North America and Southeast Asia. These groups include: (1) The Asian Red List review group, led by Kevin Laurie (Hong Kong), which is responsible for reviewing population status and trends for the three species of horseshoe crabs in Southeast Asia. The initial priority of this group is to review the status of *Tachypleus tridentatus* (the Chinese or Japanese horseshoe crab) based largely on its decline in abundance and the very pessimistic outlook for conservation of this species reported at the Hong Kong conference. (2) The *Limulus* Red List review group, led by David Smith (US), which is charged with reviewing population status and trends of *L. polyphemus* (the American horseshoe crab) in North America (including the US and Mexico). (3) The Trade, Industry and Use working group, headed by Glenn Gauvry (US), which works to assess threats to horseshoe crabs throughout their range, and to help promote industry best practices (e.g. to minimize mortality in the biomedical industry and to promote the use of bait bags and other methods that reduce the number of horseshoe crabs collected as bait). (4) The Education and Public Awareness group, led by Gary Kreamer (US), which focuses on enhancing education and public awareness about horseshoe crabs. Members of the Horseshoe Crab Specialist Group have worked with the Green Eggs and Sand program, which developed a series of education modules for primary education that focus on the ecological and biomedical importance of horseshoe crabs (O'Connell et al. 2009). The Horseshoe Crab Specialist Group promotes education and outreach at all levels, including the scientific study of horseshoe crabs and has taken a leadership position in organizing scientific meetings. Following the Hong Kong meeting (2011), the group organized a Special Session entitled "Comparative Approaches to Horseshoe Crab Ecology and Conservation in North America and Southeast Asia," at the Coastal and Estuarine Research Federation (CERF) meeting in San Diego, CA (USA) in November 2013, and the 3rd International Workshop on the Science and Conservation of Horseshoe Crabs in Sasebo, Japan in June 2015. Through these working groups, the Horseshoe Crab Specialist Group has facilitated communication among interested scientists and conservationists, empowering a strong and unified voice to address new problems that may threaten horseshoe crabs. The group's work was also facilitated by the success of a unanimous support of a resolution introduced by the Nature Society Singapore at the 2012 IUCN World Conservation Congress at Jeju, Korea for the conservation of Asian horseshoe crabs and protection of their habitats. In the following sections, we describe two newly emerging areas of concern that the group has responded to, with some degree of success.

21.5 Emerging Threats to Horseshoe Crabs

21.5.1 Importation of Asian Horseshoe Crabs to the U.S.

In late 2011, the New York Department of Environmental Conservation (NYDEC) was notified that a shipment of some 2,000 Asian horseshoe crabs originating in Thailand was being held by US Customs at JFK Airport in New York City, NY (Fig. 21.2). NYDEC sent photographs of the animals to a number of experts, who determined that the animals appeared to be immature *Tachypleus gigas*. Subsequently, information from a seafood dealer in Vietnam suggested that another New York businessman imported 7,400 kg of horseshoe crabs from Southeast Asia into

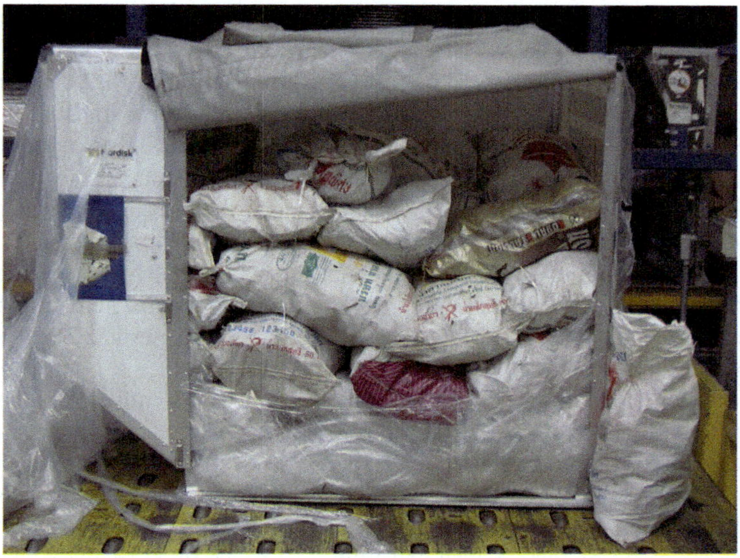

Fig. 21.2 A portion of a shipment of about 2,000 *Tachypleus gigas* from Thailand that was imported into JFK Airport, New York City for use in the bait market (Photo courtesy of Kim McKown, New York Department of Environmental Conservation)

New York State in late 2012. The Specialist Group also became aware that bait dealers from the mid-Atlantic US were reaching out to Asian suppliers to purchase more Asian horseshoe crabs (Hurdle 2013). This action did not break any existing laws; under the Lacey Act (18 USC § 42) the importation of injurious species of mammals, birds, fish (including crustaceans), amphibians, and reptiles is illegal. Because horseshoe crabs are chelicerates and not crustaceans (like shrimp, crabs or lobsters), they fall outside the scope of current Federal regulations. The Horseshoe Crab Specialist Group was greatly concerned that these international commercial trade activities could encourage further depletion of Asian horseshoe crabs.

The Specialist Group was further concerned that import of Asian horseshoe crabs could cause ecological harm in the US. In the eel and whelk fisheries, horseshoe crabs are cut into halves or quarters depending on the size of the crab and then put into a trap on the sea floor. Although the crabs are killed in this process, there is a risk of introducing parasites, pathogens or non-native (invasive) epibiont species into U.S. waters. A recent review (Botton and Shuster 2003) listed 38 species of marine organisms that encrust the shells of *L. polyphemus*. Although the organisms living on Asian horseshoe crabs have not been as well studied, barnacles, bryozoans, mollusks, and other invertebrates commonly live on their shells (Key et al. 1996; Patil and Anil 2000). The life-history strategy of many of these fouling organisms (e.g. high reproductive rate, widely dispersing larvae) means that they would have the potential to become successful invasive species if introduced to US waters. Flatworms (Platyhelminthes: Turbellaria) also infest the book gills of horseshoe crabs and are regarded by leading invertebrate pathologists as true parasites that consume host tissues (Leibovitz and Lewbart 2003). These flatworms lay eggs in tough cocoons that can easily survive even if the host crab is killed. The flatworms infesting the Asian horseshoe crabs are different species than those in North America (Kawakatsu and Sekiguchi 1988; Leibovitz and Lewbart 2003), but the possibility of survival in foreign habitats and cross-infection exists, with unknown consequences. Many other parasites and diseases of horseshoe crabs have been described, including algal and bacterial infections of the eyes and carapace, protozoan infections, and encysted stages of parasitic trematodes (Leibovitz and Lewbart 2003; Nolan and Smith 2009). The potential threats posed by the introduction of non-native parasites and diseases that may be possible due to international horseshoe crab trade, therefore, remain of high concern to the Horseshoe Crab Specialist Group.

In addition to the direct threats to American horseshoe crabs from non-native parasites and diseases, the potential for ecological damage extends further. Populations of shorebirds using the US Atlantic coast flyway are highly dependent on the horseshoe crab egg resource in Delaware Bay (Botton and Harrington 2003; Mizrahi and Peters 2009). Most notably, the Red Knot (*Calidris canutus rufa*) was recently approved by the US Fish and Wildlife Service for listing as a threatened species under the Endangered Species Act (US Fish and Wildlife Service 2014). In 1998, the Atlantic States Marine Fisheries Commission (ASMFC) in the US implemented adaptive management practices with the goal to maintain horseshoe crab populations at a level that provides sufficient eggs for migratory shorebirds (ASMFC 1998; McGowan et al. 2011). Any activity that could deplete American horseshoe crabs (such as by introducing disease or parasites), therefore, may threaten the trophic stability of Red Knots and other migratory shorebirds and interfere with US regulations designed to conserve and aid recovery of shorebird populations.

An additional point raised by the Horseshoe Crab Specialist Group was the potential for Asian horseshoe crabs to pose a threat to public health in the US. Although horseshoe crabs are commonly eaten in many countries in Southeast Asia, numerous instances of illness and even death have been caused by eating mangrove horseshoe crabs, *Carcinoscorpius rotundicauda* (reviewed by Botton and Itow 2009). The cause is contamination by a bacterial toxin called tetrodotoxin (TTX), which is a powerful neurotoxin. TTX is not destroyed by freezing or cooking and can prove fatal if consumed. *C. rotundicauda* overlaps with *T. gigas* in portions of its range (Sekiguchi 1988), and the two species are difficult to distinguish, especially as subadults. A shipment of horseshoe crabs from Thailand or Vietnam could easily contain both species. Because TTX bioaccumulates through marine food chains (Williams 2010), there is a risk that eels or whelks could become toxic if a contaminated crab were used as bait. Hence, consumers in the US (and anywhere the eels or whelks are shipped) may be at risk without directly consuming Asian horseshoe crabs.

In light of these concerns, early in 2013, the Horseshoe Crab Specialist Group consulted with the IUCN SSC and its Invasive Species Specialist Group regarding the proper evaluation of importing Asian horseshoe crabs to the US. Although IUCN guidelines state that intentional introductions need to be properly evaluated in advance, with full regard to potential impacts on biodiversity (IUCN 2000), the importing of Asian horseshoe crabs into the US took place without even a cursory risk analysis. In particular, the IUCN's guidelines for preparing risk assessments for the introduction of non-native animals and diseases both emphasize the need to identify and prioritize hazards early in the process. The Horseshoe Crab Specialist Group sent formal letters expressing concerns about Asian horseshoe crab imports to ASMFC and each of its member states and to various federal agencies including the US Fish and Wildlife Service and the National Marine Fisheries Service. Although the US Fish and Wildlife Service agreed with these concerns, the Lacey Act does not empower them to classify Asian horseshoe crabs as injurious species because they are not crustaceans. Until such time as the Lacey Act is amended, action at the federal level in the US to prohibit Asian horseshoe crab imports is not possible. At the regional level, however, action was prompt. The ASMFC reacted to ban the import and use of Asian horseshoe crabs as bait (ASMFC 2013), and as of this writing, their regional-scale resolution appears to have been effective.

21.5.2 Illegal Fishing for Horseshoe Crabs

The American horseshoe crab (*L. polyphemus*) fishery is carefully regulated in the US by an adaptive resource management program implemented by ASMFC (ASMFC 1998 and subsequent addenda; McGowan et al. 2011). All horseshoe crab fishermen must be licensed, and each member state is allocated a maximum harvest. States may choose to impose stricter regulations; for example, New Jersey presently has a total moratorium on the bait fishery for horseshoe crabs. According to various industry sources, horseshoe crabs are currently worth up to $5 USD apiece for bait. Since the supply of this valuable resource is limited, and the beaches where horseshoe crabs congregate are often rather desolate (especially at night), there is a strong incentive for illegal fishing activity (poaching).

Two widely publicized incidents in 2013 illustrate the growing problem of poaching and the difficulty of enforcing regulations. In Spring 2013, various New York newspapers including the New York Times reported that a fisherman was apprehended with illegally caught horseshoe crabs from Jamaica Bay, New York (Goodman and Nir 2013). What made the story even more compelling was that Jamaica Bay is located in a highly urbanized area, but the poaching was revealed only by a chance sighting by New York City Police using night vision equipment when they were on a routine helicopter patrol. Also in 2013, three fishermen were arrested for poaching over 1,000 horseshoe crabs from Great South Bay, New York (NYDEC 2013). Although one of the men was properly licensed, he had greatly exceeded his daily allowable catch of 30 horseshoe crabs. There are many more out of the way beaches where poaching of horseshoe crabs (and other animals) likely takes place undetected. Discussions within the horseshoe crab Specialist Group revealed that poachers were rumored to be active in other locations, including Delaware Bay and Long Island Sound, New York. Horseshoe crab poaching is lucrative and requires no specialized fishing gear, and though the penalties are strict, the probability of being apprehended is low. For example, there is a penalty of $100 for each illegally caught horseshoe crab in New York State, but it is unclear whether this is an effective deterrent. To support ongoing enforcement actions, the Horseshoe Crab Specialist Group wrote to the appropriate law enforcement organizations to emphasize the importance of horseshoe crabs and to commend them for their successful intervention in the Jamaica Bay and Great South Bay incidents.

Information on the regulation of horseshoe crab fisheries in Southeast Asia is even more limited, but poaching of horseshoe crabs is known to occur. For example, although horseshoe crabs are protected in a few provinces in China, excessive harvest of *T. tridentatus* for biomedical and human consumption has led to a substantial decline in the existing stocks (Liao and Ip 2000; Hong 2011). Incidences of illegal fishing for horseshoe crabs occasionally are also reported, e.g., a shipment of *T. gigas* was seized at a local airport (Indonesia Press Report 2012) and a truckload of same species was intercepted by district police (Jakarta Post 2013) in Indonesia under the local regulations on Preservation of Flora and Fauna. It thus appears that illegal harvesting of horseshoe crabs in some areas of Asia can be curtailed as regulations for the protection of horseshoe crabs are in place in some countries. However, enhanced law enforcement efforts are necessary, along with a wider education of the responsible officials as well as the public so that they are aware of the conservation importance of these animals.

21.6 Summary and Conclusions

The formation of the Horseshoe Crab Specialist Group in 2012 was a necessary first step to formally evaluate the IUCN Red List status of horseshoe crabs in North America and Southeast Asia. The American horseshoe crab population in Delaware Bay shows indications of increasing due in part to regional management through the ASMFC. The redirection of fishing effort from the mid-Atlantic to New England waters, however, may negatively affect the smaller, more isolated populations in that area (Rutecki et al. 2004; Smith et al. 2009). Thus, although *L. polyphemus* may currently be a species of 'least concern' overall, the status and trends for populations should be carefully monitored, particularly at the local level (Widener and Barlow 1999; Rutecki et al. 2004; Smith et al. 2009). In contrast, Asian horseshoe crabs are currently listed as 'data deficient' which hampers efforts to regulate domestic or export fisheries or to protect essential spawning and nursery habitats for these species. Updating the Red List status for the declining Chinese horseshoe crab, *T. tridentatus*, therefore, is a high priority for the Horseshoe Crab Specialist Group.

The threat from importing of Asian horseshoe crabs into US waters served to demonstrate the effectiveness of the Horseshoe Crab Specialist Group, with the backing of the IUCN, to yield positive results for conservation. The Horseshoe Crab Specialist Group brought to the attention of regulators the potential ecological and human health risks associated with introduction of non-native horseshoe crab species to the US, and the use of Asian horseshoe crabs by the US bait industry has since been halted (ASMFC 2013). In the US, federal regulations prohibiting international horseshoe crab trade (Sect. 21.5.1) will require amending the existing regulation that is specific to crustaceans (Lacey Act). In the meantime, illegal fishing for horseshoe crabs in the US and Asia is likely to continue as long as there is a strong demand for and a limited supply of these species.

Since the first international horseshoe crab meeting in New York in 2007, international participation in and awareness of research, conservation, and management of horseshoe crabs has significantly increased. Cultural and language barriers were overcome to enhance communication and enable collaborative interactions and emerging opportunities for data and concept sharing in all areas of science and conservation. With the creation of the Horseshoe Crab Specialist Group and expanded collaborative efforts through the IUCN, we are at a turning point in horseshoe crab science and conservation that will frame the future of research and management actions. By organizing scientific meetings and promoting education and awareness of horseshoe crabs, the Horseshoe Crab Specialist Group is working to gain greater public and decision-maker support for the ongoing conservation of these remarkable and valuable animals.

References

Atlantic States Marine Fisheries Commission (ASMFC) (1998) Interstate fishery management plan for horseshoe crab. Atlantic States Marine Fisheries Commission, Fishery Management report no. 32. Washington, DC

Atlantic States Marine Fisheries Commission (ASMFC) (2013) ASMFC approves resolution to ban the import and use of Asian horseshoe crabs as bait. Press release, 13 Feb 2013. http://www.asmfc.org/uploads/file/pr08HSC_AsianCrabResolution.pdf. Accessed 11 July 2014

Berkson J, Shuster CN Jr (1999) The horseshoe crab: the battle for a true multiple use resource. Fisheries 24:6–10

Botton ML (2001) The conservation of horseshoe crabs: what can we learn from the Japanese experience? In: Tanacredi JT (ed) *Limulus* in the limelight. Kluwer Academic/Plenum Publ., New York, pp 41–51

Botton ML, Harrington BA (2003) Synchronies in migration: shorebirds, horseshoe crabs, and Delaware Bay. In: Shuster CN Jr, Barlow RB, Brockmann HJ (eds) The American horseshoe crab. Harvard University Press, Cambridge, pp 5–32

Botton ML, Itow T (2009) The effects of water quality on horseshoe crab embryos and larvae. In: Tanacredi JT, Botton ML, Smith DR (eds) Biology and conservation of horseshoe crabs. Springer, New York, pp 439–454

Botton ML, Ropes JW (1987) The horseshoe crab, *Limulus polyphemus*, fishery and resource in the United States. Mar Fish Rev 49:57–61

Botton ML, Shuster CN Jr (2003) Horseshoe crabs in a food web: who eats whom? In: Shuster CN Jr, Barlow RB, Brockmann HJ (eds) The American horseshoe crab. Harvard University Press, Cambridge, pp 133–153

Botton ML, Shuster CN Jr, Sekiguchi K et al (1996) Amplexus and mating behavior in the Japanese horseshoe crab, *Tachypleus tridentatus*. Zool Sci 13:151–159

Cartwright-Taylor L, Bing YP, Hsu CC et al (2011) Distribution and abundance of horseshoe crabs *Tachypleus gigas* and *Carcinoscorpius rotundicauda* around the main island of Singapore. Aquat Biol 13:127–136

Chen CP, Yeh HY, Lin PF (2004) Conservation of the horseshoe crab at Kinmen, Taiwan: strategies and practices. Biodivers Conserv 13:1889–1904

Chen CP, Hsieh HL, Chen A et al (2009) The conservation network of horseshoe crab *Tachypleus tridentatus* in Taiwan. In: Tanacredi JT, Botton ML, Smith DR (eds) Biology and conservation of horseshoe crabs. Springer, New York, pp 543–557

Chiu HMC (2003) A review of the studies and conservation of horseshoe crabs in Hong Kong. In: Shin PKS (ed) Turning the tides. A Festschrift in Honour of Professor Brian Norton. The Marine Biological Association of Hong Kong, Hong Kong, pp 341–351

Goodman JD, Nir SM (2013) Poachers are elusive catch in city waters. New York Times, 3 July 2013. http://www.nytimes.com/2013/07/04/nyregion/poachers-are-elusive-catch-in-city-waters.html?pagewanted=all&module=Search&mabReward=relbias%3Ar&_r=0. Accessed 11 July 2014

Hong SG (2011) Biology of horseshoe crabs, *Tachypleus tridentatus*. Xiamen University Press, Xiamen, China

Hurdle J (2013) Import ban sought on Asian crabs. New York Times, 25 Feb 2013. http://green.blogs.nytimes.com/2013/02/25/import-ban-sought-on-asian-crabs/?_php=true&_type=blogs&module=Search&mabReward=relbias%3Aw&_r=0. Accessed 3 Sept 2014

Indonesia Press Report (2012) Horseshoe crab seizure–Sulawesi, Indonesia–May 2012. http://www.bkipm.kkp.go.id/bkipm/event/read/807/penggagalan-upaya-penyelundupan-jenis-ketam-tapak-kuda-(-tachipleus-gigas-)-.html. Accessed 31 Aug 2014

Itow T, Loveland RE, Botton ML (1998a) Developmental abnormalities in horseshoe crab embryos caused by exposure to heavy metals. Arch Environ Contam Toxicol 35:33–40

Itow T, Igarashi T, Botton ML et al (1998b) Heavy metals inhibit limb regeneration in horseshoe crab larvae. Arch Environ Contam Toxicol 35:457–463

IUCN (2000) Guidelines for the prevention of biodiversity loss caused by alien invasive species. SSC Invasive Species Specialist Group. http://intranet.iucn.org/webfiles/doc/SSC/SSCwebsite/Policy_statements/IUCN_Guidelines_for_the_Prevention_of_Biodiversity_Loss_caused_by_Alien_Invasive_Species.pdf. Accessed 11 July 2014

IUCN (2001) IUCN Red List categories and criteria: Version 3.1. IUCN Species Survival Commission. IUCN, Gland/Cambridge

IUCN Standards and Petitions Subcommittee (2014) Guidelines for using the IUCN Red List categories and criteria. Version 11. Prepared by the Standards and Petitions Subcommittee. http://www.iucnredlist.org/documents/RedListGuidelines.pdf. Accessed 11 July 2014

Jakarta Post (2013) Attempt to smuggle rare crabs foiled in Jambi. http://www.thejakartapost.com/news/2013/02/16/attempt-smuggle-rare-crabs-foiled-jambi.html. Accessed 11 July 2014

Kawakatsu M, Sekiguchi K (1988) Redescription of *Ectoplana limuli* (Ijima et Kaburaki, 1916) and *Ectoplana undata*, 1983 (Turbellaria, Tricladidia, Maricola) collected from three species of Asian horseshoe crabs. Jobu J Manag Inf Sci 5:57–94

Key MM, Jeffries WB, Voris HK et al (1996) Epizoic bryozoans, horseshoe crabs, and other mobile benthic substrates. Bull Mar Sci 58:368–384

Leibovitz L, Lewbart GA (2003) Diseases and symbionts: vulnerability despite tough shells. In: Shuster CN Jr, Barlow RB, Brockmann HJ (eds) The American horseshoe crab. Harvard University Press, Cambridge, pp 245–275

Levin J, Hochstein HD, Novitsky TJ (2003) Clotting cells and *Limulus* amoebocytes lysate: an amazing analytical tool. In: Shuster CN Jr, Barlow RB, Brockmann HJ (eds) The American horseshoe crab. Harvard University Press, Cambridge, pp 310–340

Liao YY, Ip FL (2000) Resources of Chinese horseshoe crabs under threat. China Aquac 10:10–11 (in Chinese)

Liao YY, Li XM (2001) Present situation of horseshoe crab resources in the sea area of China and tactics of preservation. Resour Sci 23:53–57 (in Chinese, with English summary)

Liao YY, Hong SG, Li XM (2001) A survey on the horseshoe crabs in north of South China Sea. Acta Zool Sin 47:108–111 (in Chinese, with English abstract)

McGowan CP, Smith DR, Sweka JA et al (2011) Multispecies modeling for adaptive management of horseshoe crabs and red knots in the Delaware Bay. Nat Resour Model 24:117–156

Mishra JK (2009) Horseshoe crabs, their eco-biological status along the northeast coast of India and the necessity for ecological conservation. In: Tanacredi JT, Botton ML, Smith DR (eds) Biology and conservation of horseshoe crabs. Springer, New York, pp 89–96

Mizrahi DS, Peters KA (2009) Relationships between sandpipers and horseshoe crabs in Delaware Bay: a synthesis. In: Tanacredi JT, Botton ML, Smith DR (eds) Biology and conservation of horseshoe crabs. Springer, New York, pp 65–87

Morton B, Lee CN (2011) Spatial and temporal distributions of juvenile horseshoe crabs (Arthropoda: Chelicerate) approaching extirpation along the northwestern shoreline of the new territories of Hong Kong. J Nat Hist 45:227–251

Myers JP (1986) Sex and gluttony on Delaware Bay. Nat Hist 95(5):68–77

New York Department of Environmental Conservation (NYDEC) (2013) ECOs arrests lead to felony charges against horseshoe crabbers. Press release, 11 July 2013. http://www.dec.ny.gov/press/92301.html Accessed 11 July 2014

Nolan MW, Smith SA (2009) Clinical evaluation, common diseases, and veterinary care of the horseshoe crab, *Limulus polyphemus*. In: Tanacredi JT, Botton ML, Smith DR (eds) Biology and conservation of horseshoe crabs. Springer, New York, pp 479–499

O'Connell K, Etgen C, Kreamer G et al (2009) Green eggs and sand: a collaborative effort of scientists, teaches, resource managers, and stakeholders in educating about *Limulus polyphemus*. In: Tanacredi JT, Botton ML, Smith DR (eds) Biology and conservation of horseshoe crabs. Springer, New York, pp 595–604

Patil JS, Anil AC (2000) Epibiotic community of the horseshoe crab *Tachypleus gigas*. Mar Biol 136:699–713

Rutecki D, Carmichael RH, Valiela I (2004) Magnitude of harvest of Atlantic horseshoe crabs, *Limulus polyphemus,* in Pleasant Bay, MA. Estuaries 27:179–187

Sekiguchi K (1988) Biology of horseshoe crabs. Science House, Tokyo

Shin PKS, Li HY, Cheung SG (2009) Horseshoe crabs in Hong Kong: human exploitation and current population status. In: Tanacredi JT, Botton ML, Smith DR (eds) Biology and conservation of horseshoe crabs. Springer, New York, pp 347–360

Shuster CN Jr, Botton ML, Loveland RE (2003) Horseshoe crab conservation: a coast-wide management plan. In: Shuster CN Jr, Barlow RB, Brockmann HJ (eds) The American horseshoe crab. Harvard University Press, Cambridge, pp 358–377

Smith DR, Millard MJ, Carmichael RH (2009) Comparative status and assessment of *Limulus polyphemus*, with emphasis on the New England and Delaware Bay populations. In: Tanacredi JT, Botton ML, Smith DR (eds) Biology and conservation of horseshoe crabs. Springer, New York, pp 361–386

Tanacredi JT, Botton ML, Smith DR (eds) (2009) Biology and conservation of horseshoe crabs. Springer, New York

US Fish and Wildlife Service (2014) Endangered and threatened wildlife and plants; proposed threatened status for the Rufa Red Knot (*Calidris canutus rufa*). Fed Regist 79(93):27548–27550

Widener JW, Barlow RB (1999) Decline of a horseshoe crab population on Cape Cod. Biol Bull 197:300–302

Williams BL (2010) Behavioral and chemical ecology of marine organisms with respect to tetrodotoxin. Mar Drugs 8:381–398

Chapter 22
Current Status of *Tachypleus tridentatus* in Taiwan for Red List Assessment

Hwey-Lian Hsieh and Chang-Po Chen

Abstract The horseshoe crab, *Tachypleus tridentatus*, once thrived on the west coast of Taiwan as evidenced by history, literature, colloquial language, daily usage and places named after it. After more than 15 years of efforts to conserve its populations, the horseshoe crab population remains threatened. A study on juvenile density changes in the Guningtou Horseshoe Crab Conservation Area in Kinmen Island showed that over 7 years from 2003 to 2009, the juvenile population declined rapidly from 0.30 individuals/m^2 in 2003 to between 0.02 and 0.17 individuals/m^2 in 2004–2009. Adult population surveys and scattered landing reports on fishing wharfs produced only about several tens individual catches at the most, implying a very limited adult distribution in waters surrounding Taiwan. The threats to *T. tridentatus* in Taiwan include habitat loss, habitat deterioration due to armored shorelines, pollution, *Spartina* invasion and human consumption. Conservation actions implemented include: habitat protection and improvement, population enhancement, education, and listing the horseshoe crab as a protected species. We now recommend designating new protected areas on the western and northeastern coasts of Kinmen Island as our studies show that these coasts possess beaches with the highest potential for spawning. We also strongly suggest that instars up to the third instar stage are appropriate for release, because they survive better in the field.

Keywords *Tachypleus tridentatus* • Population status • Taiwan • Red List Assessment

22.1 Introduction

The horseshoe crab, *Tachypleus tridentatus*, once thrived on the west coast of Taiwan, as attested to by evidence from history, literature, colloquial language, daily usage and places named after it (Chen et al. 2004; Chen and Chen 2011). It is well documented

H.-L. Hsieh (✉) • C.-P. Chen
Biodiversity Research Center, Academia Sinica, Taipei 115, Taiwan, Republic of China
e-mail: zohl@gate.sinica.edu.tw

R.H. Carmichael et al. (eds.), *Changing Global Perspectives on Horseshoe Crab Biology, Conservation and Management*, DOI 10.1007/978-3-319-19542-1_22

that the most effective conservation of the horseshoe crabs is accomplished by designating them as protected species and establishing their critical habitats as protected areas. With the exception of Matsu Island, an offshore island of Taiwan in the northern end of the Taiwan Strait, *Tachypleus tridentatus* is not a protected species in the remaining jurisdictional areas of Taiwan. Although an intertidal area of 800 ha in Kinmen, another offshore island in the Taiwan Strait and very close to Xiamen, is designated as the Guningtou Horseshoe Crab Conservation Area and identified as a no-take zone by the Fisheries Law of Taiwan, the horseshoe crabs living outside that area are not protected by law. Unfortunately, there are no penalties for violations within the designated area, such as habitat destruction or pollution (Chen et al. 2004). Only if this species is listed as a protected species in Taiwan can its exploitation be regulated.

To include the horseshoe crab *Tachypleus tridentatus* and the other two Asian species in the IUCN (International Union for Conservation of Nature) Red List categories has been unresolved for decades because data has been deficient. With scientific and anecdotal observations revealing continuously declining in population abundance, there is deep concern that Asian horseshoe crab populations may be decimated if no effective actions are taken to save them in the Asia-Pacific region (Shin 2011; IUCN 2012). Four strategies have been developed to achieve the implementation of Asian horseshoe crab conservation. These include: (1) updating the IUCN Red List status of horseshoe crabs, (2) raising awareness of the troubling future facing Asian horseshoe crabs, (3) reducing human demand for utilization, and (4) increasing legal protection of critical habitats (Shin 2011).

To ensure the sustainability of *T. tridentatus* in Asia, horseshoe crab scientists and experts realize that *T. tridentatus* needs to be on the IUCN Red List. Therefore, we joined the Horseshoe Crab Specialist Group, which was established in June 2012 under the auspices of the IUCN Species Survival Commission, attempting to assist the Red List assessment of Asian horseshoe crabs in Taiwan.

The objectives of this study are to update the current population status and document the distribution of horseshoe crab *Tachypleus tridentatus* in the waters surrounding Taiwan. The data lend further support for including the threatened *T. tridentatus* on the IUCN Red List. The contents of this study include the crabs' historical and current geographic ranges, juvenile and adult population abundance, habitat characteristics of nursery grounds and potential spawning grounds, threats, and conservation actions currently undertaken in Taiwan.

22.2 Geographic Range

22.2.1 Juveniles

Populations of juvenile *Tachypleus tridentatus* were once distributed on the intertidal zone along the west coast of the main island of Taiwan, in which the intertidal is dominated by sand and mud flats. However, recent studies from 1998 to 2013 (Yang et al. 2007; Chen 2011, 2012) showed that only scattered populations have been observed in very few places along the west coast of Taiwan (Fig. 22.1). Over a

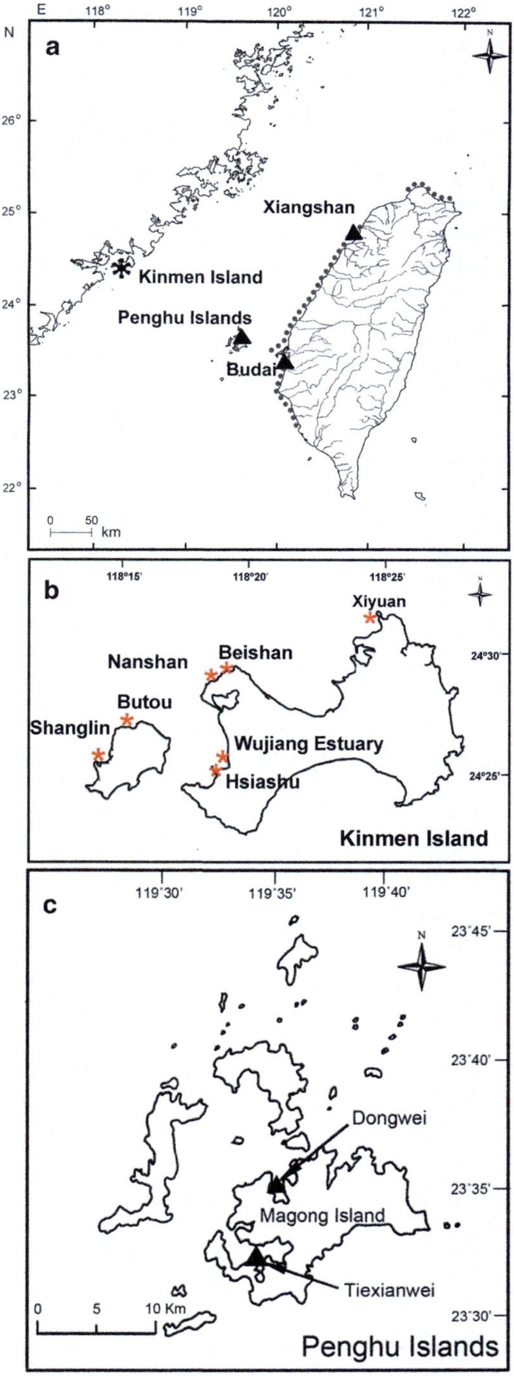

Fig. 22.1 (**a**) Historical and current distributions of juvenile *Tachypleus tridentatus* in the waters surrounding Taiwan recorded between 1998 and 2013, (**b**) Distribution on Kinmen Island, (**c**) Distribution on Penghu Islands. *Dashed lines*: locally extinct sites; *triangles*: scattered occurrence sites; *asterisk*: vitally populated sites

sampling period of several weeks, approximately 20 individuals were collected on Taiwan's offshore Penghu Islands in 2003. Because individuals were rarely found, a non-quantitative walk-through sampling was adopted. A very small population of approximately 20 individuals documented in Budai, Chiayi County between 2004 and 2005 was not found after 2006 (Yang et al. 2009). One individual with a carapace width of approximately 5 cm was reported from Xiangshan intertidal flat of Hsinchu City in 2013. Viable populations now only occur on Kinmen, an offshore island under Taiwan's jurisdiction that is very close to mainland China. The Kinmen populations have been persistently present.

22.2.2 Adults

No adults or mating pairing of *Tachypleus tridentatus* have been caught and recorded from intertidal flats of Taiwan's main island since the 1960s, nor from Kinmen Island where juvenile populations are abundant. However, over the past decade, occasional catches of adult from scattered wharfs were recorded along the western and northeastern coasts and offshore islands of Taiwan region (Fig. 22.2). These records were actual catch seen or reported by news media on those ports.

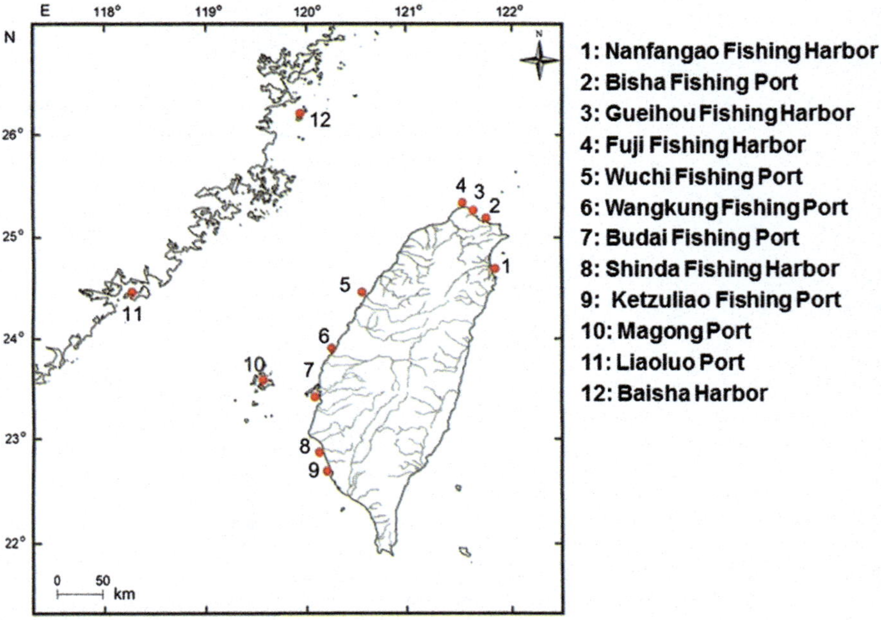

Fig. 22.2 Locations of wharfs in Taiwan region where adult *Tachypleus tridentatus* have been brought to land in the past decade

22.3 Population Density Changes

22.3.1 *Juveniles*

Viable juvenile populations occur only on Kinmen Island, an offshore island of Taiwan very close to China that was formerly under military administration. Hsieh and Chen (2009) provided a quantitative assessment of juvenile densities at sites Beishan, Nanshan, and Hsiashu on Kinmen Island. The study sites at Beishan and Nanshan are located within the Guningtou Horseshoe Crab Conservation Area. The results showed seasonal changes in juvenile abundance with the density increasing approximately tenfold from May to August, as compared to September to April (Fig. 22.3). At Beishan, Nanshan, and Hsiashu sites, the density peaked in June with ranges of 0.09–0.72, 0–0.31, and 0.20–1.90 individuals/m^2, respectively. Overall, juvenile abundance was highest at Hsiashu.

Kinmen's juvenile populations showed a marked diminishing trend. Using the same sampling method, the juvenile population declined rapidly and significantly at the Nanshan and Hsiashu sites from 2003 to 2009 (Fig. 22.4, Table 22.1). The juvenile abundance at Hsiashu dropped from an average of 1.02 individuals/m^2 in 2003 to an average of 0.18 individuals/m^2 in 2009 (adjusted $r=0.85$, $b=-0.18$, $p=0.01$) while that at Nanshan decreased from 0.18 individuals/m^2 in 2003 to 0.01 individuals/m^2 in 2009 (Spearman's correlation, $r=-0.86$, $p=0.006$). By contrast, the juvenile population at Beishan changed little and fluctuated between 0.03 individuals/m^2 in 2007 and 0.37 individuals/m^2 in 2003 (Spearman's correlation, $p=0.10$). In the Guningtou Horseshoe Crab Conservation Area where data from the two sites, Beishan and Nanshan, were combined, the juvenile density decreased from 0.30 individual/m^2 in 2003 to a low level that ranged between 0.02 and 0.17 individuals/m^2 from 2004 to 2009 (Table 22.1).

22.3.2 *Adults*

The adult population was surveyed using drift gillnets for three overnights around Penghu Island in November 2008. We used this fishing gear because bottom-trawling was prohibited by Taiwan's law. Only two sub-adult males were caught, with carapace width of 16.3 and 18.7 cm and body wet weight of 0.35 and 0.6 kg, respectively (Table 22.2). A separate survey conducted for 1 day by Kinmen Fisheries Research Institute using bottom trawl nets in 2008 caught 12 individuals (Table 22.2). Based on these data and the scattered reports that the crabs were brought to land on fishing wharfs (maximum four or five individuals ever reported from each report), we concluded that the adult population in the waters surrounding Taiwan is limited.

According to the data provided by the Kinmen Fisheries Research Institute, the Institute purchased 57 individuals from local fishermen in 2006 and 335 individuals

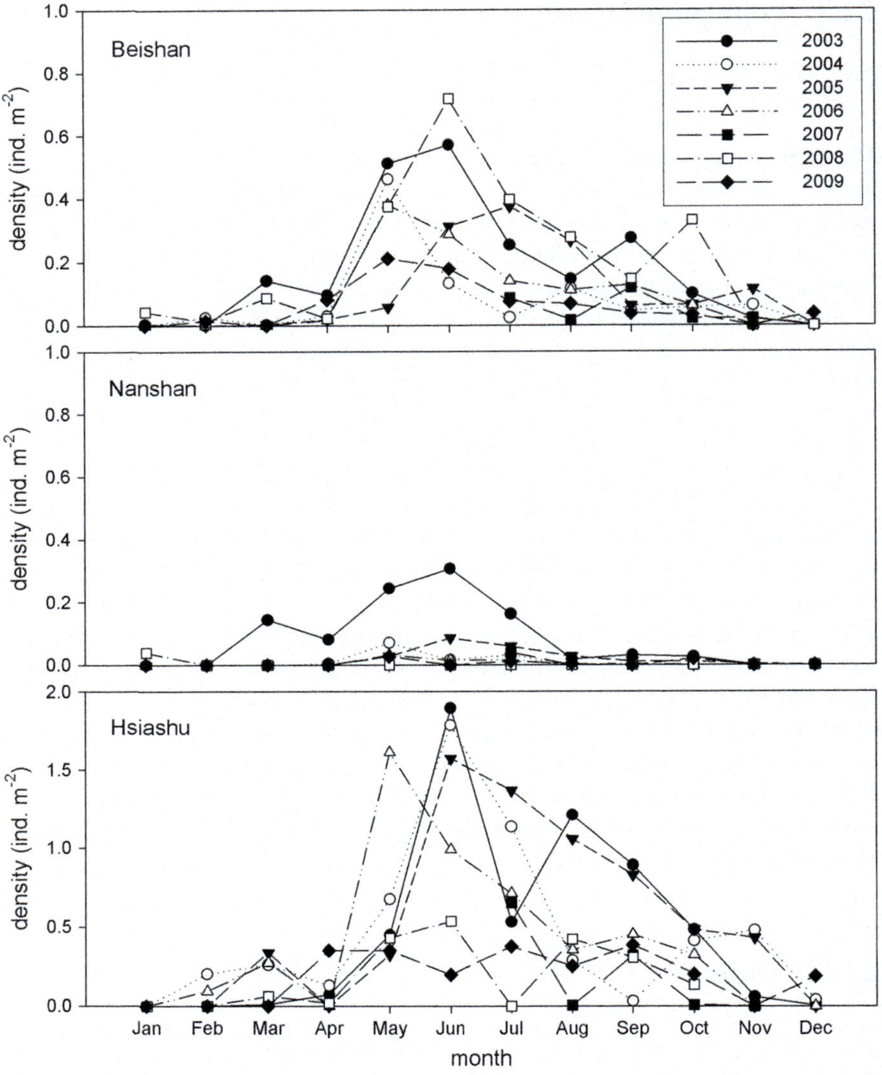

Fig. 22.3 Monthly and yearly changes in density of the juvenile *Tachypleus tridentatus* at the Beishan, Nanshan, and Hsiashu sites on Kinmen Island from 2003 to 2009. Data points are the monthly mean densities of all plots on a given transect and month at each site. Sites Beishan and Nanshan are located in Guningtou Horseshoe Crab Conservation Area

in 2007; however, no adult horseshoe crabs were offered by local fishermen in 2008. Data from the Penghu Fisheries Research Institute showed a purchase of 209 individuals in 2008. Such information may reflect policy differences among local government departments related to the purchase of adult horseshoe crabs. As we had no way to know where the fishing areas were and what fishing methods were used, it is

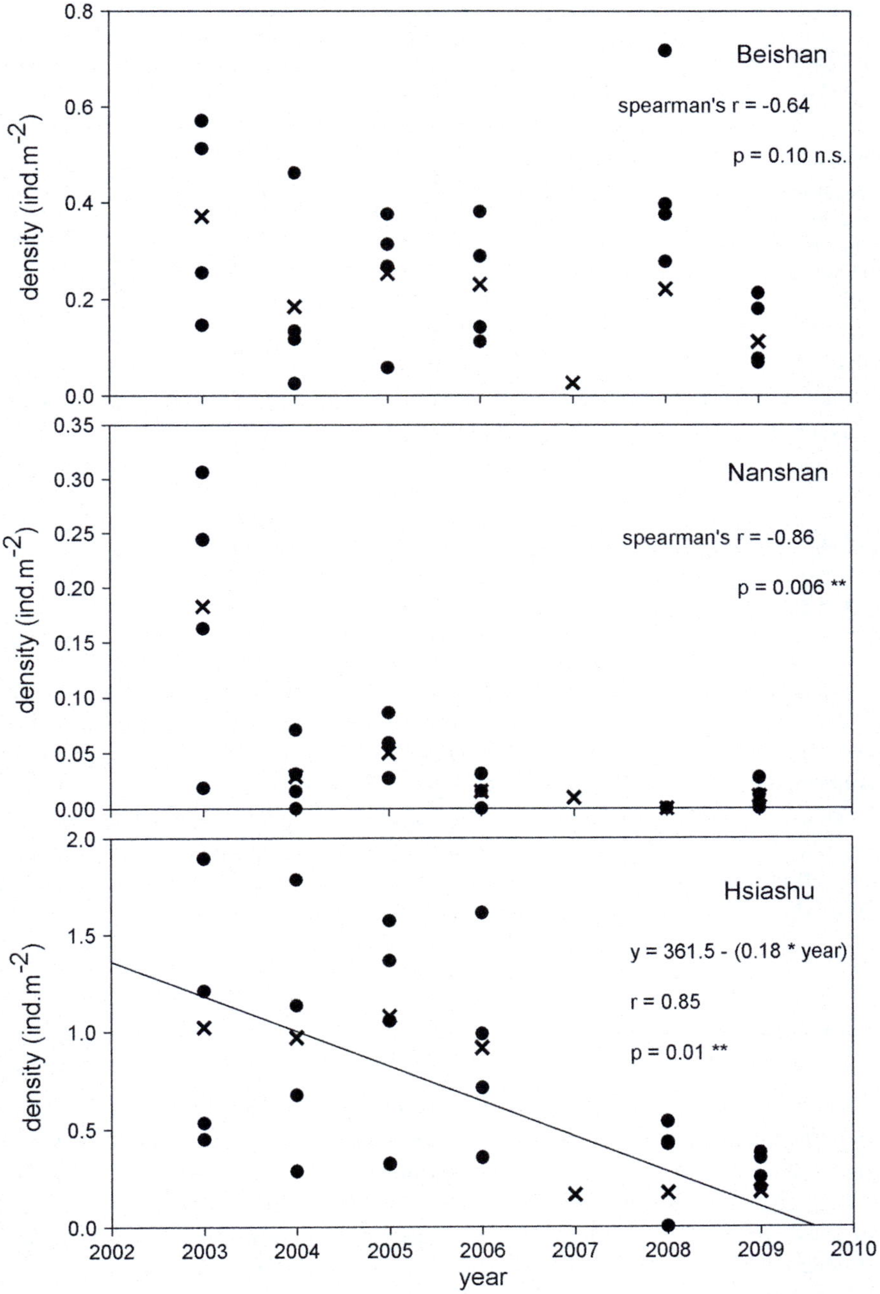

Fig. 22.4 Yearly density changes of the juvenile *Tachypleus tridentatus* at the Beishan, Nanshan, and Hsiashu sites on Kinmen Island from 2003 to 2009. Data points are monthly densities (*solid circles*) with means (*crosses*). Months refer to sampling months from May to August

Table 22.1 Changes in juvenile horseshoe crab population density inside Guningtou Horseshoe Crab Conservation Area on Kinmen Island from 2003 to 2009. Densities are the means of numbers of individuals sampled from all plots, each with 2×4 m, from May to August. The Beishan and Nanshan sites are located in this conservation area

	2003	2004	2005	2006	2007	2008	2009
Year	n=200	n=312	n=312	n=312	n=156	n=312	n=312
Population density (mean ± SE, individual/ m^2)	0.30 ± 0.06	0.12 ± 0.05	0.17 ± 0.05	0.14 ± 0.05	0.02 ± 0.02	0.13 ± 0.09	0.07 ± 0.02

difficult to estimate the adult population size based on this information. We suspected that these adults were obtained from the waters between Kinmen Island of Taiwan and Xiamen of mainland China.

22.4 Habitat and Ecology

22.4.1 Spawning Ground

Tachypleus tridentatus require various habitats for different life stages. In the Taiwan region, almost all natural spawning grounds have been degraded or lost. Potential spawning grounds for *Tachypleus tridentatus* have been experimentally confirmed (Chen et al. 2004; Hsieh and Chen 2009). In the case of Kinmen, they are located at high tidal zones and require coarse sand sediment (1.5 mm diameter grain size) with 15 % saturated water content in the top 10–45 cm when tides recede, 2.4–5.0 m tidal amplitude and 0.09–0.12 slopes (Chen et al. 2004; Hsieh and Chen 2009, the authors' unpublished data) (Fig. 22.5).

22.4.2 Nursery Ground

Suitable nursery grounds are distributed in intertidal mudflats. Taiwan's main island and offshore islands still possess some such mudflats. In the case of substrata examined at the Beishan, Nanshan and Hsiashu sites, nursery grounds possessed the following properties (Hsieh and Chen 2009): grain size ranging from 0.08 to 0.41 mm, silt/clay content from 15.1 to 45.8 %, water content from 19.1 to 28.6 %, total organic carbon content from 0.24 to 0.43 %, total nitrogen content from 0.05 to 0.07 %, chlorophyll *a* biomass from 1.9 to 2.6 μg/cm^2 and a sediment sorting coefficient of 1.87–2.76.

Table 22.2 Adult population surveys by fishing boats in Penghu Island and Kinmen Island in 2008

Collection date	Location	Type of fishing gear	Length of fishing gear × number of sets	Fishing gear deployed duration	Total number/sex/ prosomal width/ weight caught	Data source
2008 November 19	Dachang, Penghu	Drift gillnet	70 m × 2	1 day	0	This study
2008 November 21	Qinglou, Penghu	Drift gillnet	70 m × 2	1 day	0	This study
2008 November 21	Shagang, Penghu	Drift gillnet	70 m × 2	1 day	0	This study
2008 November 21	Anzhai, Penghu	Drift gillnet	70 m × 2	1 day	2 subadult males:	This study
					18.7 cm/ 0.6 kg;	
					16.3 cm/ 0.35 kg	
2008 November 22	Chikan, Penghu	Drift gillnet	70 m × 3	1 day	0	This study
2008 June 1	Liaoluowuan, Kinmen	Bottom trawl	–	–	12/4 pairs, 1 male; 3 subadults <20 cm	Kinmen County Fisheries Research Institute

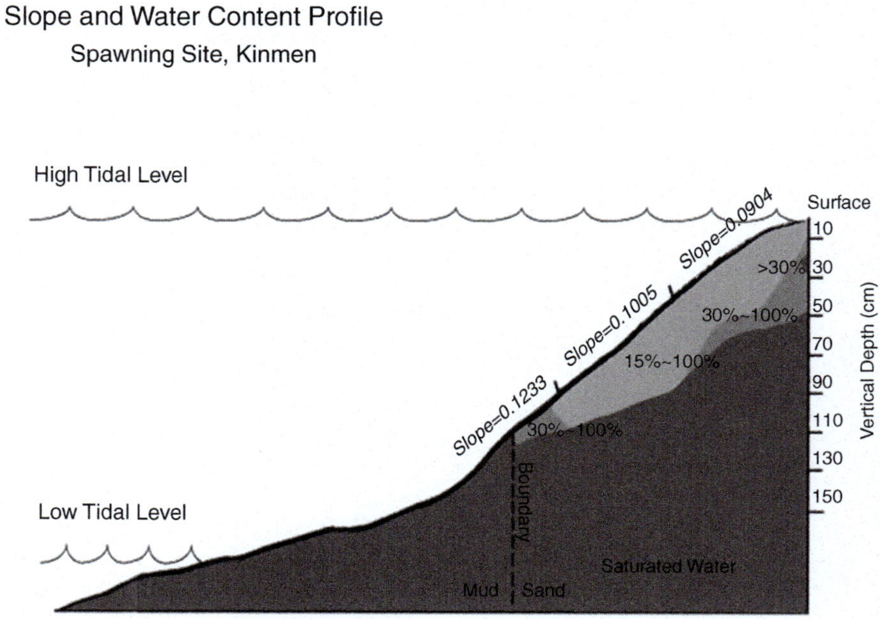

Fig. 22.5 Schematic profile showing slope and water content (%) of optimal spawning site for *Tachypleus tridentatus* in Kinmen

22.5 Threats

The major threat to *Tachypleus tridentatus* on Kinmen Island of Taiwan is habitat loss due to construction of a commercial port on nursery and spawning grounds. This port belonging to the Kinmen Passenger Service Center is located in a headland bay. The construction has reclaimed the intertidal area and dredged the subtidal area, destroying the ecological integrity of the whole bay. The horseshoe crabs are also facing habitat deterioration due to construction of dikes close to their spawning grounds. On Kinmen Island, a traditional method of harvesting oysters involves cultivation onto rock plates deployed as protruding arrays on intertidal flats. However, extensively distributed arrays become obstacles for adult horseshoe crabs when they move toward high tidal zones for spawning (Chen 2009). In addition, in recent years their nursery grounds have been lost to the invasive *Spartina alterniflora*. Human consumption has not been a major threat to horseshoe crabs in the Taiwan region since there are so few adults and the unpredictable catch cannot meet the demand of restaurants. However, there are still a few reports of horseshoe crab consumption on Kinmen Island and Taiwan's main island.

22.6 Conservation Measures

There have been persistent efforts to implement *T. tridentatus* conservation measures with a focus on the wise-use of this precious species (Chen et al. 2009):

1. Habitat protection and improvement
   This approach is to increase the supply of horseshoe crabs. We have identified and then demonstrated how to restore potential spawning and nursery grounds in the Taiwan region (Chen et al. 2004; Hsieh and Chen 2009). Currently, there is only one protected area located on Kinmen that has been set for conserving horseshoe crabs. We have initiated procedures aimed at designating more protected areas for horseshoe crab conservation. In order to achieve this goal, we consider three kinds of habitats covering the whole life cycle of horseshoe crabs, including spawning grounds, juvenile nursing grounds and adult habitats. Since adults are distributed in deep channels of 20–30 m depth (Sekiguchi 1988, and according to local fishermen's experiences), the distance between deep channels where adults live and their potential spawning grounds affects the chance that adults will reach the beaches to spawn. The shorter the distance, the higher the chance they will succeed. Based on this assumption, we measured the nearest distance between a channel of 20 m depth and a natural spawning ground at the Guningtou Horseshoe Crab Conservation Area and found that it was 5 km. Then, we circled high tidal zones of sandy beaches using this radius to identify potential spawning grounds. Using this method to assess possible spawning sites, we proposed that the western and northeastern coasts appeared to be the most suitable sites on Kinmen Island (Fig. 22.6; Chen 2009, 2011). We recommended

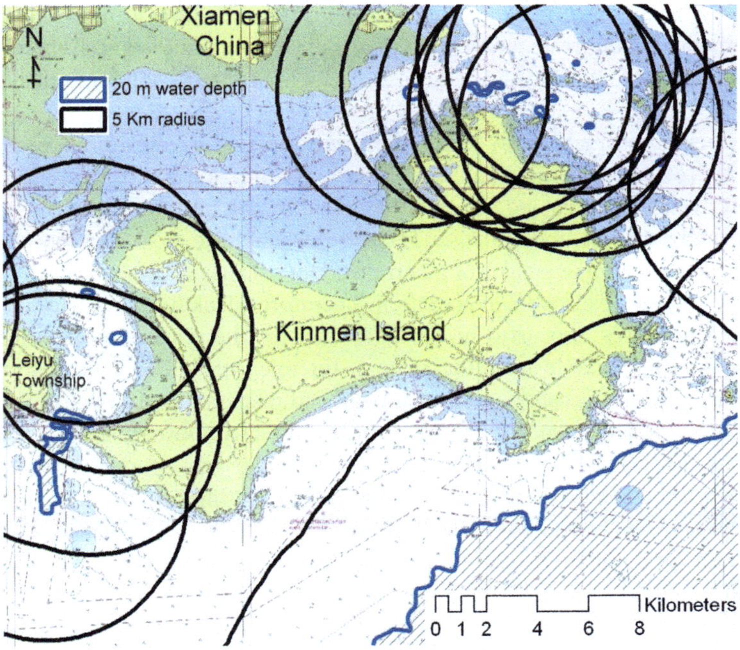

Fig. 22.6 Distribution of potential spawning grounds on the western and northeastern coasts of Kinmen Island. These proposed spawning grounds are located in the high tidal zones of sandy beaches within circled areas (*black circles*) delineated by a radius of 5 km from deep water channels located at 20 m depth (*patches with diagonal lines*)

designation of new conservation areas on these coasts. In addition, before the new conservation areas are established, any coastal development on these sections should be closely evaluated.

2. Population enhancement
This approach aims to increase the supply of live horseshoe crabs by stocking with juveniles cultivated in the laboratory to restore wild populations. We are working on a protocol to standardize the release of laboratory-cultivated juveniles to their natural habitats. Guidelines for the release include: instar stage, the point in a tidal cycle and circadian rhythm, location on the intertidal flat, and release method. With respect to the release locations, protected areas should be given priority to ensure that a healthy population can be established following a long-term release program.

Based on our laboratory cultures, the hatching rate of the first instar can be as high as 80–90 %. The first instar does not feed and molts into the second instar approximately 3 months later. The molting rate of the second to the third instar is very low, less than 10 %. This indicates that cultivating conditions for the second instar is crucial. Molting and growth at this point need to be thoroughly studied. After the third instar stage, juveniles possess better foraging and defense abilities and capacity to adapt to environmental changes, thus achieving better survival in natural habitats than the first and second instars. In laboratory rearing conditions, with much care, it took approximately 6 years for the horseshoe crab to grow from a fertilized egg to the 14th instar. Such slow growth rate makes it impractical to cultivate and release juveniles at later instar stages. Therefore, we recommend the third instar stage as the most appropriate stage for release (Chen et al. 2014).

Stocking protocols also need to take into account the substratum characteristics for released juveniles. Intertidal flats consisting of a mix of mud and sand are ideal places for releasing because the juveniles can bury immediately into mud and sand substrata upon release. The third instars have weak locomotion ability and cannot swim against strong currents. Only if they touch sediment and bury into it can they escape from predation. Therefore, slack tides with no strong currents and waves appear to be an ideal time for the release. In addition, trays appear to be suitable handling devices for releasing. Use of trays to transfer juveniles allows juveniles to crawl away and disperse freely instead of putting them into water or on the sediment surface directly. Development of a reliable trace-marker technique is also essential for assessing stocking performance. Tests will be needed to understand mark retention times for subsequent, correct readings (Chen et al. 2014).

In the past, hundreds and thousands of the first and/or second instars had been released into the wild on Kinmen Island, Taiwan, and in Xiamen and Quanzhou, Guangdong, China. However, the effectiveness of such release programs has not been evaluated. The measures just described allow us to improve the juveniles' survival rates after they are released to their natural habitats. More importantly, those measures can evaluate the effectiveness of culture and release programs.

3. Education
   This approach attempts to decrease the demand for and increase the supply of horseshoe crabs. Since 1996 we have delivered numerous lectures on horseshoe crab biology and ecology to the public and students, explaining their contributions to humans and the urgent need for conservation. In addition, we devoted effort to empowering local NGOs that promote conservation planning. We have started a "Blue Palm" project, a nation-wide campaign teaching the public to rear baby horseshoe crabs at home. Recently, we initiated and promoted Horseshoe Crab Conservation Day on Qixi, Chinese Valentine's Day, across the Taiwan Strait in 2010. Since then, conservation activities have continued. Three-way communications through web-cams have been initiated between young students and researchers from Taiwan, Hong Kong, and mainland China in 2013 to share experiences with horseshoe crab conservation work. A mini-film "Caring for Horseshoe Crabs on Qixi" has been made and posted on YouTube to enlighten the general public about the science, the crabs' significance to humans, and conservation of horseshoe crabs (Chen and Hsieh 2014). We also have made efforts to encourage people on the use of substitutes for the horseshoe crab's amebocyte lysate (LAL/TAL) through public education.
4. To list horseshoe crabs as protected species
   Although efforts to have horseshoe crabs designated as a protected species failed twice in Taiwan in 2008 and 2010, we continue to petition to have horseshoe crabs designated as a protected species in Taiwan. At the same time, we serve in the Horseshoe Crab Specialist Group approved by IUCN Species Survival Commission to assist with Red List assessment and conservation education regarding Asian horseshoe crabs.

22.7 Conclusion

Conserving the horseshoe crab is important not only because the species is so ancient that it is a living fossil, but also because human health benefits depend on the medical application of LAL/TAL. For *Tachypleus tridentatus*, adult populations are considered locally extinct from the main island of Taiwan; whereas the juvenile population is still alive on Kinmen Island. However, the juvenile population is rapidly declining, even in the Guningtou Horseshoe Crab Conservation Area of Kinmen. Given that their habitats are continuously threatened due to loss or degradation and the species is not protected by law in Taiwan, nor is its protection enforced in China, we urge that *Tachypleus tridentatus* should be categorized as a threatened species on the IUCN Red List.

References

Chang PJ (2012) Resource of juvenile horseshoe crab, *Tachypleus tridentatus*, from intertidal zone at Guningtou Conservation Area and Wuchiang Creek in Kinmen islands. Master's thesis, National Kaohsiung Marine University, Taiwan (in Chinese with an English abstract)

Chen CP (2009) Reappraisal of horseshoe crab conservation area on Kinmen. Final report to Kinmen County Government (in Chinese)

Chen CY (2011) Keynote report on Taiwan's horseshoe crabs – population crisis and conservation strategy. In: Chang CW (ed) Epic of horseshoe crabs. National Museum of Marine Biology and Aquarium, Pingdong County, pp 71–82 (in Chinese)

Chen CP, Chen YH (2011) Epic of horseshoe crabs – the case of *Tachypleus tridentatus*. In: Chang CW (ed) Epic of horseshoe crabs. National Museum of Marine Biology and Aquarium, Pingdong County, pp 5–12

Chen CP, Hsieh HL (2014) Caring horseshoe crab on Qixi. https://www.youtube.com/watch?v=jQEjpe_UWM4. Accessed 11 June 2014

Chen CP, Yeh HY, Lin PF (2004) Conservation of the horseshoe crab at Kinmen, Taiwan: strategies and practices. Biodivers Conserv 13:1889–1904

Chen CP, Hsieh HL, Chen A et al (2009) The conservation network of horseshoe crab *Tachypleus tridentatus* in Taiwan. In: Tanacredi JT, Botton ML, Smith DR (eds) Biology and conservation of horseshoe crabs. Springer, New York, pp 543–557

Chen CP, Hsieh HL, Huang TS et al (2014) A review of artificial breeding and releasing programs of the horseshoe crab *Tachypleus tridentatus* at the two sites across the Taiwan Strait. In: The across-strait conference in stocking and releasing of fishery species, Taipei, 13–14 May 2014 (in Chinese)

Hsieh HL, Chen CP (2009) Conservation program for the Asian horseshoe crab *Tachypleus tridentatus* in Taiwan: characterizing the microhabitat of nursery grounds and restoring spawning grounds. In: Tanacredi JT, Botton ML, Smith DR (eds) Biology and conservation of horseshoe crabs. Springer, New York, pp 417–438

IUCN (2012) M034: the conservation of Asian horseshoe crabs. https://portals.iucn.org/docs/2012congress/motions/es/M-034-2012-SP.pdf. Accessed 11 June 2014

Sekiguchi K (1988) Biology of horseshoe crabs. Science House, Tokyo

Shin P (2011) International workshop on the science and conservation of Asian horseshoe crabs: summary and outline of strategies for conservation. Paper presented at the International workshop on the science and conservation of Asian horseshoe crabs, Hong Kong, 13–16 June 2011

Yang MC, Chen CA, Hsieh HL et al (2007) Population subdivision of the tri-spine horseshoe crab, *Tachypleus tridentatus*, in Taiwan Strait. Zool Sci 24:219–224

Yang MC, Chen CP, Hsieh HL et al (2009) Phylogeography, demographic history, and conservation of horseshoe crab, *Tachypleus tridentatus*, in the South and East China Seaboards. In: Tanacredi JT, Botton ML, Smith DR (eds) Biology and conservation of horseshoe crabs. Springer, New York, pp 163–181

Chapter 23
Status and Threat Perceptions of the Indian Horseshoe Crabs Along the Northeast Coast of Bay of Bengal, India

Jayant K. Mishra, Anita Mishra, and Yasmin

Abstract The Indian subcontinent is surrounded by one of the oldest seas, the Bay of Bengal, at its eastern side extending over a total length of 2,090 km from the northern state of West Bengal to the southernmost tip of India and the Arabian Sea on the other side. On the eastern side of the Bay of Bengal lies Bangladesh, Myanmar and Andaman and Nicobar group of islands that surrounds Andaman Sea at its eastern side. The Bay of Bengal is unique in nature as several major river estuaries are found along its coastline, and it supports a vivid biodiversity and several unique ecosystems including mangroves, grassland, brackish water, etc. Among scores of biological resources, two horseshoe crab species *Tachypleus gigas* and *Carcinoscorpius rotundicauda* are quite prominent along the coasts of Bay of Bengal. As in other places, both these ecosystems and the biological resources including living fossils like horseshoe crabs are challenged with several threats. In the present chapter, threat perceptions to these two horseshoe crab species along the northeast coast of the Bay of Bengal with particular reference to the Gangetic Sundarban (Mangrove forest) delta are discussed. It was observed that the main threat to these two species is the destruction of their respective breeding habitat, particularly nesting zones constituting the intertidal sand for *T. gigas* and mangrove associated mudflats in case of *C. rotundicauda*.

Keywords Breeding beach • Environmental degradation • *T. gigas* • *C. rotundicauda* • Nesting population

J.K. Mishra (✉) • A. Mishra • Yasmin
Department of Ocean Studies and Marine Biology, Pondicherry University (Brookshabad Campus), Port Blair 744 112, Andaman & Nicobar Islands, India
e-mail: jayant64@yahoo.com

R.H. Carmichael et al. (eds.), *Changing Global Perspectives on Horseshoe Crab Biology, Conservation and Management*, DOI 10.1007/978-3-319-19542-1_23

23.1 Introduction

The pristine coastline of the northeast Bay of Bengal, stretching over a length of about 2,090 km, is endowed with several river estuaries, coastal lagoons, creeks, mangrove and grassland ecosystems. These environments serve as a natural incubating habitat for the spawning activity of several migratory marine animals, including living fossils such as olive ridley turtles and horseshoe crabs, particularly in the adjacent sandy intertidal zones of river estuaries. The breeding migrations of these animals along the coastal beaches are phenomenal, during which both the brooders and breeding pairs crawl up to a selective zone and lay the eggs by digging a nest in that zone. The fertilized eggs remain within the sandy nests until they hatch out as juveniles and migrate to the sea for their nourishment and growth. However, these zones are under tremendous environmental threat and undergoing rapid degradation due to several natural and anthropogenic activities. This is becoming a matter of concern for conservationists. Although several measures are being taken to save these animals, including classification of both olive ridleys and horseshoe crabs as protected species status under the Indian Environmental Act, several other precautionary measures need to be taken in terms of protecting the animals and environment types (niche/ecosystems) from destruction. It is thus binding on our part not only to protect these breeding beaches (nature's incubators) and help these living fossils to sustain forever, but also to conserve the broader coastal environment.

Indian species of horseshoe crabs, which include *Tachypleus gigas*, *Carcinoscorpius rotundicauda* and Sundarban sp. found along the northeast coast of Bay of Bengal, East coast of India (Itow et al. 2004; Mishra 2009) are under threat. In recent times, it has been observed that the migrations of breeding populations (pairs in amplexus) to the breeding beaches along the coasts of Odisha and West Bengal are in decline, as recorded by the first author since 1988. This is affecting the density of fertilized eggs in the nests (under incubation) along the breeding zones, which in turn has an adverse effect on the generation of newborn trilobites. As recorded by the author, there has been a sharp decline in the movement of number of newborn trilobites from the nests to the sea with the receding tides at Balaramgari, a major breeding place for *T. gigas*. The major cause of population decline is mostly attributed to the loss of breeding habitat for the visiting pairs of horseshoe crabs for egg laying activity.

It is well documented that all the four extant horseshoe crab species have their own preference of breeding habitats comprising muddy, sandy, coarse sand and gravel (Sekiguchi et al. 1977; Sekiguchi 1988; Brockmann 2004; Chen et al. 2004; Mishra 2009; Weber and Carter 2009) in the intertidal environment, where they undergo amplexus and release gametes for fertilization and incubation until they attain trilobite stage ready to hatch out and swim to the nourishing ground. In case of Indian species, they exhibit a very clear preference for their breeding habitats; *C. rotundicauda* breeds in the clay mud flats associated with mangrove forests, whereas *T. gigas* prefers the sandy zones with specific sand grain size of 63–125 μm (Mishra 2009). This preference for breeding habitat was established when both the species were encountered during the present investigation along the coasts of Sagar island, West Bengal, which provides contrasting habitat types like mud flats and sandy zones (Fig. 23.1).

As evident from the report of Chen et al. (2004), *T. tridentatus* along the coast of Taiwan has a suitable sediment type both for the purpose of spawning and nursing. This phenomenon is also documented in case of both the Indian species (Mishra 2009). But these critical habitats are under threat and subjected to destruction due to several natural and man made activities. The growing trend in the loss of breeding grounds (beaches) is also observed in case of all the four extant species of horseshoe crabs in the world, *Limulus polyphemus*, *Tachypleus tridentatus*, *T. gigas* and *Carcinoscorpius rotundicauda*, which is a major concern from the viewpoint of conservation (Berkson et al. 2009). Though several measures are now underway to protect these animals by conserving these breeding beaches etc. (Chen et al. 2004) in a regional basis, a comprehensive assessment of the causes of loss of spawning habitats lying in the interface of the land and sea must be carried out in all nations where horseshoe crabs reside. Similar concern has been put forth by some pioneers in the field of horseshoe crab research that the habitat is the basic requirement for the existence of all the horseshoe crab species and should be protected (Sekiguchi and Shuster 2009; Shuster and Sekiguchi 2009).

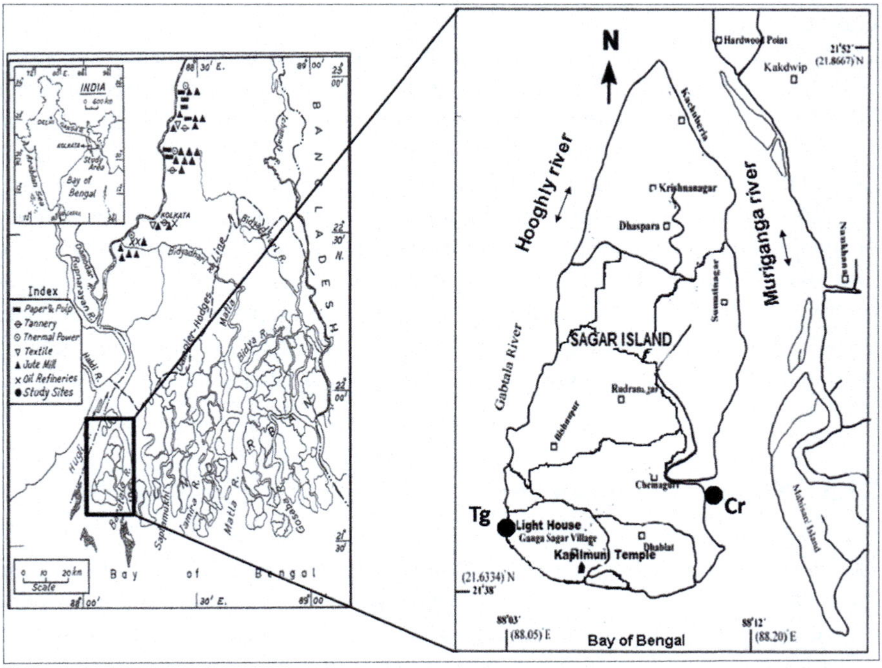

Fig. 23.1 Location of study area along the coast of Sagar Island in the Gangetic Delta, NE coast of Bay of Bengal

It is also a major concern that the use of these animals for several purposes including food, manure, bait and biomedical purposes may bring damage to the natural population (Berkson and Shuster 1999; Shuster 2004; Shuster et al. 2004; Berkson 2009; Kreamer and Michels 2009; Shin et al. 2009). Thus, assessment of the species-specific horseshoe crab population status and probable threats to them needs to be carried out. In the present chapter, the status of two species of horseshoe crabs, *C. rotundicauda* and *T. gigas*, which are prevalent in the Bay of Bengal region, are discussed. We also discuss the probable causes for their population depletion along the coasts of Balaramgari and Sagar Islands in the vicinity of Gangetic Sundarbans along the northeast coast of Bay of Bengal.

23.2 Horseshoe Crabs in India

Availability of horseshoe crabs along the northeast coast of India was carried out along the breeding zones at Balaramgadi (Lat. 21° 17′ N; Long. 87° 00′ E); NE coast of Odisha for *T. gigas* (Mishra 2009). The coast line of Sagar Island (Lat. 21° 63′ N; Long. 88° 21′ E) in the vicinity of Gangetic Sundarban), West Bengal (Fig. 23.1) was investigated for both *T. gigas* and *C. rotundicauda*. This island environment is the part of one of the important coastal ecosystems in the Gangetic delta dominated by rich mangroves and is a hot spot where both the species of horseshoe crabs coexist. Itow et al. (2004) reported the existence of another possible subspecies of *C. rotundicauda* (i.e. Sundarbans horseshoe crab) with some distinct structural features. The study was carried out in June, 2010 along the coast of Sagar Island; however, the horseshoe crabs at Balaramgadi in the vicinity of Budhabalanaga river estuary have been continuously monitored by the first author since 1988. During the study, breeding zones were investigated for the availability of breeding pairs of animals and also the nests with fertilized eggs. Simultaneously, availability of suitable breeding area and quality of breeding space for the migrating pairs were also taken into account. Sediment type and quality along the breeding zones were observed for both horseshoe crab species, with *T. gigas* utilizing coarse to fine sand particles both at Balaramgadi and Sagar Island, whereas *C. rotundicauda* laid its eggs in muddy substrates at Chemaguri creek at Sagar Island.

23.3 Status at Balaramgadi

At Balaramgadi, breeding activity of *T. gigas* occurs along the sandy zones of the intertidal region and the adjacent creek of the Budhabalanga river estuary. But this area is susceptible to several anthropogenic disturbances resulting from fishing-related activities as it is a major fish landing center along the coast of Odisha, located along the northeast coast of the Bay of Bengal. A typical feature associated with this breeding habitat at Balaramagadi is that the egg laying activity of breeding pairs shifts from one place to the other depending upon the seasonal shifting of sand beds from one side to other side of the estuary, related to the current pattern and wave action along the estuary. Mishra (2009) reported that the congregation of breeding pairs of *T. gigas* at Balaramgadi in the vicinity of Budhabalanga estuary is the highest, though the animals are also found at other places along the coastline, where there is a potential breeding zone available. Similarly, the number of nests with fertilized eggs was found to be high at Balaramgadi. But there is a sharp decline in the number of nests in this breeding habitat. Similarly, the total area available for egg laying process by the visiting pairs has decreased over the last couple of years. The reported decline in the sandy nesting zone at Balarmagadi (Mishra 2009) is solely attributed to the loss of breeding grounds due to coastal erosion and excess human activity leading towards the destruction of specific sandy zones required for nesting of copulating pairs. Simultaneously, occurrence of frequent tropical cyclones along the northeast coast of the Bay of Bengal is bringing damage to the coastline and large scale disturbance in the sandy nesting zone. As reported earlier, nests of *T. gigas* following full moon and new moon (highest high tides) during the peak breeding season at Balaramgadi number in the hundreds, accounting for thousands of fertilized eggs (Mishra 2009). However, *C. rotindicauda* is not found at this site.

23.4 Status at Sagar Island

On the other hand, both *C. rotundicauda* and *T. gigas* were encountered in the extreme North of Bay of Bengal, i.e. in the vicinity of Gangetic delta where river Ganges meets the Bay of Bengal in the form of different tributaries. The deltaic Ganges is an ecological hot spot with highest degree of biodiversity dominated by Sundarban (mangrove forests). Sagar Islands (Fig. 23.1) is located in the Sundarban region surrounded by two tributary rivers, the Muriganga river on the eastern side and the Hooghly river on the western side. The most phenomenal observation along the coastline of this island was the availability of both Indian horseshoe crab species, although we did not encounter the Sundarbans subspecies of *C. rotundicauda* reported by Itow et al. (2004) in the present surveys. Along the eastern coast of the island, eight breeding pairs of *T. gigas* were encountered at the adjacent Begua khali creek (Fig. 23.2a, b), which was found to be suitable for the spawning activity of the species.

The contrasting feature of this island ecosystem is that it also houses *C. rotundicauda* at its western side, where two pairs of the species were found at Chemaguri creek, dominated by mangrove associated mud flats (Fig. 23.3a, b). This is unusual in nature, as the distinct preferences for breeding habitats in *T. gigas* and *C. rotundicauda* usually result in a lack of geographic overlap, as evident by earlier studies (Chatterji et al. 1988; Sekiguchi 1988; Mishra 1994, 2009; Itow et al. 2004; Hajeb et al. 2009).

Though in the present investigation, threats from pollution were not addressed, several reports suggest that the ecosystem at Sagar Island is prone to environmental degradation from anthropogenic activities including industrial and municipal effluents (Bhattacharya et al. 2014, 2015).

Fig. 23.2 (**a**) Breeding ground of *T. gigas* along the creek at eastern side of the Sagar Island, (**b**) breeding pairs of *T. gigas* along the creek, eastern side of the Sagar Island

23.5 Threat Perception of Indian Horseshoe Crabs

The breeding habitats of Indian horseshoe crabs are under threat, and they are exposed to massive erosion leading to the destruction of these habitats. Along the east coast of the island (Fig. 23.4), stone packing is done to protect the shore from erosion. Other than this kind of habitat loss, there are no other major perceived threats to Indian horseshoe crabs. There are no pharmaceutical industries producing lysate as in the case of *L. polyphemus*, *T. tridentatus*, no food industries using *T. tridentatus*, *C. rotundicauda* and *T. gigas* as in some South Asian countries, and other forms of exploitation such as the production of manure or bait also do not occur in India (Swan 2001; Shuster 2004; Novisky 2009; Kreamer and Michels 2009; Hurton et al. 2009; Shin et al. 2009). But a downward trend in population is now emerging in both Indian species. This can be solely attributed to the loss of the breeding grounds as it was the case for the Japanese horseshoe crab, *T. tridentatus*, at its largest breeding ground at Kasaoka, Okayama prefecture along southeastern coast of Japan (Itow et al. 1991; Itow 1993; Seino et al. 2003; Shuster and Sekiguchi

Fig. 23.3 (**a**) Breeding ground of *C. rotundicauda* along Chemaguri creek, Western side of Sagar Island, (**b**) a solitary male *C. rotundicauda* along in Chamaguri creek water

Fig. 23.4 Stone packing along eastern side of Sagar Island to check erosion

2009; Tsuchiya 2009). By considering the report of Yang et al. (2009), which suggested that the habitat destruction in the intertidal region due to man-made activities may be responsible in influencing the species diversity and the extinction of some coastal species, necessary measures should be taken to conserve the intertidal habitat preferred by horseshoe crabs.

23.6 Conclusions

Though the direct threats to the Indian horseshoe crabs were found to be minimal in comparison to their counterparts elsewhere, measures should be taken to protect the species before it becomes too late by taking a cue from the Japanese horseshoe crab, *T. tridentatus*. The real issue now is the fragmentation of breeding habitat and its utilization. As it is found along the northeast coast of Bay of Bengal, the loss of breeding habitat is mainly due to the erosion of the intertidal sediment and construction of dikes, stone packing to check erosion, loss of mangroves, disturbances due to fisheries-related activities, etc. Therefore, protection of the natural breeding zone by alternate means is very much required. The identification of sensitive breeding habitats of the horseshoe crabs along the coast line and declaring these areas as protected areas will be helpful in both conservation of the environment and protection of the species.

Acknowledgements The authors would like to thank the Vice Chancellor, Pondicherry University for infrastructure support and Prof S. Sarkar, Department of Marine Sciences, University of Calcutta for providing the sketch of the study area.

References

Berkson J (2009) An integrative approach to horseshoe crab multiple use and sustainability. In: Tanacredi JT, Botton ML, Smith DR (eds) Biology and conservation of horseshoe crabs. Springer, New York, pp 387–398

Berkson J, Shuster CN Jr (1999) The horseshoe crab: the battle for a true multiple user resource. Fisheries 24:6–10

Berkson J, Chen C, Mishra JK et al (2009) A discussion of horseshoe crab management in five countries Taiwan, India, China, United States, and Mexico. In: Tanacredi JT, Botton ML, Smith DR (eds) Biology and conservation of horseshoe crabs. Springer, New York, pp 465–478

Bhattacharya BD, Hwang JS, Tseng L et al (2014) Bioaccumulation of trace elements in dominant mesozooplankton group inhabiting in the coastal regions of Indian Sundarban mangrove wetland. Mar Pollut Bull 87:345–351

Bhattacharya BD, Nayak DC, Sarkar SK et al (2015) Distribution of dissolved trace metals in coastal regions of Indian Sunderban mangrove wetland: a multivariate aproach. J Clean Prod 96:233–243

Brockmann HJ (2004) Nesting behavior a shoreline phenomenon. In: Shuster CN Jr, Barlow RB, Brockmann HJ (eds) The American horseshoe crab. Harvard University Press, Cambridge, MA, pp 33–49

Chatterji A, Vijaykumar R, Parulekar AH (1988) Growth and morphometric characteristics in the horseshoe crab, *Carcinoscorpius rotundicauda* (Latreille) from Canning (West Bengal). India Pak J Sci Ind Res 33:352–355

Chen CP, Hsin-Yi Y, Po-Fen L (2004) Conservation of the horseshoe crab at Kinmen, Taiwan: strategies and practices. Biodivers Conserv 13:1889–1904

Hajeb P, Christianus A, Ismail S et al (2009) Heavy metal concentration in horseshoe crab (*Carcinoscorpius rotundicauda* and *Tachypleus gigas*) eggs from Malaysian coastline. In: Tanacredi JT, Botton ML, Smith DR (eds) Biology and conservation of horseshoe crabs. Springer, New York, pp 455–463

Hurton L, Berkson J, Smith S (2009) The effect of hemolymph extraction volume and handling stress oh horseshoe crab mortality. In: Tanacredi JT, Botton ML, Smith DR (eds) Biology and conservation of horseshoe crabs. Springer, New York, pp 311–346

Itow T (1993) Crisis in the Seto Inland Sea: decimation of the horseshoe crab. EMECS News Lett 3:10–11

Itow T, Sugita H, Sekiguchi K (1991) A phenomenal decrease of the horseshoe crab in the Inland Sea of Japan and its causes. Bull Dept Manag Inf Sci Jobu Univ 4:29–46

Itow T, Mishra JK, Ahmed TA (2004) Horseshoe crabs (King crabs) in the Bay of Bengal, South Asia. Bull Fac Educ Shizuoka Univ Jpn Nat Sci Ser 54:13–30

Kreamer G, Michels S (2009) History of horseshoe crab harvest on Delaware Bay. In: Tanacredi JT, Botton ML, Smith DR (eds) Biology and conservation of horseshoe crabs. Springer, New York, pp 299–313

Mishra JK (1994) Horseshoe crab, an amazing creature of Orissa coast. Project Swarajya Publication, Orissa

Mishra JK (2009) Horseshoe crabs, their eco-biological status along the northeast coast of India and the necessity for ecological conservation. In: Tanacredi JT, Botton ML, Smith DR (eds) Biology and conservation of horseshoe crabs. Springer, New York, pp 89–96

Novisky JT (2009) Biomedical application of *Limulus* amoebocyte lysate. In: Tanacredi JT, Botton ML, Smith DR (eds) Biology and conservation of horseshoe crabs. Springer, New York, pp 315–329

Seino S, Uda T, Tsuchiya Y et al (2003) Conservation history of *Tachypleus tridentatus* and its spawning ground, a designated natural monument in Kasaoka Bay in Okayama Prefecture. Asian Pac Coasts 2003:1–14

Sekiguchi K (1988) Ecology. In: Sekiguchi K (ed) Biology of horseshoe crabs. Science House Co. Ltd., Tokyo, pp 50–68

Sekiguchi K, Shuster CN Jr (2009) Limits on the global distribution of horseshoe crabs (Limulacea): lessons learned from two lifetimes of observations: Asia and America. In: Tanacredi JT, Botton ML, Smith DR (eds) Biology and conservation of horseshoe crabs. Springer, New York, pp 5–24

Sekiguchi K, Nishiwaki S, Makioka T et al (1977) A study on the egg laying habits of horseshoe crabs, *Tachypleus gigas* and *Carcinoscorpius rotundicauda*, in Chonburi area of Thailand. Proc Jap Soc Syst Zool 13:39–45

Shin PKS, Li HY, Cheung SG (2009) Horseshoe crabs in Hong Kong: current population status and human exploitation. In: Tanacredi JT, Botton ML, Smith DR (eds) Biology and conservation of horseshoe crabs. Springer, New York, pp 347–360

Shuster CN Jr (2004) King crab fertilizer: a once-thriving Delaware Bay industry. In: Shuster CNJ, Barlow RB, Brockmann HJ (eds) The American horseshoe crab. Harvard University Press, Cambridge, MA, pp 341–357

Shuster CN Jr, Sekiguchi K (2009) Basic habitat requirements of the extant species of horseshoe crabs (Limulidae). In: Tanacredi JT, Botton ML, Smith DR (eds) Biology and conservation of horseshoe crabs. Springer, New York, pp 115–129

Shuster CN Jr, Botton ML, Loveland RE (2004) Horseshoe crab conservation: a coast-wide management plan. In: Shuster CN Jr, Barlow RB, Brockmann HJ (eds) The American horseshoe crab. Harvard University Press, Cambridge, MA, pp 358–377

Swan BL (2001) A unique medical product (LAL) from the horseshoe crab and monitoring the Delaware Bay horseshoe crab population. In: Tanacredi JT (ed) *Limulus* in the lime light. Kluwer Academic/Plenum, New York, pp 53–62

Tsuchiya K (2009) The history of horseshoe crab research and conservation in Japan. In: Tanacredi JT, Botton ML, Smith DR (eds) Biology and conservation of horseshoe crabs. Springer, New York, pp 559–570

Weber RG, Carter DB (2009) Distribution and development of *Limulus* egg clusters on Intertidal beaches in Delaware Bay. In: Tanacredi JT, Botton ML, Smith DR (eds) Biology and conservation of horseshoe crabs. Springer, New York, pp 249–266

Yang MC, Chen CP, Hsieh HL et al (2009) Is a living fossil about to go locally extinct? No mitochondrial genetic variation in horseshoe crab juveniles *Tachypleus tridentatus* at Haomeiliao Nature Reserve, Budai, Taiwan. Zool Studies 48(6):737

Chapter 24
Assessment and Management of North American Horseshoe Crab Populations, with Emphasis on a Multispecies Framework for Delaware Bay, U.S.A. Populations

Michael J. Millard, John A. Sweka, Conor P. McGowan, and David R. Smith

Abstract The horseshoe crab fishery on the US Atlantic coast represents a compelling fishery management story for many reasons, including ecological complexity, health and human safety ramifications, and socio-economic conflicts. Knowledge of stock status and assessment and monitoring capabilities for the species have increased greatly in the last 15 years and permitted managers to make more informed harvest recommendations. Incorporating the bioenergetics needs of migratory shorebirds, which feed on horseshoe crab eggs, into the management framework for horseshoe crabs was identified as a goal, particularly in the Delaware Bay region where the birds and horseshoe crabs exhibit an important ecological interaction. In response, significant effort was invested in studying the population dynamics, migration ecology, and the ecologic relationship of a key migratory shorebird, the Red Knot, to horseshoe crabs. A suite of models was developed that linked Red Knot populations to horseshoe crab populations through a mass gain function where female spawning crab abundance determined what proportion of the migrating Red Knot population reached a critical body mass threshold. These models were incorporated in an adaptive management framework wherein optimal harvest decisions for

The findings and conclusions in this article are those of the author(s) and do not necessarily represent the views of the U.S. Fish and Wildlife Service. Any use of trade, firm, or product names is for descriptive purposes only and does not imply endorsement by the U.S. Government.

M.J. Millard (✉) • J.A. Sweka
U.S. Fish and Wildlife Service, Northeast Fishery Center, Lamar, PA, USA
e-mail: mike_millard@fws.gov; john_sweka@fws.gov

C.P. McGowan
U.S. Geological Survey, Alabama Cooperative Fish and Wildlife Research Unit, Auburn University, Auburn, AL, USA
e-mail: cmcgowan@usgs.gov

D.R. Smith
U.S. Geological Survey, Leetown Science Center, Leetown, WV, USA
e-mail: drsmith@usgs.gov

R.H. Carmichael et al. (eds.), *Changing Global Perspectives on Horseshoe Crab Biology, Conservation and Management*, DOI 10.1007/978-3-319-19542-1_24

horseshoe crab are recommended based on several resource-based and value-based variables and thresholds. The current adaptive framework represents a true multispecies management effort where additional data over time are employed to improve the predictive models and reduce parametric uncertainty. The possibility of increasing phenologic asynchrony between the two taxa in response to climate change presents a potential challenge to their ecologic interaction in Delaware Bay.

Keywords Red knot • Multispecies management • Population modeling • Delaware Bay • Shorebird migration • Stock assessment • Phenology • Shorebird energetics

24.1 Introduction

The horseshoe crab fishery represents an interesting case study of a resource management program that has progressed from casual indifference to a sophisticated multispecies-based harvest framework, almost all within the twenty-first century. The reasons for this relatively rapid 15-year rise of the horseshoe crab in both public and scientific discussions are multifaceted and compelling. Numerous summaries of the pertinent economic, social, and ecologic issues have been published and continue to be debated today (Berkson and Shuster 1999; Walls et al. 2002; Berkson 2009; Okun 2012). With little recognition prior to the early 2000s, the horseshoe crab has supported important processes involving domestic agriculture, global fishery markets, human health and safety, and trans-continental migratory shorebird populations (Walls et al. 2002). As the various roles of the horseshoe crab have become more understood and more widely publicized, fishery scientists and managers have worked to continually improve management of the multi-user resource.

Harvest of horseshoe crabs along the Atlantic coast has dramatically fluctuated through time. Prior to 1998, harvest of horseshoe crabs was unregulated in most states. In addition to their use as bait in other fisheries and the biomedical industry, they were also commercially harvested for use as fertilizer and livestock feed from the mid-1800s through the mid-1900s. During this time period harvest ranged from approximately one to five million crabs annually (Shuster 1960, 1982; Shuster and Botton 1985; Finn et al. 1991). Harvest numbers dropped to between 250,000 and 500,000 in the 1950s (Shuster 1960), which was followed by a sharp increase in reported harvest in the late 1990s due to the rising demand for bait for the whelk *Busycon* spp fishery (Fig. 24.1). Harvest estimates prior to 1998 should be viewed with caution because harvest reporting was not mandatory prior to 1998 and under-reporting of actual harvest may have been significant.

As described in Smith et al. (2009b), the Atlantic States Marine Fisheries Commission (ASMFC) began formal management measures for horseshoe crabs in 1998 with the establishment of the Interstate Fishery Management Plan (IFMP) for Horseshoe Crab (ASMFC 1998). Specifically, the goal of the management plan is "*management of horseshoe crab populations for continued use by: current and*

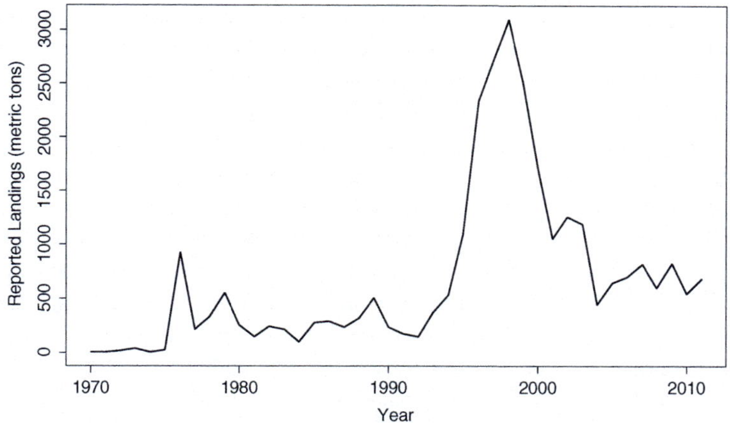

Fig. 24.1 Reported U.S. Atlantic coast horseshoe crab landings (metric tons), 1970–2011 (NMFS Commercial Landings Database August 2013)

future generations of the fishing and non-fishing public (including the biomedical industry, scientific and educational research), migratory shorebirds, and other dependent fish and wildlife (including federally listed sea turtles)." State-by-state reporting of horseshoe crab landings was mandated, with the plan requiring each State to "*continue or initiate mandatory monthly reporting of all harvest (including, but not limited to bait fisheries, bycatch, biomedical industry, and scientific and educational research harvest*" (ASMFC 1998). This IFMP mandate served to codify consistent reporting of coastwide landings data, which up to then had been inconsistent. Addendum I to the FMP established state-by-state reference period landings (RPL) for the bait fishery and implemented a 25 % reduction from each state's RPL (ASMFC 2000). Subsequent addenda to the IFMP further reduced harvest quotas, implemented seasonal bait harvest closures, and mandated male only fisheries in some or all of the four Delaware Bay states (DE, MD, NJ, and VA) (ASMFC 2004, 2006, 2008). States are allowed to implement more conservative measures than mandated by ASMFC, and New Jersey adopted a moratorium on all bait harvest in 2006. Compared to 1998 when the IFMP was established, landings in 2012 were reduced by 74 % due to the combined cumulative management actions imposed by ASMFC and states. Landings in the four Delaware Bay states decreased 79 % from 1998 to 2012.

Due to a lack of data to support traditional fishery management analyses and identification of reference points for appropriate harvest levels, the initial harvest regulations were implemented primarily with the goal of being precautionary and risk-averse. Managers recognized that management decisions needed to be more scientifically defensible and tasked their technical experts with assembling the relevant data and developing the necessary and appropriate decision support tools, particularly in the Delaware Bay region where the bulk of the harvest occurred. This approach required a greater understanding of horseshoe crab population dynamics and demographics. Stock assessments within the ASMFC technical group became progressively more insightful, time series of dependable data expanded, and initial efforts to apply traditional fishery assessment models (e.g. surplus production and catch survey) became feasible (ASMFC 2009).

Coincident with the progress towards applying the single-species approach to managing the horseshoe crab fishery was the growing need to include the ecological needs of migratory shorebirds, particularly the Red Knot *Calidrus canutus rufa,* in the fishery's management regime. During spring, northward migrating shorebirds on the Atlantic coast of North America rely heavily on horseshoe crab eggs as a food resource. Red Knots that stopover in Delaware Bay have been dietarily and physiologically linked to horseshoe crab eggs (Niles et al. 2008; Gillings et al. 2009; Fraser et al. 2010; Karpanty et al. 2011). In the early 2000s, following the large increases in horseshoe crab harvest, Red Knot populations in wintering sites in Tierra del Fuego (Argentina and Chile) and stopover sites in Delaware Bay exhibited steep apparent declines in abundance, which raised serious concerns about Red Knot population viability (Baker et al. 2004; Morrison et al. 2004; Niles et al. 2008; Gillings et al. 2009). Some researchers proposed and presented support for the hypothesis that unregulated harvests in the late 1990s caused a decline in horseshoe crab populations in Delaware Bay, which reduced spawning activity and eggs available for migrating shorebirds to consume, which then led to shorebird population collapse (Baker et al. 2004; Niles et al. 2009). As such, identification and refinement of Red Knot population dynamics, as related to the migratory energetic demands supplied during their stopover in the Delaware Bay, was needed to begin framing a multispecies approach to managing the horseshoe crab fishery. As a result, since 2000, significant effort was invested in studying Red Knot population dynamics, their migration ecology, and their ecologic relationship to horseshoe crabs (Morrison et al. 2007; Cohen et al. 2009; Gillings et al. 2009; Karpanty et al. 2011; McGowan et al. 2011a).

The growing body of information on both horseshoe crabs and Red Knots was integrated in an adaptive multispecies management framework designed to recommend harvest management strategies that were consistent with the needs for rebuilding the Red Knot population. Prior to this multispecies framework, the single-species approach for managing horseshoe crab harvest incorporated only conceptual linkages with Red Knots and was frequently mired by conflicting values among various stakeholder groups. The process of developing the adaptive management framework helped the stakeholders reduce conflict by providing the opportunity to incorporate all stakeholder values along with best available science (McGowan et al. 2009; Williams and Brown 2012). In February 2012 under Addendum VII of the Fishery Management Plan, the Atlantic States Marine Fisheries Commission (ASMFC) adopted the resultant multispecies adaptive management framework for sustainable harvest of Delaware Bay horseshoe crabs while maintaining adequate stopover habitat for migrating Red Knots, the focal shorebird species (McGowan et al. 2009; ASMFC 2012).

In the following sections, we describe and summarize the recent research on (1) horseshoe crab status and trends, as well as emerging stock assessment analyses, (2) Red Knot populations with specific focus on Delaware Bay and their interaction with horseshoe crabs, and (3) the adaptive resource management framework that provides for fishery management recommendations in a multispecies context.

24.2 Horseshoe Crab Assessment and Status Along the U.S. Atlantic Coast

Prior to the adoption of the 1998 fisheries management plan for horseshoe crabs, the status of populations along the coast was largely unknown. Quantitative assessments of the population have advanced greatly with implementation of mandatory harvest reporting and the additional fisheries independent data that have become available since 1998. The initial stock assessment (ASMFC 1999) relied on an analysis of trends in relative abundance from fisheries independent surveys (primarily trawl surveys targeting other species). A subsequent stock assessment in 2004 (ASMFC 2004) continued the analysis of trends and also added a meta-analysis (Manly 2001) to examine consistency of trends within geographic regions along the Atlantic Coast. At this time a framework was proposed to utilize a catch-survey model (Collie and Sissenwine 1983) to assess the population in the Delaware Bay region, but sufficient data for the model were not available. The next stock assessment was conducted in 2009 (ASMFC 2009), which continued the meta-analysis of regional trends in fisheries independent data sources and also added analysis of trends with autoregressive integrated moving average (ARIMA) models (Pennington 1986). By 2009, enough data were available to fit the catch-survey model and a surplus production model (Prager 1994) for the Delaware Bay region. These efforts represented the very first estimates of population status of the Delaware Bay relative to fisheries management benchmarks and provided the first quantitative insight into exploitation rates.

Box 24.1: Descriptions of Fisheries Models Applied to Horseshoe Crab Assessment in the 2009 ASMFC Stock Assessment (ASMFC 2009)

Surplus Production Model – A surplus production model is a simple population model that assumes the population grows in a logistic manner and that population growth is maximized at half of the carrying capacity. A surplus production model assesses a population in its entirety with no age-, size-, or stage-structure. It requires a time series of harvest and catch-per-effort data from a fishery independent survey. Results of a surplus production model give a time series of relative biomass and relative fishing mortality.

Catch-Survey Model – A catch survey model is a two-stage stock assessment model with pre-recruits and fully recruited animals. The recruit stage consists of animals that recruit just before or during the current time step of the model and the fully recruited animals consist of all other individuals. The model requires time series of harvest, and relative abundance of pre-recruits and fully recruited animals. Results from a catch-survey model give estimates of abundance and fishing mortality.

24.2.1 Regional Trends

Assessment of horseshoe crabs has adopted a regional approach as information from tagging (ASMFC 2009, 2013) and genetic analysis (King et al. 2003, 2005) found evidence for population structure. These regions include: New England (Rhode Island and areas north of Cape Cod), New York (embayments of Long Island and Long Island Sound), Delaware Bay (areas from the New Jersey Coast southward to Virginia), and the Southeast (areas south of Virginia). The most recent stock assessment for horseshoe crabs was conducted in 2013 (ASMFC 2013) and was an update of the 2009 stock assessment (ASMFC 2009). The 2013 stock assessment update relied primarily on the use of ARIMA models. Time series of horseshoe crab relative abundance indices, like other fisheries data, reflect two sources of variation: (1) measurement error arising from sampling variation; and (2) process error arising from actual changes in population abundance (Pennington 1986). The ARIMA approach filters the measurement error from the process variability and derives fitted estimates of abundance over the entire time series whose variance is less than the variance of the observed time series (Pennington 1986). An extension of the ARIMA models was used to infer population status relative to an index-based reference point (Helser and Hayes 1995) in the 2013 stock assessment. An index-based reference point is a reference point equivalent to some value within the observed range of values for the relative abundance index. This bootstrapping method of Helser and Hayes (1995) gave the probability of a fitted index value of a particular year (e.g. the terminal year of the survey) being less than the reference point. Chosen reference points were the 25th percentile (Q_{25}) of the fitted abundance index and the 1998 fitted abundance index. The 25th percentile was chosen because it provides a reasonable reference point for comparison for data with relatively high and low abundance over a range of years, and the 1998 fitted abundance index was chosen because that was the year when the fisheries management plan was adopted by ASMFC.

The 2013 stock assessment fit ARIMA models to a total of 39 horseshoe crab survey (primarily trawl surveys) indices of relative abundance (ASFMC 2013). Of these, 9 surveys were from the New England region, 6 from the New York Region, 19 from the Delaware Bay region, and 5 from the Southeast region (Table 24.1). Trends in the fitted abundance indices for these surveys showed much variation among surveys. Surveys with time series extending back into the mid-1990s generally showed decreasing trends through the early 2000s, but showed mixed results from the mid 2000s through the end of their time series in 2011 or 2012. For example, the Southeast Area Monitoring and Assessment Program (SEAMAP) trawl survey increased, the Delaware Bay 30 ft trawl survey remained stable, and the New York Peconic Bay trawl survey continued to decrease (Fig. 24.2).

With the exception of the New England Region, typically less than half of the surveys were less than their reference points with a probability greater than 0.50. For surveys which had adequate ARIMA model fits (normal distributions of residuals), 9 out of 33 surveys coast-wide (27 %) had 2011 or 2012 indices that were less

Table 24.1 Fisheries-independent surveys to which ARIMA models were fit in the 2013 stock assessment (ASMFC 2013)

Survey	Years of data	P (<1998)	P (< Q_{25})
New England Region			
Massachusetts Inshore Bottom Trawl (fall)	1978–2012	0.91	0.72
Massachusetts Inshore Bottom Trawl (spring)	1978–2012	0.96	0.55
New Hampshire Spawning Survey (spring)	2001–2012		0.34
New Hampshire Spawning Survey (spring – summer)	2001–2009		0.47
Rhode Island – Marine Research Inc.	1988–2012	0.87	0.53
Rhode Island – Marine Research Inc. Providence River Impingement	1992–2012	0.86	0.52
Rhode Island – Stout Survey	1975–2002	0.01	0.25
Rhode Island DFW Trawl	1998–2012	0.18	0.69
University of Rhode Island – Graduate School of Oceanography	1959–2012	1.00	1.00
New York Region			
CT Long Island Sound Trawl (fall)	1992–2012	0.22	0.11
CT Long Island Sound Trawl (spring)	1992–2012	0.11	0.08
NY Peconic Bay Trawl Survey	1987–2012	1.00	0.93
NY Western Long Island Beach Seine – Jamaica Bay	1987–2012	0.03	0.01
NY Western Long Island Beach Seine – Little Neck Bay	1987–2012	0.93	0.44
NY Western Long Island Beach Seine – Manhasset Bay	1987–2012	0.82	0.24
Delaware Bay Region			
Delaware 16 ft trawl (Juvenile)	1992–2012	0.42	0.26
Delaware 16 ft trawl (YOY)	1992–2012	0.38	0.04
Delaware 30 ft trawl (all HSC)	1990–2012	0.76	0.20
Delaware 30 ft trawl (female)	1990–2012	1.00	0.61
Delaware 30 ft trawl (male)	1990–2012	0.77	0.21
Delaware Bay Spawning Survey (female)	1999–2012		0.54
Delaware Bay Spawning Survey (male)	1999–2012		0.05
Maryland Coastal Bay	1990–2012	0.15	0.08
NJ Delaware Bay Trawl (female)	1998–2012	0.99	0.19
NJ Delaware Bay Trawl (male)	1998–2012	0.08	0.02
NJ Delaware Bay Trawl (all HSC)	1998–2012	0.09	0.04
NJ Delaware Bay Trawl (Juvenile)	1998–2012	0.15	0.08
NJ Ocean Trawl	1988–2012	0.87	0.28
NJ Surf Clam Dredge	1998–2012	0.00	0.00
NMFS bottom trawl survey (fall)	1988–2008	0.05	0.14
NMFS bottom trawl survey (spring)	1988–2008	0.17	0.05
Virginia Tech Trawl (all HSC)	2002–2011		0.10
Virginia Tech Trawl (female)	2002–2011		0.19
Virginia Tech Trawl (male)	2002–2011		0.06

(continued)

Table 24.1 (continued)

Survey	Years of data	P (<1998)	P (< Q_{25})
Southeast Region			
Florida Seahorse Key (Gulf) Spawning Survey[a]	1993–2010		0.00
Georgia Shrimp Trawl	1999–2012		0.06
NC Pamlico Sound Neuse River Gill Net	2001–2012		0.00
SEAMAP Trawl Survey	1995–2012	0.00	0.00
South Carolina Trawl	1995–2012	0.29	0.69

P (<1998) is the probability that the terminal year of the survey was less than the 1998 fitted index value and P (Q_{25}) is the probability that the terminal year of the survey was less than the 25th percentile of fitted index values. Some surveys did not have an associated P (<1998) because they began after 1998. See ASMFC (2013) for a full description of each of the surveys

[a]The Florida Seahorse Key (Gulf) Spawning survey was not conducted in 1998 or 1999, and a fitted value for P(<1998) could not be calculated

than the Q_{25} reference point and 13 out of 24 surveys (54 %) were less than the 1998 reference point. The number of surveys available to compare to the 1998 reference point was less than the number available to compare to the Q_{25} reference point because several surveys were not initiated until after 1998. In the New England region six out of seven surveys (86 %) were less than the Q_{25} reference point and five out of six (83 %) were less than the 1998 reference point. In the New York region, one out of five surveys (20 %) was less than the Q_{25} and three out of five (60 %) were below the 1998 reference point. Within the Delaware Bay region, 2 out of 16 (13 %) had 2011 or 2012 indices that were less than Q_{25} and 4 out of 11 (36 %) were less than the 1998 reference point. In the Southeast region, only one out of five surveys (20 %) was less than the Q_{25} reference point (Table 24.2).

The differences among regions in the proportion of surveys that were below their reference points in the 2013 stock assessment emphasizes the importance of management, regulation, and monitoring or horseshoe crabs on a regional or localized scale. In general, abundance indices suggest increases in the Southeast region, possible increases or at least stabilization in the Delaware Bay region, and continued population declines in the New York and New England region. Although harvest of horseshoe crabs for the bait industry has declined coast-wide since the late 1990s, the current level of harvest in the New York and New England regions may still be in excess of what those populations can sustain. The data and models necessary to quantitatively determine sustainable levels of harvest (e.g. maximum sustainable yield) in any of the regions are not yet available.

24.2.2 *Biomedical Harvest*

Despite the advances in assessment of horseshoe crab along the Atlantic Coast over the years, accounting for the mortality induced by the biomedical industry during the bleeding process remains problematic. The biomedical industry returns many of

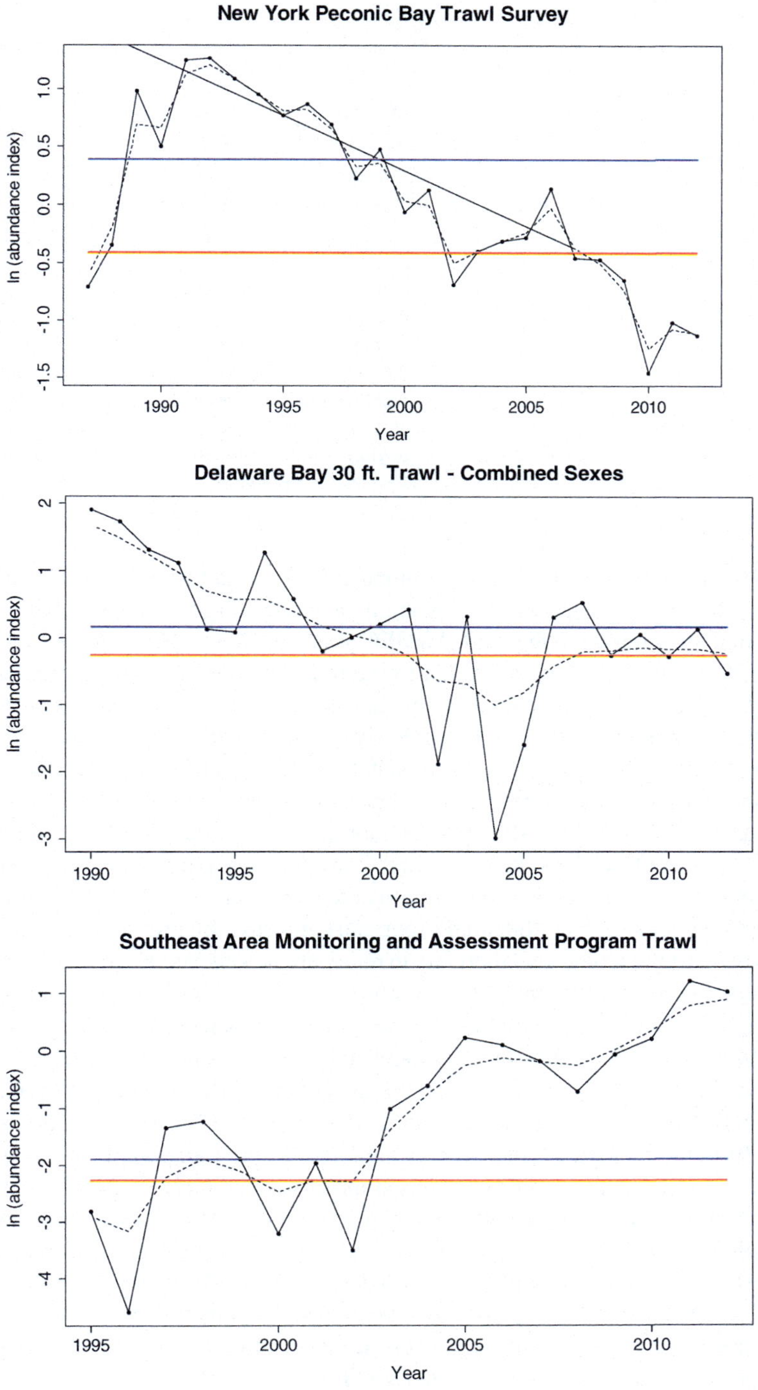

Fig. 24.2 Examples of autoregressive integrated moving average (ARIMA) model fits to horseshoe crab indices of abundance. The *solid lines* are the observed abundance indices, the *dashed line* is the ARIMA fit, the *blue horizontal lines* represent the 1998 reference points, and the *red horizontal lines* represent the 25th percentile reference points

Table 24.2 Summary of ARIMA model results from the 2013 ASMFC horseshoe crab assessment (ASMFC 2013)

Region	P (<1998) >0.50	P (<Q_{25}) >0.50
New England	5 out of 6	6 out of 7
New York	3 out of 5	1 out of 5
Delaware Bay	4 out of 11	2 out of 16
Southeast	0 out of 2	1 out of 5
Coast-wide	12 out of 24	10 out of 33

The values represent the number of surveys with terminal year having a greater than 0.50 probability of being less than the reference point (i.e. likely less than the reference point). Time series were only included in this summary if the terminal year was 2011 or 2012 and residuals from ARIMA model fits were normally distributed. Those that ended earlier are not included. Also, those surveys that did not begin until after 1998 were not included in the P ($i_f < i_{1998}$) >0.50 summary

the bled crabs to the wild following blood collection and also transfers some bled crabs to the bait industry where they are then accounted for in the bait harvest. For those that are released to the wild, ASMFC assumes 15 % suffer mortality (ASMFC 2013). This mortality rate is applied to the total number of crabs reported as bled and released by the biomedical companies coast-wide. The coast-wide estimated mortality of crabs by the biomedical industry has increased 78 %, from approximately 46,000 individuals in 2004 to approximately 80,000 in 2012. In 2012, 10 % of the total coast-wide reported mortality of horseshoe crabs came from the biomedical industry (ASMFC 2013). Although the mortality due to the biomedical industry is small compared to bait industry, we consider it significant for quantitative assessments. The limited number of biomedical companies along the Atlantic coast and policies regarding confidential proprietary information prevent the inclusion of company-specific biomedical mortality in regional assessments. The use of biomedical mortality in coast-wide analyses is permissible, but analysis on a regional scale would violate data confidentiality with individual companies. This policy prevented the application of the catch-survey model and the surplus production model, developed during the 2009 stock assessment in the Delaware Bay region (ASMFC 2009), from being included in the 2013 stock assessment (ASMFC 2013). While the mortality from the biomedical industry was ignored during initial applications of these models, including this mortality in current models would increase their applicability. Because of the data confidentiality policy precluding use of the biomedical mortality, the utility of these models was compromised such that they were not included in the 2013 assessment (ASMFC 2013). To date, no biomedical company has granted ASMFC permission to disclose individual company data. The advancement of horseshoe crabs assessment on a regional scale would benefit greatly with the availability of regional biomedical mortality data (Novitsky 2015).

24.3 Migratory Red Knots and Their Relationship to Horseshoe Crabs in Delaware Bay

Red Knots are migratory shorebirds, and the Atlantic flyway population in the western hemisphere is classified as a distinct subspecies *C. c. rufa*. The *rufa* population historically spent the majority of the non-breeding season in the Tierra del Fuego region of Argentina and Chile (Baker et al. 2013). Additional birds were known to overwinter in northern Brazil and the southeastern U.S., primarily in Florida (Baker et al. 2013). From March through May each year, the birds migrate from their wintering locations to the breeding grounds in the central Arctic regions of Canada, periodically stopping at several sites along the way (Baker et al. 2013). In Delaware Bay, Red Knots and a variety of other migratory shorebird species gather in vast numbers to exploit seasonally available horseshoe crab eggs deposited on the beach by spawning female crabs (Gillings et al. 2009; Karpanty et al. 2011). Researchers have shown that Red Knot behavior and spatial distribution is strongly associated with horseshoe crab spawning and egg availability in the bay (Gillings et al. 2009; Cohen et al. 2009; Fraser et al. 2010; Karpanty et al. 2011).

Using physiological models and flight and fuel requirements, Baker et al. (2004) proposed that Red Knots departing Delaware Bay and migrating to the arctic need to weigh at least 180 g at the time of departure in order to survive the transition to the arctic (Kvist et al. 2001, as cited by Baker et al. 2004). Studies of the Icelandic sub-species of Red Knot, *C. c. islandica,* indicate that in years of extreme cold weather late in the spring, birds that arrive in the Arctic breeding grounds at low weights have lower survival rates (Morrison 2006; Morrison et al. 2007). Studies have reported that the proportion of birds in Delaware Bay achieving the 180 g mass threshold greatly declined between 2000 and 2008 following the heavy, unregulated horseshoe crab harvests in the late 1990s (Baker et al. 2004; Atkinson et al. 2007; Niles et al. 2009). Baker et al. (2004) further proposed that the primary reason for the *rufa* Red Knot population decline was that too few birds were achieving that mass threshold during stopover in Delaware Bay and were therefore suffering mortality during their subsequent migratory flight to the Arctic.

More recent research by McGowan et al. (2011a) strongly support the hypothesis that Red Knot mass gain during stopover in Delaware Bay was closely associated with adult female horseshoe crab abundance. Using capture-mark-recapture/resight data from Delaware Bay collected between 1998 and 2008, they developed a multi-state mark recapture/resight model that tracked bird arrival, departure, annual survival, and used a multi-state structure to establish mass classes for birds in the Bay, i.e. birds above and below 180 g. The analysis estimated the probability of transition between mass classes as a function of horseshoe crab abundance covariates (McGowan et al. 2011a). The analysis demonstrated a strong association between estimated adult female horseshoe crab abundance and the probability of transitioning from the lower mass class to the higher mass class. However, McGowan et al. (2011a) reported weak associations between departure mass class and subsequent annual survival, i.e., survival of lighter weight birds was slightly lower than that of heavy birds, and all birds survived greater than 90 % annually on average across years. They found that annual adult survival was closely associated with snow depth in the arctic in early June, where survival rates were higher in years with more snow on the ground on June 10th (McGowan et al. 2011a).

Recent Red Knot mass data suggest improved stopover conditions; the proportion of birds achieving the 180 g threshold between May 26th and 28th has steadily increased from 2006 to 2010 (Dey et al. 2011). As horseshoe crab eggs have increased on the beaches of Delaware Bay, Red Knot mass at the end of the stop over season has improved (Dey et al. 2011). However, the metric of proportion of birds over the 180 g threshold at the end of the stopover season has limited inferential capacity since birds that achieved that mass threshold may have already departed the Bay or may not be actively foraging. Cannon netting (trapping) efforts tend to target foraging flocks because their movement patterns are more predictable, but cannon netting can lead to biased catches and limited inferences (Clark 1980; Pienkowski and Dick 1976). Additionally McGowan et al. (2011a) found very limited survival consequences of the mass threshold in their multistate survival analysis. McGowan et al. (2011b) demonstrated through simulation modeling that if underweight birds fail to breed, significant population decline is possible, but there is a paucity of data on *rufa* Red Knot reproductive biology (Baker et al. 2013).

24.3.1 Red Knot Population Estimates and Trends

Infrequent and unstandardized surveys in the 1980s and 1990s indicated that there may have been over 120,000 Red Knots using Delaware Bay in the early 1990s (Niles et al. 2008). Aerial surveys are riddled with error and uncertainty (Pollock and Kendall 1987; Frederick et al. 2003) and have extensive limitations as a shorebird monitoring method. However, in the 1980s and 1990s there were years with greater than 70,000 Red Knots counted in Delaware Bay (Clark et al. 1993; Niles et al. 2008), and counts in the Tierra del Fuego wintering area regularly encountered more than 50,000 birds (Morrison et al. 2004; Niles et al. 2008). Between 2000 and 2008, aerial counts of stopover sites in Delaware Bay were variable but generally declined and have since stabilized (Niles et al. 2009). Between 2009 and 2011, peak aerial counts in Delaware Bay showed signs of increasing (Dey et al. 2011). Mark-recapture based estimates of abundance for Red Knots that stopover in Delaware Bay indicate that their population abundance has been very stable since 2011 (Lyons et al. in review, see below). The Tierra del Fuego populations exhibited significant declines in the early 2000s and subsequently stabilized at approximately 10,000–15,000 birds with annual variation (Dey et al. 2011; Baker et al. 2013).

24.3.2 Population Projection Models

Baker et al. (2004) presented the first population projection model, and they predicted rapid and continued decline for *rufa* Red Knots in the 2000s and the potential for extinction as early as 2010. Baker et al. (2004) estimated survival rates of ~0.50 for Red Knots that did not make the 180 g mass threshold during stopover and included very low annual average survival rates in their projections. While Red Knots in Delaware Bay did decline significantly in the early 2000s, by 2010 the population had evidently stabilized and possibly begun to increase (Dey et al. 2011). The results of the Baker et al. (2004) model have been used to support efforts to curb and, in some locales, eliminate horseshoe crab harvest (Niles et al. 2009).

As part of a larger adaptive management effort, McGowan et al. (2011b) developed a simulation model to predict Red Knot and horseshoe crab abundance. Their model co-simulated both populations and incorporated horseshoe crab-dependent mass gain dynamics for Red Knots in the Delaware Bay region to evaluate the effects of horseshoe crab harvest actions/strategies on population viability for the birds (McGowan et al. 2011b). The horseshoe crab components of the model built upon the Sweka et al. (2007) population model and added dual sex components to allow for sex specific harvest evaluations. The males and females were linked via a sex ratio fertility function (based on Caswell 2001) in which fertilization rates declined if horseshoe crab operational sex ratio fell below the critical 2:1 male to female ratio. McGowan et al. (2011b) established the critical threshold for the sex ratio linked fertility function relying on behavioral ecology studies (Brockmann 2003), and data from un-harvested populations in Taunton Bay, ME and Seahorse Key, FL (McGowan et al. 2009, 2011b).

The McGowan et al. (2011b) model linked Red Knot populations to horseshoe crab populations through a mass gain function where female spawning crab abundance determined what proportion of the migrating Red Knot population reached the 180 g mass threshold (based on the results from McGowan et al. 2011a). In the simulation model differential survival could be applied to those that reached the mass threshold and those that did not (McGowan et al. 2011b). McGowan et al. (2011b) ran a set of simulations that included vast differences in survival of heavy and light birds (the "Baker type model," following Baker et al. 2004) and slight differences between the groups (the "weak survival effect" model, reflecting the results of McGowan et al. 2011a). McGowan et al. (2011b) reported that even with model uncertainty, a threshold based restriction on harvest, whereby crabs could only be harvested if Red Knot abundance exceeded a pre-specified threshold, allowed both populations to rebound and increase after several years. Further, McGowan et al. (2011b) assessed that the Baker type model, with significant survival consequences for light mass birds, was consistent with the observed population count data for Red Knots in Delaware Bay between 1998 and 2008. They also concluded that the observed population counts were within the 95 % confidence interval of the abundance predictions under the weak survival effect model and that model comparisons to the observed data were inconclusive (McGowan et al. 2011b).

24.3.3 Alternative Theories for Declines

The population decline in Red Knots was quickly attributed to the horseshoe crab harvest and a decline in horseshoe crab egg availability on Delaware Bay beaches without consideration of alternative hypotheses (Niles et al. 2009). Some available data support this hypothesis (Niles et al. 2009), but other data are inconclusive or even contradictory (e.g., McGowan et al. 2011a; Karpanty et al. 2011; Schwarzer et al. 2012). In light of these data, alternative causal mechanisms for the decline have been proposed with varying degrees of formality and analytical support. For example, Robinson et al. (2003) reported that birds arriving late in Delaware Bay were particularly susceptible to mass gain stress during stopover, and they suggested that birds may be arriving in the bay later in the season due to problems in the wintering grounds or at previous stop in the migration flyway. Robinson et al. offered explanations as to why birds' arrival might be delayed but found no conclusive evidence to support the explanations.

Fraser et al. (2012) proposed that Red Knot population collapse may be explained by a cessation or alteration of lemming cycles in the Arctic breeding grounds. Lemming cycles are driven by snow fall cycles whereby, in years of heavy snow in the Arctic, lemming overwinter survival is higher, and the populations exhibit spring and summer booms following years of heavy winter snows. Fraser et al. proposed that in years with lemming population booms, Arctic predators (e.g., Arctic fox, snowy owl) meet all of their energetic needs consuming super-abundant lemmings, and therefore predation pressures on Red Knot nests and fledglings are greatly reduced. Snow fall fluctuates cyclically in a 3–4 year pattern, and alterations in those patterns have been associated with shorebird population decline in Norway and other sites in the Arctic. Fraser et al. (2012) presented limited evidence of historic Red Knot population cycles at approximately 4 year intervals and showed evidence that the cycles ended in the late 1990s, coinciding with Knot population collapse in the early 2000s. The evidence was inconclusive because Fraser et al. (2012) did not have enough successive years of population monitoring data to detect a cyclical pattern with statistical significance, but the limited data do support the hypothesis. It is also noteworthy that the results of the McGowan et al. (2011a) multistate survival analysis show that Red Knot survival was positively correlated with snow depth in early June; this result is consistent with the lemming hypothesis.

24.3.4 Ongoing Monitoring and Research

Significant uncertainty remains regarding Red Knot population dynamics and the effect of horseshoe crab abundance on Red Knot population viability. Extensive monitoring efforts persist in Delaware Bay as part of ongoing management decision making and scientific research. Annually, the states of Delaware and New Jersey run banding and resighting programs to enhance the extensive database. Those data are used to update survival analyses, study spatial ecology within the Bay, estimate stopover duration, and estimate stopover arrival/departure timing. Perhaps most importantly, the mark-recapture and resight data are being used to estimate annual abundance with significantly more precision and less error and detection bias than aerial surveys (Lyons et al. in review). The estimation methods use mark-recapture data and marked to unmarked ratio data in a hierarchical model that combines the traditional Jolly-Seber model, with a binomial detection model (Lyons et al. in review). The method estimates the "super population size" accounting for the flow through nature of a fly-way population that temporarily stops at a staging site in temporal disunity (i.e., they don't all arrive and depart in unison). Preliminary results estimated super population abundance at approximately 45,000 Red Knots in 2011–2013 (J.E. Lyons, USFWS, personal communication).

24.4 The Adaptive Management Framework for Red Knots and Horseshoe Crabs

Adaptive management is increasingly used for difficult natural resource decisions characterized by recurrent decisions and reducible ecological uncertainty (Walters 1986; McCarthy and Possingham 2007; Martin et al. 2009; Runge 2011; Williams and Brown 2012). Adaptive management draws on principles of structured decision making (Conroy and Peterson 2013) and proceeds by the following steps during a set-up phase: development of the framework proceeded by articulating the management objectives, identifying the feasible alternative management actions, building models to predict the multiple species' population status given the alternative actions, and using mathematical routines to identify the optimal actions–in the sense of best meeting the management objectives. After the set-up phase, adaptive management is implemented iteratively until revision of the set-up is triggered by periodic review or onset of substantial new knowledge. Fully incorporating uncertainty in the predictive models and a well-designed monitoring program are keys to reducing uncertainty and improving future decisions.

24.4.1 Problem Definition and Management Objectives

The ASMFC is the decision maker in this context, and the decision problem was how best to set harvest regulations. Stakeholder meetings were held to discuss and develop the framework (McGowan et al. 2009). The management objective was to sustainably harvest horseshoe crabs, which could include moratoria, given objectives to provide adequate stopover habitat for migrating shorebirds. The objective was reframed into a quantitative format to support predictive modelling and optimization. After deliberation, the consensus objective or 'reward' function was

$$R_t = v_{Ft} 2H_{Ft} + v_{Mt} H_{Mt} \qquad (24.1)$$

where R_t is the reward in year *t*, H_F and H_M are number of female and male horseshoe crabs harvested in year *t*, and ν_{Ft} and ν_{Mt} are value functions determined by the level of female and male horseshoe crab harvest. The value function for females takes the value of 1 if Red Knot abundance exceeds 45,000 (i.e. 22,500 female Red Knot) or female horseshoe crab abundance exceeds 80 % of carrying capacity (i.e. 11 million, based on an age structured population model developed by Sweka et al. (2007)), and takes the value of zero otherwise. The Red Knot threshold of 45,000 was chosen to reflect historic abundances from aerial counts observed in Delaware Bay and is thought to be consistent with a recovering population (McGowan et al. 2009). The horseshoe crab threshold of 80 % carrying capacity is thought to be consistent with adequate egg availability (Smith 2007; Sweka et al. 2007). The value function for males is related to sex ratio, so that the function takes the value of zero when operational sex ratio (sex ratio on the spawning beach, expressed as males:females) is below 2:1, takes the value of 1 when operational sex ratio exceeds 3:1, and changes linearly in between those thresholds. The purpose of the value function for males is to value full fertilization and genetic diversity. The female value is multiplied by 2 because mature females are larger and considered better bait in the eel and conch fisheries; therefore females have at least twice the market value. These value functions and thresholds were developed through consultations and public meetings with managers and stakeholders with additional input from experts on horseshoe crab population ecology, specifically with respect to sex ratio and carrying capacity.

24.4.2 Alternative Management Actions

Because the decision problem addressed how to sustainably harvest horseshoe crabs, the management alternatives were expressed in terms of harvest packages (McGowan et al. 2009). Harvest packages included sex-specific harvest quotas due to male and female crabs having different market value and different contributions to population dynamics. The packages included (1) full moratorium on both sexes, (2) 250,000 males and 0 females, (3) 500,000 males and 0 females, (4) 280,000 males and 140,000 females, (5) 420,000 males and 210,000 females. In addition to the harvest quotas set through adaptive management, there are harvest controls that are constant among packages, such as season closures, trip limits, and a marine refuge off the mouth of Delaware Bay.

24.4.3 Predictive Models

To compare and select among the alternative harvest packages to implement at any given time, predictive models were necessary to express the likely outcome of implementing a package on each species' populations. Stochastic population models were developed for horseshoe crabs (Sweka et al. 2007) and Red Knots (McGowan et al. 2011b), and these models were linked to predict the outcome of implementing harvest packages on horseshoe crabs and Red Knots (McGowan et al. 2009). Uncertainty was incorporated in model parameters and by using a set of three models to represent different levels of functional dependency between horseshoe crabs and Red Knots. Once integrated, the suite of models linked horseshoe crab spawning to Red Knot weight gain and vital rates, i.e., adult survival and fecundity (McGowan et al. 2011b).

24.4.3.1 The Horseshoe Crab Population Model

The model to predict horseshoe crab abundance included age-specific survival, a sex-specific partial recruitment schedule starting at age 9, sex-specific harvest, sex-ratio specific fertilization/fecundity, and density-dependent egg mortality (Sweka et al. 2007; McGowan et al. 2009, 2011b). The horseshoe crab population model was (Sweka et al. 2007):

$$N_{i,j+1,t+1} = \left(N_{i,j,t} - H_{i,j,t}\right)S_{i,j,t}, \tag{24.2}$$

where $N_{i,j,t}$ was pre-harvest abundance, $H_{i,j,t}$ was harvest, and $S_{i,j,t}$ was survival of the ith sex, j^{th} age, and t^{th} time. Harvest quota was distributed proportionately by age among adults (i.e., $H_{i,j,t} = 0$ for subadults). Age-specific probability of recruitment to sexual maturity starting at age 9 was 0.2, 0.45, 0.99, and 1 for females and 0.36 and 1 for males (Smith et al. 2009a). Based on a maximum age for horseshoe crabs of 20 years, starting at age 18 survival rates were 0.08 to simulate senescence (Sweka et al. 2007). The number of age 1 offspring was a product of the number of fertilized eggs per female, egg to hatch survival (i.e., 1- egg mortality), and first year survival. Egg mortality was a non-decreasing function of spawning female abundance due to density-dependent nest disturbance (Smith 2007; Sweka et al. 2007). The number of fertilized eggs was modeled as a product of eggs per female and fertilization rate (McGowan et al. 2009). Fertilization rate declined linearly from one to zero as the population sex ratio (males:females) declined below 1.06. Note that for a population sex ratio of 1.06, the sex ratio during spawning, i.e., the operational sex ratio, is expected to be two males per female (McGowan et al. 2009). Other vital rates and variances were taken from Sweka et al. (2007).

24.4.3.2 The Red Knot Population Models

The set of models to predict Red Knot abundance was based on an age-structured Red Knot population model with differently parameterized relationships between horseshoe crab egg availability and Red Knot weight gain, survival, and fecundity (Sweka et al. 2007; McGowan et al. 2011b). Horseshoe crab eggs that are available for shorebird consumption result from density-dependent exhumation of nests and transportation by waves and swash. These exhumed eggs will not develop due to desiccation or low oxygen (Penn and Brockmann 1994; Nordstrom et al. 2006; Smith 2007; Jackson et al. 2008; Weber and Carter 2009), and shorebird consumption of these non-viable eggs is not thought to limit horseshoe crab population growth. Multi-state mark-recapture modeling was used to parameterize the relationship between Red Knot weight and horseshoe crab spawning abundance during the shorebird stopover period and to estimate the probability of gaining a threshold weight (McGowan et al. 2011a). We used a threshold weight of >180 g to categorize Red Knots into two weight classes (*heavy* and *light*) characterized by different survival and/or reproductive rates (McGowan et al. 2011b). The Red Knot population model was (McGowan et al. 2011b):

$$N_{t+1} = N_t\left[P_t S^H_{a,t} + (1-P_t) S^L_{a,t}\right] + N_{t-1}\left[P_{t-1} F^H_{t-1} + (1-P_{t-1}) F^L_{t-1}\right] S_{f,t-1} S_{j,t}, \quad (24.3)$$

where N_t was abundance of adults, P_t was the probability of gaining the threshold weight, $S^H{}_{a,t}$ and $S^L{}_{a,t}$ were adult survival rates for heavy and light birds, $F^H{}_t$ and $F^L{}_t$ were fecundity rates for heavy and light birds, and $S_{j,t}$ and $S_{f,t}$ were survival rates for juvenile (first and second-year birds). Unlike the adult birds, juvenile birds do not migrate through Delaware Bay during the horseshoe crab spawning period, and thus mass state of juveniles was not modeled.

Three Red Knot population models were parameterized to represent the range of hypothesized effects of weight gain on Red Knot survival and fecundity. The No Effects Model hypothesized equal survival and fecundity for heavy and light birds. The Moderate Effects model hypothesized weight-class specific survival with estimates taken from McGowan et al. (2011b). The Strong Effects model hypothesized weight-class specific survival with estimates taken from Baker et al. (2004). Fecundity was weight-specific in the Moderate and Strong Effects Models with parameter estimates taken from McGowan et al. (2011b). Variances in the vital rates were from McGowan et al. (2011b) and McGowan et al. (2011a). Model weights reflect the relative degrees of faith in each of the three models (e.g. Williams et al. 2002). The initial model weights were set based on a consensus process of members of the ASMFC technical working group to 0.2, 0.4, and 0.4 for the No Effects, Moderate Effects, and Strong Effects Models, respectively (McGowan et al. 2009).

24.4.4 Model Optimization

Adaptive stochastic dynamic programming, implemented within the software ASDP (Lubow 2001; Williams et al. 2002; Martin et al. 2009), was used to develop optimal harvest recommendations from the predetermined set of possible harvest policy options based on the reward function (Eq. 24.1) that included both harvest value for horseshoe crabs and Red Knot abundance. The optimization results in a decision matrix used to identify the optimal harvest package for the following year given the most recent estimates of horseshoe crab and Red Knot abundance.

The technical aspects of optimization start with the algorithm within ASDP, derived from Bellman's equation (Miranda and Fackler 2002; Williams et al. 2002), as given by:

$$V_t\left(\underline{N}_t, \underline{w}_t\right) = \max_{A_t}\left\{R\left(\underline{N}_t,\ \underline{w}_t,\ A_t\right) + \delta \cdot \mathrm{E}\left[V_{t+1}\left(\underline{N}_{t+1}, \underline{w}_t + 1 \middle| \underline{N}_t, \underline{w}_t, At\right)\right]\right\}, \quad (24.4)$$

where V_t was the maximized utility that depends on the state variables (i.e., the vectors of abundances, $\underline{N}_t$ and model weights, w_t), the reward function (R) was maximized over all harvest policy options (A_t), δ is a discount factor, and the expectation operator, E[·], is an average utility taken across the uncertainty in the predictive models. The recursive algorithm works by starting at a terminal time T, finding V_T, and then working backward to the current time. The discount factor (δ) was set to one in our application. The algorithm was implemented within ASDP with an arbitrarily large time horizon (T = 1,000 years), as recommended by Lubow (2001), to effectively approximate an infinite time horizon. The model weights at t + 1 are updated using Bayes' Theorem based on the previous time weights (w_t), model predictions of abundances at t + 1, and monitoring observations of abundances at t + 1.

24.4.5 *Monitoring*

Several requirements are met by the targeted monitoring programs that estimate horseshoe crab and Red Knot abundance each year. First, the most recent estimates are compared to the ASDP decision matrix to look up the optimal harvest package for the following year. Second, the estimates are compared to model predictions to update model weights using Bayes' Theorem; the updated weights reduce model uncertainty and are incorporated into the optimization analysis (Eq. 24.4). Third, the estimates are used to improve the predictive models directly and reduce parametric uncertainty.

Currently, horseshoe crab abundance is estimated by a trawl survey and Red Knot abundance by mark-recapture sampling. The sampling designs and the monitoring program, in general, are under review to ensure that the required information is available to support the adaptive management framework.

24.4.6 *Operational Adaptive Management*

The ASMFC sets harvest regulations for horseshoe crabs annually. The adaptive management framework allows harvest policy to be responsive to abundance of horseshoe crabs and Red Knots. If the abundances remain low, then harvest will remain low or decrease. Conversely, if abundances increase and management objectives are being achieved, harvest can increase. To ensure that the framework is serving its intended purpose, the ASMFC will periodically review all components of the adaptive management framework, typically on a 3–5 year cycle.

24.5 The Horseshoe Crab-Red Knot Alliance: Progress with an Uncertain Future

As prefaced by Smith et al. (2009b), the North American horseshoe crab population in the Delaware Bay region is now being managed with an explicit link to the status of a migratory shorebird population. The implementation of this multispecies management framework is a credit to the both the State and Federal fishery managers within the ASMFC, as well as the scientists who developed the protocol. If left in place, the utility of the adaptive management framework process should continue to increase over time, as both quantitative parameters and value-based thresholds points are refined. This collaborative resolution to a 15-year-long, often controversial process of managing harvest of horseshoe crabs with shorebird needs in mind can serve as a useful model for other multispecies management scenarios where the link between the species can be quantitatively characterized.

While the adaptive framework represents a significant advancement in managing the fishery in the mid-Atlantic, many variables remain that could alter the nature of both the exploitation of horseshoe crabs and the ecology of their interaction with the Red Knot. The use of bait bags can decrease the demand for horseshoe crabs, whereas advances in the development of alternative baits could potentially eliminate the bait harvest altogether (Ferrari and Targett 2003; Fisher and Fisher 2006; Gauvry 2009). We discussed the increasing trend in horseshoe crabs collected in the biomedical industry. A greater awareness of best management practices in handling and processing animals in the biomedical industry may ameliorate the concomitant trend in mortality. Moreover, adoption of a pyrogen test which employs synthetic coagulation enzymes (Ding and Ho 2001) and other alternative endotoxin assays (Novitsky 2009) would significantly decrease the dependence of the pharmaceutical industry and medical community on horseshoe crabs. On the other hand, anecdotal evidence suggests that the *Limulus* amebocyte lysate industry in Asia may be experiencing a period of rapid growth. Societal and technological pressures for prosecuting the North American *Limulus* fishery are likely to evolve in the foreseeable future.

Despite implementation of the adaptive management process, attaining the value-based population benchmarks inherent in the framework, particularly for Red Knot, is not guaranteed. As can be expected for any global migrant, the Red Knot population faces many threats outside of Delaware Bay (Niles et al. 2008). While the adaptive management framework represents the best available science for predicting the effects of horseshoe crab harvest on the population dynamics of the *rufa* Red Knot, the nature of their intertwined ecology while in Delaware Bay is likely not static. Geologically speaking, the relationship between horseshoe crabs and Red Knots in Delaware Bay is a relatively recent phenomenon, evolving during the glacial retreat and sea level rise which conspired 7,000 years ago to form the current approximate physiognomy of Delaware Bay (Delaware DNREC 2005). Phenological synchrony is an essential element of the ecological interplay between the two species, where the timing of arrival of Red Knots in Delaware Bay has generally overlapped with the bulk of the spawning of horseshoe crabs, thereby maximizing the availability of horseshoe crab eggs for meeting the energetic demands of migration. Emerging stressors, primarily climate change, are known to impact the phenology in many taxa including migratory birds (Cotton 2003; Visser and Both 2005; Both et al. 2006). The accelerated effects of climate change observed in Arctic regions present particularly acute challenges for Arctic breeding species (Liebezeit et al. 2014), with some evidence that long-distance migrants lack the flexibility to adapt adequately (Both and Visser 2001). The effect of climate change on the breeding phenology of the *rufa* Red Knot, which breeds in the Canadian Arctic, remains unknown. Given the highly disparate nature of the two species and their life histories and the environmental pressures on the respective phenologies of Red Knots and horseshoe crabs, their individual responses to such pressures are likely not consistent in direction or magnitude. Gradually increasing phenologic asynchrony remains a possible factor in the future utility of the adaptive framework as well as perhaps the success of *rufa* Red Knot population as a whole. The environmental generalist strategy of the horseshoe crab (Sekiguchi and Shuster 2009) gives hope that, even in the face of accelerating climate change, they will remain an important component of the Atlantic coast biota.

References

Atkinson PW, Baker AJ, Bennet KA et al (2007) Rates of mass gain and energy deposition in Red Knot on their final spring staging site is both time- and condition-dependent. J Appl Ecol 44:885–895

Atlantic States Marine Fisheries Commission (ASMFC) (1998) Interstate fishery management plan for horseshoe crab. Fishery Management Report No. 32 of the Atlantic States Marine Fisheries Commission. Washington, DC

Atlantic States Marine Fisheries Commission (ASMFC) (1999) Horseshoe crab stock assessment report for peer review. Stock Assessment Report No. 98-01 (Supplement) of the Atlantic States Marine Fisheries Commission. Washington, DC

Atlantic States Marine Fisheries Commission (ASMFC) (2000) Addendum 1 to the interstate fishery management plan for horseshoe crab. Fishery Management Report No. 32a of the Atlantic States Marine Fisheries Commission. Washington, DC

Atlantic States Marine Fisheries Commission (ASMFC) (2004) Addendum 3 to the interstate fishery management plan for horseshoe crab. Fishery Management Report No. 32c of the Atlantic States Marine Fisheries Commission. Washington, DC

Atlantic States Marine Fisheries Commission (ASMFC) (2006) Addendum 4 to the interstate fishery management plan for horseshoe crab. Fishery Management Report No. 32d of the Atlantic States Marine Fisheries Commission. Washington, DC

Atlantic States Marine Fisheries Commission (ASMFC) (2008) Addendum 5 to the interstate fishery management plan for horseshoe crab. Fishery Management Report No. 32e of the Atlantic States Marine Fisheries Commission. Washington, DC

Atlantic States Marine Fisheries Commission (ASMFC) (2009) Horseshoe crab stock assessment for peer review. Stock Assessment Report No. 09-02 (Supplement A) of the Atlantic States Marine Fisheries Commission. Washington, DC

Atlantic States Marine Fisheries Commission (ASMFC) (2012) Addendum VII to the fishery management plan for horseshoe crab. Fishery Management Report No. 32g of the Atlantic States Marine Fisheries Commission. Washington, DC

Atlantic States Marine Fisheries Commission (ASMFC) (2013) Horseshoe crab stock assessment update. Atlantic States Marine Fisheries Commission, Arlington

Baker AJ, Gonzalez PM, Piersma T et al (2004) Rapid population decline in Red Knots: fitness consequences of decreased refuelling rates and late arrival in Delaware Bay. Proc R Soc Lond B 271:875–882

Baker AJ, Gonzalez PM, Morrison RIG et al (2013) Red Knot (*Calidris canutus*), the birds of North America online (Poole A, ed). Cornell Lab of Ornithology, Ithaca. doi:10.2173/bna.563. http://bna.birds.cornell.edu/bna/species/563. Accessed 10 Sep 2014

Berkson J (2009) An integrative approach to horseshoe crab multiple use and sustainability. In: Tanacredi JT, Botton ML, Smith DR (eds) Biology and conservation of horseshoe crabs. Springer, New York, pp 387–398

Berkson J, Shuster CN Jr (1999) The horseshoe crab: the battle for a true multiple-use resource. Fisheries 24:6–10

Both C, Visser ME (2001) Adjustment to climate change is constrained by arrival date in a long-distance migrant bird. Nature 411:296–298

Both CS, Bouwhuis S, Lessells CM et al (2006) Climate change and population declines in a long-distance migratory bird. Nature 441:81–83

Brockmann HJ (2003) Male competition and satellite behavior. In: Shuster CN Jr, Barlow RB, Brockmann HJ (eds) The American horseshoe crab. Harvard University Press, Cambridge, MA, pp 50–82

Caswell H (2001) Matrix population models. Wiley, New York

Clark NA (1980) A simple rapid method of moving a set cannon net with minimal disruption. Wader Study Group Bull 28:32

Clark KE, Niles LJ, Burger J (1993) Abundance and distribution of migrant shorebirds in Delaware Bay. Condor 95(3):694–705

Cohen JB, Karpanty SM, Fraser JD et al (2009) Residence probability and population size of Red Knots during spring stopover in the mid-Atlantic region of the United States. J Wildl Manag 73(6):939–945

Collie JS, Sissenwine MP (1983) Estimating population size from relative abundance data measured with error. Can J Fish Aquat Sci 40:1871–1879

Conroy MJ, Peterson JT (2013) Decision making in natural resource management: a structured, adaptive approach. Wiley-Blackwell, New York

Cotton PA (2003) Avian migration phenology and global climate change. Proc Natl Acad Sci U S A 100(21):12219–12222

Delaware Department of Natural Resources and Environmental Control (DNRE) (2005) Delaware Bay and Estuary assessment report. Doc. No. 40-01-01/05/02/01. http://www.dnrec.delaware.gov/WholeBasin/Documents/DelawareBayAssessmentPages.pdf. Accessed 20 July 2014

Dey AD, Niles LJ, Sitters HP et al (2011) Update to the status of the Red Knot (*Calidris canutus*) in the western hemisphere, April 2011. Western Hemisphere Reserve Network

Ding JL, Ho B (2001) A new era in pyrogen testing. Trends Biotechnol 19(8):277–281

Ferrari KM, Targett NM (2003) Chemical attractants in horseshoe crab, *Limulus polyphemus,* eggs: the potential for an artificial bait. J Chem Ecol 29(2):477–496

Finn JJ, Shuster CN Jr, Swan BL (1991) *Limulus* spawning activity on Delaware Bay shores 1990. Finn-Tech Industries, Cape May Court House

Fisher RA, Fisher DL (2006) The use of bait bags to reduce the need for horseshoe crab as bait in the Virginia whelk fishery. VSG-06-12. VIMS Marine Resource Report No. 2006-10. Sea Grant Communications, Virginia Institute of Marine Science. http://web.vims.edu/adv/fisheries/MRR2006_10.pdf. Accessed 20 July 2014

Fraser J, Karpanty SM, Cohen JB (2010) Shorebirds forage disproportionately in horseshoe crab nest depressions. Waterbirds 33:96–100

Fraser JD, Karpanty SM, Cohen JB et al (2012) The Red Knot (*Calidris canutus rufa*) decline in the western hemisphere: is there a lemming connection? Can J Zool 91(1):13–16

Frederick PC, Hylton B, Heath JA et al (2003) Accuracy and variation in estimates of large numbers of birds by individual observers using an aerial survey simulator. J Field Ornithol 74(3):281–287

Gauvry G (2009) Community building: an integrated approach to horseshoe crab conservation. In: Tanacredi JT, Botton ML, Smith DR (eds) Biology and conservation of horseshoe crabs. Springer, New York, pp 605–612

Gillings SP, Atkinson W, Baker AJ et al (2009) Staging behavior in Red Knot in Delaware Bay: implications for monitoring mass and population size. Auk 126:54–63

Helser TE, Hayes DB (1995) Providing quantitative management advice from stock abundance indices based on research surveys. Fish Bull 93:290–298

Jackson NL, Smith DR, Nordstrom KF (2008) Physical and chemical changes in the foreshore of an estuarine beach: implications for viability and development of horseshoe crab *Limulus polyphemus* eggs. Mar Ecol Prog Ser 355:209–218

Karpanty SM, Cohen J, Fraser JD et al (2011) Evidence of sufficient horseshoe crab eggs for a reduced Red Knot population during spring migration stopover in Delaware Bay USA. J Wildl Manag 75:984–994

King TL, Eackles MS, Spidle AP et al (2003) Microsatellite DNA reveals regional differentiation among populations of the horseshoe crab (*Limulus polyphemus*). Final report to the Atlantic States Marine Fisheries Commission

King TL, Eackles MS, Spidle AP et al (2005) Regional differentiation and sexbiased dispersal among populations of the horseshoe crab (*Limulus polyphemus*). Trans Am Fish Soc 134:441–465

Kvist A, Lindstrom A, Green M et al (2001) Carrying large fuel loads during sustained flight is cheaper than expected. Nature 413:730–732

Liebezeit JR, Gurney KEB, Budde M et al (2014) Phenological advancement in arctic bird species: relative importance of snow melt and ecological factors. Polar Biol. doi:10.1007/s00300-014-1522-x

Lubow B (2001) Adaptive stochastic dynamic programming (ASDP): version 3.2. Colorado State University, Fort Collins

Manly BFJ (2001) Statistics for environmental science and management. Chapman and Hall, London

Martin J, Runge MC, Nichols JD et al (2009) Structured decision making as a conceptual framework to identify thresholds for conservation and management. Ecol Appl 19:1079–1090

McCarthy MA, Possingham HP (2007) Active adaptive management for conservation. Conserv Biol 21:956–963

McGowan CP, Smith DR, Nichols JD et al (2009) A framework for the adaptive management of horseshoe crab harvests in the Delaware Bay constrained by Red Knot conservation. Atlantic States Marine Fisheries Commission, Washington, DC

McGowan CP, Hines JE, Nichols JD et al (2011a) Demographic consequences of migratory stopover: linking Red Knot survival to horseshoe crab spawning abundance. Ecosphere 2(6):69

McGowan CP, Smith DR, Sweka JA et al (2011b) Multispecies modeling for adaptive management of horseshoe crabs and Red Knots in the Delaware Bay. Nat Resour Model 24(1):117–156

Miranda MJ, Fackler PL (2002) Applied computational economics and finance. MIT Press, Cambridge

Morrison RIG (2006) Body transformations, condition, and survival in Red Knots *Calidris canutus* travelling to breed at Alert, Ellesmere Island, Canada. Ardea 94:607–618

Morrison RIG, Ross RK, Niles LJ (2004) Declines in wintering populations of Red Knots in southern South America. Condor 106:60–70

Morrison RIG, Davidson NC, Wilson JR (2007) Survival of the fattest body stores on migration and survival in Red Knots *Calidris canutus islandica*. J Avian Biol 38:479–487

National Marine Fisheries Service (NMFS) (2013) Commercial fisheries landings database. http://www.st.nmfs.noaa.gov/commercial-fisheries/commercial-landings/annual-landings/index. Accessed 22 Aug 2014

Niles LJ, Sitters HP, Dey AD et al (2008) Status of the Red Knot in the western hemisphere. Studies Avian Biol No. 36

Niles LJ, Bart J, Sitters HP et al (2009) Effects of horseshoe crab harvest in Delaware Bay on Red Knots: are harvest restrictions working? Bioscience 59:153–164

Nordstrom KF, Jackson NL, Smith DS et al (2006) Transport of horseshoe crab eggs by waves and swash on an estuarine beach: implications for foraging shorebirds. Estuar Coast Shelf Sci 70:438–448

Novitsky TJ (2009) Biomedical applications of *Limulus* amebocyte lysate. In: Tanacredi JT, Botton ML, Smith DR (eds) Biology and conservation of horseshoe crabs. Springer, New York, pp 315–329

Novitsky TJ (2015) Biomedical implications for managing the *Limulus polyphemus* harvest along the northeast coast of the United States. In: Carmichael RH, Botton M, Shin PKS, Cheung SG (eds) Changing global perspectives on horseshoe crab biology, conservation and management. Springer, Cham, pp 483–500

Okun SB (2012) Mating in the moonlight: the battle to save the American horseshoe crab. Ocean Coastal Law J 18(1):195–221

Penn D, Brockmann HJ (1994) Nest-site selection in the horseshoe crab, *Limulus polyphemus*. Biol Bull 187:373–384

Pennington M (1986) Some statistical techniques for estimating abundance indices from trawl surveys. Fish Bull 84:519–525

Pienkowski MW, Dick WJA (1976) Some biases in cannon-and mist-netted samples of wader populations. Ring Migrat 1(2):105–107

Pollock KH, Kendall WL (1987) Visibility bias in aerial surveys: a review of estimation procedures. J Wildl Manag 51(2):502–510

Prager MH (1994) A suite of extensions to a nonequilibrium surplus-production model. Fish Bull 92:374–389

Robinson RA, Atkinson PW, Clark NA (2003) Arrival and weight gain of red knots *Calidris canutus*, ruddy rurnstone *Arenaria interpres* and Sanderling *Calidris alba* staging in Delaware Bay in spring. BTO Research Report No. 307. http://www.bto.org/sites/default/files/u196/downloads/rr307.pdf. Accessed 10 Sep 2014

Runge MC (2011) An introduction to adaptive management for threatened and endangered species. J Fish Wildl Manag 2:220–233

Schwarzer AC, Collazo JA, Niles LJ et al (2012) Annual survival of Red Knots (*Calidris canutus rufa*) wintering in Florida. Auk 129(4):725–733

Sekiguchi K, Shuster CN Jr (2009) Limits on the global distribution of horseshoe crabs (Limulacea): lessons learned from two lifetimes of observations: Asia and America. In: Tanacredi JT, Botton ML, Smith DR (eds) Biology and conservation of horseshoe crabs. Springer, New York, pp 5–24

Shuster CN (1960) Distribution of horseshoe "crabs" in Delaware Bay. Univ Delaware, Estuar Bull 5(2):1–7

Shuster CN (1982) A pictorial review of the natural history and ecology of the horseshoe crab, *Limulus polyphemus*, with reference to other Limulidae. In: Bonaventura J, Bonaventura C, Tesh S (eds) Physiology and biology of horseshoe crabs: studies on normal and environmentally stressed animals. Alan R. Liss, New York, pp 1–52

Shuster CN, Botton ML (1985) A contribution to the population biology of horseshoe crabs, *Limulus polyphemus* (L.), in Delaware Bay. Estuaries 8(4):363–372

Smith DR (2007) Effect of horseshoe crab spawning density on nest disturbance and exhumation of eggs: a simulation study. Est Coasts 30:287–295

Smith DR, Mandt MT, MacDonald PDM (2009a) Proximate causes of sexual size dimorphism in horseshoe crabs (*Limulus polyphemus*) of the Delaware Bay. J Shellfish Res 28:405–417

Smith DR, Millard MJ, Carmichael RH (2009b) Comparative status and assessment of *Limulus polyphemus* with emphasis on the New England and Delaware Bay populations. In: Tanacredi JT, Botton ML, Smith DR (eds) Biology and conservation of horseshoe crabs. Springer, New York, pp 361–386

Sweka JA, Smith DR, Millard MJ (2007) An age-structured population model for horseshoe crabs in the Delaware Bay area to assess harvest and egg availability for shorebirds. Est Coasts 30:277–286

Visser ME, Both C (2005) Shifts in phenology due to global climate change: the need for a yardstick. Proc R Soc Biol 272(1581):2561–2569

Walls EA, Berkson JM, Smith SA (2002) The horseshoe crab, *Limulus polyphemus*: 200 million years of existence, 100 years of study. Rev Fish Sci 10:39–73

Walters CJ (1986) Adaptive management of renewable resources. Blackburn Press, Caldwell

Weber RG, Carter DB (2009) Distribution and development of *Limulus* egg clusters on intertidal beaches in Delaware Bay. In: Tanacredi JT, Botton ML, Smith DR (eds) Biology and conservation of horseshoe crabs. Springer, New York, pp 249–266

Williams BK, Brown ED (2012) Adaptive management: the U.S. Department of the Interior applications guide. Adaptive Management Working Group, U.S. Department of the Interior, Washington, DC

Williams BK, Nichols JD, Conroy MJ (2002) Analysis and management of animal populations. Academic, San Diego

Chapter 25
The Mismanagement of *Limulus polyphemus* in Long Island Sound, U.S.A.: What Are the Characteristics of a Population in Decline?

Mark A. Beekey and Jennifer H. Mattei

Abstract Over the past 15 years, horseshoe crabs in Connecticut have gone from being considered a nuisance species to a species of Greatest Conservation Need in 2015. This has happened through first, its discovery as an economically important species, second through research of its ecological role in coastal estuaries, and third, through education of the public concerning its role in the environment and their own health. To manage horseshoe crab populations successfully requires long term monitoring, research and education. The use of annual or biannual trawl data trends to assess the success of management decisions is limited due to the high variance of counts by location, season, and year, and is prone to misinterpretation. Trawling also damages the very habitat that the target species needs to survive. What is required to reach the goal of a sustainably managed population is the determination of local population dynamics and then modifying management practices to maximize reproductive success of the species. In Long Island Sound (LIS), these basic requirements for sustainable management have not been available until recently. The two states that manage horseshoe crabs in Long Island Sound (Connecticut and New York) have drastically different harvest regulations and management plans that are clearly not working. The LIS horseshoe crab population has declined and remains low as shown by both the long term trawl data and our long term mark-recapture data. After 13 years of a classic ecological study we found the Long Island Sound horseshoe crab population to be one discrete management unit where at least 3 % of the recaptures were found to cross the Sound. The LIS population is aging with low recruitment of newly molted adults and is reproducing well below its maximum rate. We observed very low spawning densities, increasing numbers of single females on the beach, and less than 6 % of polyandrous mating behavior. These basic population trends over the length of this study indicate the current harvest quotas and management techniques are not sustainable. We recommend at

M.A. Beekey (✉) • J.H. Mattei
Department of Biology, Sacred Heart University,
5151 Park Avenue, Fairfield, CT 06825, USA
e-mail: beekeym@sacredheart.edu

R.H. Carmichael et al. (eds.), *Changing Global Perspectives on Horseshoe Crab Biology, Conservation and Management*, DOI 10.1007/978-3-319-19542-1_25

the very least the implementation of a unified management plan for Long Island Sound with one shared harvest quota for the entire LIS population. Second, we suggest that the number of no-harvest zones should be increased on both sides of the Sound and a ban of the harvest of spawning females should be added to existing harvest regulations. Ultimately, we would recommend the establishment of multiple Marine Protected Areas (MPAs) within LIS where the entire MPA is off limits to commercial or recreational fishing of any type.

Keywords Long Island Sound • Population dynamics • Mark-recapture data • Harvest regulations • Home range • Condition • Prosomal width • Operational sex ratios

25.1 Introduction

Historically, horseshoe crab (*Limulus polyphemus*) populations within US waters seemed boundless. With wide-ranging populations and very high fecundity, sustainable management plans were not deemed necessary. It was not until the 1990s that the *Limulus* populations experienced a notable decline, most rapidly in the mid-Atlantic region (Millard et al. 2015). Within Long Island Sound (LIS), the horseshoe crab had been harvested and used as bait for decades (Berkson and Shuster 1999). Harvest quotas of this species have been implemented in the states of Connecticut and New York, over the past 15 years, in compliance with the Atlantic States Marine Fisheries Commission (ASMFC) mandate to produce a Fishery Management Plan for *Limulus*. In Connecticut, the harvest quotas are rarely reached through the self-reporting system that has been implemented. Yet, the population in Long Island Sound continues to decline (Fig. 25.1).

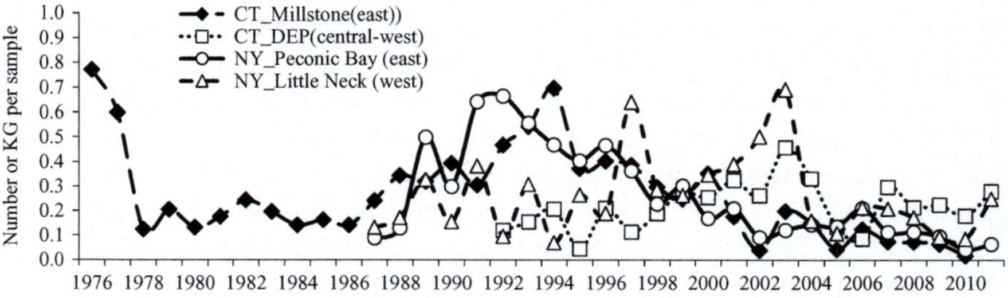

Fig. 25.1 Horseshoe crab indices of relative abundance in Long Island Sound (LIS). The number or kg of horseshoe crabs per sample (tow or seine) is plotted over time. The Millstone Power Station Trawl Monitoring Program (Dominion Nuclear Connecticut) records counts per tow and is conducted bi-weekly year round at fixed stations (total scheduled tows = 234–243) in the vicinity of the power station (Niantic River, Niantic Bay, and Jordon Cove) in eastern LIS. CT_DEP indices are based on central and western CTDEEP Long Island Sound Trawl Survey stations and represent kg/tow. NYSDEC Peconic Bay and Little Neck Bay indices are based on seine surveys and represent number horseshoe crabs per seine. Data obtained from Penny Howell, CTDEEP, Connecticut Compliance Report for Horseshoe Crab, March 2014

It is more than likely that the legal harvest of horseshoe crabs is only one source of mortality in this population. Some of the greatest losses to these economically important populations are the illegal, unregulated, and unreported (IUU) take of horseshoe crabs (Vincent and Harris 2014). Thousands of horseshoe crabs can get caught in water intake pipes used to cool shoreline power plants and industrial complexes (Applied Biology 1980). They also become bycatch during bottom trawling operations. Bottom trawling and dredging for shellfish indiscriminately pulls up many of the organisms in the benthic area and destroys the habitat as it goes. The ASMFC supports trawling for "stock assessment" purposes, yet this very activity causes harm directly or indirectly to the species that are being sustainably managed and to many species that are not currently under a management plan but are clearly part of the food web. Other causes of horseshoe crab population decline would include the loss of eggs, young of the year and subsequent juvenile life stages (at least 8–10 years of development). Causes of mortality include predation and a long list of destructive human activities: e.g., habitat degradation, pollution, invasive species, and climate change to name the most damaging (Vincent and Harris 2014).

Successful fishery/wildlife management plans would include an understanding of horseshoe crab population dynamics and life history characteristics, which can vary depending on the local environment (Walls et al. 2002). A sustainable management plan would include regulations that will maximize reproductive output and juvenile survival as well as maintain ecological services that adult horseshoe crabs provide in estuaries. Previous studies have documented that the low density of horseshoe crab populations in Long Island Sound limits egg predation by shorebirds (Beekey et al. 2013) and contributes to changes in mating behavior such as a decline in polyandry and the presence of single females on the spawning beaches (Mattei et al. 2010).

Many questions remain unanswered about general life history and population characteristics of horseshoe crabs that reside in LIS that would aid management strategies:

1. Do horseshoe crabs exhibit site fidelity within and across spawning seasons; this is important in informing conservation decisions
2. Do horseshoe crabs have home ranges and if so what is their general size; this is important to know when establishing no-harvest zones
3. Do horseshoe crabs cross the Sound from Connecticut to the north shore of Long Island, NY; this knowledge would help define the size of the population management unit
4. Do LIS horseshoe crabs leave LIS and if so, where do they go; the rates of immigration and emigration are important in population analyses
5. What is the ratio of new recruits into the breeding population using shell condition; this is an indication of an expanding or declining population and survival rates of juvenile horseshoe crabs
6. Do LIS horseshoe crabs exhibit similar population characteristics to other populations of horseshoe crabs further north in New England or further south in Delaware Bay?

Here, we present an analysis of a long-term data set collected by researchers and a large group of well-trained citizen-scientists associated with Project *Limulus* (www.projectlimulus.org) to provide answers for these general questions about the population characteristics of horseshoe crabs in Long Island Sound. We examine current management strategies and recommend changes based on the results of this long-term study. It is important for the Department of Energy and Environmental Protection and Department of Environmental Conservation agencies of Connecticut and New York State, respectively, to manage the Long Island Sound population of horseshoe crabs cooperatively to achieve a sustainable fishery; sustainably managing Atlantic coastal fisheries was the original vision of the ASMFC when it was established in the 1940s and remains its mission today.

25.1.1 Horseshoe Crab Management History

The Atlantic States Marine Fisheries Commission (ASMFC) was founded with the goal for the Atlantic states to cooperatively and sustainably manage their shared migratory fishery resources with assistance from a number of federal agencies.

> An Interstate Compact, ratified by the member states and approved by the U.S. Congress in 1942, affirmed the states' commitment to cooperative stewardship in promoting and protecting Atlantic coastal fishery resources. (asmfc.org)

Since 1998, representatives from ASMFC member states have overseen the management of the horseshoe crab fishery (ASMFC 1998). The ASMFC produced an interstate Fishery Management Plan (FMP) for horseshoe crabs with the long term goals of maintaining sustainable levels of spawning adults required to ensure the horseshoe crabs' continued role in the ecology of coastal ecosystems as well as providing for continued use by humans in the biomedical and bait industries. However, the horseshoe crab population data available to the ASMFC members was limited and flawed (Walls et al. 2002). Over the past 16 years, seven addendums have been added to the FMP for horseshoe crabs including state-by-state harvest quotas, monitoring requirements and management recommendations that in some cases recommended harvest closures. Recently, the ASMFC adopted an Adaptive Resource Management Framework (ARM) in 2012 (implemented in 2013) for horseshoe crabs originating in Delaware Bay. The ARM framework will utilize a suite of multispecies models including the status of *Limulus* and Red Knot (*Calidris canutus rufa*) populations to derive an annual allowable harvest that will be allocated among the four bay-area states (Delaware, New Jersey, Maryland, and Virginia; Millard et al. 2015). Obviously, the ARM model requires ongoing and accurate horseshoe crab stock assessments.

The most recent coast-wide stock assessment conducted in 2013 indicates that in general, horseshoe crab abundance has increased along the Delaware Bay and Southeastern US regions while declining across New England (Maine, New

Hampshire, Massachusetts, Rhode Island) and New York regions (New York State and Connecticut, see Fig. 25.1). Declines in the New England region had been evident in the 2004 Stock Assessment and the 2013 Stock Assessment confirmed the continued decline (ASMFC 2013). Trends of the New York Region surveys are variable. In western Long Island Sound, there was evidence for increasing or stable trends, while in eastern Long Island Sound/Peconic Bay, there has been a steady decline from peak levels during the mid-1990s with current levels slightly lower than levels in the mid-1980s (Smith et al. 2009; CTDEEP 2014). Harvest regulations put in place as a result of the ASMFC management plan have been successful as evident from the increasing horseshoe crab population in Delaware Bay where regulations reduced harvest by 85 % from 2000 to 2006 (Smith et al. 2009). Additionally, horseshoe crab populations in Delaware Bay were protected by the establishment of the Carl N. Shuster, Jr. Horseshoe Crab Reserve in federal waters off the mouth of the Delaware Bay estuary where horseshoe crab harvest is prohibited.

However, reduced harvest in Delaware Bay was accompanied by an increase in harvest across other regions. The New England and New York region states along with the State of Virginia saw horseshoe crab landings increase significantly after 2005 (for review, see Smith et al. 2009). This pattern of redirected harvest has raised concerns for the sustainability of populations across New England and New York regions. Numerous studies have been carried out over recent years that have revealed the vulnerability of horseshoe crab populations across New England to over-harvesting (Gibson and Olszewski 2001; Carmichael et al. 2003, 2004; Rutecki et al. 2004; James-Pirri et al. 2005; Grady and Valiela 2006; James-Pirri 2010; Schaller et al. 2010). Taken together as an aggregate, one clear conclusion from these studies is that the New England region is composed of a series of coastal embayments that support horseshoe crab populations with different characteristics and limited dispersal of adults and larvae among them. These studies helped trigger more specific regulations at the local level. For example, harvesting horseshoe crabs for bait was prohibited in Pleasant Bay, MA, and harvesting was completely banned within the Monomoy National Wildlife Refuge and Cape Cod National Seashore. In Rhode Island, regulations were implemented that prohibited harvesting from 48 h preceding and following the full and new moons from May to July (Smith et al. 2009).

The horseshoe crab population in the Delaware Bay region is dominated by one large intermixing population and management is tailored to the inter-specific relationship between horseshoe crabs and migratory shorebirds (King et al. 2005; Swan 2005; Smith et al. 2009). In contrast, the New England region is comprised of many embayment-specific populations, and consequently management regulations are more localized as a result of embayment-specific population assessments. Regulations tailored to specific population and habitat characteristics are important as variation in these characteristics can affect recovery time of an overharvested population, and measures employed may need years of assessment before discovering what strategy will speed population recovery (Beekey et al. 2013).

25.1.2 Management of Horseshoe Crabs in Long Island Sound

The Long Island Sound estuary (Fig. 25.2), with over 600 miles of coastline and 1,320 mile^2 of open water, is currently managed separately by the states of Connecticut (CT) and New York (NY). Their horseshoe crab management strategies and regulations are different although both states rely on self-reporting by harvesters. New York implemented an annual harvest quota of 150,000 horseshoe crabs in 2012, approximately 41 % of the quota allowed under ASMFC regulations (Table 25.1). From 2008 to 2013, NY harvested an average of 147,450 horseshoe crabs. This is the second largest average harvest among all states between 2008 and 2012, representing 48 % of the coastal harvest outside of the four Delaware Bay region states. The bulk of the harvest for NY occurs along the southern shore of Long Island (57.1 % in 2013; NYSDEC 2013). Harvesters are required to obtain permits and submit monthly harvesting reports to the New York State Department of Environmental Conservation (NYSDEC) from August through April. During

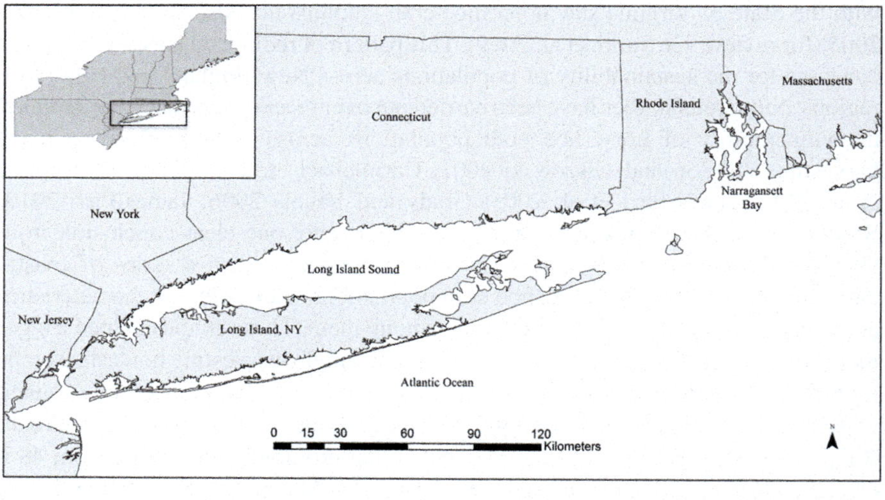

Fig. 25.2 *Inset*: Location of study area (*boxed*) within the northeastern United States. Our study area (*boxed*) includes the entire coast of Connecticut (CT), part of New York with coastline on Long Island Sound and the northern coast of Long Island, New York (NY)

Table 25.1 Reported commercial bait landings of horseshoe crabs for Connecticut and New York from 2008 to 2012

	Addendum IV quota	State quota 2012	2008	2009	2010	2011	2012	2013
CT	48,689	48,689	32,565	27,065	30,036	24,466	18,958	19,645
NY	366,272	150,000	148,719	123,653	124,808	146,995	178,899	161,623

Adapted from ASMFC (2013), CTDEEP (2014), NYDEC (2013)

May, June, and July, permittees are required to submit harvesting reports on a weekly basis. Harvesting is permitted by hand harvest, pound net, trap net, gill net, otter trawl, seine or dredge. Dredges are not permitted in the Atlantic Ocean except during the months of September and October. Horseshoe crab harvesting is currently prohibited at two beaches on the north shore of Long Island at West Meadow Beach in Stony Brook and Cedar Beach in Mount Sinai as well as on the southern shore of Long Island within the Fire Island National Seashore managed by the National Park Service.

Connecticut's annual harvest quota is set by ASMFC at 48,689 animals, although their 6-year average harvest from 2008 to 2013 has been only 52 % of their allowed quota (Table 25.1) representing approximately 9 % of the coastal harvest not including the four Delaware Bay Region states (IUCN Red List Assessment, in review). In 2013, 66 % of harvested horseshoe crabs were collected west of the Connecticut River to the NY border (CTDEEP 2014). Connecticut prohibits all harvesting and landing from July 8 through May 21. During the open period from May 22 through July 7 harvest and landing is closed from 6:00 pm on any Friday through 6:00 pm on the following Sunday. Both a permit and an endorsement letter are required for harvest. Endorsement letters can only be obtained by those on record for possessing a harvest permit between 1999 and 2006. There is a 500-crab limit per 24-h period and harvesters are only allowed to hand-harvest spawning crabs up on beaches. In 2007, CT established three no harvest zones at Menunketesuck Island in Westbrook, Sandy Point in West Haven, and Milford Point, Milford, CT in order to maximize horseshoe crab egg availability for shorebirds utilizing those areas.

New York State and Connecticut clearly have differences between their horseshoe crab harvest regulations with respect to harvest quotas (3:1 difference), allowed manner of harvest (multiple methods vs. hand harvesting alone), harvesting pressure (NY – unevenly split between the southern and northern shore of Long Island versus 100 % of CT harvest within Long Island Sound), open seasons (year round vs. late May through July) although both States have started to implement similar management strategies with the establishment of protected areas. However, both states lack basic population data on horseshoe crabs within their boundaries. While numerous studies have been conducted in New England and Delaware Bay examining population trends (size, mating patterns, movement patterns, spawning densities, egg densities) that enhance management strategies (for review see Smith et al. 2009), various population indices for horseshoe crab populations in Long Island Sound are lacking. With recent stock assessments indicating declines in abundance of horseshoe crabs in the New York Region (ASMFC 2013), it is imperative that we assess the horseshoe crab populations in this region and determine whether the different management plans of CT and NY are sustainable over the long term.

First, spatial-temporal data leading to knowledge of resource use and home range sizes are necessary for informing conservation efforts. Clearly, developing a sustainable horseshoe crab management plan for Long Island Sound is dependent upon understanding whether Long Island Sound is one large intermixed population similar to Delaware Bay or representative of the New England Region where horseshoe crabs are arranged in smaller, localized and separate management units. Second, it is important that we understand how the population is changing over time. Changes in various indices such as physical size, sex ratios, mating patterns, and condition of crabs can supply management agencies with crucial information that can help determine whether populations are being harvested in potentially damaging manners (e.g. selectively harvesting large females or taking every spawning pair off of the beaches). Also, if populations are rapidly aging with little or no recruitment of newly molted adults, and if population densities are low, as measured by beach spawning counts, or indicated by numbers of single females coming up to spawn but cannot find mates, then management regulations need to be changed to protect the long term viability of the species in a given region.

25.2 Methods

It is difficult to study migration patterns of horseshoe crabs in a body of water with over 600 miles of coastline. Following the lead of federal researchers in Delaware Bay who employed fishermen to tag adult horseshoe crabs caught in trawls and state biologists who trained volunteers to census spawning crabs, we started a community research program in 2000, Project *Limulus* (www.projectlimulus.org) to train and organize volunteers to help gather information along the coastline of the Sound. This education/research endeavor utilizes citizen scientists from non-profit environmental organizations, local aquariums and zoos, K-12 school groups, along with undergraduate/graduate research assistants to tag crabs and collect population data mainly along the coastline of Connecticut, but also in the Sound bordered by the States of New York and Rhode Island. All Project *Limulus* volunteers were trained to tag horseshoe crabs and record specific data for each horseshoe crab they tagged on a common data sheet that was returned to Sacred Heart University (SHU) via mail or e-mail. Individuals and/or groups of volunteers tagged crabs on private and public beaches depending upon access. For data quality control, every April or early May we would meet with group leaders and review the protocols, data sheets and data collection methods and distribute the tags. Each group or conservation organization therefore, had a trained group leader that was mainly responsible for training new volunteers on their beach and returning the data sheets. Members of our research team at SHU would also perform site visits periodically during the tagging season. Tagging commenced each year in May as soon as the crabs started arriving on shore to spawn and ended when all of the tags were utilized or when the crabs stopped spawning. Tagging occurred during day and night hours and on low and high tides. Typically all crabs encountered were tagged until the tags were used up or when volunteers ran out of time. The majority of beaches where tags were deployed had more than 100 horseshoe crabs tagged per season, therefore the samples of tagged individuals were a representative sample of the spawning population at each location. Volunteers were trained to tag animals as they were found regardless of sex or current mating status. From 2000 to 2006, we tagged horseshoe crabs using yellow Floy Cinch-tags (model FT-4, 8″: http://www.floytag.com) with unique identification numbers and a SHU phone number for reporting tag information. Cinch tags were attached to the crab using a #1 Yellow scratch awl (Challenge Sailcloth, Inc. Vernon-Rockville, CT) to make a hole through the lower right or left rear of the prosoma. The cinch tags were strung through the hole and cinched tightly across the posterior edge of the prosoma. In 2007, we started utilizing white disc tags issued by the U.S. Fish and Wildlife Service (USFWS). Disc tags were attached to the crab using a #2 Yellow scratch awl with a #8 stainless steel washer affixed to the awl to act as a stop to prevent the hole from getting too large. The awl was gently twisted into the lower rear of either the right or left posterior side of the prosoma. Once a hole was made, the tree pin of the disc tag was firmly inserted into the hole. Each disc tag has a unique tag number and is imprinted with a toll free number and mailing address to report recapture information to the USFWS office in Annapolis, Maryland. Mortality due to tagging is highly unlikely within and across spawning seasons (Mattei et al. 2011).

The tag number, date, sex (based on the morphology of the pedipalps), prosomal width (greatest width in cm), mating status, location of tagging and release, and any additional remarks (e.g. damage) were recorded for each tagged crab. Mating status was recorded for each crab as follows: single male, single female, pair (female and amplexed male), or F+N where F is equal to a female and N equals the total number of males mating with her including the amplexed male (e.g. F + 2 equals a female with one amplexed male and one satellite male). Starting in 2012, the condition for each tagged crab was determined based on previous studies (Brockmann and Penn 1992; Penn and Brockmann 1995; Brockmann 1996). Horseshoe crabs have a terminal molt between the ages of 8–10 years old and live approximately 18–20 years based on record tag returns. The carapace of *Limulus* erodes over time (Brockmann and Penn 1992), and the dorsal prosomal surface changes in color from a light tan with fringing hair in newly molted adults to black and hairless in older adults. We scored carapace condition based on a scale of 1–3, where:

- Condition 1 was for a newly molted, glossy prosoma, fringed with hairs;
- Condition 2 was for an intermediate prosoma (i.e., showing scratches and other signs of wear, very little fringing hair and with some gray areas), and;
- Condition 3 was for a heavily worn, no hair, mostly blackened and pitted prosoma.

For analysis of this data, we assumed Condition 1 animals to be from 0 to 3 years past their terminal molt, Condition 2 adults were 3–6 years past their final molt and Condition 3 adults were older than 6 years past their final molt. For the Chi-square analysis we assumed an expected distribution of age classes to be one-third each. A growing population would have relatively more Condition 1 crabs. We obtained recapture data from a variety of sources but the majority of recapture data was reported by individuals associated with Project *Limulus* during tagging sessions. Additional information on recaptured crabs was obtained through the general public who reported information throughout the years via phone, e-mail, or by reporting the tag information to the USFWS horseshoe crab tag database via the toll-free number.

We report the results from Milford Pt. and Sandy Pt. because the data was gathered by researchers and their supervised students from these two locations. We also show results from the combined data from all beaches in CT that includes the data from trained volunteers. Not all recapture reports called in by the public were complete; however, whenever possible, the data was used to determine differences between tag and recapture location, determine straight-line distance traveled between release and recapture location, time between release and recapture. The data were analyzed using regression analysis and one-way analysis of variance (ANOVA) in SPSS software.

25.3 Results

25.3.1 Spatial and Temporal Patterns

A total of 84,432 crabs were tagged from 2000 to 2013. Over that same period, 18,525 crabs were recaptured (Fig. 25.3) of which 72 % were live recaptures, 13 % reported as dead and 15 % with no information on their status. Over 600 volunteers associated with Project *Limulus* placed the majority of the tags deployed from 2006 to 2013 on crabs. The decrease in number of tags placed on crabs in 2012 and 2013 represents a decrease in the number of tags allotted to Project *Limulus* by the USFWS, not a decrease in volunteer participation. At the conclusion of the 2013 tagging season, our overall recapture rate was 22 %. In all, tags were put out on 241 different beaches with 75 % of the beaches along Connecticut's coastline, 21 %

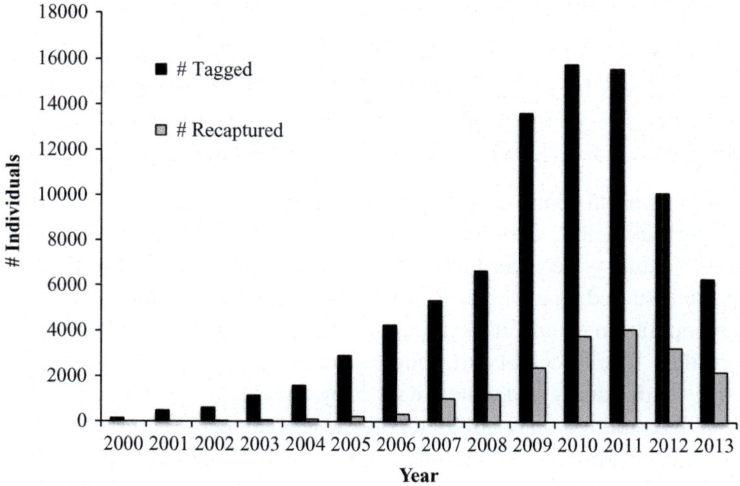

Fig. 25.3 Number of crabs tagged and recaptured by year from 2000 to 2013. Over 84,000 crabs were tagged during this time period, and over 18,000 recapture reports were generated

along the mainland coastline of New York State and the north shore of Long Island, and 4 % of the sites within Rhode Island. Project *Limulus* stopped tagging along the north shore of Long Island in 2008 with the start of a cooperative tagging study jointly administered by NYSDEC and Cornell Cooperative Extension of Suffolk County; however, tagging has continued along New York State's western shoreline of Long Island Sound around the town of Rye in Westchester County (Table 25.2).

The percentage of crabs recaptured within the same year as they were tagged ranged from a low of 25 % in 2001 to a high of 78.4 % in 2002 (Table 25.3). During 2010 and 2011 when the greatest number of crabs were tagged, 52.7 % and 60.3 % were recaptured within the same spawning season. Overall, 58.7 % of the 18,525 recaptured crabs were recaptured within the first year of release. This percentage decreased exponentially over time (Fig. 25.4). Nearly 98 % of all recaptured horseshoe crabs were found within 4 years from when they were first tagged. Only three individuals (0.02 %) were recaptured 8 years after they were tagged.

Within a single spawning season, 73 % of recaptured crabs (N = 10,650) were recaptured at the same beach as they were tagged; 85 % were recaptured within the same city (different beaches) as they were tagged; and 98 % were recaptured within the same state as they were tagged. On average, crabs were recaptured 2.64 ± 10.3 (S.D.) km away from their original tag site. After one spawning season, 22 % of recaptured crabs (N = 7,488) were reported from the same beach where they were tagged; 36 % were recaptured within the same city; and 83 % were recaptured within the same state. After the first spawning season had passed, crabs were recaptured on average 12.6 ± 22.6 (S.D.) km from where they were tagged. Twenty percent of the recaptured crabs after one season were more than 20 km from where they were originally tagged after the initial spawning season (Fig. 25.5).

For those crabs tagged in Milford (N = 1,073) and New Haven (N = 852), CT from 2000 to 2013, 84–87 % of crabs, respectively, were recaptured within 16–18 km of the tagging beach after the first spawning season (Fig. 25.6a, b). While separated from the northern shoreline of Long Island by up to 33 km, crabs tagged at Milford and New Haven, CT have been recaptured across the sound near Port Jefferson along the north shore of Long Island, NY (Fig. 25.6a, b). Close to 3.5 and 2 % of crabs tagged in Milford and New Haven, CT, respectively, were recaptured across the Sound. Three crabs originally tagged at Milford and only one crab originally tagged at New Haven were recaptured on the south shore (i.e. Atlantic Ocean side) of Long Island. No recaptures of crabs tagged at Milford or New Haven have been reported from Rhode Island or Massachusetts.

For horseshoe crabs originally tagged in Stamford and Greenwich, CT 67 % were recaptured within 16 km of their original tag site (Fig. 25.7a). Nearly 16 % of all crabs tagged in Greenwich/Stamford were recaptured in NY either along the north shore of Long Island or NY's southwestern shoreline of Long Island Sound (Fig. 25.7a). Approximately 30 % of recaptured horseshoe crabs (N = 1,424) originally tagged in Groton and Stonington, CT were found in Rhode Island and <1 % were recaptured in Massachusetts (Fig. 25.7b). On average, crabs were recaptured 8 km away from where they were tagged. Nearly 3 % of crabs tagged in either Groton or Stonington were recaptured on New York State's Fishers Island and close

Table 25.2 Summary of the cumulative number of crabs tagged, cumulative number of recaptures, and recapture rate from 2000 to 2013

	2000	2001	2002	2003	2004	2005	2006	2007	2008	2009	2010	2011	2012	2013
Cumulative # crabs tagged	354	831	1,436	2,584	4,172	7,077	11,327	16,659	23,303	36,920	52,681	68,253	78,343	84,644
Cumulative # recaptures reported	7	12	49	112	208	431	751	1,762	2,956	5,347	9,123	13,207	16,464	18,651
Recapture rate (%)	2.0	1.4	3.4	4.3	5	6	6.6	10.6	12.7	14.5	17.3	19.4	21.0	22.0

Table 25.3 Summary of the temporal distribution of recaptured crabs by year tagged

		Total	# Tagged individuals recaptured (percentage) by year													
Year	# Tagged	Recaps	2000	2001	2002	2003	2004	2005	2006	2007	2008	2009	2010	2011	2012	2013
2000	142	8	3	1	3						1					
			(37.5)	(12.5)	(37.5)						(12.5)					
2001	477	16		4	5	1	1	3	2							
				(25.0)	(31.3)	(6.3)	(6.3)	(18.8)	(12.5)							
2002	605	37			29	4	1	2	1							
					(78.4)	(10.8)	(2.7)	(5.4)	(2.7)							
2003	1,148	91				58	16	10	4	2		1				
						(63.7)	(17.6)	(11.0)	(4.4)	(2.2)		(1.1)				
2004	1,588	146					78	37	15	5	3	6			2	
							(53.4)	(25.3)	(10.3)	(3.4)	(2.1)	(4.1)			(1.4)	
2005	2,905	277						170	37	37	11	13	8		1	
								(61.4)	(13.4	(13.4)	(4.0)	(4.7)	(2.9)		(0.4)	
2006	4,250	408							259	75	29	18	19	6		2
									(63.5)	(18.4)	(7.1)	(4.4)	(4.7)	(1.5)		(0.5)
2007	5,332	1,163								892	134	53	54	15	14	1
										(76.7)	(11.5)	(4.6)	(4.6)	(1.3)	(1.2)	(0.1)
2008	6,644	1,827									1015	413	228	100	43	28
											(5.6)	(22.6)	(12.5)	(5.5)	(2.4)	(1.5)
2009	1,3617	4,072										1882	1130	561	310	188
												(46.2)	(27.8)	(13.8)	(7.6)	(4.6)

The total number of tags deployed each year of the study are presented along with the total number of recaptures from that tagging year grouped by the number of tagged individuals recaptured during the initial and subsequent years (percentage in parentheses)

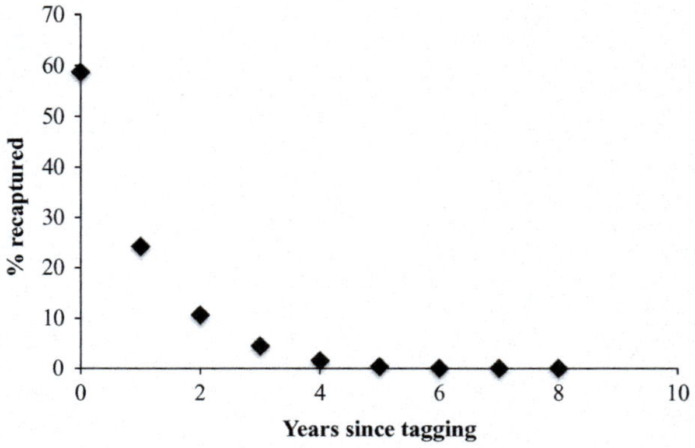

Fig. 25.4 Percentage of all crabs recaptured as a function of year since originally being tagged. The longest period of time elapsed from the date of tagging to recapture was a female crab tagged on 6/1/2000 and recaptured on 7/26/2008

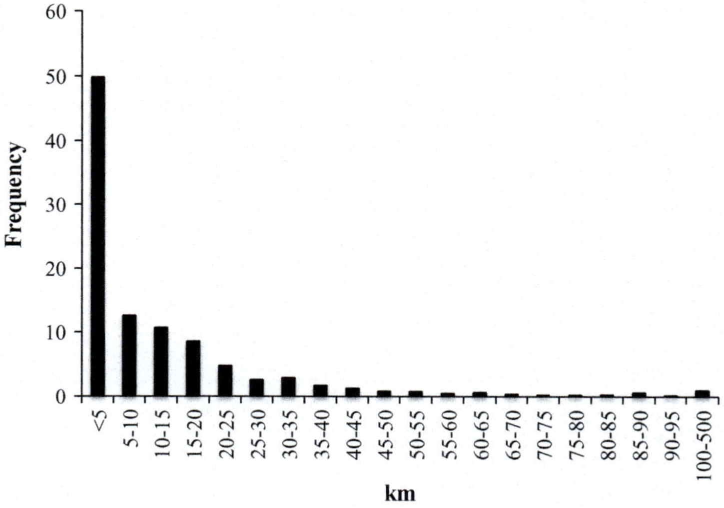

Fig. 25.5 The frequency of straight line distance (km) traveled by crabs tagged from 2000 to 2013 and recaptured after at least 1 year had passed since their initial tagging (N=7488)

to 1 % of the crabs were recaptured on the north shore of Long Island near Orient Point (northeasterly tip of Long Island).

25.3.2 Population Characteristics

Female horseshoe crabs averaged 25.4 ± 2.4 cm SD (N = 14,020) in prosomal width. Prosomal width of males averaged 19.6 ± 1.8 cm SD (N = 24,050). Overall there has been no significant change in prosomal width for females ($R^2 = 0.03$, $p > 0.05$) since 2000. However, male prosomal width has decreased significantly since 2000

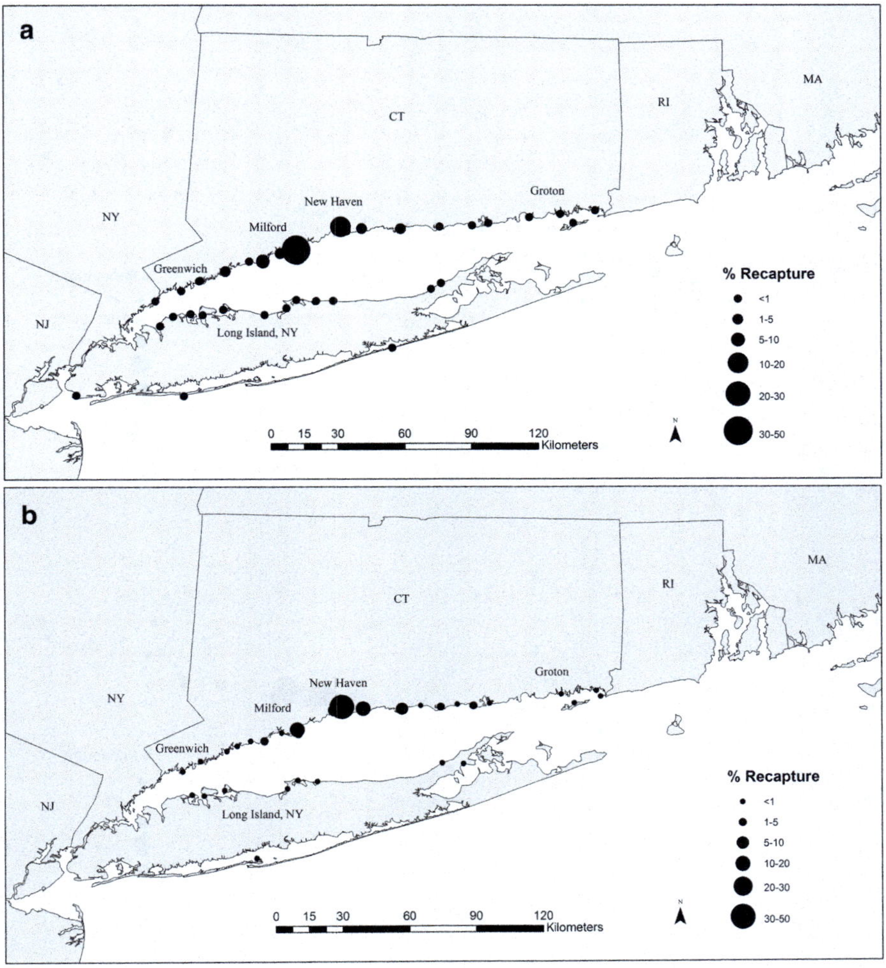

Fig. 25.6 Recapture location and frequency (%) of crabs originally tagged at (**a**) Milford, CT (*top*) and (**b**) New Haven, CT (*bottom*) and recaptured. The most eastern point–Stonington, western–Rye, NY and across the Sound on eight beaches of the North Shore of Long Island, NY

($R^2=0.296$, $p<0.05$) (Fig. 25.8). Male to female sex ratios on beaches where crabs are tagged decreased on average from 2006 to 2013 although not significantly ($R^2=0.062$, $p>0.05$) (Fig. 25.9a). At our two main research sites (Milford Point, MP and Sandy Point, SP) where we have focused our research efforts using research assistants, operational sex ratios on spawning beaches, while not significant, have decreased on average over time (MP, $R^2=0.045$, $p>0.05$; SP, $R^2=0.081$, $p>0.05$) (Fig. 25.9b).

Mating status data collected while tagging from 2007 to 2013 indicates the number of single females observed on beaches across CT generally increased ($R^2=0.436$, $p>0.05$) while the number of females with one male (paired females)

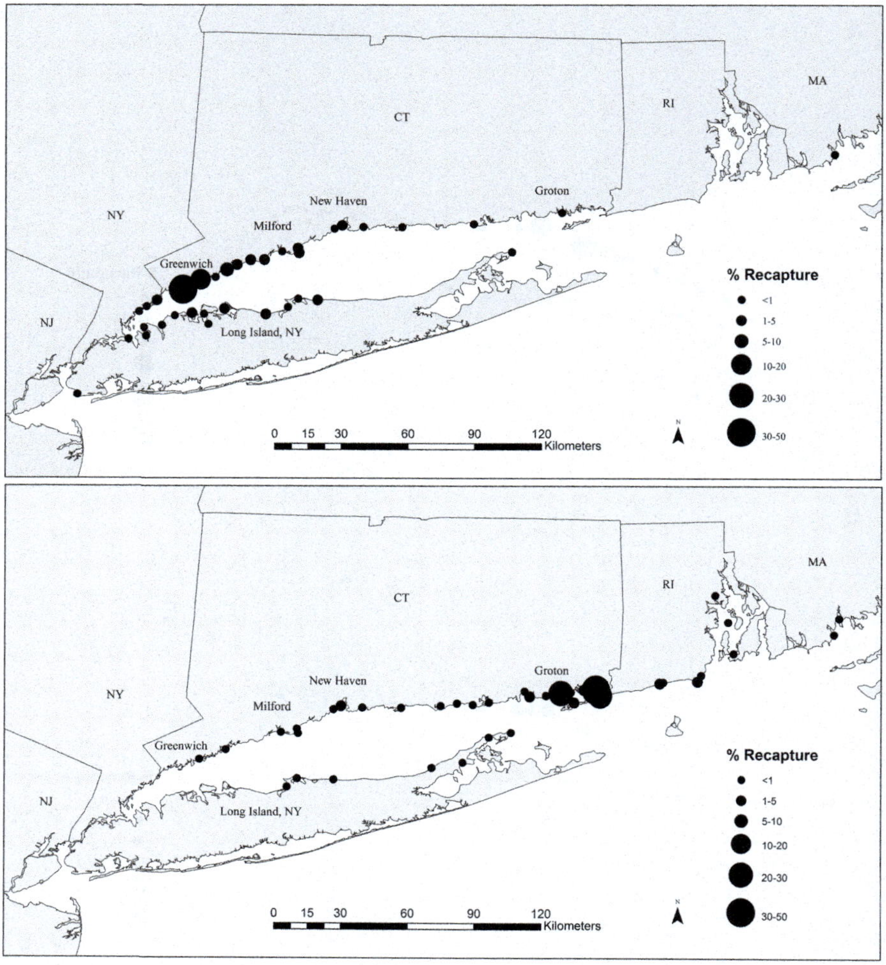

Fig. 25.7 Recapture location and frequency (%) of crabs originally tagged at (**a**) Greenwich/Stamford, CT (*top*) and (**b**) Groton/Stonington, CT (*bottom*)

($R^2=0.382$, $p>0.05$) and females with multiple males (one amplexed male and one or more satellite males) ($R^2=0.205$, $p>0.05$) decreased (Fig. 25.10a). At Milford Point (Fig. 25.10b), the number of single females has risen steadily since 2007 ($R^2=0.522$, $p>0.05$) while the number of paired females ($R^2=0.586$, $p<0.05$) and females with multiple males ($R^2=0.205$, $p>0.05$) have decreased. In contrast, the number of single females at Sandy Point has decreased slightly since 2007 ($R^2=0.097$, $p>0.05$), while the number of females with multiple mates has slightly increased ($R^2=0.441$, $p>0.05$). Paired females have remained relatively stable ($R^2=0.098$, $p>0.05$) (Fig. 25.10c).

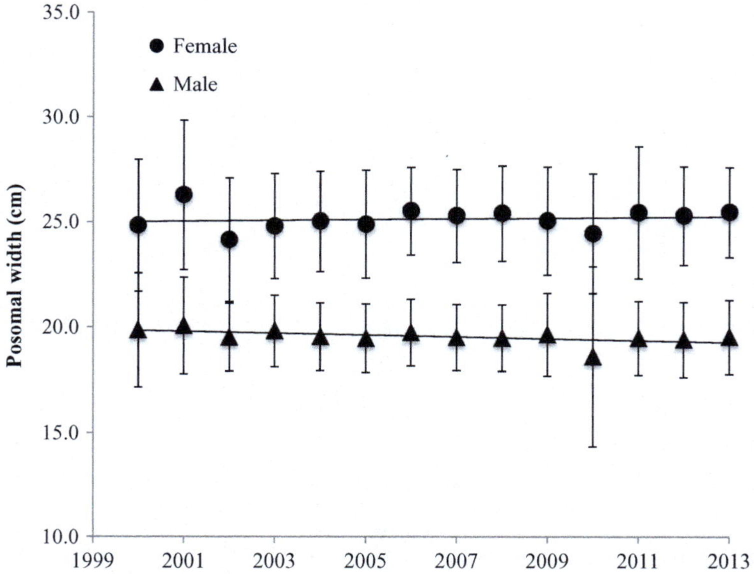

Fig. 25.8 Prosomal width (cm) by sex and year from 2000 to 2013. Regression equations as follows: Females: $y = 0.02x - 18.41$, $R^2 = 0.030$, $p > 0.05$; Males: $y = -0.04x + 104.07$, $R^2 = 0.296$, $p < 0.05$

Condition data was collected in 2012 and 2013 on a sample of adult spawning horseshoe crabs (N=9,908, N=6,165, respectively). Overall, for the combined sexes Condition 1 (newly molted adults) were significantly less abundant than Condition 2 and 3 in both years, 2012 (Fig. 25.11a, $\chi^2 = 1923.9$ $p < 0.00001$) and for 2013 (Fig. 25.11b, N=6,165, $\chi^2 = 977.6$, $p < 0.00001$). Condition 1 females without any mating scar across the opisthosoma were never observed. The number of Condition 1 females and males (new recruits) remained relatively stable from 2012 to 2013 while the number of Condition 2 ("middle aged") females and males decreased by 2.5 % and 4.3 % respectively (Fig. 25.12a). Condition 3 females and males increased on average by 2.4 %. There were fewer Condition 1 males than females in both 2012 and 2013. While the percentage of Condition 1 females and males is higher at Milford Point than the overall pattern, the decrease in the percentage of condition 2 females and males was higher than the overall average at 9.5 % and 23.7 %, respectively (Fig. 25.12b). The increase in number of Condition 3 females and males at Milford Point from 2012 to 2013 was also greater compared with the pattern observed overall in LIS. The percentage of Condition 1 females at Sandy Point was higher than the overall pattern in 2012 and 2013 (Fig. 25.12c). Also, the percentage of Condition 3 males at Sandy Point decreased from 2012 to 2013.

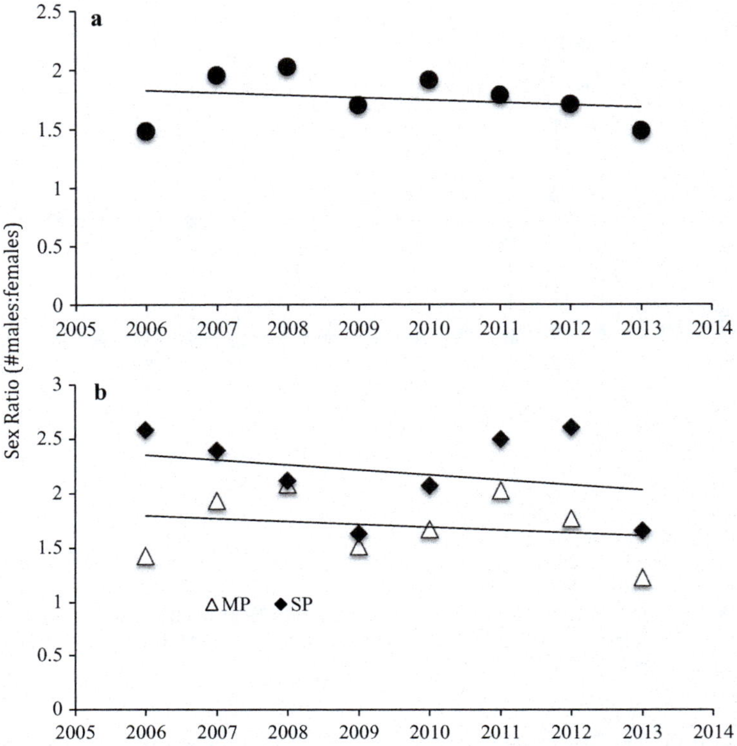

Fig. 25.9 Sex ratio (number males:females) by year from 2006 to 2013. (**a**) Overall sex ratios ($y = -0.02x + 43.58$, $R^2 = 0.062$, $p = 0.552$) (*top panel*), (**b**) Sex ratios for Milford (MP) ($y = -0.03x + 54.33$, $R^2 = 0.045$, $p = 0.613$) and Sandy Point (SP) ($y = -0.05 + 93.26$, $R^2 = 0.081$, $p = 0.495$) (*bottom panel*)

25.4 Discussion

25.4.1 Spatial and Temporal Patterns

Discernable patterns obtained from recapture data are largely due to the amount of temporal and spatial surveillance that occurs post tagging as well as accessibility to various beaches where horseshoe crabs spawn. Immediate recaptures or recaptures within a few days are more likely to occur given that volunteers are actively tagging during the initial period of the spawning season until they run out of tags. This creates a bias in favor of short-term recoveries. However, even this short term recapture data is important because it demonstrates the idea that females come to shore multiple times to lay eggs and males do also. The data reveals that some females make their nests across several beaches in a particular area during one spawning season. Accessibility creates a bias with respect to the ability to detect movement patterns since volunteers and the general public are not searching every kilometer of

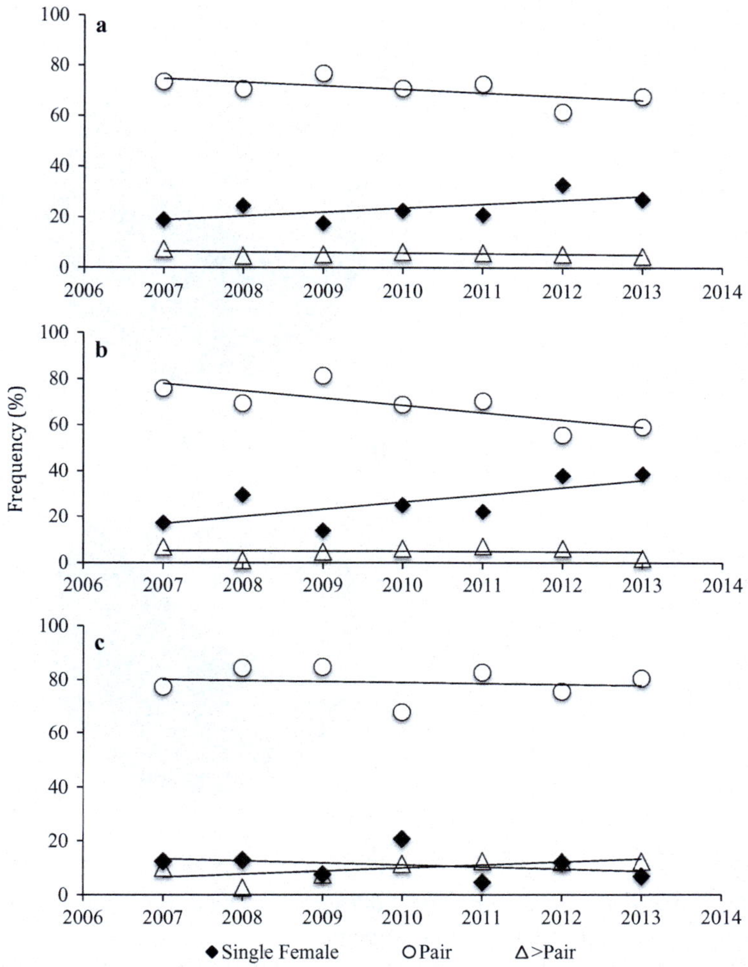

Fig. 25.10 Mating patterns for females tagged from 2007 to 2013. Paired females (P) (*circle*), single females (SF) (*diamond*), and females with multiple mates including amplexed male and additional satellite males (FM) (*triangles*) are plotted for (**a**) Females tagged at all sites (P : $y = -1.39x + 2862.5$, $R^2=0.38$, $p>0.05$; SF : $1.59x - 3167.4$, $R^2=0.43$, $p>0.05$; FM : $y = -0.1985x + 404.86$, $R^2=0.205$, $p>0.05$); (**b**) Females tagged at Milford Point (P : $y = -3.1423x + 6384.7$, $R^2=0.586$, $p<0.05$; SF : $3.2191x - 6443.9$, $R^2=0.52$, $p>0.05$; FM : $y = -0.0768x + 159.16$, $R^2=0.004$, $p>0.05$); and (**c**) Females tagged at Sandy Point, New Haven (P : $y = -0.3499x + 782.4$, $R^2=0.015$, $p>0.05$; SF : $-0.7661x + 1550.9$, $R^2=0.097$, $p>0.05$; FM : $y = 1.116x - 2233.3$, $R^2=0.440$, $p>0.05$)

the coastline where crabs can reappear after initial tagging. However, even with these biases, mark- recapture data can provide general information on spatial and temporal patterns with respect to horseshoe crab movement, particularly when large numbers of tags are deployed over a 10–14 year study period.

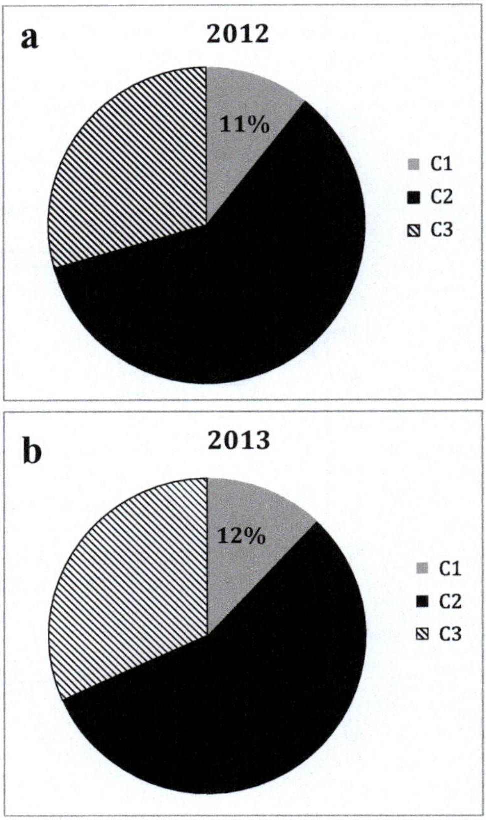

Fig. 25.11 Total frequency (%) of Condition 1, 2 and 3 for (**a**) 2012 (N = 9,908, $\chi^2 = 1923.9$ $p < 0.00001$) and (**b**) 2013 (N = 6,165, $\chi^2 = 977.6$, $p < 0.00001$)

1) Do horseshoe crabs exhibit site fidelity within and across spawning seasons; this is important in informing conservation decisions?

Within a spawning season, horseshoe crabs generally remain at the initial spawning beach where they were tagged over the course of successive tides. For instance, one female tagged at Milford Point on May 26th, 2005 was repeatedly sighted at the same beach over the next 6 days before disappearing. Horseshoe crabs that were recaptured at different beaches from where they were originally tagged within a spawning season were typically found within 2 km. This is likely because spawning horseshoe crabs never leave the shore zone to any great distance (Swan 2005). This within spawning season movement pattern has been observed in other populations along the Atlantic coast (Plum Island Sound, MA – Baptist et al. 1957; Cape Cod, MA – Widener and Barlow 1999; James-Pirri et al. 2005; Delaware Bay – Swan 2005; Great Bay Estuary, ME – Schaller et al. 2010). Thus within a spawning season there is strong evidence for moderate site fidelity.

2) Do horseshoe crabs have home ranges and if so what is their general size; this is important to know when establishing no-harvest zones?

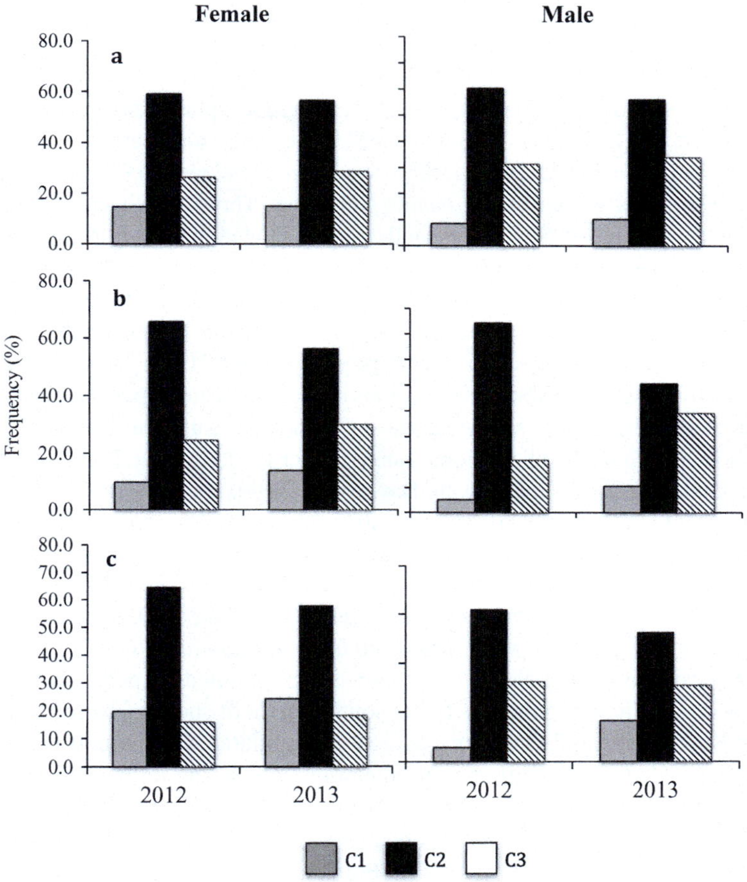

Fig. 25.12 Frequency (%) of crab conditions (1, 2, and 3) of tagged female and male crabs from (**a**) all tagging sites, (**b**) Milford Point, and (**c**) Sandy Point

After one season, horseshoe crabs in Long Island Sound (LIS) disperse more widely and are frequently recaptured on beaches other than the one they were originally tagged on. Of all crabs recaptured in subsequent spawning seasons (N=7,488), only 22 % were recaptured on the same beach as they were tagged compared to 73 % within a single spawning season. Thirty six percent of all recaptured crabs were resighted within the city limits of their original tagging location (approx. 3 km) compared to 85 % of crabs within a spawning season. However, given the differences in dispersal distances across spawning seasons, long distance dispersal was not common.

3) Do horseshoe crabs cross the Sound from Connecticut to the north shore of Long Island, NY; this knowledge would help define the size of the population management unit?

Of the 17,086 recaptured crabs originally tagged along the CT coastline, only 2.2 % were recaptured on the northern shoreline of Long Island, New York indicating that they had crossed the Sound. Three percent (N=541) were recaptured outside of Long Island Sound. The majority of these crabs were recaptured along the coastline of Narragansett Bay, Rhode Island (N=511) or the southeastern coastline of Massachusetts (N=12). Only 12 of the crabs originally tagged in CT were recaptured along the southern shore of Long Island, New York. Three crabs were reported from the Atlantic coastline of New Jersey, and one crab tagged in CT was recaptured along the Atlantic coastline of Maryland near Ocean City. On average, crabs were recaptured within 12 km of their original spawning site over subsequent spawning seasons. Greenwich/Stamford crabs were routinely recaptured on the north shore of Long Island. Clearly, crabs in the narrower western portion of Long Island Sound move more readily between CT and NY than crabs in the wider central portion of LIS. While more extensive than within spawning season, these data again reinforce the notion that horseshoe crabs have localized home ranges. Over the long term, horseshoe crabs in Long Island Sound, like in Delaware Bay (Swan 2005), are one large and interbreeding population. In fact, a genetic study of horseshoe crabs in Long Island Sound found no evidence of subpopulation structure, no evidence of inbreeding depression and that the population was in Hardy-Weinberg equilibrium (Kasinak et al. 2011).

4) Do LIS horseshoe crabs leave LIS and if so, where do they go; the rates of immigration and emigration are important in population analyses?

The recapture data indicate that there is very limited dispersal between LIS horseshoe crabs and populations to the south (southern shore of New York's Long Island, New Jersey, and Delaware) as well as populations to the north around and above Cape Cod, Massachusetts. Narragansett Bay is the only embayment with which Long Island Sound crabs are likely to intermix with, and even then, the data suggests it is mostly a result of dispersal between crabs in the easternmost portion of Long Island Sound near Groton, CT (Fig. 25.6b).

Earlier studies have demonstrated that crabs could survive for up to 5 years after they started spawning (Shuster 1950; Baptist et al. 1957; Ropes 1961). Botton and Ropes (1988) later concluded that adult crabs could live up to at least 8 years after their terminal molt. While tag loss has been demonstrated to affect tag recovery over time (Butler 2012), tagging studies can offer insight into horseshoe crab life spans once they have undergone a terminal molt and reached sexual maturity. Swan's tagging study in the Delaware Bay recorded recaptures up to 10 years post tagging (2005). In this study, only 2 % of all recaptures were resighted 4 or more years after their initial tagging (Fig. 25.4). Three crabs were recaptured 8 years after their initial tagging (Table 25.3) indicating potential life spans as outlined in previous studies.

25.4.2 Population Characteristics

One concern that has arisen with the harvesting of horseshoe crabs is that large females are preferentially targeted. Harvesters prefer females for use as bait (ASMFC 1998) because they can be cut up to bait as many as four pots, whereas cut males typically bait one to two pots (Manion et al. 2000). Eel fishermen also prefer egg-laden females due to the presumed chemical attractants found in the eggs (Ferrari and Targett 2003). The biomedical industry preferentially harvests female crabs with the assumption that they furnish more lysate (USFWS 2002). Reductions in the harvest quotas for Delaware Bay have redirected harvest pressures from the Delaware Bay to other regions (Smith et al. 2009). If there were preferential harvest of females in Long Island Sound, one would expect to see a decrease in female size and/or an increase in the operational sex ratio during spawning season when the majority of the crabs were tagged. Our long-term size data collected since 2000 indicates no significant trend in female or male size since 2000 (Fig. 25.7). Mean female ($\overline{X} = 25.4$ cm, range 12–41 cm) and male ($\overline{X} = 19.6$ cm, range 10–32 cm) prosomal widths for LIS horseshoe crabs are within the range of widths reported for Delaware Bay (females: $\overline{X} = 26.7$ cm, range = 22.5–27.2 cm; males $\overline{X} = 20.7$ cm, range = 17.5–24.5 cm) but on average larger than crabs from Massachusetts (females: $\overline{X} = 25.3$ cm, range = 12.6–29.5 cm; males $\overline{X} = 17.5$ cm, range = 9.7–25.0 cm) females and Rhode Island (females: $\overline{X} = 24.0$ cm, range = 20.1–30.0 cm; males $\overline{X} = 18.6$ cm, range = 15.9–22.4 cm) (Graham et al. 2009).

Since 2007, overall sex ratios (males:females) in Long Island Sound have decreased from 1:2 to 1:1.5 (Fig. 25.9). In Pleasant Bay, MA and Delaware Bay, the sex ratio of the entire population (not just while spawning) is skewed towards males (Carmichael et al. 2003; Smith et al. 2006). On spawning beaches, operational sex ratios are expected to be more skewed towards males due to their reproductive behavior and population demographics. Males typically spend more time on spawning beaches than females (Brockmann and Penn 1992). One male attaches to the female in amplexus while unattached satellite males surround the pair during fertilization in 40–50 % of the spawnings in Delaware Bay, while in LIS satellite behavior is only seen in 5–10 % of the spawning females depending on the beach (Mattei et al. 2010). In Cape Cod, typical spawning sex ratios range from one female to every two to three males (James-Pirri et al. 2005) while in Delaware Bay, spawning sex ratios ranged from 1:1.5 to 1:5 (Shuster and Botton 1985; Swan et al. 1996; Smith et al. 2002). Harvesting can increase the sex ratio (more males per female) on spawning beaches over time due to preferential selection of females (James-Pirri et al. 2005; Smith et al. 2009). In 2001 in Pleasant Bay, MA, an area exploited by the bait and biomedical fisheries, spawning sex ratios were 1:9 (F:M) (James-Pirri et al. 2005). Interestingly in Nauset Estuary, MA, a population that has not been exploited by harvesting, the spawning sex ratio was 1:1.6 (James-Pirri et al. 2005). Our data suggest that the number of males on spawning beaches is decreasing over time, which does not indicate preferential harvest of females. This decrease in males on spawning beaches corresponds with shifts in mating patterns we have also observed since 2007.

Since 2007, the frequency of single females observed on spawning beaches across the CT coastline of Long Island Sound has increased with a corresponding decrease in the number of females with one amplexed male and females with one amplexed male and one or more satellite male (Fig. 25.9). One might suggest that this pattern could be confounded by the types of females volunteers or researchers chose to tag (nesting or moving). However, all volunteers and researchers were instructed to tag every horseshoe crab they encountered whether nesting or moving. From 2007 to 2013, the overall percentage of females tagged that were paired with one male decreased from 73 % to 67 % while the number of single females on spawning beaches increased from 19 % to 27 %. We posit that this trend along with the decreasing sex ratios (F:M) is due to a gradual decrease in the number of males within the Long Island Sound population. Further evidence for this supposition comes from our recent assessments of crab condition.

5) What is the ratio of new recruits into the breeding population using shell condition; this is an indication of an expanding or declining population and evidence for survival rates of juvenile horseshoe crabs?

The condition data indicate that the recruitment of newly molted adults into the LIS population is low (11–12 %, Fig. 25.11a, b). This may indicate either egg and/or juvenile survivorship is very low in the Sound and that more research is needed in this area. This is yet another indication of a population in decline.

Female recruitment to the spawning population was higher than male recruitment over the past 2 years. The percentage of newly molted females (Condition 1, Fig. 25.12) was greater than the percentage of newly molted males in 2012 and 2013. In addition, the frequency of Condition 2 and 3 males is greater than the frequency of Condition 2 and 3 females. Based on these data, it appears that the male portion of the population is "aging" more rapidly than the female portion of the population, and it is not replacing itself. Previous studies have demonstrated that older males are less likely to pair with females (Brockman and Penn 1992; Penn and Brockmann 1995; Brockmann 2002; Duffy et al. 2006). Younger males (Condition 1) may be less likely to be covered with epibionts, less deterioration or fouling of the eyes, lighter and less pitted prosomas, less damage to claws used to clasp females thus, were more likely to pair and remain paired (Brockmann and Smith 2009). In a previous study, we explicitly examined male condition in relation to whether or not males were paired with a female or were a satellite male. Our data revealed that male condition played little role in determining whether a male was more likely to be found amplexed to a female or in a satellite position (Mattei et al. 2010). Interestingly, the percentage of unattached males in good condition was greater than the number of amplexed males. We attributed this behavior to low spawning density. Density plays an important role in determining alternative reproductive tactics in other species (see Courchamp et al. 1999; Stephens et al. 1999; Brockmann 2001; Gascoigne et al. 2009 for examples). We considered the presence of relatively high numbers of single females and the low level of polygynandrous behavior as evidence that horseshoe crabs along the CT coastline are not able to maximize their reproductive effort and have difficulty finding mates. This is yet more evidence of a population in decline.

6) Do LIS horseshoe crabs exhibit similar population characteristics to other populations of horseshoe crabs further north in New England or further south in Delaware Bay?

The population characteristics of horseshoe crabs in Long Island Sound are representative of both Delaware Bay and New England populations. Long Island

Sound is similar to Delaware Bay in that it is one large interbreeding population. Yet this population is discrete like those in New England in that LIS horseshoe crabs do not readily migrate out of the Sound nor have we detected any significant migration into the Sound from other populations based on the USFWS horseshoe crab tagging Program (Sheila Eyler, USFWS, personal communication). Horseshoe crab prosomal widths are more similar to Delaware Bay than populations in the New England region. LIS spawning sex ratios are at odds with both Delaware Bay and more northerly populations. Mating patterns (frequency of pairs versus females with multiple satellites) in Long Island Sound are more similar to New England populations of horseshoe crabs compared to patterns observed in Delaware Bay (Mattei et al. 2010). Based on our tagging data, we conclude that while state and federal stock assessments (ASMFC 2013; CTDEEP 2014) have classified the western Long Island Sound horseshoe crab population as relatively stable, the population density is so low that we are beginning to see a serious shift in population characteristics that are indicative of significant population decline in the future. Further decreases in population density due to over-harvesting or habitat loss could push the LIS sound population over the threshold and start a long term decline in the horseshoe crab population where recovery would be a significant challenge. Continued decrease in the number of males within the population could significantly reduce reproductive success. Given the discreteness of this population in the sense that there is limited migration into and out of Long Island Sound, the likelihood of recovery based on adult dispersal from more robust populations to the south is unlikely. Recovery based on migration of individuals from more northern populations is even less likely given that these populations are even less dense than the Long Island Sound population (Mattei et al. 2010). Finally, recovery of depleted populations is further constrained by the limited dispersal abilities of horseshoe crab larvae as demonstrated by Botton and Loveland (2003).

25.5 Management Recommendations

While CT and NY share jurisdiction over Long Island Sound, their different horseshoe crab management strategies and regulations pose a conservation heresy. In 2013, Connecticut's reported harvest was 19,645 crabs (CTDEEP 2014), while New York State's harvest of horseshoe crabs from Long Island Sound waters was estimated around 69,336 individuals (NYSDEC 2013). The combined horseshoe crab harvest from Long Island Sound (89,000) in 2013 was 12 % less than the number of crabs harvested in Delaware Bay (100,255) (ASMFC FMP review 2013). Considering that Delaware Bay's reported spawning indices (# spawning females/ m^2) are up to 400 times greater than reported spawning indices in Long Island Sound (Mattei et al. 2010), we conclude that the current level of horseshoe crab harvest in Long Island Sound is not sustainable. Our data on population characteristics supports this conclusion in that we have observed an overall increase in the number of single females and a corresponding decrease in the number of paired females and females with multiple mates since 2007. Furthermore, the lack of recruitment into the spawning population (especially for males) does not bode well for the long-term success of the horseshoe crabs in LIS. If the ultimate goal is to increase horseshoe crab abundance in Long Island Sound, we suggest that (1) Connecticut and New York focus on a unified management plan for horseshoe crabs in LIS with consistent harvesting regulations, seasonal closures, and a reduction in the quota for allowable harvest of horseshoe crabs residing within the boundaries of the Sound; and (2) an increase in the number of protected areas where harvesting is prohibited at all times. Spacing of these protected areas could be matched to existing high quality spawning locales and placed no more than 16–18 km apart from each other to encompass the average home range of LIS horseshoe crabs. However, both of these suggestions would require long-term implementation to produce results. To increase horseshoe crab abundance to levels seen before the 1980s, we suggest implementing a ban on the take of females which would allow for an immediate increase in egg deposition, as seen in Delaware Bay and is the common and accepted management practice of other economically important species including lobster and blue crab (ASMFC 2013). Connecticut and New York should abide by the vision of the ASMFC and cooperatively manage this species as well as others by setting up at least two Marine Protected Areas that remain completely undisturbed on both sides of the Sound. The economic value and conservation success of MPAs is well known (Davies et al. 2012). Connecticut and New York State should take the lead on the sustainable management of their shared natural resources.

Acknowledgements We would like to thank the hundreds of Project *Limulus* volunteers of all ages that have donated thousands of hours of their time to help understand the ecology and conservation of the horseshoe crab in Long Island Sound. Numerous conservation organizations have helped organize volunteers, conduct tagging sessions, collate data, and assist with our educational outreach. In particular, we would like to thank the staff and members of The Maritime Aquarium, Branford Land Trust, SoundWaters, Denison Pequotsepos Nature Center, CT Audubon Society, The Nature Conservancy, National Audubon and many others. The tagging aspect of the project would not have been possible without the assistance of Sheila Eyler and her staff at the US Fish & Wildlife Service, Maryland Fishery Resources Office. We would like to particularly recognize Adam Rudman, Project *Limulus* outreach coordinator who has devoted the past 5 years of his life to this project to make it a huge success. We would also like to thank the countless undergraduate and graduate students who have worked on these projects and kept the energy level high as well as the quality of data collected in excellent condition. Finally, we would like to thank David Smith, Mark Botton, and one other reviewer for comments that significantly improved this chapter.

References

Applied Biology Inc., Ray L. Lyerly & Associates (1980) Biological and environmental studies at the Florida Power & Light Company Cape Canaveral Plant and the Orlando Utilities Commission Indian River Plant, Volume II, Part I: Biological studies. Applied Biology, Inc, Atlanta; Ray L. Lyerly & Assoc., Dunedin

ASMFC (1998) Interstate fishery management plan for horseshoe crab. Atlantic States Marine Fisheries Commission, Fishery Management Report No. 32. Washington, DC

ASMFC (2013) Review of the Atlantic States Marine Fisheries Commission fishery management plan for horseshoe crab 2012 fishing year. Atlantic States Marine Fisheries Commission, Washington, DC

Baptist JP, Smith PR, Ropes JW (1957) Migrations of the horseshoe crab, *Limulus polyphemus* in Plum Island Sound, Massachusetts. US Fish and Wildlife Service Special Science Report – Fisheries No. 220:1–15

Beekey MA, Mattei JH, Pierce BJ (2013) Horseshoe crab eggs: a rare resource for predators in Long Island Sound. J Exp Mar Biol Ecol 439:152–159

Berkson J, Shuster CN Jr (1999) The horseshoe crab: the battle for a true multiple-use resource. Fisheries 24:6–10

Botton ML, Loveland RE (2003) Abundance and dispersal potential of horseshoe crab (*Limulus polyphemus*) larvae in the Delaware Estuary. Estuaries 26:1472–1479

Botton M, Ropes J (1988) An indirect method for estimating longevity of the horseshoe crab (*Limulus polyphemus*) based on epifaunal slipper shells (*Crepidula fornicata*). J Shellfish Res 7:407–412

Brockmann HJ (1996) Satellite male groups in horseshoe crabs, *Limulus polyphemus*. Ethology 102:1–21

Brockmann HJ (2001) The evolution of alternative strategies and tactics. Adv Study Behav 30:1–51

Brockmann HJ (2002) An experimental approach to altering mating tactics in male horseshoe crabs (*Limulus polyphemus*). Behav Ecol 13:232–238

Brockmann HJ, Penn D (1992) Male mating tactics in the horseshoe crab, *Limulus polyphemus*. Anim Behav 44:653–665

Brockmann HJ, Smith MD (2009) Reproductive competition and sexual selection in horseshoe crabs. In: Tanacredi JT, Botton ML, Smith DR (eds) Biology and conservation of horseshoe crabs. Springer, New York, pp 199–221

Butler CA (2012) Estimation of survival of the Atlantic horseshoe crab, *Limulus polyphemus*, by analysis of mark-recapture data with tag loss. Master's thesis, Auburn University

Carmichael RH, Rutecki D, Valiela I (2003) Abundance and population structure of the Atlantic horseshoe crab, *Limulus polyphemus*, in Pleasant Bay, Cape Cod. Mar Ecol Prog Ser 246:225–239

Carmichael RH, Rutecki D, Annett B et al (2004) Position of horseshoe crabs in estuarine food webs: N and C Stable isotopic study of foraging ranges and diet composition. J Exp Mar Biol Ecol 299:231–253

Courchamp F, Clutton-Brock T, Grenfell B (1999) Inverse density dependence and the Allee effect. Trends Ecol Evol 14:405–410

CTDEEP (2014) Compliance report to the Atlantic States Marine Fisheries Commission. Connecticut Department of Energy and Environmental Protection, Hartford

Davies TK, Martin S, Mees C et al (2012) A review of the conservation benefits of marine protected areas for pelagic species associated with fisheries. International Seafood Sustainability Foundation, McLean. ISSF Technical Report 2012–02

Duffy EE, Penn DJ, Botton ML et al (2006) Eye and clasper damage influence male mating tactics in the horseshoe crab, *Limulus polyphemus*. J Ethol 24:67–74

Ferrari KM, Targett NM (2003) Chemical attractants in horseshoe crab, *Limulus polyphemus*, eggs: the potential for an artificial bait. J Chem Ecol 29:477–496

Gascoigne J, Berec L, Gregory S et al (2009) Dangerously few liaisons: a review of a mate-finding Allee effects. Popul Ecol 51:355–372

Gibson M, Olszewski S (2001) Stock status of horseshoe crabs in Rhode Island in 2000 with recommendations for management. Rhode Island Division of Fish and Wildlife Research Reference Document 01/01

Grady SP, Valiela I (2006) Stage-structured matrix modeling and suggestions for management of Atlantic horseshoe crab, *Limulus polyphemus*, populations on Cape Cod, Massachusetts. Estuar Coast 29:685–698

Graham LJ, Botton ML, Hata D et al (2009) Prosomal-width-to-weight relationships in American horseshoe crabs (*Limulus polyphemus*): examining conversion factors used to estimate landings. Fish Bull 107:235–243

James-Pirri MJ (2010) Seasonal movement of the American horseshoe crab *Limulus polyphemus* in a semi-enclosed bay on Cape Cod, Massachusetts (USA) as determined by acoustic telemetry. Curr Zool 56:575–586

James-Pirri MJ, Tuxbury K, Marino S et al (2005) Spawning densities, egg densities, size structure, and movement patters of spawning horseshoe crabs, *Limulus polyphemus*, within four coastal embayments on Cape Cod, Massachusetts. Estuaries 28:296–313

Kasinak J, Bartholomew K, Beekey MA et al (2011) Movement patterns and population of the American horseshoe crab in relation to Long Island Sound conservation strategies. In: Proceedings of the 2010 Biennial Long Island Sound Research Conference. University of Connecticut, Stamford, 29–30 Oct 2010

King TL, Eackles MS, Spidle AP et al (2005) Regional differentiation and sex-biased dispersal among populations of the horseshoe crab (*Limulus polyphemus*). Trans Am Fish Soc 134:441–465

Manion MM, West RA, Unsworth RE (2000) Economic assessment of the Atlantic coast horseshoe crab fishery. Division of Economics, U.S. Fish and Wildlife Service, Arlington

Mattei JH, Beekey MA, Rudman A et al (2010) Reproductive behavior in horseshoe crabs: does density matter? Curr Zool 56:634–642

Mattei JH, Beekey MA, Potter HR et al (2011) Estimation of short-term tag-induced mortality in horseshoe crabs *Limulus polyphemus*. Trans Am Fish Soc 140:954–958

Millard MJ, Sweka JA, McGowan CP et al (2015) Assessment and management of North American horseshoe crab populations, with emphasis on a multispecies framework for Delaware Bay, U.S.A. populations. In: Carmichael RH, Botton M, Shin PKS, Cheung SG (eds) Changing global perspectives on horseshoe crab biology, conservation and management. Springer, New York, pp 433–461

NYSDEC (2013) 2013 compliance report to the Atlantic States Marine Fisheries Commission. New York State Department of Environmental Conservation, Albany

Penn D, Brockmann HJ (1995) Age-biased stranding and righting in horseshoe crabs, *Limulus polyphemus*. Biol Bull 187:373–384

Ropes JW (1961) Longevity of the horseshoe crab, *Limulus polyphemus* (L.). Trans Am Fish Soc 90:79–80

Rutecki D, Carmichael RH, Valiela I (2004) Magnitude of harvest of Atlantic horseshoe crabs, *Limulus polyphemus*, in Pleasant Bay, MA. Estuaries 27:179–187

Schaller SY, Chabot CC, Watson WH (2010) Seasonal movements of American horseshoe crabs, *Limulus polyphemus*, in the Great Bay Estuary, New Hampshire (USA). Curr Zool 56:587–598

Shuster CN Jr. (1950) Observations on the natural history of the American horseshoe crab, *Limulus polyphemus*. Woods Hole Oceanographic Institution Contribution Mo. 564, Woods Hole, MA

Shuster CN Jr, Botton ML (1985) A contribution to the population biology of horseshoe crabs Limulus *polyphemus* (L.), in Delaware Bay. Estuaries 8:363–372

Smith DR, Pooler PS, Swan BL et al (2002) Spatial and temporal distribution of horseshoe crab (*Limulus polyphemus*) spawning in Delaware Bay: implications for monitoring. Estuaries 25:115–125

Smith DR, Millard MJ, Eyler S (2006) Abundance of adult horseshoe crabs in Delaware Bay estimated from a large-scale mark-recapture study. Fish Bull 104:456–464

Smith DR, Millard MJ, Carmichael RH (2009) Comparative status and assessment of *Limulus polyphemus* with emphasis on the New England and Delaware Bay populations. In: Tanacredi JT, Botton ML, Smith DR (eds) Biology and conservation of horseshoe crabs. Springer, New York, pp 361–386

Stephens PA, Sutherland WJ, Freckleton RP (1999) Consequences of the Allee effect for behavior, ecology, and conservation. Trends Ecol Evol 14:401–405

Swan BL (2005) Migrations of adult horseshoe crabs, *Limulus polyphemus*, in the middle Atlantic bight: a 17-year tagging study. Estuaries 28:28–40

Swan BL, Hall WR, Shuster CN Jr. (1996) Annual survey of horseshoe crab spawning activity along the shores of Delaware Bay: 1990–1995 Summary. In Farrell J, Martin C (eds)

Proceedings of the horseshoe crab forum: status of the resource University of Delaware Sea Grant College Program, Lewes, 1996

United States Fish and Wildlife Service (USFWS) (2002) Monomoy National Wildlife Refuge compatibility determination draft. Eastern Massachusetts National Wildlife Refuge Complex, Sudbury

Vincent ACJ, Harris JM (2014) Boundless no more. Science 346(6208):420–421

Walls EA, Berkson JM, Smith SA (2002) The horseshoe crab, Limulus polyphemus: 200 million years of existence, 100 years of study. Rev Fish Sci 10:39–73

Widener JW, Barlow RB (1999) Decline of a horseshoe crab population on Cape Cod. Biol Bull 197:300–302

Chapter 26
Horseshoe Crabs in Modern Day Biotechnological Applications

A.P. Das, B. Bal, and P.S. Mahapatra

Abstract Horseshoe crabs are an important source of novel bioactive compounds. These species are the source of novel antimicrobial agents, peptides with impending pharmaceutical applications, nutritional supplements, biosensing probes, and other biotechnological implications. Many new bioactive molecules, including an antimicrobial and anticarcinogenic agent with therapeutic applications have been reported from these animals. Even if only a few products are presently on the industrial scale, a number of novel compounds are now under clinical trials and research. Despite the fact that horseshoe crabs offer an enormously lucrative source of novel bioactive compounds, they also face great challenges for their existence. This chapter highlights the advances in the exploitation of horseshoe crabs in modern day biotechnological applications, their survival and conservation.

Keywords Bioactive compounds • Antimicrobial • Anticarcinogenic • Biosensing • Conservation

26.1 Introduction

Horseshoe crabs have endured relatively unaffected for the past 400 million years and now are at the focus of competing resource uses. Over the decades, horseshoe crabs have been extensively investigated by the scientific community and are vital to diverse industrial applications. The bait industry in the US collects horseshoe crabs to catch eel and conch and have recently taken some 2.5 million crabs yearly (Smith et al. 2009). In earlier decades, horseshoe crabs were used as fertilizer and livestock feed (Shuster et al. 2004). Decreases in horseshoe crab populations have

A.P. Das (✉) • B. Bal • P.S. Mahapatra
Bioengineering Laboratory, Centre of Biotechnology, Siksha O Anusandhan University, Khandagiri Square, Bhubaneswar 751003, Odisha, India
e-mail: alokdas@soauniversity.ac.in

R.H. Carmichael et al. (eds.), *Changing Global Perspectives on Horseshoe Crab Biology, Conservation and Management*, DOI 10.1007/978-3-319-19542-1_26

been attributed to overharvesting for bait in fisheries, ecological degradation and environmental pollution (Shuster and Botton 1985; Loveland et al. 1996; Botton 2009). Scientists have used horseshoe crabs for research on vision, neurobiology, molting behavior, cellular phagocytosis, and the developmental biology (Shuster et al. 2004).

During the preceding two decades, numerous bioactive compounds with unique utility, including clotting factors, proteinase inhibitors lectins, tachyplesin, defensins, and tachystatins have been obtained from horseshoe crabs. These species possess recognized innate immune resistance against bacterial pathogens. The amebocyte is the sole type of cell that circulates in the horseshoe crab hemolymph and plays a major role in the defense mechanism. Amebocytes secrete these molecules in response to endotoxin that involves protein cross linking. The amebocytes contain intracellular granules of clotting factors, which are sensitive to a minute quantity of endotoxin. The simplicity, accuracy, and sensitivity of its antimicrobial reaction have, made the horseshoe crab an outstanding animal model for biomedical investigation. The hemolymph of horseshoe crabs presents a large catalogue of evolutionarily conserved proteins that are enormously responsive to pathogens. Numerous hemagglutinins from horseshoe crab have been well studied (Srimal et al. 1985), but their site of production and principle still remain the subject of research. Conventional systems to ascertain endotoxin rely on the response between endotoxin and the *Limulus* amebocyte lysate (LAL), which is an aqueous extract of blood cells (Miao 2013). A minute quantity of endotoxin can activate the LAL response, leading to the formation of clotting. The key application of LAL at present is in the monitoring of endotoxins in health care and biological products (Das et al. 2014). LAL is employed in the detection of endotoxemia in association with cirrhosis, meningitis, cancer, dental problems, gonorrhea, eye disease, boutonneuse fever (caused by tick-transmitted *Rickettsia*), and water-quality analysis (Rudloe 1983; Novitsky 1984; Walls et al. 2002), as well as urinary tract infections. In addition novel applications for LAL include the detection of bacterially contaminated food and dairy products, including packed stuffs (Novitsky 1984).

Biochemical characteristics and morphological stability of horseshoe crabs have remarkable significance with regard to their existence and population. Kwan et al. (2014) investigated the hemolymph quality as an indicator of health status in juvenile Chinese horseshoe crabs *Tachypleus tridentatus* in laboratory conditions. The experimental results suggested that in terms of the hemolymph quality, the crabs reared in the laboratory were considerably deprived in contrast to that of the crabs collected from the field. Alteration in hemolymph components possibly will be a potential indicator of the health status of juvenile horseshoe crabs.

Sexual size dimorphism results from sexually antagonistic selection favouring different optimal body sizes for males and females. The evolution of independent size optima is constrained by genetic correlations between the sexes, as well as physiological, developmental and phylogenetic constraints (Fairbairn et al. 2007).

Horseshoe crabs are an appealing species for the examination of sexual size dimorphism because of their unusual evolutionary history and strategy. Smith and Brockmann (2014) studied the evolution and maintenance of sexual size dimorphism in horseshoe crabs and exemplify the significance of examining several hypotheses for both males and females in understanding the decisive features underlying the extensive pattern of sexual size dimorphism.

Nossa et al. (2014) applied a novel technique of combined de novo assembly and genetic mapping to investigate the chromosome-scale genome association of the Atlantic horseshoe crab, *Limulus polyphemus*. The results suggested a relationship with other metazoan genomes, that the *L. polyphemus* genome conserves ancestral bilaterian linkage groups, and that a general ancestor of current horseshoe crabs underwent one or more ancient whole genome replication 300 million years ago, by extensive chromosome fusion.

Limulus amebocyte lysate (LAL) has been extensively used for more than a decade for the recognition of endotoxin in biological fluids and biomedical devices (Das et al. 2014). However, the restricted availability of horseshoe crabs has urged for an alternative bio-monitoring system. Recombinant Factor C (rFC) is an endotoxin inducible coagulation enzyme present in horseshoe crabs that forms the basis of a new micro-enzymatic technique for high-throughput detection of endotoxin (Ding and Bow 2001). Recently technology has been developed to reduce the quantity of LAL required for endotoxin monitoring. For example, the biosensing of endotoxin using US Food and Drug Administration (FDA) licensed cartridges uses 20 times less LAL than usual assays.

The populations of horseshoe crabs are declining globally as a result of habitat destruction due to land renovation, coastal expansion, seashore alteration, and excavation of sea beds, as well as utilization and trading, particularly for the commercial production of amebocyte lysate. However, as the information on population status is limited and scattered, the International Union for Conservation of Nature (IUCN) has classified the Chinese horseshoe crab *Tachypleus tridentatus* as data deficient. The government of India has included the horseshoe crab under the Schedule IV of the Wildlife Protection Act of 1972. To support community based horseshoe crab conservation initiatives, the Horseshoe Crab Conservation Fund was established by the Ecological Research and Development Group (ERDG) in 2004, and the fund will offer US $5,000 grants to support community based projects on the conservation of this animal. The long term survival of the world's four horseshoe crab species relies upon the conservation of their spawning grounds. The society based "Horseshoe Crab Sanctuary Program" set up in 1999 by ERDG intended to encourage coastal communities to declare their shared habitat a horseshoe crab conservation area or sanctuary.

This chapter reviews the advances in exploitation of horseshoe crabs in modern day biotechnological applications, including potential antimicrobial and anticarcinogenic agents with wide utilization in biomedical industry and their protection and conservation strategies.

26.2 Antimicrobial Activity

The emergence of antibiotic-resistant microorganisms has become an alarming health crisis. To surmount this, the development of antimicrobial agents with novel mechanistic approaches is an immediate concern. Antimicrobial peptides are a chief constituent of the defense mechanism in aquatic invertebrates, and their hemolymph encloses biologically vigorous compounds such as clotting factors, lectins, and antimicrobial proteins. The entire system constitutes an effective self-protection mechanism against pathogenic microbes. Antimicrobial compounds such as Tachyplesin I, Tachyplesin II, Tachyplesin III, Tachylectin-4, Big defensin and Polyphemusin I and II have been purified and reported from horseshoe crabs worldwide (Morita et al. 1985; Muta et al. 1990; Murakami et al. 1991; Kawabata et al. 1997; Morvan et al. 1997; Tincu and Taylor 2004; Armstrong 2010). Big defensin is a bioactive substance containing 79 amino acid residues, of which the C-terminal 37 residues contain a sequence which is interrelated to mammalian neutrophil and present in the hemocytes of *Tachypleus tridentatus* that hinders the development of both gram positive and negative bacteria, along with fungi. The horseshoe crab hemolymph carries three abundant proteins, hemocyanin, C-reactive protein, and a2-macroglobulin, and another type of granular cell, accounting for 99 % of the total hemocytes (Armstrong 1991; Toh et al. 1991). The granular cells are exceptionally receptive to endotoxin, and the cells release constituents in response to lipopolysaccharide (Muta et al. 1990; Kawabata et al. 1996). This reaction is vital for the host defense by destroying invading microbes. The hemocytes contain huge amount of small granules that particularly accumulate proteins and defense molecules, together with serine protease zymogens, a clottable protein that contributes in the coagulation cascade, antibacterial peptides, protease inhibitors and lectins. Andrä et al. (2004) reported cyclic cationic antimicrobial compounds of different sizes, depending on the *Limulus* anti-lipopolysaccharide factor (LALF), and their interface with lipopolysaccharide (LPS). As per their reports, all peptides repressed the LPS induced cytokine generation in human mononuclear cells and the *Limulus* amebocyte lysate in an amount dependent system, with the peptide involving the absolute LPS-binding loop of the LALF being the most efficient. Even though the protein exhibits no enzymatic action, it promotes the coagulation progression after it is added to *Limulus* lysate. As a membrane constituent, the endotoxin-binding protein might act as an endotoxin receptor and mediate the characteristic degranulation of amebocytes exposed to microbial endotoxin. The epithelial immunity, a cytolytic property of cuticular secretion of *Limulus polyphemus* was reported by Harrington and Armstrong (2000, 2003). They described a viscous secretion, proposed to be the product of hypodermal glands, which covers the entire surface of the organism's carapace and suggested that this secretion protects the animal from sessile colonizing organisms, or epibionts (organisms that attach themselves to the external surfaces of hosts). This secretion is referred to as dermal exudate (Harrington et al. 2008). Table 26.1 shows various peptides and proteins in horseshoe crabs with antimicrobial and anticarcinogenic activity.

Table 26.1 Antimicrobial and anticarcinogenic activity of horseshoe crabs

Horseshoe crab	Peptides/proteins	Applications	References
Limulus polyphemus	Tachyplesin II	Antibacterial activity	Saito et al. (1997)
	Polyphemusin I & II	Antimicrobial activity	Miyata et al. (1989)
	CRP and CrO	Protease inhibitory activity	Muta et al. (1990)
	Cyclic cationic antimicrobial peptides	Antimicrobial drug screening	Tincu and Taylor (2004)
			Du et al. (2011)
	Factor C-derived peptides	Neutralization of lipopolysaccharide	Andrä et al. (2004)
	Hemocyanin	Neutralization of endotoxin	Tan et al. (2000)
	Serine-protease	Phenoloxidase activity	Decker et al. (2001)
	Plasma membrane protein	Engulfment of endotoxin	Yabusaki and Aoyagi (2012)
			Liang et al. (1980)
	Limulus clotting Factor C	Protease inhibitory activity	Minetti et al. (1991)
			Urbanova et al. (2014)
	CLP19	LPS-neutralization	Ren et al. (2010)
Carcinoscorpius rotundicauda	Haemagglutinin	Antibacterial activity	Subita et al. (1985)
	Tachyplesin	Antimicrobial activity	Muta et al. (1990)
	C-reactive protein 2	Agglutination activity	Tincu and Taylor (2004)
Tachypleus gigas	Coagulogen	Antibacterial activity	Muta et al. (1990)
	Tachyplesins	Growth inhibitor	Miyata et al. (1989)
	Tachyplesin III	Antimicrobial activity	Tincu and Taylor (2004)
Tachypleus tridentatus	tCRPs	Antimicrobial activity	Ohashi et al. (1984)
	RGD-Tachyplesin	Antibacterial activity	Tincu and Taylor (2004)
	Tachyplesin	Antitumor activity	Tan et al. (2000)
	Tachyplesin-1, a disulfide stabilized β-hairpin antimicrobial peptide	Broad spectrum bactericidal activity with hemolytic property	Chen et al. (2001)
			Ouyang et al. (2002)
			Saravanan et al. (2012)

26.3 Biosensing Activity

Endotoxins are pathogenic microbial contaminants, and rapid and sensitive recognition of endotoxin is vital to check for endotoxemia. The LAL technique has been extensively used for the last 40 years for the sensing of this contaminant in different drugs, biomedical devices and biological samples (Das et al. 2014). The LAL assay represents a cascade of enzymes that are activated in the presence of minute quantity of endotoxin, resulting in a gel formation at the end of the reaction. The Factor C, existing as a zymogen, acts as the primer of this gel clot mechanism. Factor C is the ideal sensor, which alerts the horseshoe crab of the occurrence of endotoxin. The hemostatic end-point captures the attacker, destroys it and prevents further contagion. Conversely, as an endotoxin sensing system, variations in the sensitivity and accuracy of LAL and the declining availability of horseshoe crabs are raising challenges to its commercial manufacture (Ding and Ho 2010). Thus, from *in vitro* studies to manufacturing relevance, studies of Factor C have initiated an innovative period in endotoxin research.

LAL testing for endotoxin uses three major techniques, the gel clot method, the turbidimetric method and the chromogenic method. The gel clot technique is the most extensively applied method for qualitative testing, providing a simple positive/negative result. In the turbidimetric technique, the amount of insoluble coagulin increases, which results in an increase in turbidity. Kinetic turbidimetric methods offer an economical quantitative choice for water and parenteral fluids. Endpoint chromogenic method offers less product interference and more importantly is a quantitative test. The kinetic chromogenic LAL provides greater sensitivity together with low product interference from proteins, vaccines and other biologicals. A rapid and sensitive technique of endotoxin detection based on the use of stress-responsive magnetoelastic sensors that monitor the change in viscosity of the *Limulus* amebocyte lysate (LAL) assay was reported by Ong et al. (2006). These types of sensors are magnetic and wireless, which determine constraints of attention by tracking the alternation in their resonance activities. When dipped inside fluid, their resonance frequencies and amplitudes are a response to the water viscosity, separately of solution color, thus assembling them a perfect technique for monitoring the gel formation linked with LAL analysis. Recently a screen-printed endotoxin sensor based on amperometry using a novel paminophenol conjugated substrate for a *Limulus* amebocyte lysate protease reaction has been reported (Inoue et al. 2013). In this investigation they have reported a sensor for *in vitro* protease detection via amperometry. Endotoxins provoke the cascade activation of zymogens in LAL, which results in the hydrolysis of Boc-Leu-Gly-Arg-pAP (LGR-pAP) as a novel substrate for a clotting enzyme to pAP. Produced pAP is oxidized at an electrode held at +0.30 V (Inoue et al. 2008), and the resulting oxidation current was screened to compute the endotoxin in the sample. Table 26.2 shows reports on biosensing applications of horseshoe crabs.

Table 26.2 Biosensing applications of horseshoe crabs

Horseshoe crab	Techniques	Application	Reference
Limulus polyphemus	Gel-clot assay	Endotoxin detection	Ong et al. (2006)
Limulus polyphemus	Low-frequency piezoelectric tuning forks (QTFs) for endotoxin detection	Endotoxin detection	Chałupniak et al. (2014)
Limulus polyphemus	Stress-responsive magnetoelastic sensors	Endotoxin detection	Ong et al. (2006)
Limulus polyphemus	Electrochemical endotoxin sensor	Fabrication and characterization of rhTLR4/MD-2 immobilized gold electrodes	Yeo et al. (2011)
Limulus polyphemus	Electrochemical impedance spectroscopy	Endotoxins and other contaminants of microbiological origin	Heras et al. (2010)
Limulus polyphemus	Screen-printed endotoxin sensor	Amperometric detection	Inoue et al. (2013)
Limulus polyphemus	Laser scattering photometry	Bacterial endotoxins	Obata et al. (2008)

26.4 Anticarcinogenic Activity

Cancer is a terrible disease, and fighting this illness is of enormous significance to community health. The medical treatments of solid tumors, including head and neck cancers, gliomas and colorectal cancers are still comparatively unsuccessful (Chen et al. 2001; Ouyang et al. 2002). Surgery remains the main medical approach for the cure of tumors. There is great need for the exploration of novel compounds with anticancer activity, as the treatment of cancer through the current generation of anticancer drugs is frequently inadequate due to cytotoxicity in non-targeted healthy cells. New bioactive substances are becoming an established resource of sensible anticancer compounds. The exploration of anticarcinogenic properties of bioactive molecules from oceanic flora and fauna is an expanding area of biomedical investigation.

Phagocytic activity of horseshoe crab amebocytes is well documented among various pathogenic strains (Coates et al. 2012). Likewise, tachyplesin, an antimicrobial substance found in hemolymph from horseshoe crabs was established to be effective against proliferation of human hepatoma SMMC-7721 cells. It was reported to be able to reverse the malignant morphological and ultrastructural characteristics along with noticeable alteration in the actions of differentiation associated with enzymes (Nakamura et al. 1988; Ouyang et al. 2002). Conjugated RGD-Tachyplesin is an effective antitumor agent against cultured tumor of the endothelial cells and reduced the colony formation of TSU prostate cancer cells

(Chen et al. 2001). Tachyplesin at a concentration of 2.0 mg/L was found to be effective against human gastric carcinoma cell line BGC-823 cells. A marked decrease in alteration of malignant morphological and ultrastructural characteristics of carcinoma cell lines along with induction of differentiation in carcinoma cells was observed (Li et al. 2000).

Tachyplesin interacts through the lipids in the plasma membrane of eukaryotic cells (Park et al. 1992; Katsu et al. 1993; Chen et al. 2001). Essentially, as it can act jointly with the prokaryotic cells, it is likely that tachyplesin may be capable to also interact with the mitochondrial membrane of eukaryotes. These are structurally similar since mitochondria are extensively assumed to have developed from prokaryotic cells that have recognized a symbiotic relationship with the prehistoric eukaryotic cell (Gray et al. 1999). Recent studies have indicated that mitochondria play a critical role in regulating apoptosis in eukaryotic cells (Brenner and Kroemer 2000). The disruption of mitochondrial activity is marked by the release of proteins that usually are sequestered by this organelle. The discharge of compounds, such as cytochrome c and Samc, triggers caspases that, in response, activates the apoptotic cascade (Du et al. 2000). The effects of tachyplesin on the proliferation and differentiation of human hepatocellular carcinoma SMMC-7721 cells were investigated by Ouyang et al. (2002). The activity was measured with trypan blue dye exclusion test and Hematoxylin and Eosin (HE) staining. The structural morphology of the cells was investigated by light microscopy and transmission electron microscopy. Results showed that there was significant cell growth inhibition after treatment with tachyplesin at a concentration of 3.0 mg/L. The proliferation of SMMC-7721 cells was inhibited to 55.57 %, and the maximum cell mitotic index declined by 43.68 %. Thus, tachyplesin can efficiently reduce the propagation of human hepatoma SMMC-7721 cells and the levels of tumor-associated antigens.

26.5 Conclusion

Horseshoe crabs have been living on earth for over 400 million years without major morphological modifications and survive catastrophes like ice ages and asteroid collision. Their future, however, is threatened by anthropogenic activities. A decline in the horseshoe crab population was reported and associated with factors such as small population density, low genetic diversity, restricted connectivity between populations, short endurance rates of eggs, and rise in sea-level due to climate change.

Horseshoe crabs are primitive invertebrates that carry potent immune protection against invading pathogens. Ever since the FDA's authorization, the use of horseshoe crabs for LAL production by the pharmaceutical industry has increased progressively. The number of crabs being bled increased from 130,000 in 1989 to 260,000 in 1997. By 2010, over half a million crabs were bled annually. Horseshoe crab blood is estimated a value of $15,000 per quart, according to the National Oceanic and Atmospheric Administration.

The simplicity, accuracy and sensitivity of its antimicrobial and anticarcinogenic reaction have made the horseshoe crab an exceptional species for biotechnology research. The expediency of this crab is demonstrated by its widespread application for responsive sensing of endotoxins, uropathogens, bacteriuria, fungal infections and sepsis in the health care industry. Horseshoe crabs present a huge range of peptides and other cellular bioactive molecules, which are highly sensitive to pathogens. Horseshoe crabs occupy an imperative biological function in the food chain, and a reduction in their population will impact other species, particularly coastal birds and turtles. The proper care of the oceanic environment and effective resource management is required to conserve these species in the long run.

To maintain sustainable horseshoe crab populations, it is necessary to discourage activities causing degradation of natural habitats and re-establish disturbed ones. An integrated management with the incorporation of societal involvement, education and better awareness is essential in the conservation of horseshoe crabs.

References

Andrä J, Lamatab M, Tejadab GM et al (2004) Cyclic antimicrobial peptides based on *Limulus* anti-lipopolysaccharide factor for neutralization of lipopolysaccharide. Biochem Pharmacol 68:1297–1307

Armstrong PB (1991) Cellular and humoral immunity in the horseshoe crab, *Limulus polyphemus*. In: Gupta AP (ed) Immunology of insects and other arthropods. CRC Press, London, pp 4–17

Armstrong PB (2010) Role of α2-macroglobulin in the immune responses of invertebrates. Invertebr Surviv J 7:165–180

Botton ML (2009) The ecological importance of horseshoe crabs in estuarine and coastal communities: a review and speculative summary. In: Tanacredi JT, Botton ML, Smith DR (eds) Biology and conservation of horseshoe crabs. Springer, New York, pp 45–63

Brenner C, Kroemer G (2000) Mitochondria: the death signal integrators. Science 289:1150–1151

Chalupniak A, Waszczuk K, Halubek-Gluchowska K et al (2014) Application of quartz tuning forks for detection of endotoxins and Gram-negative bacterial cells by monitoring of *Limulus* amebocyte lysate coagulation. Biosens Bioelectron 58:132–137

Chen Y, Xu X, Hong S et al (2001) RGD-Tachyplesin inhibits tumor growth. Cancer Res 61:2434–2438

Coates CJ, Whalley T, Nairn J (2012) Phagocytic activity of *Limulus polyphemus* amebocytes *in vitro*. J Invertebr Pathol 111:205–210

Das AP, Kumar PS, Swain S (2014) Recent advances in biosensor based endotoxin detection. Biosens Bioelectron 51:62–75

Decker H, Ryan M, Jaenicke E, Terwilligeri N (2001) SDS-induced phenoloxidase activity of hemocyanins from *Limulus polyphemus*, *Eurypelma californicum*, and *Cancer magister*. J Biol Chem 276:17796–17799

Ding JL, Bow H (2001) A new era in pyrogen testing. Trends Biotechnol 19:277–281

Ding JL, Ho B (2010) Endotoxin detection – from *Limulus* amebocyte lysate to recombinant factor C. In: Wang XY, Quinn PJ (eds) Endotoxins: structure, function and recognition. Springer, New York, pp 187–208

Du C, Fang M, Li Y et al (2000) Smac, a mitochondrial protein that promotes cytochrome *c-dependent* caspase activation by eliminating IAP inhibition. Cell 102:33–42

Du R, Ho B, Ding JL (2011) Application of cell-free hemolymph of horseshoe crab in antimicrobial drug screening. Curr Pharm Des 17:1234–1239

Fairbairn DJ, Blanckenhorn WU, Székely T (2007) Sex, size, and gender roles: evolutionary studies of sexual size dimorphism. Oxford University Press, Oxford

Gray MW, Burger G, Lang BF (1999) Mitochondrial evolution. Science 283:1476–1481

Harrington JM, Armstrong PB (2000) Initial characterization of a potential anti-fouling system in the American horseshoe crab, *Limulus polyphemus*. Biol Bull 199:189–190

Harrington JM, Armstrong PB (2003) A liposome-permeating activity from the surface of the carapace of the American horseshoe crab, *Limulus polyphemus*. Biol Bull 205:205–206

Harrington JM, Leippe M, Armstrong PB (2008) Epithelial immunity in a marine invertebrate: a cytolytic activity from a cuticular secretion of the American horseshoe crab, *Limulus polyphemus*. Mar Biol 153:1165–1171

Heras JY, Pallarola D, Battaglini F (2010) Electronic tongue for simultaneous detection of endotoxins and other contaminants of microbiological origin. Biosens Bioelectron 25:2470–2476

Inoue KY, Yasukawa TS, Hitoshi M, Tomokazu (2008) Cell-based electrochemical assay for endotoxin using a secreted alkaline phosphatase reporter system. Electrochemistry 76(8):525–528

Inoue KY, Takano S, Takahashi S et al (2013) A screen-printed endotoxin sensor based on amperometry using a novel p-aminophenol conjugated substrate for a *Limulus* amebocyte lysate protease reaction. Analyst 138(21):6523–6531

Katsu T, Nakao S, Iwanaga S (1993) Mode of action of an anti-microbial peptide, tachyplesin I, on biomembranes. Biol Pharm Bull 16:178–181

Kawabata S, Muta T, Iwanaga S (1996) Clotting cascade and defense molecules found in hemolymph of horseshoe crab. In: Söderhäll K, Iwanaga S, Vasta GR (eds) New directions in invertebrate immunology. SOS Publications, Fair Haven, pp 255–283

Kawabata A, Saifeddine M, Al-ani B et al (1997) Protease-activated receptors: development of agonists selective for receptors triggered by either thrombin (PAR1) or trypsin (PAR2). Proc West Pharmacol Soc 40:49–51

Kwan BKY, Chan AKY, Cheung SG et al (2014) Hemolymph quality as indicator of health status in juvenile Chinese horseshoe crab *Tachypleus tridentatus* (Xiphosura) under laboratory culture. J Exp Mar Biol Ecol 457:135–142

Li QF, Ou-Yang GL, Li CY et al (2000) Effects of tachyplesin on the morphology and ultrastructure of human gastric carcinoma cell line BGC-823. World J Gastroenterol 6:676–680

Liang SM, Sakmar TP, Liu TY (1980) Studies on *Limulus* amoebocyte lysate. III. Purification of an endotoxin binding protein from Limulus amoebocyte membranes. J Biol Chem 255:5546–5590

Loveland RE, Botton ML, Shuster CN Jr (1996) Life history of the American horseshoe crab (*Limulus polyphemus* L.) in Delaware Bay and its importance as a commercial resource. In: Farrell J, Martin C (eds) Proceedings of the horseshoe crab forum: status of the resource. University of Delaware Sea Grant College Program, Lewes, pp 15–22

Miao P (2013) Electrochemical sensing strategies for the detection of endotoxin: a review. RSC Adv 3:9606–9617

Minetti CA, Lin YA, Cislo T et al (1991) Purification and characterization of an endotoxin-binding protein with protease inhibitory activity from *Limulus* amebocytes. J Biol Chem 266:20773–20780

Miyata T, Tokunaga F, Yoneya T et al (1989) Antimicrobial peptides, isolated from horseshoe crab hemocytes, tachyplesin II, and polyphemusins I and II: chemical structures and biological activity. J Biochem 106:663–668

Morita T, Ohtsubo S, Nakamura T et al (1985) Isolation and biological activities of *Limulus* anticoagulant (anti-LPS factor) which interacts with lipopolysaccharide (LPS). J Biochem 97:1611–1620

Morvan A, Iwanaga S, Comps M et al (1997) *In vitro* activity of the *Limulus* antimicrobial peptide tachyplesin I on marine bivalve pathogens. J Invertebr Pathol 69:177–182

Murakami MT, Niwa FT, Miyata T et al (1991) Direct virus inactivation of tachyplesin I and its isopeptides from horseshoe crab hemocytes. Chemotherapy 37:327–334
Muta T, Fujimoto T, Nakajima H et al (1990) Tachyplesins isolated from hemocytes of southeast Asian horseshoe crabs (*Carcinoscorpius rotundicauda and Tachypleus gigas*): identification of a new tachyplesin, tachyplesin III, and a processing intermediate of its precursor. J Biochem 108:261–266
Nakamura T, Furunaka F, Miyata T et al (1988) Tachyplesin, a class of antimicrobial peptide from the hemocytes of the horseshoe crab (*Tachypleus tridentatus*). Isolation and chemical structure. J Biol Chem 263(32):16709–16713
Nossa CW, Paul H, Jia-Xing Y et al (2014) Joint assembly and genetic mapping of the Atlantic horseshoe crab genome reveals ancient whole genome duplication. GigaScience 3:9
Novitsky TJ (1984) Discovery to commercialization: the blood of the horseshoe crab. Oceanus 27:13–18
Obata T, Nomura M, Kase Y et al (2008) Early detection of the *Limulus* amebocyte lysate reaction evoked by endotoxins. Anal Biochem 373:281–286
Ohashi K, Niwa M, Nakamura T et al (1984) Anti-LPS factor in the horseshoe crab, *Tachypleus tridentatus*: its hemolytic activity on the red blood cell sensitized with lipopolysaccharide. FEBS Lett 176:207–210
Ong KG, Leland JM, Zeng K et al (2006) A rapid highly-sensitive endotoxin detection system. Biosens Bioelectron 21:2270–2274
Ouyang G, Li Q, Peng X et al (2002) Effects of tachyplesin on proliferation and differentiation of human hepatocellular carcinoma SMMC-7721 cells. World J Gastroenterol 86:1053–1058
Park NG, Lee S, Oishi O et al (1992) Conformation of tachyplesin I from *Tachypleus tridentatus* when interacting with lipid matrices. Biochemistry 31:12241–12247
Ren J, Gao H, Tang M et al (2010) Lipopolysaccharide (LPS) detoxification of analogue peptides derived from *Limulus* anti-LPS factor. Peptides 31:1853–1859
Rudloe A (1983) The effect of heavy bleeding on mortality of the horseshoe crab, *Limulus polyphemus*, in the natural environment. J Invertebr Pathol 42:167–176
Saito T, Hatada M, Iwanaga S et al (1997) A newly identified horseshoe crab lectin with binding specificity to O-antigen of bacterial lipopolysaccharides. J Biol Chem 272:30703–30708
Saravanan R, Mohanram H, Joshi M et al (2012) Structure, activity and interactions of the cysteine deleted analog of tachyplesin-1 with lipopolysaccharide micelle: mechanistic insights into outer-membrane permeabilization and endotoxin neutralization. Biochim Biophys Acta 1818:1613–1624
Schuster CN Jr, Botton ML (1985) A contribution to the population biology of horseshoe crabs, *Limulus polyphemus* (L.), in Delaware Bay. Estuaries 8(4):363–372
Shuster CN Jr, Barlow RB, Brockmann HJ (eds) (2004) The American horseshoe crab. Harvard University Press, Cambridge, MA
Smith MD, Brockmann HJ (2014) The evolution and maintenance of sexual size dimorphism in horseshoe crabs: an evaluation of six functional hypotheses. Anim Behav 97:127–139
Smith DR, Millard MJ, Carmichael RH (2009) Comparative status and assessment of *Limulus polyphemus*, with emphasis on the New England and Delaware Bay populations. In: Tanacredi JT, Botton ML, Smith DR (eds) Biology and conservation of horseshoe crabs. Springer, New York, pp 361–386
Srimal S, Miyata T, Kawabata S et al (1985) The complete amino acid sequence of coagulogen isolated from southeast Asian horseshoe crab, *Carcinoscorpius rotundicauda*. J Biochem 98:305–318
Subita S, Toshiyuki M, Shun-ichiro K et al (1985) The complete amino acid sequence of coagulogen isolated from Southeast Asian horseshoe crab, *Carcinoscorpius rotundicauda*. J Biochem 98:305–318
Tan NS, Patricia ML, Yau YH et al (2000) Definition of endotoxin binding sites in horseshoe crab Factor C recombinant sushi proteins and neutralization of endotoxin by sushi peptides. FASEB J 14:1801–1813

Tincu JA, Taylor SW (2004) Antimicrobial peptides from marine invertebrates. Antimicrob Agents Chemother 48:3645–3654

Toh Y, Mizutani A, Tokunaga F et al (1991) Morphology of the granular hemocytes of the Japanese horseshoe crab *Tachypleus tridentatus* and immunocytochemical localization of clotting factors and antimicrobial substances. Cell Tissue Res 266:137–147

Urbanova V, Hartmann D, Grunclova L et al (2014) IrFC–an Ixodes ricinus injury-responsive molecule related to *Limulus* Factor C. Dev Comp Immunol 46:439–447

Walls EA, Berkson J, Smith SA (2002) The horseshoe crab, *Limulus polyphemus*: 200 million years of existence, 100 years of study. Rev Fish Sci 10(1):39–73

Yabusaki K, Aoyagi H (2012) Simplified preparation of crude and functional coagulogen by thermal inactivation of serine proteases in *Limulus* amebocyte lysate and its application for rapid endotoxin determination. J Biosci Bioeng 113:406–411

Yeo TY, Choi JS, Lee BK et al (2011) Electrochemical endotoxin sensors based on TLR4/MD-2 complexes immobilized on gold electrodes. Biosens Bioelectron 28:139–145

Chapter 27
Current Horseshoe Crab Harvesting Practices Cannot Support Global Demand for TAL/LAL: The Pharmaceutical and Medical Device Industries' Role in the Sustainability of Horseshoe Crabs

Glenn Gauvry

Abstract Globally, injectable pharmaceuticals and parenterals, implantable medical devices and similar therapeutic products are safety tested for endotoxin contamination using indicator reagents derived from the blood of horseshoe crabs. *Tachypleus* and *Limulus* Amoebocyte Lysate (TAL and LAL, respectively) are used in quality control laboratories for testing of raw materials, intermediates and final products to protect public health. As the demand for human and animal drugs and devices increases, so does the demand for TAL and LAL. In the United States, the harvesting of horseshoe crabs (*Limulus polyphemus*) is regulated and monitored. Harvesting limits have been exceeded in recent years and may lead to enforcement of harvesting limits to the American LAL industry. Throughout most of Asia, the harvesting of horseshoe crabs is poorly regulated and unmonitored. All indicators point to a declining population. If the *Tachypleus* population cannot support TAL production for the Asia and Pacific pharmaceutical and medical device industries, companies will entirely depend on LAL for testing. The already limited LAL supply is not sufficient to support this increased global demand and will likely result in increased cost and constrain the availability of pharmaceutical and medical device products. As consumers of horseshoe crab-based products, the pharmaceutical and medical device industries can act to help protect horseshoe crab species and, in turn, protect the long-term availability of endotoxin testing products. Global supply chain initiatives that integrate social and environmental considerations into the qualification and choice of vendors, use of best practices for TAL and LAL production, and developing alternative testing methods will promote a more sustainable TAL/LAL industry that can include reduction of animal-based testing and global conservation of horseshoe crab species.

G. Gauvry (✉)
Ecological Research & Development Group Inc.,
190 Main Street, Little Creek, Dover, DE 19901, USA
e-mail: kabutogani@comcast.net

R.H. Carmichael et al. (eds.), *Changing Global Perspectives on Horseshoe Crab Biology, Conservation and Management*, DOI 10.1007/978-3-319-19542-1_27

Keywords Endotoxin detection • LAL/TAL • Conservation • Pharmaceuticals • Parenterals • Vaccines • Healthcare • Biomedical harvest • Protecting health

27.1 Introduction

The healthcare industry, specifically the manufacturers and users of endotoxin detection products derived from horseshoe crabs, are in a unique position to affect the sustainability of horseshoe crabs as a natural resource that is in decline throughout much of Asia and only marginally stable along the Atlantic coast of the United States, where *Limulus* amoebocyte lysate (LAL) is derived (Fig. 27.1).

The horseshoe crab species are slow to mature and suffer high mortality during growth to adulthood. Horseshoe crabs take 10–12 years to reach maturity. During early developmental years, the juveniles are predated upon by a host of marine creatures. Only 3 out of 100,000 survive their first year (Botton et al. 2003; Carmichael et al. 2003).

Once they reach adulthood, horseshoe crabs have few predators other than humans. If not harvested for bait, human consumption, scientific research or biomedical bleeding, horseshoe crabs are expected to live 10 years or more after reaching maturity (Ropes 1961; Botton and Ropes 1988). It is only when they become adults that they begin their annual spring migration from the shelter of deeper waters to their spawning beaches. Adult horseshoe crabs annually congregate on spawning beaches where females burrow into the sand and lay masses of eggs, which are then fertilized by the males around her. Females lay as many as 25 egg clusters each year or around 100,000 eggs (Shuster and Botton 1985; Leschen et al. 2006). It is during spawning when horseshoe crabs are most vulnerable to harvest for TAL/LAL, bait for several different fisheries, human consumption and folk remedies in certain parts

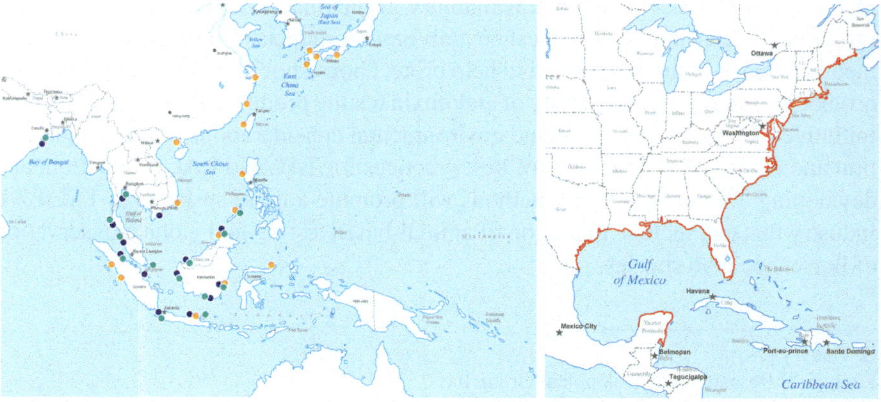

Fig. 27.1 Geographic ranges of *Carcinoscorpius rotundicauda* (*green*), *Tachypleus tridentatus* (*orange*), *Tachypleus gigas* (*blue*), and *Limulus polyphemus* (*red*)

of the world. It is this often unregulated harvesting pressure, coupled with loss of habitat, which poses the greatest threat to the future survival of these remarkable animals.

The Ecological Research & Development Group (ERDG) a nonprofit organization founded in 1995, whose mission is the conservation of the world's four horseshoe crab species, has partnered with members of the endotoxin detection industry with the belief that the industry, and their customers, have the potential to make a significant difference in the conservation of the world's four horseshoe crab species in ways not possible through traditional conservation strategies. Presently, the growth of the global healthcare industry is entirely dependent upon the harvest and collection of blood from live horseshoe crabs to produce TAL/LAL. Although direct mortality of horseshoe crabs due to LAL production is estimated to be relatively low, 8–15 % (Rudloe 1983; Walls and Berkson 2003), the mortality associated with TAL production is 100 % because after bleeding, the animals are sold to secondary markets for food and chitin production. The survival of this biomedical industry is dependent on the survival of three of the world's four horseshoe crabs species. Hence, any factors that affect horseshoe crab survival, even those external to the industry, have potential to affect the sustainability of the industry, giving industry members a unique motivation to broadly support horseshoe crab conservation.

Few people understand how deeply the TAL/LAL industry affects the lives of nearly every man, woman, child and domestic animal in the world, who are dependent upon medical service for their health. The safety of much of the world's pharmaceutical and medical devices must be tested for the presence of life-threatening endotoxins prior to public use, and the most reliable endotoxin detection test currently available is TAL/LAL. There is no indication that the world's human and animal population will become less dependent on medical services in the years to come. In fact, as our global population expands, ages, and medical advancements improve and/or prolong life, we expect to become more, not less reliant upon endotoxin detection methodologies, which currently means TAL or LAL. It is questionable whether current harvesting levels for TAL/LAL can be sustained, much less meet the projected future demands of this rapidly growing market, particularly if Asian horseshoe crab species are harvested to functional extinction.

27.2 Projected TAL/LAL Market Growth

Demand for TAL/LAL to test biomedical devices and vaccines is likely to rise significantly in the next two decades. Approximately 25 % of the medical device market is currently dependent upon TAL/LAL for endotoxin detection. Although the medical device market is about half the size of the world's pharmaceutical market (~$350 billion USD annually), it is showing rapid growth, driven in part by greater life expectancy and an aging population. Although vaccines are a small segment of the global pharmaceutical market, they represent the fastest growing segment of this industry and are heavily reliant upon endotoxin detection. According to the World Health Organization (WHO) the vaccine market's projected growth over the next 15 years is 10–15 % annually, vs. 5–7 % for pharmaceuticals. And these figures do not include the rapidly growing veterinary vaccine market. According to WHO, in 2007 global vaccine sales were $19 billion USD; by 2010 they had grown to $28 billion USD and are projected to reach $100 billion USD by 2025.

Much of this growth will occur in emerging markets, in countries such as China, India and Brazil (IMS 2012). An emerging market is defined as the percentage of growth into the $5,000 USD or above income bracket. These individuals are assumed to have access to medical services previously unavailable to them. China's income growth, by this definition, dwarfs all other emerging markets combined (IMS 2012).

The size of the pharmaceutical market in 2006 was $658 billion USD with emerging markets representing 14 %. By 2011 it had grown to $956 billion USD with emerging markets representing 20 %. The projected growth for 2016 is $1,175–1,205 billion USD with emerging markets representing 30 % (IMS 2012). Here again these emerging markets represent many of the same Asian countries that are experiencing rapid growth in the vaccine market.

Currently, the medical device market is dominated by U.S. companies making up approximately 45 % of total revenue, with Europe coming in second at approximately 25 %. However, the Asian market is growing rapidly. Countries such as China, India and Brazil are responsible for producing more than 86 % of the world's vaccines by volume, with growth projected to 87 % by 2020 (Visiongain 2010). China, the world's largest vaccine producer, produces more than one billion vaccine doses annually and is preparing to take on a larger role in the international vaccine market. With the world's largest population and 17 million newborn babies each year (Central Intelligence Agency 2014), growing demand is not surprising. Considering the global use of vaccines alone illustrates how reliant consumers are on horseshoe crabs via the endotoxin detection industry, and in emerging markets a large percentage of endotoxin detection is dependant upon TAL, derived from the two species of horseshoe crabs (*Tachypleus tridentatus* and *T. gigas*) that are in serous decline.

There is a growing trend to outsource product manufacturing to emerging markets, which may make good economic sense, if it were not for the fact that a good portion of these emerging markets are dependent upon the blood of the two Asian horseshoe crab species that are in serious decline.

While not all pharmaceutical products are tested with TAL/LAL, any product or device that comes in contact with blood, lymph, spinal fluids or mucus membranes (i.e., products that do not pass through the body via an oral pathway) is tested for the presence of endotoxins. There is also a considerable amount of endotoxin detection utilized in the research and development of new products. For every 250 drugs that make it to preclinical testing, only one may make it to FDA approval.

Unfortunately, it is difficult to obtain reliable data on the size of the TAL/LAL industry itself, in terms of annual revenue. However, the size of the TAL/LAL market is not really important, but rather the size of the market they serve, for it is growth on the consumer side that places pressure on the horseshoe crab resource needed for endotoxin detection.

27.3 Meeting Growing Demand

Can the world's horseshoe crab population meet the growing demands of the global healthcare industry? Although we lack definitive data on the size and health of the Asia horseshoe crab population, by all accounts, it is diminishing (IUCN SSC Horseshoe Crab Specialist Group, personal communication). Yet the Asian healthcare industry is growing at an unprecedented level, and is largely dependent upon TAL derived from *Tachypleus tridentatus* or *T. gigas* for endotoxin detection.

In the United States, home to *Limulus*, the harvest of horseshoe crabs for the LAL industry is regulated by the Atlantic States Marine Fisheries Commission (ASMFC), which is tasked with implementing the Interstate Fishery Management Plan (FMP) for the horseshoe crab (ASMFC 1998). In 1998, the FMP established a biomedical mortality threshold, which could trigger action by the Management Board if exceeded (ASMFC 2013). With the United States horseshoe crab population just beginning to show signs of recovery after years of decline, it is questionable whether the US horseshoe crab population can manage the growth of the LAL consumer market, much less the TAL consumer market, if there were to be an issue of supply and demand. Is it wise to have the global healthcare industry so directly tied to a diminishing and/or marginally stable resource?

The challenges for the three Asian species are much more complex; their ranges are extensive (Fig. 27.1) and encompass a host of countries with very different social, economic and environmental priorities. We have a considerable amount of information about the species in areas where the population has declined, but much less information about the species in areas where the population appears reasonably healthy. It is these areas of relative health that are of major concern, for as the need for TAL grows, and the resources in traditional harvesting areas decline, these healthy, often unregulated populations will be the first to be exploited, if they are not being exploited already.

27.4 Fisheries Management

As the result of serious decline in the American horseshoe crab population during the 1990s, the ASMFC, a governmental regulatory body represented by each of the states along the East Coast of the United States, implemented an Interstate FMP. This plan, focused on the sustainability of the species, has been revised seven times since its inception, to keep pace with changing conditions in the horseshoe crab population. Built into this plan is a biomedical exemption based on an estimated bleed and release mortality of 57,500 horseshoe crabs, which is calculated at 15 % of the total biomedical harvest.

To meet the projected healthcare industry demand, it is likely that the allowable mortality threshold set for *Limulus polyphemus* will have to be raised or the level of estimated mortality associated with biomedical harvest reduced. LAL producers

Table 27.1 Number of horseshoe crabs harvested for LAL production and the estimated number lost due to the bleeding process (assuming 15 % mortality) each year from 2005 to 2012

	2005	2006	2007	2008	2009	2010	2011	2012
Harvested for LAL	323,149	367,914	500,251	511,478	512,552	548,751	628,476	611,827
Total mortality	44,830	49,182	63,432	63,285	60,642	75,428	80,827	79,786

Source: ASMFC

have been able to manage market growth by improving the industries harvesting, handling and release protocols, as well as through advancements in manufacturing and testing methodologies. With production waste now at a minimum and testing methodologies fine-tuned, LAL producers will likely need to harvest more animals to keep pace with global demand. Under current regulations, however, there is no further allowance for increased harvest. Every year since 2005, biomedical harvest for LAL production has increased, and every year since 2007, LAL producers have exceeded the mortality threshold set by the ASMFC (Table 27.1). A review process is underway to determine what if anything should be done to enforce the proposed mortality limit.

Two of the three Asian horseshoe crab species are used in the production of TAL, *Tachypleus tridentatus* and *Tachypleus gigas*. TAL is only produced in China and used primarily in China, some in Russia, very little in India and hardly any in Japan, South Korea and South East Asia. Chinese multi-national pharmaceutical and medical device companies that manufacture in China, yet sell worldwide, purchase and test their end products with LAL. However, with the world's largest population of over 1.3 billion people, one fifth of the earth's population, and the largest aging population in the world, China by itself, is a sizable healthcare market.

Unfortunately, we do not have reliable horseshoe crab harvesting figures for China's TAL industry, nor knowledge of how it may be regulated. We are only just beginning to gather data on where, outside of China, *Tachypleus* is harvested for TAL.

Tachypleus is generally harvested in pairs, and bled to death, then sold to a secondary market for human consumption and chitin, resulting in 100 % mortality. The limited available harvesting numbers suggest a drop in abundance from 600,000 pairs 5–10 years ago, to less than 100,000 pairs currently. With the three Asian horseshoe crab species trending towards decline, how will TAL producers keep pace with their countries growing health care market?

The demand for horseshoe crabs throughout its Asian spawning range is placing significant stress on the species and hindering conservation efforts. Although we do not yet have sound data on the health of the three Asian horseshoe crab species, we do know that the horseshoe crab populations in Japan, Taiwan, China, Hong Kong and Singapore, once vibrant, are now endangered.

27.5 Conclusion

Without a regulatory process on TAL similar to what is in place for LAL, the current rate of harvest for *Tachypleus* is unsustainable. Failure to address this problem will continue to add to the decline of Asian horseshoe crab species, and shift the harvest pressure onto the US LAL market to absorb the residual demand. As a result of sound US management policies implemented since 1999, *Limulus* is beginning to recover. It is questionable, however, whether the LAL industry can offset a decline in TAL availability in a rapidly growing healthcare market, even if regulations are loosened to allow greater harvest and, in turn, greater mortality of *Limulus polyphemus* stocks.

Investment in the development and utilization of alternative endotoxin detection methodologies that reduce or avoid reliance on live horseshoe crabs is the ideal solution to ensure the sustainability of the endotoxin detection industry, their consumer supply line, and horseshoe crab populations world-wide. The question is whether the pharmaceutical and medical device industry will drive this transition. By choosing vendors who use best practices for TAL and LAL production, supporting global supply chain initiatives that integrate social and environmental considerations into the qualification and choice of vendors, as well as offering or developing alternative testing methods, industry leaders have the potential to drive the TAL/LAL industry toward the reduction of animal-based testing and make a significant contribution to global conservation of the horseshoe crab species. Recommended best practices include moist and temperature-controlled transport, prescreening for injured crabs, separating crabs to avoid re-bleeding, aseptic environments and sterile needle use, and returning bled crabs to their point of origin. In 2014, ERDG launched a new section on their website called "Protecting Health." The web content is designed to delineate the problems outlined in this paper, raise awareness of the possible implications for global health industry, and provide tools to promote communication regarding how the biomedical industry can become a driving force behind the conservation of the world's four horseshoe crab species.

References

ASMFC (Atlantic States Marine Fisheries Commission) (1998) Interstate fishery management plan for horseshoe crab. Atlantic States Marine Fisheries Commission, Fishery Management Report No 32. Washington, DC

ASMFC (Atlantic States Marine Fisheries Commission) (2013) Horseshoe crab stock assessment update 2.2

Botton ML, Ropes JW (1988) An indirect method for estimating longevity of the horseshoe crab (*Limulus polyphemus*) based on epifaunal slipper shells (*Crepidula fornicata*). J Shellfish Res 7:407–412

Botton ML, Loveland RE, Tiwari A (2003) Distribution, abundance, and survivorship of young-of-the-year in a commercially exploited population of horseshoe crabs *Limulus polyphemus*. Mar Ecol Prog Ser 265:175–184

Carmichael RH, Rutecki D, Valiela I (2003) Abundance and population structure of the Atlantic horseshoe crab *Limulus polyphemus* in Pleasant Bay, Cape Cod. Mar Ecol Prog Ser 246:225–239

Central Intelligence Agency (2014) China. In The World Factbook. https://www.cia.gov/library/publications/the-world-factbook/geos/br.html. Accessed Sept 2013

IMS (2012) Institute for Healthcare Informatics, IMS Market Prognosis, May 2012, Economic Intelligence Unit, Jan 2012

Leschen AS, Grady SP, Valiela I (2006) Fecundity and spawning of the Atlantic horseshoe crab, *Limulus polyphemus*, in Pleasant Bay, Cape Cod, Massachusetts, USA. Mar Ecol 27:54–65

Ropes JW (1961) Longevity of the horseshoe crab, *Limulus polyphemus* (L.). Trans Am Fish Soc 90:79–80

Rudloe A (1983) The effect of heavy bleeding on mortality of the horseshoe crab, *Limulus polyphemus*, in the natural environment. J Invertebr Pathol 42:167–176

Shuster CN Jr, Botton ML (1985) A contribution to the population biology of horseshoe crabs, *Limulus polyphemus* (L) in Delaware Bay. Estuaries 8:363–372

Visiongain (2010) Vaccines manufacturing technology and services: commercial prospects 2010–2025. The Global Vaccine Market, London, 2009–2025

Walls EA, Berkson J (2003) Effects of blood extraction on horseshoe crabs (*Limulus polyphemus*). Fish Bull 101:457–459

Chapter 28
Biomedical Implications for Managing the *Limulus polyphemus* Harvest Along the Northeast Coast of the United States

Thomas J. Novitsky

Abstract North American horseshoe crabs (*Limulus polyphemus*) have been continuously harvested in Massachusetts for the production of *Limulus* amebocyte lysate (LAL); since the mid 1970s as bait for American eel (*Anguilla rostrata*); and since the mid 1990s as bait for whelk (*Busycon* ssp.). Harvest regulations were promulgated by the Atlantic States Marine Fisheries Commission (ASMFC) in 1998 over concern for the observed decline of the Red Knot (*Calidris canutus*), a correlation between its feeding behavior (horseshoe crab eggs during the spring migration), and a decline in spawning horseshoe crabs likely due to harvesting for bait. Generally, horseshoe crab harvest for the production of LAL (biomedical use) is exempted from harvest regulations, since the animals bled for LAL production are returned alive to the ocean, and mortality is considered low. However, evidence is accumulating that mortality of bled horseshoe crabs is higher than originally thought (29 % vs 15 %); that females may have an impaired ability to spawn following bleeding and release; and that bled crabs become disoriented and debilitated for various lengths of time following capture, handling, bleeding, and release. This paper reviews the latest evidence for mortality and impairment of horseshoe crabs bled for biomedical use, especially in Massachusetts where horseshoe crab populations in small bays and inlets are particularly vulnerable, and since State regulations allow for using bled crabs as bait rather than returning to the site of capture. A novel management plan is proposed that can satisfy all affected parties as well as insure a continued supply of horseshoe crabs for the vitally important production of LAL. This plan may also serve as a model for other areas of the Atlantic coast where biomedical harvest occurs.

Keywords Limulus amebocyte lysate (LAL) • Biomedical • Atlantic States Marine Fisheries Commission (ASMFC) • Massachusetts • Best Management Practices (BMP) • Mortality • Sublethal effects • Artificial bait • Alternative assays • Food and Drug Administration (FDA) • Open reporting • Dual use

---

T.J. Novitsky (✉)
Horseshoe Crab Conservation Association, P.O. Box 2334, Orleans, 02653-2334 MA, USA
e-mail: tjnovitsky@gmail.com

R.H. Carmichael et al. (eds.), *Changing Global Perspectives on Horseshoe Crab Biology, Conservation and Management*, DOI 10.1007/978-3-319-19542-1_28

28.1 Introduction

The American horseshoe crab, *Limulus polyphemus*, has long been of interest to researchers due to the properties of its blood (hemolymph) and blood cells (amebocytes) (Howell 1885; Loeb 1902). In addition to the interest generated in studying a "living fossil" since the horseshoe crab was recognizable in the fossil record back some 445 mya, and nearly identical in gross morphology to fossils from 250 mya (Rudkin et al. 2008), use of *Limulus* as a research tool was facilitated by its abundance, size, ease of collection, suitability to be maintained in aquaria, and, perhaps most importantly, by the design of its circulatory system and large blood volume. These attributes most likely contributed to the decision by Dr. Frederik Bang in 1953 to use the horseshoe crab as a surrogate for studying the clotting mechanisms of human blood (Bang 1953). Subsequent discoveries by Bang and colleague Dr. Jack Levin (Levin and Bang 1968) led to an assay for endotoxin (pyrogen) employing an amebocyte extract, *Limulus* amebocyte lysate (LAL).

In October 1978, a symposium on the "Biomedical Applications of the Horseshoe Crab (Limulidae)" was held at the Marine Biological Laboratory (MBL) in Woods Hole, Massachusetts (Cohen et al. 1979). This symposium's location, timing, and subject matter reflected an incredible growth in interest in the horseshoe crab, especially LAL. At the time of the symposium, it had been only just over 1 year since the United States Food and Drug Administration (FDA) licensed the first manufacturer, Associates of Cape Cod, Inc, also located in Woods Hole, MA to produce LAL for use to detect pyrogen (endotoxin) in injectable pharmaceuticals (Novitsky 1991, 2009; Levin et al. 2003). Although the symposium covered all aspects of horseshoe crab biology, the majority of presentations related to the biomedical use of the recent commercial availability of LAL. It is notable, even at this early stage of the LAL industry, that concern over the survival of the species was raised. Anne Rudloe (1979) who would later be contracted by the FDA to conduct a study examining the mortality associated with horseshoe crabs bled to produce LAL stated in her presentation: "The emerging importance of *Limulus* to biomedical research, as the source of *Limulus* lysate, requires more complete knowledge of the biology of the species, so that it can be wisely managed as a natural resource." She went on to say: "There has developed in recent years a substantial harvest of horseshoe crabs for this purpose. Such harvesting *has most often occurred on breeding beaches*." In another paper, Sydney R. Galler (1979), Assistant Secretary for Environmental Affairs, US Department of Commerce, called on the conference participants as well as the scientific community to develop a practical plan for the conservation and protection of the horseshoe crab in parallel with their efforts to disseminate information about biomedical applications. Thus from the very beginning of the LAL industry, population depletion was a concern. To be fair, Jack Levin released the horseshoe crabs he used for his research immediately following bleeding and delighted in watching them swim away from the dock near the MBL (personal communication). Likewise, Associates of Cape Cod, Inc. instituted a policy of immediate release followed by the rest of the industry. This practice also became part of the

FDA's LAL licensing requirements (Levin et al. 2003) until the FDA relinquished authority for the general handling of horseshoe crabs prior to bleeding and the disposition of horseshoe crabs following bleeding to the ASMFC in 2002 (Horseshoe crab technical committee, Atlantic States Marine Fisheries Commission Draft, March 5–6, 2002). Biomedical use, i.e. the production of LAL, continues to be regulated by the FDA but only with respect to the quality of the reagent, not to the harvest or final disposition of the horseshoe crabs used as a raw material.

It was not until 1998 with the institution of a Horseshoe Crab Working Group organized under the auspices of the Atlantic States Marine Fisheries Commission (ASMFC) that coast-wide regulations for the harvest of horseshoe crabs was instituted (ASMFC 1998). Prior to 1998, certain states (Rhode Island, Massachusetts, New York, and Florida) had no catch limit and no reporting, but required a license. Virginia also had no catch limit but required a license and prohibited capture by trawl and dredge. Other Atlantic states (New Hampshire, Delaware, Maryland, and South Carolina) had various regulations covering limits as well as other aspects of capture (ASMFC 1998). Also, up until 1998 regulation in federal waters varied along the coast. For example, harvest in certain National Wildlife Refuges and National Seashores was allowed either by precedent, i.e. horseshoe crab harvest had occurred historically, or as was the case in the Monomoy National Wildlife Refuge (NWR), allowed by permit (Compatibility Determination Eastern Massachusetts National Wildlife Refuge Complex 2002; James-Pirri 2002). Prior to 1998, commercial horseshoe crab harvesting had been ongoing for many years on Monomoy NWR. Officially, however, commercial harvest of horseshoe crabs was never allowed on refuge land or waters due to the fact that a compatibility determination was never completed by the United States Fish and Wildlife Service, which had authority over the refuge. However, after discussions with the refuge staff, Jay Harrington was issued a 'Special Use Permit' in 1991 to legally harvest horseshoe crabs in "closed areas" of the refuge, specifically for the biomedical company, Associates of Cape Cod, Inc. Permits were renewed annually, and a scheduled compatibility review conducted in 1994 reconfirmed that the harvest of horseshoe crabs for biomedical use was compatible. However in 1999, due to public concern over reports of excessive horseshoe crab harvesting in the refuge, it was determined that the Monomoy NWR did not have the authority to regulate intertidal areas within the refuge boundary. As a result, in 2000, the United States Fish and Wildlife Service refused to renew a harvest permit for Jay Harrington (Compatibility Determination Eastern Massachusetts National Wildlife Refuge Complex 2002). This action was accompanied by a near simultaneous denial for harvest by the National Park Service in the nearby waters of the Cape Cod National Seashore. Facing the loss of a significant portion of their raw material supply, Associates of Cape Cod, Inc. and Jay Harrington, whose very livelihood mainly depended on harvesting horseshoe crabs, brought legal action against the United States Department of Interior (DOI) and were granted an injunction to continue harvesting in these areas for the 2000 and 2001 seasons (Associates of Cape Cod, Inc. 2000). The DOI prevailed, however, and in May of 2001 the government's win resulted in the prohibition of horseshoe crab harvest for any reason in the National Seashore and in the Monomoy National

Wildlife Refuge (actually a ban was instituted for the refuge until a new compatibility study could be completed) (Compatibility Determination Eastern Massachusetts National Wildlife Refuge Complex 2002). The ruling eventually resulted in a ban for the harvesting of horseshoe crabs in all federal waters (James-Pirri 2012).

Coast-wide harvest regulations begun in 1998 were promulgated not from a concern for the future of the increasingly important LAL (biomedical) industry, but from an environmental concern over a potential lack of food (horseshoe crab eggs) for migratory shorebirds, in particular the Red Knot (*Calidris canutus*), that stopped in the Delaware Bay during migration to their Arctic nesting grounds. This migration synchronized perfectly with the annual horseshoe crab spawning event, and the birds needed this critical stopover and the excellent nutrition from horseshoe crab eggs for a successful migration (Mizrahi and Peters 2009).

To further protect the LAL industry that used far fewer crabs than the bait industry, and since mortality from bleeding was considered insignificant (most bled crabs were returned to their environment alive), the biomedical industry was exempt from restrictions on harvesting horseshoe crabs with the exception of a requirement to report the number of horseshoe crabs bled (ASMFC 1998; Novitsky 2009). The extremely small number of horseshoe crabs harvested specifically for research is considered inconsequential and taking for research purposes is completely exempt.

Currently harvest restrictions in the US seem to have stabilized horseshoe crab populations, although in the Northeast where there are many small populations restricted to bays and inlets, the ASMFC and State-specific limits may not be sufficient (Smith et al. 2009). The 'stock status determination' of the ASMFC concluded "…the evidence from New York and New England suggest that the current harvest within these regions is not sustainable." In Wellfleet Harbor, for example, over-fishing for bait has resulted in a horseshoe crab decline when compared to other nearby areas (James-Pirri 2002; James-Pirri et al. 2005). The horseshoe crab harvest is thought to have degraded the Wellfleet Harbor's bottom, which is thought to adversely affect the shellfish harvest (Town of Wellfleet, The Wellfleet Shellfish Advisory Board Letter to Paul J. Diodati, Director of Massachusetts DMF, 2012). In addition, there is no longer a biomedical harvest in the bay since the females are too small (personal communication, Town of Wellfleet, The Wellfleet Shellfish Advisory Board 2012 Letter to Paul J. Diodati, Director of Massachusetts DMF). Other environmental factors on Cape Cod, Massachusetts, in particular poorly located shoreline stabilization and beach nourishment projects, may also reduce horseshoe crab populations by their impact on critical spawning habitat (Fabie 2009).

Other examples abound. In a relatively compact area near Mashnee Dike on the upper reaches of Buzzards Bay Massachusetts, a small but stable breeding population that had been studied for years by scientists from the Marine Biological Laboratory in Woods Hole was virtually wiped out using a hand harvest during spawning (Widener and Barlow 1999). A similar situation was found at Stage Harbor (Widener and Barlow 1999). Two other surveyed bays in Massachusetts, Wellfleet Harbor and Cape Cod Bay have also shown serious declines related to spawning indices (Faherty 2012). This type of extermination of local populations most likely repeats itself in many other bays and inlets in the Northeast even though state-wide harvest quotas initially set by the ASMFC are never reached, i.e. demand for bait seems to be satisfied (ASMFC 2013). Basically the horseshoe crab is an easy fishing target during its most vulnerable spawning period. Crabs come into shallow water on sandy beaches that are well known to fishermen and where they are easily captured by hand/rake from small boats. This type of harvest should raise an immediate concern for the continued survival of this species (Fabie 2009). Harvest numbers from the other two Northeastern states, Maine and New Hampshire, while having small populations of horseshoe crabs and regulated bait harvests, are considered *de minimis* by the ASMFC Interstate Fisheries Management Fisheries Program Charter. *De minimis* is defined as "a situation in which, under existing condition of the stock and scope of the fishery, conservation, and enforcement actions taken by an individual state would be expected to contribute insignificantly to a coastwide conservation program required by a Fishery Management Plan or amendment." Maine and New Hampshire reported to the ASMFC bait harvest totals from 1998 to 2003 and 1998 to 2002 respectively until they were granted *de minimis* status. Application for this status can be made if average horseshoe crab bait landings for two consecutive years constitute less than one percent of the coastwide total. Once granted, ASMFC member states are no longer required to report their harvests as long as the conditions for maintaining their status continue (ASMFC 2001).

Although it was recognized that there was mortality associated with biomedical use, these limits were regarded as insignificant and initially based on the 10–15 % reported by Rudloe (1983). Several recent papers have reported mortalities between 10 and 30 %. Most of these studies attempted to control conditions for bleeding and handling crabs and some have attempted to mimic conditions currently employed by the LAL industry (Walls and Berkson 2003; Hurton and Berkson 2006; Leschen and Correia 2010; Anderson et al. 2013). Due to the variations in methodology, it is difficult to compare these studies. It would be best to copy the tagging and release modality used originally by Rudloe (1983), i.e. tagging experimental (bled) and controls (unbled) horseshoe crabs then releasing them back into the ocean environment and subsequently calculating mortality based on the numbers of recovered tagged animals. However, with Rudloe's (1983) method, it is extremely difficult to recover sufficient animals to reach statistical significance without initially using a very large number of animals. Rudloe (1983) for example, barely achieved statistical significance with her 2-year study. Unfortunately nearly all of the mortality studies following Rudloe's study (1983) have used ponds or aquaria for observing bled horseshoe crabs and controls. These studies, of course, can produce their own experimental bias, and the ones showing highest mortalities are often criticized by the LAL industry by employing the unproven argument that bled horseshoe crabs released to their "natural" environment would fare much better than those released to "artificial" environments. For now the ASMFC "accepts" mortality for bled horseshoe crabs released to the environment as 15–30 % (ASMFC 2013).

In addition to mortality, the vitality of crabs released following bleeding for biomedical use may now be cause for concern. The strongest circumstantial evidence for this may be the extremely skewed ratios of female to male crabs in Pleasant Bay, an area in Massachusetts that has been traditionally harvested for biomedical use (Carmichael et al. 2003) and since 2006 has been set aside solely for biomedical harvest, i.e. no bait harvest allowed (James-Pirri et al. 2005; Leschen et al. 2005; Massachusetts Division of Marine Fisheries 2006; James-Pirri 2012). Recently, direct experimental evidence has been published (see Sect. 28.4, below) that supports an adverse effect on bled crabs that have been released (Kurz and James-Pirri 2002; Leschen and Correia 2010; James-Pirri et al. 2012; Anderson et al. 2013).

Today, there are four companies operating in the United States that produce LAL from *Limulus polyphemus* harvested from various locations along the Atlantic Coast. It should be noted that there exists a similar industry in Southeast Asia where other species of horseshoe crabs, namely *Tachypleus tridentatus*, *Tachypleus gigas*, and *Carcinoscorpius rotundicauda* are, or can potentially be used to make an LAL equivalent, *Tachypleus* amebocyte lysate (TAL) and *Carcinoscorpius* amebocyte lysate (CAL).

This paper reviews the current status of the horseshoe crab harvest along the Atlantic coast of the United States compared to that in the Northeast, in particular Massachusetts, with respect to all stakeholders. In the Northeast especially, the concept of "sustainable yield" may need to be replaced with "population recovery" if a slow but certain decline of the horseshoe crab is to be reversed.

28.2 Effect of Bleeding on *Limulus* Mortality

Several studies followed Rudloe's (1983) initial mortality assessment, many of which attempted to more closely mimic conditions used by LAL manufacturers (Kurz and James-Pirri 2002; Walls and Berkson 2003; Leschen and Correia 2010). These additional studies, while not strictly comparable due to the different methods used, found mortalities ranging from 5 to 30 %. Despite one recent (Leschen and Correia 2010), very carefully controlled study conducted at a functioning biomedical facility and using only female horseshoe crabs (females are preferred by the biomedical industry due to their larger size and hence larger hemolymph content) that found a 29 % mortality, the ASMFC has retained their initial estimate of 15 % mortality for bled crabs not returned to their native environment (ASMFC 2013). In 1998 the ASMFC Fisheries Management Board established a mortality threshold for bled and returned horseshoe crabs of 57,500. This figure is based on the 15 % mortality estimate (ASMFC 1998) and has been exceeded in 2007 and every year since (ASMFC 2013). With an increase in crabs bled for biomedical comes an increase in mortality. Although mortality following bleeding is only an estimate, additionally biomedical harvesters are required to report actual mortality from capture to return (ASMFC 2013). In 2012 estimated mortality (15 %) was 79,786, well above the threshold of 57,500 (ASMFC 2013). In 2011, 545,164 horseshoe crabs were harvested for biomedical use compared with 650,539 for bait (ASMFC 2013). This represents 45.6 % of the total harvest. If horseshoe crabs harvested for biomedical use and subsequently used for bait are included, the total number for biomedical becomes 52.6 %. Actual mortality, i.e., not estimates from numbers of bled and released horseshoe crabs, have gone from 4,391 in 2004 to 9,665 in 2010, a 220 % increase.

At their summer meeting on 4 Aug 2011, the ASMFC Horseshoe Crab Management Board approved formation of an ad-hoc working group, made up of technical committee members and biomedical representatives, to develop best management practices (BMP) to minimize coast-wide mortality from the practice of collecting horseshoe crab blood for biomedical use (ASMFC 2011b). In October the newly formed Horseshoe Crab Biomedical Ad-hoc Working Group issued a report with recommended best management practices (BMP) (ASMFC 2011a). Many of the recommendations, e.g. keeping transit times to a minimum, keeping crabs cool and moist during transit and storage between collection and return, and returning crabs to the ocean as soon as practical following bleedings, were already standard practice for most of the biomedical industry. BMPs were never formalized, and the BMPs listed by the ad-hoc group were not made mandatory nor was there any reporting requirement. While it would have been too soon to attribute a positive effect due to the implementation of BMP, harvest-related mortality actually fell to 6,917 in 2011 (ASMFC 2012). Mortality fell again in 2012 to 6,819 when biomedical users should have been employing the BMP. However, the lowest mortality number was recorded in 2008 (2,973) well before the BMP recommendations were made. It is therefore too early to tell whether BMP as currently recommended will significantly affect mortality.

Despite the variability of mortality rates from more recent studies, and a biomedical industry that may now be the major user of the horseshoe crab, the ASMFC Horseshoe Crab Stock Assessment Subcommittee continues to endorse the 15 % mortality rate despite the 85 % increase in the biomedical harvest since 2004 with a corresponding increase in mortality of 75 % (ASMFC 2013).

While the Massachusetts bait harvest as a percentage of the coast-wide harvest has remained fairly consistent–ranging between 9 and 21 % with no clear trend up or down (Table 28.1), there is no clear idea of the number of horseshoe crabs harvested for 'biomedical use only' in the state. Unlike the fishing industry, the biomedical industry argues it would hurt competition if its use numbers were released. Due to this industry secrecy and its acceptance by the ASMFC, critical review outside the jurisdiction of the ASMFC has been impossible. However, from the data publically available, a comparison made between horseshoe crabs counted against the bait quota and horseshoe crabs harvested only for biomedical use, a range of 12.7–23.3 % from 2004 to 2012 was observed, with no clear trend apparent (Table 28.2). ASMFC reports also indicate that biomedical use (all manufacturers) has increased nearly twofold, from 343,126 crabs in 2004 to 611,827 crabs in 2012 (ASMFC 2013). If the total bait harvest is compared to all horseshoe crabs bled, the numbers have also been steadily increasing–from a low of 42 % in 2005 to over 83 % in 2012 (Table 28.3). Massachusetts however, is currently the only state that *consistently* allows crabs bled for biomedical use to be subsequently used as bait (dual-use) and be counted against the bait quota as allowed by the ASMFC since 2004 (Marin Hawk, Fishery Management Plan Coordinator ASMFC, personal

Table 28.1 Massachusetts (MA) bait harvest compared with coast-wide bait harvest 1998–2012 (ASMFC 2013)

Year	MA bait	Coast-wide bait	% MA bait
1998	400,000	2,743,585	14.6
1999	545,715	2,600,914	21.0
2000	272,930	1,656,967	16.5
2001	134,143	1,013,697	13.2
2002	138,613	1,265,926	10.9
2003	125,364	1,052,493	11.9
2004	69,436	681,323	10.2
2005	73,740	769,323	9.6
2006	171,906	840,944	20.4
2007	150,829	827,554	18.2
2008	103,963	660,983	15.7
2009	98,332	817,265	12.0
2010	54,782	605,511	9.0
2011	67,087	662,622	10.1
2012	106,821	729,100	14.7

Table 28.2 Coast-wide comparison of number of bait horseshoe crabs bled vs. number of biomedical-only horseshoe crabs bled (ASMFC 2013)

Year	Bait	Biomedical	Total	% Bait bled
2004	50,366	275,194	325,560	18.3
2005	39,429	270,496	309,925	12.7
2006	58,625	296,958	355,583	16.5
2007	71,379	398,844	470,223	15.2
2008	87,864	402,080	489,994	21.9
2009	110,350	362,291	472,641	23.3
2010	66,047	438,417	504,464	13.1
2011	83,312	492,734	576,046	14.5
2012	81,030	485,965	566,995	14.3

communication). The Massachusetts Department of Marine Fisheries (DMF) believes this dual-use exemption results in a significant reduction in the total number of horseshoe crabs harvested in the state (Massachusetts 2008 compliance report to the Atlantic States Marine Fisheries Commission). Thus the 81,030 crabs most recently reported by the ASMFC as "number of bait crabs bled" coast-wide in 2012, may therefore be due in large part to the Massachusetts harvest (ASMFC 2013). The number of bait crab landings reported for Massachusetts for the same year was 106,821 (number includes bait crabs bled for biomedical). Thus, if the assumption that the Massachusetts biomedical industry is using most of the total number of horseshoe crabs counted against the bait quota as reported for the entire biomedical industry, then up to 80 % of the Massachusetts bait harvest may be attributed to the biomedical use. If one adds in the numbers harvested for biomedical use only, the

Table 28.3 Coast-wide bait harvest vs. coast-wide biomedical harvest, 2004–2012 (ASMFC 2013)

Year	Total bait[a]	Biomedical[b]	% Biomedical
2004	681,323	343,126	50.4
2005	769,429	323,149	42.0
2006	840,944	367,914	43.8
2007	827,554	500,251	60.4
2008	660,983	511,478	77.4
2009	817,265	512,552	62.7
2010	605,511	548,751	90.6
2011	662,622	628,476	94.8
2012	729,100	611,827	83.9

[a]Includes crabs harvested for biomedical use but counted against bait quota
[b]Includes all biomedical crabs, i.e. biomedical use only plus those counted against bait quota

Massachusetts industry may easily be the largest user of crabs in the state. It is unfortunate that only estimates of the numbers of horseshoe crabs used by region can be made due to the secrecy surrounding the biomedical industry and their operations. If the percent of biomedical crabs bled actually is mainly attributable to the Massachusetts harvest, the Massachusetts DMF may be correct in assuming their "dual-use" policy helps reduce the total number of crabs harvested in their state. However, without better disclosure, there can be no independent verification. In any case, if the coast-wide number of horseshoe crabs harvested for bait is compared to those harvested solely for biomedical use and the biomedical crabs counted against the bait quota, the biomedical industry as a whole is now using a significant percentage of the total horseshoe crab harvest in the United States–as much as 94.8 % in 2011 (Table 28.3).

28.3 Harvest Effects on Spawning *Limulus*, Especially Females

Female horseshoe crabs have been preferentially used for bait and biomedical bleeding, albeit for different reasons. Fishing lore attributes females more attractive as bait than males, most likely due to the presence of eggs and/or a female specific pheromone thought to attract predators as well as males prior to and during spawning. For the biomedical industry, sexually mature females are preferred as the larger females yield more blood than the smaller males for the same amount of work. It is thought that females have an additional molt compared to males before their final, i.e. terminal ecdysis, and hence grow larger (Shuster and Sekiguchi 2003). Data collected from the Massachusetts bait harvest yearly since 2000 confirm this size difference (Table 28.4). Calculated average prosomal width (mm) of harvested females vs. males was 251.9 ± 8.2 vs. 197.6 ± 2.7 respectively.

Table 28.4 Relative size (prosoma width) difference between sexes from Massachusetts bait harvest 2000–2012 (ASMFC 2013)

Year	Female/male
2000	1.32
2001	1.32
2002	1.32
2003	1.29
2004	1.26
2005	1.33
2006	1.28
2007	1.29
2008	1.26
2009	1.26
2010	1.22
2011	1.22
2012	1.20

Table 28.5 Female to male ratios from Massachusetts (MA) bait harvest 2003–2011 ((ASMFC 2012); 2013 data from Vincent Malkoski, Senior Marine Fisheries Biologist, Massachusetts Division of Marine Fisheries, personal communication)

Year	% Female	% Unclassified[a]
2003	51.4	0
2004	55.9	0.06
2005	50.4	0
2006	49.5	0.05
2007	48.8	0.06
2008	52.8	0.02
2009	53.2	0.08
2010	62.0	10.7
2011[b]	NR[c]	100
2012	52.8	29.8
2013	58.9	33.4

[a]Unclassified count not used to calculate % Female
[b]Preliminary data
[c]Not Reported by MA to ASMFC

Massachusetts and Rhode Island bait harvest ratios (% Females) are shown in Tables 28.5 and 28.6 respectively. The greater percentage of females harvested, albeit small, supports the claim that females are preferred as bait over males. Also, beginning in 2010, bait harvest percentage in Massachusetts of "unclassified" spiked well above 0.1 % (the 100 % figure reported for 2011 is an anomaly due to the fact that no numbers according to sex were reported) and was much higher compared to percentages reported by the biomedical industry over the same time period (Table 28.7). This most likely reflects a more careful reporting by the biomedical industry but may be related to the dual use of horseshoe crabs harvested for biomedical then used as bait rather than being released following bleeding. Horseshoe crabs used for this purpose are typically marketed by bait dealers and not by fishermen directly, so counts may be missed due to lack of a clear reporting authority. However, while sex ratio data is unavailable for industry-wide biomedical, there is

Table 28.6 Female to male ratios from Rhode Island bait harvest 2009–2011 (ASMFC 2012)

Year	% Female	% Unclassified[a]
2009	58.2	9.8
2010	56.2	11.5
2011[b]	51.4	0

[a]Unclassified count not used to calculate % Female
[b]Preliminary data

Table 28.7 Female to male ratios from Massachusetts biomedical harvest 2008–2013 (Massachusetts Compliance Report to ASMFC 2008–2012; 2013 data from Vincent Malkoski, Senior Marine Fisheries Biologist, Massachusetts Division of Marine Fisheries, personal communication)

Year	% Female	% Unclassified
2008	64	1.5
2009	66	<1
2010	62	10
2011	NR	NR
2012	72	1
2013	66	0

NR not reported

still a greater female horseshoe crabs preference for biomedical use when compared to those harvested for bait in Massachusetts (Table 28.7).

Spawning surveys in Massachusetts also indicate a disproportional effect on female horseshoe crab abundance. Massachusetts DMF reported that 48 % of all 2008 surveys in the state recorded no female crabs. In addition, only 12 % of these surveys recorded more than 10 females (Massachusetts Compliance Report to ASMFC 2012). Surveys of Wellfleet and Cape Cod Bay indicated an even smaller fraction of female crabs on average compared with those for other embayments in the state (Faherty 2012).

A feature unique to the horseshoe crab harvest in Massachusetts is a provision for biomedical harvest only since 2006 in Pleasant Bay (Massachusetts Division of Marine Fisheries 2006). Horseshoe crabs for bait and biomedical have been harvested in Pleasant Bay for over 30 years (Leschen and Correia 2010). Sex ratios for this area differ dramatically from other areas in Massachusetts as well as other areas along the Atlantic coast in that they are skewed toward males (James-Pirri 2012). Data from the 1950s indicate a female:male ratio of 1:2.5, which was similar to that observed in Cape Cod Bay (1:2.4). More recent data find ratios of 1:5.8 (2000–2001) and 1:9.5 (2008) compared with 1:2.9 and 1:1.7 for Pleasant Bay and Cape Cod Bay respectively for the same time periods (Massachusetts Division of Marine Fisheries 2008; James-Pirri 2012). These differences suggest a higher mortality and/or morbidity for female horseshoe crabs bled for biomedical use and released. Even before Pleasant Bay was closed to harvesting for bait, mortality associated with the biomedical harvest (assuming 10–15 % of crabs bled and released) was higher simply due to the fact that the biomedical harvest was up to 25 times greater (Rutecki et al. 2004). These observations, i.e. a selective effect on females, and overall mortality, should be given priority for further investigation. As an example, biomedical horseshoe crab harvest in the Monomoy NWR was allowed by permit from 1991 until 2001 when populations were reevaluated and the practice of biomedical harvest only, even when the bled horseshoe crabs were supposedly returned to the area where they were captured, was deemed to be detrimental (Compatibility Determination Eastern Massachusetts National Wildlife Refuge Complex 2002).

Extensively harvested areas in Massachusetts also indicate an impact on spawning horseshoe crabs. Spawning indices, summarized by the Massachusetts Audubon Society from their annual surveys indicate a decline in spawning females in Wellfleet Bay and Cape Cod Bay relative to other areas, e.g. Pleasant Bay where regular surveys are conducted (Faherty 2012; James-Pirri 2012).

One especially dramatic example showed not only a decline in spawning, but the elimination of nearly an entire local population in the Mashnee Dike area of Buzzards Bay, MA. This particular population had been studied for years by vision researcher Robert Barlow at the MBL but was essentially decimated in a relatively short period of time with a 95 % reduction of the population between 1984 and 1999 (Widener and Barlow 1999).

28.4 Non-lethal Effects of Bleeding on *Limulus*

Traditionally the biomedical industry has bled crabs only once per year. Early studies indicated that it took at least a week for the crab to regain blood volume and several weeks to regain baseline amebocyte counts (Anderson et al. 2013). Since it is impractical to maintain crabs in holding ponds until they regain blood volumes and cell count, a practice of one bleed per year became the norm. Although bled crabs were seldom tagged, a fresh scar or needle puncture mark on the arthrodial membrane was quite apparent so that even if a bled crab was recaptured in the same year, a trained technician could avoid a second bleeding if a scar was in evidence. However, there is no provision in the proposed BMPs for preventing crabs being bled twice or more during a single season, and the effect on crab mortality is unknown. Likewise, due to the design of the horseshoe crab's circulatory system (open, i.e. no separate veins with capillaries connected to the arteries to circulate hemolymph back to the cardiac sinus), once the cardiac sinus (large tubular heart) and 11 major arteries (Shuster 2003) are empty, the blood flow slows to a drip or stops completely. It has been estimated that no more than 30 % of the entire blood of an individual crab is ever removed using a gravity flow as opposed to vacuum aspiration (Novitsky 1991). This type of bleeding, i.e. using gravity flow, appears to have become an industry standard (Levin et al. 2003), but due to the secrecy associated with the biomedical industry and a lack of a provision in the BMPs, it remains unclear whether this method is used universally. While equations have been developed to calculate the amount of blood that can be removed from an individual crab

(Hurton et al. 2005), use of this for relating blood volume removed to mortality or other physiological effects is unreliable as the amount of blood removed from an individual crab is extremely variable for reasons unknown (personal observation).

In one study conducted on female crabs using actual bleeding protocols employed by the biomedical manufacturer in Massachusetts (with some additional modifications to comply with the proposed BMP, i.e. employing a refrigerated truck for crab transport, loading fewer crabs per transport container to reduce crushing, and keeping the crabs moist by covering the transport containers with wet burlap), a significant female mortality following bleeding from 22.5 to 29.8 % was found compared to the 3 % for unbled controls (Leschen and Correia 2010).

In another study that examined selected hemolymph components of biomedically bled vs. wild caught female horseshoe crabs (using the same crabs as the aforementioned study), it was found that a number of constituents differed between bled and unbled crabs, in particular, protein concentration (James-Pirri et al. 2012). These authors concluded that the lower protein concentrations found in bled horseshoe crabs and prolonged biomedical bleeding may impact crab physiology. In a study examining specific effects of bleeding on behavior and physiology, results suggested that biomedical bleeding may decrease female horseshoe crab fitness (Anderson et al. 2013). These authors found reduced activity in horseshoe crabs following bleeding measured by several parameters. They also found unexpected mortality in one group. Physiological parameters were also affected, e.g. hemocyanin content. The authors do point out that thermal stress in combination with bleeding may have also affected activity level and physiology. Indeed, Coates et al. (2012) found horseshoe crabs (unbled) were adversely affected by extended captivity, and the effects were exacerbated with increasing temperature. While the maximum temperature tested in the Coates et al. (2012) study was only 23 °C (a maximum of 38 °C was reached with one group in the Anderson et al. (2013) study), captivity in the Coates et al. (2012) study covered 56 days, a much longer time than Anderson et al. (2013). Also, the thermal death point measured in horseshoe crabs collected from Woods Hole, Massachusetts was found to be 41 °C (Mayer 1914). It is unlikely therefore that temperature accounted for the mortality observed by Anderson et al. (2013).

28.5 Summary and Recommendations

Despite state-by-state quotas imposed by the ASMFC and additional harvest restrictions imposed by individual states, horseshoe crabs, especially females, continue to decline in the Northeast, in particular Massachusetts (Massachusetts Division of Marine Fisheries 2009, 2010, 2011, 2012). In New England, horseshoe crab habitat is comprised of a series of coastal embayments with limited movement of populations between them. Because of this, horseshoe crab populations would benefit from more local management, e.g., embayment by embayment, rather than overall harvest quota/restrictions for the entire State (Smith et al. 2009). Likewise, harvesting females during spawning, where populations spawn in relatively small and easily accessible harbors, bays and inlets, may explain the decline in females as well as the overall number of horseshoe crabs. Where numbers are not necessarily declining, e.g., Pleasant Bay, due to the return of horseshoe crabs following bleeding and a lack of a bait harvest, the ratio of spawning females to males is highly skewed when compared to other areas, even where both bait and biomedical harvest is permitted. This latter phenomenon may in fact be related to yearly, preferential bleeding of females. Female horseshoe crabs captured during the act of spawning may be at additional risk. This may be particularly true in Pleasant Bay. The Massachusetts DMF should set a priority to review the current data in Pleasant Bay and if females are at particular risk due to biomedical use, institute additional restrictions in this unique area, an area that has perhaps been studied with respect to horseshoe crabs more than any other. Grady and Valiela (2007) using a matrix modeling method on data collected from horseshoe crab populations on Cape Cod, including Pleasant Bay, concluded that low harvest pressure and restricting harvest to sexually mature individuals would be sufficient to sustain populations. However, the biomedical industry already accepts only larger, sexually mature animals, and low harvest pressure may be impossible given the fact that ASMFC quotas have not been reached in years, and small isolated populations, without additional protection, will continue to be threatened. Thus, the single most effective step to ensure sustainability of the horseshoe crab harvest in the Northeast, particularly Massachusetts, is to completely restrict all commercial harvest during the spawning season. This restriction would include no harvest for a time prior to spawning when the horseshoe crab is coming out of its winter inactivity and beginning to feed and move toward spawning beaches, as well as during the actual spawning period. Current restrictions that cover only small periods around the monthly tides do not protect the horseshoe crabs that remain near the breeding beaches waiting for the next tidal cycle. To accomplish this and to facilitate enforcement, a no harvest period from January 1 through July 31 is suggested. Crabs captured after July 31 by trawling in deeper water would provide sufficient numbers of animals to satisfy both the bait and biomedical industries.

While the fate of horseshoe crabs used for bait is clear, there is much uncertainty surrounding both the mortality and subsequent fitness of horseshoe crabs bled for biomedical use. Thus, an immediate updating of BMPs to remove ambiguity and subsequent inclusion in an enforceable regulation is imperative. In Leschen and Correia's (2010) study for example, the authors' needed to secure a refrigerated truck for horseshoe crab transport and insist on packing fewer horseshoe crabs per container than were currently being used in the biomedical facility.

The practice of "dual-use," i.e., allowing biomedical horseshoe crabs to be used for bait following bleeding, should be completely eliminated. While the concept of dual-use would suggest a decrease in the total number of horseshoe crabs harvested, this is not evident from harvest data. In fact, in Massachusetts, the biomedical harvest may be close to equal that of the bait harvest. Dual-use may also encourage and sustain large-scale commercial horseshoe crab operations, since a more sophisticated operation is required to supply both industries. Of course dual-use horseshoe crabs command a higher price as they are "sold twice."

"Open" reporting needs to be applied to the biomedical industry. If small commercial fishermen and bait dealers are required to report their catches, sex ratios, and other information, there can be no valid reason for biomedical manufacturers to be exempt from full disclosure. Since the entire bait industry is several orders of magnitude smaller by revenue than the biomedical industry, claiming confidentiality due to company size, as is the case in Massachusetts, would be more appropriately applied to fishermen rather than the biomedical industry.

The ASMFC should officially encourage development and use of an artificial bait, especially one that does not contain horseshoe crabs specifically sacrificed for this purpose. An artificial bait currently exists (University of Delaware 2013). There is also bait that can be made from the cell-free hemolymph that results from the production of LAL. This "waste" product, when added to a binder has be shown to be an effective replacement bait (Novitsky et al. 2002).

Finally, in the age of genetic engineering, there is no reason why a synthetic replacement for LAL cannot be designed. In fact, one LAL manufacturer currently sells a synthetic substitute along with a traditional LAL (Lonza 2014). This synthetic substitute was invented by scientists using one of the horseshoe crab genes responsible for the main enzyme component of LAL to engineer a reagent produced in yeast (Ding et al. 1977). According to the FDA however, this synthetic reagent is not by definition, LAL, i.e. a lysate (L) of *Limulus* (L) amebocytes (A) and thus cannot be licensed. The major users of LAL, the pharmaceutical industry (various lots of intravenous solutions, biologics, and medical devices are required to be tested with FDA-licensed LAL prior to release for distribution and use) do not have a choice of using a synthetic substitute until the FDA changes regulations. It is interesting to note that LAL may be one of only a few diagnostic reagents (if not the only one) that is regulated on its composition (LAL) rather than what it detects (endotoxin). As endotoxin has been standardized as to its toxicity (pyrogenic dose in humans), and an official reference standard is commercially available and accepted by several pharmacopeias and the FDA, any reagent that can accurately and routinely detect the pyrogenic dose of endotoxin, i.e. by testing with the reference standard, should be a ready substitute for LAL. The PyroGene™ synthetic reagent already does this (Lonza 2014), as do some other endotoxin tests currently under development or that have been described in the literature, such as the in vitro pyrogen test (Daneshian et al. 2006). Thus all those concerned with horseshoe crab conservation, especially state agencies responsible for the regulation of horseshoe crab harvest and the ASMFC, should actively encourage the FDA to allow LAL substitutes as long as the substitutes can be properly validated (i.e., shown to detect a pyrogenic level of endotoxin in an actual pharmaceutical drug and device).

References

Anderson RL, Watson WH III, Chabot CC (2013) Sublethal behavioral and physiological effects of the biomedical bleeding process on the American horseshoe crab. Biol Bull 225:137–151

Associates of Cape Cod, Inc. and Jay Harrington v. Bruce Babbitt (2000) Civ. Act. No. 00-10549-RWZ

ASMFC (Atlantic States Marine Fisheries Commission) Horseshoe Crab Plan Development Team (1998) Interstate fisheries management plan for horseshoe crabs. Fishery management report no. 32:1–58. Washington, DC http://www.asmfc.org/uploads/file/hscFMP.pdf. Accessed Dec 1998

ASMFC (Atlantic States Marine Fisheries Commission) Horseshoe Crab Plan Development Team (2001) Addendum I to the interstate fisheries management plan for horseshoe crabs. Fishery management report no. 32a. Washington, DC http://www.asmfc.org/uploads/file/hscAddendumI.pdf. Accessed Apr 2001

ASMFC (Atlantic States Marine Fisheries Commission) (2011a) Horseshoe Crab Biomedical Ad-hoc Working Group report. http://www.asmfc.org/uploads/file/biomedAdHocWGReport_Oct2011.pdf. Accessed 5 Oct 2011

ASMFC (Atlantic States Marine Fisheries Commission) (2011b) Summer meeting summary. http://www.asmfc.org/uploads/file/Meetings/Summer2011/2011SummerMeetingSummary.pdf. Accessed 4 Aug 2011

ASMFC (Atlantic States Marine Fisheries Commission) Horseshoe Crab Review Team (2012) Review of the fishery management plan in 2011 for horseshoe crab (*Limulus polyphemus*). http://www.asmfc.org/uploads/file/hscFMPReview2012.pdf. Accessed May 2012

ASMFC (Atlantic States Marine Fisheries Commission) Horseshoe Crab Stock Assessment Committee (2013) Atlantic States Marine Fisheries Commission 2013 horseshoe crab stock assessment update. http://www.asmfc.org/uploads/file/52a88db80213HSC_StockAssessmentUpdate.pdf. Accessed 5 Aug 2013

Bang FB (1953) The toxic effect of a marine bacterium on *Limulus* and the formation of blood clots. Biol Bull 105:447–448

Carmichael RH, Rutecki D, Valiela I (2003) Abundance and population structure of the Atlantic horseshoe crab *Limulus polyphemus* in Pleasant Bay, Cape Cod. Mar Ecol Prog Ser 246:225–239

Coates CJ, Bradford EL, Krome CA et al (2012) Effect of temperature on biochemical and cellular properties on captive *Limulus polyphemus*. Aquaculture 334:30–38

Cohen E, Bang FB, Levin J, Machalonis JJ, Pistole TG, Prendergast RA, Shuster C, Watson SW (eds) (1979) Biomedical applications of the horseshoe crab (Limulidae). Alan R. Liss, Inc., New York

Compatibility Determination Eastern Massachusetts National Wildlife Refuge Complex (2002) http://www.fws.gov/northeast/planning/Compatibility%20Determinations/2002%20Horseshoe%20Crab%20Harvesting%20at%20Monomoy%20NWR.pdf. Accessed 27 Nov 2014

Daneshian M, Guenther A, Wendel A et al (2006) *In vitro* pyrogen test for toxic or immunomodulatory drugs. J Immunol Methods 313:169–175

Ding JL, Chai C, Pui AWM et al (1977) Expression of full length and deletion homologues of *Carcinoscorpius rotundicauda* Factor C in *Saccharomyces cerevisiae*: immunoreactivity and endotoxin binding. J Endotoxin Res 4:33–43

Fabie M (2009) Potential threats to horseshoe crabs on Cape Cod, Massachusetts. Dissertation, Duke University

Faherty M (2012) Summary of Wellfleet Harbor horseshoe crab data, 2000–2012. http://www.massaudubon.org/content/download/11507/185193/file/Summary-of-Wellfleet-Harbor-Horseshoe-Crab-Data-2001-2012.pdf. Accessed 30 July 2014

Galler SR (1979) *Limulus polyphemus*, a target of opportunity. In: Cohen E, Bang FB, Levin J, Machalonis JJ, Pistole TG, Prendergast RA, Shuster C, Watson SW (eds) Biomedical applications of the horseshoe crab (Limulidae). Alan R. Liss, Inc., New York, pp 107–108

Grady SP, Valiela I (2007) Stage-structured matrix modeling and suggestions for management of Atlantic horseshoe crab, *Limulus polyphemus*, populations on Cape Cod, Massachusetts. Estuar Coasts 29:685–698

Howell WH (1885) Observations upon the chemical composition and the coagulation of the blood of *Limulus polyphemus*, *Callinectes hastatus*, and *Cucumaria* sp. Johns Hopkins Circ 5:4

Hurton L, Berkson J (2006) Potential causes of mortality for horseshoe crabs (*Limulus polyphemus*) during the biomedical bleeding process. Fish Bull 104:293–298

Hurton L, Berkson J, Smith S (2005) Estimation of total hemolymph volume in the horseshoe crab *Limulus polyphemus*. Mar Freshw Behav Physiol 38:139–147

James-Pirri MJ (2002) Population demographics and spawning densities of the horseshoe crab, *Limulus polyphemus*, within Cape Cod National Seashore, Cape Cod Bay, and Monomoy National Wildlife Refuge, Massachusetts. Final Report to National Park Service, Boston, MA

James-Pirri MJ (2012) Assessment of spawning horseshoe crabs (*Limulus polyphemus*) at Cape Cod National Seashore, 2008–2009. Natural Resource Technical Report NPS/CACO/NRTR – 2012/573, National Park Service, Fort Collins, Colorado

James-Pirri MJ, Tuxbury K, Marino S et al (2005) Spawning densities, egg densities, size structure, and movement patterns of spawning horseshoe crabs, *Limulus polyphemus*, within four coastal embayments on Cape Cod, Massachusetts. Estuaries 28:296–313

James-Pirri MJ, Veillette PA, Leschen AS (2012) Selected hemolymph constituents of captive, biomedically bled, and wild caught adult female American horseshoe crabs (*Limulus polyphemus*). Mar Freshw Behav Physiol 45:281–289

Kurz W, James-Pirri MJ (2002) The impact of biomedical bleeding on horseshoe crab, *Limulus polyphemus*, movement patterns on Cape Cod, Massachusetts. Mar Freshw Behav Physiol 35:261–268

Leschen AS, Correia SJ (2010) Mortality of female horseshoe crabs (*Limulus polyphemus*) from biomedical bleeding and handling: implications for fisheries management. Mar Freshw Behav Physiol 43(2):135–147

Leschen AS, Grady SP, Valiela I (2005) Fecundity and spawning of the Atlantic horseshoe crab, *Limulus polyphemus*, in Pleasant Bay, Cape Cod, Massachusetts, USA. Mar Ecol 27:54–65

Levin J, Bang FB (1968) Clottable protein in *Limulus*: its localization and kinetics of its coagulation by endotoxin. Thromb Diath Haemorrh 19:186–197

Levin J, Hochstein HD, Novitsky TJ (2003) Clotting cells and *Limulus* amebocyte lysate: an amazing analytical tool. In: Shuster CN, Barlow RB, Brockmann JH (eds) The American horseshoe crab. Harvard University Press, Cambridge, pp 310–340

Loeb L (1902) The blood cells and inflammatory process of *Limulus*. J Med Res 2:145–158

Lonza (2014) PyroGene™ recombinant factor C endotoxin detection assay. http://bio.lonza.com/go/literature/44453. Accessed 30 Nov 2014

Massachusetts Division of Marine Fisheries (2006) Marine Fisheries Advisory. Massachusetts Marine Fisheries Advisory Commission approves immediate closure of horseshoe crab bait fishery in the Pleasant Bay/Little Pleasant Bay Complex. http://www.mass.gov/eea/docs/dfg/dmf/marinefisheriesnotices/2006/hc-bait-fishery-closure-071406.pdf. Accessed 14 Jul 2006

Massachusetts Division of Marine Fisheries (2008) Massachusetts 2008 compliance report to the Atlantic States Marine Fisheries Commission–Horseshoe Crab. http://www.edc.uri.edu/nrs/classes/NRS555/assets/readings_2011/MADMF_2008_Leschen_asmfc__MA_hcrab_report.pdf. Accessed 22 Jul 2014

Massachusetts Division of Marine Fisheries (2009) Massachusetts 2009 compliance report to the Atlantic States Marine Fisheries Commission–Horseshoe Crab. http://www.mass.gov/eea/docs/dgf/dmf/programsandprojects/2009-asmfc-hcrab-report.pdf. Accessed 22 Jul 2014

Massachusetts Division of Marine Fisheries (2010) Massachusetts 2010 compliance report to the Atlantic States Marine Fisheries Commission–Horseshoe Crab. http://www.mass.gov/eea/docs/dgf/dmf/programsandprojects/2010-asmfc-hcrab-report.pdf. Accessed 22 Jul 2014

Massachusetts Division of Marine Fisheries (2011) Massachusetts 2011 compliance report to the Atlantic States Marine Fisheries Commission–Horseshoe Crab. http://www.mass.gov/eea/docs/dgf/dmf/programsandprojects/2011-asmfc-hcrab-report.pdf. Accessed on 22 Jul 2014

Massachusetts Division of Marine Fisheries (2012) Massachusetts 2012 compliance report to the Atlantic States Marine Fisheries Commission–Horseshoe Crab. http://www.mass.gov/eea/docs/dgf/dmf/programsandprojects/2012-asmfc-hcrab-report.pdf. Accessed on 22 Jul 2014

Mayer AG (1914) The effects of temperature on tropical marine animals. Carnegie Inst Washington Publ 183:1–25

Mizrahi DS, Peters KA (2009) Relationship between sandpipers and horseshoe crab in Delaware Bay: a synthesis. In: Tanacredi JT, Botton ML, Smith D (eds) Biology and conservation of horseshoe crabs. Springer, New York, pp 65–87

Novitsky TJ (1991) Discovery to commercialization: the blood of the horseshoe crab. Oceanus 27:13–18

Novitsky TJ (2009) Biomedical applications of *Limulus* amebocyte lysate. In: Tanacredi JT, Botton ML, Smith D (eds) Biology and conservation of horseshoe crabs. Springer, New York, pp 315–329

Novitsky TJ, Dawson ME, Paus EJ (2002) Artificial bait. US Patent 6,391,295, 21 May 2002

Rudkin DM, Young GA, Nowlan GS (2008) The oldest horseshoe crab: a new xiphosurid from late Ordovician Konsevat-Lagerstatten deposits Manitoba, Canada. Palaeontology 51:1–9

Rudloe A (1979) *Limulus polyphemus*: a review of the ecologically significant literature. In: Cohen E, Bang FB, Levin J, Machalonis JJ, Pistole TG, Prendergast RA, Shuster C, Watson SW (eds) Biomedical applications of the horseshoe crab (Limulidae). Alan R. Liss, Inc., New York, pp 27–35

Rudloe A (1983) The effect of heavy bleeding on mortality of the horseshoe crab, *Limulus polyphemus*, in the natural environment. J Invertebr Pathol 42:167–176

Rutecki D, Carmichael RH, Valiela I (2004) Magnitude of harvest of Atlantic horseshoe crabs, *Limulus polyphemus*, in Pleasant Bay, Massachusetts. Estuaries 27:179–187

Shuster CN (2003) A blue blood: the circulatory system. In: Shuster CN, Barlow RB, Brockmann JH (eds) The American horseshoe crab. Harvard University Press, Cambridge, pp 276–287

Shuster CN, Sekiguchi K (2003) Growing up takes about ten years and eighteen stages. In: Shuster CN, Barlow RB, Brockmann JH (eds) The American horseshoe crab. Harvard University Press, Cambridge, pp 103–132

Smith DR, Millard MJ, Carmichael RH (2009) Comparative status and assessment of *Limulus polyphemus* with emphasis on the New England and Delaware Bay populations. In: Tanacredi JT, Botton ML, Smith D (eds) Biology and conservation of horseshoe crabs. Springer, New York, pp 361–386

University of Delaware (2013) Bait alternative. http://www.udel.edu/udaily/2013/may/artificial-bait-052913.html. Accessed 29 May 2013

Walls EA, Berkson J (2003) Effects of blood extraction on horseshoe crabs (*Limulus polyphemus*). Fish Bull 101:457–459

Widener JW, Barlow RB (1999) Decline of a horseshoe crab population on Cape Cod. Biol Bull 197(2):300–302

Chapter 29
Biotechnology Efforts to Conserve Horseshoe Crabs Through the Development of Recombinant Factor C-Based Endotoxin Test

Peng Li, Bow Ho, and Jeak Ling Ding

Abstract During Gram-negative bacterial infection, endotoxin released from the outer membrane of the bacteria interacts with host sensor/receptor proteins to drive inflammatory responses. In severe persistent infection, septic shock ensues. Therefore endotoxin, which is ubiquitous, poses a threat to the biotechnology and healthcare industries. The quality assurance for injectable drugs and medical devices started with the slow, less efficient and expensive pyrogen test using rabbits. In 1970, the US Food and Drug Administration (FDA) approved the *Limulus* amoebocyte lysate (LAL) for testing endotoxin contamination. However, the LAL test requires the harvesting and bleeding of the horseshoe crab. Problems with the specificity of LAL to endotoxin and the potential for negative effects on horseshoe crab populations have called for an alternative reliable endotoxin test. Here, we review biotechnology efforts into recombinant Factor C (rFC), the endotoxin-inducible sensor protein in LAL, which establishes the foundation of a synthetic and rapid detection of endotoxin. Incidentally, in its "Guidance for Industry–pyrogen and endotoxin testings", the US FDA has listed the rFC-based PyroGene® assay as an alternative endotoxin testing method. The industry has been slow to take the lead in using this highly sensitive test, which has been well-validated by the manufacturer.

P. Li
Centre for Biomedical and Life Sciences, Singapore Polytechnic,
500 Dover Road, Singapore 139651, Singapore

B. Ho (✉)
Department of Microbiology, Yong Loo Lin School of Medicine, National University of Singapore, 5 Science Drive 2, Singapore 117545, Singapore
e-mail: michob@nus.edu.sg

J.L. Ding (✉)
Department of Biological Sciences, Faculty of Science, National University of Singapore,
14 Science Drive 4, Singapore 117543, Singapore
e-mail: dbsdjl@nus.edu.sg

R.H. Carmichael et al. (eds.), *Changing Global Perspectives on Horseshoe Crab Biology, Conservation and Management*, DOI 10.1007/978-3-319-19542-1_29

It is noteworthy that the rFC satisfies the typical analytical characteristics used in method validation (USP chapter <1225>). Thus, the sooner rFC can be accepted to replace the traditional LAL assay, the quicker and surer it will be to conserve horseshoe crabs, a threatened “living fossil”.

Keywords Conservation • Limulus amoebocyte lysate (LAL) test • Biotechnology • Recombinant Factor C (rFC) • Endotoxin/pyrogen assay

29.1 Introduction

In the process of infection and invasion in the host, the Gram-negative bacteria (GNB) release endotoxin, also known as lipopolysaccharide or exogenous biological pyrogen. The endotoxin is normally anchored on the bacterial outer membrane. The presence of endotoxin in the bloodstream of the host causes inflammation and can lead to sepsis. Endotoxin is recognized by receptors (e.g. CD14, TLR4) on the immune responsive cells, and this activates the endotoxin-signaling pathway. Endotoxin-stimulated human macrophages release inflammatory chemokines and cytokines. However, excessive inflammation causes septic shock, multiple organ failure and death (Bone 1996; McCormick et al. 2001). Thus, the ubiquitous endotoxin is a bane to the medical and healthcare industries as indicated by the wide application of the endotoxin test (Table 29.1). The US Food and Drug Administration (FDA) has set an endotoxin upper limit to 5.0 endotoxin units (EU)/kg body weight/hour per injection to ensure safe injectable drugs (Petsch and Anspach 2000). Hence, the biotechnology industry and pharmaceutical companies need a sensitive and specific endotoxin test to assure the endotoxin-free/endotoxin-low injectables and medical devices to pass US FDA guidelines prior to commercialization of parenteral products.

Table 29.1 The application of endotoxin detection assay in the biomedical industry

Application	Reference
Antibiotics product	Yamamoto et al. (2000)
Cardiac regeneration	Soncin et al. (2009)
Meningitis	Ichinohe et al. (1995)
Nanoparticles formulation	Dobrovolskaia et al. (2010)
Packaging components	Madsen et al. (2014)
Radiopharmaceuticals	Weatherman et al. (2013)
Sterile compound preparation	Cooper et al. (2010)
Urethritis	Prior and Spagna (1980)
Vaccine	Geier and Geier (2002)
Water contamination	Zhang et al. (2014)

29.2 Endotoxin: The Nemesis

Endotoxin or lipopolysaccharide (LPS) is also referred to as exogenous biological pyrogen because of its toxic and pyrogenic (fever-inducing) properties against the mammalian hosts. The membrane of an *E. coli* contains approximately 2×10^6 endotoxin molecules, constituting about 20 femtograms (Minabe et al. 1994). The endotoxin molecule comprises covalently linked fragments of the O-specific antigen, the outer and inner core oligosaccharides and lipid A (Ulmer et al. 2002).

Lipid A is the most structurally-conserved moiety of the endotoxin molecule (Frecer et al. 2000a, b). It is the bioactive center, as almost all endotoxin-induced biological reactions in the mammalian cells are lipid A-dependent. Direct evidence comes from studies of free native lipid A (Galanos et al. 1992) and synthetic lipid A (Takayama et al. 1983), both of which elicit full pyrogenic activity. Lipid A is made up of a phosphorylated β1,6-linked D-glucosamine disaccharide that consists of variable numbers of asymmetrically placed amide or ester-linked acyl chains. This structure is the minimal necessity for the bioactivity of lipid A for it to induce cytokine production (Rietschel and Brade 1992) and inflammatory response. The distinctive structure of lipid A most likely suggests its important roles in the outer membrane of the bacteria as it confers resistance against host phospholipases, and hence, survival to the GNB. Endotoxins are uniquely thermo-stable and fairly insensitive to pH variations. The removal of endotoxin requires treatment at 200 °C for more than 2 h, or short durations in high concentrations of acids or bases, such as in pH 1.0 citric acid for 3 min (Minabe et al. 1994). Endotoxin is ubiquitous and indomitable in nature, existing in water, soil and the human gut, which hosts a wide variety of microbial commensals, including GNB. However, intravenous introduction of the endotoxin triggers a sequence of medical problems in the host. Thus, understanding the chemistry of endotoxin and defining its bioactive domain (Goh et al. 2002a, b; Frecer et al. 2004) has aided the design and development of better and more effective diagnostics and potential therapeutics against endotoxin.

29.3 Horseshoe Crab, a "Living Fossil": Nature's Gift/Foe Against Pyrogen?

The horseshoe crab, commonly referred to as *Limulus*, has been considered a "living fossil." It belongs to an ancient protostome in the order, Xiphosura. Studies on the horseshoe crab have enlightened us on its evolutionary biology and palaeontology. There are four types of horseshoe crabs in the world: *Limulus polyphemus* (in the eastern coast of North and Central America); *Tachypleus tridentatus* (in East Asia) and *Tachypleus gigas* and *Carcinoscorpius rotundicauda* in Southeast Asia. These four extant species differ in morphology, size and habitat, especially, that of the *C. rotundicauda*–the smallest member, which thrives in estuarine swamps, while the larger *Tachypleus* and *Limulus* inhabit the seas.

Invertebrates offer an ideal model to study innate immunity in isolation since they lack adaptive immunity. Compared with other invertebrate models, the horseshoe crab has unique advantages. Firstly, it harbors a powerful innate immune system. As a mud-dwelling species, the *C. rotundicauda* co-habits with myriads of microbial pathogens, including Gram-negative and Gram-positive bacteria, and fungi. Throughout evolution, it has developed a plethora of formidable innate immune mechanisms, which have protected it from the extremely challenging microbes, and therefore it has conceivably co-evolved an excellent frontline immune defense system, thus providing a unique animal model for the study of innate immunity. Secondly, the horseshoe crab has a relatively large body size with substantial volume of tissues for biochemical and physiological studies. These studies are less feasible when using small invertebrates like the *Drosophila*. Thirdly, although the genomic DNA sequences of the horseshoe crab are still unavailable, the recent launch of the expressed sequence tag (EST) clusters (Ding et al. 2005) annotating frontline host defense, signaling, apoptosis and stress response genes of the horseshoe crab, has significantly facilitated the innate immune response study at the molecular level. During the last few decades, the horseshoe crab has been an ideal animal model for worldwide biotechnological research and innovations on the premise that alternative environmentally-friendly diagnostics and potential therapeutics were achieved in ways consistent with conservation of the species.

29.4 Factor C, an Endotoxin-Sensor in the Horseshoe Crab Blood

In an effort to produce LAL from *ex vivo* sources, researchers have attempted for several decades, but in vain, to aquaculture (Kropach 1979) and to establish a cell line from the horseshoe crab blood amoebocytes (Sanborn et al. 1959; Pearson and Woodland 1979; Ding et al. 1988). Parallel attempts were made to purify and characterize the LAL active enzymes to better understand how the blood clots in the presence of GNB (Muta et al. 1995; Ding et al. 2005). These efforts were made with a long-term goal of cloning the Factor C cDNA to synthesize an alternate pyrogen detection system (see Sect. 29.5).

The horseshoe crab relies on many of its innate immune proteins that are involved in different frontline defense mechanisms to stave off an infection (Ho et al. 1994; Yeo et al. 1996). Some examples include the complement cascade proteins like C3 (Ng et al. 2004; Tan et al. 2005; Zhu et al. 2005), melanization reaction in a phenoloxidase-induced production of anti-microbial compounds from a protein like hemocyanin (Nagai et al. 2001; Jiang et al. 2007; Du et al. 2010, 2011), lectins (Zhu et al. 2006; Li et al. 2007; Ng et al. 2007), antimicrobial peptide defense proteins, like tachyplesin (Nakamura et al. 1988a; Yeo et al. 1993), and an exceptionally sensitive blood coagulation cascade sensor protein, Factor C, that is triggered by picogram levels of endotoxin (Muta et al. 1991; Ding et al. 1993a).

The horseshoe crab hemolymph harbors mostly one kind of blood cells called amoebocytes, which are very sensitive to bacterial endotoxin. The amoebocytes release two types of granular components (large and small granules) into the plasma to participate in host defense via blood coagulation (Iwanaga and Lee 2005) and incapacitate the infecting microbe (Ding et al. 1988, 1993b). Figure 29.1 summarizes the defense proteins present in the large and small granules in the amoebocytes. Amongst all of these defense proteins, Factor C, which is present in the large granules, initiates the endotoxin-sensitive coagulation cascade and has been proposed to function in frontline defense against the invading GNB.

In the past decades, the coagulation cascade has been characterized (Iwanaga et al. 1994; Ding et al. 1995). This cascade consists of three serine protease zymogens: Factor C, Factor B, proclotting enzyme, and additionally, a clottable protein, coagulogen (Ho et al. 1993; Muta and Iwanaga 1996). As a biosensor of endotoxin, Factor C initiates the coagulation cascade reaction. Factor C in *C. rotundicauda* amoebocyte lysate (CAL) is capable of detecting trace levels of endotoxin (Navas et al. 1990). Ho (1983) further enhanced the sensitivity of endotoxin detection by spiking the test sample with known levels of endotoxin. Bacterial endotoxin induces the amoebocytes to aggregate and degranulate. Presumably, horseshoe crabs use the endotoxin-induced coagulation cascade as an important defense mechanism against the invasion of GNB (Armstrong and Rickles 1982). Thus, the horseshoe crab is heavily dependent on the coagulation cascade for its survival in the harsh environment. Nature seems to have provided the horseshoe crab with a powerful defense, which humans have exploited as a resource for LAL-endotoxin test (Levin et al. 1970) during the last >40 years.

Further search for other potential endotoxin-sensitive proteins in the horseshoe crab blood led to more research on additional serine protease cascades, which contribute to the fundamental biological processes in both the invertebrates and vertebrates.

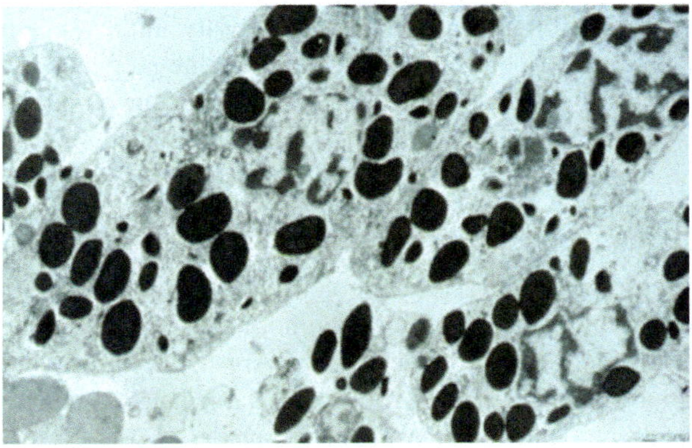

Fig. 29.1 The horseshoe crab amoebocytes. There are two types of electron dense granules, the large and small granules. Antimicrobial peptides are present in the small granules while Factor C and other coagulation cascade enzymes are present in large granules

Serine proteases play significant roles in signaling pathways involved in development, immunity, and hemostasis. However, no homologues of horseshoe crab Factor C have been found in mammals or insects besides the C-terminal serine protease domain sequence of the human α-thrombin, which bears 36.7 % similarity. A similarity between the Factor C-driven serine protease cascade and the *Drosophila* dorsoventral-determination serine protease cascade has been reported (Gay and Keith 1992)–the horseshoe crab clotting enzyme and the 36 *Drosophila* serine proteases including Snake and Easter, show significant sequence homology (Smith and DeLotto 1992; Ding et al. 1993a). Horseshoe crab Factor B, proclotting enzyme and coagulogen also exhibit a similar primary structure to the counterparts in *Drosophila* (Snake, Easter and Spätzle) (Wang et al. 2003), but there is no homologue of Factor C identified in the *Drosophila*. Thus, amongst the best studied invertebrate model organisms, only the horseshoe crab Factor C (a serine protease) appears to display unique endotoxin-binding properties. Is this an unfortunate endowment on the horseshoe crab, which is being exploited and endangered by humans?

29.5 Molecular Cloning of Factor C for Pyrogen Detection Alternative: From LAL to rFC-Based PyroGene Test

Recently, genetic engineering of Factor C in bacteria (Roopashree et al. 1995, 1997a), yeast (Roopashree et al. 1996; Pui et al. 1997) insect cell line (Wang et al. 2001) and mammalian cells (Roopashree et al. 1997b) expressed recombinant Factor C (rFC) capable of agglutinating GNB and binding endotoxin. Subsequent cloning and molecular manipulations of the Factor C cDNA in the baculoviral-insect cell (*Spodoptera frugiperda*) system expressed rFC with an outstanding endotoxin-detection sensitivity at 0.001 EU/mL (Ding and Ho 2001). The rFC is a biologically functional synthetic biosensor for bacterial endotoxin. It remains as a zymogen until it binds to endotoxin, where it undergoes a conformational transformation to exhibit its full enzymatic activity, indicating the presence of endotoxin in a sample. Thus a rapid, sensitive one-step diagnostic assay for endotoxin has been developed (Fig. 29.2). In this assay, the stimulated rFC enzyme hydrolyses a synthetic substrate to produce a product, which quantifies the amount of endotoxin in the sample.

An ELISA-based colorimetric assay is formulated using Boc-Val-Pro-Arg-pNA (pNA, p-nitroanilide) as a chromogenic substrate that is hydrolysed by endotoxin-induced rFC to produce a measurable colorimetric product (Ding and Ho 2001). Alternatively, in a fluorimetric assay, the endotoxin-activated rFC hydrolyses a fluorogenic substrate such as Boc-Val-Pro-Arg-MCA (Boc, butoxy-carbonyl; MCA, 7-amido-4-methylcoumarin), yielding a fluorimetric product, measurable at an excitation at 380 nm and an emission at 460 nm.

Compared with the commercial LAL, which relies on all the coagulation cascade components to be functional to form a final reliable coagulation product, the rFC-based pyrogen assay has lower background and selective response to bacterial

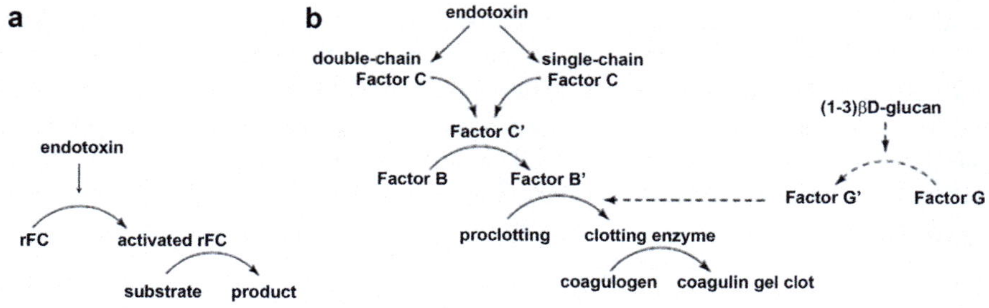

Fig. 29.2 Recombinant Factor C (rFC) pyrogen assay. (**a**) The novel single-step activation of rFC which hydrolyses a substrate to produce colored (colorimetric assay) result by colorimetric reaction, or fluorescence (fluorimetric assay) result by fluorimetric reaction, (**b**) The traditional Factor C-initiated coagulation cascade in the LAL assay (Adapted from (Ding and Ho 2001) with permission from Elsevier)

endotoxin (Ding and Ho 2001). In addition, the rFC assay, which involves only a single endotoxin-sensitive proenzyme, does not suffer from any potential false positive due to the fungal toxin, (1-3)β-D-glucan-activated Factor G pathway (see Fig. 29.2b). Furthermore, the microfluorimetric or microcolorimetric format integrated into the rFC assay has made it possible for rapid high-throughput screens of endotoxin in large quantities of samples (Ding and Ho 2001). Since 2000, the rFC clone has been outlicensed to Lonza Inc., which has incorporated it into a sensitive pyrogen-detection assay (PyroGene kit, Lonza Inc, http://www.lonza.com/products-services/pharma-biotech/endotoxin-detection/endotoxin-detection-assays.aspx).

29.6 How Can Biotechnology Circumvent the Need to Harvest and Bleed the Endangered Horseshoe Crab?

The rapid and specific detection of endotoxin is urgently needed for the development and production of clinical drugs and medical devices where state-of-the-art pyrogen-free facilities are required. It is necessary that every major pharmaceutical company employs efficient pyrogen testing for quality control and process endorsement. In biotechnology, where the purified plasmid DNAs from bacterial expressing-host cells, such as *E. coli* are used as diagnostics and therapeutics, the medical regulatory agencies are obligated to set strict rules to guard against endotoxin contamination. The biotechnological and pharmaceutical companies and healthcare industry have relied on LAL for quality assurance of parenteral products for more than 40 years, although there are problems related with the lack of specificity of the LAL assay. Recent programs to protect the dwindling population of *Limulus* have not eased these problems (Gauvry 2009; Ding and Ho 2010). In fact, harvesting of the horseshoe crab has intensified for both its blood and, in some areas, for cuisine (Liao and Li 2001; Berkson 2009). The rFC has offered several biotechnological applications–the horseshoe crab has contributed a reliable replacement for the

LAL-based pyrogen assay as well as a novel pyrogen-removal technology (Ding et al. 2001). Before it is too late, we must conserve this species without further harvesting and exploitation. Because of the presence of multiple endotoxin-binding sites on each molecule, the rFC is extremely sensitive to endotoxin and is, hence, the ideal replacement for the LAL assay. The replacement of the LAL assay with the rFC assay will avert the need to collect *Limulus* and thus, help to protect this endangered "living fossil".

29.7 Perpetuity of the "Living Fossil" Depends on a Change in Human Mindset

Fifteen years passed between the FDA's approval of the LAL test (1970; http://en.wikipedia.org/wiki/Limulus_amebocyte_lysate) and the start of basic research (mid-1980s) on the coagulation enzymes (Ding et al. 1988; Nakamura et al. 1988b), cloning and developing the rFC into a synthetic pyrogen test (Muta et al. 1991; Navas 1991; Roopashree et al. 1996; Ding et al. 1997; Pui et al. 1997; Tan et al. 2000; Ding and Ho 2001; Wang et al. 2002, 2003). In 1997, the environmentally-cognizant "BBC Tomorrow's World" program drew the attention of the world to the genetic engineering of rFC, with the promise of an alternate and even better pyrogen test without ever having to harvest the horseshoe crab again! This year (2015) will mark the passing of another 15 years since the rFC clone was licensed (2000) to Lonza Inc. The rFC-based PyroGene waited for 12 years for the US FDA to list it as an alternative valid test method for endotoxin– "Guidance for Industry – Pyrogen and Endotoxins Testing: Questions and Answers"; www.fda.gov/drugs/guidancecomplianceregulatoryinformation/guidances/ucm314718.htm.

Environmental concerns and conservation issues have loomed over the industry since the days of the rabbit pyrogen test, which predated the LAL test. There has been concern regarding survival of horseshoe crabs after they have been bled. Studies have shown that casualties may be as high as 30 % (Rudloe 1983; Thompson 1998; Walls and Berkson 2003). Lonza Inc (USA) and Hyglos Inc (Germany) reached an agreement to co-commercialize the rFC-based endotoxin test kits: PyroGene and Endozyme (http://www.hyglos.de/en/news-events/news-events/news/article/press-release-lonza-and-hyglos-gmbh-reach-agreement-for-endotoxin-detection-patents.html). One of the companies' heads stated: "*…we believe that finally, the sustainable animal-free alternative to LAL will be the future of endotoxin detection*". Various companies also market forms of LPS-removal kits. All of these businesses arising from research on the horseshoe crab are potentially good news for the "living fossil"; would this species now need to be further exploited for the industry? A concerted drive by: (1) the US and European Pharmacopoeias to approve, recognize, adopt, and mandate the use of the PyroGene and Endozyme assays worldwide to replace the LAL test; (2) the IUCN's efforts and new government policies to prohibit the harvesting of horseshoe crabs (for blood or for food); and (3) a paradigm shift and a *bona fide* change in the mindset of ordinary citizens and Pharmaceutical companies and Biotech industry towards a sincere sentiment for the conservation of the horseshoe crab will help ensure the perpetuity of this 500 million year-old "living fossil".

29.8 Conclusion

The applications of the rFC in endotoxin detection have been supported by research work around the world. For many years, scientists have envisaged that research findings on the horseshoe crab innate immunity will benefit the biotechnology, pharmaceutical and human healthcare industries. There is no need for continuous harvesting from dwindling populations of horseshoe crab when there are readily available alternative tests that are more consistent and have a better endotoxin-detection limit compared to LAL.

References

Armstrong PB, Rickles FR (1982) Endotoxin-induced degranulation of the *Limulus* amebocyte. Exp Cell Res 140(1):15–24

Berkson J (2009) An integrative approach to horseshoe crab multiple use and sustainability. In: Tanacredi JT, Botton ML, Smith DR (eds) Biology and conservation of horseshoe crabs. Springer, New York, pp 387–398

Bone RC (1996) The sepsis syndrome: definition and general approach to management. Clin Chest Med 17(2):175–181

Cooper JF, Latta KS, Smith D (2010) Automated endotoxin testing program for high-risk-level compounded sterile preparations at an institutional compounding pharmacy. Am J Health-Syst Pharm 67(4):280–286

Ding JL, Ho B (2001) A new era in pyrogen testing. Trends Biotechnol 19(8):277–281

Ding JL, Ho B (2010) Endotoxin detection – from Limulus amebocyte lysate to recombinant factor C. Sub-Cell Biochem 53:187–208

Ding JL, Kim JC, Ho B (1988) Pokeweed mitogen stimulates DNA synthesis in cultured amoebocytes of *Carcinoscorpius rotundicauda*. Cytobios 55(222–223):147–154

Ding JL, Navas MA 3rd, Ho B (1993a) Two forms of factor C from the amoebocytes of *Carcinoscorpius rotundicauda*: purification and characterisation. Biochim Biophys Acta 1202(1):149–156

Ding JL, Sababathy TK, Ho B (1993b) Morphological changes of *Carcinoscorpius rotundicauda* amoebocyte and *Escherichia coli* during their interaction. Cytobios 75(300):21–32

Ding JL, Navas MA 3rd, Ho B (1995) Molecular cloning and sequence analysis of factor C cDNA from the Singapore horseshoe crab, *Carcinoscorpius rotundicauda*. Mol Mar Biol Biotechnol 4(1):90–103

Ding JL, Chai C, Pui AWM et al (1997) Expression of full length and deletion homologues of *Carcinoscorpius rotundicauda* Factor C in *Saccharomyces cerevisiae*: immunoreactivity and endotoxin binding. J Endotoxin Res 4:33–43

Ding JL, Tan KC, Thangamani S et al (2005) Spatial and temporal coordination of expression of immune response genes during *Pseudomonas* infection of horseshoe crab, *Carcinoscorpius rotundicauda*. Genes Immunol 6(7):557–574

Ding JL, Zhu Y, Ho B (2001) High-performance affinity capture-removal of bacterial pyrogen from solutions. J Chromatogr B Biomed Sci Appl 759(2):237–246

Dobrovolskaia MA, Neun BW, Clogston JD et al (2010) Ambiguities in applying traditional *Limulus* amebocyte lysate tests to quantify endotoxin in nanoparticle formulations. Nanomedicine 5(4):555–562

Du R, Ho B, Ding JL (2010) Rapid reprogramming of haemoglobin structure-function exposes multiple dual-antimicrobial potencies. EMBO J 29(3):632–642

Du R, Ho B, Ding JL (2011) Application of cell-free hemolymph of horseshoe crab in antimicrobial drug screening. Curr Pharm Design 17(13):1234–1239

Frecer V, Ho B, Ding JL (2000a) Interpretation of biological activity data of bacterial endotoxins by simple molecular models of mechanism of action. Eur J Biochem 267(3):837–852

Frecer V, Ho B, Ding JL (2000b) Molecular dynamics study on lipid A from *Escherichia coli*: insights into its mechanism of biological action. Biochem Biophys Acta 1466(1–2):87–104

Frecer V, Ho B, Ding JL (2004) De novo design of potent antimicrobial peptides. Antimicrob Agents Chemother 48(9):3349–3357

Galanos C, Freudenberg MA, Katschinski T, Salomao R, Mossmann H, Kumazawa Y (1992) Tumor necrosis factor and host response to endotoxin. In: Morrison DC, Ryan JL (eds) Bacterial endotoxic lipopolysaccharides, vol 11, Immunpharmacology and pathophysiology. Boca Raton/CRC Press, Florida, pp 75–102

Gauvry G (2009) Community building: an integrated approach to horseshoe crab conservation. In: Tanacredi JT, Botton ML, Smith DR (eds) Biology and conservation of horseshoe crabs. Springer, New York, pp 605–612

Gay NJ, Keith FJ (1992) Regulation of translation and proteolysis during the development of embryonic dorso-ventral polarity in *Drosophila*: homology of easter proteinase with *Limulus* proclotting enzyme and translational activation of Toll receptor synthesis. Biochim Biophys Acta 1132(3):290–296

Geier DA, Geier MR (2002) Clinical implications of endotoxin concentrations in vaccines. Ann Pharmacother 36(5):776–780

Goh YY, Frecer V, Ho B et al (2002a) Rational design of green fluorescent protein mutants as biosensor for bacterial endotoxin. Protein Eng 15(6):493–502

Goh YY, Ho B, Ding JL (2002b) A novel fluorescent protein-based biosensor for gram-negative bacteria. Appl Environ Microbiol 68(12):6343–6352

Ho B (1983) An improved *Limulus* gelation assay. Microbios Letts 24:81–84

Ho B, Kim JC, Ding JL (1993) Electrophoretic analysis of endotoxin-activated gelation reaction of *Carcinoscorpius rotundicauda* amoebocyte lysate. Biochem Mol Biol Int 29(4):687–694

Ho B, Yeo DS, Ding JL (1994) A tetrodotoxin neutralizing system in the haemolymph of the horseshoe crab, *Carcinoscorpius rotundicauda*. Toxicon 32(7):755–762

Ichinohe S, Inada K, Nemoto T et al (1995) Usefulness of endotoxin-specific *Limulus* test for the measurement of endotoxin in cerebrospinal fluid in diagnosis of bacterial meningitis. Kansenshogaku zasshi 69(11):1227–1234

Iwanaga S, Lee BL (2005) Recent advances in the innate immunity of invertebrate animals. J Biochem Mol Biol 38(2):128–150

Iwanaga S, Muta T, Shigenaga T et al (1994) Role of hemocyte-derived granular components in invertebrate defense. Ann NY Acad Sci 712:102–116

Jiang N, Tan NS, Ho B et al (2007) Respiratory protein-generated reactive oxygen species as an antimicrobial strategy. Nat Immunol 8(10):1114–1122

Kropach C (1979) Observations on the potential of *Limulus* aquaculture in Israel. In: Cohen E, Bang FB (eds) Biomedical applications of the horseshoe crab (Limulidae): proceedings of a conference held at Marine Biological Laboratory, Woods Hole, Massachusetts. Alan R. Liss Inc, New York

Levin J, Tomasulo PA, Oser RS (1970) Detection of endotoxin in human blood and demonstration of an inhibitor. J Lab Clin Med 75(6):903–911

Li Y, Ng PM, Ho B et al (2007) A female-specific pentraxin, CrOctin, bridges pattern recognition receptors to bacterial phosphoethanolamine. Eur J Immunol 37(12):3477–3488
Liao Y, Li X (2001) Poisoning from eating horseshoe crab and its prevention and treatment. Wei sheng yan jiu (J Hyg Res) 30(2):122–124
Madsen AM, Tendal K, Frederiksen MW (2014) Attempts to reduce exposure to fungi, beta-glucan, bacteria, endotoxin and dust in vegetable greenhouses and a packaging unit. Sci Total Environ 468–469:1112–1121
McCormick JK, Yarwood JM, Schlievert PM (2001) Toxic shock syndrome and bacterial superantigens: an update. Annu Rev Microbiol 55:77–104
Minabe M, Takeuchi K, Kumada H et al (1994) The effect of root conditioning with minocycline HCl in removing endotoxin from the roots of periodontally-involved teeth. J Periodontol 65(5):387–392
Muta T, Iwanaga S (1996) The role of hemolymph coagulation in innate immunity. Curr Opin Immunol 8(1):41–47
Muta T, Miyata T, Misumi Y et al (1991) *Limulus* factor C: an endotoxin-sensitive serine protease zymogen with a mosaic structure of complement-like, epidermal growth factor-like, and lectin-like domains. J Biol Chem 266(10):6554–6561
Muta T, Seki N, Takaki Y et al (1995) Purified horseshoe crab factor G: reconstitution and characterization of the (1-->3)-beta-D-glucan-sensitive serine protease cascade. J Biol Chem 270(2):892–897
Nagai T, Osaki T, Kawabata S (2001) Functional conversion of hemocyanin to phenoloxidase by horseshoe crab antimicrobial peptides. J Biol Chem 276(29):27166–27170
Nakamura T, Furunaka H, Miyata T et al (1988a) Tachyplesin, a class of antimicrobial peptide from the hemocytes of the horseshoe crab (*Tachypleus tridentatus*): isolation and chemical structure. J Biol Chem 263(32):16709–16713
Nakamura T, Tokunaga F, Morita T et al (1988b) Intracellular serine-protease zymogen, factor C, from horseshoe crab hemocytes. Its activation by synthetic lipid A analogues and acidic phospholipids. Eur J Biochem 176(1):89–94
Navas MA (1991) Purification, cDNA cloning and characterization of Factor C from amoeboctyes of the horseshoe crab, *Carcinoscorpius rotundicauda*. National University of Singapore, Singapore
Navas MA 3rd, Ding JL, Ho B (1990) Inactivation of factor C by dimethyl sulfoxide inhibits coagulation of the *Carcinoscorpius* amoebocyte lysate. Biochem Int 21(5):805–813
Ng PM, Jin Z, Tan SS et al (2004) C-reactive protein: a predominant LPS-binding acute phase protein responsive to *Pseudomonas* infection. J Endotoxin Res 10(3):163–174
Ng PM, Le Saux A, Lee CM et al (2007) C-reactive protein collaborates with plasma lectins to boost immune response against bacteria. EMBO J 26(14):3431–3440
Pearson FC, Woodland E (1979) The *in vitro* cultivation of *Limulus* amoebocytes. Progr Clin Biol Res 29:93–102
Petsch D, Anspach FB (2000) Endotoxin removal from protein solutions. J Biotechnol 76(2–3):97–119
Prior RB, Spagna VA (1980) Comparative evaluation of the tube and microdilution *Limulus* lysate techniques for rapid presumptive diagnosis of gonococcal urethritis in men. J Clin Microbiol 11(4):340–342
Pui AWM, Ho B, Ding JL (1997) Yeast recombinant Factor C from horseshoe crab binds endotoxin and causes bacteriostasis. J Endotoxin Res 4:391–400
Rietschel ET, Brade H (1992) Bacterial endotoxins. Sci Am 267(2):54–61
Roopashree SD, Chai C, Ho B et al (1995) Expression of *Carcinoscorpius rotundicauda* factor C cDNA. Biochem Mol Biol Int 35(4):841–849
Roopashree SD, Ho B, Ding JL (1996) Expression of *Carcinoscorpius rotundicauda* Factor C in *Pichia pastoris*. Mol Mar Biol Biotechnol 5(4):334–343
Roopashree SD, Ho B, Ding JL (1997a) The cys-rich and EGF-like domains of *C. rotundicauda* Factor C yields soluble fusion protein with GFP. Biotechnol Lett 19(11):1147–1150
Roopashree SD, Ho B, Ding JL (1997b) Recombinant COS-1 cells express *Carcinoscorpius rotundicauda* Factor C. Biotechnol Lett 19:357–361

Rudloe A (1983) The effect of heavy bleeding on mortality of the horseshoe crab, *Limulus polyphemus*, in the natural environment. J Invertebr Pathol 42(2):167–176
Sanborn RC, Haskell JA, Fisher FMJ (1959) Cells of *Limulus polyphemus* in tissue culture. Biol Bull 117:399
Smith CL, DeLotto R (1992) A common domain within the proenzyme regions of the *Drosophila* snake and easter proteins and *Tachypleus* proclotting enzyme defines a new subfamily of serine proteases. Protein Sci 1(9):1225–1226
Soncin S, Lo Cicero V, Astori G et al (2009) A practical approach for the validation of sterility, endotoxin and potency testing of bone marrow mononucleated cells used in cardiac regeneration in compliance with good manufacturing practice. J Transl Med 7:78
Takayama K, Qureshi N, Mascagni P et al (1983) Fatty acyl derivatives of glucosamine 1-phosphate in *Escherichia coli* and their relation to lipid A: complete structure of A diacyl GlcN-1-P found in a phosphatidylglycerol-deficient mutant. J Biol Chem 258(12):7379–7385
Tan NS, Ho B, Ding JL (2000) High-affinity LPS binding domain(s) in recombinant factor C of a horseshoe crab neutralizes LPS-induced lethality. Faseb J 14(7):859–870
Tan SS, Ng PM, Ho B, Ding JL (2005) The antimicrobial properties of C-reactive protein (CRP). J Endotoxin Res 11(4):249–256
Thompson M (1998) Assessments of the population biology and critical habitat for the horseshoe crab, *Limulus polyphemus*, in the South Atlantic Bight. M.S. thesis, University of Charleston
Ulmer AJ, Rietschel ET, Zähringer U et al (2002) Lipopolysaccharide: structure, bioactivity, receptors, and signal transduction. Trends Glycosci Glyc 14:53–88
Walls EA, Berkson J (2003) Effects of blood extraction on horseshoe crabs (*Limulus polyphemus*). Fish Bull 101(2):457–459
Wang J, Ho B, Ding JL (2001) Functional expression of full length *Limulus* Factor C in stably transformed Sf9 cells. Biotechnol Lett 23:71–76
Wang J, Tan NS, Ho B et al (2002) Modular arrangement and secretion of a multidomain serine protease: evidence for involvement of proline-rich region and N-glycans in the secretion pathway. J Biol Chem 277(39):36363–36372
Wang L, Ho B, Ding JL (2003) Transcriptional regulation of *Limulus* factor C: repression of an NFkappaB motif modulates its responsiveness to bacterial lipopolysaccharide. J Biol Chem 278(49):49428–49437
Weatherman KD, Augustine S, Christoff J et al (2013) Establishing benchmark rates of microbial and bacterial endotoxin contamination for radiopharmaceuticals compounded in commercial nuclear pharmacy settings. Int J Pharm Compd 17(2):168–174
Yamamoto A, Ochiai M, Fujiwara H et al (2000) Evaluation of the applicability of the bacterial endotoxin test to antibiotic products. Biologicals 28(3):155–167
Yeo DS, Ding JL, Ho B (1993) An antimicrobial factor from the plasma of the horseshoe crab, *Carcinoscorpius rotundicauda*. Microbios Lett 73:45–58
Yeo DS, Ding JL, Ho B (1996) Neuroblastoma cell culture assay shows that *Carcinoscorpius rotundicauda* haemolymph neutralizes tetrodotoxin. Toxicon 34(9):1054–1057
Zhang C, Liu WJ, Zhang ML et al (2014) Review on characteristics and detecting assay of bacterial endotoxin contamination in water environment. Huan jing ke xue = Huanjing kexue / [bian ji, Zhongguo ke xue yuan huan jing ke xue wei yuan hui "Huan jing ke xue" bian ji wei yuan hui] 35(4):1597–1601
Zhu Y, Thangamani S, Ho B et al (2005) The ancient origin of the complement system. EMBO J 24(2):382–394
Zhu Y, Ng PM, Wang L et al (2006) Diversity in lectins enables immune recognition and differentiation of wide spectrum of pathogens. Int Immunol 18(12):1671–1680

Chapter 30
Development of Aquaculture Methods to Enhance Horseshoe Crab Populations: An Example from Delaware Bay, U.S.A.

Brenda J. Landau, David R. Jones, Chester B. Zarnoch, and Mark L. Botton

Abstract The New Jersey Aquaculture Innovation Center (AIC) at Rutgers University has instituted a program to enhance the population of the American horseshoe crab *Limulus polyphemus* in the Delaware Bay Estuary (DBE) by using aquaculture at the earliest stages in its life history, when losses are greatest and production costs are minimal. Located on the Cape May Canal, the AIC pumps provide both raw and treated (sand-filtered, UV-sterilized) DBE water throughout the facility. Eggs of *L. polyphemus* were collected from local, sub-optimal nesting beaches impacted by coastal development and rising sea level, and hatched in jars normally used for trout eggs. A hatching system was built that uses recirculating treated seawater and a header tank to provide constant, independently controllable flow through 12 jars. An experiment performed using the hatching system tested the effects of egg stocking density and seawater flow rates on hatching success. The results showed highest hatching success at high egg densities with low flow rates. Hatchlings were reared in downweller silos in a raceway tank with flow-through, raw seawater. Half of the hatchlings were allowed to feed on organic matter in the incoming raw seawater while the others were given a supplement of *Artemia* nauplii. Statistically, there was no difference in growth or development between the treatments. After 10 weeks, most hatchlings were at the 3rd instar stage. This

B.J. Landau (✉)
Pendoley Environmental Pty Ltd, 12A Pitt Way, Booragoon, WA 6154, Australia
e-mail: brenda.landau@penv.com.au

D.R. Jones
The New Jersey Aquaculture Innovation Center at Rutgers University,
3920 Bayshore Road, Cape May, NJ 08204, USA

C.B. Zarnoch
Baruch College, City University of New York,
17 Lexington Avenue, New York, NY 10010, USA

M.L. Botton
Department of Natural Sciences, Fordham University,
113 West 60th Street, New York, NY 10023, USA

R.H. Carmichael et al. (eds.), *Changing Global Perspectives on Horseshoe Crab Biology, Conservation and Management*, DOI 10.1007/978-3-319-19542-1_30

suggests that aquaculture of the early stages of *L. polyphemus*, utilizing natural estuarine DBE water, maximizes the impact to the future population by protecting eggs and early instars, the most vulnerable life stages, at minimal cost.

Keywords *Limulus* aquaculture • Delaware Bay • Hatching methods • Hatching success • Stocking density • Flow rate • *Limulus* eggs • Supplemental feeding • Growth rate • Developmental index • Population enhancement

30.1 Introduction

The many challenges to the recovery of natural populations of the American horseshoe crab, *Limulus polyphemus*, in Delaware Bay (DBE; NJ, USA) and elsewhere have led to increased interest in aquaculture production of this species. There have been efforts to raise horseshoe crabs in laboratory culture, but these have typically been small-scale, research-oriented projects, not aimed at stock enhancement (e.g. Mishra 2009; Schreibman and Zarnoch 2009; Tsuchiya 2009). Additionally, these studies vary significantly in species studied, culture conditions, and reported response variables, making it difficult to advance the field from these efforts (Carmichael and Brush 2012). Thus, there is a need to more rigorously develop culture methodologies and test culture variables to support future aquaculture programs. Among the most important needs to improve culture efficiency is a detailed description of the design and performance of culture systems (Schreibman and Zarnoch 2009; Carmichael and Brush 2012) and development of feeds for the early instars (Schreibman and Zarnoch 2009; Tzafrir-Prag et al. 2010; Carmichael and Brush 2012; Hu et al. 2013).

Schreibman and Zarnoch (2009) used closed recirculating aquaculture systems with artificial seawater for over 7 years to produce juvenile *Limulus*. Their culture systems included egg hatching and juvenile grow-out systems. The egg hatching system used commercially-available McDonald hatching jars to culture *Limulus* from egg to 1st instar. The 1st instars were then transferred to downwellers, typically used in bivalve production, for 2–3 months of growth. After molting to the 2nd instar stage, juvenile *Limulus* were fed live brine shrimp nauplii (*Artemia* sp.). The instars were held in the recirculating downweller system for 2–3 months and grown to 0.6 cm in prosomal width. They were then transferred to a recirculating system with a layer of sand on the bottom of the tank and fed chopped clams. Although these systems were successfully used to culture *Limulus*, Schreibman and Zarnoch (2009) suggested that land-based systems that use natural seawater may improve efficiency of large-scale production because of non-costs for artificial sea-salts and the potential for suspended organic particulates in the seston to serve as a food source for the early instars (Gaines et al. 2002; Carmichael et al. 2009).

The best use of aquaculture may be to enhance survival of *Limulus* in its early, most vulnerable stages. A general rule of thumb is that, in the wild, there is an order of magnitude loss in numbers with each molt in the first year of life, from the first through fourth instars (Botton et al. 2003a). Therefore, potential impact of enhancement is maximized by culturing the vulnerable eggs and early post-hatch animals, while minimizing costs and space requirements as compared to those for the culture of larger animals. This concept was developed and tested at the New Jersey Aquaculture Innovation Center (AIC) in Cape May, New Jersey in 2012 (Fig. 30.1).

The aquaculture concept was designed to minimize costs by:

- culturing the smallest stages that would require the least space and food
- utilizing the population's natural spawning effort in Delaware Bay
- using ambient temperature
- using natural seawater
- using sunk costs (the AIC already has flow-through seawater [both raw and treated] systems, infrastructure and staff), and
- releasing the animals before over-wintering costs (heat, food, etc.) are incurred.

It is important to note that the AIC is located on Cape May Canal, and its seawater source is the same water as for the natural spawning effort in the Delaware Bay Estuary. While not investigated in this study, using the same water may also limit any adverse biological impacts from a non-natural source of water to, for example, imprinting signature, biochemical cues, and nutrition. The AIC pumps move raw (2 mm-filtered), ambient temperature seawater (salinity, 28–31) from Cape May Canal through the facility. Some of that water can be further treated (sand filtered, UV-irradiated, and heated) before being pumped through the facility; hereafter referred to as "treated seawater."

The *Limulus* culture system has two components, an egg hatching system and a post-hatch grow-out system. The designs for both are based on the systems used to hatch eggs and culture hatchlings as described by Schreibman and Zarnoch (2009). The initial stocking density of the hatching jars was 400 eggs/jar, and the flow rate was approximately 4 L/min. The grow-out system was used to culture the *Limulus* from 1st instar to 3rd/4th instar and was designed to utilize existing infrastructure normally employed to culture small oyster seed (post-metamorphosed larvae).

The current study was designed to build off of Schreibman and Zarnoch's (2009) efforts and to investigate the following questions regarding culture systems and large-scale production:

1. If we increase the stocking density from 400 eggs/jar, do we need to increase the flow rate to maintain the health of the culture?
2. If the hatched trilobites are grown out in raw seawater, do they need to be fed or can they survive and grow on the suspended particulate material in the raw seawater?
3. What is the overall capacity of this system?
4. Could this approach be easily replicated by other groups interested in horseshoe crab aquaculture?

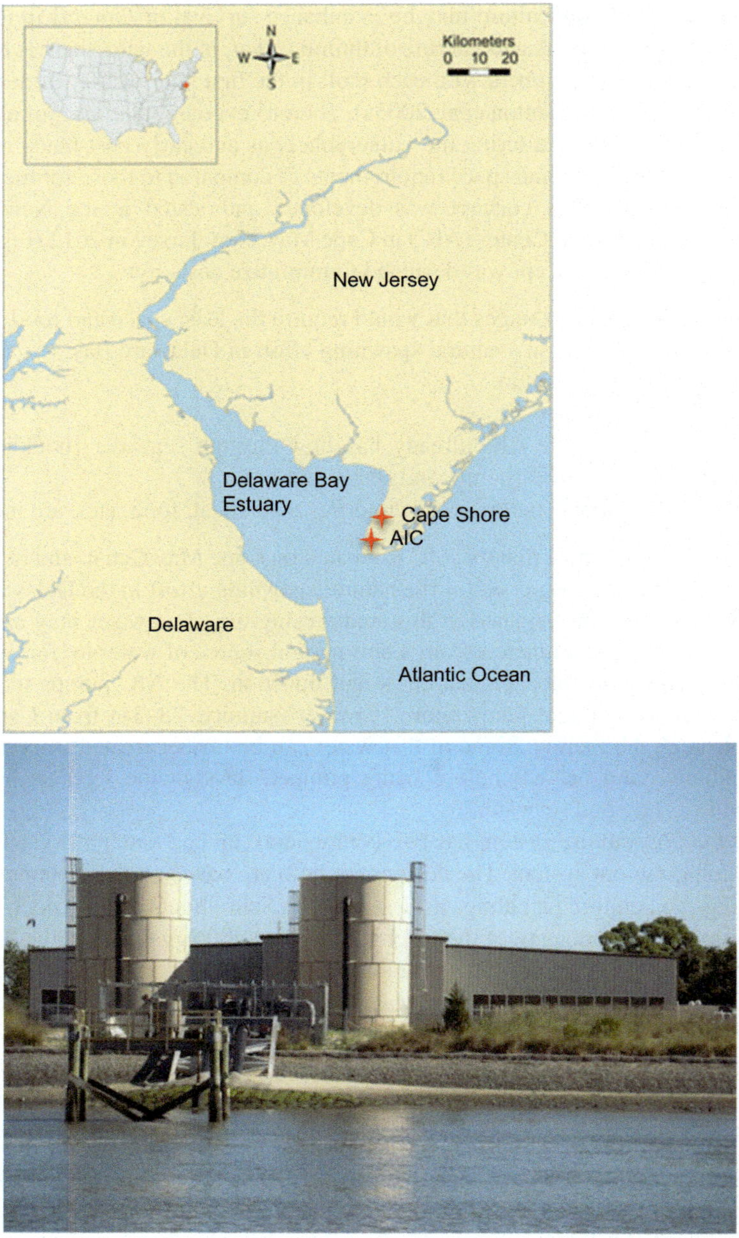

Fig. 30.1 Locations of *Limulus* collection beaches at the New Jersey Aquaculture Innovation Center (AIC) and Cape Shore, NJ on the Delaware Bay Estuary, USA. The photograph shows the AIC in Cape May, NJ, USA where the egg hatching and supplemental feeding experiments were conducted

Our culture systems were designed to perform experiments and collect data to address these questions. In addition to presenting the results of these experiments, we provide detailed descriptions of the culture systems to inform those interested in repeating and/or extending these results.

30.2 Methods and Materials

30.2.1 Egg Collection

Two collection sites on the New Jersey shore of Delaware Bay were selected, both for ease of access. The Cape Shore beach (https://goo.gl/maps/bT7XH; lat 39.074379, long -74.912000) is adjacent to the Cape Shore facilities of Rutgers University, and the AIC "beach" (https://goo.gl/maps/56ED4; lat 38.967651, long −74.942725), is a recent (since 2006) feature of accumulated sand beneath the AIC intake pipes along the Cape May Canal (Fig. 30.1).

Eggs from the Cape Shore location were collected using the method described by Brockmann (2003). There was a spawning event expected the night of the full moon on 4 June 2012. An hour before the nocturnal high tide, active nests were marked. About 2 h later, eggs from four different females were harvested from the marked nests. Females usually deposit a consecution of eggs along the beach slope, and we were careful to only take one of each female's clutches. Using this method we find that nests are easy to excavate, enough time has elapsed for fertilization to be completed, and the date of fertilization is certain.

Because the AIC beach is briefly intertidal, i.e., exposed only for a short time at low tide, a different approach to egg collection was taken. Authors Jones and Landau were onsite during the day at the AIC and able to frequently monitor any nesting activity occurring a short walk from the building. Nesting females were observed and a nest site was staked and excavated during low tide when the nest site was exposed.

The eggs collected from the Cape Shore and AIC beaches were cleaned of sand, weighed wet, pooled, and mixed such that a random sample of eggs would be 80 % from Cape Shore and 20 % from the AIC. No pre-selection of eggs, for example by color, was done. This was intentional, anticipating a future, large-scale implementation of the hatching system that would minimize both labor costs and the handling of eggs.

30.2.2 Egg Hatching System

The egg hatching system is a closed recirculating system. It is comprised of hatching jars to host the eggs, a shallow trough in which the hatching jars rest, and a water circulation system that was designed to allow control of the flow rate to individual hatching jars so that the consequences of flow rate to hatching success at different stocking densities could be investigated. We had three objectives: (1) model our system after that used by Schreibman and Zarnoch (2009); (2) build something affordable, modular and easy to use; and (3) build something with enduring practical value for both research and aquaculture. General features of the hatching system design include:

- constructed of low-cost materials
- unencumbered access to electricity, treated seawater, and drainage
- high visibility of the eggs and water volume for ease of monitoring
- easy access to the components for adjustments and cleaning, and
- fail-safe feature that protects eggs from desiccation if a pump fails or if there is an unintended draw-down of the sump.

The various components, overall footprint, and water circulation of the egg hatching system are illustrated in Figs. 30.2, 30.3, 30.4, and 30.5.

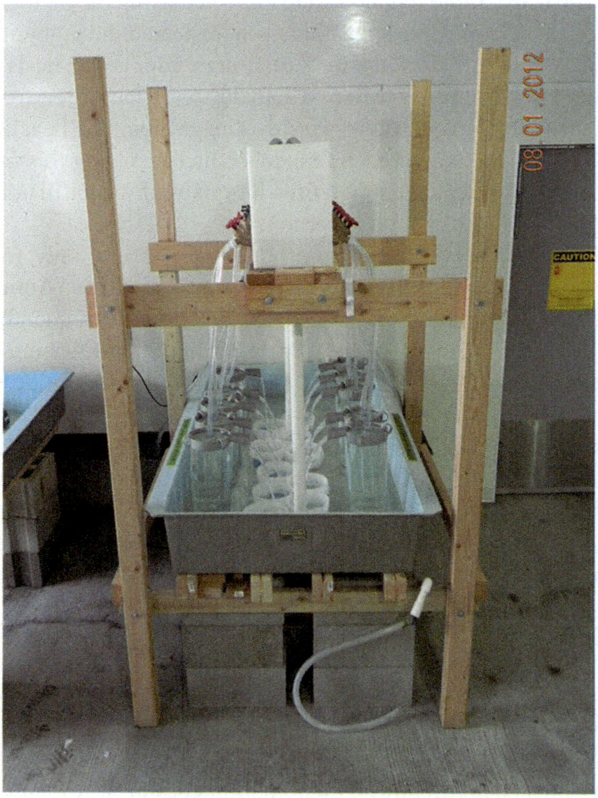

Fig. 30.2 Egg hatching system, showing header tank, hatching jars and catch tubs; sump not shown

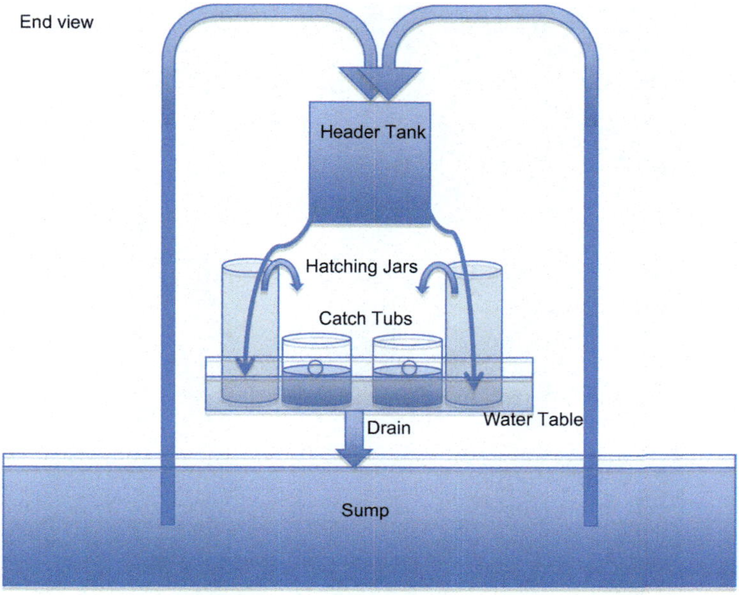

Fig. 30.3 Egg hatching system schematic, end view. Blue = treated DBE seawater. *Arrows* show direction of flow

30.2.2.1 Hatching Jars

The system was designed for 12 McDonald jars that are commercially available (Pentair/Aquatic Ecosystems, Apopka, Florida, USA). Each jar is 45.7 cm tall with a diameter of 15.8 cm. They are transparent and hold 6 L. A mesh screen at the top retains the eggs while water flows through the jar. The jars are designed to hold up to 100,000 trout eggs with a flow rate of 3.785 L/min to ensure the eggs are properly rotated. Water was delivered to the hatching jars from the header tank through 1.27 cm i.d. clear vinyl hose. Water enters the jars at the top, but a central pipe delivers it to the bottom creating vertical flow up to and out of the top of the jar. A standard plastic faucet valve controlled flow to each hatching jar, and individual jar flow rate was adjustable from 0 to 18.9 L/min. The McDonald jars, two rows of six, stood in a shallow (0.15 m deep × 0.9 m wide × 1.8 m long), fiberglass trough with a drain at one end that was positioned over the sump.

30.2.2.2 Water Component

A common volume of treated seawater was recirculated through the egg hatching system. This seawater passed through a sand filter (grain size, 0.45–0.55 mm in diameter; nominal particle retention, 20–100 μm) and was then exposed to ultraviolet light. Most of the volume resided in the sump, an open tank that was

Fig. 30.4 Sump features: (1) hoses to header tank where water flows back by gravity through hatching jars to the sump; (2) two submersible pumps return water to header tank; (3) standpipe maintains water level in sump, and can be rotated by hand to (4) floor drain

approximately 0.6 m deep × 0.6 m wide × 3 m long and held about 1,135 L. Two submersible pumps returned water from the sump through 1.9 cm plastic hoses to the header tank. Water in the header tank flowed by gravity through hoses into individual hatching jars at independent rates determined by the faucet valve opening dedicated to each jar. Water overflowing from the top of the hatching jars pooled in the shallow water table then drained into the sump, etc.

The level of the pooled water in the shallow trough can be controlled by standpipe height such that it can serve as a water bath (as needed) to moderate temperature fluctuations in the hatching jars.

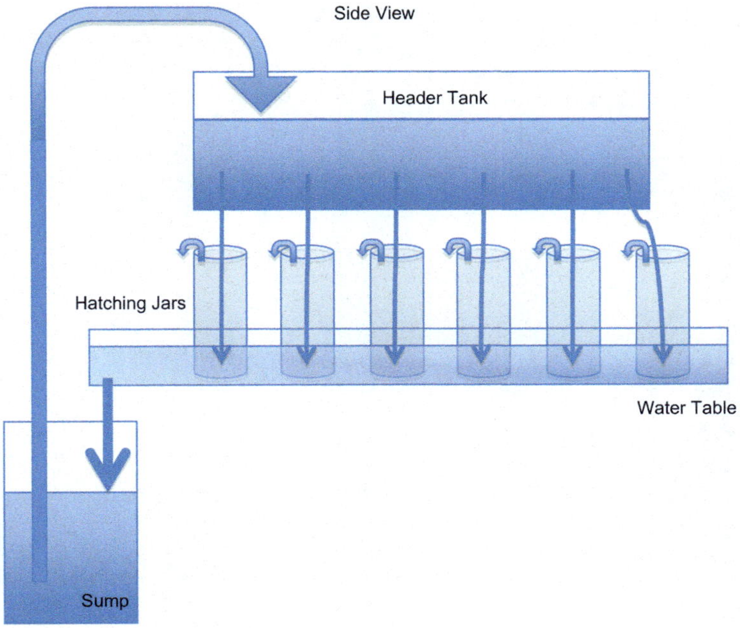

Fig. 30.5 Side view of water circulation through the egg hatching system. Blue = treated DBE seawater. *Arrows* show direction of flow

The resident volume was not heated or further treated because we were interested in system performance under ambient conditions (23–28 °C). If desired, and at greater expense, the water volume in the sump could be further heated, cooled, aerated or filtered.

30.2.2.3 Catch Tubs

Overflow from the McDonald jars was directed into "catch tubs". These tubs were made from disposable plastic 2 quart paint buckets with several 2.5 cm diameter drain holes punched through the sides at least 2.5 cm above the bottom. These buckets were lined with 3.8 L, fine-mesh bags (sold as paint strainers) with an integral elastic band edging that held them in place. Any material leaving the McDonald jars in the overflow water was caught in the mesh liner and held in these catch tubs for collection or disposal. If the mesh screen lid of the hatching jar was released while water circulated through the jar, swimming trilobites entrained in the vertical flow out of the jar were caught in the catch basin (Fig. 30.6), while the denser eggs remained near the bottom of the jar. Once the screen lid was again secured, the paint strainer was easily removed from the catch tub and its contents rinsed into a beaker; this is how animals were collected for counting and removal to the grow-out system.

Fig. 30.6 Hatching jars overflow into catch tubs lined with fine-mesh bags to collect outflow. The screen lid is clipped to the top of the hatching jar with PVC rings to retain eggs and hatchlings despite flow. Drain holes in sides of catch tubs maintain water to their level, the weight of which keeps the tubs in place

30.2.2.4 System Management

The water in the egg hatching system is a common volume so if infection or other undesirable contamination shows up in one jar, immediate action needs to be taken or all jars could become infected. The primary management option for contamination was to flush the jars (as described below) then replace the entire water volume. Daily management practice included careful visual inspection of the eggs and the system, and taking water quality measurements. Temperature, salinity, and dissolved oxygen (DO) were measured using a hand-held digital meter with a submersible probe (YSI 85; YSI Inc., Yellow Springs, Ohio, USA). In addition, a water sample was tested using a commercial kit (LaMotte Co., Chestertown, MD, USA) to monitor the total available nitrogen (TAN) in the system. The management goal was to keep TAN below 0.05 mg/L.

At least once a day, the entire water volume of the header tank and sump was replaced. The system was designed to make this easy thereby increasing the likelihood that it was actually done. To change the water, the pumps were shut off (plugs pulled from wall sockets) and the external standpipe on the sump tank was rotated (by hand) to empty into the floor drain. This quickly drained the header and sump tanks. Once empty, the header tank and sump could be hosed out and the vinyl tubing from the header to each jar could be cleaned more thoroughly as needed. The hatching jars would retain 6 L of water, but could be tipped with the mesh in-place to drain almost that entire volume. Once clean, the system was quickly refilled from the treated seawater input. One trained person could quickly and handily complete this process in 30 min.

Moreover, the system could be run in "flushing" mode as follows. The sump's standpipe was rotated below the sump tank height. A slow continuous input of fresh, treated seawater to the header tank would flush through the hatching jars before mixing into the resident volume in the sump, thereby diluting waste from the egg cultures in the jars, before flowing out the sump through the lowered standpipe; the lower the standpipe height, the faster the turnover of the resident volume and flushing of the system. Other management options to improve water quality include increasing the size of the reservoir and/or chilling the water volume.

30.2.2.5 Fail Safes

Staffing at the AIC during these experiments was for an 8 h workday, including weekends. Often, staff were around for 12 h, so the culture systems were usually unattended for 12 h every day. The system was designed with fail-safe features to minimize desiccation of the eggs in the event of a power failure or other unplanned event, especially when no one was on-site. Two submersible pumps were used to return water from the sump to the header tank so that flow would be maintained if one pump failed. The hatching jars empty at the top, so if the system experienced an unintended drain-down, a jar would retain 6 L of water. The mesh screen lids for the jars were secured with multiple clips made from PVC pipe that protected the contents should a jar unintentionally tip. Finally, the catch basins could be left in place to catch any escapees, while the placement of their drain holes ensured that the basins would not overflow yet would retain water to protect escapees.

30.2.3 Grow-Out System

The grow-out system is a flow-through system designed to make use of existing materials, infrastructure and the availability of raw DBE seawater at the AIC. The physical set-up for this system (Fig. 30.7) was based on the culture system described by Schreibman and Zarnoch (2009). As with the egg hatching system, the culture vessels, in this case downweller silos normally used in shellfish culture, were designed to be independent of each other. For example, an individual silo could be removed for either cleaning or to work-over the animals for data collection without interrupting the functioning of the other silos. The basic components of the system are a hosting tank, silos, and air lifts connected to a manifold that distributes air pressure equally to all ports. Both raw seawater and air were continuously delivered to the system.

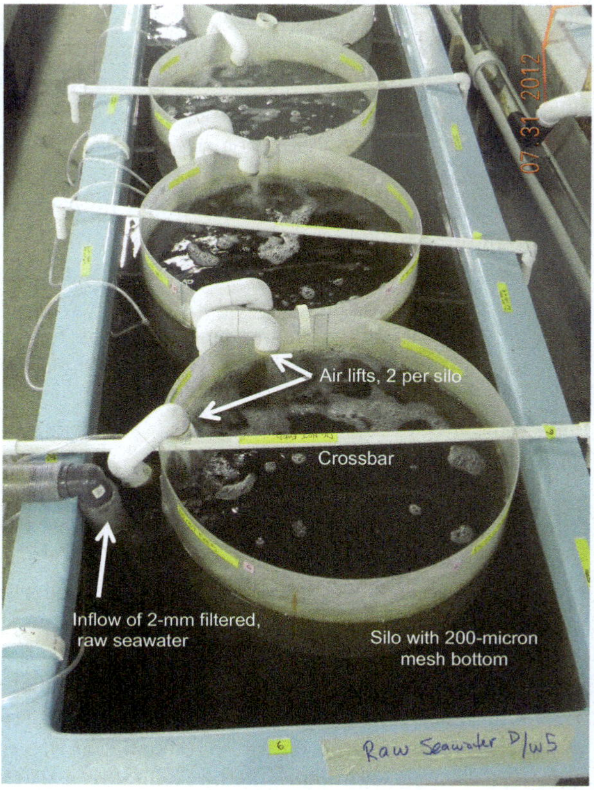

Fig. 30.7 The grow-out system comprised of silos suspended from a crossbar in a tank with flow-through raw seawater and two air lifts that circulate water vertically through each silo

30.2.3.1 Tank

A rectangular tank, 0.76 m wide by 0.61 m deep by 7.3 m long, was used to host the downweller silos. The water level was kept at a depth of about 0.45 m; therefore, the tank water volume was 2,548 L.

30.2.3.2 Silos

We used downweller silos typically employed in the grow-out of juvenile bivalves. They are designed to have water delivered to the top of the silo that leaves from the bottom. The cylindrical silos are 60 cm in diameter and 46 cm deep. A rod was slipped through two holes on opposite sides at the top of the silo and hung from the tank rim, suspending the silo in the tank (Fig. 30.7). As suspended, each silo held an average water depth of 25 cm for an average culture volume of about 70 L. The clearance between the silos and the bottom of the tank was about 20 cm. The mesh bottom of the silos had openings of 200 μm.

30.2.3.3 Air Lifts

Two air lifts constructed of 2.5 cm PVC pipe were hung on each silo so that water from below and outside the silo was pumped into the silo at the top, creating a vertical circulation in the silo. Each lift was attached to a port in the air supply manifold by aquarium tubing. The manifold supplied air at a constant pressure and valves on the ports allowed the rate of pumping by each air lift to be adjusted. Typical flow rates were about 4 L/min.

30.2.3.4 Culture Water

The inflow and outflow for raw DBE seawater were at opposite ends of the culture tank. The outflow was directed into an AIC floor drain that empties back into the Cape May Canal. A standpipe at the outflow set the water depth in the tank. The standpipe was collared, with holes at the base of the collar, so that the pull of water to the outflow came from the bottom of the tank. Raw seawater was delivered to the tank through a 5 cm i.d. hose at a rate of about 86 L/min. At this rate, it took about 89 min to replace 95 % of the water volume. This rate allowed continuous replacement of water in the tank while minimizing turbulence and re-suspension of sediments off the bottom of the tank.

30.2.3.5 Food Source

The AIC routinely grows out oyster seed (juveniles 2–12 mm) in a similar system. No supplemental food is added and the seed oysters do grow. Therefore, it was a reasonable assumption that the raw DBE seawater flowing through the AIC contained a natural food supply of phytoplankton and other suspended organic material that could be utilized by early *Limulus* instars. In treatments where feeding was supplemented (see below), it was with newly hatched brine shrimp (Pentair/Aquatic Ecosystems). Brine shrimp were used because they have been successfully used as a food source in other studies and because they are easy to produce.

30.2.3.6 System Management

The grow-out system was designed for ease of maintenance. The raw DBE seawater carries a sediment load that settled out of suspension when it entered the grow-out tank, so the tank was "dropped" daily. First, the inflow was deferred and the standpipe removed, which quickly emptied the tank. The accumulated sediment could be quickly hosed out. At the AIC, we have the luxury of a gravity-fed seawater supply that behaves like pressurized freshwater in a hose. Modifying the nozzle to produce a gentle spray allowed us to wash down the hatchlings with seawater. Once the cleaning was done, the inflow of raw DBE seawater was restored and the standpipe replaced; the tank quickly refilled. This procedure can (and should) be done quickly so that the *Limulus* hatchlings can remain in place in their silos without the stress of exposure or excessive handling.

Because the grow-out system was flow-through and the accumulated sediment was removed daily, concern for contamination, invasion or infection by undesirable biologics was minimal. The air lifts ensured that the water in the silos was highly oxygenated (DO measurements were always at saturation). The rough filtration at 2 mm (done at the intake for the raw seawater system) ensured that large predators or competitors were not introduced to the tank. Daily temperature and salinity readings were taken as part of the routine check of all of the animal culture systems at the AIC. Unlike in our closed systems, we had no cause to think that TAN could accumulate; therefore, no routine water quality tests for nitrogen species were done. Because the approach is minimalist, the main management options to correct water quality issues for this kind of system are a complete water change with a tank/silo cleaning and/or a reduction in stocking density.

30.2.3.7 Fail Safes

If the mesh in a downweller silo clogged without correction, the silo would overflow, allowing *Limulus* to escape into the tank where they could eventually flow out the standpipe into the floor drain thence back into the Cape May Canal unharmed. If the system was cleaned daily and lightly monitored, silos did not clog and overflow. An unplanned release due to clogged mesh did happen after the completion of the supplemental food experiment.

30.2.4 Context of Culture Experiments

Aquaculturists are interested in the capacity of their systems so they can adjust stocking density accordingly. Knowing the capacity is a way to maximize the use of the system and to determine the costs for programs that use those systems. For the egg hatching system, we wanted to find out if the flow rate of water through the system had to be increased if the egg stocking density was increased in order to maintain adequate viability and hatching success.

A major cost associated with the aquaculture of animals is feed. An attractive feature of the culture of early life stages of *Limulus* is that food costs are minimal; they do not begin feeding until the first post-hatch molt (Brown and Clapper 1981). Because the flow of raw DBE seawater through the AIC facility is a sunk cost, we wanted to investigate the necessity of having to feed post-hatch *Limulus* cultured in 2 mm-filtered raw DBE seawater in a flow-through system. Typical of molting animals, growth of *Limulus* is a step function. Furthermore, for the early molts of *Limulus*, growth steps or stages are also developmental. For example, the post-hatch trilobite has a tail stub, the next instar has a longer tail that is fused to the

Table 30.1 Experimental design for investigating the effects of egg stocking density and flow rate on hatching success, including the hatching success results

	Treatment 1 (Control)	Treatment 2	Treatment 3
Stocking density	400 eggs/jar	4,000 eggs/jar	4,000 eggs/jar
Flow rate	4 L/min	4 L/min	12 L/min
No. of replicates	3	3	3
Proportion hatched	0.7517	0.8129	0.6996

body, and the next instar has a tail that articulates. Therefore, a supplemental feeding experiment was designed that compared growth and development of post-hatch *Limulus*.

30.2.4.1 Stocking Density Experiment

The experimental design had three treatments (Table 30.1). Treatment 1 was the control; it had the same stocking density (400 eggs/jar) and flow rate (4 L/min) described by Schreibman and Zarnoch (2009). Treatment 2 had the same flow rate but 10 times the number of eggs (4,000). This number was chosen because it is close to the average of 3,650 eggs per clutch reported by Shuster and Sekiguchi (2003). Both the stocking density and the flow rate were increased for Treatment 3, to 4,000 eggs/jar and a flow rate of 12 L/min. There were three replicates of each treatment.

Eggs were counted out for the three replicates with an initial stocking density of 400 eggs/jar. The larger stocking density (4,000 eggs/jar) was apportioned by weight, using an empirically determined average egg weight.

Every day thereafter, for 21 days, hatched trilobites were collected from each jar and tallied. This data was used to determine the overall hatch rate for each treatment. Hatched trilobites were transferred to the grow-out system.

30.2.4.2 Supplemental Feeding Experiment

The supplemental feeding experiment included two treatments with three replicates each. Treatment 1 (T1) involved no supplemental feeding. Treatment 2 (T2) involved a supplemental feeding of newly hatched brine shrimp to each replicate silo at 1.5 % (wet weight) of the total silo biomass once each week. The three silos for T1 were positioned nearest the inflow of raw DBE seawater. They were labelled so that their positions could be retained throughout the experiment. The three silos for T2 were positioned down-flow of those for T1. They were also labelled so that their positions could be retained throughout the experiment (Fig. 30.8).

The animals for this experiment were collected from the egg hatching experiment. The majority of these larvae had hatched on Day 16 post-collection, so they were 1 week old at the start of the supplemental food experiment. To randomize

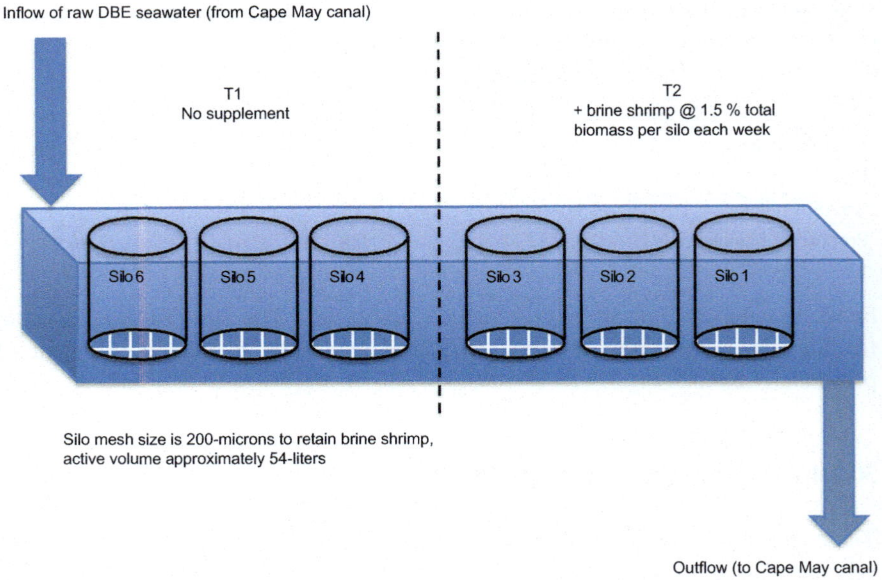

Fig. 30.8 Design of the supplemental feeding experiment

variation, the hatchlings were pooled, weighed wet and divided equally by weight into six aliquots, one for each treatment silo. Based on the results of the egg hatching experiment, there were roughly 1,000 hatchlings in each silo. The approximate liquid volume of the silos, in-system, was 70 L; therefore, the initial stocking density was about 14 animals/L.

The 200 μm mesh size for the bottom of the silos was chosen because it would retain the brine shrimp. All six experimental silos (T1 and T2) were of the same size and had the same size mesh bottom.

The supplemental feeding treatment (T2) silos were fed newly hatched brine shrimp nauplii at the prescribed amount once each week. Addition of the brine shrimp would trigger a feeding response behavior in the hatchlings; they became noticeably more active. The hatchlings could be observed feeding on the brine shrimp and would take on the bright orange hue of the brine shrimp after feeding on them.

For weeks 1–3, the starting biomass (weight of the initial stock) was used to determine the weight of brine shrimp to add to the T2 silos instead of handling the animals to get actual weights. This was because the animals were deemed too delicate to safely handle and we wanted to avoid excessive handling of the animals. Starting in week 4, the animals appeared robust enough to collect, clean and weigh. During this procedure, the contents of each silo were poured onto a screen and gently washed with treated seawater before being emptied into a mesh bag. Most of the excess water was allowed to drip off the bag before the bag and contents were weighed. Non-*Limulus* material that could easily be picked out of the culture was removed at this time.

Flow rates for the air lifts were set so that water circulation in the silo was not so vigorous that the animals had to work to maintain their position in the water column. Behavioral observations of individual *Limulus* in each silo were used to set the air lift flow. Our reasoning was that the hatchlings would grow best if they were not expending superfluous energy.

Data for hatchlings in each silo were collected once every week for the next 10 weeks as follows. Hatchlings were suspended in the water column by gently swirling the water in the silo by hand before scooping a sample out with a beaker. From this sample, 100 animals were haphazardly selected and sorted by instar stage. The number of individuals in each instar group was counted (n) and the group was weighed wet (g). After data collection, the animals were returned to their respective silos.

30.3 Results and Discussion

30.3.1 Stocking Density Experiment

The proportion of hatched eggs for each treatment was determined as the sum of the hatched larvae from each replicate (3) divided by 1,200 for Treatment 1 and by 12,000 for Treatments 2 and 3 (Table 30.1). Treatment 2 had the highest hatch rate, 81.3 %, and Treatment 3 had the lowest hatch rate, 70.0 %. Treatment 1 had an intermediate hatch rate of 75.2 %.

Hatching rates were compared following the method of Pazer and Swanson (1972). By this method, $\chi^2 = 417.954$ ($P < 0.0001$); therefore, the null hypothesis of no difference was rejected, i.e., there is a difference between these hatching rates.

Because there was a difference, we used a Z-test to see if there was a difference in hatching success between the two best treatments (Treatment 1 and Treatment 2). We found $Z = 4.71046$ ($P < 0.0001$), so the null hypothesis was rejected, i.e., Treatment 2 was significantly better than Treatment 1.

In a natural intertidal nest, waste material from the eggs would be flushed on the tidal cycle. New water would also replenish oxygen around the eggs. Additionally, agitation of eggs with sand has been shown to enhance hatching success (Ehlinger and Tankersley 2003). Thus, it was reasonable to hypothesize that a tenfold increase in stocking density would require an increased flow rate. Our results suggest otherwise and invite study about the potential use of hatching jars in large-scale aquaculture production of *Limulus*.

Besides the removal of waste material from and delivery of oxygen to the developing eggs, there may also be antibiotic properties in the spawned material (Loveland 2007) surrounding the eggs. Development may also be harmed if the egg is moved too much at key times during development, as seen in other in-beach nesters—the green sea turtle (Parmenter 1980) and the diamondback terrapin (J. Rokita pers. obs.). Until the effect of excessive agitation on *Limulus* eggs is studied, lower flow rates may be desirable, with concomitant attention paid to maintaining water quality as stocking density increases. It could be that the higher stocking density would minimize agitation due to water flow through the hatching jar. These and other issues should be of concern for an aquaculture system designed to replace the natural nest processes.

30.3.2 Supplemental Feeding Experiment

Measurements performed each week on the grow-out system allowed for determination of an average weight for each stage, and because the sample from each silo was random, the percentage at each instar stage could be applied to the silo population as a whole (Table 30.2). We did not perform total counts for the silos, and so we could not quantify survival. Throughout the course of this study, we did not observe any dead *Limulus* in the samples counted or any evidence of dead *Limulus* in the silos during routine care and observation.

The data generated from this experiment were used to compare growth and development between the two treatments as a way to investigate the value of supplemental feeding during the grow-out of post-hatch *Limulus*. An index was created to compare development rates between Treatment 1 (no supplement) and Treatment 2 (supplement), which we refer to as the Instar Developmental Index (IDI). The equation for the IDI was

$$\mathrm{IDI} = \left(\%\mathrm{1st\,instar} \times 1\right) + \left(\%\mathrm{2nd\,instar} \times 2\right) + \left(\%\mathrm{3rd\,instar} \times 3\right) + \left(\%\mathrm{4th\,instar} \times 4\right).$$

The IDI for T1 and T2 was calculated weekly (Fig. 30.9). The differences in the paired (by week) IDI were compared using the non-parametric Wilcoxon test with the null hypothesis of no difference between the treatment groups. We found $t=16.00$ ($P=0.2754$); therefore, the null hypothesis was not rejected, i.e., there was no statistically significant difference in development between the treatments over the experimental period.

Mean sample weights (Fig. 30.10) were used to compare growth between the two treatments. First, regression was used to derive linear equations for mean weight as a function of time defined as week number from 1 to 10. The derived equations were

- $\mathrm{Mean\ weightT1} = 1.511 + 0.159 \times \mathrm{week\ number}$
- $\mathrm{Mean\ weightT2} = 1.603 + 0.147 \times \mathrm{week\ number}$.

Slopes (growth rates) were significantly different from zero (ANOVA; $P=0.010$ for T1 and $P=0.0004$ for T2); therefore, the slopes were compared. There was no difference between the slopes, thus no difference in the rate of mean weight change between the treatments ($t=0.230$, $P=0.823$).

In week one, 100 % of the hatchlings in the samples from both treatment groups were 1st or 2nd instars. By week ten, 87 % (73–100 %) of the hatchlings in the samples from Treatment 1 were 3rd or 4th instars compared to 46 % (11–74 %) from Treatment 2 (Table 30.2).

Table 30.2 Pooled data from the Supplemental Feeding Experiment: Treatment 1 (T1; raw seawater) and Treatment 2 (T2; *Artemia* nauplii supplement)

	% 1st instar		% 2nd instar		% 3rd instar		% 4th instar	
Week	T1	T2	T1	T2	T1	T2	T1	T2
1	24.00±5.20	23.67±5.77	76.00±5.20	76.33±5.77				
2	6.00±1.00	5.67±1.53	94.00±1.00	94.33±1.53				
3		0.67±0.58	97.67±3.21	97.33±1.53	2.33±3.21	2.00±1.00		
4			99.67±0.58	100.00±0.00	0.33±0.58			
5			99.33±0.58	98.33±2.89	0.67±0.58	1.67±2.89		
6			99.33±0.58	91.33±11.72	0.67±0.58	8.67±11.72		
7	0.33±0.58	0.33±0.58	99.33±1.15	89.33±12.22	0.33±0.58	10.33±12.34		
8			94.33±4.16	80.33±19.14	5.67±4.16	19.67±19.14		
9			55.67±10.50	57.33±27.47	44.33±10.50	38.67±24.09		4.00±5.29
10			13.00±13.53	54.00±38.02	57.33±40.82	35.33±21.96	29.67±51.38	10.67±12.66

Average proportion (%) and standard deviation reported for weeks 1–10 by instar stage

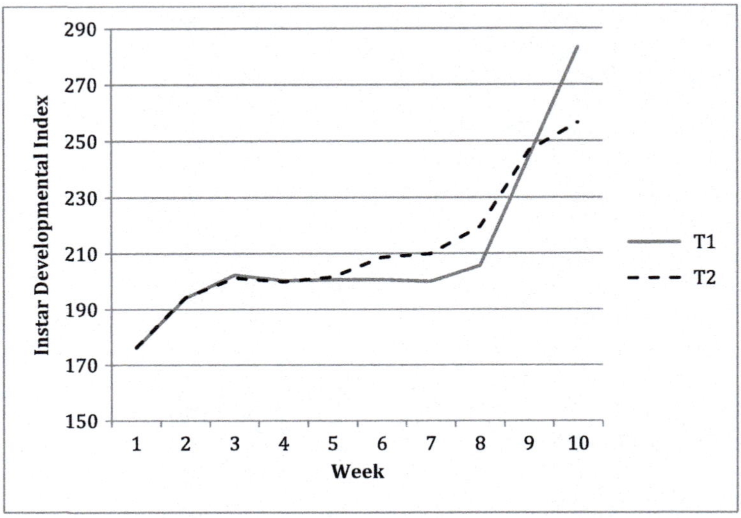

Fig. 30.9 Weekly values of the Instar Developmental Index for treatments 1 (non-supplemented) and 2 (supplemented)

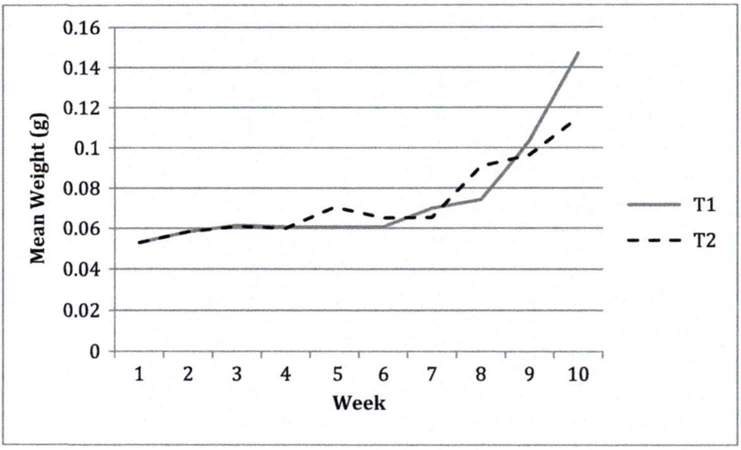

Fig. 30.10 Mean weight (g) of sample for treatments 1 (non-supplemented) and 2 (supplemented)

30.3.3 System Potential

The MacDonald jars are designed to host 100,000 trout eggs. Trout eggs range from 3.5 to 4.5 mm in diameter (http://www.softmilkingegg.com/Fish-egg-sizes.html) and the average diameter of *Limulus* embryos immediately prior to hatching is 3.6 mm (Shuster and Sekiguchi 2003). Given that the flow rate in the hatching jars may not need to be increased as stocking density increases, and the similarity in size to trout eggs, the capacity of the egg hatching system might well be 100,000 eggs per jar. If so, then each hatching system like the one we designed can produce 5,760,000 hatchlings/year based on the following parameters:

- 12 McDonald jars each stocked at 100,000 eggs
- An 80 % hatching success rate
- An average hatching period of 21 days, thus six hatching periods between mid-May and mid-September.

If the trilobites hatched from eggs in this system, they (ca. 80 %) were immediately released to Delaware Bay, the differential compared to the putative natural survival (one out of ten) puts eight times more trilobites into the population. And if these eggs were collected from sub-optimal locations (for hatching), the impact would be greater. Currently, it is impractical to culture horseshoe crabs through the adult stages, which is the stage required for the pharmaceutical and bait industries, due to their slow maturation time (Shuster and Sekiguchi 2003).

The impact of a *Limulus* aquaculture and stocking program on local populations could be assessed through large releases of hatchlings into areas of restricted circulation that can be routinely sampled over time (every year for a decade or more). More expensive technologies, based on genetics or a mark and recapture program could also be employed. The lack of a definitive way to assess the impact on the population, however, should not impede the use of aquaculture to release more early instars into the population since the costs are not onerous.

The potential for maintaining *Limulus* beyond the 2nd and 3rd instars can be assessed after the grow-out system is better studied. For example, survival rates for each instar stage need to be determined and stocking density must be optimized. For now, the use of the two systems described here for egg hatching and grow-out provides a sound aquaculture base for enhancing the natural population of *Limulus*. In conjunction with that effort, it can provide a system and a source of animals for research and, with experience gained, it can underpin the extension of culture into overwintering and for older instars for scientific research.

30.3.4 Future Research

Several ideas for future research come to mind from the design and use of the egg hatching and grow-out systems. Amongst the obvious are experiments to test the capacity of the systems, for example, the maximum stocking densities of both the jars and the silos, and the effects of water quality parameters like temperature or residence time on the outcomes.

The quality and quantity of the food found in the raw seawater needs to be assessed, and quick, standardized methodologies for characterizing this food supply with respect to early stage *Limulus* need to be developed. This would allow for dynamic management decisions to be made about flow rates to increase or decrease the residence time of water volumes and the food contained therein, to concentrate particulate organic matter, or the need for supplemental feeding.

In the supplemented silos, *Limulus* gorged themselves on brine shrimp. If you are what you eat, then the nutritional content of brine shrimp, especially if eaten preferentially, may displace more nutritious choices and unbalance the *Limulus* diet. Thus, the nutritional diversity of potential food supplements should be investigated (Carmichael and Brush 2012). If brine shrimp remain a preferred food, the nutritional content can be varied by choice of strain and/or the food that brine shrimp eat. Production costs should be included in any investigation of food supplements.

The lower hatching success associated with increased flow rate should be investigated. It is possible that physical deformities in trilobites are related to the level of agitation during egg incubation. This could easily be tested using the current system. The types and timing of deformities could be informative to *Limulus* biology.

The design of the egg hatching system, particularly the use of independent jars, allows for studies to explore how egg variables may influence hatching success. For example, one may compare hatching rates for eggs from different environments such as those from optimal vs. sub-optimal nesting sites, or for eggs collected from the same nesting site in different months, or for eggs laid by the same female but from successive clutches, etc.

The behavior of the hatchlings in the grow-out system could be monitored in relation to photoperiod, tidal cycles, and interactions among individuals (i.e. grooming, competition) and then related to growth.

Finally, the nutritional impacts of a natural food supply can be studied, as the raw seawater used in the grow-out system is a reasonable proxy for what is naturally available. Future efforts will need to characterize the seston (i.e., chlorophyll-a, organic content, C:N) and quantify feeding rates in order to improve the efficiency of feeding early instars using raw seawater.

30.4 Summary

We have demonstrated that aquaculture of *Limulus polyphemus* can be performed in a cost-effective way if the youngest stages are cultured, the natural spawning cycle is utilized, the sunk costs of the AIC are employed, local (DBE) seawater is used at ambient temperatures, and the hatchlings are released before feeding or over-wintering become necessary. The impact is further enhanced by protecting the most vulnerable stages from predation and by collecting eggs from sub-optimal locations.

The results from the egg hatching experiment suggest that the hatching jars can be stocked at high densities with comparatively low flow rate so long as water quality is maintained. The results from the supplemental feeding experiment suggest that, if coincident with the natural spawning season, early post-hatch stages (instars) can be reared in coarsely strained raw seawater at ambient conditions without supplemental feeding up to the 3rd instar stage. The standard diet for early instars has been *Artemia* nauplii as previous research has demonstrated good growth performance (Mishra 2009; Schreibmen and Zarnoch 2009; Hu et al. 2013); however, the cost and labor associated with using this feed may be prohibitive for large-scale aquaculture production. The results presented here show similar growth between early instars fed *Artemia* nauplii and those surviving on suspended organic material from raw seawater. This is a significant advancement in *Limulus* culture since it will improve economic efficiency. The result is also consistent with previous studies that suggest that early instars feed upon microalgae and other detrital material in the field (Gaines et al. 2002; Botton et al. 2003b; Carmichael et al. 2009). Moreover both hatching and grow-out systems were designed to be constructed and operated using easily available, mostly non-specialized materials, and to make their use and upkeep easy.

Acknowledgments We thank Dr. Matthew Landau, Richard Stockton College of New Jersey, who designed the instar developmental index and did all of the statistical analyses. We also thank Drs. Margaret Brennan-Tonetta and Bradley Hillman from the Agriculture Experimental Station at Rutgers University for program support. A DuPont Clear Into the Future grant partially funded this project.

References

Botton ML, Loveland RE, Tiwari A (2003a) Recruitment and survival of young-of-the-year in a commercially exploited population of horseshoe crabs (*Limulus polyphemus*). Mar Ecol Prog Ser 265:175–184

Botton ML, Shuster CN Jr, Keinath J (2003b) Horseshoe crabs in a food web: who eats whom? In: Shuster CN Jr, Barlow RB, Brockmann HJ (eds) The American horseshoe crab. Harvard University Press, Cambridge, MA, pp 133–153

Brockmann HJ (2003) Nesting behavior: a shoreline phenomenon. In: Shuster CN Jr, Barlow RB, Brockmann HJ (eds) The American horseshoe crab. Harvard University Press, Cambridge, MA, pp 333–347

Brown GG, Clapper DL (1981) Procedures for maintaining adults, collecting gametes, and culturing embryos and juveniles of the horseshoe crab, *Limulus polyphemus* (L.). In: Hinegardner R, Atz J, Fay R, Fingerman M, Josephson R, Meinkoth N (eds) Laboratory animal management, marine invertebrates. National Academy Press, Washington, DC, pp 268–290

Carmichael RH, Brush E (2012) Three decades of horseshoe crab rearing: a review of conditions for captive growth and survival. Rev Aquac 4:32–43

Carmichael RH, Gaines E, Sheller Z et al (2009) Diet composition of juvenile horseshoe crabs: implications for growth and survival of natural and cultured stocks. In: Tanacredi JT, Botton ML, Smith DR (eds) Biology and conservation of horseshoe crabs. Springer, New York, pp 521–534

Ehlinger GS, Tankersley RA (2003) Larval hatching in the horseshoe crab, *Limulus polyphemus*: facilitation by environmental cues. J Exp Mar Biol Ecol 292:199–212

Gaines EF, Carmichael RH, Grady SP et al (2002) Stable isotopic evidence for changing nutritional sources of juvenile horseshoe crabs. Biol Bull 203:228–230

Hu M, Wang Y, Cheung SG et al (2013) Comparison of different frozen natural foods on survival and growth of juvenile Chinese horseshoe crab *Tachypleus tridentatus* (Leach, 1819): implications on laboratory culture. Aquac Res 44:567–573

Loveland RE (2007) Comment made at the 2007 international horseshoe crab conference. Dowling College, New York

Mishra JK (2009) Larval culture of *Tachypleus gigas* and its molting behavior under laboratory conditions. In: Tanacredi JT, Botton ML, Smith DR (eds) Biology and conservation of horseshoe crabs. Springer, New York, pp 513–519

Parmenter CJ (1980) Incubation of the eggs of the green sea turtle, *Chelonia mydas*, in Torres Strait, Australia: the effect of movement on the hatchability. Aust Wildl Res 7:487–491

Pazer HL, Swanson LA (1972) Modern methods for statistical analysis. Intext Educational Publishers, Scranton

Schreibman MP, Zarnoch CB (2009) Aquaculture methods and early growth of juvenile horseshoe crabs (*Limulus Polyphemus*). In: Tanacredi JT, Botton ML, Smith DR (eds) Biology and conservation of horseshoe crabs. Springer, New York, pp 501–511

Shuster CN Jr, Sekiguchi K (2003) Growing up takes about ten years and eighteen stages. In: Shuster CN Jr, Barlow RB, Brockmann HJ (eds) The American horseshoe crab. Harvard University Press, Cambridge, MA, pp 103–130

Tsuchiya K (2009) The history of horseshoe crab research and conservation in Japan. In: Tanacredi JT, Botton ML, Smith DR (eds) Biology and conservation of horseshoe crabs. Springer, New York, pp 559–570

Tzafrir-Prag T, Schreibman MP, Lupatsch I et al (2010) Preliminary studies of energy and protein requirements of Atlantic horseshoe crabs, *Limulus polyphemus*, grown in captivity. J World Aquac Soc 41:874–883

Chapter 31
Horseshoe Crab Research in Urban Estuaries: Challenges and Opportunities

Jennifer H. Mattei, Mark L. Botton, Mark A. Beekey, and Christina P. Colón

Abstract Horseshoe crabs rely on estuaries for food resources, places to spawn and for larvae and juveniles to develop and grow. Many of these estuaries are becoming increasingly urbanized and dominated by human activity. An urban estuary is characterized by armored shorelines, high nutrient loads, large fluctuations in algal and bacteria populations, increased levels of pollutants like heavy metals and pesticides, and seasonally low oxygen levels and pH. While urban estuaries are challenging for horseshoe crab survival and to researchers trying to study them, there are also opportunities for involving the public in research and increasing public awareness of the importance of the conservation and survival of horseshoe crabs. Two recent studies in New York and Connecticut have involved citizen scientists to tag and gather valuable data on horseshoe crab population dynamics. It has been discovered that there is very low recruitment of new adults into the two urban estuaries studied and that spawning populations of horseshoe crabs are relatively low when compared to less disturbed more rural estuaries.

Keywords Urban estuary • Eutrophication • Acidification • Armored shoreline • Jamaica Bay NY • New Haven CT • Citizen science • Civic Engagement • Volunteer-based projects • Carapace condition

J.H. Mattei (✉) • M.A. Beekey
Department of Biology, Sacred Heart University,
5151 Park Avenue, Fairfield, CT 06825, USA
e-mail: matteij@sacredheart.edu

M.L. Botton
Department of Natural Sciences, Fordham University,
113 West 60th Street, New York, NY 10023, USA

C.P. Colón
Department of Biological Sciences, Kingsborough Community College,
2001 Oriental Boulevard, Brooklyn, NY 11235, USA

R.H. Carmichael et al. (eds.), *Changing Global Perspectives on Horseshoe Crab Biology, Conservation and Management*, DOI 10.1007/978-3-319-19542-1_31

31.1 Introduction

Both horseshoe crabs and human beings are estuarine-dependent species. Horseshoe crabs use estuaries as spawning and feeding habitat for adults (Brockmann 2003; Botton 2009), a site for egg development (Botton et al. 2010), and as juvenile nursery areas (Botton et al. 2003; Carmichael et al. 2003; Burton et al. 2009). Humans depend on estuaries for many ecosystem services including transportation, recreation, food and water resources, waste disposal, industry, energy needs and community development (Mattei 2005; Barbier et al. 2011). With some 123.3 million people in the US (39 % of the total population) living in coastal counties (National Ocean Service 2014) and well over three billion people worldwide found living within 200 km of a coastline (Creel 2003; Valiela 2009), it was inevitable that horseshoe crabs and humans would come into conflict over the use of the coastal zone. Sea walls, bulkheads, groins, and jetties, ubiquitous coastal protection measures used in virtually all of our urban estuaries, may impede or preclude access by horseshoe crabs to critical beach spawning habitats (Botton 2001; Chiu and Morton 2003; Jackson and Nordstrom 2009; Tsuchiya 2009; Jackson et al. 2010). Water quality in urban estuaries is commonly affected by municipal, industrial, and non-point source pollution carried by stormwater runoff and ground water discharges. Pollutants often include multiple biotic stressors such as heavy metals, pesticides, pharmaceuticals, plastics, and raw sewage. Pollution can change nutrient cycling which may result in large fluctuations in algal and bacterial populations and seasonal decline in dissolved oxygen levels, i.e. hypoxia (Varekamp et al. 2014). These common conditions of urban seas may have adverse effects on the developmental success of horseshoe crab eggs, growth and survival of juveniles, and survival of adults (Botton and Itow 2009; Beekey and Mattei 2015).

Much of the literature on American horseshoe crabs is based on studies conducted in less urban, i.e., less disturbed habitats, such as lower Delaware Bay (Shuster and Botton 1985; Smith et al. 2002), Cape Cod (Barlow et al. 1986), or the Gulf Coast of Florida (Rudloe 1980; Brockmann and Johnson 2011), with far fewer studies from urban sites (e.g. Jamaica Bay, New York, Botton et al. 2006a and New Haven Harbor, Connecticut, Mattei et al. 2010; Beekey et al. 2013). With respect to many biotic and abiotic characteristics, urban estuaries significantly differ with the intensity of coastal development and accompanying issues pertaining to habitat and water quality, access to research sites, and anthropogenic influences on research sites. Urban estuaries present challenges to both horseshoe crabs and those who study them, but we suggest that they offer opportunities that in many cases can enhance data collection and ultimately conservation of a species under threat living close to our cities. Here, we discuss the challenges and opportunities that urban estuaries present for conducting research on horseshoe crabs.

31.2 Urban Estuaries

By definition, urban ecology concerns the study of organisms and their interactions in mainly terrestrial sites or within watersheds based on the predominance of buildings and infrastructure, coupled with dense human populations (McDonnell and Pickett 1990; Mayer et al. 2010). In contrast, most ecological studies are concentrated in less urban or rural areas, where sites have little or no human infrastructure and low human population density. Comparative studies are based on a gradient of human impact. Due to the very nature of ecosystems having diffuse boundaries, very few to no habitats would be considered pristine or without some human influence (Ellis et al. 2013). Many papers have been published based on comparing and contrasting ecological processes on this urban-rural gradient (Niemela et al. 2011). It would logically follow then that urban estuaries and seas are defined by the comparatively high level of human use and development of the surrounding areas and would also warrant study and comparison to less urbanized estuaries. When studying horseshoe crabs or other urban estuarine biota, many factors can lead to changes in population dynamics, species behavior, health and interactions with other species that may differ from their counterparts in rural estuaries (Mattei et al. 2010).

Urban estuaries differ substantially from more rural estuaries by the type of human disturbance and the level of disturbance. For example, as human populations grow in cities along the coast, the volume of nitrogenous waste and carbon load increases proportionately. Discharge from wastewater treatment facilities supplies over 53,000 tons per year of nitrogenous waste products to Long Island Sound (Lopez et al. 2014). In fact, the watershed for the Sound includes five states, all of which compounds the waste problem. Other sources of nitrogenous waste or excess nitrogen include: effluent from industrial facilities, fertilizers from farms, lawns, and gardens, as well as waste produced by domesticated animals (from cows to cats and dogs); all eventually ending up in our urban seas like Jamaica Bay (JB) and Long Island Sound (LIS). One of the more biologically deleterious effects of nitrogenous waste and macronutrient loading is hypoxia. Nitrogenous waste has a fertilizing effect on JB and LIS waters, causing algal blooms. As the algae die and sink to the bottom, bacteria decompose the organic matter and deplete the oxygen within the bottom two thirds of the water column. Surface waters have dissolved oxygen because of the close contact with the atmosphere. Hypoxia usually occurs in the western half of LIS including New Haven Harbor (NHH) and in the upper eastern sections of JB during the months of July through October (Wallace et al. 2014). Most of the organisms (fish, crabs, shellfish, and other invertebrates) either die or leave the affected area (Lopez et al. 2014; Varekamp et al. 2014). Remarkably, horseshoe crab adults and juveniles can survive very low levels of dissolved oxygen, and large die-offs in Western LIS have not been reported (Botton et al. 2010, Penny Howell, Connecticut Department of Energy and Environmental Protection, pers. comm.). In addition to hypoxia and its negative impact on the biodiversity of LIS, pathogens are also carried into LIS from untreated waste water (Lopez et al. 2014). Eutrophication and increases in green algal blooms may also lead to disease outbreaks in horseshoe crab populations (Nolan et al. 2009; Braverman et al. 2012). Various heavy metals in differing types and amounts are also present in urban estuaries (Varekamp et al. 2014). These pollutants may be tolerated by horseshoe crabs, but it certainly can decrease the quality of the habitat and quantity of food resources (Varekamp et al. 2014).

Carbon dioxide (CO_2), another byproduct of eutrophication and decomposition can substantially reduce the pH of the urban estuary (Wallace et al. 2014). This acidification is in addition to general ocean acidification from increasing atmospheric CO_2 (Doney et al. 2009). In a recent study of Northeastern urban estuaries including JB and LIS, waters that exhibit hypoxia also during the late summer and fall experience a drop in pH from 8.2 to 7.2 on average over a 2 year study period (Wallace et al. 2014). The acidification is from the carbonic acid produced by bacteria during decomposition of organic matter and from atmospheric CO_2 (Feely et al. 2010). It was also recently discovered that aragonite, a polymorph of calcium carbonate, which is important to shellfish development, is also under-saturated in JB and NHH (Wallace et al. 2014). This will directly impact shellfish reproduction and survival, which in turn will affect the availability of food resources for horseshoe crabs (Botton 2009; Wallace et al. 2014). Calcium carbonate has not been found in large quantities in the exterior shell of horseshoe crabs; however, aragonite minerals may be important in the formation of the interior core of the horseshoe crab shell during growth and development (Chen et al. 2008). The effects of under saturation of these compounds in estuaries on *Limulus* development have not been reported and are in need of further study.

Urban estuaries have high sediment loads, algal and microbe densities, which results in loss of water clarity and light penetration (Valiela 2009). Male horseshoe crabs usually find females during mating season by sight (Barlow and Powers 2003; Brockmann and Smith 2009). The eutrophication of urban estuaries could negatively affect horseshoe crabs by reducing visibility. In low density populations in murky urban waters, males may be unable to see females, which could leave some females single, resulting in lower reproductive output of the population (Mattei et al. 2010; Beekey and Mattei 2015).

Urban estuaries, adjoining harbors and beaches experience artificial light pollution. A survey of nightly light pollution in coastal areas in 2010 revealed that Asia and North America had the second and third largest areas affected, respectively, worldwide (Davies et al. 2014). Known and potential impacts on estuarine biota include disruption of reproductive cycles, decreasing survival and changing predator-prey interactions (Becker et al. 2013; Davies et al. 2014). How light pollution may affect horseshoe crab behavior and survival is unknown. However, there is some evidence that changing the light exposure may affect either circadian or circatidal rhythms in some animals during lab experiments (Chabot and Watson 2010).

Physical alterations of urban estuaries are extensive (Defeo et al. 2009). Often the natural habitats like oyster reefs and beds, eel grass and saltmarsh grasses are removed or buried with sediments and organic matter from human activity that includes dredging and dumping (Weigold and Pillsbury 2014). To prevent erosion, the shoreline is often armored. Seawalls, bulkheads, large rip-rap, breakwaters and groins may hold sediments in the short run, but in the long run numerous problems have been observed. An armored shoreline cuts the connection between the land and the sea that many organisms rely on to survive including shorebirds and horseshoe crabs (Dugan et al. 2008). The armoring reflects the wave energy back onto the beach and causes vertical erosion, loss of beach sediments, organic material and intertidal habitats. Loss of spawning beaches from bulkhead expansion along many urban coastlines in Japan and other countries is directly correlated with the decline of Asian horseshoe crab species (Shuster and Sekiguchi 2009).

In spite of the challenges imposed by sharing the estuarine habitat with humans, horseshoe crabs are remarkably resilient animals that appear to thrive or are at least tolerant of our polluted urban estuaries. Studies of horseshoe crabs in urban estuaries have provided important insights into possible reasons for the evolutionary success of the group. For example, although horseshoe crabs obviously did not evolve in the presence of man-made pollutants except in the last 100 years, fossil evidence suggests that they have always tracked the presence of shallow-water marine habitats (Rudkin and Young 2009; Kin et al. 2013) and have possibly spawned and deposited eggs intertidally (Diedrich 2011) as they do today. Horseshoe crab spawning occurs in habitats that require their eggs, embryos and larvae to tolerate wide swings in environmental conditions (e.g. Botton et al. 2010). Horseshoe crab eggs, embryos and larvae have been shown to be highly tolerant of the typical organic and inorganic pollutants found in urban estuaries, which is consistent with their overall physiological plasticity (Botton and Itow 2009). They have biochemical defense mechanisms (e.g. stress proteins or Heat Shock Proteins) that help guard against damage to physiologically important proteins that might be caused by extremes in temperature (Botton et al. 2006b) and salinity (Greene et al. 2011). So, perhaps part of the explanation for pollution tolerance in horseshoe crabs is related to these kinds of adaptations to survive environmental fluctuations in general.

All coastal estuaries will be affected in one way or another by sea level rise and other factors associated with global climate change. During our current age of human domination, known as the Anthropocene, species that are generalists, which have a wide tolerance to temperature changes, salinity changes, and have variable diets, will have a good chance for survival. However, there may be a tipping point in urban estuaries, where multiple factors become too harsh for horseshoe crabs to survive. As this occurs, the results would be a reduction in the horseshoe crab population size with a patchy distribution similar to what we observe for *Tachypleus* in Southeast Asia (Nishida and Koike 2009; Shin et al. 2009; Yang et al. 2009).

Each summer our urban estuaries become 'hot, sour, and breathless' (i.e. increase in temperature, lower pH, and hypoxic), and adding each of the additional stressors discussed above could lead to synergistic effects that cause losses in biodiversity. The need for management of human derived pollution could never be more imperative (Gruber 2011; Wallace et al. 2014).

31.3 Study Areas

Jamaica Bay (Fig. 31.1) is located almost entirely within the New York City limits. It occupies an area of 101 km^2 and has a mean depth of 4.0 m. The estuary is highly eutrophic and receives the majority of its freshwater and pollutant inputs from water treatment plants, storm sewers, and combined sewer overflows (O'Shea and Brosnan 2000; Beck et al. 2009). Jamaica Bay is experiencing rapid loss of wetlands and beaches, attributed to global sea level rise and local processes including land subsidence (Gornitz et al. 2002; Hartig et al. 2002). The hydrodynamics within Jamaica Bay have been affected by extensive dredging for land reclamation, including the construction of Floyd Bennett Field and JFK International Airport, and the dredging of navigational channels. Other large-scale disturbances to Jamaica Bay include extensive bulkheading throughout the bay, and the construction of several sanitary landfills along the northern margin. Despite these major perturbations, Jamaica Bay is one of the most important spawning locations for horseshoe crabs in New York State's marine district, with especially high densities occurring at Plumb Beach, Dead Horse Bay, and Big Egg (Sclafani et al. 2009). Smaller, more isolated patches of sand that are interspersed within regions of armored shorelines can also attract high densities of spawning crabs (Botton et al. 2006a).

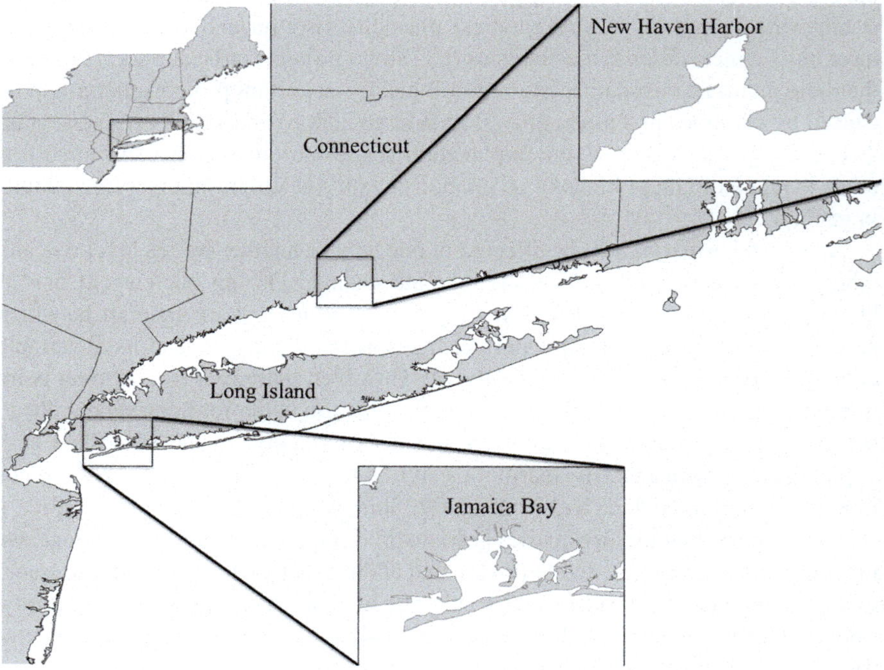

Fig. 31.1 A map of two urban estuaries, Jamaica Bay, New York and New Haven Harbor, Connecticut

New Haven Harbor, located on the north shore of Long Island Sound (Fig. 31.1), is an active port that has been heavily impacted by many decades of industrial activity along the shoreline and within the watershed (Rozan and Benoit 2001). In addition to being an active port receiving many shipments of steel, lumber, oil, salt and sand; on the east side of the Harbor is a large, oil burning, electrical power plant, and on the west side is a sewage treatment plant that empties directly into the harbor (Spiegel 2006). The navigational channel is periodically dredged at ca. 15 year intervals to maintain a depth of approximately 10 m, but water depths elsewhere in the harbor are typically 1–7 m (Rozan and Benoit 2001). Most of the horseshoe crab spawning activity within New Haven Harbor occurs at Sandy Point, a long sand spit in West Haven, Connecticut that juts out into the middle of the Harbor (Beekey et al. 2013). Spawning takes place intermittently and at lower densities within the inner harbor in patches of sand that occur within broader areas of rubble fill and salt marsh (Beekey and Mattei 2008; Botton et al. 2009).

31.4 Opportunities for Citizen Science and Research in Urban Estuaries

The study of horseshoe crabs in urban estuaries provides excellent opportunities for public education and engagement with the local marine environment. The opportunity for ecological education of the public is one of numerous ecosystem services that urban estuaries provide (Haase et al. 2014). For example, horseshoe crab and shorebird viewing is a popular guided activity at the Jamaica Bay Wildlife Refuge, part of the Gateway National Recreation Area (NRA; Riepe 2001). At Plumb Beach, Brooklyn, another unit of Gateway NRA, faculty and students from Kingsborough Community College (KCC) and Fordham University partner with dozens of volunteers from the New York City Audubon Society in the annual high tide census of spawning horseshoe crabs (Colón and Rowden 2014). This is part of a larger survey effort in New York State designed to provide simultaneous coverage of spawning activity on a total of 12 evening high tides bracketing the new and full moons in May and June (Sclafani et al. 2009). Personnel first receive training from an experienced marine educator on how to conduct the survey; this is held in conjunction with the annual Eco-Festival at KCC, which also features lectures and exhibits about the local marine environment. Hands-on training continues in the field, where volunteers are shown how to distinguish sexes, tag animals, etc. Some students and volunteers participate in other aspects of the project, including core sampling to determine egg densities, measuring and determining carapace condition of adult crabs, and surveying the intertidal sand flats for the presence of juvenile crabs. This citizen science project has enormous value to the students and volunteers, many of which have little previous awareness of the ecological and biomedical importance of horseshoe crabs and are often unaware of the rich diversity of flora and fauna in their local marine habitats (Colón and Rowden 2014).

In 2014, the horseshoe crab research was used as the basis of a pilot program at Kingsborough Community College to add Citizen Science to the spectrum of Civic Engagement opportunities for students. Students who participated in the training and surveys were awarded credit towards their Civic Engagement graduation requirement. Students who participate in the summer research are funded through several grants including Bridges to the Baccalaureate through the National Institutes of Health (NIH) and C-STEP (Collegiate Science and Technology and Entry Program). These opportunities represent an entry point into scientific research and environmental activism for many students who have never explored the outdoors and were unaware of many local and global biodiversity and conservation issues. Raised either in an urban landscape where parks are places to avoid, or arriving from overseas and unaware of the natural landscape of New York Harbor, many Kingsborough students who participate in this research are deeply impacted by their first visit to an urban estuary and their first encounter with a live horseshoe crab.

To understand the horseshoe crab population dynamics within a single urban estuary like New Haven Harbor (NHH), faculty and students at Sacred Heart University (SHU) attempted to recruit people who frequent the beach to report basic information on spawning horseshoe crabs. From 1998 to 2000 the majority of people encountered on the beach were either repulsed by the sight of horseshoe crabs or were afraid of them because their telson looked like a stinger and their claws looked dangerous. However, to answer very basic questions about how far horseshoe crabs travel during their adult life span and if spawning adults tagged in NHH come back regularly to the same beach, it would be helpful if people who visit the beach frequently during the spawning season have a way to report their observations. In fact, with the millions of people who visit Long Island Sound annually, volunteers could be recruited from all around the Sound. It was not known at the start of the study if horseshoe crabs could cross the Sound on a regular basis or if they stayed within a specific estuary on the Connecticut coast. Evidence from studies in deep embayments in New England revealed that animals rarely leave (James-Pirri 2010). Yet other studies have evidence of horseshoe crabs traveling quite broadly over their 18–20 year life span (Swan 2005). So what would be the pattern in a large urban estuary?

In order to recruit volunteers to help, first an education program was needed so people would learn that horseshoe crabs are benign and are beneficial to humans and to the estuary they inhabit. Letters and emails were sent to local non-profit environmental organizations asking if they wanted to join a newly established conservation program called Project *Limulus* even though at the time it consisted of one professor and two undergraduate research assistants. Faculty and undergraduate students offered free lectures and beach walks/talks to their members. The staff members at the Connecticut Audubon Society, The Maritime Aquarium and The Nature Conservancy were trained to tag, record data and lead their own groups of volunteers out to local beaches during the spawning season. It was discovered that members of conservation organizations wanted to donate their time and be active participants in conservation activities rather than always be asked to donate money. After several years, Project *Limulus* had hundreds of volunteers (Mattei and Beekey

2008). The research and outreach programs were supported by Connecticut Sea Grant, Disney Worldwide Conservation Fund, the National Fish & Wildlife Foundation, the College of Arts & Sciences, Undergraduate Research Initiative at SHU and other local granting agencies. The outreach program was expanded to include local schools and teachers modifying the program to include children from pre-K to high school (Figs. 31.2 and 31.3). This also involved training students who were interested in a career in education. From 2007 to 2014 Beekey and Mattei and other Project *Limulus* staff delivered over 300 lectures, dozens of newspaper stories, three different spots on National Public Radio and on several occasions, local television news stories and the Aqua Kids TV show in 2009 and 2014. In 2013:

- Fifteen Project *Limulus* information sessions ran that targeted the general public, and 13 outreach programs were offered through the context of various non-profit organizations (see Table 31.1). Over 750 adults and 525 children participated in these outreach events.
- Project *Limulus* staff traveled to 15 schools ranging from 3rd to 12th grade (See Table 31.2). A total of 1,062 students participated in these outreach events along with 85 teachers and an additional 197 adults (parents of students who participated in the school field trips or in-school presentations).
- Outreach events were conducted for some summer regional enrichment programs targeting underprivileged children for a total of 160 students, 19 adults, and 21 teachers including 48 students involved with the Horizons National Youth Program run through Sacred Heart University (http://www.sacredheart.edu/academics/universitycollegepart-timenon-creditprograms/horizonsnationaly-

Fig. 31.2 Project *Limulus* Outreach Coordinator, Adam Rudman, with three undergraduate Biology students (Fordham and Sacred Heart University) are teaching the 2nd Grade class of Unquowa School in Fairfield, Connecticut the economic and ecological value of horseshoe crabs in Long Island Sound (Photo by: T. Deer-Mirek)

Fig. 31.3 Pre-school children, like 4-year old Alice Young, learn early that horseshoe crabs are harmless through public education programs like Project *Limulus* (Photo by: T. Deer-Mirek)

Table 31.1 Project *Limulus* outreach events in 2013

Group name	No. of kids	No. of adults
Ash Creek Nature Center	15	30
Beardsley Zoo	27	30
Bruce Museum	103	213
Connecticut Audubon	16	35
Darien Nature Center	19	13
Denison Pequotsepos Nature Center	65	121
Menunkatuck Audubon	51	146
Mystic Aquarium	50	66
Nature Conservancy	25	45
Private Group	26	12
Sherwood Island Nature Center	10	2
SoundWaters	100	20
CT Surfriders	20	20
Total	527	753

outhprogram/), 32 students from the Yale Exploration program (http://explo.org/360/yale), and 80 students participating in the Inter-district Aquaculture Program run through the Center for Creative Youth (Fig. 31.4; http://www.crec.org/magnetschools/schools/ccy/).

- Egg hatching kits were distributed to 22 classrooms from seven different school districts with a combined total of 790 students ranging from K-12th grade. All of the kits were successfully returned and all of the hatched horseshoe crabs were released back into the water on the beach where they were originally collected from. Overall our participating teachers had a 60 % hatching success rate that is similar to what we observe in laboratory conditions.

Table 31.2 K-12 Project *Limulus* Educational Outreach 2013

School/group name	Town	State	No. of students	Grades	No. of adults	No. of teachers
Madison Public School District	Madison	CT	281w		85	5
City Hill Middle School	Naugatuck	CT	120	8th	5	5
Harborside Middle School	Milford	CT	192	6th	29	27
Hopkins School	New Haven	CT	8	11–12th	1	1
Metropolitan Business Center	New Haven	CT	13	9–12th	2	2
Naugatuck High School	Naugatuck	CT	22	11–12th	3	3
Our Lady of Assumption School	Fairfield	CT	145	K-8th	26	21
Rye Country Day School	Rye	NY	30	11–12th	2	1
Sayless School	Baltic	CT	25	8th	1	1
Shelton Intermediate	Shelton	CT			1	1
St. Luke's School	New Canaan	CT	44	5th	6	6
St. Thomas School	Stratford	CT	17	1st	8	1
Unquowa School	Fairfield	CT	40	2nd grade, PreK4	16	3
ACES	Hamden	CT	25	9–12th	2	2
John G. Prendergast School	Ansonia	CT	100	9–12th	10	6
Total			1,062		197	85

An overlooked advantage of working in urban estuaries is their accessibility to students and volunteers, compared with more remote locations such as Delaware Bay. Jamaica Bay and New Haven Harbor are within the reach of millions of people by public transportation. Moreover, locations such as Jamaica Bay and New Haven Harbor provide a mosaic of habitat types in close proximity to each other, enabling visitors and volunteers to easily contrast horseshoe crab activity on relatively "natural" beaches and highly degraded habitats with bulkheads, rubble fill, or other kinds of human alterations.

31.4.1 Examining Recruitment in Horseshoe Crab Populations Using Volunteer-Based Projects

As a specific example of the kind of useful data that can be collected by students and citizen scientists, we have measured and determined carapace condition of adult horseshoe crabs in Jamaica Bay and New Haven Harbor for several years as an indicator of the "health" of the population (Duffy et al. 2006; Shuster 2009).

Fig. 31.4 High school students from Bridgeport, Connecticut help undergraduate research assistants from Sacred Heart University tag and collect data on spawning horseshoe crabs such as size, sex, and condition (Photo by: A. Rudman)

Although there are no known methods for determining the chronological age of adult crabs, the assignment of a "relative" age can be done quite easily by examining the condition of the carapace. We have used a simple scoring procedure (after Brockmann 2002; Duffy et al. 2006; Shuster 2009), which can easily be demonstrated to volunteers in the field. A score of 1 is given for a "young" or newly molted animal, which has a lustrous shell with sharp spines, typically with few if any attached invertebrates, and hairs are present along the inner edge of the prosoma (Fig. 31.5a). A score of 2 is used for a shell in intermediate condition ("middle age"), one that has scratches and other signs of wear, with some blackened areas, often with some coverage by barnacles, mussels (*Mytilus*), slipper shells (*Crepidula*), bryozoans, and other invertebrates. A score of 3 is assigned to "old age" crabs that have a thin, pitted, and heavily blackened shell, often with partial to complete rotting of the lateral eyes (Fig. 31.5b, Duffy et al. 2006). A classic model of population growth that reflects an increasing or expanding population would include greater numbers of younger adults (i.e. Condition 1) entering the population than older or aging adults (i.e. Condition 2 and 3). Equal numbers of young, middle aged and older adults would indicate a stable population. The characteristic configuration for relative age classes of a population in decline has disproportionately greater numbers of older adults (e.g. Pew Research Center 2014).

Using groups of volunteers, we measured and assessed carapace condition for horseshoe crabs from both the Jamaica Bay and New Haven Harbor populations in 2012 and 2013. Using multiple teams of three (one person to gather the crabs, the second to measure and determine carapace condition, and the third to record data),

Fig. 31.5 Contrasting pictures of Condition 1 and 3 adult horseshoe crabs. (**a**) A newly molted Condition 1 adult female. Her shell is shiny, golden brown, with the hinged edge lined with small hairs. (**b**) An old, Condition 3 horseshoe crab with dull, black, pitted shell and damaged dark colored eyes. No hairs are present (Photo by: J. Mattei)

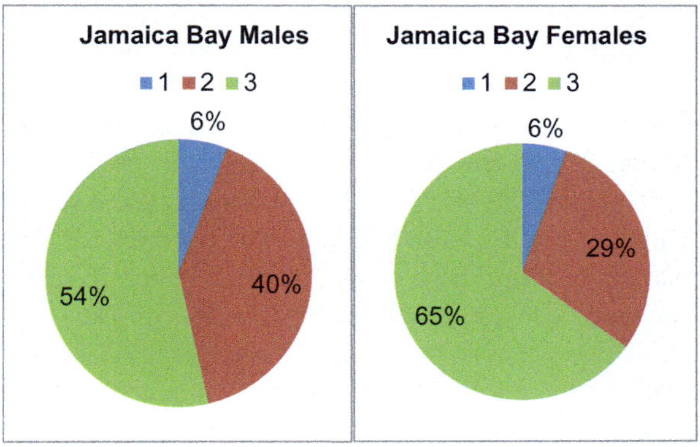

Fig. 31.6 Carapace condition of a sample of adult horseshoe crabs separated by sex from Jamaica Bay (N=84 females and 177 males) during the 2013 spawning season. Carapace condition was scored as (1) newly molted, (2) middle aged and (3) oldest (See text for description)

we can easily assess several hundred crabs on a single high tide, giving a robust sample for statistical analysis. Both Jamaica Bay and New Haven Harbor populations are dominated by condition 2 and 3 crabs (Figs. 31.6 and 31.7, respectively), suggesting that there has been little recent recruitment of new adults into the population. For JB in 2013, fewer than 6 % of the male and female crabs were judged to be new recruits (Condition 1, Fig. 31.6). This could have important demographic consequences, since alternative male mating tactics (attached vs. satellite) and the risk of beach stranding are known to be condition-dependent (Penn and Brockmann 1995; Duffy et al. 2006). With the help of volunteers in NHH, we were able to document variation in recruitment from year to year. The total spawning adults in 2012 that were recorded as condition 1 was 9 %, and in 2013 19 % were newly molted (Fig. 31.7). These percentages need to be followed for several more

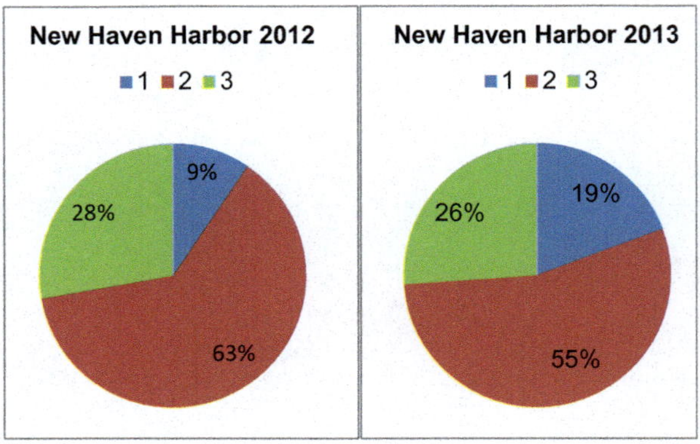

Fig. 31.7 Carapace condition of total adult spawning horseshoe crabs sampled from New Haven Harbor during the 2012 (N=988) and 2013 (N=993) spawning seasons. Carapace condition was scored as (1) newly molted, (2) middle aged and (3) oldest (see text for description)

years across more than one beach, but this preliminary data demonstrates that recruitment of young adults is low and that the population is declining. Data from our volunteers has redirected our efforts to investigate why more juvenile horseshoe crabs are not entering the spawning adult population.

Through our outreach programs that include training sessions, volunteers learn to tag horseshoe crabs as well as how to conduct a spawning census in order to estimate density of spawning females on a beach (tag data reported in Beekey and Mattei 2015). By modifying the protocol used in other states (James-Pirri et al. 2005; Smith and Michels 2006) volunteers count spawning adults on a pre-measured length of beach, walking at high tide covering the first 3 m into the water from the waterline. From these counts over a number of years we have discovered that the spawning density is well below estimates on Delaware Bay beaches in New Jersey and Delaware (0.03 +/− 0.04 females/m^2 in NHH and ~1.0 females/m^2 in DE) and that there is a lot of variation between counts, which was correlated with day vs. night, tidal height, moon phase, water temperature, wind speed, and from year to year (Mattei et al. 2010). While it is difficult to calculate actual population size in a large open body of water, with the help of volunteers one can obtain an estimate of spawning density on many different beaches and population trends over multiple years.

31.5 Summary and Conclusions

Horseshoe crab populations inhabiting urban estuaries are subjected to many different stressors, including pollution and shoreline development. Yet, these resilient animals are able to thrive in urban habitats such as New Haven Harbor and Jamaica Bay. Much of research on horseshoe crab ecology has come from studies in less developed embayments, such as Cape Cod and Delaware Bay. Mating behaviors, food web relationships, and many other aspects of ecology and behavior may prove to be significantly different in the urban settings. To begin to explore these differences in greater detail and to gain insights into how horseshoe crabs can survive in urban estuaries, we have found that help of citizen scientists drawn from nearby communities can be invaluable. In the long run, the conservation of horseshoe crabs depends on greater public awareness of the importance of these animals.

Acknowledgments The authors would like to thank all the volunteers of all ages that have donated hundreds of hours of time to help understand the ecology and conservation of the horseshoe crab in our urban seas. Numerous conservation organizations have helped with turning the image of the horseshoe crab from the 'cockroach of the sea' to the 'living fossil' that entangles our own health and the health of our estuaries in its own well-being. In particular, we would like to thank the staff and members of The Maritime Aquarium, Connecticut Audubon Society, New York City Audubon Society, The Nature Conservancy and many others. The tagging aspect of the project would not have been possible without the assistance of Sheila Eyler and her staff at the U.S. Fish & Wildlife Service, Maryland Fishery Resources Office. The Jamaica Bay component of this research was supported by grant 2R25GM0600309 of the Bridge Program of NIGMS under the supervision of Dr. Edward Catapane, grant 0537121091 of the CSTEP Program of NYSED, and grant R/CMB-39 from the New York Sea Grant Program. We thank the New York City Department of Parks, New York State Department of Environmental Conservation, Connecticut Department of Energy and Environmental Protection and the National Park Service – Gateway National Recreation Area for issuing the necessary permits to conduct research at Plumb Beach, Jamaica Bay and in Long Island Sound. The work in Long Island Sound was supported by grants from Disney Worldwide Conservation Fund, Connecticut Sea Grant, National Fish and Wildlife Foundation, and the College of Arts & Sciences, Sacred Heart University, Undergraduate Research Initiative program. We would also like to thank the many undergraduate and graduate students who have worked on these projects who with great enthusiasm help to manage our volunteers and data collection.

References

Barbier EB, Hacker SD, Kennedy C et al (2011) The value of estuarine and coastal ecosystem services. Ecol Monogr 81:169–193

Barlow RB, Powers MK (2003) Seeing at night and finding mates: the role of vision. In: Shuster CN Jr, Barlow RB, Brockman HJ (eds) The American horseshoe crab. Harvard University Press, Cambridge, pp 83–102

Barlow RB, Powers MK, Howard H et al (1986) Migration of *Limulus* for mating: relation to lunar phase, tide height, and sunlight. Biol Bull 171:310–329

Beck AJ, Cochran JK, Sañudo-Wilhemy SA (2009) Temporal trends of dissolved trace metals in Jamaica Bay, NY: importance of wastewater input and submarine groundwater discharge in an urban estuary. Estuar Coasts 32:535–550

Becker A, Whitfield AK, Cowley PD et al (2013) Potential effects of artificial light associated with anthropogenic infrastructure on the abundance and foraging behaviour of estuary associated fishes. J Appl Ecol 50:43–50

Beekey MA, Mattei JH (2008) What long term mark/recapture studies reveal about horseshoe crab population dynamics in Long Island Sound. In: Proceedings of the 2008 Biennial Long Island Sound research conference, Connecticut College, Groton, Connecticut, 30–31 October, 2008

Beekey MA, Mattei JH, Pierce BJ (2013) Horseshoe crab eggs: a rare resource for predators in Long Island Sound. J Exp Mar Biol Ecol 439:152–159

Beekey MA, Mattei JH (2015) The mismanagement of *Limulus polyphemus* in Long Island Sound: what are the characteristics of a population in decline. In: Carmichael RH, Botton M, Shin PKS, Cheung SG (eds) Changing global perspectives on horseshoe crab biology, conservation and management. Springer, New York, pp 537–555

Botton ML (2001) The conservation of horseshoe crabs: what can we learn from the Japanese experience? In: Tanacredi JT (ed) *Limulus* in the limelight. Kluwer Academic/Plenum Publishers, New York, pp 41–51

Botton ML (2009) The ecological importance of horseshoe crabs in estuarine and coastal communities: a review and speculative summary. In: Tanacredi JT, Botton ML, Smith DR (eds) Biology and conservation of horseshoe crabs. Springer, New York, pp 45–63

Botton ML, Itow T (2009) The effects of water quality on horseshoe crab embryos and larvae. In: Tanacredi JT, Botton ML, Smith DR (eds) Biology and conservation of horseshoe crabs. Springer, New York, pp 439–454

Botton ML, Loveland RE, Tiwari A (2003) Distribution, abundance, and survivorship of young of the year in a commercially exploited population of horseshoe crabs *Limulus polyphemus*. Mar Ecol Prog Ser 265:175–184

Botton ML, Loveland RE, Tanacredi JT et al (2006a) Horseshoe crabs (*Limulus polyphemus*) in an urban estuary (Jamaica Bay, New York), and the potential for ecological restoration. Estuar Coasts 29:820–830

Botton ML, Pogorzelska M, Smoral L et al (2006b) Thermal biology of horseshoe crab embryos and larvae: a role for heat shock proteins. J Exp Mar Biol Ecol 336:65–73

Botton ML, Mattei J, Robinson GS et al (2009) Horseshoe crabs in New Haven Harbor: an initial study of reproductive biology and beach quality. In: Tanacredi JT, Botton ML, Smith DR (eds) Biology and conservation of horseshoe crabs. Springer, New York, p 613

Botton ML, Tankersley RA, Loveland RE (2010) Developmental ecology of the American horseshoe crab, *Limulus polyphemus*. Curr Zool 56:550–562

Braverman H, Leibovitz L, Lewbart GA (2012) Green algal infection of American horseshoe crab (*Limulus polyphemus*) exoskeletal structures. J Invertebr Pathol 111:90–93

Brockmann HJ (2002) An experimental approach to altering mating tactics in male horseshoe crabs (*Limulus polyphemus*). Behav Ecol 13:232–238

Brockmann HJ (2003) Nesting behavior, a shoreline phenomenon. In: Shuster CN Jr, Barlow RB, Brockmann HJ (eds) The American horseshoe crab. Harvard University Press, Cambridge, pp 33–49

Brockmann HJ, Johnson SL (2011) A long-term study of spawning activity in a Florida Gulf Coast population of horseshoe crabs (*Limulus polyphemus*). Estuar Coasts 34:1049–1067

Brockmann HJ, Smith MD (2009) Reproductive competition and sexual selection in horseshoe crabs. In: Tanacredi JT, Botton ML, Smith DR (eds) Biology and conservation of horseshoe crabs. Springer, New York, pp 199–221

Burton WH, Kelley FS, Franks EA (2009) Distribution of juvenile horseshoe crabs in subtidal habitats of Delaware Bay using a suction-dredge sampling device. In: Tanacredi JT, Botton ML, Smith DR (eds) Biology and conservation of horseshoe crabs. Springer, New York, pp 285–293

Carmichael RH, Rutecki D, Valiela I (2003) Abundance and population structure of the Atlantic horseshoe crab *Limulus polyphemus* in Pleasant Bay, Cape Cod. Mar Ecol Prog Ser 246:225–239

Chabot CC, Watson WH III (2010) Circatidal rhythms of locomotion in the American horseshoe crab *Limulus polyphemus*: underlying mechanisms and cues that influence them. Curr Zool 56:499–517

Chen PY, Lin AYM, Lin YS et al (2008) Structure and mechanical properties of selected biological materials. J Mech Behav Biomed Mater 1:208–226

Chiu HMC, Morton B (2003) The status of horseshoe crabs in Hong Kong. In: Morton B (ed) Perspectives on marine environment change in Hong Kong and southern China. Hong Kong University Press, Hong Kong, pp 741–756

Colón CP, Rowden J (2014) Blurring the roles of scientist and activist through citizen science. In: Finley A (ed) Civic learning and teaching: a bridge to civic life and a life of learning. Bringing Theory to Practice, Washington, DC, pp 45–52

Creel L (2003) Ripple effects: population and coastal regions. Population Reference Bureau, Measure Communication, Washington, DC. www.prb.org

Davies TW, Duffy JP, Bennie J et al (2014) The nature, extent, and ecological implications of marine light pollution. Front Ecol 12:347–355

Defeo O, McLachlan A, Schoeman DS et al (2009) Threats to sandy beach ecosystems: a review. Estuar Coast Shelf Sci 81:1–12

Diedrich CG (2011) Middle Triassic horseshoe crab reproduction areas on intertidal flats of Europe with evidence of predation by archosaurs. Biol J Linn Soc 103:76–105

Doney SC, Fabry VJ, Feely RA et al (2009) Ocean acidification: the other CO_2 problem. Ann Rev Mar Sci 1:169–192

Duffy EE, Penn D, Botton ML et al (2006) Eye and clasper damage influence male mating tactics in the horseshoe crab, *Limulus polyphemus*. J Ethol 24:67–74

Dugan JE, Hubbard DM, Rodil IF et al (2008) Ecological effects of coastal armoring on sandy beaches. Mar Ecol 29:160–170

Ellis EC, Kaplan JO, Fuller DQ et al (2013) Used planet: a global history. Proc Natl Acad Sci U S A 110:7978–7985

Feely RA, Alin SR, Newton J et al (2010) The combined effects of ocean acidification, mixing, and respiration on pH and carbonate saturation in an urbanized estuary. Estuar Coast Shelf Sci 88:442–449

Gornitz V, Couch S, Hartig EK (2002) Impacts of sea level rise in the New York City metropolitan area. Glob Planet Change 32:61–88

Greene M, Hamilton MG, Botton ML (2011) Physiological responses of horseshoe crab (*Limulus polyphemus*) embryos to osmotic stress, and a possible role for stress proteins (HSP's). Mar Biol 158:1691–1698

Gruber N (2011) Warming up, turning sour, losing breath: ocean biogeochemistry under global change. Phil Trans R Soc A 369:1980–1996

Haase D, Larondelle N, Andersson E et al (2014) A quantitative review of urban ecosystem service assessments: concepts, models, and implementation. Ambio 43(4):413–433

Hartig EK, Gornitz V, Kolker A et al (2002) Anthropogenic and climate-change impacts on salt marshes of Jamaica Bay, New York City. Wetlands 22:71–89

Jackson NL, Nordstrom KF (2009) Strategies to conserve and enhance sandy barrier habitat for horseshoe crabs (*Limulus polyphemus*) on developed shorelines in Delaware Bay, USA. In: Tanacredi JT, Botton ML, Smith DR (eds) Biology and conservation of horseshoe crabs. Springer, New York, pp 399–416

Jackson NL, Nordstrom KF, Smith DR (2010) Armoring of estuarine shorelines and implications for horseshoe crabs on developed shorelines in Delaware Bay. In: Shipman H, Dethier MN, Gelfenbaum G, Fresh KL, Dinicola RS (eds) Puget Sound shorelines and the impacts of armoring – Proceedings of a state of the science workshop, May 2009. US Geological Survey Scientific Investigations Report 2010-5254, pp 195–202

James-Pirri MJ (2010) Seasonal movement of the American horseshoe crab *Limulus polyphemus* in a semi-enclosed bay on Cape Cod, Massachusetts (USA) as determined by acoustic telemetry. Curr Zool 56:575–586

James-Pirri MJ, Tuxbury K, Marino S et al (2005) Spawning densities, egg densities, size structure, and movement patterns of spawning horseshoe crabs, *Limulus polyphemus*, within four coastal embayments on Cape Cod, Massachusetts. Estuaries 28:296–313

Kin A, Gruszczynski M, Martill M et al (2013) Palaeoenvironment and taphonomy of a Late Jurassic (Late Tithonian) Lagerstätte from central Poland. Lethaia 46:71–81

Lopez G, Carey D, Carlton JT et al (2014) Biology and ecology of Long Island Sound. In: Latimer JS, Tedesco MA, Swanson RL, Yarish C, Stacey P, Garza C (eds) Long Island Sound: prospects for the urban sea. Springer, New York, pp 285–480

Mattei JH (2005) The challenge of environmental protection. In: Rose GL (ed) Public policy in Connecticut: challenges and perspectives. Sacred Heart University Press, Fairfield, pp 66–85

Mattei J, Beekey M (2008) The horseshoe crab conundrum. In: Van Patten M (ed) Wrack Lines, 8:1, Spring/Summer, Groton, pp 2–7

Mattei JH, Beekey MA, Rudman A et al (2010) Reproductive behavior in horseshoe crabs: does density matter? Curr Zool 56:634–642

Mayer PM, Grimm NB, Lepczyk C et al (2010) Urban ecosystems research joins mainstream ecology. Nature 467:153

McDonnell MJ, Pickett STA (1990) Ecosystem structure and function along urban-rural gradients: an unexploited opportunity for ecology. Ecology 71:1232–1237

National Ocean Service (2014) What percentage of the American population lives near the coast? http://oceanservice.noaa.gov/facts/population.html. Accessed 2 June 2014

Niemela J, Breuste JH, Guntenspergen G, McIntyre NE, Elmqvist T, James P (eds) (2011) Urban ecology: patterns, processes, and applications. Oxford University Press, New York

Nishida S, Koike H (2009) Genetic structure of Japanese populations of *Tachypleus tridentatus* by mtDNA AT-rich region sequence analysis. In: Tanacredi JT, Botton ML, Smith DR (eds) Biology and conservation of horseshoe crabs. Springer, New York, pp 183–196

Nolan M, Smith SA, Smith DR (2009) Clinical evaluation, common diseases, and veterinary care of the horseshoe crab. In: Tanacredi JT, Botton ML, Smith DR (eds) Biology and conservation of horseshoe crabs. Springer, New York, pp 479–499

O'Shea ML, Brosnan TM (2000) Trends in indicators of eutrophication in western Long Island Sound and the Hudson-Raritan estuary. Estuaries 23:877–901

Penn D, Brockmann HJ (1995) Age-biased stranding and righting in male horseshoe crabs, *Limulus polyphemus*. Anim Behav 49:1531–1539

Pew Research Center (2014) Population change in the U.S. and the world from 1950 to 2050. http://www.pewglobal.org/2014/01/30/chapter-4-population-change-in-the-u-s-and-the-world-from-1950-to-2050/. Accessed 2 June 2014

Riepe D (2001) Horseshoe crabs: an ancient wonder of New York and a great topic for environmental education. In: Tanacredi JT (ed) *Limulus* in the limelight – a species 350 million years in the making and in peril. Kluwer Academic/Plenum Publishers, New York, pp 131–134

Rozan TF, Benoit G (2001) Mass balance of heavy metals in New Haven Harbor, Connecticut: predominance of nonpoint sources. Limnol Oceanogr 46:2032–2049

Rudkin DM, Young GA (2009) Horseshoe crabs – an ancient ancestry revealed. In: Tanacredi JT, Botton ML, Smith DR (eds) Biology and conservation of horseshoe crabs. Springer, New York, pp 25–44

Rudloe A (1980) The breeding behavior and patterns of movement of horseshoe crab, *Limulus polyphemus*, in vicinity of breeding beaches in Apalachee, Florida. Estuaries 3:177–183

Sclafani M, Brousseau L, McKown K et al (2009) T-3-1 Study 5: Horseshoe crab spawning activity survey final report. NY State DEC Technical Report. http://www.NYhorseshoecrab.org. Accessed 2 June 2014

Shin PKS, Li HY, Cheung SG (2009) Horseshoe crabs in Hong Kong: current population status and human exploitation. In: Tanacredi JT, Botton ML, Smith DR (eds) Biology and conservation of horseshoe crabs. Springer, New York, pp 347–360

Shuster CN Jr (2009) Public participation in studies on horseshoe crab populations. In: Tanacredi JT, Botton ML, Smith DR (eds) Biology and conservation of horseshoe crabs. Springer, New York, pp 585–594

Shuster CN Jr, Botton ML (1985) A contribution to the population biology of horseshoe crabs, *Limulus polyphemus* (L.), in Delaware Bay. Estuaries 8:363–372

Shuster CN Jr, Sekiguchi K (2009) Basic habitat requirements of the extant species of horseshoe crabs (Limulacea). In: Tanacredi JT, Botton ML, Smith DR (eds) Biology and conservation of horseshoe crabs. Springer, New York, pp 115–162

Smith DR, Michels SF (2006) Seeing the elephant: importance of spatial and temporal coverage in a large-scale volunteer-based program to monitor horseshoe crabs. Fisheries 31(10):485–491

Smith DR, Pooler PS, Loveland RE et al (2002) Horseshoe crab (*Limulus polyphemus*) reproductive activity on Delaware Bay beaches: interaction with beach characteristics. J Coast Res 18:730–740

Spiegel JE (2006) Who is minding the state ports? The New York Times. http://www.nytimes.com/2006/03/05/nyregion/nyregionspecial2/05ctports.html?pagewanted=print&_r=0. Accessed 5 Mar 2006

Swan BL (2005) Migrations of adult horseshoe crabs, *Limulus polyphemus*, in the Middle Atlantic Bight: a 17-year tagging study. Estuaries 18:28–40

Tsuchiya K (2009) The history of horseshoe crab research and conservation in Japan. In: Tanacredi JT, Botton ML, Smith DR (eds) Biology and conservation of horseshoe crabs. Springer, New York, pp 559–570

Valiela I (2009) Global coastal change. Blackwell Publishing, Malden

Varekamp JC, McElroy AE, Mullaney JR et al (2014) Metals, organic compounds, and nutrients in Long Island Sound: sources, magnitudes, trends, and impacts. In: Latimer JS, Tedesco MA, Swanson RL, Yarish C, Stacey P, Garza C (eds) Long Island Sound: prospects for the urban sea. Springer, New York, pp 203–283

Wallace RB, Baumann H, Grear JS et al (2014) Coastal ocean acidification: the other eutrophication problem. Estuar Coast Shelf Sci 148:1–13

Weigold ME, Pillsbury E (2014) Long Island Sound: a socioeconomic perspective. In: Latimer JS, Tedesco MA, Swanson RL, Yarish C, Stacey P, Garza C (eds) Long Island Sound: prospects for the urban sea. Springer, New York, pp 1–46

Yang M-C, Chen C-P, Hsieh H-L et al (2009) Phylogeography, demographic history, and reserves network of horseshoe crab, *Tachypleus tridentatus*, in the south and east China seaboards. In: Tanacredi JT, Botton ML, Smith DR (eds) Biology and conservation of horseshoe crabs. Springer, New York, pp 163–181

Chapter 32
Green Eggs & Sand, Team Limulus, and More: Educating for Horseshoe Crab Conservation in the United States

Gary Kreamer and Sharon W. Kreamer

Abstract Efforts to educate people about the American horseshoe crab, *Limulus polyphemus*, have expanded greatly in the United States in recent years. This article provides highlights of five high profile programs, including: Ecological Research & Development Group's (ERDG) *Horseshoe Crab Conservation Network*™, the *Green Eggs & Sand* project, *Raising Horseshoe Crabs in the Classroom*, *Team Limulus* and *Crab App*. These programs represent a variety of approaches taken to increase public awareness, promote stewardship, and deepen understanding of the unique life history, ecological connections and human uses of this ancient species. The article includes profiles of programs targeting bayshore homeowners, teachers, school children and volunteer monitoring groups. Selected links and references to additional resources for horseshoe crab education are also provided.

Keywords ERDG • *Limulus polyphemus* • Raising horseshoe crabs • Molts • Education • Conservation • Sanctuary • Monitoring • Apps • Curriculum

32.1 Introduction

Over the last 15 years, interest in the American horseshoe crab has inspired increasingly widespread and varied efforts to promote awareness and conservation of this ancient animal. In just about every Atlantic coast state where horseshoe crabs come into shore and spawn, educational programs have sprung up to enlighten people about this species. This article cannot presume to cover them all. What it will do is highlight five examples of *Limulus* education programs representative of the varied

G. Kreamer (✉)
Delaware Division of Fish and Wildlife,
4876 Hay Point Landing Road, Smyrna, DE 19977, USA
e-mail: gary.kreamer@state.de.us

S.W. Kreamer
The Tatnall School, 1501 Barley Mill Road, Wilmington, DE 19807, USA

R.H. Carmichael et al. (eds.), *Changing Global Perspectives on Horseshoe Crab Biology, Conservation and Management*, DOI 10.1007/978-3-319-19542-1_32

approaches taken, with the hope that – in profiling these programs for those who work with horseshoe crabs in other parts of the world – pathways may emerge for adapting, applying and/or expanding those approaches to new audiences, thus broadening the impacts of education for conservation of this important group of animals.

32.2 Ecological Research & Development Group's (ERDG) *Horseshoe Crab Conservation Network*™

One could hardly present an overview of horseshoe crab conservation through education in the United States without highlighting the exceptional body of work done through the non-profit ERDG over the past two decades. As outlined by Gauvry (2009), the essence of ERDG's varied initiatives boils down to one common denominator – establishing a compassionate connection to an animal – the horseshoe crab – that for much of human history has gone unnoticed or been viewed as a trash species or commodity. In establishing that connection to this animal, seeds of awareness are planted in people, leading them to want to learn more, and with that knowledge, inspiring an interest in taking action for conservation. Building on that premise, ERDG's *Horseshoe Crab Conservation Network*™ offers a multifaceted, cross cultural suite of initiatives designed to inform and engage individuals, communities, organizations and scientists around the conservation of the world's four horseshoe crab species. Key components of that effort are summarized in the paragraphs to follow.

32.2.1 Just flip 'em*!* ®

One of the earliest directed forays into horseshoe crab education for conservation in the US occurred in the late 1990s when ERDG introduced its *Just flip 'em*! ® campaign (ERDG 2006). At the time, public awareness of horseshoe crabs along the Atlantic coast, and even at its Delaware Bay spawning epicenter, was relatively low, with many people living along spawning beaches viewing the animals as a nuisance or, at best, with utter indifference. It was not uncommon at the time for communities to enlist contractors to come in with heavy equipment to bury or haul away horseshoe crabs stranded in spawning to keep their beaches looking (and smelling) cleaner. In the time since, *Just flip 'em*! ® has worked wonders in mitigating that need and changing that mindset through a simple, but effective message: if one sees a live horseshoe crab flipped over on its back along the beach – rather than walk by and ignore it – take a moment to gently turn it over and give it the opportunity to continue its life.

The rationale for this derives from the scientifically established observation that when horseshoe crabs are flipped over on their backs along spawning shorelines, a certain percentage of them (10–15 % or so) will not be able to right themselves, resulting in death by desiccation, predation by gulls, or other factors (Botton and Loveland 1989). But the intent behind *Just flip 'em*! ® runs deeper than the number of crabs or percentage of the population that might be saved by this action. The larger goal lies in getting people to overcome their indifference to this animal (Fig. 32.1). In flipping over a stranded horseshoe crab on the beach, an individual makes a conscious compassionate connection to the animal. From that grows an interest in learning more, and ultimately, taking action in transferring that knowledge and passion to others.

Consider the person who is flipping horseshoe crabs during their morning walk along the beach (as has become the habit of many Delaware Bay shore residents in recent years). Along the way, that individual crosses path with a neighbor or visitor who sees her/him flipping horseshoe crabs and asks "why?". At that point that individual makes the transition from crab flipper to crab educator. In this way, a chain of awareness, linked to conservation action, grows person to person, expanding out to the community and beyond. Over the years, numerous wonderful examples of such impacts have emerged from the *Just flip 'em*! ® program, including the woman, who during her morning walks along her Delaware Bayshore backyard, puts a stone in her pocket for every 100 crabs she flips, and in this way documented rescue of more than 20,000 crabs during a single spawning season! Then there's the story of Jasper the Golden Retriever – made famous in a recent poster sponsored by Charles

Fig. 32.1 ERDG's *Just flip 'em*! ® program

Fig. 32.2 Jasper's discovery poster (Produced by Charles River Laboratories in collaboration with ERDG)

River Laboratories (see Fig. 32.2) for the dog's ability to find dying crabs stranded high on the beach that his human crab-flipping companion had overlooked (Charles River is one of four companies in the United States that bleeds horseshoe crabs for human health biomedical benefits and is currently working with ERDG in promoting an *ICare* campaign for horseshoe crab awareness). In recent years, compassionate connecting to horseshoe crabs has expanded to include an annual blessing of the horseshoe crabs event that brings hundreds of Buddhists from across the Northeast to a Delaware Bay beach to pay homage to this ancient mariner (MacMillan 2012).

32.2.2 Back Yard Stewardship™

Residents in many of these same beaches where *Just flip 'em*! ® has become a popular activity can further their conservation commitment by banding together – through ERDG's *Back Yard Stewardship*™ program – in declaring their shared habitats as horseshoe crab sanctuaries (ERDG 2009). This program was founded with the idea that the future survival of the world's four horseshoe crab species will ultimately depend upon preservation of spawning habitat – a challenging prospect in light of the ever-increasing human density along the same beaches horseshoe crabs rely on for spawning. Since the program was launched in 1999, ten beaches on Delaware Bay (seven on the Delaware side and three in New Jersey), encompassing over 15 miles of some of the most productive horseshoe crab spawning habitats in the world, have been protected through designation as horseshoe crab sanctuaries. Efforts are also underway with citizen advocates in New York, Florida and the Yucatan to enlist community support for establishing sanctuaries in other parts of *Limulus*' range where horseshoe crab spawning occurs along human occupied beaches.

As described by ERDG, creating a horseshoe crab sanctuary is, at its simplest, a community's expression of how they view their beach and the natural resources it encompasses. Communities vary widely in how this designation is manifested (Gauvry 2009). Many communities erect signs to alert visitors to the significance of their beach to horseshoe crab spawning. Some have incorporated the horseshoe crab into their town logo, symbol or slogan and a few even hold festivals to celebrate the annual spawning ritual. Others engage more quietly, enjoying the resource and educating their friends, neighbors and visitors about it as opportunities arise.

32.2.3 ERDG's Horseshoe Crab Website

Not everyone lives on or visits beaches where horseshoe crabs spawn, and even for those that do, there needs to be a place where people can satisfy their interest in learning more about this animal. The information-rich, user-friendly and ever-evolving ERDG website fulfills that need and then some. In addition to being the most frequented horseshoe crab website on the internet, www.horseshoecrab.org (ERDG 2008) serves as the umbrella for ERDG's global horseshoe crab conservation network. As such, it provides community interaction, cultural exchange, education resources and peer-reviewed information on the world's four horseshoe crab species to over 300,000 visitors annually and 1,500 linked sites worldwide. The success of this site can be measured by the tens of thousands of students, educators and researchers from around the world who use it annually. Major topics addressed on the web site include natural history, anatomy, global distribution, evolution, medical uses, conservation, and news (the latter featuring postings of the latest articles and news releases about horseshoe crabs from around the world). The depth and breadth, user-friendliness, and accuracy of information about horseshoe crabs available on this site are unparalleled.

Efforts to improve the website and update the information it provides are ongoing. A key recent addition is development of a comprehensive data base for horseshoe crab research (ERDG 2012), providing citations, listings (and a growing number of pdf downloads) for over a thousand publications on various aspects of horseshoe crab research. The ultimate goal of providing the world with the most comprehensive source of peer-reviewed information on the four horseshoe crab species drives these efforts. The user-friendly data base interface offers visitors the ability to search by topic, resource type and key words, along with access to easy download and upload of resources. This same approach will be applied to a recently conceived (in development) "Teacher Toolbox" section of the website, that will be dedicated to exchange of lesson plans, posters, PowerPoints, booklets, student research submissions, and various other outreach and education resources from around the world. One of the latest website enhancements provides important new information relating to the biomedical use of horseshoe crabs, and conservation practices critical to that use. For social media enthusiasts, ERDG's popular Facebook page (https://www.facebook.com/horseshoecrab.erdg) provides an engaging venue for more casual and immediate global sharing of horseshoe crab information, stories, photos, and videos (ERDG 2014a).

32.2.4 Young Voices™ Horseshoe Crab & The Arts™

The final component of ERDG's educational repertoire cultivates horseshoe crab awareness and education via a whole other arena – engagement of young people in sharing their heartfelt expression for horseshoe crabs through poems, stories, images and even musical compositions. Since the inception of the *Young Voices™ Horseshoe Crab & The Arts™* program in 1999 (ERDG 2014b), thousands of students from various locations around the world have contributed to this collective voice, exploring, expressing and nurturing the compassionate heart – not only in themselves through the power of expression and creativity they bring to, and take from, their work, but also in adults who view these works and are moved in their own hearts by the messages they conveyed and emotions they evoke. Figure 32.3 offers a montage of a few examples of the many wonderful works contributed by these young artists.

32.3 *Green Eggs & Sand*

In 1999, at a time when schools in Delaware, Maryland and New Jersey were teaching middle and high school students ecology via lessons on Australian kangaroo rats, mid-western prairie dogs, and South American tropical rain forests, an important ecological and economical drama was occurring right in their own backyards. A major conflict was being played out in the media, courts and politics between the rights of fishermen to harvest horseshoe crabs for bait and the need to conserve horseshoe crabs for providing a superabundance of eggs as a vital food resource for migratory shorebirds.

Consequently, a group of environmental education specialists from those three states, along with a prominent horseshoe crab video documentarian, Michael Oates (2014) put their collective heads and resources together and decided to do something to change that. After considerable thought, planning, and a great deal of effort, a program emerged from all this that came to be called *Green Eggs & Sand.*

From the outset, the goal of *Green Eggs & Sand* (GE&S) was to provide an intensive teacher workshop experience and comprehensive classroom curriculum for immersing Mid-Atlantic educators and their students in the horseshoe crab/migratory shorebird phenomenon and the contentious management controversy sur-

Fig. 32.3 Montage of selected young voices horseshoe crab in the arts competition winners

rounding it. At the first International Symposium on the Science and Conservation of Horseshoe Crabs on Long Island in 2007, members of the GE&S team delivered a detailed overview of the history, philosophy and underpinnings of that approach (O'Connell et al. 2009). A key focus of that presentation, and a big reason for the success of GE&S, can be described in one word – collaboration. By working together across state and agency lines, costs in staff time and funding are spread and shared (so no one program or state's budget is too burdened), at the same time that diverse talents and access to resources are synergized. Equally vital has been the collaboration fostered in bringing scientists and educators together through this project. Over the 15 years of presenting GE&S workshops in various Atlantic Coast venues, more than 100 expert presenters – including horseshoe crab and shorebird researchers, biomedical scientists, fisheries managers, commercial fishermen, and representatives of the conservation community – have shared their time, talents and perspectives on these issues with participants. The contributions of these experts in providing real data, video lessons and scientific ground-truthing to the curriculum have likewise been significant.

Building on that foundation of collaboration, GE&S continues to expand its audience, focus and resources in enhancing education about *Limulus* in numerous arenas. Educator workshops remain at the heart of that approach, with interest in the program leading to expansion of these opportunities well beyond the original Mid-Atlantic region focus. As such, three intensive weekend-long workshops are now offered each spring in tune with the full and new moon peak spawning times for horseshoe crab spawning in various locations, including: a late April/early May session (hosted by the University of Georgia's Marine Education Center and Aquarium on Skidaway Island), a mid-to-late May session on Delaware Bay (alternating between venues in Delaware and New Jersey), and a "northern" workshop in June (including sessions in recent years hosted by partners in New York, Massachusetts, Maine and Connecticut). These workshops (TYDB 2014) feature a mix of field activities (including at least one horseshoe crab spawning observation trip), numerous expert and stakeholder seminars, and demonstration activities from each of the four GE&S curriculum modules. Since the inaugural Delaware Bay workshop in 2000, 36 GE&S workshops have been presented, reaching over 1,000 educator/scientist participants, encompassing 25 states and 5 foreign nationalities. Educator surveys over the years document use of the GE&S experience and resources in reaching more than 100,000 students coast wide with teachings on various aspects of horseshoe crab biology, ecology, and connections to humans and the environment.

In 2010, as a consequence of the expanding geographic interest in horseshoe crabs, and to address the ever-changing and growing body of knowledge around horseshoe crabs, shorebirds and human use of the resource, the GE&S coordination team initiated a major update and revision to the then decade-old curriculum. A cadre of past participant educators were recruited from up and down the Atlantic coast and assigned to teams for review, update and enhancements to lessons in the first three modules of the curriculum. Over the course of several writing workshops held during 2010 and 2011, each of these teams collaborated in completing substantive rewrites/revisions of all lessons from the Horseshoe Crab, Shorebird and Human Connections modules (Fig. 32.4). Gaps in the curriculum were also identified, resulting in development of several new lessons for each of those modules, including many featuring strong interdisciplinary connections and/or offering an engaging STEM (Science, Technology, Engineering and Math) focus. This included new lessons providing in-depth exploration of horseshoe crab anatomy and vision, the tidal cycle (as it relates to horseshoe crab spawning), shorebird sampling methodologies, webquest information gathering and reporting challenges, and (thanks to the generous support of Lonza USA Inc.) simulation of the gel-clot test for biomedical use of horseshoe crab blood product. In the horseshoe crab module, new video segments were also produced to further improve its content, including four clips highlighting the functional anatomy of *Limulus* (inspired and directed by Dr. Carl N. Shuster Jr., one of world's foremost authorities on horseshoe crabs), and a six-segment companion piece featuring Dr. Jane Brockmann (an expert on the spawning behavior of *Limulus*) demonstrating field research projects to engage student observation of *Limulus* spawning activities (Brockmann 2012). As a standing policy of the GE&S team, the new curriculum-on-DVD is only distributed to educators participating in a full-weekend workshop.

Figure 32.5 illustrates a piece of one of the new GE&S lessons developed for the horseshoe crab biology and ecology module – involving use of molt specimens as a tool for studying finer points of horseshoe crab anatomy. Inspired by a short paper

Fig. 32.4 Screenshot of *Humans and Horseshoe Crabs* module menu from *Green Eggs & Sand* curriculum

contributed by Dr. Shuster some 50 years ago (Shuster and Horrell 1966), this hands-on lab-based activity engages students in rotating through a series of eight stations, each challenging the student to observe particular aspects of the horseshoe crab dorsal, ventral and even internal shell anatomy. A favorite station for students is one where they sex specimens by observation of gonopore structure on the molt operculum. As they move through the respective stations, students answer questions about what they observe via a two-page worksheet. Copies of the station cards, worksheet and answer key are provided with the lesson plan, along with an interactive PowerPoint for assessing student learning. Once the molt study kits are assembled for conducting this lab, they can be brought out year after year for repeated use with new groups of teachers or students. Thanks to the savvy of Delaware Aquatic Resources Education (ARE) program assistant Anthony Jackson, a process for presoaking molts in a 50/50 solution of glycerin and water imparts long-term flexibility to the molt specimens for enhancing ease of manipulation of various structures during the labs. In late summer and early fall, horseshoe crab molts of various sizes can pile up in large numbers along many Delaware Bay and coastal Atlantic beaches, enabling GE&S coordinators to collect sufficient quantities to distribute to teachers as part of the wealth of take-home materials provided through a GE&S workshop.

As GE&S looks to the future, the goal of providing quality workshops at venues up and down the Atlantic coast to equip and empower legions of new classroom

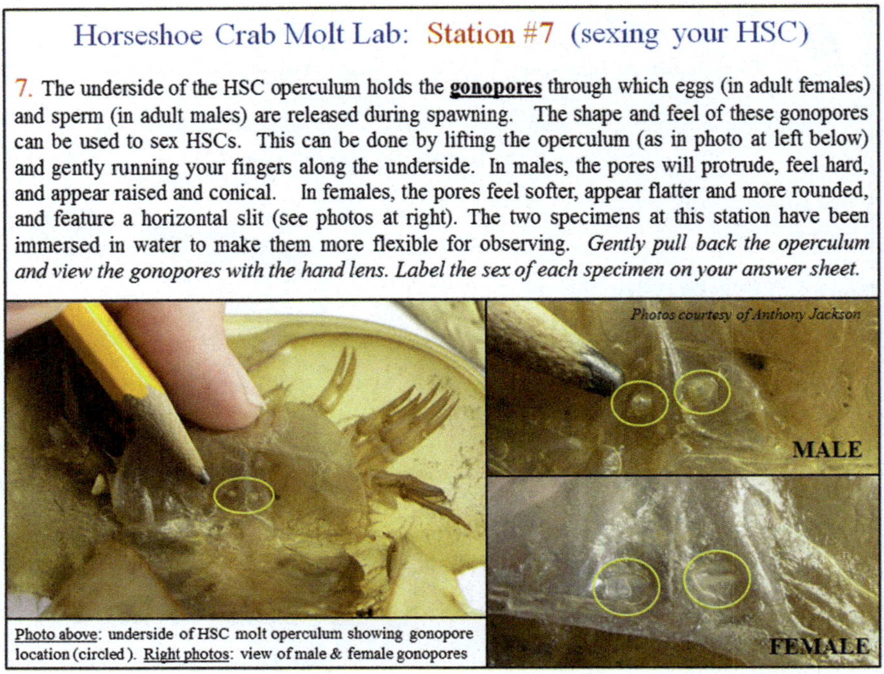

Fig. 32.5 Horseshoe crab molt anatomy lesson components (From *Green Eggs & Sand* curriculum)

teachers, curriculum coordinators and non-formal educators with knowledge and resources for educating about the wonders of horseshoe crabs will continue to be given high priority. Efforts to obtain the funding required to appropriately update the fourth and final video-driven management module of the curriculum – which serves as a wonderful case study of the chronology, complexities and challenges inherent in managing the multiple-use horseshoe crab resource – is also a key area for future focus. The GE&S team is also open to opportunities and inquiries on adapting the GE&S approach and materials to educating about horseshoe crabs in other parts of the world.

32.4 *Raising Horseshoe Crabs in the Classroom*

One of the tried-and-true ways of getting people connected to caring about a little-known or unappreciated animal is to put them in touch with the living creature in its early life stages, and that has certainly proven true with horseshoe crabs. Back in the mid-1990s and early on in the emerging horseshoe crab controversy (and in the principal author's work with horseshoe crab education), this approach was put to the test via construction of a simple three-window rectangular viewing box, each window containing a small aquarium featuring horseshoe crabs eggs, larvae and hatchlings at various stages of development. The box was taken to workshops, conferences and special event venues around Delaware, and everywhere it went, it attracted interest and engagement of people far surpassing the large number of fancy, high-quality information displays surrounding it. In addition to the attraction of viewing these tiny, elegant, seldom-seen life forms in a whole different light – relative to the somewhat coarse and beat-up versions of adults most people were accustomed to seeing spawning on bay beaches – people were amazed to hear that it took as long as it did (8–10 years) for those little creatures to become those large adults on the beach. At the time such awareness carried a valuable message in getting the public to appreciate the challenge inherent in sustaining populations of a slow-to-mature species like *Limulus* in the face of the heavy harvest pressure it was experiencing.

The success of that venture prompted Delaware ARE educators to initiate workshops for teachers demonstrating how to hatch and grow horseshoe crab eggs in the classroom. Although other staff and program priorities made this project short-lived in Delaware, the idea and approach were passed along to education colleagues in Maryland's Department of Natural Resources ARE Program, who have since adapted, enhanced and expanded it (MDNR 2014). The ambitious product of their efforts – *Raising Horseshoe Crabs in the Classroom* – engages teachers and students in raising newly-hatched horseshoe crabs in the classroom and monitoring conditions for their growth and development. Since its launching in 1999, this program has engaged more than 20,000 Maryland students in exploring the wonders of the horseshoe crab life cycle (Fig. 32.6).

As with *Green Eggs & Sand*, teacher attendance at a *Raising Horseshoe Crabs in the Classroom* workshop is a prerequisite for receiving the materials and accompanying lesson plans. At these workshops, in addition to a supply of pre-collected eggs or larvae, teachers receive a 10-gal aquarium, filter, dipnet, sand, salt, hydrometer, thermometer and water quality test kits. A Teachers' Guide, including instructions, lesson plans, data collection procedures and forms, rearing protocols, and other resources is also provided. The protocols enable schools participating in the project to learn the most up-to-date techniques for raising horseshoe crabs in the classroom, as well as procedures for monitoring salinity, nutrient levels and other water quality indicators that might impact rearing success. As part of their obligations, participating teachers are required to record and submit a monthly data sheet containing water quality and horseshoe crab maintenance information.

At the end of each academic year, schools may participate in one of four horseshoe crab release events on Chesapeake Bay. These events offer an opportunity for students to celebrate what they have learned, release horseshoe crabs they have raised, and participate in a host of hands-on field activities. These events mark the culmination of months of study for students involved in the program. Ada Stambaugh, a high school teacher participant in the program since 2001, described the benefits of her students work in the program as follows: "Several years ago I challenged my students to think of what horseshoe crabs might run into as food in the Bay possibly working toward the continental shelf. So they felt we should feed

Fig. 32.6 Scenes from Maryland's raising horseshoe crabs in the classroom program

them a variety of things that could be found pretty easily in the grocery store. We tried many things to see what they might enjoy the most and scallops were one of those things. We also feed them in smaller amounts and not as often: chicken liver/hearts, frozen clams (large stone clams), flounder, cod, rock fish, squid, and raw whole shrimp cut into pieces … I have really found this project helpful in making my students think and solve real life problems."

A key aspect of this program is that the number of young horseshoe crabs reared and released is a very small percentage relative to the wild populations, and are thus not intended to provide population restoration outcomes. This contrasts with horseshoe crab culture programs going on in Asia – including excellent initiatives in Hong Kong (Cheung et al. 2011), Taiwan (Chen et al. 2009) and Japan (Tsuchiya 2009) – where horseshoe crab populations are in serious decline, and captive breeding and release programs are a vital component to species survival. But comparable rewards and benefits in terms of enhanced student appreciation of the need for conservation of this remarkable animal are certainly seen through Maryland's program and reinforce that key idea articulated earlier in this article – planting seeds of connectivity in young people for horseshoe crabs that they will carry with them for the rest of their lives.

32.5 *Team Limulus*

Another unique example of student involvement in the research and conservation of horseshoe crabs is seen in *Team Limulus* at the Wheeler School, located in Providence, Rhode Island. Founded in 2008, *Team Limulus* is a group of high school and middle school students (ages 12–18) who are committed to monitoring a local population of horseshoe crabs and to educating other students and local communities about the importance of these animals (The Wheeler School 2011). Membership in the team is not limited to a particular grade, but rather based on the interest of students. As an indication of the impact such participation has on students, more than a dozen *Team Limulus*/Wheeler School graduates have returned at the end of the spring semester of college to continue participation in the program.

Student members of *Team Limulus* (Fig. 32.7) along with a faculty advisor, conduct spawning surveys (James-Pirri 2011, 2012a) at Gaspee Point Beach in Warwick, RI during the months of April, May, and June. The team also tags horseshoe crabs as part of the US Fish and Wildlife Service's Cooperative Horseshoe Crab Tagging Program (DCIB 2014a, b). Tagging data provide information about abundance, distribution, and migration patterns of populations along the Atlantic Coast. State agencies use this data to manage populations, so the students are collecting and contributing meaningful and significant data. The spawning survey data collected by *Team Limulus* are uploaded to two larger research data bases: one maintained by researcher Mary Jane James-Pirri of the Graduate School of Oceanography at the University of Rhode Island (James-Pirri 2012a, b) and the other by researchers Jennifer Mattei and Mark Beekey in the Department of Biology at Sacred Heart University in Fairfield, Connecticut (SHU 2014), the latter of whom also manage *Project Limulus*, a community-based research program within Long Island Sound (see Mattei et al. 2015 for more information on that effort). A Rhode Island Environmental advocacy group, *Save the Bay*, has also used *Team Limulus*' spawn-

Fig. 32.7 *Team Limulus* students at work: collecting morphometric data while tagging horseshoe crabs at Gaspee Point Beach, Warwick, RI (*Photo left*) and presenting their research at Massachusetts Audubon Wellfleet Bay Wildlife Sanctuary in Wellfleet, MA (*Photo right*)

ing survey data to support lobbying for greater harvesting restriction of horseshoe crabs during the spawning season in Narragansett Bay, RI.

32.5.1 Student Presentations

Beyond the field work, student members of *Team Limulus* also organize, analyze, and present their findings from the spawning and tagging data at local conferences. Students presented their research from 2009 to 2010 at a Massachusetts Audubon Society conference on horseshoe crabs (Fig. 32.6) and as a keynote speaker group for an evening talk as part of a summer series at the Wellfleet Bay Audubon Sanctuary in Massachusetts. Additionally, *Team Limulus* students share their passion for horseshoe crabs with younger students by guest teaching in science classes in both the Wheeler School elementary and middle school divisions. As part of those efforts, the high school students design and present hands-on activities with *Limulus* molts to engage the younger students in discovering the unique features of horseshoe crabs and helping them develop a love and respect for the organism.

32.5.2 High School Biology Level Activities

At Wheeler, interest and involvement in *Team Limulus* is also stimulated through use of elements of the *Green Eggs & Sand* in various classes. This includes engaging students in the Advanced Placement (AP) Biology class in running the LAL (*Limulus* Amoebocyte Lysate) lab during their microbiology and immunology unit, as a practical, hands-on application of the concepts they are learning in class. In this lab, student use LAL to test for the presence of bacterial endotoxin in their saliva and various water sources. This lab activity allows students to gain an awareness of the vital role that horseshoe crabs play in human health. As a culmination to the ecology unit, AP students participate in another *Green Eggs & Sand* activity: *Webquest*: *the horseshoe crab controversy* (Fig. 32.8). Here students learn about the challenges and difficulties involved in resource management of a fishery species. Teams of students are assigned different roles and must research their stakeholder's position online in preparation for a town hall style meeting. Stakeholder roles include: watermen, biomedical researchers, birders/ecotourists, environmentalists, beachfront home owners, and natural resource managers. The end goal is for each stakeholder group to knowledgably and confidently present their viewpoint, back it up with data, answer questions successfully, and ultimately work together to create a management solution. Students truly enjoy this activity and the opportunity to role play an actual real-world scenario. It is an excellent learning experience because students come away from it with the realization that there is no easy solution that will satisfy all the stakeholders' needs in this complex conservation and management issue.

Fig. 32.8 AP Biology students participate in a mock town meeting from the *Green Eggs & Sand* curriculum

32.6 *Crab App* Smart Phone Technology

One of the newer tools available to groups such as *Team Limulus* for use in their horseshoe crab monitoring efforts is the aptly titled "*Crab App*." Over the last decade in the United States, increasing focus has been given to citizen science, often referred to as "public participation in scientific research" (Mattei and Beekey 2008; Hand 2010). At the same time, increasing use of smartphone technology has opened a new door equipping the ordinary citizen to collect data that can then be used in scientific research. *Crab App* combines these two trends in offering a free I-Phone app that not only educates the user about horseshoe crabs, but allows a person who sees/finds a horseshoe crab on a beach to upload meaningful data to an online data base. Dr. Matthew Sclafani, of Cornell Cooperative Extension, Suffolk County, Marine Program in New York, and Jared Lamenzo, of The Wildlab (2014) are the creators of this application. The goal of this app is to enable citizen scientists to count horseshoe crabs, monitor populations and report data in a meaningful way that can then be easily accessed and utilized by researchers. The App identifies location using GPS and has an information button that teaches the user how to distinguish between male and female crabs and provides other useful information (Fig. 32.9). Crab App version 1.1.0 is available at no cost and can be downloaded from Itunes (Apple, Inc. 2014). Another more recent example of this approach is the

Fig. 32.9 Screen shots of homepage and data collection pages of *Crab App*

My HSC App Android mobile application. Designed by a team of scientists from Malaysia, and downloadable from Google Play (2014), this app provides a valuable, user-friendly tool for inputting information on the global distribution of all four horseshoe crab species.

32.7 Conclusions

A wealth and variety of efforts are being directed towards educating about horseshoe crabs in the United States (Fig. 32.10). It is interesting to contemplate how substantially and quickly this point of focus has emerged and magnified. Two decades ago, one would be hard-pressed to name a single program dedicated to horseshoe crab education in this part of the world. Today, interest in *Limulus* and the conservation of their estuarine habitats appears to be growing exponentially, with new video programs, books, website components, social media networks, celebratory festivals, and educational resource materials exploding on the scene each and every year. It is essential to find ways to tap into and translate this vein of human interest and fascination to address serious challenges in the conservation of horseshoe crabs in parts of the world where resources are lacking, and priorities conflicting. A recently-convened Education Working Group of the International Union for the Conservation of Nature's (IUCN) Horseshoe Crab Science Specialist Group (2013) is poised to tackle that challenge. It is hoped that, through that network – in sharing the above and other initiatives from the United States, along with the advanced education programs underway in Japan, Taiwan, Hong Kong and Singapore – further development of approaches and pathways to help address that goal will be realized.

Selected Horseshoe Crab Education Websites from the United States

Ecological Research Development Group: www.horseshoecrab.org

ERDG's research data base: http://horseshoecrab.org/research/

Green Eggs and Sand workshop information: www.tydb.org/greeneggs_sand.php

Maryland horseshoe crab information site: www.dnr.state.md.us/education/horseshoecrab

University of Delaware horseshoe crab site: www.ceoe.udel.edu/horseshoecrab

Team Limulus: www.wheelerschool.org/podium/default.aspx?t=127732&rc=1

US Fish and Wildlife Service horseshoe crab tag reporting site: http://www.fws.gov/crabtag/

Project Limulus: www.sacredheart.edu/pages/13692_project_limulus.cfm

The Horseshoe Crab App: Wildlab: www.thewildlab.org/species/view/Horseshoe_Crab

Horseshoe crab App in Itunes: http://itunes.apple.com/us/app/crab-app/id399402040?mt=8

New York Horseshoe Crab Monitoring Network: www.nyhorseshoecrab.org

Marine Biological Lab: www.mbl.edu/marine_org/images/animals/Limulus/index.html

Green Eggs and Sand, Team Limulus, and more presentation at 2011 Hong Kong workshop
www.cityu.edu.hk/bch/iwscahc2011/Download/O-20%20Gary%20Kreamer.pdf

Fig. 32.10 Selected websites featuring horseshoe crab education information and resources from the US

References

Apple Inc (2014) Crab App. Mediated Spaces, Inc. https://itunes.apple.com/us/app/crab-app/id399402040?mt=8. Accessed 21 July 2014

Botton M, Loveland R (1989) Reproductive risk: high mortality associated with spawning by horseshoe crabs (*Limulus polyphemus*) in Delaware Bay, USA. Mar Biol 101:143–151

Brockmann J (2012) Lesson 9. Field projects and data analysis. In: Module 1: the horseshoe crab, green eggs & sand (curriculum DVD). The Green Eggs and Sand Partnership

Chen CP, Hsieh HL, Chen A et al (2009) The conservation network of horseshoe crab *Tachypleus tridentatus* in Taiwan. In: Tanacredi JT, Botton ML, Smith DR (eds) Biology and conservation of horseshoe crabs. Springer, New York, pp 543–557

Cheung J, Lau T, Cheung SG et al (2011) Conservation in action: a rearing and releasing programme of juvenile horseshoe crabs in Hong Kong schools. http://www.cityu.edu.hk/bch/iwscahc2011/Download/O-17%20Joe%20Cheung.pdf. Accessed 23 July 2014

DCIB (Delaware Center for the Inland Bays) (2014a) Horseshoe crab resighting form. http://www.fws.gov/northeast/marylandfisheries/Fish%20Facts/horseshoe%20crab%20fact.html. Accessed 21 July 2014

DCIB (Delaware Center for the Inland Bays) (2014b) Horseshoe crab tagging protocol. http://www.inlandbays.org/wp-content/documents/tagging_procedure.pdf. Accessed 21 July 2014

ERDG (Ecological Research & Development Group Inc.) (2006) Just flip 'em! http://www.horseshoecrab.org/act/flipem.html. Accessed 24 July 2014

ERDG (Ecological Research & Development Group Inc.) (2008) The horseshoe crab. http://www.horseshoecrab.org/. Accessed 10 Feb 2014

ERDG (Ecological Research & Development Group Inc.) (2009) Backyard Stewardship™: coastal communities define their shared habitat as a horseshoe crab sanctuary. http://www.horseshoecrab.org/act/sanctuary.html. Accessed 23 July 2014

ERDG (Ecological Research & Development Group Inc.) (2012) Horseshoe crab research data base. http://horseshoecrab.org/research/. Accessed 23 July 2014

ERDG (Ecological Research & Development Group Inc.) (2014a) Ecological Research & Development Group Inc. (ERDG) community. https://www.facebook.com/horseshoecrab.erdg. Accessed 23 July 2014

ERDG (Ecological Research & Development Group Inc.) (2014b) Young voices: horseshoe crab in the arts. http://www.horseshoecrab.org/act/contest.html. Accessed 23 July 2014

Gauvry G (2009) Community building: an integrated approach to horseshoe crab conservation. In: Tanacredi JT, Botton ML, Smith DR (eds) Biology and conservation of horseshoe crabs. Springer, New York, pp 605–612

Google Play (2014) My HSC App. https://play.google.com/store/apps/details?id=com.myhsc&hl=en. Accessed 23 Nov 2014

Hand E (2010) Citizen science: people power. Nature 466(7307):685–687. doi:10.1038/466685a

IUCN SSC Horseshoe Crab Specialist Group (2013) Annual report of the Horseshoe Crab Specialist Group for 2012. http://horseshoecrab.org/press/2013/02/Annual-Report-of-HSC-SSG-2012.pdf. Accessed 23 July 2014

SHU (Sacred Heart University) (2014) Project *Limulus*. http://www.sacredheart.edu/academics/collegeofartssciences/academicdepartments/biology/projectlimulus/. Accessed 21 July 2014

James-Pirri MJ (2011) 2011 spawning survey schedule: Gaspee Point Beach, Warwick, RI. Accessed 21 July 2014

James-Pirri MJ (2012a) Horseshoe crab spawning survey instructions. http://www.gso.uri.edu/mjjp/HSC_survey_instructions.pdf. Accessed 21 July 2014

James-Pirri MJ (2012b) Mary Jane James-Pirri – GSO, URI Homepage. http://www.gso.uri.edu/mjjp/. Accessed 21 July 2014

MacMillan M (2012) Hundreds pray for horseshoe crabs. Cape Gazette.com http://capegazette.villagesoup.com/p/hundreds-pray-for-horseshoe-crabs/836460, http://files.meetup.com/1400103/HSC%20Blessings.pdf. Accessed 23 July 2014

Mattei JH, Beekey MA (2008) The horseshoe crab conundrum; can we harvest and conserve? Wrack Lines 8(1):2–7

Mattei JH, Botton ML, Beekey MA et al (2015) Horseshoe crab research in urban estuaries: challenges and opportunities. In: Carmichael RH, Botton M, Shin PKS, Cheung SG (eds) Changing global perspectives on horseshoe crab biology, conservation and management. Springer, New York, pp 557–574

MDNR (Maryland Department of Natural Resources) (2014) Raising horseshoe crabs in the classroom. http://www.dnr.state.md.us/education/are/GES/horseshoecrab/raising.asp. Accessed 23 July 2014

O'Connell K, Etgen C, Kreamer G et al (2009) *Green eggs and sand*: a collaborative effort of scientists, teachers, resource managers and stakeholders in educating about *Limulus polyphemus*. In: Tanacredi JT, Botton ML, Smith DR (eds) Biology and conservation of horseshoe crabs. Springer, New York, pp 595–604

Oates M (2014) Michael Oates 302 stories. http://vimeo.com/user15002426. Accessed 23 July 2014

Shuster CN Jr, Horrell HC (1966) *Limulus* exoskeleton as a teaching aid. Turtox News 44(1):40–41

The Wheeler School (2011) Team *Limulus* 2011: student horseshoe crab research. http://www.wheelerschool.org/podium/default.aspx?t=127732. Accessed 21 July 2014

The Wildlab (2014) Homepage. www.thewildlab.org. Accessed 23 July 2014

Tsuchiya K (2009) The history of horseshoe crab research and conservation in Japan. In: Tanacredi JT, Botton ML, Smith DR (eds) Biology and conservation of horseshoe crabs. Springer, New York, pp 559–570

TYDB (Thank You Delaware Bay) (2014) General green eggs and sand. http://tydb.mobiusnm.com/general-green-eggs-and-sand. Accessed 23 July 2014

Chapter 33 Building Local Community Awareness of Horseshoe Crabs in Kujukushima, Nagasaki, Japan

Chihiro Nishimura and Chikako Iwaoka

Abstract Kujukushima is one of the several habitats of the horseshoe crab *Tachypleus tridentatus* in Japan. However, this is not widely known by local residents. This article introduces networking activities concerning horseshoe crabs in Kujukushima, Nagasaki, Japan.

The Kujukushima Aquarium was reopened in July 2009. It averages about 400,000 visitors a year with approximately 30 % of them being children. It is a community-based aquarium. Previously, there was only one horseshoe crab exhibition tank, but in the newly renovated aquarium, the horseshoe crab exhibition area is much expanded. Exhibitions include living horseshoe crabs (adults, juveniles, eggs), information on their life history and ecology, and activities on community awareness building and origami (related to marine life) for children. The horseshoe crab laboratory is set up behind the main exhibition room.

The Kujukushima Aquarium is collaborating with the University of Nagasaki. University students, who are studying regional policy with a focus on eco-tourism, participated in the design of the new aquarium. The collaborative work aims to build community awareness and provide quality education on the importance of horseshoe crab conservation. In the aquarium, the students participate as volunteers in the horseshoe crab exhibition and in programs for children once a month in the children's educational playroom.

Former students have organized and established the Kujukushima Ramblers' Club for building community awareness and conservation of horseshoe crabs. They started up the project, the *Kabuto-ganikki* (Horseshoe Crab Diary), and garnered support from the Kao Corporation in 2009 and 2011 as a part of their corporate social responsibility program (CSR). Such kind of cooperation between the aquarium

C. Nishimura (✉)
Department of Regional Policy, Faculty of Economics, University of Nagasaki, 123 Kawashimo, Sasebo, Nagasaki 858-8580, Japan
e-mail: cnishimu@sun.ac.jp

C. Iwaoka
Kujukushima Aquarium, 1008 Kashimae, Sasebo, Nagasaki 858-0922, Japan

R.H. Carmichael et al. (eds.), *Changing Global Perspectives on Horseshoe Crab Biology, Conservation and Management*, DOI 10.1007/978-3-319-19542-1_33

and the university serves as a good model to fulfill education goals for both the university and the community.

Keywords Kujukushima • *Tachypleus tridentatus* • Networking activities • Building community awareness

33.1 Introduction

Nagasaki Prefecture is not large in area (4,094 km^2), but its coastline is quite complicated, having what is called a "saw-tooth" coastline. Because of this, Nagasaki Prefecture has the longest coastline (4,137 km) in all the administrative divisions of Japan. The notable feature of horseshoe crab habitats in Nagasaki is that many small habitats are scattered along this complicated coastline, and there may be many undiscovered habitats where surveys have not yet been conducted. Therefore, there is a possibility of finding new habitats and spawning sites in the future.

Figure 33.1 shows the distributions of horseshoe crabs in Japan and Nagasaki Prefecture. Habitats in the Seto Inland Sea have almost disappeared, leaving behind Kyushu Island being the main horseshoe crab habitat in Japan. Conservation activities in Nagasaki Prefecture, therefore, are especially important for the survival of horseshoe crabs in Japan.

At the New York symposium in 2007, Iwaoka reported on public awareness activities, including cooperation with local fishermen, exhibitions and environmental education activities in the old aquarium, lectures at elementary and junior high schools, field excursions, the use of mass media, and collaboration with the local

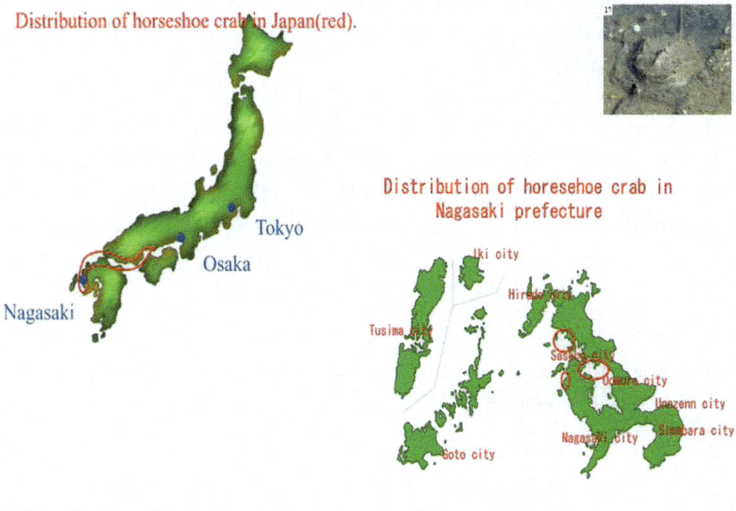

Fig. 33.1 Distribution of horseshoe crabs in Japan and Nagasaki Prefecture

government (Iwaoka and Okayama 2009). In the next symposium in Hong Kong in 2011, we reported on a new network for building community awareness concerning horseshoe crabs in Kujukushima, Nagasaki, Japan. This new network was established between the aquarium and the university with university students studying regional policy helping out in various activities. After these two symposia, we have extended our community-based conservation work to local residents. The aim of this article is to introduce and describe the network for the conservation of horseshoe crabs in Kujukushima.

33.2 Kujukushima and the Aquarium

The Kujukushima Islands are located within the Saikai National Park and consist of many small islands which have saw-tooth coastlines (Fig. 33.1). Though "Kujukushima" means "99 islands" in Japanese, there are in fact 208 islands in the area. The Kujukushima Islands have relatively long natural coastlines forming various habitats such as rocky beaches, sandy beaches, tidelands and algal beds, which provide diverse environments for the survival of the horseshoe crab.

The aquarium is located in Sasebo City, Nagasaki Prefecture (Fig. 33.2). Not only the breeding and exhibition of flora and fauna but also public awareness programs and investigation and research activities by local residents are emphasized. For example, the biggest horseshoe crab in Japan was found here. It was dried and preserved, and is now displayed in the aquarium. An American horseshoe crab,

Fig. 33.2 Saikai National Park Kujukushima Aquarium "Umikirara"

Limulus polyphemus, long considered to be an alien species, was found by local residents in a supermarket parking lot. It was delivered to the aquarium by local police. Staff specialists working at the aquarium provide environmental education on horseshoe crabs and tidal flat observation for local resident groups, school groups, etc. Through these activities, we hope that more information on horseshoe crabs from this region will be collected and their conservation value will receive more recognition.

The Kujukushima Aquarium opened at the Saikai Pearl Sea Resort in 1994 and was renovated in 2009. It is a compound type facility having an exhibition pavilion containing a model ship, an IMAX dome theater, and an aquarium. The aquarium operates under the concept of community-based aquarium. Through investigation and research, the aquarium collects information on local flora and fauna with an aim to raise the interest and concern of local residents on them.

Before the renovation, there was only one tank for horseshoe crabs, and no place was available for conservation activities. The new aquarium possesses live display of various life stages of horseshoe crabs (adults, juveniles, eggs) with information on the life history and ecology being shown. Community awareness building, and marine life related activities are organized for children (Fig. 33.3). The horseshoe crab laboratory is set up behind the exhibition area. There are two tanks that are filled with eggs and juveniles. The backyard, where breeding, investigation, and research are undertaking has been enhanced as well. The latest findings of horse-

Fig. 33.3 Horseshoe crab section in Kujukushima Aquarium before and after reopening

shoe crabs provided by staff or local residents can be written directly on a map of the Kujukushima Sea area. With such efforts, we believe that the awareness of the conservation of horseshoe crabs among children of Sasebo will be growing.

33.3 Cooperation with the University Laboratory

33.3.1 The University Laboratory

The University of Nagasaki (Sasebo Campus) is only an 8 min drive from the Kujukushima Aquarium. Students from the Department of Regional Policy, Faculty of Economics studying building community awareness participated in the planning and execution for the expansion of the horseshoe crab section of the aquarium. Figure 33.4 shows the conceptual diagram of their study which adopts a trans-scientific approach and is largely field-based with emphasis on evidence-based policy and practice. Another feature is a focus on the "glocal (global and local)" perspective. Kujukujshima is one of the field sites of their study and typical activities are shown in Fig. 33.5. These activities include walking in the countryside (Fig. 33.5a) and sea kayaking in the Kujukushima Sea (Fig. 33.5b). These activities are compulsory for all laboratory students. The students shop in the local fish market (Fig. 33.5c) and do their cooking of the seafood (Fig. 33.5d) caught in the Kujukushima Sea. This program is now adapted for children. The visitor experience in the aquarium was measured using psychological and physiological parameters

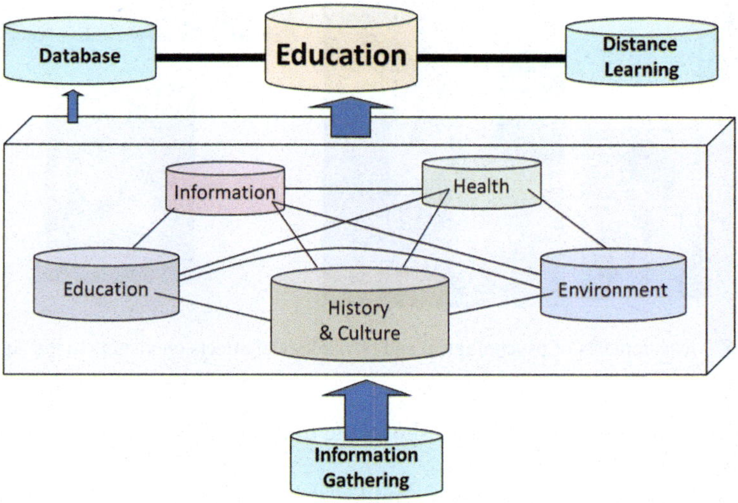

Fig. 33.4 Conceptual diagram of the study in the Department of Regional Policy, Faculty of Economics, The University of Nagasaki

Fig. 33.5 Typical activities for the university laboratory around Kujukushima

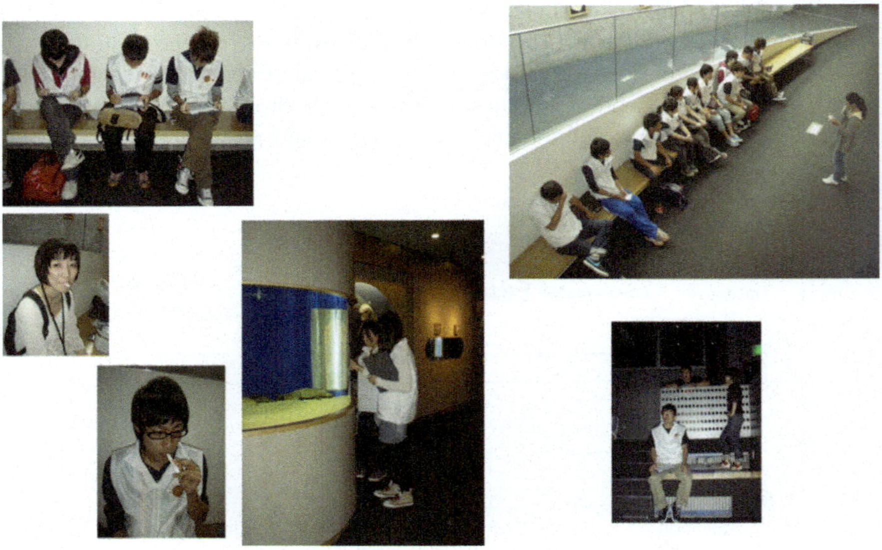

Fig. 33.6 Measurements of psychological and physiological effects on visitors to the aquarium

(Fig. 33.6). The photo at the center shows the training of the guides for introducing horseshoe crabs to visitors.

The students are also involved in guided tours for introducing horseshoe crabs to visitors using exhibition panels and videos, and they assist in horseshoe crab origami activities (they can complete an origami horseshoe crab within 3 min!) and a program for children once a month in the children's educational playroom.

The aims of these activities are to educate the visitors and enhance their community awareness of the conservation of horseshoe crabs.

The university laboratory students held various activities in the aquarium that formed part of their research for the graduation reports. For example, a female student was involved in the production and delivery of a picture story based on a folk tale of the horseshoe crab (Fig. 33.7a). Some students took charge of the exhibition (Fig. 33.7b) or children's educational programs, which included card-making on Mother's Day, a quiz about Kujukushima and its associated flora and fauna, a picture story show, a short play, and Christmas-tree making (Fig. 33.7c). A female student (Fig. 33.7d) researched on the degree of recognition of the concept of this aquarium by local residents. Two university students participated in a Chinese intensive program in the university and published in Taiwan a picture book of horseshoe crabs (Fig. 33.7e, f). The male student (Fig. 33.7f) took charge of storytelling in Chinese and Japanese at the children's educational playroom and at an elementary school.

33.3.2 *Kujukushima Ramblers' Club*

Kujukushima Ramblers' Club is a citizens' group established in 2001 by students graduated from the University of Nagasaki (Sasebo Campus). When the club was just established, activities only included walking and sea kayaking around Kujukushima but later were extended to building community awareness and horseshoe crabs conservation. Students and graduates of this university started the *Kabuto-ganikki* (horseshoe crab diary) project in 2009. *Kabutogani* means

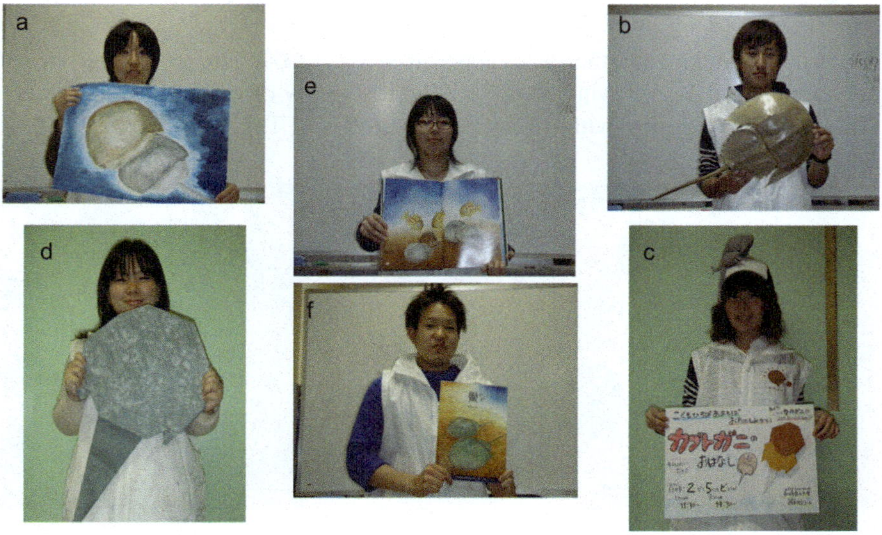

Fig. 33.7 Various activities planned and led by University students

horseshoe crab and *nikki* means diary in English. The combination in one word means "diary or record of horseshoe crabs." This club was supported by Kao Corporation in 2009 and 2011 as a part of their corporate social responsibility (CSR).

33.3.3 Educational Programs for Children

Programs for children are offered in the children's educational playroom and the horseshoe crab section in the Kujukushima Aquarium. Figure 33.8 shows the participation of the children in making a very big origami horseshoe crab with the help of the university students in June of 2011. The size of that origami horseshoe crab is even bigger than the largest biological specimen collected in Japan (Fig. 33.9). The contents and time periods for exhibitions in the horseshoe crab section in the aquarium are shown in Table 33.1.

According to tradition, a pair of horseshoe crabs brings luck. In Japan, they are often mentioned in speeches in wedding ceremonies because, traditionally speaking, people believe that a pair of horseshoe crabs stays together forever. In 2011, The Kujukushima Ramblers' Club held an event called "Pictures with Horseshoe Crabs," and couples and families who visited the aquarium took pictures with preserved horseshoe crabs on St. Valentine's Day (Fig. 33.10). After obtaining permission, pictures were exhibited in the horseshoe crab section in the aquarium a month later.

33.4 Evaluation and Future Prospects

The Kujukushima Aquarium was reopened in July 2009. This community-based aquarium was designed with the help of the University of Nagasaki students who studied regional policy with a focus on eco-tourism. The aims are to build the

Fig. 33.8 Horseshoe crab origami in the children's educational playroom at Kujukushima Aquarium

Fig. 33.9 This giant origami horseshoe crab is bigger than the largest specimen of horseshoe crab ever found in Japan

Table 33.1 Exhibition list for children in the horseshoe crab section in the aquarium

1. Habitat (2009.07–2009.08)
2. Natural monument (2009.08–2009.10)
3. Conservation (2009.10–2009.11)
4. Events (2009.11–2009.12)
5. Exchanges (2009.12–2010.01)
6. Kujukushima Ramblers' club (2010.01–2010.03)
7. Ecotourism (2010.03–2010.04)
8. Community building in Japan (2010.03–2010.05)
9. Education for children (2010.05–2010.09)
10. Story around the world (2010.09–2011.02)
11. Public awareness (2011.02–2011.06)

community awareness and provide the general public education on horseshoe crab conservation. Figure 33.11 shows the mode of cooperation between the aquarium and the university. Their activities reflect the community-based concept of the Kujukushima Aquarium. The exhibition presented by university students catches visitors' interest and is very popular. Moreover, various advices from the university especially with regard to the building of community awareness provide much support to the aquarium staff and strengthen the recognition of the importance of horseshoe crabs among local residents and visitors. In addition to field surveys in collaboration with the university, a program that makes the horseshoe crab a motif is offered in the children's educational playroom. It seems that this may be a chance to successfully instill a concern for valuable living organisms such as horseshoe crabs in the next generation. The aquarium intends to strengthen the cooperation

Fig. 33.10 "Pictures with horseshoe crabs" on Valentine's Day

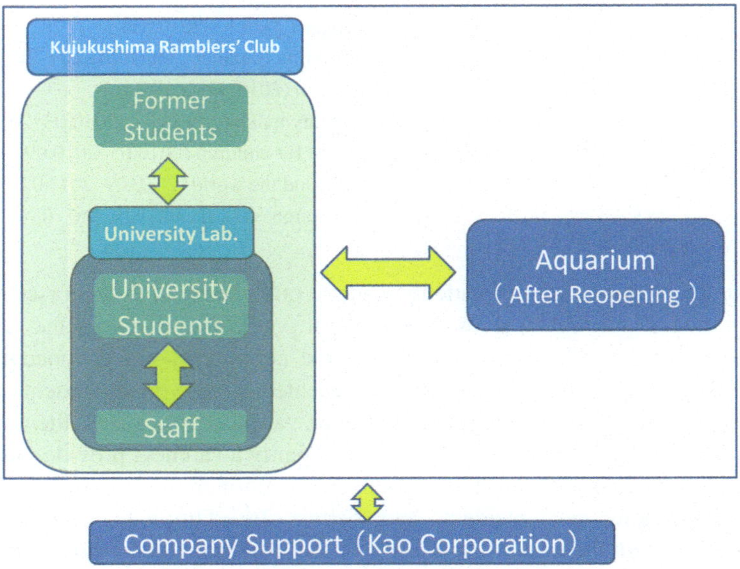

Fig. 33.11 The model of cooperation between the University of Nagasaki and the Kujukushima Aquarium

with local residents and local educational institutions on the conservation of horseshoe crabs and to continue to provide community-based activities in the future. On the other hand, the personal development of the university students was observed with abilities to take action, and skills in thinking and teamwork much enhanced. It suggests that such kind of cooperation between the aquarium and the university serves as a good model to fulfill education goals for both the university and the community.

33.5 Conclusion

As mentioned above, such cooperation on horseshoe crabs is underway in Kujukushima, and new efforts have been made from a local perspective. We would like to extend our effort to regional collaborations and establish a network in East Asia on the conservation of horseshoe crabs. It is well known that Japan has had a deep connection with China and Taiwan in a range of areas including natural, historical, cultural, and economic. A typical example is the establishment of friendship cities between Japan and China such as between Hirado City, Japan and Nan'an City, China, and between Nagasaki City, Japan and Fuzhou City, China. We wish to study the conservation of horseshoe crabs from a trans-scientific approach with all of our Asian friends.

Acknowledgments These activities were supported by the Kao Community Museum Project of the Kao Corporation. Mr. S.E. Rife provided helpful advices on the manuscript. The authors thank Kao Corporation and Mr. S.E. Rife for their support.

Reference

Iwaoka C, Okayama T (2009) Public awareness and community-based conservation for the horseshoe crab at Saikai National Park in Nagasaki Prefecture, Japan. In: Tanacredi JT, Botton ML, Smith DR (eds) Biology and conservation of horseshoe crabs. Springer, New York, pp 571–581

Chapter 34
Young Voices: Through the Arts, Future Environmental Stewards Have a Global Voice

Glenn Gauvry and Ruth H. Carmichael

Abstract Young Voices, Horseshoe Crabs and the Arts™, is a global outreach campaign directed toward students from pre-K though high school, but shared with people of all ages. The program has been operated by the Ecological Research and Development Group, Inc. since inception in 1999. During the past 15 years, the program has grown from a few hundred to participation by many thousands around the world. We celebrate these young environmental stewards whose compassion for the conservation of this remarkable species, expressed through art, has elevated and transformed our perception of this often-misunderstood mariner. To the artists and educators who have made this program possible, we give our heart-felt thanks.

Keywords Conservation • Education • ERDG • Student • Art • School

34.1 Introduction

The Ecological Research & Development Group (ERDG), founded in 1995, is a non-profit wildlife conservation organization whose primary focus is the conservation of the world's four remaining horseshoe crab species. ERDG's philosophy is that the foundation of a successful conservation program is effective public education and coastal community support, combined with conservation initiatives that preserve essential habitat and minimize human impact.

To support their mission, ERDG has an annual juried arts competition to which students from around the world submit poems, tales, and images in appreciation and celebration of the remarkable horseshoe crab. The competition is open to all grade levels (pre-K–12). Students may enter through their school or individually.

G. Gauvry (✉)
Ecological Research & Development Group, Inc., 190 Main Street, Little Creek, Dover, DE 19901, USA
e-mail: kabutogani@comcast.net

R.H. Carmichael
Dauphin Island Sea Lab, University of South Alabama, Dauphin Island, AL 36528, USA

Department of Marine Sciences, University of South Alabama, Mobile, AL 36688, USA

R.H. Carmichael et al. (eds.), *Changing Global Perspectives on Horseshoe Crab Biology, Conservation and Management*, DOI 10.1007/978-3-319-19542-1_34

34.2 Approach

Each year 30–50 student visual and language artworks are selected from several hundred submittals to appear in an annual anthology, "Horseshoe Crabs & the Arts™", which is posted on the ERDG web site (ERDG 2015), where all works are publically available and can be viewed anytime. The reach of these young voices is further extended by ERDG's global social media contacts by posting on the ERDG Facebook page (ERDG-Facebook 2015). The works also have been reproduced for a traveling art exhibition, which has been hosted in art and nature centers as well as at public conservation events throughout the United States, Japan, Hong Kong, and Taiwan.

34.3 Results

In 2014 alone, the Young Voices program received 450 entries from 5 countries (Japan, Hong Kong, Norway, Mexico and the USA), of which 35 were selected from each grade level that entered. Here we share some of the most compelling works among the winning entries from the past 15 years, since the program began in 1999. Through the arts, our future environmental stewards truly have a voice, which is heard globally.

Order of images:

Donald Hudson 2005, 5th grade, Spruance Elementary School, Philadelphia, PA, USA, medium: collage

Faith Schaffer 2009, 10th grade, Haverford High School, Havertown, PA, USA, medium: watercolor

Anneliese Olbrich 2013, 8th grade, Rodney Thompson Middle School, Stafford, VA. USA, medium: watercolor

Aneet Mishra 2010, 8th grade, Kendriya Vidyalaya Port Blair, Andamans, India, medium: oil paint

Carlos Bonilla 2005, 5th grade, Spruance Elementary School, Philadelphia, PA, USA, medium: collage

Madison McGlade 2005, 5th grade, Our Lady Star of the Sea, Cape May, NJ, USA, medium: watercolor

Mary Shoup 2010, 8th grade, Rodney Thompson Middle School, Stafford, VA, USA, medium: computer graphics

Renee McGee 2012, 8th grade, Rodney Thompson Middle School, Stafford, VA, USA, medium: ink drawing

Rosetta Silva 2014, 3rd grade, Field Street Elementary School, Carney's Point, NJ, USA, medium: water color

Ruby Tik Kei Ng 2014, 6th grade, Tsung Tsin Christian Academy, Cheung Sha Wan, Hong Kong, medium: colored pencil

References

ERDG (2015) Ecological Research and Development Group. http://horseshoecrab.org. Accessed 16 Apr 2015

ERDG-Facebook (2015) Ecological Research and Development Group-Facebook. https://www.facebook.com/horseshoecrab.erdg. Accessed 16 Apr 2015

Index

[*]Due to frequent use of this term, page numbers refer to the first occurrence in each chapter

R.H. Carmichael et al. (eds.), *Changing Global Perspectives on Horseshoe Crab Biology, Conservation and Management*, DOI 10.1007/978-3-319-19542-1

The manufacturer's authorised representative in the EU is Springer Nature Customer Service Centre GmbH, Europaplatz 3, 69115 Heidelberg, Germany. If you have any concerns regarding our products, please contact ProductSafety@springernature.com

Printed and bound by CPI Group (UK) Ltd, Croydon, CR0 4YY
30/09/2026
02225563-0018